Corollary (Buffer Flexibility): *Flexibility reduces the amount of variability buffering required in a production system.*

Law (Conservation of Material): *In a stable system, over the long run, the rate out of a system will equal the rate in, less any yield loss, plus any parts production within the system.*

Law (Capacity): *In steady state, all plants will release work at an average rate that is strictly less than the average capacity.*

Law (Utilization): *If a station increases utilization without making any other changes, average WIP and cycle time will increase in a highly nonlinear fashion.*

Law (Process Batching): *In stations with batch operations or with significant changeover times:*

1. *The minimum process batch size that yields a stable system may be greater than one.*
2. *As process batch size becomes large, cycle time grows proportionally with batch size.*
3. *Cycle time at the station will be minimized for some process batch size, which may be greater than one.*

Law (Move Batching): *Cycle times over a segment of a routing are roughly proportional to the transfer batch sizes used over that segment, provided there is no waiting for the conveyance device.*

Law (Assembly Operations): *The performance of an assembly station is degraded by increasing any of the following:*

1. *Number of components being assembled.*
2. *Variability of component arrivals.*
3. *Lack of coordination between component arrivals.*

Definition (Station Cycle Time): *The average cycle time at a station is made up of the following components:*

$$\text{Cycle time} = \text{move time} + \text{queue time} + \text{setup time} + \text{process time}$$
$$+ \text{wait-to-batch time} + \text{wait-in-batch time}$$
$$+ \text{wait-to-match time}$$

Definition (Line Cycle Time): *The average cycle time in a line is equal to the sum of the cycle times at the individual stations, less any time that overlaps two or more stations.*

Law (Rework): *For a given throughput level, rework increases both the mean and standard deviation of the cycle time of a process.*

Law (Lead Time): *The manufacturing lead time for a routing that yields a given service level is an increasing function of both the mean and standard deviation of the cycle time of the routing.*

Law (CONWIP Efficiency): *For a given level of throughput, a push system will have more WIP on average than an equivalent CONWIP system.*

Law (CONWIP Robustness): *A CONWIP system is more robust to errors in WIP level than a pure push system is to errors in release rate.*

Law (Self-Interest): *People, not organizations, are self-optimizing.*

Law (Individuality): *People are different.*

Law (Advocacy): *For almost any program, there exists a champion who can make it work—at least for a while.*

Law (Burnout): *People get burned out.*

Law (Responsibility): *Responsibility without commensurate authority is demoralizing and counterproductive.*

FACTORY PHYSICS
Foundations of Manufacturing Management
SECOND EDITION

Wallace J. Hopp
Northwestern University

Mark L. Spearman
Georgia Institute of Technology

Boston   Burr Ridge, IL   Dubuque, IA   Madison, WI   New York   San Francisco   St. Louis
Bangkok   Bogotá   Caracas   Lisbon   London   Madrid
Mexico City   Milan   New Delhi   Seoul   Singapore   Sydney   Taipei   Toronto

McGraw-Hill Higher Education

A Division of The **McGraw-Hill** *Companies*

FACTORY PHYSICS: FOUNDATIONS OF MANUFACTURING MANAGEMENT
Published by Irwin/McGraw-Hill, an imprint of The McGraw-Hill Companies, Inc., 1221 Avenue of the
Americas, New York, NY 10020. Copyright © 2001, 1999, 1995, by The McGraw-Hill Companies, Inc. All
rights reserved. No part of this publication may be reproduced or distributed in any form or by any means, or
stored in a database or retrieval system, without the prior written consent of The McGraw-Hill Companies,
Inc., including, but not limited to, in any network or other electronic storage or transmission, or broadcast for
distance learning.

1 2 3 4 5 6 7 8 9 0 DOC/DOC 0 9 8 7 6 5 4 3 2 1 0

ISBN 0-256-24795-1

Publisher: Jeffrey J. Shelstad
Executive editor: Richard Hercher
Developmental editor: Gail Korosa
Marketing manager: Zina Craft
Project manager: Kimberly D. Hooker
Production supervisor: Kari Geltemeyer
Coordinator freelance design: Mary Christianson
Supplement coordinator: Becky Szura
New media: Edward Przyzycki
Freelance cover designer: Larry Didona Design Images
Cover photographs: Wright Brothers © Corbis
Compositor: Techsetters, Inc.
Typeface: 10/12 Times Roman
Printer: R. R. Donnelley & Sons Company

Library of Congress Cataloging-in-Publication Data

Hopp, Wallace J.
    Factory physics : foundations of manufacturing management / Wallace J. Hopp, Mark
  L. Spearman.
        p. cm.
    Includes bibliographical references and index.
    ISBN 0-256-24795-1
    1. Factory management. 2. Production management. I. Spearman, Mark L. II. Title.
  TS155.H679   2000
    658.5–dc21   99-086385

www.mhhe.com

To Melanie, Elliott, and Clara
W.J.H.

To Blair, my best friend and spiritual companion who has always been there to lift me
    up when I have fallen,
to Jacob, who has taught me to trust in the Lord and in whom I have seen a mighty
    work,
to William, who has a tender heart for God,
to Rebekah in whom God has graciously blessed me, and
To him who is able to keep you from falling and to present you before his glorious
    presence without fault and with great joy
to the only God our Savior be glory, majesty, power and authority, through Jesus
    Christ our Lord, before all ages, now and forevermore! Amen.

—Jude 24–25

M.L.S.

Origins of Factory Physics

In 1988 we were working as consultants at the IBM raw card plant in Austin, Texas, helping to devise more effective production control procedures. Each time we suggested a particular course of action, our clients would, quite reasonably, ask us to explain *why* such a thing would work. Being professors, we responded by immediately launching into theoretical lectures, replete with outlandish metaphors and impromptu graphs. After several semicoherent presentations, our sponsor, Jack Fisher, suggested we organize the essentials of what we were saying into a formal one-day course.

We did our best to put together a structured description of basic plant behavior. While doing this, we realized that certain very fundamental relations—for example, the relation between throughput and WIP, and several other basic results of Part II of this book—were not well known and were not covered in any standard operations management text. Our six offerings of the course at IBM were well received by audiences ranging from machine operators to mid-level managers. During one class, a participant observed, "Why, this is like physics of the factory!" Since both of us have bachelor's degrees in physics and keep a soft spot in our hearts for the subject, the name stuck. Factory physics was born.

Buoyed by the success of the IBM course, we developed a two-day industry course on short-cycle manufacturing, using factory physics as the organizing framework. Our focus on cycle time reduction forced us to strengthen the link between fundamental relations and practical improvement policies. Teaching to managers and engineers from a variety of industries helped us extend our coverage to more general production environments.

In 1990, Northwestern University launched the Master of Management in Manufacturing (MMM) program, for which we were asked to design and teach courses in management science and operations management. By this time we had enough confidence in factory physics to forgo traditional problem-based and anecdote-based approaches to these subjects. Instead, we concentrated on building intuition about basic manufacturing behavior as a means for identifying areas of leverage and comparing alternate control policies. For completeness and historical perspective, we added coverage of conventional topics, which became the basis for Part I of this book. We received enthusiastic support from the MMM students for the factory physics approach. Also, because they had substantial and varied industry experience, they constructively challenged our ideas and helped us sharpen our presentation.

In 1993, after having taught the MMM courses and the industry short course several times, we began writing out our approach in book form. This proved to be a slow process because it revealed a number of gaps between our presentation of concepts and their implementation in practice. Several times we had to step back and draw upon our own research and that of many others, to develop practical discussions of key manufacturing management problem areas. This became Part III of this book.

Factory physics has grown a great deal since the days of our terse tutorials at IBM and will undoubtedly continue to expand and mature. Indeed, this second edition contains several new developments and changes of presentation from the first edition. But while details will change, we are confident that the fundamental insight behind factory physics—that there are principles governing the behavior of manufacturing systems, and understanding them can improve management practice—will remain the same.

Intended Audience

Factory Physics is intended for three principal academic audiences:

1. *Manufacturing management students* in a core manufacturing operations course.
2. *MBA students* in a second operations management course following a general survey course.
3. *BS and MS industrial engineering students* in a production control course.

We also hope that practicing manufacturing managers will find this book a useful training reference and source of practical ideas.

How to Use this Book

After a brief introductory chapter, the book is organized into three parts: Part I, The Lessons of History; Part II, Factory Physics; and Part III, Principles in Practice. In our own teaching, we generally cover Parts I, II, and III in order, but vary the selection of specific topics depending on the course. Regardless of the audience, we try to cover Part II completely, as it represents the core of the factory physics approach. Because it makes extensive use of pull production systems, we make sure to cover Chapter 4 on "The JIT Revolution" prior to beginning Part II. Finally, to provide an integrated framework for carrying the factory physics concepts into the real world, we regard Chapter 13, "A Pull Planning Framework," as extremely important. Beyond this, the individual instructor can select historical topics from Part I, applied topics from Part III, or additional topics from supplementary readings to meet the needs of a specific audience.

The instructor is also faced with the choice of how much mathematical depth to use. To assist readers who want general concepts with minimal mathematics, we have set off certain sections as *Technical Notes*. These sections, which are labeled and indented in the text, present justification, examples, or methodologies that rely on mathematics (although nothing higher than simple calculus). These sections can be skipped completely without loss of continuity.

In teaching this material to both engineering and management students, we have found, not surprisingly, that management students are less interested in the mathematical aspects of factory physics than are engineering students. However, we have not found management students to be averse to mathematics; it is math without a concrete purpose to which they object. When faced with quantitative developments of core manufacturing ideas, these students not only are capable of grasping the math, but also are able to appreciate the practical consequences of the theory.

New to the Second Edition

The basic structure of the second edition is the same as that of the first. Aside from moving Chapter 12 on Total Quality Manufacturing from Part III to Part II, where it has been adapted to highlight the importance of quality to the science of factory physics, the basic content and placement of the chapters are unchanged. However, a number of enhancements have been made, including the following:

- *More problems.* The number of exercises at the end of each chapter has been increased to offer the reader a wider range of practice problems.
- *More examples.* Almost all models are motivated with a practical application before the development of any mathematics. Frequently, these applications are then used as examples to illustrate how the model is used.
- *Web support.* Powerpoint presentations, case materials, spreadsheets, derivations, and a solutions manual are now available on the Web. These are constantly being updated as more material becomes available. Go to http://www.mhhe.com/pom under Text Support for our web site.
- *Inventory management.* The development of inventory models in Chapter 2 has been enhanced to frame historical results in terms of modern theory and to provide the reader with the most sophisticated tools available. Excel spreadsheets and inventory function add-ins are available over the Web to facilitate the more complex inventory calculations.
- *Enterprise resources planning.* Chapters 3 and 5 describe how materials requirements planning (MRP) has evolved into enterprise resources planning (ERP) and gives an outline of a typical ERP structure. We also describe why ERP is not the final solution to the production planning problem.
- *People in production systems.* Chapter 7 now includes some laws concerning the behavior of production lines in which personnel capacity is an important constraint along with equipment capacity.
- *Variability pooling.* Chapter 8 introduces the fundamental idea that variability from independent sources can be reduced by combining the sources. This basic idea is used throughout the book to understand disparate practices, such as how safety stock can be reduced by stocking generic parts, how finished goods inventories can be reduced by "assembling to order," and how elements of push and pull can be combined in the same system.
- *Systems with blocking.* Chapter 8 now includes analytic models for evaluating performance of lines with finite, as well as infinite, buffers between stations. Such models can be used to represent kanban systems or systems with physical limitations of interstation inventory. A spreadsheet for examining the tradeoffs of additional WIP buffers, decreasing variability, and increasing capacity is available on the Web.
- *Sharper variability results.* Several of the laws in Chapter 9, The Corrupting Influence of Variability, have been restated in clearer terms; and some important new laws, corollaries, and definitions have been introduced. The result is a more complete science of how variability degrades performance in a production system.
- *Optimal batch sizes.* Chapters 9 and 15 extend the factory physics analysis of the effects of batching to a normative method for setting batch sizes to minimize cycle times in multiproduct systems with setups and discuss implications for production scheduling.

- *General CONWIP line models.* Chapter 10 now includes an analytic procedure for computing the throughput of a CONWIP line with general processing times. Previously, only the case with balanced exponential stations (the practical worst case) was analyzed explicitly. These new models are easy to implement in a spreadsheet (available on the Web) and are useful for examining inventory, capacity, and variability tradeoffs in CONWIP lines.

- *Quality control charts.* The quality discussion of Chapter 12 now includes an overview of statistical process control (SPC).

- *Forecasting.* The section on forecasting has been expanded into a separate section of Chapter 13. The treatment of time series models has been moved into this section from an appendix and now includes discussion of forecasting under conditions of seasonal demand.

- *Capacitated material requirements planning.* The MRP-C methodology for scheduling production releases with explicit consideration of capacity constraints has been extended to consider material availability constraints as well.

- *Supply chain management.* The treatment of inventory management is extended to the contemporary subject of supply chain management. Chapter 17 now deals with this important subject from the perspective of multiechelon inventory systems. It also discusses the "bullwhip effect" as a means for understanding some of the complexities involved in managing and designing supply chains.

W.J.H.
M.L.S.

Since our thinking has been influenced by too many people to allow us to mention them all by name, we offer our gratitude (and apologies) to all those with whom we have discussed factory physics over the years. In addition, we acknowledge the following specific contributions.

We thank the key people who helped us shape our ideas on factory physics: Jack Fisher of IBM, who originated this project by first suggesting that we organize our thoughts on the laws of plant behavior into a consistent format; Joe Foster, former adviser who got us started at IBM; Dave Woodruff, former student and lunch companion extraordinaire, who played a key role in the original IBM study and the early discussions (arguments) in which we developed the core concepts of factory physics; Souvik Banerjee, Sergio Chayet, Karen Donohue, Izak Duenyas, Silke Kröckel, Melanie Roof, Esma Senturk-Gel, Valerie Tardif, and Rachel Zhang, former students and valued friends who collaborated on our industry projects and upon whose research portions of this book are based; Yehuda Bassok, John Buzacott, Eric Denardo, Bryan Deuermeyer, Steve Graves, Uday Karmarkar, Steve Mitchell, George Shantikumar, Rajan Suri, Joe Thomas, Michael Zazanis, and Paul Zipkin, colleagues whose wise counsel and stimulating conversation produced important insights in this book. We also acknowledge the National Science Foundation, whose consistent support made much of our own research possible.

We are grateful to those who patiently tested this book (or portions of it) in the classroom and provided us with essential feedback that helped eliminate many errors and rough spots: Karla Bourland (Dartmouth), Izak Duenyas (Michigan), Paul Griffin (Georgia Tech), Steve Hackman (Georgia Tech), Michael Harrison (Stanford), Phil Jones (Iowa), S. Rajagopalan (USC), Jeff Smith (Texas A&M), Marty Wortman (Texas). We thank the many students who had to put up with typo-ridden drafts during the testing process, especially our own students in Northwestern's Master of Management in Manufacturing program, in BS/MS-level industrial engineering courses at Northwestern and Texas A&M, and in MBA courses in Northwestern's Kellogg Graduate School of Management.

We give special thanks to the reviewers of the original manuscript, Suleyman Tefekci (University of Florida), Steve Nahmias (Santa Clara University), David Lewis (University of Massachusetts, Lowell), Jeffrey L. Rummel (University of Connecticut), Pankaj Chandra (McGill University), Aleda Roth (University of North Carolina, Chapel Hill), K. Roscoe Davis (University of Georgia), and especially Michael H. Rothkopf (Rutgers University), whose thoughtful comments greatly improved the quality of our ideas and presentation. We also thank Mark Bielak who assisted us in our first attempt to write fiction.

In addition to those who helped us produce the first edition, many of whom also helped us on the second edition, we are grateful to individuals who had particular influence on the revision. We acknowledge the people whose ideas and suggestions helped us deepen our understanding of factory physics: Jeff Alden (General Motors), John Bartholdi (Georgia Tech), Corey Billington (Hewlett-Packard), Dennis E. Blumenfeld (General Motors), Sunil Chopra (Northwestern University), Mark Daskin (Northwestern University), Greg Diehl (Network Dynamics), John Fowler (Arizona State University), Rob Herman (Alcoa), Jonathan M. Heuberger (DuPont Pharmaceuticals), Sayed Iravani (Northwestern University), Tom Knight (Alcoa), Hau Lee (Stanford University), Leon McGinnis (Georgia Tech), John Mittenthal (University of Alabama), Lee Schwarz (Purdue University), Alexander Shapiro (Georgia Tech), Kalyan Singhal (University of Baltimore), Tom Tirpak (Motorola), Mark Van Oyen (Loyola University), Jan Van Mieghem (Northwestern University), Joe Velez (Alcoa), William White (Bell & Howell), Eitan Zemel (New York University), and Paul Zipkin (Duke University).

We would like to thank particularly the reviewers of the first edition whose suggestions helped shape this revision. Their comments on how the material was used in the classroom and how specific parts of the book were perceived by their students were extremely valuable to us in preparing this new edition: Diane Bailey (University of Southern California), Charles Bartlett (Polytechnic University), Guillermo Gallego (Columbia University), Marius Solomon (Northeastern University), M. M. Srinivasan (University of Tennessee), Ronald S. Tibben-Lembke (University of Nevada, Reno), and Rachel Zhang (University of Michigan).

Finally, we thank the editorial staff at Irwin: Dick Hercher, Executive Editor, who kept us going by believing in this project for years on the basis of all talk and no writing; Gail Korosa, Senior Developmental Editor, who recruited the talented team of reviewers and applied polite pressure for us to meet deadlines, and Kimberly Hooker, Project Manager, who built a book from a manuscript.

B R I E F   C O N T E N T S

CONTENTS

PART III

PRINCIPLES IN PRACTICE

13 A Pull Planning Framework 408

0   FACTORY PHYSICS?

Perfection of means and confusion of goals seem to characterize our age.
Albert Einstein

0.1   The Short Answer

What is factory physics, and why should one study it?

Briefly, **factory physics** is a *systematic description of the underlying behavior of manufacturing systems*. Understanding it enables managers and engineers to work with the natural tendencies of manufacturing systems to

1. Identify opportunities for improving existing systems.
2. Design effective new systems.
3. Make the tradeoffs needed to coordinate policies from disparate areas.

0.2   The Long Answer

The above definition of factory physics is concise, but leaves a great deal unsaid. To provide a more precise description of what this book is all about, we need to describe our focus and scope, define more carefully the meaning and purpose of factory physics, and place these in context by identifying the manufacturing environments on which we will concentrate.

0.2.1   Focus: Manufacturing Management

To answer the question of why one should study factory physics, we must begin by answering the question of why one should study manufacturing at all. After all, one frequently hears that the United States is moving to a service economy, in which the manufacturing sector will represent an ever-shrinking component. On the surface this appears to be true: Manufacturing employed on the order of 50 percent of the workforce in 1950, but only about 20 percent by 1985. To some, this indicates a trend in manufacturing that parallels the experience in agriculture earlier in the century. In 1929, agriculture employed 29 percent of the workforce; by 1985, it employed only three percent. During this time there was a shift away from low-productivity, low-pay jobs in agriculture and toward higher-productivity, higher-pay jobs in manufacturing, resulting in a dramatic increase in the overall standard of living. Similarly, proponents of this analogy argue, we are currently shifting from a manufacturing-based workforce to an even more productive service-based workforce, and we can expect even higher living standards.

However, as Cohen and Zysman point out in their elegant and well-documented book *Manufacturing Matters: The Myth of the Post-Industrial Economy* (1987), there is a fundamental flaw in this analogy. Agriculture was *automated,* while manufacturing, at least partially, is being moved *offshore*—moved abroad. Although the number of agricultural jobs declined, due to a dramatic increase in productivity, American agricultural output did not decline after 1929. As a result, most of the jobs that are *tightly linked* to agriculture (truckers, vets, crop dusters, tractor repairers, mortgage appraisers, fertilizer sales representatives, blight insurers, agronomists, chemists, food processing workers, etc.) were not lost. When these tightly linked jobs are considered, Cohen and Zysman estimate that the number of jobs currently dependent on agricultural production is not three million, as one would obtain by looking at an SIC (standard industrial classification) count, but rather something on the order of six to eight million. That is, two or three times as many workers are employed in jobs tightly linked to agriculture as are employed directly in agriculture itself.

Cohen and Zysman extend this linkage argument to manufacturing by observing that many jobs normally thought of as being in the service sector (design and engineering services, payroll, inventory and accounting services, financing and insuring, repair and maintenance of plant and machinery, training and recruiting, testing services and labs, industrial waste disposal, engineering support services, trucking of semifinished goods, etc.) depend on manufacturing for their existence. If the number of manufacturing jobs declines due to an increase in productivity, many of these tightly linked jobs will be retained.

But if American manufacturing declines by being moved offshore, many tightly linked jobs will shift overseas as well. There are currently about 21 million people employed directly in manufacturing. Therefore, if a similar multiplier to that estimated by Cohen and Zysman for agriculture applies, there are some 20 to 40 million tightly linked jobs that depend on manufacturing. This implies that over half of the jobs in America are strongly tied to manufacturing. Even without considering the indirect effects (e.g., unemployed or underemployed workers buy fewer pizzas and attend fewer symphonies) of losing a significant portion of the manufacturing jobs in this country, the potential economic consequences of moving manufacturing offshore are enormous.

During the 1980s when we began work on the first edition of this book, there were many signs that American manufacturing was not robust. Productivity growth relative to that in other industrialized countries had slowed dramatically. Shares of domestic firms in several important markets (e.g., automobiles, consumer electronics, machine tools) had declined alarmingly. As a result of rising imports, America had become the world's largest debtor nation, mounting huge trade deficits with other manufacturing powers, such as Japan. The fraction of American patents granted to foreign inventors had doubled over the previous two decades. These and many other trends seemed to indicate that American manufacturing was in real trouble.

The reasons for this decline were complex and controversial, as we will discuss further in Part I. Moreover, in many regards, American manufacturing made a recovery in the 1990s as net income of manufacturers rose almost 65 percent in constant dollars from 1985 to 1994 (Department of Commerce 1997). But one conclusion stands out as obvious—global competition has intensified greatly since World War II, particularly since the 1980s, due to the recovery of economies devastated by the war. Japanese, European, and Pacific Rim firms have emerged as strong competitors to the once-dominant American manufacturing sector. Because they have more options, customers have become increasingly demanding. It is no longer possible to offer products, as Henry Ford once did, in "any color as long as it's black." Customers expect variety, reasonable price, high quality, comprehensive service, and responsive delivery. Therefore, from now on, in good economic times and bad, only those firms that can keep pace along all these dimensions will survive.

Although speaking of manufacturing as a monolithic whole may continue to make for good political rhetoric, the reality is that the rise or fall of the American manufacturing sector will occur one firm at a time. Certainly a host of general policies, from tax codes to educational initiatives, can help the entire sector somewhat; the ultimate success of each individual firm is fundamentally determined by the *effectiveness of its management*. Hence, quite literally, our economy, and our very way of life in the future, depends on how well American manufacturing managers adapt to the new globally competitive environment and evolve their firms to keep pace.

0.2.2 Scope: Operations

Given that the study of manufacturing is worthwhile, how should we study it? Our focus on management naturally leads us to adopt the high-level orientation of **"big M" manufacturing,** which includes product design, process development, plant design, capacity management, product distribution, plant scheduling, quality control, workforce organization, equipment maintenance, strategic planning, supply chain management, interplant coordination, as well as direct production—**"little m" manufacturing**—functions such as cutting, shaping, grinding, and assembly.

Of course, no single book can possibly cover all big M manufacturing. Even if one could, such a broad survey would necessarily be shallow. To achieve the depth needed to promote real understanding, we must narrow our scope. However, to preserve the "big picture" management view, we cannot restrict it too much; highly detailed treatment of narrow topics (e.g., the physics of metal cutting) would constitute such a narrow viewpoint that, while important, would hardly be suitable for identifying effective management policies. The middle ground, which represents a balance between high-level integration and low-level details, is the operations viewpoint.

In a broad sense, the term **operations** refers to the application of resources (capital, materials, technology, and human skills and knowledge) to the production of goods and services. Clearly, all organizations involve operations. Factories produce physical goods. Hospitals produce surgical and other medical procedures. Banks produce checking account transactions and other financial services. Restaurants produce food and perhaps entertainment. And so on.

The term *operations* also refers to a specific function in an organization, distinct from other functions such as product design, accounting, marketing, finance, human resources, and information systems. Historically, people involved in the operations function are housed in departments with names like production control, manufacturing engineering, industrial engineering, and planning, and are responsible for the activities directly related to the production of goods and services. These typically include plant scheduling, inventory control, quality assurance, workforce scheduling, materials management, equipment maintenance, capacity planning, and whatever else it takes to get product out the door.

In this book, we view operations in the broad sense rather than as a specific function. We seek to give general managers the insight necessary to sift through myriad details in a production system and identify effective policies. The operations view focuses on the *flow of material* through a plant, and thereby places clear emphasis on most of the key measures by which manufacturing managers are evaluated (throughput, customer service, quality, cost, investment in equipment and materials, labor costs, efficiency, etc.). Furthermore, by avoiding the need for detailed descriptions of products or processes, this view concentrates on *generic manufacturing behavior,* which makes it applicable to a wide range of specific environments.

The operations view provides a unifying thread that runs through all the various big-M manufacturing issues. For instance, operations and product design are linked in that a product's design determines how it must flow through a plant and how difficult it will be to make. Adopting an operations viewpoint in the design process therefore promotes **design for manufacturability.** In the same fashion, operations and strategic planning are closely tied, since strategic decisions determine the number and types of products to be produced, the size of the manufacturing facilities, the degree of vertical integration, and many other factors that affect what goes on inside the plant. Embedding a concern for operations in strategic decision making is essential for ensuring feasible plans. Other manufacturing functions have analogous relationships to operations, and hence can be coordinated with the actual production process by addressing them from an operations viewpoint.

The traditional field in which operations are studied is called **operations management (OM).** However, OM is broader than the scope of this book, since it encompasses operations in service, as well as manufacturing, organizations. Just as our operations scope covers only part of (big M) manufacturing, our manufacturing focus includes only part of operations management. In short, the scope of this book can be envisioned as the intersection between OM and manufacturing, as illustrated in Figure 0.1.

The operations view of manufacturing may seem a somewhat technical perspective for a management book. This is not accidental. Some degree of technicality is required just to accurately describe manufacturing behavior in operations terms. More importantly, however, is the reality that in today's environment, *manufacturing itself is technical.* Intense global competition is relentlessly raising market standards, causing seemingly small details to take on large strategic importance. For example, quality acceptable to customers in the 1970s may have been possible with relatively unsophisticated systems. But to meet customer expectations and comply with standards common

FIGURE 0.1

Manufacturing and operations management

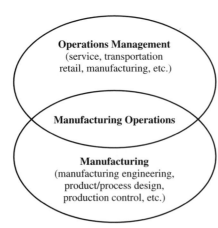

for vendor certification today is virtually impossible without rigorous quality systems in place. Similarly, it was not so long ago when customer service could be ensured by maintaining large inventories. Today, rapid technological change and smaller profit margins make such a strategy uneconomical—literally forcing companies into the tighter control systems necessary to run with low-inventory levels. These shifts are making operations a more integral, and more technical, component of running a manufacturing business.

The trends of the 1990s may make it appear that the importance of operations is a new phenomenon. But, as we will discuss in greater depth in Part I, low-level operations details have *always* had strategic consequences for manufacturing firms. A recent reminder of this fact was the experience of Japan in the 1970s and 1980s. As Chapter 4 describes, Japanese firms, particularly Toyota, were able to carry out a strategy of low-cost, small-lot production only through intense attention to minute details on the factory floor (e.g., die changing, statistical process control, material flow control) over an extended time. The net result was an enormously effective competitive weapon that permitted Toyota to rise from obscurity to a position as a worldwide automotive leader.

Today, the importance of operations to the health, and even viability, of manufacturing firms is greater than ever due to global competition in the following three dimensions:

1. **Cost.** This is the traditional dimension of competition that has always been the domain of operations management. Efficient utilization of labor, material, and equipment is essential to keeping costs competitive. We should note, however, that from the customer standpoint it is **unit cost** (total cost divided by total volume) that matters, implying that both cost reduction and volume enhancement are worthy OM objectives.

2. **Quality.** The 1980s brought widespread recognition in America that quality is a key competitive weapon. Of course, *external* quality, that seen by the customer, has always been a concern in manufacturing. The quality revolution of the 1980s served to focus attention on *internal* quality at each step in the manufacturing process, and its relationship to customer satisfaction. Facets of operations management, such as statistical process control, human factors, and material flow control, have loomed large in this context as components of **total quality management (TQM)** strategies.

3. **Speed.** While cost and quality remain critical, the 1990s can be dubbed the *decade of speed*. Rapid development of new products, coupled with quick customer delivery, are pillars of the **time-based competition (TBC)** strategies that have been adopted by leading firms in many industries. Bringing new products to market swiftly requires both performance of development tasks in parallel and the ability to efficiently ramp up production. Responsive delivery, without inefficient excess inventory, requires short manufacturing cycle times, reliable processes, and effective integration of disparate functions (e.g., sales and manufacturing). These issues are central to operations management, and they arise repeatedly throughout this book.

These three dimensions are broadly applicable to most manufacturing industries, but their relative importance obviously varies from one firm to another. A manufacturer of a commodity (baking soda, machine screws, resistors) depends critically on efficiency, since low cost is a condition for survival. A manufacturer of premium goods (luxury automobiles, expensive watches, leatherbound books) relies on quality to retain its market. A manufacturer of a high-technology product (computers, patent-protected pharmaceuticals, high-end consumer electronics) requires speed of introduction to be competitive and to maximally exploit potential profit during the limited economic lifetime of the product. Clearly, the management challenges in these varying environments are different. Since operations are integral to cost, quality, and speed, however, operations management has a key strategic role in each.

0.2.3   Method: Factory Physics

So far, we have determined that the focus of this book is manufacturing management, and the scope is operations. The question now becomes, How can managers use an operations viewpoint to identify a sensible combination of policies that are both effective now and flexible enough to adapt to future needs?

In our opinion, some conventional approaches to manufacturing management fall short:

1. *Management by imitation* is not the answer. Watching the competition can provide a company with a valuable source of benchmarking and may help it to avoid getting stuck in established modes of thinking. But imitation cannot provide the impetus for a truly significant competitive edge. Bold new ideas must come from within, not without.

2. *Management by buzzword* is not the answer. Manufacturing firms have become inundated with a wave of "revolutions" in recent years. MRP, JIT, TQM, BPR, TBC (and even a few without three-letter acronyms) have swept through the manufacturing community accompanied by soaring rhetoric and passionate emotion, but with little concrete detail. As we will observe in Part I, these movements have contained many valuable insights. However, they are very dangerous as management systems because it is far too easy for managers to become attached to catchy slogans and trendy buzzwords and lose sight of the fundamental objectives of the business. The result can be very poor decisions for the long run.

3. *Management by consultant* is, at best, only a partial solution. A good consultant can make an objective evaluation of a firm's policies and provide a source of new ideas. However, as an outsider, the consultant is not in a position to obtain the support of key people so critical to implementing new management systems. Additionally, a consultant can never have the intimate familiarity with the business that an insider has, and is therefore likely to push generic solutions, rather than customized methods that match the specific needs of the firm. No matter how good an off-the-shelf technology (e.g., scheduling tools, optical scanners, AGVs, robots) is, the manufacturing *system* must be ultimately designed in-house, if it is to be effective as a whole.

So, what is the answer? In our view, the answer is not *what to do* about manufacturing problems but rather *how to think* about them. Each manufacturing environment is unique. No single set of procedures can work well under all conditions. Therefore, effective manufacturing managers of the future will have to rely on a solid understanding of their systems to enable them to identify leverage points, creatively leapfrog the competition, and engender an environment of continual improvement. For the student of manufacturing management, this is something of a "good news–bad news" message. The bad news is that manufacturing managers will need to know more about the fundamentals of manufacturing than ever before. The good news is that the manager who has developed these skills will be increasingly valuable in industry.

From an operations viewpoint, there are behavioral tendencies shared by virtually all manufacturing enterprises. We feel that these can be organized into a body of knowledge to serve as a manufacturing manager's knowledge base, just as the field of medicine serves as a physician's knowledge base. In this book, we employ a spirit of rational inquiry to seek a **science of manufacturing** by establishing basic concepts as building blocks, stating fundamental principles as "manufacturing laws," and identifying general insights from specific practices. Our primary goal is to provide the reader with an organized framework from which to evaluate management practices and develop useful intuition about manufacturing systems. Our secondary goal is to encourage others to push the science of manufacturing even further, developing new and better structures than we can offer at this time.

We use the term **factory physics** to distinguish our long-term emphasis on general principles from the short-term fixation on specific techniques inherent in the buzzword approach. We emphatically stress that factory physics is *not factory magic*. Rather, it is a discipline based on the scientific method that has several features in common with the field of physics:

1. *Problem-solving framework.* Just as there are few easy solutions in physics, there are few in manufacturing management. Physics offers rational approaches for understanding nature. An understanding of basic physics is critical to the engineer in building or designing a complex system. Likewise, factory physics provides a context for understanding manufacturing operations that allows the manufacturing manager or engineer to pose and solve the right problems.

2. *Technical approach.* Physics is generally viewed as a hard, technical subject. But, as we noted, OM is a hard technical subject as well. A presentation of OM without some technical content is like a newspaper description of an engineering feat without any physical description—it sounds interesting but the reader cannot tell how it is actually done. Such an approach might be legitimate as a *survey* of operations management, but is not suited to developing the skills needed by manufacturing managers and engineers.

3. *Role of intuition.* Physicists generally have well-developed intuition about the physical world. Even before writing any mathematical equations to represent a system, a physicist forms a qualitative feel for the important parameters and their relationships. Analogously, to make good decisions, a manager needs sound intuition about system behavior and the consequences of various actions. Thus, while we will spend a fair amount of time developing concepts with mathematical models, our real concern is not the analyses themselves, but rather the general intuition we can draw from them.

In the spirit of factory physics, we can summarize the key skills that will be required by the manager of the future as falling into three distinct categories: **basics, intuition, and synthesis.**[1] The relation of these to operations management and their role in this book are as follows:

1. **Basics.** The language and elementary concepts for describing manufacturing systems are essential prerequisites for any manufacturing manager. Although many basics of relevance to the manufacturing manager (e.g., elementary mathematics, statistics, physics of manufacturing processes) are outside the realm of OM and therefore the scope of this text, we do present a number of basic concepts integral to OM, dealing with variability, reliability, behavior of queuing systems, and so on. These are introduced as needed in Part II. We also cull valuable basic concepts from traditional OM practices in the historical survey of Part I.

2. **Intuition.** The single most important skill of a manufacturing manager is intuition regarding the behavior of manufacturing systems. Solid intuition enables a manager to identify leverage points in a plant, evaluate the impacts of proposed changes, and coordinate improvement efforts. We therefore devote the bulk of Part II to developing intuition about key types of manufacturing behavior.

---

[1]While these categories may be new for a manufacturing book, they are hardly revolutionary. The *Trivium,* which constituted the basis for a liberal education in the Middle Ages and consisted of grammar (the basic rules), logic (rational relationships), and rhetoric (fitting it all together), is virtually identical to our structure.

3. **Synthesis.** Close behind intuition on the list of important skills for a manufacturing manager is the ability to bring together the disparate components of a system into an effective whole. In part, this is related to the ability to understand tradeoffs and focus on critical parameters. But it also depends on the capacity to step back and view the system from a holistic perspective. We discuss a formal method for problem solving based on this view—the **systems approach**—in Chapter 6. A good manufacturing manager also considers improvements based on many different approaches (e.g., process changes, logistics changes, personnel policy changes) and is sensitive to the effects of changes in one area or another. In Part III, we present a production planning hierarchy that integrates potentially disjoint decisions, and we describe the interfaces between different functions. Often, the "biggest bang for the buck" lies at the interfaces, so we highlight them wherever possible throughout Parts II and III.

0.2.4 Perspective: Flow Lines

To use the factory physics method to study manufacturing management from an operations standpoint, we must select a primary perspective through which to view manufacturing systems. Without this, environmental differences will tend to obscure common underlying behavior and make development of a science of manufacturing impossible. The reason is that even when we adopt an operations view and ignore the low-level differences in products and processes, manufacturing environments vary greatly with respect to their **process structure,** that is, the manner in which material moves through the plant. For instance, the continuous flow nature of a chemical plant behaves very differently and hence presents a very different management picture than does the one-at-a-time artisan environment of a custom machine shop. Hayes and Wheelwright (1979) classify manufacturing environments by process structure into four categories (see Figure 0.2) which can be summarized as follows:

1. **Job shops.** Small lots are produced with a high variety of routings through the plant. Flow through the plant is jumbled, setups are common, and the environment has more of an atmosphere of project work than pacing. For example, a commercial printer, where each job has unique requirements, will generally be structured as a job shop.

2. **Disconnected flow lines.** Product batches are produced on a limited number of identifiable routings (i.e., paths through the plant). Although routings are distinct, individual stations within lines are not connected by a paced material handling system, so that inventories can build up between stations. The majority of manufacturing systems in industry resemble the disconnected flow line environment to some extent. For example, a heavy equipment (e.g., tank car) manufacturer will use well-defined assembly lines but, because of the scale and complexity of the processes at each station, generally will not automate and pace movement between stations.

3. **Connected flow lines.** This is the classic moving assembly line made famous by Henry Ford. Product is fabricated and assembled along a rigid routing connected by a paced material handling system. Automobiles, where frames travel along a moving assembly line between stations at which components are attached, are the classic application of the connected flow line. But, despite the familiarity and historic appeal of this type of system, automatic assembly lines are actually much less common than disconnected flow lines in industry.

4. **Continuous flow processes.** Continuous product (food, chemicals, oil, roofing materials, fiberglass insulation, etc.) flows automatically down a fixed routing. Many food processing plants, such as sugar refineries, make use of continuous flow to achieve high efficiency and product uniformity.

FIGURE 0.2

The product process matrix
(Source: Hayes and Wheelwright 1979)

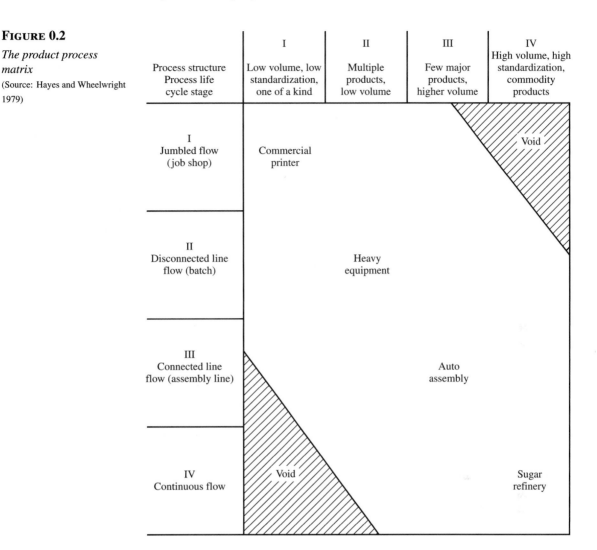

Process structure Process life cycle stage	I Low volume, low standardization, one of a kind	II Multiple products, low volume	III Few major products, higher volume	IV High volume, high standardization, commodity products
I Jumbled flow (job shop)	Commercial printer			Void
II Disconnected line flow (batch)		Heavy equipment		
III Connected line flow (assembly line)			Auto assembly	
IV Continuous flow	Void			Sugar refinery

These environments are suited to different types of products. Because a job shop provides maximum flexibility, it is well suited to low-volume, highly customized products. However, because a job shop is not very efficient on a unit cost basis, it is unattractive for higher-volume products. Therefore, most discrete parts manufacturing plants make at least partial use of some kind of flow line. The decision of how much to automate and pace the line depends on whether the volume and expected economic life justify the necessary capital investment. In continuous product manufacturing, the analogous decision is how far to move from "bench-top" batch production toward a continuous flow process.

Figure 0.2 presents an often-cited **product process matrix** that relates process structure to product type. The basic message of this figure is that higher volumes tend to go hand in hand with smoother-flow process structures. This suggests that the appropriate manufacturing environment may depend on the stage of the product in its life cycle. Newly introduced products are typically produced in small volumes and are subject to design tinkering during a start-up phase, which makes them well suited to the flexibility provided by a job shop environment. As the product progresses through growth and maturation phases, volumes justify a shift to a more efficient (disconnected) flow line. If the product matures into a commodity (i.e., instead of declining out of the market), even greater standardization of flow, in an automated assembly line or continuous flow line, may be justified. This evolution can be viewed as traversing the diagonal of the product process matrix in Figure 0.2 from the upper left to the lower right over the life of the product.

While the product process matrix is useful for characterizing differences in process structures and their relationship to product requirements, it presents only part of the picture. If manufacturing strategy were simply a matter of noting the type of product and selecting the appropriate process from such a matrix, we wouldn't need a science of manufacturing (or highly trained manufacturing managers). But, as we have stressed, customers today want it all: variety, low cost, high quality, and quick responsive delivery. A major challenge facing modern manufacturing firms is how to structure the environment so that it attains the speed and low cost of the high-volume flow lines while retaining the flexibility and customization potential of a low-volume job shop, all within an atmosphere of continually improving quality.

In this book, we select as our perspective discrete parts production on disconnected flow lines. We do this in part because such environments are most prevalent in industry. Additionally, the flow line perspective enables us to identify concepts for "unjumbling" flow and improving efficiency in job shop environments. Finally, flow lines provide a logical link between discrete parts production and continuous flow processes, and hence a vehicle for looking to continuous systems as a source of ideas for smoothing flow and improving cost efficiency. Thus, the disconnected flow line perspective serves as the foundation upon which to build a problem-solving framework that is applicable across a broad range of manufacturing environments.

0.3   An Overview of the Book

The remainder of this book is divided into three major parts:

Part I, The Lessons of History, provides a history of manufacturing in America, along with a review of traditional operations management techniques, including inventory control models, material requirements planning (MRP), and just-in-time (JIT). In reviewing each of these, we identify the essential insights that are necessary components of the science of manufacturing. Part I concludes with a critical review of why these historical techniques are, by themselves, inadequate for the future needs of manufacturing.

Part II, Factory Physics, presents the core concepts of the book. We begin with the basic structure of the science of manufacturing and a discussion of the systems approach to problem solving. Then we examine key behavioral tendencies of manufacturing plants, starting with basic relationships between measures (e.g., throughput, inventory, and cycle time) and working up to the subtle influences of variability. We also examine the science behind some popular Japanese techniques by comparing push and pull production systems. For clarity, the main conclusions are stated as "manufacturing laws," although, as we will discuss, some of these laws are true laws that always hold, while others are useful generalities that hold most of the time. We include in Part II a brief discussion of critical human issues in manufacturing systems to emphasize the essential point that manufacturing is more than just machinery and logistics—it is people, too. We also identify key links between logistics and quality, to provide some science behind TQM practices.

Part III, Principles in Practice, treats specific manufacturing management issues in detail. By applying the lessons of Part I and the laws of Part II, we contrast and compare different approaches to problems commonly encountered in running a manufacturing facility. These include shop floor control, sequencing and scheduling, long-range aggregate planning, workforce planning, capacity management, and coordination of planning and control across levels in a hierarchical system. The focus is on choosing the right measures and controls and providing a framework within which to build solutions. We illustrate problem-solving procedures by providing explicit "how to" instructions for selected problems. The purpose of these detailed solutions is not so much to provide user-ready tools, but rather to help the reader visualize how general concepts of Part II can be applied to specific problems.

This three-part approach roughly parallels the three categories of skills required by manufacturing managers and engineers: basics, intuition, and synthesis. Part I concentrates on basics, by providing a historical perspective and introducing traditional terms and techniques. Part II focuses on intuition, by describing fundamental behavior of manufacturing systems. Part III addresses synthesis, by developing a framework for integrating disparate manufacturing planning problems. A manufacturing professional with mastery of these three skills can identify the essential problems in a factory *and* do something about them.

And now, on to *Factory Physics.*

I   THE LESSONS OF HISTORY

Those who cannot remember the past are condemned to repeat it.
George Santayana

1 MANUFACTURING IN AMERICA

What has been will be again, what has been done will be done again; there is nothing new under the sun.

Ecclesiastes

1.1 Introduction

A fundamental premise of this book is that to manage something effectively, one must first understand it. But manufacturing systems are complex entities that can be viewed in many ways,[1] many of which are integral to sound managerial insight. A particularly important perspective, which provides an organizing framework for all others, is that of history.

A sense of history is fundamental to manufacturing managers for two main reasons. First, in manufacturing, as in all walks of life, the ultimate test of an idea is the test of time. Since short-term success may be the result of luck or exogenous circumstances, we can only identify concepts of lasting value by taking the long-term view. Second, because the requirements for success in business change over time, it is critical for managers to make decisions with the future in mind. One of the very best tools for consistently anticipating the future is a sound appreciation of the past.

The history of American manufacturing, which follows its rise from meager colonial beginnings to undisputed worldwide leadership by mid-20th century, through a period of serious decline in the 1970s and 1980s, and into a revitalization in the complex global environment of the 1990s, is a fascinating story. Sadly, we have neither the space nor the expertise to offer comprehensive coverage here. Instead, we highlight major events and trends with emphasis on themes that will be crucial later in the book. We hope the reader will be sufficiently interested in these historical issues to pursue more basic sources. The following are attractive starting points. Wren (1987) provides an excellent general overview from a management perspective. Boorstin in *The Americans* trilogy (1958, 1965, 1973) offers a number of highly readable insights into American business

---

[1] For example, to a mechanical engineer a manufacturing system is a set of physical processes for altering material, to an operations manager it is a logistical network of product flows, to an organization behavior specialist it is a community of people with shared concerns, to an accountant it is a collection of interrelated cash flows, and so on.

in a cultural context. Chandler (1977, 1990) gives a towering treatment of the evolution of large-scale management in America, as well as Germany and Great Britain. We have drawn heavily on these works, and their references, in what follows.

1.2   The American Experience

In many ways, America began with a clean slate. A vast, wide-open continent offered unparalleled resources and unlimited opportunities for development. Unshackled by the traditions of the Old World, Americans were free to write their own rules. Government, law, cultural practices, and social mores were all choices to be made as part of the grand American experiment.

Naturally, these choices reflected the times in which they were made. In 1776, antimonarchist sentiment, which would soon fuel the French Revolution, was on the rise in both the Old World and the New. America chose democracy. In 1776, Scotsman Adam Smith (1723–1790) proclaimed the end of the old mercantilist system and the beginnings of modern capitalism in his *Wealth of Nations,* in which he articulated the benefits of the division of labor and explained the workings of the "invisible hand" of capitalism.[2] America chose the free market system. In 1776, James Watt (1736–1819) sold his first steam engine in England and began the first industrial revolution in earnest. America embraced the new factory system, evolved a unique style of manufacturing, and eventually led the transportation and communications breakthroughs that sparked the second industrial revolution. In 1776, English common law was the standard for the civilized world. America adapted this tradition, borrowed from Roman law and the *Code Napoléon,* and rapidly became the most litigious country in the world.[3]

In almost all cases, Americans did not invent revolutionary concepts from scratch. Rather, they borrowed freely (and even stole) ideas from the Old World and adapted them to the New. Because the needs of the New World were different, because they were not bound by Old World customs and traditions, and, quite frankly, because they were naive, the social and economic institutions that resulted were uniquely American.

The very fact that America had the opportunity to create itself has done much to shape its national identity. Unlike the countries of the Old World, which coalesced as nations long after they had acquired a national spirit, the United States of America achieved nationhood as a composite of colonies with little sense of common identity. Hence, Americans actively sought an identity in the form of cultural symbols. The strongest and most uniquely American cultural icon was that of the rugged individualist seeking freedom on the frontier. This spawned the wild comic legends about Davy Crockett and Mike Fink and later played a large part in transforming Abraham Lincoln into a revered national icon as the "rail splitter" president. Even after the frontier was gone, the myth of the frontier lived on in popular literature and cinema about the cowboys, ranchers, gunfighters, and prospectors of the Old West.

In more recent times, the myth of the frontier evolved into the myth of the self-made person, which has roots stretching back to the aphorisms of Benjamin Franklin (1706–1790) and the essays of Ralph Waldo Emerson (1803–1882), and which found fertile ground in the Protestant work ethic. This myth made heroes out of successful industrialists of the 19th century (e.g., Carnegie, Rockefeller, Morgan) and provided

---

[2]It is not coincidence that Henry Ford, one of the men most visibly associated with capitalism, would write a book 150 years after Smith's and with the penultimate chapter entitled "The Wealth of Nations."

[3]Two-thirds of the world's lawyers practice in the United States where there are 1,000 lawyers to every 100 engineers. Japan, on the other hand, has 1,000 engineers to every 100 lawyers (Lamm 1988, 17).

cultural support for the unvarnished pursuit of wealth by the corporate raiders of the 1980s. The terms that referred to the players in the takeover games of that "decade of greed"—*gunslinger, white knight, masters of the universe*—were not accidental. Nor is the fact that marketing and finance have consistently been more popular in American business schools than operations management. The perception has been that in finance and marketing, one can do something "big" or "bold" by starting daring new ventures or launching exciting new products, while in operations management one can only struggle to save a few pennies on the cost side—necessary, perhaps, but not very exciting. Attention to detail may be a virtue in Europe or Japan, where resource limits have long been a fact of life; it is decidedly dull in the land of the cowboy.

A third cultural force permeating the American identity is an underlying faith in the *scientific method.* From the period of the Enlightenment, which in America took the form of the popular science of Franklin and then the pragmatic inventions of Whitney, Bell, Eastman, Edison, and others, Americans have always embraced the rational, reductionist, analytical approach of science. The first uniquely American management system became known as **scientific management.**[4] The notion of "managing by the numbers" has deep roots in our cultural propensity for things scientific.

The *reductionist* method favored by scientists analyzes systems by breaking them down into their component parts and studying each one. This was a fundamental tenet of scientific management, which worked to improve overall efficiency by decomposing work into specific tasks and then improving the efficiency of each task. Today's industrial engineers and operations researchers still use this approach almost exclusively and are very much a product of the scientific management movement.

While reductionism can be an extremely profitable paradigm for analyzing complex systems—and certainly Western science has attained many triumphs via this approach—it is not the only valid perspective. Indeed, as has become obvious from the huge gap between academic research and actual practice in industry, too much emphasis on individual components can lead to a loss of perspective for the overall system.

In contrast to the reductionism of the West, Far Eastern societies seem to maintain a more **holistic** or **systems** perspective. In this approach, individual components are viewed much more in terms of their interactions with other subsystems and in the light of the overall goals of the system. This systems perspective undoubtedly influenced the development of just-in-time (JIT) systems in Japan, as we will discuss more thoroughly in Chapter 4.

The difference between the reductionist and holistic perspectives is starkly illustrated by the differing responses taken by the Americans and the Japanese to the problem of setups in manufacturing operations. Setup time is the time required for changeover of a machine from making one product to making another. In the American industrial engineering/operations research literature, for decades, setup times were regarded as constraints, leading to the development of all sorts of complex mathematical models for determining "optimal" lot sizes that would balance setup costs against inventory carrying costs. This view made perfect sense from a reductionist perspective, in which the setups were a given for the subsystem under consideration. In contrast, the Japanese, looking at manufacturing systems in the more holistic sense, recognized that setup times were not a given—they could be reduced. Moreover, from a systems perspective, there was clear value in reducing setup times. Clever use of jigs, fixtures, off-cycle preparations, and the like, which became known as *single minute exchange of die,* or SMED (Shingo 1985), enabled some Japanese factories to realize significantly shorter setup times than those

---

[4]This is in spite of the fact that its developer, Frederick W. Taylor, himself preferred the terms *task management* or the *Taylor system.*

in comparable American plants. In particular, the Japanese automobile industry became among the most productive in the world. These plants became simpler to manage and more flexible than their American counterparts.

Of course, the Japanese system had its weak points as well. Its convoluted pricing and distribution systems made Japanese electronic devices cheaper in New York than in Tokyo. Competition was tightly regulated by a traditional corporate network that kept out newcomers and led to bad investments. Strong profits of the 1980s were plowed into overvalued stocks and real estate. When the bubble burst in the 1990s, Japan found itself mired in an extended recession that precipitated the "Asian crisis" throughout the Pacific Rim. But Japanese workers in many industries remain productive, their investment rate is high, and personal debt is low. These sound economic basics make it very likely that Japan will continue to be a strong source of competition well into the 21st century.

1.3   The First Industrial Revolution

Prior to the first industrial revolution, production was small-scale, for limited markets, and labor- rather than capital-intensive. Work was carried out under two systems, the **domestic system** and **craft guilds.** In the domestic system, material was "put out" by merchants to homes where people performed the necessary operations. For instance, in the textile industry, different families spun, bleached, and dyed material, with merchants paying them on a piecework basis. In the craft guilds, work was passed from one shop to another. For example, leather was tanned by a tanner, passed to curriers, then passed to shoemakers and saddlers. The result was separate markets for the material at each step of the process.

The first industrial revolution began in England during the mid-18th century in the textile industry. This revolution, which dramatically changed manufacturing practices and the very course of human existence, was stimulated by several innovations that helped mechanize many of the traditional manual operations. Among the more prominent technological advances were the *flying shuttle* developed by John Kay in 1733, the *spinning jenny* invented by James Hargreaves in 1765 (Jenny was Mrs. Hargreaves), and the *water frame* developed by Richard Arkwright in 1769. By facilitating the substitution of capital for labor, these innovations generated economies of scale that made mass production in centralized locations attractive for the first time.

The single most important innovation of the first industrial revolution, however, was the steam engine, developed by James Watt in 1765 and first installed by John Wilkinson in his iron works in 1776. In 1781 Watt developed the technology for transforming the up-and-down motion of the drive beam to rotary motion. This made steam practical as a power source for a host of applications, including factories, ships, trains, and mines. Steam opened up far greater freedom of location and industrial organization by freeing manufacturers from their reliance on water power. It also provided cheaper power, which led to lower production costs, lower prices, and greatly expanded markets.

It has been said that Adam Smith and James Watt did more to change the world around them than anyone else in their period of history. Smith told us why the modern factory system, with its division of labor and "invisible hand" of capitalism, was desirable. Watt, with his engines (and the well-organized factory in which he, his partner Matthew Boulton and their sons built them), showed us how to do it. Many features of modern life, including widespread employment in large-scale factories, mass production of inexpensive goods, the rise of big business, the existence of a professional managerial class, and others, are direct consequences of their contributions.

1.3.1 The Industrial Revolution in America

England had a decided technological edge over America throughout the 18th century, and protected her competitive advantage by prohibiting export of models, plans, or people that could reveal the technologies upon which her industrial strength was based. It was not until the 1790s that a technologically advanced textile mill appeared in America—and that was the result of an early case of industrial espionage!

Boorstin (1965, 27) reports that Americans made numerous attempts to invent machinery like that in use in England during the later years of the 18th century, going so far as to organize state lotteries to raise prize money for enticing inventors. When these efforts failed repeatedly, Americans tried to import or copy English machines. Tench Coxe, a Philadelphian, managed to get a set of brass models made of Arkwright's machinery; but British customs officers discovered them on the dock and foiled his attempt. America finally succeeded in its efforts when Samuel Slater (1768–1835)—who had been apprenticed at the age of 14 to Jedediah Strutt, the partner of Richard Arkwright (1732–1792)—disguised himself as a farmer and left England secretly, without even telling his mother, to avoid the English law prohibiting departure of anyone with technical knowledge. Using the promise of a partnership, Moses Brown (for whom Brown University was named), who owned a small textile operation in Rhode Island with his son-in-law William Almy, enticed Slater to share his illegally transported technical knowledge. With Brown and Almy's capital and Slater's phenomenal memory, they built a cotton-spinning frame and in 1793 established the first modern textile mill in America at Pawtucket, Rhode Island.

The *Rhode Island system,* as the management system used by the Almy, Brown, and Slater partnership became known, closely resembled the British system on which it was founded. Focusing only on spinning fine yarn, Slater and his associates relied little on vertical integration and much on direct personal supervision of their operations. However, by the 1820s, the American textile industry would acquire a distinctly different character from that of the English by consolidating many previously disparate operations under a single roof. This was catalyzed by two factors.

First, America, unlike England, had no strong tradition of craft guilds. In England, distinct stages of production (e.g., spinning, weaving, dying, printing, in cotton textile manufacture) were carried out by different artisans who regarded themselves as engaged in distinct occupations. Specialized traders dealt in yarn, woven goods, and dyestuffs. These groups all had vested interests in not centralizing or simplifying production. In contrast, America relied primarily on the domestic system for textile production throughout its colonial period. Americans of this time either spun and wove for themselves or purchased imported woolens and cottons. Even in the latter half of the 18th century, a large proportion of American manufacturing was carried out by village artisans without guild affiliation. As a result, there were no organized constituencies to block the move toward integration of the manufacturing process.

Second, America, unlike England, still had large untapped sources of water power in the late 18th and early 19th centuries. Thus, the steam engine did not replace water power in America on a widespread basis until the Civil War. With large sources of water power, it was desirable to centralize manufacturing operations. This is precisely what Francis Cabot Lowell (1775–1817) did. After smuggling plans for a power loom out of Britain (Chandler 1977, 58), he and his associates built the famous cotton textile factories at Waltham and Lowell, Massachusetts, in 1814 and 1821. By using a single source of water power to drive all the steps necessary to manufacture cotton cloth, they established an early example of a modern integrated factory system. Ironically, because steam facilitated power generation in smaller units, its earlier introduction in England served to keep the production process smaller and more fragmented in England than in water-reliant America.

The result was that Americans, faced with a fundamentally different environment than that of the technologically and economically superior British firms, responded by innovating. These steps toward vertical integration in the early-19th-century textile industry were harbingers of a powerful trend that would ultimately make America the land of big business. The seeds of the enormous integrated mass production facilities that would become the norm in the 20th century were planted early in our history.

1.3.2 The American System of Manufacturing

Vertical integration was the first step in a distinctively American style of manufacturing. The second and more fundamental step was the production of interchangeable parts in the manufacture of complex multipart products. By the mid-19th century it was clear that the Americans were evolving an entirely new approach to manufacturing. The 1851 Crystal Palace Exhibition in London saw the first use of the term *American system of manufacturing* to describe the display of American products, such as the locks of Alfred Hobbs, the repeating pistol of Samuel Colt, and the mechanical reaper of Cyrus McCormick, all produced using the method of interchangeable parts.

The concept of interchangeable parts did not originate in America. The Arsenal of Venice was using some standard parts in the manufacture of warships as early as 1436. French gunsmith Honore LeBlanc had shown Thomas Jefferson musket components manufactured using interchangeable parts in 1785; but the French had abandoned his approach in favor of traditional craft methods (Mumford 1934, Singer 1958). It fell to two New Englanders, Eli Whitney (1765–1825) and Simeon North, to prove the feasibility of interchangeable parts as a sound industrial practice. At Jefferson's urging, Whitney was contracted to produce 10,000 muskets for the American government in 1801. Although it took him until 1809 to deliver the last musket, and he made only $2,500 on the job, he established beyond dispute the workability of what he called his "Uniformity System." North, a scythe manufacturer, confirmed the practicality of the concept and devised new methods for implementing it, through a series of contracts between 1799 and 1813 to produce pistols with interchangeable parts for the War Department. The inspiration of Jefferson and the ideas of Whitney and North were realized on a large scale for the first time at the Springfield Armory between 1815 and 1825, under the direction of Colonel Roswell Lee.

Prior to the innovation of interchangeable parts, the making of a complex machine was carried out in its entirety by an artisan, who fabricated and fitted each required piece. Under Whitney's uniformity system, the individual parts were mass-produced to tolerances tight enough to enable their use in any finished product. The division of labor called for by Adam Smith could now be carried out to an extent never before achievable, with individual workers producing single parts rather than completed products. The highly skilled artisan was no longer necessary.

It is difficult to overstate the importance of the idea of interchangeable parts, which Boorstein (1965) calls "the greatest skill-saving innovation in human history." Imagine producing personal computers under the skilled artisan system! The artisan would first have to fabricate a silicon wafer and then turn it into the needed chips. Then the printed-circuit boards would have to be produced, not to mention all the components that go into them. The disk drives, monitor, power supply, and so forth—all would have to be fabricated. Finally, all the components would be assembled in a handmade plastic case. Even if such a feat could be achieved, personal computers would cost millions of dollars and would hardly be "personal." Without exaggeration, our modern way of life depends on and evolved from the innovation of interchangeable parts. Undoubtedly, the Whitney and North contracts were among the most productive uses of federal funds to stimulate technological development in all of American history.

The American system of manufacturing, emphasizing mass production through use of vertical integration and interchangeable parts, started two important trends that impacted the nature of manufacturing management in this country to the present.

First, the concept of interchangeable parts greatly reduced the need for specialized skills on the part of workers. Whitney stated his aim as to "substitute correct and effective operations of machinery for that skill of the artist which is acquired only by long practice and experience, a species of skill which is not possessed in this country to any considerable extent" (Boorstein 1965, 33). Under the American system, workers without specialized skills could make complex products. An immediate result was a difference in worker wages between England and America. In the 1820s, unskilled laborers' wages in America were one-third or one-half higher than those in England, while highly-skilled workers in America were only slightly better paid than in England. Clearly, America placed a lower premium on specialized skills than other countries from a very early point in her history. Workers, like parts, were interchangeable. This early rise of the undifferentiated worker contributed to the rocky history of labor relations in America. It also paved the way for the sharp distinction between planning (by management) and execution (by workers) under the principles of scientific management in the early 20th century.

Second, by embedding specialization in machinery instead of people, the American system placed a greater premium on general intelligence than on specialized training. In England, unskilled meant unspecialized; but the American system broke down the distinction between skilled and unskilled. Moreover, machinery, techniques, and products were constantly changing, so that open-mindedness and versatility became more important than manual dexterity or task-specific knowledge. A liberal education was useful in the New World in a way that it had never been in the Old World, where an education was primarily a mark of refinement. This trend would greatly influence the American system of education. It also very likely prepared the way for the rise of the professional manager, who is assumed able to manage any operation without detailed knowledge of its specifics.

1.4   The Second Industrial Revolution

In spite of the notable advances in the textile industry by Slater in the 1790s and the practical demonstration of the uniformity system by Whitney, North, and Lee in the early 1800s, most industry in pre-1840 America was small, family-owned, and technologically primitive. Before the 1830s, coal was not widely available, so most industry relied on water power. Seasonal variations in the power supply, due to drought or ice, plus the lack of a reliable all-weather transportation network made full-time, year-round production impractical for many manufacturers. Workers were recruited seasonally from the local farm population, and goods were sold locally or through the traditional merchant network established to sell British goods in America. The class of permanent industrial workers was small, and the class of industrial managers almost nonexistent. Prior to 1840, there were almost no manufacturing enterprises sophisticated enough to require anything more than traditional methods of direct factory management by the owners.

Before the Civil War, large factories were the exception rather than the rule. In 1832, Secretary of the Treasury Louis McLane conducted a survey of manufacturing in

10 states and found only 36 enterprises with 250 or more workers, of which 31 were textile factories. The vast majority of enterprises had assets of only a few thousand dollars, had fewer than a dozen employees, and relied on water power (Chandler 1977, 60–61). The Springfield Armory, often cited as the most modern plant of its time—it used interchangeable parts, division of labor, cost accounting techniques, uniform standards, inspection/control procedures, and advanced metalworking methods—rarely had more than 250 employees.

The spread of the factory system was limited by the dependence on water power until the opening of the anthracite coal fields in eastern Pennsylvania in the 1830s. From 1840, anthracite-fueled blast furnaces began providing an inexpensive supply of pig iron for the first time. The availability of energy and raw material prompted a variety of industries (e.g., makers of watches, clocks, safes, locks, pistols) to build large factories using the method of interchangeable parts. In the late 1840s, newly invented technologies (e.g., sewing machines and reapers) also began production using the interchangeable-parts method.

However, even with the availability of coal, large-scale production facilities did not immediately arise. The modern integrated industrial enterprise was not the consequence of the technological and energy innovations of the first industrial revolution. The mass *production* characteristic of large-scale manufacturing required coordination of a mass *distribution* system to facilitate the flow of materials and goods through the economy. Thus, the second industrial revolution was catalyzed by innovations in transportation and communication—railroad, steamship, and telegraph—that occurred between 1850 and 1880. Breakthroughs in distribution technology in turn prompted a revolution in mass production technology in the 1880s and 1890s, including the Bonsack machine for cigarettes, the "automatic-line" canning process for foods, practical implementation of the Bessemer steel process and electrolytic aluminum refining, and many others. During this time, America visibly led the way in mass production and distribution innovations and, as a result, by World War II had more large-scale business enterprises than the rest of the world combined.

1.4.1   The Role of the Railroads

Railroads were the spark that ignited the second industrial revolution for three reasons:

1. They were America's first big business, and hence the first place where large-scale management hierarchies and modern accounting practices were needed.

2. Their construction (and that of the telegraph system at the same time) created a large market for mass-produced products, such as iron rails, wheels, and spikes, as well as basic commodities such as wood, glass, upholstery, and copper wire.

3. They connected the country, providing reliable all-weather transportation for factory goods and creating mass markets for products.

Colonel John Stevens received the first railroad charter in America from the New Jersey legislature in 1815 but, because of funding problems, did not build the 23-mile-long Camden and Amboy Railroad until 1830. In 1850 there were 9,000 miles of track extending as far as Ohio (Stover 1961, 29). By 1865 there were 35,085 miles of railroad in the United States, only 3,272 of which were west of the Mississippi. By 1890, the total had reached 199,876 miles, 72,473 of which were west of the Mississippi. Unlike in the Old World and in the eastern United States, where railroads connected established population centers, western railroads were generally built in sparsely populated areas, with lines running from "Nowhere-in-Particular to Nowhere-at-All" in the anticipation of development.

The capital required to build a railroad was far greater than that required to build a textile mill or metalworking enterprise. A single individual or small group of associates was rarely able to own a railroad. Moreover, because of the complexity and distributed nature of its operations, the many stockholders or their representatives could not directly manage a railroad. For the first time, a new class of salaried employees—middle managers—emerged in American business. Out of necessity the railroads became the birthplace of the first administrative hierarchies, in which managers managed other managers.

A pioneer of methods for managing the newly emerging structures was Daniel Craig McCallum (1815–1878). Working for the New York and Erie Railroad Company in the 1850s, he developed principles of management and a formal organization chart to convey lines of authority, communication, and division of labor (Chandler 1977, 101). Henry Varnum Poor, editor of the *American Railroad Journal,* widely publicized McCallum's work in his writings and sold lithographs of his organization chart for $1 each. Although the Erie line was taken over by financiers with little concern for efficiency (i.e., the infamous Jay Gould and his associates), Poor's publicity efforts ensured that McCallum's ideas had a major impact on railroad management in America.

Because of their complexity and reliance on a hierarchy of managers, railroads required large amounts of data and new types of analysis. In response to this need, innovators like J. Edgar Thomson of the Pennsylvania Railroad and Albert Fink of the Louisville & Nashville invented many of the basic techniques of modern accounting during the 1850s and 1860s. Specific contributions included introduction of standardized ratios (e.g., the ratio between a railroad's operating revenues and its expenditures, called the *operating ratio*), capital accounting procedures (e.g., renewal accounting), and unit cost measures (e.g., cost per ton-mile). Again, Henry Varnum Poor publicized the new accounting techniques and they rapidly became standard industry practice.

In addition to being the first big businesses, the railroads, along with the telegraph, paved the way for future big businesses by creating a mass distribution network and thereby making mass markets possible. As the transportation and communication systems improved, commodity dealers, purchasing agricultural products from farmers and selling to processors and wholesalers, began to appear in the 1850s and 1860s. By the 1870s and 1880s, mass retailers, such as department stores and mail-order houses, followed suit.

1.4.2   Mass Retailers

The phenomenal growth of these mass retailers provided a need for further advances in the management of operations. For example, Sears and Roebuck's sales grew from $138,000 in 1891 to $37,789,000 in 1905 (Chandler 1977, 231). Otto Doering developed a system for handling the huge volume of orders at Sears in the early years of the 20th century, a system which used machinery to convey paperwork and transport items in the warehouse. But the key to his process was a complex and rigid scheduling system that gave departments a 15-minute window in which to deliver items for a particular order. Departments that failed to meet the schedule were fined 50 cents per item. Legend has it that Henry Ford visited and studied this state-of-the-art mail-order facility before building his first plant (Drucker 1954, 30).

The mass distribution systems of the retailers and mail-order houses also produced important contributions to the development of accounting practices. Because of their high volumes and low margins, these enterprises had to be extremely cost-conscious. Analogous to the use of operating ratios by the railroads, retailers used gross margins (sales receipts less cost of goods sold and operating expenses). But since retailers, like the railroads, were single-activity firms, they developed specific measures of process efficiency unique to their type of business. Whereas the railroads concentrated on cost per ton-mile, the retailers focused on inventory turns or "stockturn" (the ratio of annual sales to average on-hand inventory). Marshall Field was tracking inventory turns as early as 1870 (Johnson and Kaplan 1987, 41), and maintained an average of between five and six turns during the 1870s and 1880s (Chandler 1977, 223), numbers that equal or better the performance of some retail operations today.

It is important to understand the difference between the environment in which American retailers flourished and the environment prevalent in the Old World. In Europe and Japan, goods were sold to populations in established centers with strong word-of-mouth contacts. Under such conditions, advertising was largely a luxury. Americans, on the other hand, marketed their goods to a sparse and fluctuating population scattered across a vast continent. Advertising was the life blood of firms like Sears and Roebuck. Very early on, marketing was more important in the New World than in the Old. Later on, the role of marketing in manufacturing would be further reinforced when makers of new technologies (sewing machines, typewriters, agricultural equipment) found they could not count on wholesalers or other intermediaries to provide the specialized services necessary to sell their products, and formed their own sales organizations.

1.4.3 Andrew Carnegie and Scale

Following the lead of the railroads, other industries began the trend toward big business through horizontal and vertical integration. In horizontal integration, a firm bought up competitors in the same line of business (steel, oil, etc.). In vertical integration, firms subsumed their sources of raw material and users of the product. For instance, in the steel industry, vertical integration took place when the steel mill owners purchased mining and ore production facilities on the upstream end and rolling mills and fabrication facilities on the downstream end.

In many respects, modern factory management first appeared in the metal making and working industries. Prior to the 1850s, the American iron and steel industry was fragmented into separate companies that performed the smelting, rolling, forging, and fabrication operations. In the 1850s and 1860s, in response to the tremendous growth of railroads, several large integrated rail mills appeared in which blast furnaces and shaping mills were contained in a single works. Nevertheless, in 1868, America was still a minor player in steel, producing only 8,500 tons compared with Britain's production of 110,000 tons.

In 1872, Andrew Carnegie (1835–1919) turned his hand to the steel industry. Carnegie had worked for J. Edgar Thompson on the Pennsylvania Railroad, rising from telegraph operator to division superintendent, and had a sound appreciation for the accounting and management methods of the railroad industry. He combined the new Bessemer process for making steel with the management methods of McCallum and Thompson, and he brought the industry to previously unimagined levels of integration and efficiency. Carnegie expressed his respect for his railroad mentors by naming his first integrated steel operation the Edgar Thompson Works. The goal of the E. T. Works was "a large and regular output," accomplished through the use of the largest and most technologically advanced blast furnaces in the world. More importantly, the E. T. Works took full advantage of integration by maintaining a continuous work flow—it was the first steel mill whose layout was dictated by material flow. By relentlessly exploiting his scale advantages and increasing velocity of throughput, Carnegie quickly became the most efficient steel producer in the world.

Carnegie further increased the scale of his operations by integrating vertically into iron and coal mines and other steel-related operations to improve flow even more. The effect was dramatic. By 1879, American steel production nearly equaled that of Britain. And by 1902, America produced 9,138,000 tons, compared with 1,826,000 for Britain.

Carnegie also put the cost accounting skills acquired from his railroad experience to good use. A stickler for accurate costing—one of his favorite dictums was, "Watch the costs and the profits will take care of themselves"—he instituted a strict accounting system. By doggedly focusing on unit cost, he became the low-cost producer of steel and was able to undercut competitors who had a less precise grasp of their costs. He used this information to his advantage, raising prices along with his competition during periods of prosperity and relentlessly cutting prices during recessions.

In addition to graphically illustrating the benefits from scale economies and high throughput, Carnegie's was a classic story of an entrepreneur who made use of minute data and prudent attention to operating details to gain a significant strategic advantage in the marketplace. He focused solely on steel and knew his business thoroughly, saying

> I believe the true road to preeminent success in any line is to make yourself master in that line. I have no faith in the policy of scattering one's resources, and in my experience I have rarely if ever met a man who achieved preeminence in money-making—certainly never one in manufacturing—who was interested in many concerns. The men who have succeeded are men who have chosen one line and stuck to it. (Carnegie 1920, 177)

Aside from representing one of the largest fortunes the world had known, Carnegie's success had substantial social benefit. When Carnegie started in the steel business in the 1870s, iron rails cost $100 per ton; by the late 1890s they sold for $12 per ton (Chandler 1984, 485).

1.4.4   Henry Ford and Speed

By the beginning of the 20th century, integration, vertical and horizontal, had already made America the land of big business. High-volume production was commonplace in process industries such as steel, aluminum, oil, chemicals, food, and tobacco. Mass production of mechanical products such as sewing machines, typewriters, reapers, and industrial machinery, based on new methods for fabricating and assembling interchangeable metal parts, was in full swing. However, it remained for Henry Ford (1863–1947) to make high-speed mass production of complex mechanical products possible with his famous innovation, the moving assembly line.

Like Carnegie, Ford recognized the importance of throughput velocity. In an effort to speed production, Ford abandoned the practice of skilled workers assembling substantial subassemblies and workers gathering around a static chassis to complete assembly. Instead, he sought to bring the product to the worker in a nonstop, continuous stream. Much has been made of the use of the moving assembly line, first used at Ford's Highland Park plant in 1913. However, as Ford noted, the principle was more important than the technology:

> The thing is to keep everything in motion and take the work to the man and not the man to the work. That is the real principle of our production, and conveyors are only one of many means to an end. (Ford 1926, 103)

After Ford, mass production became almost synonymous with assembly-line production.

Ford had signaled his strategy to provide cheap, reliable transportation early on with the Model N, introduced in 1906 for $600. This price made it competitive with much less sophisticated motorized buggies and far less expensive than other four-cylinder automobiles, all of which cost more than $1,000. In 1908, Ford followed with the legendary Model T touring car, originally priced at $850. By focusing on continual improvement of a single model and pushing his mass production techniques to new limits at his Highland Park plant, Ford reduced labor time to produce the Model T from 12.5 to 1.5 hours, and he brought prices down to $360 by 1916 and $290 by the 1920s. Ford sold 730,041 Model T's in fiscal year 1916/17, roughly one-third of the American automobile market. By the early 1920s, Ford Motor Company commanded two-thirds of the American automobile market.

Henry Ford also made his share of mistakes. He stubbornly held to the belief in a perfectible product and never appreciated the need for constant attention to the process of bringing new products to market. His famous statement that "the customer can have any color car as long as it's black" equated mass production with product uniformity. He failed to see the potential for producing a variety of end products from a common set of standardized parts. Moreover, his management style was that of a dictatorial owner. He never learned to trust his managerial hierarchy to make decisions of importance. Peter Drucker (1954) points to Henry's desire to "manage without managers" as the fundamental cause of Ford's precipitous decline in market share (from more than 60 percent down to 20 percent) between the early 1920s and World War II.

But Henry Ford's spectacular successes were not merely a result of luck or timing. The one insight he had that drove him to new and innovative manufacturing methods was his appreciation of the strategic importance of speed. Ford knew that high throughput and low inventories would enable him to keep his costs low enough to maintain an edge on his competition and to price his product so as to be available to a large segment of the public. It was his focus on speed that motivated his moving assembly line. But his concern for speed extended far beyond the production line. In 1926, he claimed, "Our finished inventory is all in transit. So is most of our raw material inventory." He boasted that his company could take ore from a mine and produce an automobile in 81 hours. Even allowing for storage of iron ore in winter and other inventory stocking, he claimed an average cycle time of not more than five days. Given this, it is little wonder that Taiichi Ohno, the originator of just-in-time systems, of whom we will have more to say in Chapter 4, was an unabashed admirer of Ford.

The insight that speed is critical, to both cost and throughput, was not in itself responsible for Ford's success. Rather, it was his attention to the details of implementing this insight that set him apart from the competition. The moving assembly line was just one technological innovation that helped him achieve his goal of unimpeded flow of materials through the entire system. He used many of the methods of the newly emerging discipline of scientific management (although Ford had evidently never heard of its founder, Frederick Taylor) to break down and refine the individual tasks in the assembly process. His 1926 book is filled with detailed stories of technical innovations—in glass making, linen manufacture, synthetic steering wheels, artificial leather, heat treating of steel, spindle screwdrivers, casting bronze bushings, automatic lathes, broaching machines, making of springs—that evidence his attention to details and appreciation of their importance. For all his shortcomings and idiosyncrasies, Henry Ford knew his business and used his intimacy with small issues to make a big imprint on the history of manufacturing in America.

1.5   Scientific Management

Although management has been practiced since ancient times (Peter Drucker credits the Egyptians who built the pyramids with being the greatest managers of all time), management *as a discipline* dates back to the late 19th century. Important as they were, the practical experiences and rules of thumb offered by such visionaries as Machiavelli did not make management a field because they did not result from a systematized method of critical scrutiny. Only when managers began to observe their practices in the light of the rational, deductive approach of scientific inquiry could management be termed a discipline and gain some of the respectability accorded to other disciplines using the scientific method, such as medicine and engineering. Not surprisingly, the first proponents of a scientific approach to management were engineers. By seeking to introduce a management focus into the professional fabric of engineering, they sought to give it some of engineering's effectiveness and respectability.

Scientific observation of work goes back at least as far as Leonardo da Vinci, who measured the amount of earth a man could shovel more than 450 years ago (Consiglio 1969). However, as long as manufacturing was carried out in small facilities amenable to direct supervision, there was little incentive to develop systematic work management procedures. It was the rise of the large integrated business enterprise in the late 19th and early 20th centuries that caused manufacturing to become so complex as to demand more sophisticated control techniques. Since the United States led the drive toward increased manufacturing scale, it was inevitable that it would also lead the accompanying managerial revolution.

Still, before American management writers developed their ideas in response to the second industrial revolution, a few British writers had anticipated the systematizing of management in response to the first industrial revolution. One such visionary was Charles Babbage (1792–1871). A British eccentric of incredibly wide-ranging interests, he demonstrated the first mechanical calculator, which he called a "difference machine," complete with a punch card input system and external memory storage, in 1822. He turned his attention to factory management in his 1832 book *On the Economy of Machinery and Manufactures,* in which he elaborated on Adam Smith's principle of division of labor and described how various tasks in a factory could be divided among different types of workers. Using a pin factory as an example, he described the detailed tasks required in pin manufacture and measured the times and resources required for each. He suggested a profit-sharing scheme in which workers derive a share of their wages in proportion to factory profits. Novel as his ideas were, though, Babbage was a writer, not a practitioner. He measured work rates for descriptive purposes only; he never sought to improve efficiency. He never developed his computer to commercial reality, and his management ideas were never implemented.

The earliest American writings on the problem of factory management appear to be a series of letters to the editor of the *American Machinist* by James Waring See, writing under the name of "Chordal," beginning in 1877 and published in book form in 1880 (Muhs, Wrege, Murtuza 1981). See advocated high wages to attract quality workers, standardization of tools, good "housekeeping" practices in the shop, well-defined job descriptions, and clear lines of authority. But perhaps because his book *(Extracts from Chordal's Letters)* did not sound like a book on business or because he did not interact with other pioneers in the area, See was not widely recognized or cited in future work on management as a formal discipline.

The notion that management could be made into a profession began to surface during the period when engineering became recognized as a profession. The American Society of Civil Engineers was formed in 1852, the American Institute of Mining Engineers in 1871, and, most importantly for the future of management, the American Society of Mechanical Engineers (ASME) in 1880. ASME quickly became the forum for debate of issues related to factory operation and management. In 1886, Henry Towne (1844–1924), engineer, cofounder of Yale Lock Company, and president of Yale and Towne Manufacturing Company, presented a paper entitled "The Engineer as an Economist"

(Towne 1886). In it, he held that "the matter of shop management is of equal importance with that of engineering … and the *management of works* has become a matter of such great and far-reaching importance as perhaps to justify its classification also as one of the modern arts." Towne also called for ASME to create an "Economic Section" to provide a "medium for the interchange" of experiences related to shop management. Although ASME did not form a Management Division until 1920, Towne and others kept shop management issues in prominence at society meetings.

1.5.1   Frederick W. Taylor

It is easy in hindsight to give credit to many individuals for seeking to rationalize the practice of management. But until Frederick W. Taylor (1856–1915), no one generated the sustained interest, active following, and systematic framework necessary to plausibly proclaim management as a discipline. It was Taylor who persistently and vocally called for the use of science in management. It was Taylor who presented his ideas as a coherent system in both his publications and his many oral presentations. It was Taylor who, with the help of his associates, implemented his system in many plants. And it is Taylor who lies buried under the epithet "father of scientific management."

Although he came from a well-to-do family, had attended the prestigious Exeter Academy, and had been admitted to Harvard, Taylor chose instead to apprentice as a machinist; and he rose rapidly from laborer to chief engineer at Midvale Steel Company between 1878 and 1884. An engineer to the core, he earned a degree in mechanical engineering from Stevens Institute on a correspondence basis while working full-time. He developed several inventions for which he received patents. The most important of these, high-speed steel (which enables a cutting tool to remain hard at red heat), would have been sufficient to guarantee him a place in history even without his involvement in scientific management.

But Taylor's engineering accomplishments pale in comparison to his contributions to management. Drucker (1954) wrote that Taylor's system "may well be the most powerful as well as the most lasting contribution America has made to Western thought since the Federalist Papers." Lenin, hardly a fan of American business, was an ardent admirer of Taylor. In addition to being known as the father of scientific management, he is claimed as the "father of industrial engineering" (Emerson and Naehring 1988).

But what were Taylor's ideas that accord him such a lofty position in the history of management? On the surface, Taylor was an almost fanatic champion of efficiency. Boorstein (1973, 363) calls him the "Apostle of the American Gospel of Efficiency." The core of his management system consisted of breaking down the production process into its component parts and improving the efficiency of each. In essence, Taylor was trying to do for work units what Whitney had done for material units: standardize them and make them interchangeable. Work standards, which he applied to activities ranging from shoveling coal to precision machining, represented the work rate that should be attainable by a "first-class man."

But Taylor did more than merely measure and compare the rates at which men worked. What made Taylor's work scientific was his relentless search for the best way to do tasks. Rules of thumb, tradition, standard practices were anathema to him. Manual tasks were honed to maximum efficiency by examining each component separately and eliminating all false, slow, and useless movements. Mechanical work was accelerated through the use of jigs, fixtures, and other devices, many invented by Taylor himself. The "standard" was the rate at which a "first-class" man could work *using the "best" procedure.*

With a faith in the scientific method that was singularly American, Taylor sought the same level of predictability and precision for manual tasks that he achieved with the "feed and speed" formulas he developed for metal cutting. The following formula for the time required to haul material with a wheelbarrow B is typical (Taylor 1903, 1431):

$$B = \left\{ p + [a + 0.51 + (0.0048)\text{distance hauled}]\frac{27}{L} \right\} 1.27$$

Here p represents the time loosening one cubic yard with the pick, a represents the time filling a barrow with any material, L represents the load of a barrow in cubic feet, and all times are in minutes and distances in feet.

Although Taylor was never able to extend his "science of shoveling" (as his opponents derisively termed his work) into a broader theory of work, it was not for lack of trying. He hired an associate, Sanford Thompson, to conduct extensive work measurement experiments. While he was never able to reduce broad categories of work to formulas, Taylor remained confident that this was possible:

> After a few years, say three, four or five years more, someone will be ready to publish the first book giving the laws of the movements of men in the machine shop—all the laws, not only a few of them. Let me predict, just as sure as the sun shines, that is going to come in every trade.[5]

Once the standard for a particular task had been scientifically established, it remained to motivate the workers to achieve it. Taylor advocated all three basic categories of worker motivation:

1. *The "carrot."* Taylor proposed a "differential piece rate" system, in which workers would be paid a low rate for the first increment of work and a substantially higher rate for the next increment. The idea was to give a significant reward to workers who met the standard relative to those who did not.

2. *The "stick."* Although he tried fining workers for failure to achieve the standard, Taylor ultimately rejected this approach. A worker who is unable to meet the standard should be reassigned to a task to which he is more suited and a worker who refuses to meet the standard ("a bird that can sing and won't sing") should be discharged.

3. *Factory ethos.* Taylor felt that a *mental revolution,* in which management and labor recognize their common purpose, was necessary in order for scientific management to work. For the workers this meant leaving the design of their work to management and realizing that they would share in the rewards of efficiency gains via the piece rate system. The result, he felt, would be that both productivity and wages would rise, workers would be happy, and there would be no need for labor unions. Unfortunately, when piecework systems resulted in wages that were considered too high, it was a common practice for employers to reduce the rate or increase the standard.

Beyond time studies and incentive systems, Taylor's engineering outlook led him to the conclusion that management authority should emanate from expertise rather than power. In sharp contrast to the militaristic unity-of-command character of traditional management, Taylor proposed a system of "functional foremanship" in which the traditional single foreman is replaced by eight different supervisors, each with responsibility for specific functions. These included the *inspector,* responsible for quality of work; the *gang boss,* responsible for machine setup and motion efficiency; the *speed boss,* responsible for machine speeds and tool choices; the *repair boss,* responsible for machine

---

[5] Abstract of an address given by Taylor before the Cleveland Advertising Club, March 3, 1915, and repeated the next day. It was his last public appearance. Reprinted in Shafritz and Ott 1990, 69–80.

maintenance and repair; the *order of work* or *route clerk,* responsible for routing and scheduling work; the *instruction card foreman,* responsible for overseeing the process of instructing bosses and workers in the details of their work; the *time and cost clerk,* responsible for sending instruction cards to the men and seeing that they record time and cost of their work; and the *shop disciplinarian,* who takes care of discipline in the case of "insubordination or impudence, repeated failure to do their duty, lateness or unexcused absence."

Finally, to complete his management system, Taylor recognized that he required an accounting system. Lacking personal expertise in financial matters, he borrowed and adapted a bookkeeping system from Manufacturing Investment Company, while working there as general manager from 1890 to 1893. This system was developed by William D. Basley, who had worked as the accountant for the New York and Northern Railroad, but was transferred to the Manufacturing Investment Company, also owned by the owners of the railroad, in 1892. Taylor, like Carnegie before him, successfully applied railroad accounting methods to manufacturing.

To Taylor, scientific management was not simply time and motion study, a wage incentive system, an organizational strategy, and an accounting system. It was a philosophy, which he distilled to four principles. Although worded in various ways in his writings, these are concisely stated as (Taylor 1911, 130)

1. The development of a true science.
2. The scientific selection of the worker.
3. His scientific education and development.
4. Intimate friendly cooperation between management and the men.

The first principle, by which Taylor meant that it was the managers' job to pursue a scientific basis for running their business, was the foundation of scientific management. The second and third principles paved the way for the activities of personnel and industrial engineering departments for years to come. However, in Taylor's time there was considerably more science in the writing about selection and education of workers than there was in practice. The fourth principle was Taylor's justification for his belief that trade unions were not necessary. Because increased efficiency would lead to greater surplus, which would be shared by management and labor (an assumption that organized labor did not accept), workers should welcome the new system and work in concert with management to achieve its potential. Taylor felt that workers would cooperate if offered higher pay for greater efficiency, and he actively opposed the rate-cutting practices by which companies would redefine work standards if the resulting pay rates were too high. But he had little sympathy for the reluctance of workers to be subjected to stopwatch studies or to give up their familiar practices in favor of new ones. As a result, Taylor never enjoyed good relations with labor.

1.5.2   Planning versus Doing

What Taylor meant in his fourth principle by "intimate friendly cooperation" was a clear separation of the jobs of management from those of the workers. Managers should do the planning—design the job, set the pace, rhythm, and motions—and workers should work. In Taylor's mind, this was simply a matter of matching each group to the work for which it was best qualified.

In concept, Taylor's views on this issue represented a fundamental observation: that planning and doing are distinct activities. Drucker describes this as one of Taylor's most valuable insights, "a greater contribution to America's industrial rise than stopwatch or time and motion study. On it rests the entire structure of modern management" (Drucker 1954, 284). Clearly Drucker's *management by objectives* would be meaningless without the realization that management will be easier and more productive if managers plan their activities before undertaking them.

But Taylor went further than distinguishing the activities of planning and doing. He placed them in entirely separate jobs. All planning activities rested with management. Even management was separated according to planning and doing. For instance, the gang boss had charge of all work up to the time that the piece was placed in the machine (planning), and the speed boss had charge of choosing the tools and overseeing the piece in the machine (doing). The workers were expected to carry out their tasks in the manner determined by management (scientifically, of course) as best. In essence, this is the military system; officers plan and take responsibility, enlisted men do the work but are not held responsible.[6] Taylor was adamant about assigning workers to tasks for which they were suited; evidently he did not feel they were suited to planning.

But, as Drucker (1954, 284) points out, planning and doing are actually two parts of the same job. Someone who plans without even a shred of doing "dreams rather than performs," and someone who works without any planning at all cannot accomplish even the most mechanical and repetitive task. Although it is clear that workers *do* plan in practice, the tradition of scientific management has clearly discouraged American workers from thinking creatively about their work and American managers from expecting them to. Juran (1992, 365) contends that the removal of responsibility for planning by workers had a negative effect on quality and resulted in reliance by American firms on inspection for quality assurance.

In contrast, the Japanese, with their quality circles, suggestion programs, and empowerment of workers to shut down lines when problems occur, have legitimized planning on the part of the workers. On the management side, the Japanese requirement that future managers and engineers begin their careers on the shop floor has also helped remove the barrier between planning and doing. "Quality at the source" programs are much more natural in this environment, so it is not surprising that the Japanese appreciated the ideas of quality prophets, such as Deming and Juran, long before the Americans did.

Taylor's error with regard to the separation of planning and doing lay in extending a valuable conceptual insight to an inappropriate practice. He made the same error by extending his reduction of work tasks to their simplest components from the planning stage to the execution stage. The fact that it is effective to analyze work broken down into its elemental motions does not necessarily imply that it is effective to carry it out in this way. Simplified tasks could improve productivity in the short term, but the benefits are less clear in the long term. The reason is that simple repetitive tasks do not make for satisfying work, and therefore, long-term motivation is difficult. Furthermore, by encouraging workers to concentrate on motions instead of on jobs, scientific management had the unintended result of making workers inflexible. As the pace of change in technology and the marketplace accelerated, this lack of flexibility became a clear competitive burden. The Japanese, with their holistic perspective and worker empowerment practices, have consciously encouraged their workforce to be more adaptable.

By making planning the explicit duty of management and by emphasizing the need for quantification, scientific management has played a large role in spawning and shaping

[6]Taylor's functional management represented a break with the traditional management notion of a single line of authority, which the proponents of scientific management called "military" or "driver" or "Marquis of Queensberry" management (see, e.g., L. Gilbreth 1914). However, he adhered to, even strengthened, the militaristic centralization of responsibility with management.

the fields of industrial engineering, operations research, and management science. The reductionist framework established by scientific management is behind the traditional emphasis by the industrial engineers on line balancing and machine utilization. It is also at the root of the decades-long fascination by operations researchers with simplistic scheduling problems, an obsession that produced 30 years of literature and virtually no applications (Dudek, Panwalker, and Smith 1992). The flaw in these approaches is not the analytic techniques themselves, but the lack of an objective that is consistent with the overall system objective. Taylorism spawned powerful tools but not a framework in which those tools could achieve their full potential.

1.5.3 Other Pioneers of Scientific Management

Taylor's position in history is in no small part due to the legions of followers he inspired. One of his earliest collaborators was Henry Gantt (1861–1919), who worked with Taylor at Midvale Steel, Simond's Rolling Machine, and Bethlehem Steel. Gantt is best remembered for the Gantt chart used in project management. But he was also an ardent efficiency advocate and a successful scientific management consultant. Although Gantt was considered by Taylor as one of his true disciples, Gantt disagreed with Taylor on several points. Most importantly, Gantt preferred a "task work with a bonus" system, in which workers were guaranteed their day's rate but received a bonus for completing a job within the set time, to Taylor's differential piece rate system. Gantt was also less sanguine than Taylor about the prospects for setting truly fair standards, and therefore he developed explicit procedures for enabling workers to protest or revise the standards.

Others in Taylor's immediate circle of followers were Carl Barth (1860–1939), Taylor's mathematician and developer of special-purpose slide rules for setting "feeds and speeds" for metal cutting; Morris Cook (1872–1960), who applied Taylor's ideas both in industry and as Director of Public Works in Philadelphia; and Horace Hathaway (1878–1944), who personally directed the installation of scientific management at Tabor Manufacturing Company and wrote extensively on scientific management in the technical literature.

Also adding energy to the movement and luster to Taylor's reputation were less orthodox proponents of scientific management, with some of whom Taylor quarreled bitterly. Most prominent among these were Harrington Emerson (1853–1931) and Frank Gilbreth (1868–1924). Emerson, who had become a champion of efficiency independently of Taylor and had reorganized the workshops of the Santa Fe Railroad, testified during the hearings of the Interstate Commerce Commission concerning a proposed railroad rate hike in 1910–1911 that scientific management could save "a million dollars a day." Because he was the only "efficiency engineer" with firsthand experience in the railroad industry, his statement carried enormous weight and served to emblazon scientific management on the national consciousness. Later in his career, Emerson became particularly interested in the selection and training of employees. He is also credited with originating the term *dispatching* in reference to shop floor control (Emerson 1913), a phrase which undoubtedly derives from his railroad experience.

Frank Gilbreth had a somewhat similar background to that of Taylor. Although he had passed the qualifying exams for MIT, Gilbreth became an apprentice bricklayer instead. Outraged at the inefficiency of bricklaying, in which a bricklayer had to lift his own body weight each time he bent over and picked up a brick, he invented a movable scaffold to maintain bricks at the proper level. Gilbreth was consumed by the quest for efficiency. He extended Taylor's time study to what he called *motion study,* in which he made detailed analyses of the motions involved in bricklaying in the search for a more efficient procedure. He was the first to apply the motion picture camera to the task of analyzing motions, and he categorized the elements of human motions into 18 basic components, or therbligs (Gilbreth spelled backward, sort of). That he was successful was evidenced by the fact that he rose to become one of the most prominent builders in the country. Although Taylor feuded with him concerning some of his work for nonbuilders, he gave Gilbreth's work on bricklaying extensive coverage in his 1911 book, *The Principles of Scientific Management*.

1.5.4   The Science in Scientific Management

Scientific management has been both venerated and vilified. It has generated both proponents and opponents who have made important contributions to our understanding and practice of management. One can argue that it is the root of a host of management-related fields, ranging from organization theory to operations research. But in the final analysis, it is the basic realization that management can be approached scientifically that is the primary contribution of scientific management. This is an insight we will never lose, an insight so basic that, like the concept of interchangeable parts, once it has been achieved it is difficult to picture life without it. Others intimated it; Taylor, by sheer perseverance, drove it into the consciousness of our culture. As a result, scientific management deserves to be classed as the first management *system*. It represents the starting point for all other systems. When Taylor began the search for a management system, he made it possible to envision management as a profession.

It is, however, ironic that scientific management's legacy is the application of the scientific method to management, because in retrospect we see that scientific management itself was far from scientific. Taylor's *Principles of Scientific Management* is a book of advocacy, not science. While Taylor argued for his own differential piece rate in theory, he actually used Gantt's more practical system at Bethlehem Steel. His famous story of Schmidt, a first-class man who excelled under the differential piece rate, has been accused of having so many inconsistencies that it must have been contrived (Wrege and Perroni 1947). Taylor's work measurement studies were often carelessly done, and there is no evidence that he used any scientific criteria to select workers. Despite using the word *scientific* with numbing frequency, Taylor subjected very few of his conjectures to anything like the scrutiny demanded by the scientific method.

Thus, while scientific management fostered quantification of management, it did little to place it in a real scientific framework. Still, to give Taylor his due, by sheer force of conviction, he tapped into the underlying American faith in science and changed our view of management forever. It remains for us to realize the full potential of this view.

1.6   The Rise of the Modern Manufacturing Organization

By the end of World War I, scientific management had firmly taken hold, and the main pieces of the American system of manufacturing were in place. Large-scale, vertically integrated organizations making use of mass production techniques were the norm. Although family control of large manufacturing enterprises was still common, salaried managers ran the day-to-day operations within centralized departmental hierarchies. These organizations had essentially fully exploited the potential economies of scale for producing a single product. Further organizational growth would require taking advantage of economies of *scope* (i.e., sharing production and distribution resources across multiple products). As a result, development of institutional structures and management procedures for controlling the resulting organizations was the main theme of American manufacturing history during the interwar period.

1.6.1 Du Pont, Sloan, and Structure

The classic story of growth through diversification is that of General Motors (GM). Formed in 1908 when William C. Durant (1861–1947) consolidated his own Buick Motor Company with the Cadillac, Oldsmobile, and Oakland companies, GM rapidly became an industrial giant. The flamboyant but erratic Durant was far more interested in acquisition than in organization, and he continued to buy up units (including Chevrolet Motor Company) to the point where, by 1920, GM was the fifth largest industrial enterprise in America. But it was an empire without structure. Lacking corporate offices, demand forecasting, and coordination of production, the corporation encountered financial difficulties whenever sales slowed. Du Pont Company came to Durant's aid more than once by investing heavily in GM and finally forced him out in 1920 (Bryant and Dethloff 1990).

Pierre Du Pont (1870–1954) came out of semiretirement to succeed Durant as president with the hope of making the Du Pont Company's GM investments profitable. A more capable successor could not possibly have been found. In 1902, he and his cousins Alfred and Coleman had purchased control of E. I. du Pont de Nemours & Company, a collection of single-function explosives manufacturers, and had consolidated it into a centrally governed, multidepartmental, integrated organization (Chandler and Salsbury 1971). Well aware of scientific management principles,[7] Du Pont and his associates installed Taylor's manufacturing control techniques and accounting system, and introduced psychological testing for personnel selection. Perhaps Du Pont's most influential innovation, however, was the refined use of return on investment (ROI) to evaluate the relative performance of departments. By 1917, Du Pont Powder Company stood as the first modern American manufacturing corporation.[8]

When he moved to General Motors, Du Pont quickly identified Alfred P. Sloan (1875–1966) as his main collaborator and set out to reorganize the company. Du Pont and Sloan agreed that GM's activities were too numerous, scattered, and varied to be amenable to the centralized organization in use at Du Pont Powder Company. With Du Pont's support, Sloan crafted a plan to structure the company as a collection of autonomous operating divisions coordinated (but not run) by a strong general office. The various divisions were carefully targeted at specific markets (e.g., Cadillac at the high-priced market, Chevrolet at the low end to compete directly with Ford, and Buick and Oldsmobile in the middle; Pontiac was introduced between Chevrolet and Oldsmobile in the mid-1920s) in accordance with Sloan's goal of "a car for every purse and purpose" (Cray 1979). Under Sloan's reorganization, GM's general office borrowed ROI methods from Du Pont Powder Company for evaluating units, and also developed sophisticated new procedures for demand forecasting, inventory tracking, and market share estimation.

---

[7] A. J. Moxham and Coleman du Pont had hired Frederick Taylor as a consultant at Steel Motor Company, and were instrumental in implementing Taylor's system when they later joined Du Pont as executives.

[8] The other candidate for the first modern manufacturing corporation would be General Electric, formed in 1892 by the merger of Edison General Electric and Thomson-Houston Electric, both of which were themselves products of mergers. To manage this first major consolidation of machinery-making companies, GE set up a modern structure of top and middle management patterned after that used by the railroads. However, its financial measures were not as sophisticated as those used by Du Pont and, unlike in the modern American corporation, a board of directors dominated by outside financiers held considerable veto power (Chandler 1977).

These techniques gradually became standard throughout American industry and are still used in modified form today.

Sloan's strategy was stunningly effective. In 1921, GM was a distant second with 12.3 percent of the automotive market to Ford's 55.7 percent. With its targeted product lines and regular introduction of new models, GM increased its share to 32.3 percent by 1929, while Ford, which waited until 1927 to replace the Model T with the Model A, fell to 31.3 percent. By 1940, Ford, which was still run by Henry, his son Edsel, and a tiny group of executives, was in serious trouble, having fallen to 18.9 percent and third place behind Chrysler's 23.7 percent share and far behind GM's 47.5 percent (Chandler 1990). Only a massive reorganization by Henry Ford II, beginning in 1945 and following the GM model, saved Ford from extinction.

In addition to forging hugely successful firms, Pierre Du Pont and Alfred Sloan shaped the American manufacturing corporation of the 20th century. While exhibiting many variations, all large industrial enterprises in the 20th century have used one of two basic structures. The centralized, functional department organization developed at Du Pont is used predominantly by firms with a single line of products in a single market. The multidivisional, decentralized structure developed at GM is the rule for firms with several product lines or markets. The environment in which we practice manufacturing today owes its existence to the efforts of these two innovators and their many associates.

1.6.2   Hawthorne and the Human Element

As industrial organizations grew larger and more technologically complex, the role of the worker took on increased importance. Indeed, the primary goals of scientific management—motivating workers and matching workers to tasks—were essentially behavioral. However, Taylor, being the true engineer, seemed to believe that human beings could be optimized in the same sense as a metal-cutting machine. For example, he observed that because a worker "strains every nerve to secure victory for his side" in a baseball game (Taylor 1911, 13), he or she should be capable of similar exertion at work. Despite the fact that he was an accomplished athlete, Taylor did not show the slightest appreciation for the psychological difference between work and play. Similarly, while he could spend countless hours studying and educating workers in the science of shoveling, he had no patience for a worker's sentimental attachment to the shovel he had handled for years. Although his writings certainly indicate a concern for the workers, Taylor never managed to understand their points of view.

In spite of Taylor's personal blind spots, scientific management served to catalyze the behavioral approach to management by systematically raising questions on authority, motivation, and training. The earliest writers in the field of industrial psychology acknowledged their debt to scientific management and framed their discussions in terms consistent with Taylor's system.

The acknowledged father of industrial psychology was Hugo Munsterberg (1863–1916). Born and educated in Germany, Munsterberg came to America and established a famous psychology laboratory at Harvard, where he studied a wide range of psychological questions in education, crime, and philosophy as well as industry. In his 1913 book *Psychology and Industrial Efficiency,* he paid tribute to scientific management and directly addressed it in three parts entitled "The Best Possible Man" (i.e., worker selection), "The Best Possible Work" (i.e., training and working conditions), and "The Best Possible Effect" (i.e., achieving management goals). Munsterberg's groundbreaking work paved the way for a steady stream of industrial psychology textbooks and a psychological testing fad shortly after World War I.

Among the Americans who led the way in the application of psychology to industry was Walter Dill Scott (1869–1955), who studied worker selection and rating for promotion (Scott 1913). A series of articles he wrote in 1910 to 1911 for *System* magazine (now *Business Week*) under the title "The Psychology of Business" were highly influential in raising awareness of the field of psychology among managers. He later turned to psychological research in advertising, defined the proper role of the newly arising personnel management function, and served as president of Northwestern University.

Lillian Gilbreth (1878–1972) was an early and visible proponent of industrial psychology from inside the ranks of scientific management. Wife of scientific management pioneer Frank Gilbreth and matriarch of the brood made famous by the book *Cheaper by the Dozen* (Gilbreth and Carey 1949), Gilbreth was one of the pioneers of the scientific management movement. In addition to collaborating with her husband on his motion studies work and carrying on this work after his death, she became one of the first advocates of psychology in management with her book *The Psychology of Management* (1914), based on her doctoral thesis in psychology at Brown University. In this book she contrasted scientific management with traditional management along various dimensions, including individuality. Her premise was that because of its emphasis on scientific selection, training, and functional foremanship, scientific management offered ample opportunity for individual development, while traditional management stifled such development by concentrating power in a central figure. Although the details of her work in psychology read today like an apology for scientific management and have largely been forgotten, Lillian Gilbreth deserves a place in management history for her early call for the humanization of the management process.

Mary Parker Follett (1868–1933) belonged chronologically to the scientific management era, but her thinking on the sociology and psychology of work was far ahead of its time. Like Lillian Gilbreth, she found in Taylor's functional foremanship a sound basis for allocating authority:

> One person should not give orders to another person, but both should agree to take their orders from the situation …We have here, I think, one of the largest contributions of scientific management; it tends to depersonalize orders. (Follett 1942, 59)

However, Follett was repelled by the relegation of the worker to simply carrying out tasks given and designated by management. She held that "not consent but participation is the right basis for all social relations" (Follett 1942, 211). By "participation," Follett meant to include the workers' ideas as well as their labor. Her rationale was that the ideas are valuable in themselves, but more importantly, the very process of participation is essential to establishing a functional work environment. Although at times her ideas sound idealistic, the depth and range of her work are astonishing and many of her insights still apply today.

A major episode in the quest to understand the human side of manufacturing was the series of studies conducted at the Western Electric Hawthorne plant in Chicago between 1924 and 1932. The studies originally began with a simple question: How does workplace illumination affect worker productivity? Under sponsorship of the National Academy of Science, a team of researchers from Massachusetts Institute of Technology observed groups of coil-winding operators under different lighting levels. They observed that productivity relative to a control group went up as illumination was increased, as had been expected. Then, in another experiment, they observed that productivity also went up when illumination was *decreased,* even to the level of moonlight (Roethlisberger and Dickson 1939).

Unable to explain the results, the original team abandoned the illumination studies and began other tests—of the effects on productivity of rest periods, length of work week, incentive plans, free lunches, and supervisory styles. In most cases, the trend was for higher-than-normal output by the groups under study.

Various experts were brought in to study the puzzling Hawthorne data, most notably George Elton Mayo (1880–1949) from Harvard. Approaching the problem from the perspective of the "psychology of the total situation," he came to the conclusion that the results were primarily due to "a remarkable change of mental attitude in the group." In the legend that subsequently grew up around the Hawthorne studies, Mayo's interpretation was reduced to the simple explanation that productivity increased as a result of the attention received by the workers under study, and this was dubbed the *Hawthorne effect.* However, in his writings, Mayo (1933, 1945) was not satisfied with this simple explanation and modified his view beyond this initial insight, arguing that work is essentially a group activity and that workers strive for a sense of belonging, not simply financial gain, in their jobs. By emphasizing the need for listening and counseling by managers in order to improve worker collaboration, the industrial psychology movement shifted the emphasis of management from technical efficiency, the focus of Taylorism, to a richer, more complex, human relations orientation.

1.6.3   Management Education

In addition to fostering the human relations perspective, the rise of the modern integrated business enterprise solidified the position of the professional managerial class. Prior to 1920, the majority of large-scale businesses were run by owner-entrepreneurs such as Carnegie, Ford, and Du Pont. Growth and integration after World War I resulted in systems too large to be run by owners (although Henry Ford tried, with disastrous results). Consequently, more and more decision-making responsibility was given to managers, middle and upper, who were without significant holdings in the firm.

In the 19th and early 20th centuries, it was not uncommon for these professional managers to be drawn from the ranks of the skilled workers (e.g., machinists). But as the modern business enterprises matured, formal university training became increasingly necessary. Many managers of this era were educated in traditional engineering disciplines (e.g., mechanical, electrical, civil, chemical). Some, however, began to seek education directly related to management, in either business schools or industrial engineering programs, both of which were emerging in the wake of the scientific management movement at the turn of the century.

The first American undergraduate business program was established in 1881 at the University of Pennsylvania's Wharton School. This was followed by schools at Chicago and Berkeley in 1898, and at Dartmouth (with the first master's level program), New York University, and Wisconsin in 1900. By 1910 there were more than a dozen separately organized schools of business at American universities, although the programs were generally small and had curricula restricted to background (e.g., economics, law, foreign languages) with anecdotes about the best industrial practices. The leading program of the time, Harvard, was organized in large part by Arch Shaw who had previously lectured at Northwestern and, as head of a Chicago publishing house, had published *Library of Factory Management.* Shaw relied heavily on outside lecturers from the scientific management movement (e.g., Frederick Taylor, Harrington Emerson, Carl Barth, Morris Cooke) and was instrumental in introducing the case method, which became Harvard's trademark and would heavily influence business education across America (Chandler 1977).

Between 1914 and 1940, American business schools grew and diversified their curricula. During this period most of the state universities introduced business programs; among them were Ohio State (1916); Alabama, Minnesota, North Carolina (1919); Virginia (1920); Indiana (1921); Kansas and Michigan (1924) (Pierson 1959). As the number of programs grew, so did the number of degrees granted: from 1,576 BAs and 110 MBAs in 1920, to 18,549 BAs and 1,139 MBAs in 1940 (Gordon and Howell 1959). At the same time, the functional areas of a business education were being standardized; by the mid-1920s, more than half of the 34 schools belonging to the American Association of Collegiate Schools of Business required students to take courses in accounting, business law, finance, statistics, and marketing. Textbooks supporting this functional orientation also began to appear (e.g., Hodge and McKinsey 1921 in accounting, Lough 1920 and Bonneville 1925 in finance, and Cherington 1920 in marketing).

American engineering schools also responded to the need for management education by introducing industrial engineering (IE) programs. Like the early business schools, the first IE departments were heavily influenced by the scientific management movement. Hugo Diemer taught the first shop management course in the mechanical engineering department of the University of Kansas in 1901 to 1902 and later went on to found the first IE curriculum at Penn State in 1908. Other engineering schools followed, and by the end of World War II there were more than 25 IE curricula in American universities. After the war, growth of the IE field tracked that of the economy; by the 1980s the number of IE programs had reached about 100 (Emerson and Naehring 1988).

The tools of industrial engineering evolved as the field grew during the interwar period. In addition to the methods of time and motion study (Gilbreth 1911; Barnes 1937), techniques of cost engineering (Fish 1915; Grant 1930), quality control (Shewhart 1931; Grant and Leavenworth 1946), and production/inventory management (Spriegel and Lansburgh 1923; Mitchell 1931; Raymond 1931; Whitin 1953) were presented in textbook form and widely introduced into industrial engineering curricula. By the end of World War II, all the major components of the IE discipline were in place, with the exception of the quantitative tools of operations research, which did not appear in a major way until after the war.

1.7   Peak, Decline, and Resurgence of American Manufacturing

Although the modern American manufacturing enterprise had largely been formed by the 1920s, the depression of the 1930s and the war of the 1940s prevented the country from reaping the full benefits of its powerful manufacturing sector. Thus, it was not until the post–World War II period, in the 1950s and 1960s, that America enjoyed a golden era of manufacturing. This era shaped the attitudes of a generation of managers, heavily influenced business and engineering schools, and set the stage for the not-so-golden era of manufacturing in the 1980s and 1990s.

1.7.1   The Golden Era

American manufacturing went into World War II in an extremely strong position, having mastered the techniques of mass production and distribution and management of large-scale enterprises. It emerged from the war in a position of undisputed global dominance. In 1945 the American industrial plant was easily the strongest in the world. The American market was eight times the size of the next-largest market in the world, giving American firms a huge scale advantage. American per capita income was eight times that of Japan in the 1950s, providing a vast source of capital, despite the fact that savings rates were lower than those in other countries. The American primary and secondary education system was the finest in the world. And with the GI Bill added to the land grant college system, America outpaced the rest of the world in higher education as well. Labor productivity (measured as gross domestic product per worker-hour) was nearly double that of any European country, and fully three times that of Germany and seven times that of Japan (Maddison 1984). With its huge domestic market, ready capital, and well-trained, productive workforce, America could produce and distribute goods at a pace and scale unthinkable to anyone else.

In contrast, the rest of the world lay virtually in ruins. The industrial plant in Europe and Japan had been physically devastated by the war. The scientific establishments of many countries were in disarray as America inherited some of their best brains. Furthermore, at the war's end, because transportation was expensive and trade policies protectionist, economies were far less global than they are today. Because the primary market for almost everything was in America, other countries would have been at a huge disadvantage even without their inferior physical plants and disrupted R&D base.

The resulting postwar boom in American manufacturing was undoubtedly exhilarating and was certainly profitable. Americans saw per capita income (in constant 1958 dollars) rise from $1 in 1950 to $3 in 1970 (U.S. Department of Commerce 1972). In 1947, the 200 largest industrial firms in America were responsible for 30 percent of the world's value added in manufacturing and 47.2 percent of total corporate manufacturing assets. By 1963, they accounted for 41 percent of value added and 56.3 percent of assets. By 1969 the top 200 American industrials accounted for 60.9 percent of the world's manufacturing assets (Chandler 1977, 482). For a while the living was easy. But as many of the baby boom generation enjoyed "Leave It to Beaver" lives in suburbia, the competitive world that would be their inheritance was being shaped as America's former enemies and allies recovered from the war.

1.7.2   Accountants Count and Salesmen Sell

During the golden era following World War II, the principal opportunities for American manufacturing firms were plainly in the areas of marketing, to develop the huge potential markets for new products, and finance, to fuel growth. As we mentioned earlier, America already had a stronger history in advertising than the Old World. Moreover, as indicated by the reliance of Du Pont and GM on financial measures to coordinate their large-scale enterprises, American manufacturers were well acquainted with the tools of finance. The manufacturing function itself became of secondary importance. American dominance in manufacturing was so formidable that eminent economist John Kenneth Galbraith proclaimed the problem of production "solved" (Galbraith 1958).

But as the manufacturing boom of the 1950s and 1960s turned into the manufacturing bust of the 1970s and 1980s, it became plain that something was wrong. The simplest explanation is that since the details of manufacturing didn't matter during the golden era, American firms became lax. Because American goods were the envy of the world, firms could largely dictate the quality specifications of their products, and managers learned to take quality for granted. Because of the American technological advantage and the lack of competition, continual improvement was unnecessary to maintain market share, and managers learned to take the status quo for granted. When foreign firms, which could not afford to take anything for granted, recovered sufficiently to present a legitimate challenge, many American firms lacked the vigor to meet it.

While this simple explanation may be accurate for some firms or industries, it does not give the whole story. The influences of the golden era on the current condition of American manufacturing are subtle and complex. Besides promoting a deemphasis on manufacturing details, the emphasis on marketing and finance in the 1950s and 1960s profoundly influenced today's American manufacturing firms. Recognizing these areas as having the greatest career potential, more and more of the "best and brightest" chose careers in marketing and finance. These became the glamour functions, while manufacturing and operations were increasingly viewed as dead-end "career breakers." This led to the simultaneous rise of the marketing and finance outlooks as dominant perspectives in American manufacturing firms. We trace some of the consequences below.

The Marketing Outlook.   With top executives and rising stars increasingly preoccupied with selling, the organizations themselves took on more of the marketing outlook. While there is nothing intrinsically wrong with the marketing outlook for the marketing department, for the firm itself it can be an overly conservative perspective. The principal task of marketing is to analyze the introduction of new products. But the products that are most amenable to analysis tend to be imitative, rather than innovative.

A good case history that illustrates the pitfalls of the marketing outlook is that of IBM and the xerography process. In the late 1950s, Haloid Company (which had introduced the first commercial xerographic copier in 1949 and later changed its name to Xerox) offered IBM the opportunity to jointly develop the first practical office copier. IBM enlisted Arthur D. Little, a Boston management consulting firm, to conduct a market study on the potential for such a product. A. D. Little, basing its conclusions on consumption of carbon paper and assessments of which offices needed to make paper copies, estimated maximum demand to be no more than 5,000 machines, far less than necessary to justify the development costs (Kearns and Nadler 1992). IBM declined the offer, and Xerox went on to make so much money that royalties to Battelle Memorial Institute, the research laboratory where the process was developed, threatened its not-for-profit status.

The conclusion is that the marketing outlook will often not justify the high-risk, high-payoff ventures associated with truly innovative new products. The Xerox machine *created* a demand for paper copies that did not previously exist. While hard to analyze, revolutionary products such as this can be enormously profitable. An overreliance on marketing may have caused large American manufacturing firms to take on fewer of these ventures than they should have. As evidence of this, consider that the last major automotive innovation to appear first on an American car was the automatic transmission in the 1940s. Four-wheel drive, four-wheel steering, turbocharging, and antilock brakes were all introduced first by foreign automakers (Dertouzos, Lester, Solow 1989, 19).

The Finance Outlook.   As noted earlier, Du Pont pioneered the use of ROI as a measure of the effectiveness of capital in a large-scale enterprise shortly after the turn of the century. However, in the 1910s, Du Pont Powder Company was primarily owned and managed by the Du Pont family; so there was no question that it was to be managed for the long-term benefit of its owners. Pierre Du Pont would never have used short-term ROI to evaluate the performance of individual managers. By the 1950s and 1960s, high-level managers were no longer owners, and the pervasiveness of the finance outlook had extended short-term ROI in the form of quarterly reports to a measure of individual performance.

An overreliance on short-term ROI discouraged managers from pursuing high-risk or long-term ventures and thus further aggravated the tendency toward the conservatism promoted by the marketing outlook. Short-term ROI can be artificially inflated for a while, possibly many years, through reduction in the investment base by forgoing process upgrades, equipment maintenance, and replacement, and by purchasing less than state-of-the-art facilities. However, in the long run, such practices can put a firm at a distinct competitive disadvantage. For instance, Dertouzos, Lester, and Solow (1989, 57) cite statistics showing that the rate of business-sector capital investment as a percentage of net output in Japan and West Germany has significantly outpaced that of America since 1965, precisely the period over which these countries significantly narrowed the productivity gap between themselves and America.

Moreover, the finance outlook, which views manufacturing management as essentially analogous to portfolio management, implies that the way to minimize risk is to diversify. The portfolio manager diversifies investments by purchasing various types of securities. The manufacturing executive diversifies by acquiring businesses outside the firm's core activities. As the rest of the world recovered from the war and began to give American firms serious competition in the 1960s, manufacturing firms increasingly turned to the financial response of diversification, almost to the point of mania in the late 1960s. In 1965 there were 2,000 mergers and acquisitions in America; by 1969 the number had risen to more than 6,000. Moreover, of the assets acquired during the 1963–1972 merger wave, nearly three-fourths were for product diversification, and one-half of these were in unrelated products (Chandler 1977). The effect was a dramatic change in the character of America's large manufacturing firms. In 1949, 70 percent of the 500 largest American firms earned 95 percent of revenues from a single business. By 1969, 70 percent of the largest firms no longer had a dominant business (Davidson 1990).

Like the marketing outlook, the finance outlook is too restrictive a perspective for the entire firm. While managers of purely financial portfolios are certainly rational in their use of diversification to achieve stable returns, manufacturing firms that use the same strategy are neglecting an important difference between portfolio and manufacturing management: Manufacturing firms influence their destinies in a far more direct way than do investors. The profitability of a manufacturing business is a function of many things, including product design, product quality, process efficiency, customer service, and so forth. When a firm moves away from its core business, there is a danger that it will fail to perform on these key measures. This can more than offset any potential advantage from diversification and can even threaten the existence of the company.

Indeed, the preponderance of statistical evidence paints a negative picture of the effectiveness of the merger-and-acquisition strategy. A detailed survey by Ravenscraft and Scherer (1987) of mergers during the 1960s and early 1970s showed that, on average, profitability and efficiency of firms decline after they are acquired. Hayes and Wheelwright (1984, 13) cite further statistics from Fruhan (1979) and *Forbes* magazine showing that highly diversified conglomerates tend to underperform relative to firms with highly focused product markets. In the realm of popular culture, books like *Barbarians at the Gate* (Burrough 1990) and *Merchants of Debt* (Anders 1992) graphically illustrate how far pure unbridled greed can take the merger-and-acquisition process from any consideration of manufacturing effectiveness. Scherer and Ross (1990, 173), in a comprehensive survey of firm structure and economic performance, sum up the effectiveness of the merger-and-acquisition approach with this statement: "The picture that emerges is a pessimistic one: widespread failure, considerable mediocrity, and occasional successes."

1.7.3   The Professional Manager

The rapid growth following World War II profoundly shaped the manufacturing manager in two additional ways. First, strong demand for managers prompted an acceleration of the promotion process, under the "fast-track manager" system. Second, unable to nurture enough managers internally, industry increasingly looked to the universities to provide professional management training. Before the war, MBA-trained managers were still a rarity; only 1,139 master's degrees in business were granted in 1940 (Gordon and Howell 1959, 21). After the war, this tripled to 3,357 in 1948 and continued growing steadily, so that by the 1980s the MBA had become the standard credential for the business executive in America. This intensified emphasis led to changes in the character of both corporations and business schools.

The Fast-Track Manager.   As Hayes and Wheelwright (1984) point out, before the war, it was traditional for managers to spend considerable time—a decade or more—in a job before being moved up the managerial ladder. After the war, however, there were simply not enough qualified people to fill the expanding need for managers. To fill the gap, business organizations identified rising stars and put them on fast tracks to executive levels. These individuals did shorter rotations through lower-level assignments—two or three years—on their way to upper-level positions. As a result, top manufacturing managers who came of age in the 1960s and 1970s were likely to have substantially less depth of experience at the operating levels than their predecessors.

Worse yet, the concept of a fast-track manager, first introduced to fill a genuine postwar need, gradually became institutionalized. Once some "stars" had moved up the promotion ladder quickly, it became impossible to convince those who followed to return to the slower, traditional pace. A bright young manager who was not promoted quickly enough would look for opportunities elsewhere. Lifelong loyalty to a firm became a thing of the past in America, and it became commonplace for top managers in one industry to have come up from the ranks of an entirely different one.[9] American business schools preached the concept of the professional manager who could manage any firm regardless of the technological or customer details, and American industry practiced it.[10] The days of Carnegie and Ford, owner-entrepreneur-managers who knew the details of their businesses from the bottom up, were gone.

Academization of Business Schools.   As business schools expanded after the war to meet the demand for professional managers, their pedagogical approaches came under increasing scrutiny. In 1959, two influential studies of American business schools, commissioned by Ford Foundation (Gordon and Howell 1959) and Carnegie Corporation (Pierson 1959), were released. These studies criticized American universities for taking an overly vocational approach to business education and called for an increase in academic standards and a broadening of emphasis to promote general knowledge, based on the "fundamental disciplines" of the behavioral sciences, economics, and mathematics and statistics. The studies advocated an interesting mix of specialization (i.e., emphasis on more sophisticated analytical techniques[11]) and generalization (i.e., development of professional managers who are prepared to deal with virtually any management problem).

Having been on the fringe of academic respectability from their inception, the business schools took the studies' recommendations seriously. They hired faculty specialists in psychology, sociology, economics, mathematics, and statistics—many without any

---

[9]For example, John Scully came from Pepsi to head Apple Computer, and Archie McCardle came from Xerox to head International Harvester.

[10]For that matter, American government practiced it. When Secretary of the Treasury Donald Regan and White House Chief of Staff James Baker exchanged jobs during the Reagan administration, there was little mention of it in the press—except to note the different management styles of the two men.

[11]Presumably this had something to do with the fact that the studies were done in the era of Sputnik—a time of widespread faith in science.

business background whatever. They revised curricula to include more courses in these basic "theoretical" subjects and reduced courses aimed at training students for specific jobs. Operations research, which had burst onto the scene with some well-publicized military successes during World War II and was developing rapidly in the 1960s with the evolution of the digital computer, was quickly absorbed into operations management. The concept of the professional manager became the ruling paradigm in American business education.

This "modernizing" of the business schools did more than produce a generation of managers long on general theories and short on specific practical skills. It eroded the business schools' traditional, albeit small, role as repositories of the best of industry practice. With specialists in psychology and mathematics pursuing narrowly focused research in arcane academic journals, it is hardly surprising that when productivity growth declined in the late 1970s and early 1980s, industry did not look to the universities for help. Instead, it turned to Japanese examples (e.g., Schonberger 1982) and anecdotal surveys of industry practice by consultants (e.g., Peters and Waterman 1982). Thus, after being educated in the "scientific" tools of management, the MBA-trained professional managers of the 1980s and 1990s were wooed by an endless stream of quick fixes for their management woes. Fads based on buzzwords, such as theory Z, management by objectives, zero-based budgeting, decentralization, quality circles, restructuring, "excellence," management by walking around, matrix management, entrepreneuring, value chain analysis, one-minute managing, just-in-time, total quality management, time-based competition, business process reengineering, and many others, came and went with numbing regularity. While many of these "theories" contain valuable insights, the sheer number of them is evidence that the fix is not quick.

The ultimate irony occurred in the 1980s when, in a desperate attempt to win back the trust of students alienated by the almost total disconnect between classroom and boardroom, many operations management courses began to teach the buzzword fads themselves. In doing so, business schools gave up their role as arbiter of what works and what does not. Instead of being trendsetters, they became trend followers.

It is apparent that business schools and corporations have swung far apart since the Ford and Carnegie studies of 1959, with industry naively relying on glib buzzword approaches and academia leaning too far toward specialized research and imitative teaching. It is time for a reappraisal of both. Business schools need to recover their foundation in practice, in order to focus their tools on problems of real industry interest instead of on abstract intellectual challenge. Industry needs to recover its appreciation of the importance of the technical details of manufacturing and develop the capacity to systematically evaluate which management practices work, instead of lurching from one bandwagon to the next. By adjusting the attitudes of both academics and practitioners, we have the potential to apply the tools and technology developed in the decades since World War II to sustain manufacturing as a solid base of the American economy well into the 21st century.

1.7.4   Recovery and Globalization of Manufacturing

In many respects, the 1990s represented a resurgence of American manufacturing after the decline of the 1970s and 1980s. In 1997, manufacturing profits were at a 40-year high, and unemployment was at its lowest level in more than two decades. Annual productivity increases in manufacturing had returned to a healthy rate above three percent. Seven years of economic growth had spurred investment in physical plant, so that nonresidential equipment owned by business nearly doubled between 1987 and 1996 (*Business Week,* June 9, 1970, 70).

Good times for American manufacturers also extended beyond the domestic market. The Institute for Management Development in Lausanne, Switzerland, ranked America as the most globally competitive nation in the world every year during the period 1993 to 1997. A 1993 survey by the Center for the Study of American Business (CSAB) at Washington University in St. Louis of 48 manufacturing executives found that 90 percent considered their firms more competitive than they had been five years earlier (Chilton 1995). Large majorities of these executives also reported that quality and product development time had improved substantially over this same period.

While encouraging, the situation in the mid-1990s was far from a return to that of the mid-1960s. The American economy was strong but hardly dominant. World-class firms in struggling economies retained their potential to offer intense competition; for example, despite improved profitability of America's "big three," Toyota is still widely regarded as the premier automaker in the world (Taylor 1997). American manufacturers remained keenly aware of competition from around the globe. The CSAB survey reported that 75 percent of manufacturing executives strongly agreed (and an additional 10 percent somewhat agreed) that the competition they faced in 1993 was much stiffer than that 10 years earlier, and large majorities agreed that even more improvements in quality and product development times would be needed in the next five years in order to keep pace.

Furthermore, some statistics gave troublesome or ambiguous signs. For instance, trade deficits in the 1990s remained at or near record levels, although the deficit as a percentage of exports fell significantly. Also, labor productivity increases that were ascribed to downsizing of the manufacturing workforce in the 1980s and 1990s may have been partially a by-product of a workforce reclassification, from permanent employees to temporary employees and consultants. Finally, the productivity increases and economic recovery did not translate to a surge in real wages. From 1970 to 1985 productivity grew at a pace of 1.9 percent per year while real wages grew 0.87 percent per year. However, from 1985 to 1996 the growth in productivity was 2.5 percent while wage growth was only 0.26 percent per year. This may have been partly due to the increasing influence of Wall Street. The bull market of the 1990s (driven at least in part by baby-boomers seeking retirement investments) encouraged analysts to look more closely than ever at anticipated earnings. This in turn motivated management to continue making sharp cuts in the workforce (both labor and middle management), using temporary help, and instituting many other productivity improvements, all while keeping wages nearly constant. Clearly, the world of manufacturing has become a very different, and much more intensely competitive, place than it was during America's golden era.

1.8   The Future

America's manufacturing future cannot help but be influenced by its past. The practices and institutions used today have evolved over the past 200 years. The influences range from the ramifications of the myth of the frontier to our love affair with finance and marketing, and they will not evaporate overnight. An appreciation of what has gone before can at least make us conscious of what we are dealing with (a brief summary of manufacturing milestones is given in Table 1.1). But history shapes only the possibilities for the future, not the future itself. It is up to the next generation of manufacturing managers to evolve the American system of manufacturing to its next level.

What will this level be? Although no one can say for sure, it is our belief that the concept of the professional manager is bankrupt. In a world of intense global competition, simply setting appropriate general guidelines is not enough. Managers need detailed knowledge about their business, knowledge that must include *technical* details.

TABLE 1.1 Milestones in the History of Manufacturing

Date	Event
4000 B.C.	Egyptians coordinate large-scale projects to build pyramids.
1500	Leonardo da Vinci systematically studies shoveling.
1733	John Kay invents flying shuttle.
1765	James Hargreaves invents spinning jenny.
1765	James Watt invents steam engine.
1776	Adam Smith publishes *Wealth of Nations,* introducing the notions of division of labor and the invisible hand of capitalism.
1776	James Watt sells first steam engine.
1781	James Watt invents system for producing rotary motion from up-and-down stroke of steam engine.
1785	Honore LeBlanc shows Thomas Jefferson interchangeable musket parts.
1793	First modern textile mill in America established in Pawtucket, RI.
1801	Eli Whitney contracted by U.S. government to produce muskets, using system of interchangeable parts.
1814	Integrated textile facility established in Waltham, MA.
1832	Charles Babbage publishes *On the Economy of Machinery and Manufactures,* dealing with organization and costing procedures for factories.
1840	Opening of anthracite coal fields in eastern Pennsylvania provides first American source of inexpensive nonwater power.
1851	Crystal Palace Exhibition in London displays "American system of manufacturing."
1854	Daniel C. McCallum develops and implements earliest large-scale organization management system at New York and Erie Railroad.
1855	Henry Bessemer patents a process for refining iron into steel that was far better suited to mass production than earlier "puddling" processes.
1869	The first transcontinental railroad, the Union Pacific–Central Pacific, is completed.
1870	Marshall Field makes use of inventory turns as a measure of retail operation performance.
1875	Andrew Carnegie opens the Edgar Thompson Steel Works in Pittsburgh, the first integrated Bessemer rail mill built from scratch and for decades the largest steel works in the world.
1877	Arthur Wellington publishes *The Economic Theory of the Location of Railways,* the first book to present methods of capital budgeting.
1880	American Society of Mechanical Engineers (ASME) founded.
1886	Charles Hall of the United States and Paul Heroult in Europe simultaneously invent electrolytic method for reducing bauxite into aluminum.
1886	Henry Towne presents paper at ASME calling for an "Economic Section" devoted to shop management.
1910	Hugo Diemer publishes *Factory Organization and Administration,* the first industrial engineering textbook.
1911	F. W. Taylor publishes *The Principles of Scientific Management.*
1913	Henry Ford introduces first moving automotive assembly line in Highland Park, MI.
1913	Ford W. Harris publishes *How Many Parts to Make at Once.*
1914	Lillian Gilbreth publishes *The Psychology of Management.*
1915	John C. L. Fish publishes *Engineering Economics: First Principles,* the first text to present discounted cash flow methods.
1916	Henri Fayol publishes first overall theory of management as *Administration industrielle et générale* (not translated into English until 1929).
1920	Alfred P. Sloan reorganizes General Motors to consist of a general office and several autonomous divisions.
1924	Hawthorne studies begin at Western Electric plant in Chicago; they continue to 1932.
1931	Walter Shewhart publishes *Economic Control of Quality of Manufactured Product,* introducing the concept of the control chart.
1945	ENIAC (Electronic Numerical Integrator and Calculator), the first fully electronic digital computer, is built at the Univeristy of Pennsylvania.
1947	Herbert Simon publishes *Administrative Behavior,* marking a change in focus of organization theory from the structure of organizations to the process of decision making.
1953	Thomson Whitin publishes *The Theory of Inventory Management,* the first book to develop a theory to underlie the practice of inventory control.
1954	Peter Drucker publishes *The Practice of Management,* introducing the concept of Management by Objectives (MBO) on a wide scale.
1964	The IBM 360 becomes the first computer based on silicon chips.
1975	Joseph Orlicky publishes *Material Requirements Planning.*
1977	Introduction of the Apple II starts the personal computer revolution in earnest.
1978	Taichi Ohno publishes *Toyota seisan hoshiki* on the Toyota production system.

Unfortunately, the rise of such monolithic software packages as Enterprise Requirements Planning (the subject of Chapter 3) which purport to encapsulate "best practices" may prove to be a giant step backward in terms of managers better understanding their practices.

In the future, survival itself is likely to depend on understanding these details. The manufacturing function is no longer a necessary evil that can be taken for granted; it is a vital strategic function. In an era when products move from cutting-edge technology to commodities in the blink of an eye, inefficient manufacturing is likely to be fatal. The economic recovery of the 1990s and the fact that several universities have initiated programs in manufacturing management that stress the technical aspects and operating details of manufacturing are encouraging signs that we are adjusting to the new era.

But change will not come uniformly to all of American manufacturing. Some firms will adapt—indeed, have already adapted—to the new globally competitive world of manufacturing; others will resist change or will continue to seek some kind of technological quick fix. American firms will not rise or fall as a group. Firms that master the intricacies of manufacturing under the new world order will thrive. Those that cling to the methods evolved under the unique, and long-gone, conditions following World War II will not. Those that continue to increase profits by squeezing their employees to increase productivity without allowing real wages to rise will also fail (it appears that the General Motors strike in the summer of 1998 was a crack in the veneer of new American juggernaut).

To make the transition to the new era of manufacturing, it is crucial to remember the lessons of history. Consistently, the key to effective manufacturing has been not technology alone, but also the organization in which the technology was used. The only way for a manufacturing firm of the future to gain a significant strategic advantage over the long term will be to focus and coordinate its manufacturing operation, in conjunction with product and market development, with customer needs. The goal of this book is to provide the manufacturing manager with the intuition and tools needed to do just this.

Discussion Points

1. Before 1900, despite its weaknesses in effective management of workers, manufacturing leadership was well provided by top management. They were technological entrepreneurs, architects of productive systems, veritable lions of industry. But when they delegated their production responsibilities to a second-level department, the factory institution never recovered its vitality. The lion was tamed. Its management systems became protective and generally were neither entrepreneurial nor strategic. Production managers since then have typically had little to do with initiating substantially new process technology—in contrast to their predecessors before 1900. Wickham Skinner (1985)
   a. Do you agree with Skinner?
   b. What structural differences between manufacturing enterprises before 1890 and after 1920 contributed to this difference in managerial orientation?
   c. Why have manufacturing managers become increasingly seen as "custodians of financial assets"? (What were the impacts on the role of manufacturing as part of a business strategy?)
   d. How is Japan (or Germany) different from (or the same as) America with regard to this trend in manufacturing leadership?
   e. Taking the structural characteristics of manufacturing enterprises (e.g., scale, complexity, pace of technological change) as given, what can be done to revitalize manufacturing leadership?

2. America's industrial rise took place following a war with its principal rival (England); Japan's rise also took place following a war with its primary rival (America). America's success could be attributed to its system (i.e., interchangeable parts and vertical integration), while Japan's success could be attributed to its system (i.e., just-in-time).
   a. What other parallels can be drawn between the manufacturing stories of America and Japan?
   b. What are key differences?
   c. What relevance do these similarities and differences have to the manufacturing manager and policy maker of today?

Study Questions

1. What events characterized the first and second industrial revolutions? What effects did these changes have on the nature of manufacturing management?

2. List three key impacts of Frederick W. Taylor's scientific management on the practice of manufacturing management in America.

3. Proponents of a service economy for America sometimes compare the recent decline in manufacturing jobs to the earlier decline in agriculture jobs. In what way are these two declines different? How might this affect the argument that a shift to a service economy will not reduce our standard of living?

4. What are some signs of the decline of American manufacturing? How long has this been going on?

5. Give a counterargument for each of the following "usual answers" as to why American manufacturing is in decline:
   a. Growth of government regulation, taxes, etc.
   b. Deterioration in the American work ethic combined with adversary relationship between labor and management.
   c. Interruptions in supply and price increases in energy since first OPEC oil shock.
   d. Massive influx of new people into workforce—teenagers, women, and minority groups—who had to be conditioned and trained.
   e. Advent of unusually high capital costs caused by high inflation.
   If the real answer is none of the above, what else is left?

6. Name two post–World War II trends in management that have contributed to the decline of American manufacturing.

7. Why was it unimportant for a manager to be terribly concerned with production details in the 1950s and early 1960s? How did this affect the nature of American business schools during this period and their impact on management practices today?

8. Give some pros and cons of the portfolio management approach to managing a complex manufacturing enterprise.

9. What caused the need for the fast-track manager in the 1950s and 1960s? What potential impacts on the perspective of management might this practice have?

10. Compare a professional manager (i.e., a manager who is allegedly capable of managing any business) to a manager of a purely financial portfolio. List some strengths and weaknesses that such a person might bring to the manufacturing environment.

11. What attitudes does a modern professional manager in America share with the early settlers of this country? What negative consequences might this have?

12. Even in circumstances where it can be documented that innovative designs have had markedly better long-term performance, why do many managers pursue *imitative* designs?

13. It has been widely claimed that many of the troubles of American manufacturing can be traced to an overreliance on short-term financial measures. Name some policies, at both the government and firm levels, that might be used to discourage this type of mind-set.

14. What essential skill does a manufacturing manager need to be able to appreciate the big picture and still pay attention to important details without becoming completely overwhelmed?

15. In very rough terms, one could attribute the success of American manufacturing to effective competition on the cost dimension (i.e., via economies of scale due to mass production), the success of German manufacturing to effective competition on the quality dimension (i.e., via a reputation for superior product design and conformance with performance specifications), and the success of Japanese manufacturing to effective competition on the time dimension (i.e., via short manufacturing cycle times and rapid introduction of new products). Of course, each newly ascendant manufacturing power had to compete on the dimensions of its predecessors as well, so Germany had to be cost-competitive and Japan used cost and quality in addition to time. Thinking in terms of this simple model, that represents global competition as a succession of new competitive dimensions, give some suggestions for what might be the next important dimension of competition.

2 INVENTORY CONTROL: FROM EOQ TO ROP

When your pills get down to four
Order more.

Anonymous, from Hadley and Whitin (1963)

2.1 Introduction

Scientific management (SM) made the modern discipline of operations management (OM) possible. Not only did SM establish management as a discipline worthy of study, but also it placed a premium on quantitative precision that made mathematics a management tool for the first time. Taylor's primitive work formulas were the precursors to a host of mathematical models designed to assist decision making at all levels of plant design and control. These models became standard subjects in business and engineering curricula, and entire academic research disciplines sprang up around various OM problem areas, including inventory control, scheduling, capacity planning, forecasting, quality control, and equipment maintenance. The models, and the SM focus that motivated them, are now part of the standard language of business.

Of the operations management subdisciplines that spawned mathematical models, none was more central to factory management, nor more typical of the American approach to OM, than that of inventory control. In this chapter, we trace the history of the mathematical modeling approach to inventory control in America. Our reasons for doing this are as follows:

1. The inventory models we discuss are among the oldest results of the OM field and are still widely used and cited. As such, they are essential components of the language of manufacturing management.

2. Inventory plays a key role in the logistical behavior of virtually all manufacturing systems. The concepts introduced in these historical models will come back in our factory physics development in Part II and our discussion of inventory management in Chapter 17.

3. These classical inventory results are central to more modern techniques of manufacturing management, such as material requirements planning (MRP), just-in-time (JIT), and time-based competition (TBC), and are therefore important as a foundation for the remainder of Part I.

We begin with the oldest, and simplest, model—the economic order quantity (EOQ), and we work our way up to the more sophisticated reorder point (ROP) models. For each model we give a motivating example, a presentation of its development, and a discussion of its underlying insight.

2.2   The Economic Order Quantity Model

One of the earliest applications of mathematics to factory management was the work of Ford W. Harris (1913) on the problem of setting manufacturing lot sizes. Although the original paper was evidently incorrectly cited for many years (see Erlenkotter 1989, 1990), Harris's EOQ model has been widely studied and is a staple of virtually every introductory production and operations management textbook.

2.2.1   Motivation

Consider the situation of MedEquip, a small manufacturer of operating-room monitoring and diagnostic equipment, which produces a variety of final products by mounting electronic components in standard metal racks. The racks are purchased from a local metalworking shop, which must set up its equipment (presses, machining stations, and welding stations) each time it produces a "run" of racks. Because of the time wasted setting up the shop, the metalworking shop can produce (and sell) the racks more cheaply if MedEquip purchases them in quantities greater than one. However, because MedEquip does not want to tie up too much of its precious cash in stores of racks, it does not want to buy in excessive quantities.

This dilemma is precisely the one studied by Harris in his paper "How Many Parts to Make at Once." He puts it thus:

> Interest on capital tied up in wages, material and overhead sets a maximum limit to the quantity of parts which can be profitably manufactured at one time; "set-up" costs on the job fix the minimum. Experience has shown one manager a way to determine the economical size of lots. (Harris 1913)

The problem Harris had in mind was that of a factory producing various products and switching between products entails a costly setup. As an example, he described a metalworking shop that produced copper connectors. Each time the shop changed from one type of connector to another, machines had to be adjusted, clerical work had to be done, and material might be wasted (e.g., copper used up as test parts in the adjustment process). Harris defined the sum of the labor and material costs to ready the shop to produce a product to be the **setup cost.** (Notice that if the connectors had been purchased, instead of manufactured, then the problem would remain similar, but setup cost would correspond to the cost of placing a purchase order.)

The basic tradeoff is the same in the MedEquip example and Harris's copper connector case. Large lots reduce setup costs by requiring less frequent changeovers. But small lots reduce inventory by bringing in product closer to the time it is used. The EOQ model was Harris's systematic approach to striking a balance between these two concerns.

2.2.2   The Model

Despite his claim in the above quote that the EOQ is based on experience, Harris was consistent with the scientific management emphasis of his day on precise mathematical

FIGURE 2.1

Inventory versus time in the EOQ model

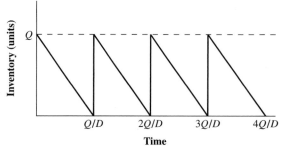

approaches to factory management. To derive a lot size formula, he made the following assumptions about the manufacturing system:[1]

1. *Production is instantaneous.* There is no capacity constraint, and the entire lot is produced simultaneously.
2. *Delivery is immediate.* There is no time lag between production and availability to satisfy demand.
3. *Demand is deterministic.* There is no uncertainty about the quantity or timing of demand.
4. *Demand is constant over time.* In fact, it can be represented as a straight line, so that if annual demand is 365 units, this translates to a daily demand of one unit.
5. *A production run incurs a fixed setup cost.* Regardless of the size of the lot or the status of the factory, the setup cost is the same.
6. *Products can be analyzed individually.* Either there is only a single product or there are no interactions (e.g., shared equipment) between products.

With these assumptions, we can use Harris's notation, with slight modifications for ease of presentation, to develop the EOQ model for computing optimal production lot sizes. The notation we will require is as follows:

D = demand rate (in units per year)

c = unit production cost, not counting setup or inventory costs (in dollars per unit)

A = fixed setup (ordering) cost to produce (purchase) a lot (in dollars)

h = holding cost (in dollars per unit per year); if the holding cost consists entirely of interest on money tied up in inventory, then $h = ic$, where i is the annual interest rate

Q = lot size (in units); this is the decision variable

For modeling purposes, Harris represented both time and product as continuous quantities. Since he assumed constant, deterministic demand, ordering Q units each time the inventory reaches zero results in an average inventory level of $Q/2$ (see Figure 2.1). The holding cost associated with this inventory is therefore $hQ/2$ per year. The setup cost is A per order, or AD/Q per year, since we must place D/Q orders per year to satisfy demand. The production cost is c per unit, or cD per year. Thus, the total (inventory,

---

[1] The reader should keep in mind that *all* models are based on simplifying assumptions of some sort. The real world is too complex to analyze directly. Good modeling assumptions are those that facilitate analysis while capturing the essence of the real problem. We will be explicit about the underlying assumptions of the models we discuss in order to allow the reader to personally gauge their reasonableness.

setup, and production) cost per year can be expressed as

$$Y(Q) = \frac{hQ}{2} + \frac{AD}{Q} + cD \tag{2.1}$$

Example:

To illustrate the nature of $Y(Q)$, let us return to the MedEquip example. Suppose that its demand for metal racks is fairly steady and predictable at $D = 1,000$ units per year. The unit cost of the racks is $c = \$250$, but the metalworking shop also charges a fixed cost of $A = \$500$ per order, to cover the cost of shutting down the shop to set up for a MedEquip run. MedEquip estimates its opportunity cost or hurdle rate for money at 10 percent per year. It also estimates that the floorspace required to store a rack costs roughly \$10 per year in annualized costs. Hence, the annual holding cost per rack is $h = (0.1)(250) + 10 = \$35$. Substituting these values into expression (2.1) yields the plots in Figure 2.2.

We can make the following observations about the cost function $Y(Q)$ from Figure 2.2:

1. The holding cost term hQ/D increases linearly in the lot size Q and eventually becomes the dominant component of total annual cost for large Q.

2. The setup cost term AD/Q diminishes quickly in Q, indicating that while increasing lot size initially generates substantial savings in setup cost, the returns from increased lot sizes decrease rapidly.

3. The unit-cost term cD does not affect the relative cost for different lot sizes, since it does not include a Q term.

4. The total annual cost $Y(Q)$ is minimized by some lot size Q. Interestingly, this minimum turns out to occur precisely at the value of Q for which the holding cost and setup cost are exactly balanced (i.e., the hQ/D and AD/Q cost curves cross).

FIGURE 2.2

Costs in the EOQ model

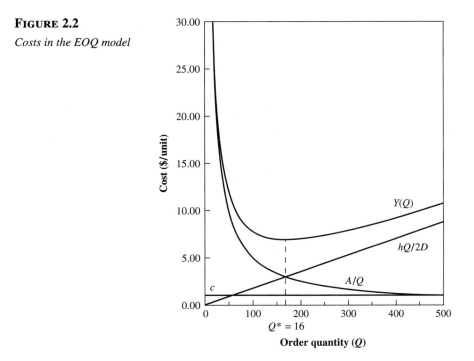

Harris wrote that finding the value of Q that minimizes $Y(Q)$ "involves higher mathematics" and simply gives the solution without further derivation. The mathematics he is referring to (calculus) does not seem quite as high today, so we will fill in some of the details he omitted in the following technical note. Those not interested in such details can skip this and subsequent technical notes without loss of continuity.

Technical Note

The standard approach for finding the minimum of an unconstrained function, such as $Y(Q)$, is to take its derivative with respect to Q, set it equal to zero, and solve the resulting equation for Q^*. This will find a point where the slope is zero (i.e., the function is flat). If the function is convex (as we will verify below), then the zero-slope point will be unique and will correspond to the minimum of $Y(Q)$.

Taking the derivative of $Y(Q)$ and setting the result equal to zero yields

$$\frac{dY(Q)}{dQ} = \frac{h}{2} - \frac{AD}{Q^2} = 0 \tag{2.2}$$

This equation represents the *first-order condition* for Q to be a minimum. The *second-order condition* makes sure that this zero-slope point corresponds to a minimum (i.e., as opposed to a maximum or a saddle point) by checking the second derivative of $Y(Q)$:

$$\frac{d^2Y(Q)}{dQ^2} = 2\frac{AD}{Q^3} \tag{2.3}$$

Since this second derivative is positive for any positive Q (that is, $Y(Q)$ is convex), it follows that solving (2.2) for Q^* (as we do in (2.4) below) does indeed minimize $Y(Q)$.

The lot size that minimizes $Y(Q)$ in cost function (2.1) is

$$Q^* = \sqrt{\frac{2AD}{h}} \tag{2.4}$$

This square root formula is the well-known **economic order quantity (EOQ),** also referred to as the **economic lot size.** Applying this formula to the example in Figure 2.2, we get

$$Q^* = \sqrt{\frac{2AD}{h}} = \sqrt{\frac{2(500)(1,000)}{35}} = 169$$

The intuition behind this result is that the large fixed cost ($500) associated with placing an order makes it attractive for MedEquip to order racks in fairly large batches (169).

2.2.3   The Key Insight of EOQ

The obvious implication of the above result is that the optimal order quantity increases with the square root of the setup cost or the demand rate and decreases with the square root of the holding cost. However, a more fundamental insight from Harris's work is the one he observed in his abstract, namely, the realization that

There is a tradeoff between lot size and inventory.

Increasing the lot size increases the average amount of inventory on hand, but reduces the frequency of ordering. By using a setup cost to penalize frequent replenishments, Harris articulated this tradeoff in clear economic terms.

The basic insight on the previous page is incontrovertible. However, the specific mathematical result (i.e., the EOQ square root formula) depends on the modeling assumptions, some of which we could certainly question (e.g., how realistic is instantaneous production?). Moreover, the usefulness of the EOQ formula for computational purposes depends on the realism of the input data. Although Harris claimed that "The set-up cost proper is generally understood" and "may, in a large factory, exceed *one dollar* per order," estimating setup costs may actually be a difficult task. As we will discuss in detail later in Parts II and III, setups in a manufacturing system have a variety of other impacts (e.g., on capacity, variability, and quality) and are therefore not easily reduced to a single invariant cost. In purchasing systems, however, where some of these other effects are not an issue and the setup cost can be cleanly interpreted as the cost of placing a purchase order, the EOQ model can be very useful.

It is worth noting that we can use the insight that there is a tradeoff between lot size and inventory without even resorting to Harris's square root formula. Since the average number of lots per year F is

$$F = \frac{D}{Q} \tag{2.5}$$

and the total inventory investment is

$$I = \frac{cQ}{2} = \frac{cD}{2F} \tag{2.6}$$

we can simply plot inventory investment I as a function of replenishment frequency F in lots per year. We do this for the MedEquip example with $D = 1,000$ and $c = \$250$ in Figure 2.3. Notice that this graph shows us that the inventory is cut in half (from \$12,500 to \$6,250) when we produce or order 20 times per year rather than 10 times per year (i.e., change the lot size from 100 to 50). However, if we replenish 30 times per year instead of 20 times per year (i.e., decrease the lot size from 50 to 33), inventory only falls from \$6,250 to \$4,125, a 34 percent decrease.

This analysis shows that there are decreasing returns to additional replenishments. If we can attach a value to these production runs or purchase orders (i.e., the setup cost A), then we can compute the optimal lot size using the EOQ formula as we did in Figure 2.2. However, if this cost is unknown, as it may well be, then the curve in Figure 2.3 at least gives us an idea of the impact on total inventory of an additional annual replenishment. Armed with this tradeoff information, a manager can select a reasonable number of changeovers or purchase orders per year and thereby specify a lot size.

FIGURE 2.3

Inventory investment versus lots per year

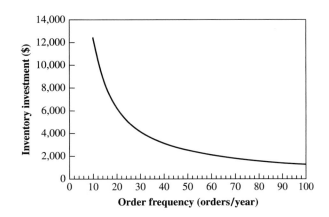

2.2.4 Sensitivity

A second insight that follows from the EOQ model is that

> Holding and setup costs are fairly insensitive to lot size.

We can see this in Figure 2.2, where the total cost only varies between seven and eight for values of Q between 96 and 306. This implies that if, for any reason, we use a lot size that is slightly different than Q^*, the increase in the holding plus setup costs will not be large. This feature was qualitatively observed by Harris in his original paper. The earliest quantitative treatment of it of which we are aware is by Brown (1967, 16).

To examine the sensitivity of the cost to lot size, we begin by substituting Q^* for Q into expression (2.1) for Y (but omitting the c term, since this is not affected by lot size), and we find that the minimum holding plus setup cost per unit is given by

$$Y^* = Y(Q^*) = \frac{hQ^*}{2} + \frac{AD}{Q^*}$$
$$= \frac{h\sqrt{2AD/h}}{2} + \frac{AD}{\sqrt{2AD/h}}$$
$$= \sqrt{2ADh} \tag{2.7}$$

Now, suppose that instead of using Q^*, we use some other arbitrary lot size Q', which might be larger or smaller than Q^*. From expression (2.1) for $Y(Q)$, we see that the annual holding plus setup cost under Q' can be written

$$Y(Q') = \frac{hQ'}{2} + \frac{AD}{Q'}$$

Hence, the ratio of the annual cost using lot size Q' to the optimal annual cost (using Q^*) is given by

$$\frac{Y(Q')}{Y^*} = \frac{hQ'/2 + AD/Q'}{\sqrt{2ADh}}$$
$$= \frac{Q'}{2}\sqrt{\frac{h^2}{2ADh}} + \frac{1}{Q'}\sqrt{\frac{A^2D^2}{2ADh}}$$
$$= \frac{Q'}{2}\sqrt{\frac{h}{2AD}} + \frac{1}{2Q'}\sqrt{\frac{2AD}{h}}$$
$$= \frac{Q'}{2Q^*} + \frac{Q^*}{2Q'}$$
$$= \frac{1}{2}\left(\frac{Q'}{Q^*} + \frac{Q^*}{Q'}\right) \tag{2.8}$$

To appreciate (2.8), suppose that $Q' = 2Q^*$, which implies that we use a lot size twice as large as optimal. Then the ratio of the resulting holding plus setup cost to the optimum is $\frac{1}{2}(2 + \frac{1}{2}) = 1.25$. That is, a 100 percent error in lot size results in a 25 percent error in cost. Notice that if $Q' = Q^*/2$, we also get an error of 25 percent in the cost function.

We can get further sensitivity insight from the EOQ model by noting that because demand is deterministic, the order interval is completely determined by the order quantity. We can express the time between orders T as

$$T = \frac{Q}{D} \tag{2.9}$$

Hence, dividing (2.4) by D, we get the following expression for the optimal order interval

$$T^* = \sqrt{\frac{2A}{hD}} \tag{2.10}$$

and by substituting (2.9) into (2.8), we get the following expression for the ratio of the cost resulting from an arbitrary order interval T' and the optimum cost:

$$\frac{\text{Annual cost under } T'}{\text{Annual cost under } T^*} = \frac{1}{2}\left(\frac{T'}{T^*} + \frac{T^*}{T'}\right) \tag{2.11}$$

Expression (2.11) is useful in multiproduct settings, where it is desirable to order such that different products are frequently replenished at the same time (e.g., to facilitate sharing of delivery trucks). A method for facilitating this that has been widely proposed in the operations research literature is to order items at intervals given by *powers of 2*. That is, make the order interval one week, two weeks, four weeks, eight weeks, etc.[2] The result is that items ordered at 2^n week intervals will be placed at the same time as orders for items with 2^k intervals for all k smaller than n (see Figure 2.4). This will facilitate sharing of trucks, consolidation of ordering effort, simplification of shipping schedules, etc.

Moreover, the sensitivity results we derived above for the EOQ model imply that the error introduced by restricting order intervals to powers of 2 will not be excessive. To see this, suppose that the optimal order interval for an item T^* lies between 2^m and 2^{m+1} for some m (see Figure 2.5). Then T^* lies either in the interval $[2^m, 2^m\sqrt{2}]$ or in the interval $[2^m\sqrt{2}, 2^{m+1}]$. All points in $[2^m, 2^m\sqrt{2}]$ are no more than $\sqrt{2}$ times as large as 2^m. Likewise, all points in the interval $[2^m\sqrt{2}, 2^{m+1}]$ are no less than 2^{m+1} divided by $\sqrt{2}$. For instance, in Figure 2.5, 2^m is within a multiplicative factor of $\sqrt{2}$ of T_1^*, and 2^{m+1} is within a multiplicative factor of $1/\sqrt{2}$ of T_2^*. Hence, the power-of-2 order interval T' must lie in the interval $[T^*/\sqrt{2}, \sqrt{2}T^*]$ around the optimal order interval T^*. Thus, the maximum error in cost will occur when $T' = \sqrt{2}T^*$, or $T' = T^*/\sqrt{2}$. From (2.11), the error from using $T' = \sqrt{2}T^*$ is

$$\frac{1}{2}\left(\sqrt{2} + \frac{1}{\sqrt{2}}\right) = 1.06$$

and is the same when $T' = T^*/\sqrt{2}$. Hence, the error in the holding plus setup cost resulting from using the optimal power-of-2 order interval instead of the optimal order interval is guaranteed to be no more than six percent. Jackson, Maxwell, and Muckstadt (1985); Roundy (1985, 1986); and Federgruen and Zheng (1992) give algorithms for

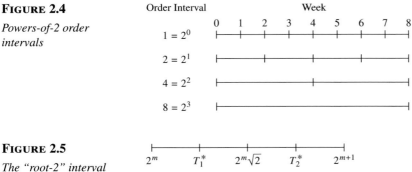

FIGURE 2.4

Powers-of-2 order intervals

FIGURE 2.5

The "root-2" interval

---

[2]To be complete, we must also consider negative powers of 2 or one-half week, one-fourth week, one-eighth week, etc. However, if we use a sufficiently small unit of time as our baseline (e.g., days instead of weeks), this will not be necessary in practice.

computing the optimal power-of-2 policy and extend the above results to more general multipart settings.

As a concrete illustration of these concepts, consider once again the MedEquip problem. We computed the optimal order quantity for racks to be $Q^* = 169$. Hence, the optimal order interval is $T^* = Q^*/D = 169/1,000 = 0.169$ year, or $0.169 \times 52 = 8.78$ weeks. Suppose further that MedEquip orders a variety of other parts from the same supplier. The unit price of $250 for racks is a delivered price, assuming an average shipping cost. However, if MedEquip combines orders for different parts, total shipping costs can be reduced. If the minimum order interval for any of the products under consideration is one week, then the order interval for racks can be rounded to the nearest power of 2 of $T = 8$ weeks or $8/52 = 0.154$ year. This implies an order quantity of $Q = TD = 0.154(1,000) = 154$. The holding plus order cost of this modified order quantity is

$$Y(Q) = \frac{hQ}{2} + \frac{AD}{Q} = \frac{35(154)}{2} + \frac{500(1,000)}{154} = \$5,942$$

The optimal annual cost (i.e., from using $Q^* = 169$) is given by

$$Y^* = \sqrt{2ADh} = \sqrt{2(500)(1,000)(35)} = \$5,916$$

So the modified order quantity results in less than a one percent increase in cost. The other parts ordered from the same supplier will have similar increases in holding plus order cost—but none of more than six percent. If these increases are offset by the reduced transportation cost, then the power-of-2 order schedule is worthwhile.

2.2.5   EOQ Extensions

Harris's original formula has been extended in a variety of ways over the years. One of the earliest extensions (Taft 1918) was to the case in which replenishment is not instantaneous; instead, there is a finite, but constant and deterministic, production rate. This model is sometimes called the **economic production lot (EPL)** model and results in a similar square root formula to the regular EOQ. Other variations of the basic EOQ include backorders (i.e., orders that are not filled immediately, but have to wait until stock is available), major and minor setups, and quantity discounts among others (see Johnson and Montgomery 1974; McClain and Thomas 1985; Plossl 1985; Silver, Pyke, and Peterson 1998).

2.3   Dynamic Lot Sizing

As we noted above, the EOQ formulation is predicated on a number of assumptions, specifically,

1. Instantaneous production.
2. Immediate delivery.
3. Deterministic demand.
4. Constant demand.
5. Known constant setup costs.
6. Single product or separable products.

We have already noted that Taft relaxed the assumption of instantaneous production. Introducing delivery delays is straightforward if delivery times are known and fixed (i.e., compute order quantities according to the EOQ formula and place the orders at times equal to desired delivery minus delivery time). If delivery times are uncertain, then a different approach is required. However, a more prevalent and important source of randomness than delivery times is in demand. The topic of relaxing the assumption of deterministic demand will be taken up in the next section on statistical inventory models. We have already discussed an approach for getting around the specification of a constant setup cost (i.e., by examining the inventory versus order frequency tradeoff). In Chapter 17 we will discuss approaches for handling multiproduct cases where parts cannot be analyzed separately. This leaves the assumption of constant demand.

2.3.1   Motivation

Consider the situation of RoadHog, Inc., which is a small manufacturer of motorcycle accessories. It makes a muffler with fins (that does little to suppress engine noise) on a line that is also used to make a variety of other products. Because it is costly to set up the line to produce the gauges, RoadHog has an incentive to produce them in batches. However, while customer demand is known over a 10-week planning horizon (because it is entered into a master production schedule and "frozen"), it is not necessarily constant from week to week. Since this violates a key assumption of the EOQ model, we need a fundamentally different model to balance the setup and holding costs.

The main historical approach to relaxing the constant-demand assumption is the Wagner–Whitin model (Wagner and Whitin 1958). This model considers the problem of determining production lot sizes when demand is deterministic but time-varying and all the other assumptions for the EOQ model are valid. The importance of this **dynamic lot-sizing** approach is that it has had a substantial impact on the literature in production control, and later influenced the development of materials requirements planning (MRP), as we will discuss in Chapter 3. For these reasons, we now present an overview of the Wagner–Whitin dynamic lot-sizing procedure.

2.3.2   Problem Formulation

When demand varies over time, a continuous time model, like the EOQ model, is awkward to specify. So, instead, we will clump demand into discrete periods, which could correspond to days, weeks, or months, depending on the system. A daily production schedule might make sense for a high-volume system with rapidly changing demand, while a monthly schedule may be adequate for a low-volume system with demand that changes more slowly.

To specify the problem and model, we will make use of the following notation, which represents the dynamic counterpart to the static notation used for the EOQ model:

t = a time period (e.g., day, week, month); we will consider $t = 1, \ldots, T$, where T represents the **planning horizon**

D_t = demand in period t (in units)

c_t = unit production cost (in dollars per unit), not counting setup or inventory costs in period t

A_t = setup (order) cost to produce (purchase) a lot in period t (in dollars)

h_t = holding cost to carry a unit of inventory from period t to period $t + 1$ (in dollars per unit per period); for example, if holding cost consists entirely of interest on money tied up in inventory, where i is the annual interest rate and periods correspond to weeks, then $h_t = ic_t/52$

I_t = inventory (in units) left over at the end of period t

Q_t = lot size (in units) in period t; there are T such decision variables, one for each period

Example:

With this notation, we can specify the RoadHog problem precisely. We suppose that the data for the next 10 weeks are as given in Table 2.1. Note that for simplicity we have assumed that the setup costs A_t, the production cost c_t, and the holding cost h_t are all constant over time, although this is not necessary for the Wagner–Whitin model. The basic problem is to satisfy all demands at minimal cost (i.e., production plus setup plus holding cost). The only controls are the production quantities Q_t. However, since all demands must be filled, only the timing of production is open to choice, not the total production quantity. Hence if the unit production cost is constant (that is, c_t does not vary with t), then production cost will be the same regardless of timing and therefore can be omitted altogether.

The simplest lot-sizing procedure one might think of is to produce exactly what is required in each period. This is called the **lot-for-lot rule,** and as we will see in Chapter 3, it can make sense in some situations. However, in this problem, the lot-for-lot rule implies that we will have to produce, and hence pay a setup cost, in every period. Table 2.2 shows the production schedule and resulting costs for this policy. Since we never carry inventory, the total cost is just that of the 10 setups, or $1,000.

Another plausible policy is to produce a fixed amount each time we perform a setup. This is known as the **fixed order quantity** lot-sizing rule. Since there are 300 units to produce, one possible fixed order quantity would be 100 units. This would require us to produce exactly three times, resulting in three setups, and would not leave any product left over at the end of period 10. Table 2.3 illustrates the production schedule and resulting costs for this policy. Notice that under this policy we frequently produce more

---

TABLE 2.1   Data for the RoadHog Dynamic Lot-Sizing Example

t	1	2	3	4	5	6	7	8	9	10
D_t	20	50	10	50	50	10	20	40	20	30
c_t	10	10	10	10	10	10	10	10	10	10
A_t	100	100	100	100	100	100	100	100	100	100
h_t	1	1	1	1	1	1	1	1	1	1

TABLE 2.2   Lot-for-Lot Solution to the RoadHog Example

t	1	2	3	4	5	6	7	8	9	10	Total
D_t	20	50	10	50	50	10	20	40	20	30	300
Q_t	20	50	10	50	50	10	20	40	20	30	300
I_t	0	0	0	0	0	0	0	0	0	0	0
Setup cost	100	100	100	100	100	100	100	100	100	100	1,000
Holding cost	0	0	0	0	0	0	0	0	0	0	0
Total cost	100	100	100	100	100	100	100	100	100	100	1,000

TABLE 2.3   Fixed Order Quantity Solution to the RoadHog Example

t	1	2	3	4	5	6	7	8	9	10	Total
D_t	20	50	10	50	50	10	20	40	20	30	300
Q_t	100	0	0	100	0	0	100	0	0	0	300
I_t	80	30	20	70	20	10	90	50	30	0	0
Setup cost	100	0	0	100	0	0	100	0	0	0	300
Holding cost	80	30	20	70	20	10	90	50	30	0	400
Total cost	180	30	20	170	20	10	190	50	30	0	700

than is required in a given period and therefore pay inventory carrying costs. However, the total inventory carrying cost is only $400, which, when added to the $300 setup cost, results in a total cost of $700. This is lower than the cost from the lot-for-lot policy. But can we do better? We will find out below by developing a procedure that is guaranteed to find the minimum setup plus inventory cost.

2.3.3   The Wagner–Whitin Procedure

A key observation for solving the dynamic lot-sizing problem is that if we produce items in period t (and incur a setup cost) for use to satisfy demand in period $t + 1$, then it cannot possibly be economical to produce in period $t + 1$ (and incur another setup cost). Either it is cheaper to produce *all* of period $t + 1$'s demand in period t, or all of it in $t + 1$; it is never cheaper to produce some in each. (Notice that we violated this property in the fixed order quantity solution given in Table 2.3.) In more general terms, we can state this result as follows:

Wagner–Whitin Property
Under an optimal lot-sizing policy either the inventory carried to period $t + 1$ from a previous period will be zero or the production quantity in period $t + 1$ will be zero.

This result greatly facilitates computation of optimal production quantities, as we will see.[3]

The Wagner–Whitin property implies that either $Q_t = 0$ or Q_t will be one of the following: D_t, $D_t + D_{t+1}$, $D_t + d_{t+1} + D_{t+2}, \ldots, D_t + D_{t+1} + \cdots + D_T$. That is, we will produce either nothing or exactly enough to satisfy demand in the current period plus some integer number of future periods. We could compute the minimum-cost production schedule by enumerating all possible combinations of periods in which production occurs. However, since we can either produce or not produce in each period, the number of such combinations is 2^{N-1}, which can be quite large if many periods are considered. To be more efficient, Wagner and Whitin suggested an algorithm that is well suited to computer implementation. We will describe this algorithm by means of the RoadHog example.

---

[3]Some pundits have noted that, while useful mathematically, in real systems the Wagner–Whitin property is either obvious or ridiculous. In essence, it states we should not produce until inventory falls to zero. If one really accepts all the modeling assumptions, particularly those of known, deterministic demand and well-defined fixed setup costs, then the property is nearly tautological. However, in real systems where uncertainty complicates things, one almost always starts production before inventory is exhausted (i.e., to provide protection against stockouts caused by random disruptions).

The Wagner–Whitin algorithm proceeds forward in time, starting with period 1 and finishing with period N. By the Wagner–Whitin property, we know that we will only produce in a period if the inventory carried to that period is zero. If this is the case, then our decision can be thought of in terms of how many periods of demand to produce. For instance, in a six-period problem, there are six possibilities for the amount we can produce in period 1, namely, D_1, $D_1 + D_2$, $D_1 + D_2 + D_3$, ..., $D_1 + D_2 + D_3 + D_4 + D_5 + D_6$. If we choose to produce $D_1 + D_2$, then inventory will run out in period 3 and so we will have to produce again in that period. In period 3, then, we will have the option of producing for period 3 only; periods 3 and 4; periods 3, 4, and 5; or periods 3, 4, 5, and 6.

Step 1

We begin the algorithm by looking at the one-period problem. That is, we act as though the world ends after one period. The optimal policy for this problem is trivial; we produce 20 units to satisfy demand in period 1, and we are done. Since there is no inventory carried from one period to another, and we are neglecting production cost, the minimum cost in the one-period problem, which we denote by Z_1^*, is

$$Z_1^* = A_1 = 100$$

As we will see as the algorithm unfolds, it is also useful to keep track of the last period in which production occurs in each problem we consider. Here, obviously, production takes place only in period 1, so the last period of production in the one-period problem, which we denote by j_1^*, is

$$j_1^* = 1$$

Step 2

In the next step of the algorithm we increase the time horizon and consider the two-period problem. Now we have two options for the production in period 2; we can cover demand in period 2 with production either in period 1 or in period 2. If we produce in period 1, we will incur a holding cost associated with carrying inventory from period 1 to period 2. If we produce in period 2, we will incur an extra setup cost in period 2. Notice also that if we produce in period 2, then the cost of satisfying previous demand (i.e., demand in period 1) is given by Z_1^*. Since we are trying to minimize cost, the optimal policy is to choose the period with the lower total cost, that is,

$$Z_2^* = \min \begin{cases} A_1 + h_1 D_2 & \text{produce in period 1} \\ Z_1^* + A_2 & \text{produce in period 2} \end{cases}$$

$$= \min \begin{cases} 100 + 1(50) = 150 \\ 100 + 100 \quad = 200 \end{cases}$$

$$= 150$$

The optimal decision is to produce for both periods 1 and 2 in period 1. Therefore, the last period in which production takes place in an optimal two-period policy is

$$j_2^* = 1$$

Step 3

Now, we proceed to the three-period problem. Ordinarily four possible production schedules would need to be considered: produce in period 1 only, produce in periods 1 and 2, produce in periods 1 and 3, or produce in periods 1, 2, and 3. However, we need to consider only three of these: one only, one and two, and one and three. This is because we only need to consider when we are going to produce the demand for period 3. We have already solved the two- and one-period problems. Note that the gain in speed grows sharply as the number of periods grows. For instance, for the 10-period problem we reduce the number of schedules we must check from 512 to 10. We will reduce these even more with the "planning horizon" result discussed later.[4]

If we decide to produce in period 3, then we know from our solution to the two-period problem that it will be optimal to produce for periods 1 and 2 in period 1.

$$
Z_3^* = \min \begin{cases} A_1 + h_1 D_2 + (h_1 + h_2)D_3 & \text{produce in period 1} \\ Z_1^* + A_2 + h_2 D_3 & \text{produce in period 2} \\ Z_2^* + A_3 & \text{produce in period 3} \end{cases}
$$

$$
= \min \begin{cases} 100 + 1(50) + (1+1)(10) = 170 \\ 100 + 100 + 1(10) \qquad\quad = 210 \\ 150 + 100 \qquad\qquad\qquad = 250 \end{cases}
$$

$$
= 170
$$

Again, it is optimal to produce everything in period 1, so

$$
j_3^* = 1
$$

Step 4

The situation changes when we move to the next step, the four-period problem. Now there are four options for the timing of production for period 4, namely, periods 1 to 4:

$$
Z_4^* = \min \begin{cases} A_1 + h_1 D_2 + (h_1 + h_2)D_3 + (h_1 + h_2 + h_3)D_4 & \text{produce in period 1} \\ Z_1^* + A_2 + h_2 D_3 + (h_2 + h_3)D_4 & \text{produce in period 2} \\ Z_2^* + A_3 + h_3 D_4 & \text{produce in period 3} \\ Z_3^* + A_4 & \text{produce in period 4} \end{cases}
$$

$$
= \min \begin{cases} 100 + 1(50) + (1+1)(10) + (1+1+1)(50) = 320 \\ 100 + 100 + 1(10) + (1+1)(50) \qquad\qquad = 310 \\ 150 + 100 + 1(50) \qquad\qquad\qquad\qquad\quad = 300 \\ 170 + 100 \qquad\qquad\qquad\qquad\qquad\qquad\quad = 270 \end{cases}
$$

$$
= 270
$$

This time, it turned out to be optimal not to produce in period 1, but rather to meet period 4's demand with production in period 4. Hence,

$$
j_4^* = 4
$$

If our planning horizon were only 4 periods, we would be done at this point. We would translate our results to a lot-sizing policy by reading the j_t^* values backward in time. The fact that $j_4^* = 4$ means that we would produce $D_4 = 50$ units in period 4. This would leave us with a three-period problem. Since $j_3^* = 1$, it would be optimal to produce $D_1 + D_2 + D_3 = 80$ units in period 1.

---

[4]This technique of solving successively longer horizon problems and using the solutions from previous steps to reduce the amount of computation in each step is known as *dynamic programming*. Dynamic programming is a form of *implicit enumeration,* which allows us to consider all possible solutions without explicitly computing the cost of each one.

Step 5 and Beyond

But our planning horizon is not 4 periods; it is 10 periods. Hence, we must continue the algorithm. However, before doing this, we will make an observation that will further reduce the computations we must make. Notice that up to this point, each step in the algorithm has increased the number of periods we must consider for the last period's production. So, by step 4, we had to consider producing for period 4 in all periods 1 through 4. It turns out that this is not always necessary.

Notice that in the four-period problem it is optimal to produce in period 4 for period 4. What this means is that the cost of setting up in period 4 is less than the cost setting up in period 1, 2, or 3 and carrying the inventory to period 4. If it weren't, then we would have chosen to produce in one of these periods. Now consider what this means for period 5. For instance, could it be cheaper to produce for period 5 in period 3 than in period 4? Production in periods 3 and 4 must be held in inventory from period 4 to period 5 and therefore incur the same carrying cost for that period. Therefore the only question is whether it is cheaper to set up in period 3 and carry inventory from period 3 to period 4 than it is to set up in period 4. But we already know the answer to this question. The fact that $j_4^* = 4$ tells us that it is cheaper to set up in period 4. Therefore, it is unnecessary to consider producing in periods 1, 2, and 3 for the demand in period 5. We need to consider only periods 4 and 5.

This reasoning can more generally be stated as follows:

Planning Horizon Property

If $j_t^* = \bar{t}$, then the last period in which production occurs in an optimal $t + 1$ period policy must be in the set $\bar{t}, \bar{t} + 1, \ldots, t + 1$.

Using this property, the calculation required to compute the minimum cost for the five-period problem is

$$Z_5^* = \min \begin{cases} Z_3^* + A_4 + h_4 D_5 & \text{produce in period 4} \\ Z_4^* + A_5 & \text{produce in period 5} \end{cases}$$

$$= \min \begin{cases} 170 + 100 + 1(50) = 320 \\ 270 + 100 \qquad\quad = 370 \end{cases}$$

$$= 320$$

Given that we are going to set up in period 4 anyway, it is cheaper to carry inventory from period 4 to period 5 than to set up again in period 5. Hence,

$$j_5^* = 4$$

We solve the remaining five periods, using the same approach, and summarize the results of these calculations in Table 2.4. Notice the blank spaces in the upper right-hand corner of this table. These are the result of our use of the planning horizon property. Without this property, we would have had to calculate values for each of these spaces.

2.3.4   Interpreting the Solution

The minimum total setup plus inventory carrying cost is given by $Z_{10} = \$580$, which we note is indeed lower than the cost achieved by either the lot-for-lot or fixed order quantity solutions we offered earlier. The optimal lot sizes are determined from the j_t^* values. Since $j_{10}^* = 8$, it is optimal to produce for periods 8, 9, and 10 in period 8. Hence, $Q_8^* = D_8 + D_9 + D_{10} = 90$. With periods 8, 9, and 10 taken care of, we are

TABLE 2.4   Solution to Wagner–Whitin Example

Last Period with Production	Planning Horizon t									
	1	**2**	**3**	**4**	**5**	**6**	**7**	**8**	**9**	**10**
1	100	150	170	320						
2		200	210	310						
3			250	300						
4				270	320	340	400	560		
5					370	380	420	540		
6						420	440	520		
7							440	480	520	610
8								500	520	580
9									580	610
10										620
Z_t^*	100	150	170	270	320	340	400	480	520	580
j_t^*	1	1	1	4	4	4	4	7	7 or 8	8

left with a seven-period problem. Since $j_7^* = 4$, it is optimal to produce for periods 4, 5, 6, and 7 in period 4. Hence, $Q_4^* = D_4 + D_5 + D_6 + D_7 = 130$. This leaves us with a three-period problem. Since $j_3^* = 1$, we should produce for periods 1, 2, and 3 in period 1, so $Q_1^* = D_1 + D_2 + D_3 = 80$.

2.3.5   Caveats

Although the calculations underlying Table 2.4 are certainly tedious to do by hand, they are not difficult for a computer. Given this, it is rather surprising that many production and operations management textbooks have omitted the Wagner–Whitin algorithm in favor of simpler heuristics that do not always give the optimal solution. Presumably, "simpler" meant both less computationally burdensome and easier to explain. Given that the algorithm is only used where production planning is computerized, the computational-burden argument is not compelling. Furthermore, the concepts underlying the algorithm are not difficult—certainly not so difficult as to prevent practitioners from using commercial software incorporating it!

However, there are more important concerns about the entire concept of "optimal" lot sizing whether one is using the Wagner–Whitin algorithm or any of the heuristic approaches that approximate it.

1. Like the EOQ model, the Wagner–Whitin model assumes setup costs known in advance of the lot-sizing procedure. But, as we noted earlier, setup costs can be very difficult to estimate in manufacturing systems. Moreover, the true cost of a setup is influenced by capacity. For instance, shutting down to change a die is very costly in terms of lost production when operating close to capacity, but not nearly as costly when there is a great deal of excess capacity. This issue cannot be addressed by any model that assumes independent setup costs. Thus, it would appear that the Wagner–Whitin model, like EOQ, is better suited to purchasing than production systems.

2. Also like the EOQ model, the Wagner–Whitin model assumes deterministic demand and deterministic production. Uncertainties, such as order cancellations, yield loss, and delivery schedule deviations are not considered. The result is that the "optimal"

production schedule given by the Wagner–Whitin algorithm will have to be adjusted to meet real conditions (e.g., reduced to accommodate leftover inventory from order cancellations or inflated for expected yield loss). The fact that these adjustments will be made on an ad hoc basis, coupled with the speculative nature of the setup costs, could make this theoretically optimal schedule perform poorly in practice.

3. Another key assumption is that of *independent products,* that is, that production for different products does not make use of common resources. This assumption is clearly violated in many instances. This can be important if some resources are highly utilized.

4. The Wagner–Whitin property leads us to the conclusion that we should produce either nothing in a period or the demand for an integer number of future periods. This property follows from (1) the fact that a fixed setup cost is incurred each time production takes place and (2) the assumption of infinite capacity. In the real world, where setups have more subtle consequences and capacity is finite, a sensible production plan may be quite different. For instance, it may be reasonable to produce according to a level production plan (i.e., produce approximately the same amount in each period), in order to achieve a degree of pacing or rhythm in the line. Wagner–Whitin, by focusing exclusively on the tradeoff between fixed and holding costs, may actually serve to steer our intuition away from realistic concerns.

2.4   Statistical Inventory Models

All the models discussed up to this point have assumed that demand is fixed and known. Although there are cases in which this assumption may approximate reality (e.g., when the schedule is literally frozen over the horizon of interest), often it does not. If demand is random, then there are two basic approaches to take:

1. Model demand as if it were deterministic for modeling purposes and then modify the solution to account for randomness.
2. Explicitly represent randomness in the model.

Neither approach is correct or incorrect in any absolute sense. The real question is, Which is more *useful?* In general, the answer depends on the circumstances. When planning is over a sufficiently long horizon to ensure that random deviations "average out," a deterministic model may work well. Also, a deterministic model with appropriate "fudge factors" to anticipate randomness, coupled with a suitably frequent regeneration cycle to get back on track, can be effective. However, to determine these fudge factors or to help design policies for dealing with time frames in which randomness is critical, a model that explicitly incorporates randomness may be more appropriate.

Historically, the operations management literature has pursued both approaches. The most prevalent deterministic model for production scheduling is materials requirements planning (MRP), the subject of Chapter 3. The most prevalent probabilistic models are the **statistical reorder point** approaches, which we examine in this section.

Statistical modeling of production and inventory control problems is not new, dating back at least to Wilson (1934). In this classic paper, Wilson breaks the inventory control problem into two distinct parts:

1. Determining the **order quantity,** or the amount of inventory that will be purchased or produced with each replenishment.
2. Determining the **reorder point,** or the inventory level at which a replenishment (purchase or production) will be triggered.

In this section, we will address this two-part problem in three stages.

First, we will consider the situation in which we are only interested in a single replenishment, so that the only issue is to determine the appropriate order quantity in the face of uncertain demand. This has traditionally been called the **news vendor model** because it could apply to a person who purchases newspapers at the beginning of the day, sells a random amount, and then must discard any leftovers.

Second, we will consider the situation in which inventory is replenished one unit at a time as random demands occur, so that the only issue is to determine the reorder point. The target inventory level we set for the system is known as a *base stock level,* and hence the resulting model is termed the **base stock model.**

Third, we will consider the situation where inventory is monitored continuously and demands occur randomly, possibly in batches. When the inventory level reaches (or goes below) r, an order of size Q is placed. After a lead time of ℓ, during which a stockout might occur, the order is received. The problem is to determine appropriate values of Q and r. The model we use to address this problem is known as the (Q, r) **model.**

These models will make use of the concepts and notation found in the field of *probability.* If it has been awhile since the reader has reviewed these, now might be a good time to peruse Appendix 2A.

2.4.1   The News Vendor Model

Consider the situation that a manufacturer of Christmas lights faces each year. Demand is somewhat unpredictable and occurs in such a short burst just prior to Christmas that if inventory is not on the shelves, sales are lost. Therefore, the decision of how many sets of lights to produce must be made prior to the holiday season. Additionally, the cost of collecting unsold inventory and holding it until next year is too high to make year-to-year storage an attractive option. Instead, any unsold sets of lights are sold after Christmas at a steep discount.

To choose an appropriate production quantity, the important pieces of information to consider are (1) anticipated demand and (2) the costs of producing too much or too little. To develop a formal model, we make the following assumptions:

1. *Products are separable.* We can consider products one at a time since there are no interactions (e.g., shared resources).

2. *Planning is done for a single period.* We can neglect future periods since the effect of the current decision on them is negligible (e.g., because inventory cannot be carried across periods).

3. *Demand is random.* We can characterize demand with a known probability distribution.

4. *Deliveries are made in advance of demand.* All stock ordered or produced is available to meet demand.

5. *Costs of overage or underage are linear.* The charges for having too much or too little inventory is proportional to the amount of the overage or underage.

We make use of these assumptions to develop a model using the following notation:

$X =$ demand (in units), a random variable

$G(x) = P(X \leq x) =$ cumulative distribution function of demand; for this model we will assume that G is a continuous distribution because it is analytically convenient, but the results are essentially the same if G is discrete (i.e., restricted to integer values), as we will note

$g(x) = \dfrac{d}{dx} G(x) =$ density function of demand

μ = mean demand (in units)

σ = standard deviation of demand (in units)

c_o = cost (in dollars) per unit left over after demand is realized

c_s = cost (in dollars) per unit of shortage

Q = production or order quantity (in units); this is the decision variable

Example:

Now consider the Christmas lights example with some numbers. Suppose that a set of lights costs $1 to make and distribute and sells for $2. Any sets not sold by Christmas will be discounted to $0.50. In terms of the above modeling notation, this means that the unit overage cost is the amount lost per excess set, or c_o = $(1 − 0.50) = $0.50. The unit shortage cost is the lost profit from a sale, or c_s = $(2 − 1) = $1. Suppose further that demand has been forecast to be 10,000 units with a standard deviation of 1,000 units and that the normal distribution is a reasonable representation of demand.

The firm could choose to produce 10,000 sets of lights. But recall that the symmetry (i.e., bell shape) of the normal distribution implies that it is equally likely for demand to be greater or less than 10,000 units. If demand is less than 10,000 units, the firm will lose c_o = $0.50 per unit of overproduction. If demand is greater than 10,000 units, the firm will lose c_s = $1 per unit of underproduction. Clearly, shortages are worse than overages. This suggests that perhaps the firm should produce more than 10,000 units. But how much more? The model we develop below is aimed at answering exactly this question.

To develop a model, observe that if we produce Q units and demand is X units, then the number of units of overage is given by

$$\text{Units over} = \max \{Q - X, 0\}$$

That is, if $Q \geq X$, then the overage is simply $Q - X$; but if $Q < X$, then there is a shortage and so the overage is zero. We can calculate the expected overage as

$$E[\text{units over}] = \int_0^\infty \max \{Q - x, 0\} g(x) \, dx$$

$$= \int_0^Q (Q - x) g(x) \, dx \tag{2.12}$$

Similarly, the number of units of shortage is given by

$$\text{Units short} = \max \{X - Q, 0\}$$

That is, if $X \geq Q$, then the shortage is simply $X - Q$; but if $X < Q$, then there is an overage and so the shortage is zero. We can calculate the expected shortage as

$$E[\text{units short}] = \int_0^\infty \max \{x - Q, 0\} g(x) \, dx$$

$$= \int_Q^\infty (x - Q) g(x) \, dx \tag{2.13}$$

Using (2.12) and (2.13), we can express the expected cost as a function of the production quantity as

$$Y(Q) = c_o \int_0^Q (Q - x) g(x) \, dx + c_s \int_Q^\infty (x - Q) g(x) \, dx \tag{2.14}$$

We will find the value of Q that minimizes this expected cost in the following technical note.

Technical Note

As we did for the EOQ model, we will find the minimum of $Y(Q)$ by taking its derivative and setting it equal to zero. To do this, however, we need to take the derivative of integrals with limits that are functions of Q. The tool we require for this is *Leibnitz's rule,* which can be written as

$$\frac{d}{dQ} \int_{a_1(Q)}^{a_2(Q)} f(x, Q)\, dx = \int_{a_1(Q)}^{a_2(Q)} \frac{\partial}{\partial Q}[f(x, Q)]\, dx + f(a_2(Q), Q)\frac{da_2(Q)}{dQ}$$
$$-f(a_1(Q), Q)\frac{da_1(Q)}{dQ}$$

Applying this to take the derivative of $Y(Q)$ and setting the result equal to zero yields

$$\frac{dY(Q)}{dQ} = c_o \int_0^Q 1 g(x)\, dx + c_s \int_Q^\infty (-1)g(x)\, dx$$
$$= c_o G(Q) - c_s[1 - G(Q)] = 0 \qquad (2.15)$$

Solving (2.15) (which we simplify below in (2.16)) for Q^* yields the production (order) quantity that minimizes $Y(Q)$.

To minimize expected overage plus shortage cost, we should choose a production or order quantity Q^* that satisfies

$$G(Q^*) = \frac{c_s}{c_o + c_s} \qquad (2.16)$$

First, note that since $G(Q^*)$ represents the probability that demand is less than or equal to Q^*, this result implies that Q^* should be chosen such that the probability of having enough stock to meet demand is $c_s/(c_o + c_s)$. Second, notice that since $G(x)$ increases in x (cumulative distribution functions are always monotonically increasing), so that anything that makes the right-hand side of (2.16) larger will result in a larger Q^*. This implies that increasing c_s will increase Q^*, while increasing c_o will decrease Q^*, as we would intuitively expect.

We can further simplify expression (2.16) if we assume that G is normal. For this case we can write

$$G(Q^*) = \Phi\left(\frac{Q^* - \mu}{\sigma}\right) = \frac{c_s}{c_o + c_s}$$

where Φ is the cumulative distribution function (cdf) of the standard normal distribution.[5] This means that

$$\frac{Q^* - \mu}{\sigma} = z$$

where z is the value in the standard normal table (see Table 1 at the end of the book) for which $\Phi(z) = c_s/(c_o + c_s)$, and hence

$$Q^* = \mu + z\sigma \qquad (2.17)$$

---

[5]We are making use of the well-known result that if X is normally distributed with mean μ and standard deviation σ, then $(X - \mu)/\sigma$ is normally distributed with mean zero and standard deviation one (i.e., the standard normal distribution).

Expression (2.17) implies that for the normal case, Q^* is an increasing function of the mean demand μ. It is also increasing in the standard deviation of demand σ, provided that z is positive. This will be the case whenever $c_s/(c_o + c_s)$ is greater than one-half, since $\Phi(0) = 0.5$ and $\Phi(z)$ is increasing in z. However, if costs are such that $c_s/(c_o + c_s)$ is less than one-half, then the optimal order size Q^* will decrease as σ increases.

Example:
Now we return to the Christmas lights example. Because demand is normally distributed, we can compute Q^* from (2.17). To do this, we must find z by computing

$$\frac{c_s}{c_o + c_s} = \frac{1}{1 + 0.5} = 0.67$$

and by looking up in a standard normal table to find that $\Phi(0.44) = 0.67$. Hence $z = 0.67$ and

$$Q^* = \mu + z\sigma = 10,000 + (0.44)1,000 = 10,440$$

Notice that this answer can be interpreted as telling us to produce 0.44 standard deviation above mean demand. Therefore, if the standard deviation of demand had been 2,000 units, instead of 1,000, the answer would have been to produce $0.44 \times 2,000 = 880$ units above mean demand, or 10,880 units.

The news vendor problem, and its intuitive critical ratio solution given in (2.16), can be extended to a variety of applications that, unlike the Christmas lights example, have more than one period. One common situation is the problem in which

1. A firm faces periodic (e.g, monthly) demands that are independent and have the same distribution $G(x)$.
2. All orders are backordered (i.e., met eventually).
3. There is no setup cost associated with producing an order.

It can be shown that an "order up to Q" policy (i.e., after each demand, produce enough to bring the inventory level up to Q) is optimal under these conditions. Moreover, the problem of finding the optimal order-up-to level Q^* can be formulated as a news vendor model (see Nahmias 1993, 291–294). The solution Q^* therefore satisfies Equation (2.16), where c_o represents the cost to hold one unit of inventory in stock for one period and c_s represents the cost of carrying a unit of backorder (i.e., an unfilled order) for one period. Similarly, under the same conditions, except that sales are lost instead of backordered, the optimal order-up-to level is found by solving (2.16) for Q^* with c_o equal to the one-period holding cost and c_s equal to the unit profit (i.e., selling price minus production cost).

We conclude this section by summarizing the basic insights from the news vendor model:

1. In an environment of uncertain demand, the appropriate production or order quantity depends on both the distribution of demand *and* the relative costs of overproducing versus underproducing.
2. If demand is normally distributed, then increasing the variability (i.e., standard deviation) of demand will increase the production or order quantity if $c_s/(c_s + c_o) > 0.5$ and decrease it if $c_s/(c_s + c_o) < 0.5$.

2.4.2   The Base Stock Model

Consider the situation facing Superior Appliance, a store that sells a particular model of refrigerator. Because space is limited and because the manufacturer makes frequent deliveries of other appliances, Superior finds it practical to order replacement refrigerators each time one is sold. In fact, it has a system that places purchase orders automatically whenever a sale is made. But because the manufacturer is slow to fill replenishment orders, the store must carry some stock in order to meet customer demands promptly. Under these conditions, the key question concerns how much stock to carry.

To answer this question, we need a model. To develop one, we make use of a continuous-time framework (e.g., like the EOQ model) and the following modeling assumptions:

1. *Products can be analyzed individually.* There are no product interactions (e.g., shared resources).
2. *Demands occur one at a time.* There are no batch orders.
3. *Unfilled demand is backordered.* There are no lost sales.
4. *Replenishment lead times are fixed and known.* There is no randomness in delivery lead times. (We will show how to relax this assumption to consider variable lead times later in this chapter.)
5. *Replenishments are ordered one at a time.* There is no setup cost or constraint on the number of orders that can be placed per year, which would motivate batch replenishment.

We will relax the last assumption in the next section on the (Q, r) model, where ordering in bulk will become a potentially attractive option.

We also make use of the following notation:

ℓ = replenishment lead time (in days), assumed constant throughout this section

X = demand during replenishment lead time (in units), a random variable

$p(x) = P(X = x)$ = probability demand during replenishment lead time equals x (probability mass function). We are assuming demand is discrete (i.e., countable), but sometimes it is convenient to approximate demand with a continuous distribution. When we do this, we assume a density function $g(x)$ in place of the probability mass function

$G(x) = P(X \leq x) = \sum_{i=0}^{x} p(i)$ = probability demand during replenishment lead time is less than or equal to x (cumulative distribution function)

$\theta = E[X]$, mean demand (in units) during lead time ℓ

σ = standard deviation of demand (in units) during lead time ℓ

h = cost to carry one unit of inventory for one year (in dollars per unit per year)

b = cost to carry one unit of backorder for one year (in dollars per unit per year)

r = reorder point (in units), which represents inventory level that triggers a replenishment order; this is the decision variable

$R = r + 1$, base stock level (in units)

$s = r - \theta$, safety stock level (in units)

$S(R)$ = fill rate (fraction of orders filled from stock) as a function of R

$B(R)$ = average number of outstanding backorders as a function of R

$I(R)$ = average on-hand inventory level (in units) as a function of R

Since we place an order when there are r units in stock and expect to incur demand for θ units while we wait for the replenishment order to arrive, $r - \theta$ is the amount of inventory we expect to have on hand when the order arrives. If $s = r - \theta > 0$, then we call this the **safety stock** for this system, since it represents inventory that protects it against stockouts due to fluctuations in either demand or deliveries. Since finding $r - \theta$ is equivalent to finding r (because θ is a constant), we can view the problem as finding the optimal base stock level ($R = r + 1$), reorder point r, or safety stock level ($s = r - \theta$).

We can approach the problem of finding an optimal base stock level in one of two ways. We can follow the procedure we have used up to now (in the EOQ, Wagner–Whitin, and news vendor models) and formulate a cost function and find the reorder point that minimizes this cost. Or we can simply specify the desired customer service level and find the smallest reorder point that attains it. We will develop both approaches below. But first we need to develop expressions for the performance measures $S(R)$, $B(R)$, and $I(R)$.

We begin by analyzing the relationship between inventory, replenishment orders, and backorders under a base stock policy. To do this, we distinguish between **on-hand inventory,** which represents physical inventory in stock (and hence can never be negative), and **inventory position,** which represents the balance of on-hand inventory, backorders, and replenishment orders and is given by

$$\text{Inventory position} = \text{on-hand inventory} - \text{backorders} + \text{orders} \qquad (2.18)$$

Under a base stock policy we place a replenishment order every time a demand occurs. Hence, at all times the following holds:

$$\text{Inventory position} = R \qquad (2.19)$$

Using (2.18) and (2.19), we can derive expressions for the performance measures.

Service Level.   Consider a specific replenishment order. Because lead times are constant, we know that all the other $R - 1$ items either in inventory or on order will be available to fill new demand before the order under consideration arrives. Therefore, the only way the order can arrive after the demand for it has occurred is if demand during the replenishment lead time is greater than or equal to R (that is, $X \geq R$). Hence, the probability that the order arrives *before* its demand (i.e., does not result in a backorder) is given by $P(X < R) = P(X \leq R - 1) = G(R - 1) = G(r)$. Since all orders are alike with regard to this calculation, the fraction of demands that are filled from stock is equal to the probability that an order arrives before the demand for it has occurred, or

$$S(R) = G(R - 1) = G(r) \qquad (2.20)$$

Hence, $G(R - 1)$ represents the fraction of demands that will be filled from stock. This is normally called the **fill rate** and represents a reasonable definition of customer service for many inventory control systems.

Backorder Level.   At any time, the number of orders is exactly equal to the number of demands that have occurred during the last ℓ time units. If we let X represent this (random) number of demands, then from (2.18) and (2.19)

$$\text{On-hand inventory} - \text{backorders} = R - X \qquad (2.21)$$

Notice that on-hand inventory and backorders can never be positive at the same time (i.e., because if we had both inventory and backorders, we would fill backorders until either stock ran out or the backorders were all filled). So, at a point where the number of outstanding orders is $X = x$, the backorder level is given by

$$\text{Backorders} = \begin{cases} 0 & \text{if } x < R \\ x - R & \text{if } x \geq R \end{cases}$$

The expected backorder level can be computed by averaging over possible values of x:

$$B(R) = \sum_{x=R}^{\infty} (x - R)p(x) \tag{2.22}$$

Expression (2.22) is a very important and useful function in the theory of inventory control. Because it measures the amount of unmet demand (backorder level), it is referred to as a *loss function*. While it can be computed in the form given in (2.22), it is frequently more convenient to write it in terms of the cumulative distribution function as follows:

$$B(R) = \theta - \sum_{x=0}^{R} [1 - G(x)] \tag{2.23}$$

This loss function will come up again in the (Q, r) model. Even simpler spreadsheet-implementable formulas for computing $B(R)$ are given in Appendix 2B for the cases where demand is Poisson-distributed and also for the case where demand is approximated by the (continuous) normal distribution.

Inventory Level.   Taking the expectation of both sides of Equation (2.21) and noting that $I(R)$ represents expected on-hand inventory, $B(R)$ represents expected backorder level, and $E[X] = \theta$ is the expected lead time demand, we get

$$I(R) = R - \theta + B(R) \tag{2.24}$$

Example:
We can now analyze the Superior Appliance example. Suppose from past experience we know that mean demand for the refrigerator under consideration is 10 units per month and replenishment lead time is one month. Therefore, mean demand during lead time is $\theta = 10$ units. Further suppose that we model demand using the Poisson distribution.[6] Specifically, for any integer values of k and x, we set

$$p(R) = \text{Prob}\{\text{demand during lead time} = R\} = \frac{\theta^R e^{-\theta}}{R!} = \frac{10^R e^{-10}}{R!}$$

and

$$G(R) = \sum_{k=0}^{R} p(k) = \sum_{k=0}^{R} \frac{10^k e^{-10}}{k!}$$

With these we can also compute the $B(r)$ function by using the formulas from Appendix 2B. We summarize the results in Table 2.5. If we want to achieve a fill rate of at least

---

[6]The Poisson distribution is a good modeling choice for demand processes where demands occur one by one and do not exhibit cyclic fluctuations. It is completely specified by only one parameter, the mean, and is therefore convenient when one lacks information concerning the variability of demand. The standard deviation of the Poisson is equal to the square root of the mean.

TABLE 2.5   Fill Rates for Various Values of *R*

R	*p(R)*	*G(R)*	*B(R)*	*R*	*p(R)*	*G(R)*	*B(R)*
0	0.000	0.000	10.000	12	0.095	0.792	0.531
1	0.000	0.000	9.000	13	0.073	0.864	0.322
2	0.002	0.003	8.001	14	0.052	0.917	0.187
3	0.008	0.010	7.003	15	0.035	0.951	0.103
4	0.019	0.029	6.014	16	0.022	0.973	0.055
5	0.038	0.067	5.043	17	0.013	0.986	0.028
6	0.063	0.130	4.110	18	0.007	0.993	0.013
7	0.090	0.220	3.240	19	0.004	0.997	0.006
8	0.113	0.333	2.460	20	0.002	0.998	0.003
9	0.125	0.458	1.793	21	0.001	0.999	0.001
10	0.125	0.583	1.251	22	0.000	0.999	0.000
11	0.114	0.697	0.834	23	0.000	1.000	0.000

90 percent, we must choose R such that $G(R-1) \geq 0.9$. From Table 2.5 we see this requires $R - 1 = 14$, or $R = 15$, which results in a 91.7 percent fill rate. Since average demand during a replenishment lead time is 10 units, this is equivalent to setting a safety stock level of $r - \theta = 14 - 10 = 4$ units. The average backorder level resulting from $R = 15$ is given by $B(15) = 0.103$. The average inventory level is given by

$$I(R) = R - \theta + B(R) = 15 - 10 + 0.103 = 5.103$$

If we were to increase the base stock level from 15 to 16, the fill rate would increase to 95.1 percent, the backorder level would fall to 0.055, and the average inventory level would increase to 6.055. Whether or not the improved customer service (as measured by fill rate and backorder level) is worth the additional inventory investment is a value judgment for Superior Appliance. One way to balance these competing issues is to use a cost optimization model, as we show below.

In general, the higher the mean demand during replenishment lead time, the higher the base stock level required to achieve a particular fill rate. This is hardly surprising, since the reorder point r must contain enough inventory to cover demand while orders are coming. If the distribution of demand during lead time is symmetric (e.g., bell-shaped), then the probability of demand exceeding θ during the lead time is one-half. Hence, any fill rate greater than one-half will require r to be greater than θ.

In addition to mean demand, the variability of the demand process affects the choice of base stock level. The higher the standard deviation of demand during a replenishment lead time, the larger r will have to be for a given fill rate. If, in the previous example, we had approximated $G(x)$ by the normal distribution with mean θ and standard deviation σ, the choice of σ would have influenced the results in Table 2.5. Choosing $\sigma = \sqrt{\theta}$ would give results similar to those generated by using the Poisson distribution for $G(x)$ (since the standard deviation is always the square root of the mean in the Poisson). Higher values of σ would have given lower fill rates for the various values of r, while lower values of σ would have resulted in higher fill rates.

The base stock model has been widely studied in the operations management literature. This is partly because it is comparatively simple to analyze, but also because it is easily extended to a range of situations. For instance, base stocks can be used to control work releases in a multistage production line. In such a system, a base stock level is established for each inventory buffer in the line (e.g., in front of the workstations). Whenever an item is removed from the buffer, a replenishment order is triggered. As we will discuss in Chapter 4, this is essentially what the Japanese kanban system does.

Finally, we consider an optimization approach to setting the base stock level. To do this, we approximate demand with a continuous distribution $G(x)$ with density $g(x)$. Then we can write the cost function consisting of the sum of inventory holding costs plus backorder costs as

$$Y(R) = \text{holding cost} + \text{backorder cost} \tag{2.25}$$
$$= hI(R) + bB(R)$$
$$= h(R - \theta + B(R)) + bB(R)$$
$$= h(R - \theta) + (b + h)B(R) \tag{2.26}$$

We compute the base stock level R that minimizes $Y(R)$ in the following technical note.

Technical Note

Treating R as a continuous variable, we can take the derivative of $Y(R)$ as follows:

$$\frac{dY(R)}{dR} = h + (b + h)\frac{dB(R)}{dR}$$

The continuous-version expression of (2.22), the backorder function, $B(R)$, is given by

$$B(R) = \int_R^\infty (x - R)g(x)\,dx \tag{2.27}$$

so $dB(R)/dR$ can be computed as

$$\frac{dB(R)}{dR} = \frac{d}{dR}\int_R^\infty (x - R)g(x)\,dx$$
$$= -\int_R^\infty g(x)\,dx$$
$$= -[1 - G(R)]$$

Setting $dY(R)/dR$ equal to zero yields

$$\frac{dY(R)}{dR} = h - (b + h)[1 - G(R)] = 0 \tag{2.28}$$

Solving (2.28) yields the optimal value of R.

The base stock level R that minimizes holding plus backorder cost is given by

$$G(R^*) = \frac{b}{b + h} \tag{2.29}$$

Notice that this formula has the same critical ratio structure that we saw in the news vendor solution given in (2.16). This implies that the optimal base stock level is the one for which the fill rate is given by $b/(b + h)$. This result makes intuitive sense, since increasing the holding cost h causes R^* to decrease, while increasing the backorder cost b causes R^* to increase. Note that when backorder and holding costs are equal, the resulting fill rate is one-half so that $R^* = \theta$, the average demand during the replenishment time, and thus there is no safety stock.

As we did for the news vendor problem, we can simplify (2.29) for the case where G is normal. Using the same arguments we used to derive expression (2.17), we can show that

$$R^* = \theta + z\sigma \qquad (2.30)$$

where z is the value from the standard normal table for which $\Phi(z) = b/(b+h)$ and μ and σ are the mean and standard deviation, respectively, of lead-time demand. Note that R^* increases in θ and also increases in σ provided that $z > 0$. This will be the case as long as $b/(b+h) > 0.5$, or equivalently $b > h$. Since carrying a unit of backorder is typically more costly than carrying a unit of inventory, it is generally the case that the optimal base stock level is an increasing function of demand variability.

Example:

Let us return to the Superior Appliance example. To approximate demand with a continuous distribution, we assume lead-time demand is normally distributed with mean $\theta = 10$ units per month and standard deviation $\sigma = \sqrt{\theta} = 3.16$ units per month. (Choosing $\sigma = \sqrt{\theta}$ makes the standard deviation the same as that for the Poisson distribution used in the earlier example.) Suppose that the wholesale cost of the refrigerators is \$750 and Superior uses an interest rate of two percent per month to charge inventory costs, so that $h = 0.02(750) = \$15$ per unit per month. Further suppose that the backorder cost is estimated to be \$25 per unit per month, because Superior typically has to offer discounts to get sales on out-of-stock items.

Then the optimal base stock level can be found from (2.30) by first computing z by calculating

$$\frac{b}{b+h} = \frac{25}{25+15} = 0.625$$

and looking up in a standard normal table to find $\Phi(0.32) = 0.625$. Hence, $z = 0.32$ and

$$R^* = \theta + z\sigma = 10 + 0.32(3.16) = 11.01 \approx 11$$

Using Table 2.5, we can compute the fill rate for this base stock level as $S(R) = G(R - 1) = G(10) = 0.583$. (Notice that even though we used a continuous model to find R^*, we used the discrete formula in Table 2.5 to compute the actual fill rate because in real life, demand for refrigerators is discrete.) This is a pretty low fill rate, which may indicate that our choice for the backorder cost b was too low.

If we were to increase the backorder cost to $b = \$200$, the critical ratio would increase to 0.93, which (because $z_{0.93} = 1.48$) would increase the optimal base stock level to $R^* = 10 + 1.48(3.16) = 14.67 \approx 15$. This is the base stock level we got in our previous analysis where we set it to achieve a fill rate of 90 percent, and we recall that the actual fill rate it achieves is 91.7 percent. We can make two observations from this. First, the actual fill rate computed from Table 2.5 using the Poisson distribution— 91.7 percent even after rounding R up to 15—is generally lower than the critical ratio in (2.29), 93 percent, because a continuous demand distribution tends to make inventory look more efficient than it really is. Second, the backorder cost necessary to get a base stock level of 15, and hence a fill rate greater than 90 percent, is very large

($200 per unit per month!), which suggests that such a high fill rate is not a economical.[7]

We conclude by noting that the primary insights from the simple base stock model are as follows:

1. Reorder points control the probability of stockouts by establishing **safety stock.**
2. The required base stock level (and hence safety stock) that achieves a given fill rate is an increasing function of the mean and (provided that unit backorder cost exceeds unit holding cost) standard deviation of the demand during replenishment lead time.
3. The "optimal" fill rate is an increasing function of the backorder cost and a decreasing function of the holding cost. Hence, if we fix the holding cost, we can use either a service constraint or a backorder cost to determine the appropriate base stock level.
4. Base stock levels in multistage production systems are very similar to kanban systems, and therefore the above insights apply to those systems as well.

2.4.3   The (Q, r) Model

Consider the situation of Jack, a maintenance manager, who must stock spare parts to facilitate equipment repairs. Demand for parts is a function of machine breakdowns and is therefore inherently unpredictable (i.e., random). But, unlike in the base stock model, suppose that the costs incurred in placing a purchase order (for parts obtained from an outside supplier) or the costs associated with setting up the production facility (for parts produced internally) are significant enough to make one-at-a-time replenishment impractical. Thus, the maintenance manager must determine not only how much stock to carry (as in the base stock model), but also how many to produce or order at a time (as in the EOQ and news vendor models). Addressing both of these issues simultaneously is the focus of the (Q, r) model.

From a modeling perspective, the assumptions underlying the (Q, r) model are identical to those of the base stock model, except that we will assume that either

1. There is a fixed cost associated with a replenishment order.   or
2. There is a constraint on the number of replenishment orders per year.

and therefore replenishment quantities greater than 1 may make sense.

The basic mechanics of the (Q, r) model are illustrated in Figure 2.6, which shows the net inventory level (on-hand inventory minus backorder level) and inventory position (net inventory plus replenishment orders) for a single product being continuously monitored. Demands occur randomly, but we assume that they arrive one at a time, which is why net inventory always drops in unit steps in Figure 2.6. When the inventory position reaches the reorder point r, a replenishment order for quantity Q is placed. (Notice that because the order is placed exactly when inventory position reaches r, inventory position

---

[7]Part of the reason that b must be so large to achieve $R = 15$ is that we are rounding to the nearest integer. If instead we always round up, which would be reasonable if we want service to be at least $b/(b + h)$, then a (still high) value of $b = \$135$ makes $b/(b + h) = 0.9$ and results in $R = 14.05$ which rounds up to 15. Since the continuous distribution is an approximation for demand anyway, it does not really matter whether a large b or an aggressive rounding procedure is used to obtain the final result. What does matter is that the user perform sensitivity analysis to understand the solution and its impacts.

FIGURE 2.6

Net inventory and inventory position versus time in the (Q, r) model with $Q = 4$, $r = 4$

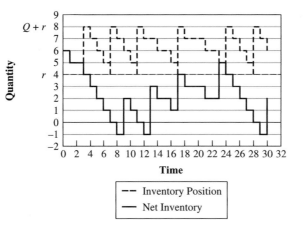

immediately jumps to $r + Q$ and hence never spends time at level r.) After a (constant) lead time of ℓ, during which stockouts might occur, the order is received. The problem is to determine appropriate values of Q and r.

As Wilson (1934) pointed out in the first formal publication on the (Q, r) model, the two controls Q and r have essentially separate purposes. As in the EOQ model, the replenishment quantity Q affects the tradeoff between production or order frequency and inventory. Larger values of Q will result in few replenishments per year but high average inventory levels. Smaller values will produce low average inventory but many replenishments per year. In contrast, the reorder point r affects the likelihood of a stockout. A high reorder point will result in high inventory but a low probability of a stockout. A low reorder point will reduce inventory at the expense of a greater likelihood of stockouts.

Depending on how costs and customer service are represented, we will see that Q and r can interact in terms of their effects on inventory, production or order frequency, and customer service. However, it is important to recognize that the two parameters generate two fundamentally different kinds of inventory. The replenishment quantity Q affects **cycle stock** (i.e., inventory that is held to avoid excessive replenishment costs). The reorder point r affects **safety stock** (i.e., inventory held to avoid stockouts). Note that under these definitions, all the inventory held in the EOQ model is cycle stock, while all the inventory held in the base stock model is safety stock. In some sense, the (Q, r) model represents the integration of these two models.

To formulate the basic (Q, r) model, we combine the costs from the EOQ and base stock models. That is, we seek values of Q and r to solve either

$$\min_{Q,r} \{\text{fixed setup cost} + \text{backorder cost} + \text{holding cost}\} \qquad (2.31)$$

or

$$\min_{Q,r} \{\text{fixed setup cost} + \text{stockout cost} + \text{holding cost}\} \qquad (2.32)$$

The difference between formulations (2.31) and (2.32) lies in how customer service is represented. Backorder cost assumes a charge per unit time a customer order is unfilled, while stockout cost assumes a fixed charge for each demand that is not filled from stock (regardless of the duration of the backorder). We will make use of both approaches in the analysis that follows.

Notation.   To develop expressions for each of these costs, we will make use of the following notation:

D = expected demand per year (in units)

ℓ = replenishment lead time (in days); initially we assume this is constant, although we will show how to incorporate variable lead times at the end of this section

X = demand during replenishment lead time (in units), a random variable

$\theta = E[X] = D\ell/365$ = expected demand during replenishment lead time (in units)

σ = standard deviation of demand during replenishment lead time (in units)

$p(x) = P(X = x)$ = probability demand during replenishment lead time equals x (probability mass function). As in the base stock model, we assume demand is discrete. But when it is convenient to approximate it with a continuous distribution, we assume the existence of a density function $g(x)$ in place of the probability mass function

$G(x) = P(X \leq x) = \sum_{i=0}^{x} p(i)$ = probability demand during replenishment lead time is less than or equal to x (cumulative distribution function)

A = setup or purchase order cost per replenishment (in dollars)

c = unit production cost (in dollars per unit)

h = annual unit holding cost (in dollars per unit per year)

k = cost per stockout (in dollars)

b = annual unit backorder cost (in dollars per unit of backorder per year); note that failure to have inventory available to fill a demand is penalized by using either k or b but not both

Q = replenishment quantity (in units); this is a decision variable

r = reorder point (in units); this is the other decision variable

$s = r - \theta$ = safety stock implied by r (in units)

$F(Q, r)$ = order frequency (replenishment orders per year) as a function of Q and r

$S(Q, r)$ = fill rate (fraction of orders filled from stock) as a function of Q and r

$B(Q, r)$ = average number of outstanding backorders as a function of Q and r

$I(Q, r)$ = average on-hand inventory level (in units) as a function of Q and r

Costs

Fixed Setup Cost.   There are two basic ways to address the desirability of having an order quantity Q greater than one. First, we could simply put a constraint on the number of replenishment orders per year. Since the number of orders per year can be computed as

$$F(Q, r) = \frac{D}{Q} \tag{2.33}$$

we can compute Q for a given order frequency F as $Q = D/F$. Alternatively, we could charge a fixed order cost A for each replenishment order that is placed. Then the annual fixed order cost becomes $F(Q, r)A = (D/Q)A$.

Stockout Cost.   As we noted earlier, there are two basic ways to penalize poor customer service. One is to charge a cost each time a demand cannot be filled from stock (i.e., a stockout occurs). The other is to charge a penalty that is proportional to the length of time a customer order waits to be filled (i.e., is backordered).

The annual stockout cost is proportional to the average number of stockouts per year, given by $D[1 - S(Q, r)]$. We can compute $S(Q, r)$ by observing from Figure 2.6 that inventory position can only take on values $r + 1, r + 2, \ldots, r + Q$ (note it cannot be equal to r since whenever it reaches r, another order of Q is placed immediately). In fact, it turns out that over the long term, inventory position is equally likely to take on any value in this range. We can exploit this fact to use our results from the base stock model in the following analysis (see Zipkin 1999 for a rigorous version of this development).

Suppose we look at the system[8] after it has been running a long time and we observe that the current inventory position is x. This means that we have inventory on hand and on order sufficient to cover the next x units of demand. So we ask the question, What is the probability that the $(x + 1)$st demand will be filled from stock? The answer to this question is precisely the same as it was for the base stock model. That is, since all outstanding orders will have arrived within the replenishment lead time, the only way the $(x + 1)$st demand can stock out is if demand during the replenishment lead time is greater than or equal to x. From our analysis of the base stock model, we know that the probability of a stockout is

$$
\begin{aligned}
P\{X \geq x\} &= 1 - P\{X < x\} \\
&= 1 - P\{X \leq x - 1\} \\
&= 1 - G(x - 1)
\end{aligned}
$$

Hence, the fill rate given an inventory position of x is one minus the probability of a stockout, or $G(x - 1)$. Since the Q possible inventory positions are equally likely, the fill rate for the entire system is computed by simply averaging the fill rates over all possible inventory positions:

$$
S(Q, r) = \frac{1}{Q} \sum_{x=r+1}^{r+Q} G(x - 1) = \frac{1}{Q}[G(r) + \cdots + G(r + Q - 1)] \tag{2.34}
$$

We can use (2.34) directly to compute the fill rate for a given (Q, r) pair. However, it is often more convenient to convert this to another form. By using the fact that the base stock backorder level function $B(R)$ can be written in terms of the cumulative distribution function as in (2.23), it is straightforward to show that the following is an equivalent expression for the fill rate in the (Q, r) model:

$$
S(Q, r) = 1 - \frac{1}{Q}[B(r) - B(r + Q)] \tag{2.35}
$$

This exact expression for $S(Q, r)$ is simple to compute in a spreadsheet, especially using the formulas given in Appendix 2B. However, it is sometimes difficult to use in analytic expressions. For this reason, various approximations have been offered. One approximation, known as the **base stock** or **type I service** approximation, is simply the (continuous demand) base stock formula for fill rate, which is given by

$$
S(Q, r) \approx G(r) \tag{2.36}
$$

From Equation (2.34) it is apparent that $G(r)$ underestimates the true fill rate. This is because the cdf $G(x)$ is an increasing function of x. Hence, we are taking the smallest

---

[8]This technique is called *conditioning* on a random event (i.e., the value of the inventory position) and is a very powerful analysis tool in the field of probability.

term in the average. However, while it can seriously underestimate the true fill rate, it is very simple to work with because it involves only r and not Q. It can be the basis of a very useful heuristic for computing good (Q, r) policies, as we will show below.

A second approximation of fill rate, known as **type II service,** is found by ignoring the second term in expression (2.35) (Nahmias 1993). This yields

$$S(Q, r) \approx 1 - \frac{B(r)}{Q} \tag{2.37}$$

Again, this approximation tends to underestimate the true fill rate, since the $B(r+Q)$ term in (2.35) is positive. However, since this approximation still involves both Q and r, it is not generally simpler to use than the exact formula. But as we will see below, it does turn out to be a useful intermediate approximation for deriving a reorder point formula.

Backorder Cost.   If, instead of penalizing stockouts with a fixed cost per stockout k, we penalize the time a backorder remains unfilled, then the annual backorder cost will be proportional to the average backorder level $B(Q, r)$. The quantity $B(Q, r)$ can be computed in a similar manner to the fill rate, by averaging the backorder level for the base stock model over all inventory positions between $r + 1$ and $r + Q$:

$$B(Q, r) = \frac{1}{Q} \sum_{x=r+1}^{r+Q} B(x) = \frac{1}{Q}[B(r + 1) + \cdots + B(r + Q)] \tag{2.38}$$

Again, this formula can be used directly or converted to simpler form for computation in a spreadsheet, as shown in Appendix 2B. As with the expression for $S(Q, r)$, it is sometimes convenient to approximate this with a simpler expression that does not involve Q. One way to do this is to use the analogous formula to the type I service formula and simply use the base stock backorder formula

$$B(Q, r) \approx B(r) \tag{2.39}$$

Notice that to make an exact analogy with the type I approximation for fill rate, we should have taken the minimum term in expression (2.38), which is $B(r + 1)$. While this would work just fine, it is a bit simpler to use $B(r)$ instead. The reason is that we typically use such an approximation when we are also approximating demand with a continuous function; under this assumption the backorder expression for the base stock model really does become $B(r)$ [instead of $B(R)$].

Holding Cost.   The last cost in problems (2.31) and (2.32) is the inventory holding cost, which can be expressed as $hI(Q, r)$. We can approximate $I(Q, r)$ by looking at the *average* net inventory and acting as though demand were deterministic, as in Figure 2.7, which depicts a system with $Q = 4$, $r = 4$, $\ell = 2$, and $\theta = 2$. Demands are perfectly regular, so that every time inventory reaches the reorder point ($r = 4$), an order is placed, which arrives two time units later. Since the order arrives just as the last demand in the replenishment cycle occurs, the lowest inventory level ever reached is $r - \theta + 1 = s + 1 = 3$. In general, under these deterministic conditions, inventory will decline from $Q + s$ to $s + 1$ over the course of each replenishment cycle. Hence, the average inventory is given by

$$I(Q, r) \approx \frac{(Q + s) + (s + 1)}{2} = \frac{Q + 1}{2} + s = \frac{Q + 1}{2} + r - \theta \tag{2.40}$$

In reality, however, demand is variable and sometimes causes backorders to occur. Since on-hand inventory cannot go below zero, the above deterministic approximation underestimates the true average inventory by the average backorder level. Hence, the exact

FIGURE 2.7

*Expected inventory versus
time in the (Q, r) model
with $Q = 4$, $r = 4$, $\theta = 2$*

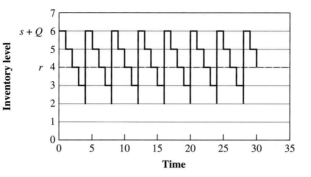

expression is

$$I(Q, r) = \frac{Q + 1}{2} + r - \theta + B(Q, r) \tag{2.41}$$

Backorder Cost Approach.   We can now make verbal formulation (2.31) into a mathematical model. The sum of setup and purchase order cost, backorder cost, and inventory carrying cost can be written as

$$Y(Q, r) = \frac{D}{Q}A + bB(Q, r) + hI(Q, r) \tag{2.42}$$

Unfortunately, there are two difficulties with the cost function $Y(Q, r)$. The first is that the cost parameters A and b are difficult to estimate in practice. In particular, the backorder cost is nearly impossible to specify, since it involves such intangibles as loss of customer goodwill and company reputation. Fortunately, however, the objective is not really to minimize this cost; it is to strike a reasonable balance between setups, service, and inventory. Using a cost function allows us to conveniently use optimization tools to derive expressions for Q and r in terms of problem parameters. But the quality of the policy must be evaluated directly in terms of the performance measures, as we will illustrate in the next example. The expressions for $B(Q, r)$ and $I(Q, r)$ involve both Q and r in complicated ways. So using exact expressions for these quantities does not lead us to simple expressions for Q and r. Therefore, to achieve tractable formulas, we approximate $B(Q, r)$ by expression (2.39) and use this in place of the true expression for $B(Q, r)$ in the formula for $I(Q, r)$ as well. With this approximation our objective function becomes

$$Y(Q, r) \approx \tilde{Y}(Q, r) = \frac{D}{Q}A + bB(r) + h\left[\frac{Q + 1}{2} + r - \theta + B(r)\right] \tag{2.43}$$

We compute the Q and r values that minimize $\tilde{Y}(Q, r)$ in the following technical note.

Technical Note

Treating Q as a continuous variable, differentiating $\tilde{Y}(Q, r)$ with respect to Q, and setting the result equal to zero yield

$$\frac{\partial \tilde{Y}(Q, r)}{\partial Q} = \frac{-DA}{Q^2} + \frac{h}{2} = 0 \tag{2.44}$$

Approximating lead-time demand with a continuous distribution with density $g(x)$, differentiating $\tilde{Y}(Q, r)$ with respect to r, and setting the result equal to zero yield

$$\frac{\partial \tilde{Y}(Q, r)}{\partial r} = (b + h)\frac{dB(r)}{dr} + h = 0 \tag{2.45}$$

Since, as in the base stock case, the continuous analog for the $B(r)$ function is

$$B(r) = \int_r^\infty (x - r)g(x)\, dx$$

we can compute the derivative of $B(r)$ as

$$\frac{dB(r)}{dr} = \frac{d}{dr}\int_r^\infty (x - r)g(x)\, dx$$

$$= -\int_r^\infty g(x)\, dx$$

$$= -[1 - G(r)]$$

and rewrite (2.45) as

$$-(b + h)[1 - G(r)] + h = 0 \tag{2.46}$$

Hence, we must solve (2.44) and (2.46) to minimize $\tilde{Y}(Q, r)$, which we do in (2.47) and (2.48).

The optimal reorder quantity Q^* and reorder point r^* are given by

$$Q^* = \sqrt{\frac{2AD}{h}} \tag{2.47}$$

$$G(r^*) = \frac{b}{b + h} \tag{2.48}$$

Notice that Q^* is given by the EOQ formula and the expression for r^* is given by the critical ratio formula for the base stock model. (The latter is not surprising, since we used a base stock approximation for the backorder level.) If we further assume that lead-time demand is normally distributed with mean θ and standard deviation σ, then we can simplify (2.48) as we did for the base stock model in (2.30) to get

$$r^* = \theta + z\sigma \tag{2.49}$$

where z is the value in the standard normal table such that $\Phi(z) = b/(b + h)$.

It is important to remember that these values for Q^* and r^* are only approximate. So we should check their performance (e.g., in terms of average inventory, fill rate, order frequency, and backorder level) by using exact formulas. If performance is not adequate, then the cost parameters can be adjusted. Typically, it makes sense to leave holding cost h alone and adjust the fixed order cost A and the backorder cost b, since these are more difficult to estimate in advance. Note that increasing A increases Q^* and hence reduces average order frequency, while increasing b increases r^* and hence reduces stockout rate and average backorder level. We illustrate this in the example on page 82.

Stockout Cost Approach.   As an alternative to the backorder cost approach, we can make verbal formulation (2.32) into a mathematical model by writing the sum of the annual setup or purchase order cost, stockout cost, and inventory carrying cost as

$$Y(Q, r) = \frac{D}{Q}A + kD[1 - S(Q, r)] + hI(Q, r) \tag{2.50}$$

As was the case for the backorder model, this cost function involves parameters that are difficult to specify. In particular, the stockout cost k is dependent on the same intangibles (lost customer goodwill and company reputation) as is the backorder cost b. Hence, again, this cost function is merely a means for deriving expressions for Q and r that reasonably balance setups, service, and inventory. It is not a performance measure in itself.

Also like the backorder model, the stockout model cost function contains expressions $S(Q, r)$ and $I(Q, r)$ that involve both Q and r and therefore do not lead to simple expressions. So we will make two levels of approximation to generate closed-form expressions for Q and r.

First, analogous to what we did in the backorder cost model above, we will assume that the effect of Q on the fill rate $S(Q, r)$ and the backorder correction factor $B(Q, r)$ in the inventory term $I(Q, r)$ can be ignored. This leads to the familiar EOQ formula for the order quantity

$$Q^* = \sqrt{\frac{2AD}{h}}$$

Second, to compute an expression for the reorder point, we make two approximations in (2.50). We replace the service $S(Q, r)$ by type II approximation (2.37) and the backorder correction term $B(Q, r)$ in the inventory term by base stock approximation (2.39). This yields the following approximate cost function

$$Y(Q, r) \approx \tilde{Y}(Q, r) = \frac{D}{Q}A + kD\frac{B(r)}{Q} + h\left[\frac{Q+1}{2} + r - \theta + B(r)\right] \quad (2.51)$$

Going through the usual optimization procedure (taking the derivative with respect to r, setting the result equal to zero, and solving for r) yields the following expression for the optimal reorder point:

$$G(r^*) = \frac{kD}{kD + hQ} \quad (2.52)$$

If we further assume that lead-time demand is normally distributed with mean θ and standard deviation σ, then we can simplify the expression for the reorder point to

$$r^* = \theta + z\sigma \quad (2.53)$$

where $\Phi(z) = kD/(kD + hQ)$.

Notice that unlike formula (2.49), expression (2.53) is sensitive to Q. Specifically, making Q larger makes the ratio $kD/(kD + hQ)$ smaller and hence reduces r^*. The reason is that larger Q values serve to increase the fill rate (because the reorder point is crossed less frequently) and hence require a smaller reorder point to achieve a given level of service.

Example:
Jack, the maintenance manager, has collected historical data that indicate one of the replacement parts he stocks has annual demand (D) of 14 units per year. The unit cost c of the part is $150, and since the firm uses an interest rate of 20 percent, the annual holding cost h has been set at $0.2(\$150) = \$30$ per year. It takes 45 days to receive a replenishment order, so average demand during a replenishment lead time is

$$\theta = \frac{14}{365} \times 45 = 1.726$$

The part is purchased from an outside supplier, and Jack estimates that the cost of time and materials required to place a purchase order A is about $15. The one remaining cost required by our model is the cost of either a backorder or stockout. Although he is very uncomfortable trying to estimate these, when pressed, Jack made a guess that the annualized cost of a backorder is about $b = \$100$ per year, and the cost per stockout event can be approximated by $k = \$40$.[9] Finally, Jack has chosen to model demands using the Poisson distribution.[10]

Regardless of whether we use the backorder cost model or the stockout cost model, the order quantity is computed by using (2.47), which yields

$$Q^* = \sqrt{\frac{2AD}{h}} = \sqrt{\frac{2(15)(14)}{30}} = 3.7 \approx 4$$

To compute the reorder point, we can use either the backorder cost or the stockout cost model. To use expression (2.49) from the normal demand version of the backorder model, we approximate the Poisson by the normal, with mean $\theta = 1.726$ and standard deviation $\sigma = \sqrt{1.726} = 1.314$. The critical ratio is given by

$$\frac{b}{b+h} = \frac{100}{100+30} = 0.769$$

and from a standard normal table, $\Phi(0.736) = 0.769$. Hence, $z = 0.736$ and

$$r^* = \theta + z\sigma = 1.726 + 0.736(1.314) = 2.693 \approx 3$$

As we noted earlier, the performance of this policy should not be judged in terms of the cost function or the approximate performance measures. Instead, it should be evaluated in terms of the exact expressions for order frequency, fill rate, backorder level, and inventory level. To compute these, we need to compute $p(r)$, $G(r)$, and $B(r)$. We do this by using the formulas from Appendix 2B for the Poisson demand case, and we summarize the results in Table 2.6. With these we can compute the order frequency, fill rate, backorder level, and average inventory level for the policy ($Q = 4$, $r = 3$) as follows:

$$F(Q,r) = \frac{D}{Q} = \frac{14}{4} = 3.5$$

$$S(Q,r) = 1 - \frac{1}{Q}[B(r) - B(r+Q)]$$

$$= 1 - \tfrac{1}{4}[B(3) - B(3+4)]$$

$$= 1 - \tfrac{1}{4}(0.140 - 0.001)$$

$$= 0.965$$

$$B(Q,r) = \frac{1}{Q}\sum_{x=r+1}^{r+Q} B(x)$$

$$= \tfrac{1}{4}[B(4) + B(5) + B(6) + B(7)]$$

$$= \tfrac{1}{4}(0.042 + 0.011 + 0.003 + 0.001)$$

$$= 0.014$$

---

[9]Notice that either approach for penalizing backorders or stockouts assumes that the cost is independent of which machine it affects. Of course, in reality, stockouts for heavily used critical machines are far more costly than stockouts affecting lightly used machines with excess capacity.

[10]The Poisson is a good assumption when demands are generated by many independent sources, such as failures of different machines. However, if demands were generated by a more regular process, such as scheduled preventive maintenance procedures, the Poisson distribution will tend to overestimate variability and lead to conservative, possibly excessive, safety stock levels.

TABLE 2.6 $p(r), G(r), B(r)$ for
$\theta = 1.726$

r	$p(r)$	$G(r)$	$B(r)$
0	0.178	0.178	1.726
1	0.307	0.485	0.904
2	0.265	0.750	0.389
3	0.153	0.903	0.140
4	0.066	0.969	0.042
5	0.023	0.991	0.011
6	0.007	0.998	0.003
7	0.002	1.000	0.001
8	0.000	1.000	0.000
9	0.000	1.000	0.000
10	0.000	1.000	0.000

$$I(Q, r) = \frac{Q+1}{2} + r - \theta + B(Q, r)$$

$$= \frac{4+1}{2} + 3 - 1.726 + 0.014$$

$$= 3.79$$

As an alternative to using the backorder cost model, we could have computed the reorder point by using expression (2.53) from the stockout cost model. The critical ratio in this formula is

$$\frac{kD}{kD + hQ} = \frac{40(14)}{40(14) + 30(4)} = 0.824$$

and from a standard normal table $\Phi(0.929) = 0.824$ so $z = 0.929$ and

$$r^* = \theta + z\sigma = 1.726 + 0.929(1.314) = 2.946 \approx 3$$

Since this policy ($Q = 4, r = 3$) is the same as that resulting from the backorder cost model, the performance measures will also be the same. So, in a practical sense, the backorder and stockout costs chosen by Jack are equivalent. In the single-product case, either model could be used—increasing either b or k will serve to increase service and decrease backorder level (at the expense of a higher inventory level). So either model can be used to generate a set of efficient solutions by varying these cost parameters. But we will see in Chapter 17 that the two models can behave differently in multiproduct systems.

The policy generated by the current cost coefficients will require placing replenishment orders three and one-half times per year, the fill rate is fairly high (96.5 percent), there will be few backorders (only 0.014 on average), and on-hand inventory will average a bit under four units (3.79). The decision maker might look at these values and feel that the policy is just fine. If not, then sensitivity analysis should be used to find variants of the solution.

For instance, suppose that the decision maker felt that three and one-half replenishment orders per year were too few and that, given the capacity of the purchasing department, $F = 7$ orders per year would be manageable. Then we could use

$Q = D/F = 14/7 = 2$. But if we stick with a reorder point of $r = 3$, then the fill rate becomes

$$S(Q, r) = 1 - \frac{1}{Q}[B(r) - B(r + Q)] = 1 - \tfrac{1}{2}(0.140 - 0.011) = 0.936$$

which may be too low for a repair part. If we increase the reorder point to $r = 4$, then the fill rate becomes

$$S(Q, r) = 1 - \frac{1}{Q}[B(r) - B(r + Q)] = 1 - \tfrac{1}{2}(0.042 - 0.003) = 0.980$$

For this new policy ($Q = 2, r = 4$) we can easily compute the backorder level and average inventory level to be

$$B(Q, r) = \frac{1}{Q}[B(5) + B(6)]$$

$$= \tfrac{1}{2}(0.011 + 0.003)$$

$$= 0.007$$

$$I(Q, r) = \frac{Q + 1}{2} + r - \theta + B(Q, r)$$

$$= \frac{2 + 1}{2} + 3 - 1.726 + 0.007$$

$$= 3.78$$

The increased reorder point has lowered the backorder rate, and the increased order frequency has reduced the average inventory level relative to the original policy of ($Q = 4, r = 3$). Of course, the cost of doing this is an additional three and one-half replenishment orders per year.

An alternate method for doing sensitivity analysis would be to modify the fixed order cost A until the order frequency $F(Q, r)$ is satisfactory and then modify the backorder cost b or the stockout cost k (depending on which model is being used) until the fill rate $S(Q, r)$ and/or the backorder level $B(Q, r)$ is acceptable. In a single-product problem like this, there is no great advantage to this approach, since we are still searching over two variables (that is, A and b or k instead of Q and r). But as we will see in Chapter 17, this approach is *much* more efficient in multiproduct problems, where one can search over a single (A, b) or (A, k) pair instead of (Q, r) values for each product. Furthermore, since expressions (2.47), (2.49), and (2.53) are simple closed-form equations involving the problem data, they are extremely simple to compute in a spreadsheet.

Modeling Lead-Time Variability.   Throughout our discussion of the base stock and (Q, r) models we have assumed that the replenishment lead time ℓ is fixed. All the variability in the system was assumed to be due to demand variability. However, in many practical situations, the lead time may also be subject to variability. For instance, a supplier of a part may sometimes be late (or early) on a delivery. The primary effect of this additional variability is to inflate the standard deviation of the demand during the replenishment lead time σ. By computing a formula for σ that considers lead-time variability, we can easily incorporate this additional source of variability into the base stock and (Q, r) models.

To develop the appropriate formula, we must introduce a bit of additional notation:

L = replenishment lead time (in number of periods), a random variable

$\ell = E[L]$ = expected replenishment lead time (in number of periods)

σ_L = standard deviation of replenishment lead time (in days)

D_t = demand on day t (in units), a random variable. We assume that demand is stationary over time, so that D_t has same distribution for each day t; we also assume daily demands are independent of one another

$d = E[D_t]$ = expected daily demand (in units)

σ_D = standard deviation of daily demand (in units)

As before, we let X represent the (random) demand during the replenishment lead time. With the above notation, this can be written as

$$X = \sum_{t=1}^{L} D_t \qquad (2.54)$$

Because daily demands are independent and identically distributed, we can compute the expected demand during the replenishment lead time as

$$E[X] = E[L]E[D_t] = \ell d = \theta \qquad (2.55)$$

which is what we have been using all along. However, variable lead times change the variance of demand during replenishment lead time. Using the standard formula for sums of independent, identically distributed random variables, we can compute

$$\mathrm{Var}(X) = E[L]\,\mathrm{Var}(D_t) + E[D_t]^2\,\mathrm{Var}(L) = \ell\sigma_D^2 + d^2\sigma_L^2 \qquad (2.56)$$

Although the "units" of (2.56) look wrong (the first term appears to have units of time while the second has units of time-squared), both terms are actually dimensionless. The reason is that L is defined as a random variable representing the *number* of periods and not the periods themselves. While the mean and variance of L do not have to be an integer, realizations of the random variable itself, do. For instance, by counting the number of days, we might observe lead times of five, six, and three days yielding a mean of 4.667. However, it is not valid to observe 5/7, 6/7, and 3/7 weeks and then compute a mean. Hence, the standard deviation of lead-time demand is

$$\sigma = \sqrt{\mathrm{Var}(X)} = \sqrt{\ell\sigma_D^2 + d^2\sigma_L^2} \qquad (2.57)$$

To get a better feel for how formula (2.57) behaves, consider the case where demand is Poisson. This implies that $\sigma_D = \sqrt{d}$, since the standard deviation is always the square root of the mean for Poisson random variables. Substituting this into (2.57) yields

$$\sigma = \sqrt{\ell d + d^2\sigma_L^2} = \sqrt{\theta + d^2\sigma_L^2} \qquad (2.58)$$

Notice that if $\sigma_L = 0$, which represents the case where the replenishment lead time is constant, then this reduces to $\sigma = \sqrt{\theta}$, which is exactly what we have been using for the Poisson demand case. If $\sigma_L > 0$, then formula (2.58) serves to inflate σ above what it would be for the constant-lead-time case.

To illustrate the use of the above formula in an inventory model, let us return to the Superior Appliance example from Section 2.4.2. There we assumed that demand for refrigerators was Poisson-distributed with a mean of 10 per month and that lead time was one month (30 days). So mean daily demand is $d = \frac{10}{30} = \frac{1}{3}$. Because we are assuming Poisson demand, we can use (2.58) to calculate σ. Using the same holding and backorder cost as in Section 2.4.2, $h = 15$ and $b = 25$, the critical ratio is $b/(h+b) = 25/(15+25) = 0.625$, so $z = 0.32$ since $\Phi(0.32) = 0.625$. The optimal base stock level (assuming the normal approximation of demand) is therefore

$$R^* = \theta + z\sigma = \theta + z\sqrt{\theta + d^2\sigma_L^2}$$

If $\sigma_L = 0$, then we get $R^* = 11.01$, which is what we got previously. If $\sigma_L = 30$ (i.e., the variability in replenishment lead time is so large that the standard deviation is equal to the mean), then we get $R^* = 13.34$. The additional 3.33 units of inventory are required to achieve the same service level in the face of the more variable demand.

Formula (2.57) or (2.58) can be used in this same fashion to inflate the reorder point in the (Q, r) model in either equation (2.49) or (2.53) to account for variable replenishment lead times.

Basic (Q, r) Insights.   Apart from all the mathematical and modeling complexity, the basic insights behind the (Q, r) model are essentially those of the EOQ and base stock models, namely that

> Cycle stock increases as replenishment frequency decreases.

and

> Safety stock provides a buffer against stockouts.

The (Q, r) model places these insights into a unified framework.

Historically, the (Q, r) model (including the special case of the base stock model, which is just a (Q, r) model with $Q = 1$) was one of the earliest attempts to explicitly model variability in the demand process and provide quantitative understanding of just how safety stock affects service level. In terms of rough intuition, this model suggests that safety stock, service level, and backorder level are primarily affected by the reorder point r, while cycle stock and order frequency are essentially functions of replenishment quantity Q. However, the mathematics of the model show that the situation is somewhat more subtle. As we saw above, the expressions for service and backorder level depend on Q as well as r. The reason is that if Q is large, so that the part is replenished infrequently in large batches, then stock level seldom reaches the reorder point and therefore has few opportunities for stockouts. If, on the other hand, Q is small, then stock level frequently falls to the reorder point and therefore has a greater chance of stocking out.

Beyond these qualitative observations, the (Q, r) model offers some quantitative insight into the factors that affect the optimal stocking policy. From approximate formulas (2.47), (2.49), and (2.53) we can draw the following conclusions.

1. Increasing the average annual demand D tends to increase the optimal order quantity Q.

2. Increasing the average demand during a replenishment lead time θ will tend to increase the optimal reorder point r. Note that increasing either the annual demand D or the replenishment lead time ℓ will serve to increase θ. The implication is that either high demand or long replenishment lead times will tend to require more inventory in stock.

3. Increasing the variability of the demand process σ will tend to increase the optimal reorder point r.[11] The key insight here is that a highly variable demand process typically requires more safety stock as protection against stockouts than does a very stable demand process.

4. Increasing the holding cost h will tend to decrease both the optimal replenishment quantity Q and reorder point r. Note that the holding cost can be increased by increasing

---

[11]Note that this is only true if the critical ratio in (2.49) or (2.53) is at least one-half. If this ratio is less than one-half, then z will be negative and the optimal order point will actually decrease in the standard deviation of lead-time demand. But this only occurs when the costs are such that it is optimal to set a relatively low fill rate for the product. So, the case where z is positive is very common in practice.

the cost of the item, the interest rate associated with inventory, or the noninterest holding costs (e.g., handling and spoilage). The point is that the more expensive it is to hold inventory, the less we should hold.

The (Q, r) model is a happy example of an approach that provides both powerful general insights and useful practical tools. As such it is a basic component of any manufacturing manager's skill set.

2.5   Conclusions

Although this chapter has covered a wide range of inventory modeling approaches, we have barely scratched the surface of this vast branch of the OM literature. The complexity and variety of inventory systems have spawned a wide array of models. Table 2.7 summarizes some of the dimensions along which these models differ and classifies the five models we have treated in this chapter (i.e., EOQ, Wagner–Whitin (WW), news vendor (NV), base stock (BS), and (Q, r)), plus the economic production lot (EPL) model that we mentioned as an EOQ extension. (Notice that some of the entries in Table 2.7 contain dashes, which indicates that the particular modeling decision has been rendered meaningless by other modeling assumptions and therefore does not apply.) The OM literature contains models representing all reasonable combinations of these dimensions, as well as models with features that go beyond them (e.g., substitution between products, explicit links between spare-parts inventory and utilization of maintenance personnel, and perishable inventories). In this book, we will return to the important subject of inventory management in Chapter 17, where we will extend some of the models of this chapter into the important practical environments of multiple products and multilevel inventory systems. The reader interested in a more comprehensive summary than we can provide in two chapters is encouraged to consult Graves, Rinnooy Kan, Zipkin (1993); Hadley and Whitin (1963); Johnson and Montgomery (1974); McClain and Thomas (1985); Nahmias (1993); Peterson and Silver (1985); Sherbrooke (1992); and Zipkin (1998).

Although some of these models require data that may be difficult or impossible to obtain, they do offer some basic insights:

1. *There is a tradeoff between setups (replenishment frequency) and inventory.* The more frequently we replenish inventory, the less *cycle stock* we will carry.
2. *There is a tradeoff between customer service and inventory.* Under conditions of random demand, higher customer service levels (i.e., fill rates) require higher levels of *safety stock*.
3. *There is a tradeoff between variability and inventory.* For a given replenishment frequency, if customer service remains fixed (at a sufficiently high level), then the higher the variability (i.e., standard deviation of demand or replenishment lead time), the more inventory we must carry.

Despite the efforts of some just-in-time advocates to deny the existence of such trade-offs, they are facts of manufacturing life. The commonly heard admonitions "Inventory is evil" or "Setups are bad" do little to guide the manager to useful policies.

In contrast, an understanding of the dynamics of inventory, replenishment frequency, and customer service enables a manager to evaluate which actions are likely to have the greatest impact. Such intuition can help address such questions as, Which setups are

TABLE 2.7   Classification of Inventory Models

Modeling Decision	Model					
	EOQ	**EPL**	**WW**	**NV**	**BS**	(Q, r)
Continuous (C) or discrete (D) time	C	C	D	D	C	C
Single (S) or multiple (M) products	S	S	S	S	S	S
Single (S) or multiple (M) periods	—	—	M	S	—	—
Backordering (B) or lost sales (L)	—	—	—	L	B	B
Setup or order cost [yes (Y) or no (N)]	Y	Y	Y	N	N	Y
Deterministic (D) or random (R) demand	D	D	D	R	R	R
Deterministic (D) or random (R) production	D	D	D	D	D	D
Constant (C) or dynamic (D) demand	C	C	D	—	C	C
Finite (F) or infinite (I) production rate	I	F	I	—	I	I
Finite (F) or infinite (I) horizon	I	I	F	F	I	I
Single (S) or multiple (M) echelons	S	S	S	S	S	S

most disruptive? How much inventory is too much? How much will an improvement in customer service cost? How much is a more reliable vendor worth? And so on. We will develop additional insights regarding inventory in Part II and will return to the practical considerations of inventory management in Chapter 17 of Part III.

The inventory models and insights discussed here also provide a framework for thinking about higher-level actions that can change the nature of these tradeoffs, such as increased system flexibility, better vendor management, and improved quality. Finding ways to alter these fundamental relationships is a key management priority that we will explore more fully in Parts II and III.

APPENDIX 2A
BASIC PROBABILITY

Random Experiments and Events

The starting point of the field of **probability** is the random experiment. A **random experiment** is any measurement or determination for which the outcome is not known in advance. Examples include measuring the hardness of a piece of bar stock, checking a circuit board for short circuits, or tossing a coin.

The set of all possible outcomes of the experiment is called the **sample space.** For example, consider the random experiment of tossing two coins. Let (a, b) denote the outcome of the experiment, where a is H if the first coin comes up heads or T if it comes up tails, with b defined similarly for the second coin. The sample space is then {(H, H), (H, T), (T, H), (T, T)}.

An **event** is a subset of the sample space. The individual elements in the sample space are called **elementary events.** A nonelementary event in our sample space is "at least one coin comes up heads," which corresponds to the set {(H, H), (H, T), (T, H)}. Events are used to make **probability statements.** For instance, we can ask, What is the probability that no tails appear?

Once the set of events has been defined, we can make statements concerning their probability.

Definitions of Probability

Over the years, three basic definitions of probability have been proposed: (1) classical or a priori probability, (2) frequency or a posteriori probability, and (3) subjective probability. The different definitions are useful for different types of experiments.

   A priori probability is appropriate when the random experiment has a sample space composed of n mutually exclusive and equally likely outcomes. Under these conditions, if event A is made up of n_A of these outcomes, we define the probability of A occurring as n_A/n. This definition is useful in describing games of chance. For example, the question regarding the probability of no tails occurring when two coins are tossed can be interpreted in this way. Clearly, all the outcomes in the sample space are mutually exclusive. If the coins are "fair," then no particular outcome is "special" and therefore cannot be more likely to occur than any other. Thus, there are four mutually exclusive and equally likely outcomes. Only one of these contains no tails. Therefore the probability of no tails is $\frac{1}{4}$, or 0.25.

   The second definition of probability, **frequency** or **a posteriori probability,** is also couched in terms of a random experiment, but *after* the experiment instead of *before* it. To describe this definition, we imagine performing a number of experiments, say N, of which M result in event E. Then we define the probability of E to be the number p to which the ratio of M/N converges as N becomes larger and larger. For instance, suppose $p = 0.75$ is the long-run fraction of good chips produced on a line in a wafer fabrication. Then we can consider p to be the probability of producing a good wafer on any given try.

   Subjective probability can be used to describe experiments that are intrinsically impossible to replicate. For instance, the probability of rain at the company picnic tomorrow is a meaningful number, but is impossible to determine experimentally since tomorrow cannot be repeated. So when the weather forecaster says that the chance of rain tomorrow is 50 percent, this number represents a purely subjective estimate of likelihood.

   Fortunately, regardless of the definition of probability used, the tools and techniques for analyzing probability problems are the same. The first step is to assign probabilities to events by means of a probability function. A **probability function** is a mathematical function that takes as input an event and produces a number between zero and one (i.e., a probability).

   For example, consider again the two-coin toss experiment. Suppose P is the corresponding probability function. Since there is nothing unique about any of the outcomes listed above, they should be equally likely. Thus, we can write

$$P\{(\text{H, T})\} = \tfrac{1}{4}$$

Also, since the events (H, T) and (T, H) are mutually exclusive, their probabilities are additive, so

$$P\{(\text{H, T}) \text{ or } (\text{T, H})\} = \tfrac{1}{4} + \tfrac{1}{4} = \tfrac{1}{2}$$

Similarly, the probability of the "sure event" (i.e., that (H, H), (H, T), (T, H), or (T, T) will occur) must be one. Probability functions provide a useful shorthand for making statements regarding random events.

Random Variables and Distribution Functions

The majority of probability results turn on the concept of a **random variable.** Unfortunately, the term *random variable* is a misnomer since it is neither random nor a variable. Like a probability function, a random variable is a function. But instead of defining probabilities to events, it assigns numbers to *outcomes* of a random experiment. This greatly simplifies notation by replacing clumsy representations of outcomes like (H, T) with numbers.

For example, a random variable for the two coin experiment can be defined as

Outcome	Value of Random Variable
(H, H)	0
(H, T)	1
(T, H)	2
(T, T)	3

A random variable for the experiment to measure the hardness of bar stock might be the output of a device that applies a known pressure to the bar and reads out the Rockwell hardness index. A random variable for the circuit-board experiment might be simply the number of short circuits.

Random variables can be either **continuous** or **discrete**. Continuous random variables assign real numbers to their associated outcomes. The hardness experiment is one such example. Discrete random variables, on the other hand, assign outcomes to integers. Examples of discrete random variables are the random variable defined above for the coin toss experiment and the number of short circuits on a circuit board.

Random variables are also useful in defining events. For instance, all the outcomes of the circuit-board experiment with no more than five short circuits constitute an event. The linkage between the event referenced by a random variable and the probability of the event is given by its associated **distribution function,** which we will denote by G. For instance, let X denote the hardness of a piece of steel with an associated distribution function G. Then the probability that the hardness is less than or equal to some value x can be written as

$$P\{X \leq x\} = G(x)$$

If the event of interest is that the hardness is in some range of values, say from x_1 to x_2, we can write

$$P\{x_1 < X \leq x_2\} = G(x_2) - G(x_1)$$

Note that since X is continuous, it can take on values with an infinite number of decimal places of accuracy. Thus, the probability of X being *exactly* any number in particular (say, $X = 500.0000\ldots$) is zero. However, we can talk about the **probability density function** f as the probability of X lying in a small interval divided by the size of the interval, so that

$$g(x) \, \Delta x = P\{x \leq X \leq x + \Delta x\}$$

Of course, to be precise, $g(x)$ is defined only in the limit as Δx goes to zero. But for practical purposes, as long as Δx is small, this expression is almost exact. For instance,

$$P[4.9999 \leq X \leq 5.0001] \approx f(5) \cdot 0.0002$$

to a high degree of accuracy.

For continuous random variables defined for positive real numbers, g and G are related by

$$G(x) = \int_0^x g(x) \, dx$$

Analogous to the probability density functions of continuous random variables, discrete random variables have **probability mass** functions. We typically denote these functions by $p(x)$ to distinguish them from density functions. For instance, in the two-coin experiment, the event of two heads coming up is the same as the event $\{X = 0\}$. Its associated probability is

$$P\{\text{two heads}\} = P\{X = 0\} = p(0) = \tfrac{1}{4}$$

Notice that, unlike in the continuous case, in the discrete case there *is* a finite probability of particular values of the random variable.

In many cases, discrete random variables are defined from zero to positive infinity. For these discrete distributions, the relationship between p and G is given by

$$G(x) = \sum_{i=0}^{x} p(i)$$

Using the distribution function G for the two-coin experiment, we can write the probability of one or fewer tails as

$$P\{\text{one or fewer tails}\} = P[X \leq 2] = G(2) = p(0) + p(1) + p(2)$$

Expectations and Moments

The probability density and mass functions can be used to compute the **expectation** of a random variable, which is also known as the **first moment, mean,** or **average** and is often denoted by μ. For a discrete random variable X defined from zero to infinity with probability mass function p, the **expected value** of X, frequently written $E[X]$, is given by

$$\mu = E[X] = p(1) + 2p(2) + 3p(3) + \cdots = \sum_{x=0}^{\infty} x p(x)$$

For a continuous random variable with density g, the expected value is defined analogously as

$$\mu = E[X] = \int_{0}^{\infty} x g(x)\, dx$$

Note that it follows from these definitions that the mean of the sum of random variables is the sum of their means. For example, if X and Y are random variables of any kind (e.g., discrete or continuous, independent or not), then

$$E[X + Y] = E[X] + E[Y]$$

In addition to computing the expectation, one can compute the expected value of virtually any function of a random variable, although only a few are commonly used. The most important function of a random variable, which measures its **dispersion** or spread, is $(X - E[X])^2$. Its expectation is called the **variance,** usually denoted as σ^2, and is given by

$$\sigma^2 = E[(X - E[X])^2] = E[X^2 - 2X E[X] - E[X]^2] = E[X^2] - E[X]^2$$
$$= \sum_{x=0}^{\infty} x^2 p(x) - \mu^2$$

for the discrete case and by

$$\sigma^2 = E[(X - E[X])^2] = E[X^2] - E[X]^2$$
$$= \int_{0}^{\infty} x^2 g(x)\, dx - \mu^2$$

for the continuous case. The **standard deviation** is defined as the square root of the variance. Note that the standard deviation has the same units as the mean and the random variable itself.

In Chapters 8 and 9, both the mean and the standard deviation are used extensively to describe many important random variables associated with manufacturing systems (e.g., capacity, cycle time, and quality).

Conditional Probability

Beyond simply characterizing the likelihood of individual events, it is often important to describe the dependence of events on one another. For example, we might ask, What is the probability that a machine is out of adjustment *given* it has produced three bad parts in a row? Questions like these are addressed via the concept of **conditional probability.**

The conditional probability that event E_1 occurs, given event E_2 has occurred, written $P[E_1|E_2]$, is defined by

$$P[E_1|E_2] = \frac{P[E_1 \text{ and } E_2]}{P[E_2]}$$

To illustrate this concept, consider the following questions related to the experiment with two coins: What is the probability of two heads, given the first coin is a head? and What is the probability of two heads, given there is at least one head?

To answer the first question, let E_1 be the event "two heads" and let E_2 be the event "the first coin is a head." Note that the event "E_1 and E_2" is equivalent to the event E_1 (the only way to have two heads *and* the first coin to be a head is to have two heads). Hence,

$$P[E_1 \text{ and } E_2] = P[E_1] = \tfrac{1}{4}$$

Since there are two ways for the first coin to be a head [(H, H) and (H, T)], the probability of E_2 is one-half, so

$$P[E_1|E_2] = \frac{P[E_1 \text{ and } E_2]}{P[E_2]}$$

$$= \frac{P[E_1]}{P[E_2]}$$

$$= \frac{\frac{1}{4}}{\frac{1}{2}} = \frac{1}{2}$$

One way to think about conditioning is that the information of knowing an event has occurred serves to reduce the "effective" sample space. In the above example, knowing that "the first coin is a head" eliminates the outcomes (T, H) and (T, T), leaving only (H, H) and (H, T). Since the event "two heads" [(H, H)] corresponds to one-half of the remaining outcomes, its probability is one-half.

To answer the second question, let E_2 be the event "at least one head." Again, the event "E_1 and E_2" is equal to the event E_1 and has probability of one-fourth. However, there are three ways to have at least one head [(H, H), (H, T), and (T, H)], so $P[E_2] = \tfrac{3}{4}$ and

$$P[E_1|E_2] = \frac{P[E_1 \text{ and } E_2]}{P[E_2]}$$

$$= \frac{P[E_1]}{P[E_2]}$$

$$= \frac{\frac{1}{4}}{\frac{3}{4}} = \frac{1}{3}$$

This time, knowing that "at least one head" occurred eliminates only the outcome (T, T), which leaves the outcome (H, H) as one of three equally likely outcomes, which therefore has a probability of one-third.

As another example, consider a random experiment involving the tossing of two dice. The sample space of the experiment is given by $\{(d_1, d_2)\}$, where $d_i = 1, 2, \ldots, 6$ is the number of dots on die i. There are 36 different points in the sample space; by symmetry, these are all equally likely.

Now let X be a random variable equal to the sum of the number of spots on the dice. Note that the number of possible values of X is 11 and that these do *not* have equal probability. To compute the probability of any particular value of X, we must count the number of ways it can result (i.e., the number of outcomes making up the event) and divide by the total number outcomes in the sample space. Thus, the probability of rolling a six is found by noting there are five outcomes that result in a 6—$\{(1, 5), (2, 4), (3, 3), (4, 2), (5, 1)\}$—out of 36 possible outcomes, so $P[X = 6] = \tfrac{5}{36}$.

Computing the conditional probability of rolling a six given that the first die is three or less is a bit more complicated. Let E_1 be the event "rolling a six" and E_2 be the event "the first die is three or less." The event corresponding to E_1 *and* E_2 corresponds to three outcomes in the sample space—$\{(1, 5), (2, 4), (3, 3)\}$—so that $P[E_1 \text{ and } E_2] = \frac{3}{36} = \frac{1}{12}$. Event E_2 corresponds to 18 outcomes in the sample space

$$\{(1, 1), (1, 2), (1, 3), (1, 4), (1, 5), (1, 6)(2, 1), (2, 2), (2, 3),$$
$$(2, 4), (2, 5), (2, 6)(3, 1), (3, 2), (3, 3), (3, 4), (3, 5), (3, 6)\}$$

so $P[E_2] = \frac{18}{36} = \frac{1}{2}$. Thus, the conditional probability of rolling a six given that the first die is three or less is

$$P[E_1|E_2] = \frac{P[E_1 \text{ and } E_2]}{P[E_2]} = \frac{\frac{1}{12}}{\frac{1}{2}} = \frac{1}{6}$$

Independent Events

Conditional probability allows us to define the notion of **stochastic independence** or, simply, independence. Two events E_1 and E_2 are defined to be **independent** if

$$P[E_1 \text{ and } E_2] = P[E_1]P[E_2]$$

Notice that this definition implies that if E_1 and E_2 are independent and $P(E_2) > 0$, then

$$P[E_1|E_2] = \frac{P[E_1 \text{ and } E_2]}{P[E_2]} = \frac{P[E_1]P[E_2]}{P[E_2]} = P[E_1]$$

Thus, events E_1 and E_2 are independent if the fact that E_2 has occurred does not influence the probability of E_1.

If two events are independent, then the random variables associated with these events are also independent. Independent random variables have some nice properties. One of the most useful is that the expected value of the product of two independent random variables is simply the product of the expected values. For instance, if X and Y are independent random variables with means of μ_x and μ_y, respectively, then

$$E[XY] = E[X]E[Y] = \mu_x\mu_y$$

This is not true in general if X and Y are not independent.

Independence also has important consequences for computing the variance of the sum of random variables. Specifically, if X and Y are independent, then

$$\text{Var}(X + Y) = \text{Var}(X) + \text{Var}(Y)$$

Again, this is not true in general if X and Y are not independent.

An important special case of this variance result occurs when random variables X_i, $i = 1, 2, \ldots, n$, are independent and identically distributed (i.e., they have the same distribution function) with mean μ and variance σ^2, and Y, another random variable, is defined as $\sum_{i=1}^{n} X_i$. Then since means are always additive, the mean of Y is given by

$$E[Y] = E\left[\sum_{i=1}^{n} X_i\right] = n\mu$$

Also, by independence, the variance of Y is given by

$$\text{Var}(Y) = \text{Var}\left(\sum_{i=1}^{n} X_i\right) = n\sigma^2$$

Note that the standard deviation of Y is therefore $\sqrt{n}\sigma$, which does not increase with the sample size n as fast as the mean. This result is important in statistical estimation, as we note later in this appendix.

Special Distributions

There are many different types of distribution functions that describe various kinds of random variables. Two of the most important for modeling production systems are the (discrete) Poisson distribution and the (continuous) normal distribution.

The Poisson Distribution.   The **Poisson distribution** describes a discrete random variable that can take on values $0, 1, 2, \ldots$. The probability mass function (pmf) is given by

$$p(k) = \frac{e^{-\mu}\mu^k}{k!} \qquad k = 0, 1, 2, \ldots$$

and the cumulative distribution function (cdf) is given by

$$G(x) = \sum_{k=0}^{x} p(x)$$

The mean (expectation) of the Poisson is μ, and the standard deviation is $\sqrt{\mu}$. Notice that this implies that the Poisson is a "one-parameter distribution" because specifying the mean automatically specifies the standard deviation.

To illustrate the use of the Poisson pmf and cdf, suppose the number of customers who place orders to a particular plant on any given day is Poisson-distributed with a mean of two. Then the probability of zero orders being placed is given by

$$p(0) = \frac{e^{-2}2^0}{0!} = e^{-2} = 0.135$$

The probability of exactly one order on a given day is

$$p(1) = \frac{e^{-2}2^1}{1!} = e^{-2} \times 2 = 0.271$$

The probability of two or more orders on a given day is one minus the probability of one or fewer orders, which is given by

$$1 - G(1) = 1 - p(0) - p(1) = 1 - 0.135 - 0.271 = 0.594$$

Part of the reason that the Poisson distribution is so important is that it arises frequently in practice. In particular, **counting processes** that are composed of a number of independent counting processes tend to look Poisson. For example, in the situation used for the numerical calculations above, the underlying counting process is the number of customers who place orders. This is made up of the sum of the separate counting processes representing the number of orders placed by individual customers. To be more specific, if we let $N(t)$ denote the total number of orders that have been placed on the plant by time t, we let $N_i(t)$ denote the number of orders placed by customer i by time t (which may or may not be Poisson), and we let M denote the total number of potential customers, then clearly

$$N(t) = N_1(t) + \cdots + N_M(t)$$

As long as M is "large enough" (say 20 or more, the exact number depends on how close the $N_i(t)$ are to Poisson) and the times between counts for processes $N_i(t)$ are independent, identically distributed, random variables for each i, then $N(t)$ will be a Poisson process. (Note that the interarrival times between orders need only be identically distributed for each given customer; they do not need to be the same for different customers. So it is entirely permissible to have customers with different rates of ordering.)

If $N(t)$ is a Poisson process with a rate of λ arrivals per unit time, then the number of arrivals in t units of time is Poisson-distributed with mean λt. That is, the probability of exactly k arrivals in an interval of length t is

$$p(k) = \frac{e^{-\lambda t}(\lambda t)^k}{k!} \qquad k = 0, 1, 2, \ldots$$

This situation arises frequently. The historical application of the Poisson process was in characterizing the number of phone calls to an exchange in a given time interval. Since callers tend to space their phone calls independently of one another, the total number of phone calls received by the exchange over an interval of time tends to look Poisson. For this same reason, many other arrival processes (e.g., customers in a bank or a restaurant, hits on a Web site, demands experienced by a retailer) are well characterized by the Poisson distribution. A related situation of importance to manufacturing is the number of failures that a machine experiences. Since complex machinery can fail for a wide variety of reasons (e.g., power loss, pump failure, jamming, loss of coolant, and component breakage) and since we do not replace *all* the components whenever one breaks, we end up with a set of components having different times to failure and different ages. Thus, we can think of the failures as "arriving" from a number of different sources. Since these different sources are often independent, the number of failures experienced during a given interval of operating time tends to look Poisson.

The Exponential Distribution

One additional important point about the Poisson distribution is that the times between arrivals in a Poisson process with arrival rate λ are **exponentially distributed.** That is, the time between the kth and $(k + 1)$st arrival is a continuous random variable with density function

$$g(t) = \lambda e^{-\lambda t} \qquad \lambda \geq 0$$

and cumulative distribution function

$$G(t) = 1 - e^{-\lambda t} \qquad \lambda \geq 0$$

The mean of the exponential is $1/\lambda$, and the standard deviation is also $1/\lambda$; so, like the Poisson, the exponential is a one-parameter distribution.

To illustrate the relationship between the Poisson and exponential distributions, let us reconsider the previous example in which we had a Poisson process with an arrival rate of two orders per day. The probability that the time until the first order is less than or equal to one day is given by the exponential cdf as

$$G(1) = 1 - e^{(-2)(1)} = 0.865$$

Notice that the probability that the first order arrives within one day is exactly the same as the probability of one or more orders on the first day. This is 1 minus the probability of zero arrivals on the first day, which can be computed using the Poisson probability mass function as

$$1 - p(0) = 1 - 0.135 = 0.865$$

We see that there is a close relationship between the Poisson (which measures the number of arrivals) and exponential (which measures times between arrivals) distributions. However, it is important to keep the two distinct, since the Poisson distribution is discrete and therefore suited to counting processes, while the exponential is continuous and therefore suited to times.

A fascinating fact about the exponential distribution is that it is the *only* continuous distribution that possesses the **memorylessness** property. This property is defined through the **failure rate function,** which is also called the **hazard rate function** and is defined for any random variable X with cdf $G(t)$ and pdf $g(t)$ as

$$h(t) = \frac{g(t)}{1 - G(t)} \tag{2.59}$$

To interpret $h(t)$, suppose that the random variable X has survived for t hours. The probability that it will not survive for an additional time dt is given by

$$\begin{aligned} P[X \in (t, t + dt)|X > t] &= \frac{P[X \in (t, t + dt), X > t]}{P[X > t]} \\ &= \frac{P[X \in (t, t + dt)]}{P[X > t]} \\ &= \frac{g(t)\,dt}{1 - G(t)} \\ &= h(t)\,dt \end{aligned}$$

Hence, if X represents a lifetime, then $h(t)$ represents the conditional density that a t-year-old item will die (fail). If X represents the time until an arrival in a counting process, then $h(t)$ represents the probability density of an arrival given that no arrivals have occurred before t.

A random variable that has $h(t)$ increasing in t is called **increasing failure rate (IFR)** and becomes more likely to fail (or otherwise end) as it ages. A random variable that has $h(t)$ decreasing in t is called **decreasing failure rate (DFR)** and becomes less likely to fail as it ages. Some random variables (e.g., the life of an item that goes through an initial burn-in period during which it grows more reliable and then eventually goes through an aging period in which it becomes less reliable) are neither IFR nor DFR.

Now let us return to the exponential distribution. The failure rate function for this distribution is

$$h(t) = \frac{g(t)}{1 - G(t)} = \frac{\lambda e^{-\lambda t}}{1 - (1 - e^{-\lambda t})} = \lambda$$

which is constant! This means that a component whose lifetime is exponentially distributed grows neither more nor less likely to fail as it ages. While this may seem remarkable, it is actually quite common because, as we noted, Poisson counting processes, and hence exponential interarrival times, occur often. For instance, as we observed, a complex machine that fails due to a variety of causes will have failure events described by a Poisson process, and hence the times until failure will be exponential.

The Normal Distribution

Another distribution that is extremely important to modeling production systems, arises in a huge number of practical situations, and underlies a good part of the field of statistics is the normal distribution. The normal is a continuous distribution that is described by two parameters, the mean μ and the standard deviation σ. The density function is given by

$$g(x) = \frac{1}{\sqrt{2\pi}\sigma} e^{-(x-\mu)^2/(2\sigma^2)}$$

The cumulative distribution function, as always, is the integral of the density function

$$G(x) = \int_{-\infty}^{x} g(y)\,dy$$

Unfortunately, it is not possible to write $G(y)$ as a simple, closed-form expression. But it is possible to "standardize" normal random variables and compute $G(x)$ from a lookup table of the standard normal distribution, as we describe below.

A **standard normal distribution** is a normal distribution with mean zero and standard deviation of one. Its density function is virtually always denoted by $\phi(z)$ and is given by

$$\phi(z) = \frac{1}{\sqrt{2\pi}} e^{-z^2/2}$$

The cumulative distribution function is denoted by $\Phi(z)$ and is given by

$$\Phi(z) = \int_{-\infty}^{z} \phi(y)\,dy$$

There is no closed-form expression for $\Phi(z)$ either, but this function is readily available in lookup tables, such as Table 1 at the end of the book and using functions built into scientific calculators and spreadsheet programs.

The reason that standard normal tables are so useful is that if a random variable X is normally distributed with mean μ and standard deviation σ, then the "standardized" random variable

$$Z = \frac{X - \mu}{\sigma}$$

is normally distributed with mean zero and standard deviation one.

To illustrate how this property can be exploited, suppose a casting process produces castings whose weights are normally distributed with mean 1,000 grams and standard deviation 150 grams.

Let X denote the (random) weight of a given casting. Then the probability that the casting will weigh less than or equal to 850 grams is

$$G(850) = P(X \leq 850) = P\left(\frac{X - 1,000}{150} \leq \frac{850 - 1,000}{150}\right) = P(Z \leq -1) = \Phi(-1)$$

From a standard normal table we find that $\Phi(-1) = 0.159$. Hence, we would expect 15.9 percent of the castings to have weights less than 850 grams. Similarly, the probability of the casting having a weight greater than 1,150 grams is

$$1 - G(1,150) = 1 - P(X \leq 1,150) = 1 - P\left(Z \leq \frac{1,150 - 1,000}{150}\right) = 1 - P(Z \leq 1) = \Phi(1)$$

From a standard normal table, $\Phi(1) = 0.841$, so $1 - \Phi(1) = 0.159$. Notice that this is the same as $\Phi(-1)$. The reason is that the standard normal distribution is symmetric (bell-shaped). Hence, the probability of a random sample one standard deviation or more below the mean is equal to the probability of a random sample one standard deviation or more above the mean.

The probability that a randomly chosen casting weighs between 850 and 1,150 grams is given by $1 - G(1,150) - G(850) = 1 - 0.159 - 0.159 = 0.682$. These kinds of calculations are central to **statistical quality control.** For instance, if we were to observe fewer than 68.2 percent of castings in the weight range between 850 grams and 1,150 grams, then this would be a sign that the process was no longer producing castings whose weights are normally distributed with mean 1,000 and standard deviation 150. This could be due to a change in either the mean or the standard deviation in the underlying process. This type of logic can be used to construct **process control charts** for monitoring the behavior of many different types of processes.

A major reason that the normal distribution is so important in practice is that it arises frequently in nature. This is due to the famous **central limit theorem,** which states (roughly) that the sum of a sufficiently large number (say, greater than 30) of random variables will be normally distributed.

To illustrate this, suppose we measure the times between arrivals of phone calls to an exchange. From our discussion of the Poisson distribution, we know that these times are likely to be exponentially distributed. The exponential is very different from the normal, as we can see from the density functions. The normal density is a symmetric, bell-shaped function with its peak at the mean value μ. The exponential density, on the other hand, is only defined above zero, takes on its maximum value at zero, and declines exponentially above zero. Also, because the exponential always has a standard deviation equal to its mean, while the normal generally has a standard deviation less than its mean, we typically say that exponential random variables are more *variable* than normal random variables. We define a measure of variability and discuss this concept in greater depth in Chapter 8.

But even though the interarrival times between calls are far from normal, the central limit theorem implies that the sum of these times will tend to look normal. That is, if we add 40 interarrival times, which would represent the time until the 40th arrival and repeat this many times to create a histogram, the result will be a bell-shaped curve indistinguishable from that of a normally distributed random variable.

The central limit theorem is fundamental to statistics because in statistics we frequently compute means from data. For instance, if we select N individuals randomly from the population of the United States and measure their heights, then letting X_i represent the (random) height of the ith individual, we see the mean height of the selected group is

$$\bar{X} = \frac{X_1 + \cdots + X_N}{N}$$

If we were to repeat this experiment over and over, we would get different values for the N heights. Hence, the average $\bar{X}$ is itself a random variable. If N is large enough, $\bar{X}$ will be normally distributed. This fact allows us to use the normal distribution to compute the probability that $\bar{X}$ lies within a given interval (i.e., a confidence interval) and make a variety of statistical tests.

Parameters and Statistics

The true probabilities of events (e.g., the probability that a machine will run without breakdown for at least 100 hours) and moments of distributions (e.g., the mean time to process a job) are

parameters of the system. These are typically known only to God. We mere humans can only compute **estimates** of the true values of parameters. This is the basic task of the field of **statistics.**

To estimate a parameter, we take a **random sample,** which represents a collection of independent, identically distributed random variables from a given **population.**[12] For instance, since we cannot measure the hardness of every point on a piece of bar stock, we take a sample of measurements to give us an indication of the true hardness.

A **statistic** is simply a function of a random sample that can be computed (i.e., it has no unknown parameters). Two common statistics (also called **estimators**) are the **sample mean** and the **sample variance** of a random variable. Consider a sample of n independent and identically distributed random variables X_i, $i = 1, 2, \ldots, n$, each with mean μ and variance σ^2. The sample mean $\bar{X}$ is given by the average of the observations, computed as

$$\bar{X} = \frac{1}{n} \sum_{i=1}^{n} X_i$$

Note that the sample mean is itself a random variable. The mean of $\bar{X}$ is also μ. Estimators, such as $\bar{X}$, whose expectation is equal to the value of the parameter being estimated are called **unbiased** estimators. Because the X_i are independent, the variance of $\bar{X}$ is given by

$$\text{Var}(\bar{X}) = \text{Var}\left(\frac{1}{n} \sum_{i=1}^{n} X_i\right) = \frac{1}{n^2} \text{Var}\left(\sum_{i=1}^{n} X_i\right) = \frac{1}{n^2} n\sigma^2 = \frac{\sigma^2}{n}$$

Hence, while the variance of any single observation is σ^2, the variance of the mean of n observations is σ^2/n (so the standard deviation is $\sigma/\sqrt{n}$). Since this variance decreases with n, the implication is that larger samples yield better (i.e., tighter) estimates of the true population mean.

This notion is formalized by the concept of a **confidence interval.** The $(1 - \alpha)$ percent confidence interval for the true mean of the population (i.e., the interval in which we expect the sample mean to lie $(1 - \alpha)$ percent of the time if we estimate it over and over) is given by

$$\bar{X} \pm \frac{z_{\alpha/2}\sigma}{\sqrt{n}}$$

where $z_{\alpha/2}$ is the value in the standard normal table such that $\Phi(z_{\alpha/2}) = 1 - \alpha/2$. Notice that as n grows larger, this interval becomes tighter, meaning that more data yield better estimates.

The above confidence interval assumes that the population variance is known with certainty. But in general the variance is also unknown and hence must itself be estimated. This is done by computing the **sample variance** s^2 which is an unbiased estimator for the true variance and is given by

$$s^2 = \frac{\sum_{i=1}^{n}(X_i - \bar{X})^2}{n - 1}$$

or, in a form that is easier to compute, by

$$s^2 = \frac{\sum_{i=1}^{n} X_i^2 - n\bar{X}^2}{n - 1}$$

The confidence interval for the population mean becomes

$$\bar{X} \pm \frac{t_{\alpha/2;n-1}s}{\sqrt{n}}$$

where $t_{\alpha/2;n-1} > z_{\alpha/2}$, so that the confidence interval is wider due to the uncertainty introduced by having to estimate the variance. However, as n grows large, $t_{\alpha/2;n-1}$ converges to $z_{\alpha/2}$; so for large sample sizes the two confidence intervals are essentially the same.

For example, suppose we wish to characterize the process times of a new machine. The first job takes 90 minutes of run time, the second job 40 minutes, and the third job 110 minutes. Based

---

[12]In a sense, the job of the field of *statistics* is the reverse of that of the field of *probability*. In statistics we use samples to estimate properties of a population. In probability we use properties of the population to describe the likelihood of samples.

on these data, we estimate the mean process time to be $\bar{X} = (90 + 40 + 110)/3 = 80$ hours. Similarly, the estimate of the variance is $s^2 = 1,300$ (so $s = \sqrt{1,300} = 36.06$). For this particular case (assuming the run times are normally distributed), it turns out that $t_{\alpha/2;n-1} = t_{0.05;2} = 2.92$, so the 90 percent confidence interval for the true mean time between outages is given by

$$\bar{X} \pm \frac{t_{\alpha/2;n-1}s}{\sqrt{n}} = 80 \pm \frac{2.92(36.06)}{\sqrt{3}} = 80 \pm 60.78$$

Not surprisingly, with only three observations, we do not have much confidence in this estimate.

In this book we are primarily interested in how systems behave as a function of their parameters (e.g., mean process time, variance of process time) and thus will assume we know these exactly. We caution the reader, however, that in practice one must use estimates of the true parameters. Often, these estimates are not very good, so collecting more data is an important part of the analysis.

APPENDIX 2B
INVENTORY FORMULAS

Poisson Demand Case

If demand during replenishment lead time is Poisson-distributed with mean θ, then the probability mass function (pmf) and cumulative distribution function (cdf) are given by $p(x)$ and $G(x)$, respectively, where

$$p(x) = \frac{e^{-\theta}\theta^r}{r!} \qquad x = 0, 1, 2, \ldots \tag{2.60}$$

$$G(x) = \sum_{k=0}^{r} p(k) \qquad x = 0, 1, 2, \ldots \tag{2.61}$$

These are the basic building blocks of all the performance measures. They can be easily entered as formulas in a spreadsheet, or in some spreadsheets they are actually built in. For example, in Excel

$$p(x) = \text{POISSON}(x, \theta, \text{ FALSE})$$

$$G(x) = \text{POISSON}(x, \theta, \text{ TRUE})$$

Here θ represents the mean, and TRUE and FALSE are used to toggle between the cdf and the pmf. We caution the reader, however, that the Poisson functions in Excel are not always stable for large x, because the formula for $p(x)$ involves the ratio of two large numbers. When θ is large (and hence the reorder point r is likely to be large), it is often safer to use the normal distribution (formulas) with mean θ and standard deviation $\sqrt{\theta}$.

By using the $G(x)$ function, it is simple to compute the fill rate for the base stock model with base stock level R as

$$S(R) = G(R - 1) \tag{2.62}$$

Next we compute the loss function $B(R)$, which represents the average backorder level in a base stock model with base stock level R. Alternatively, $B(R)$ can be interpreted as the expected amount by which lead-time demand exceeds R. It can be written in various forms, including

$$B(R) = \sum_{x=R}^{\infty} (x - R)p(x)$$

$$= \theta - \sum_{x=0}^{R-1} [1 - G(x)]$$

$$= \theta p(R) + [\theta - R](1 - G(R)) \tag{2.63}$$

The last form is the most convenient for use in spreadsheets, since it can be computed without the use of any sums but is correct only for the Poisson case.

Using $B(R)$, we can compute the average inventory level $I(R)$ for the base stock model as a function of the base stock level R as

$$I(R) = R - \theta + B(R) \tag{2.64}$$

Now we turn to the performance measures for the (Q, r) model under the assumption of Poisson demand. As we observed in Section 2.4.3, the inventory position in the (Q, r) model is uniformly spread over the values between $r + 1$ and $r + Q$, which enables us to compute the fill rate by averaging the base stock fill rates for these levels as follows:

$$
\begin{aligned}
S(Q, r) &= \frac{1}{Q} \sum_{x=r+1}^{r+Q} G(x - 1) \\
&= \frac{1}{Q} \sum_{x=r}^{r+Q-1} G(x) \\
&= 1 - \frac{1}{Q} [B(r) - B(r + Q)]
\end{aligned} \tag{2.65}
$$

The last form, which expresses the fill rate in terms of the $B(x)$ function, is the most convenient for use in a spreadsheet, since it does not require compution of a sum.

We can use the same type of argument to compute the backorder level for the (Q, r) model as the average of the backorder levels of the base stock model over the inventory positions from $r + 1$ to $r + Q$:

$$B(Q, r) = \frac{1}{Q} \sum_{x=r+1}^{r+Q} B(x) \tag{2.66}$$

However, we can write this in a simpler form by defining the following function:

$$
\begin{aligned}
\beta(x) &= \sum_{k=x+1}^{\infty} B(k) \\
&= \tfrac{1}{2} \{ [(x - \theta)^2 + x][1 - G(x)] - \theta(x - \theta) p(x) \}
\end{aligned} \tag{2.67}
$$

with the last expression holding only for the Poisson case. The function $\beta(x)$ is sometimes referred to as a **second-order loss function,** since it represents the sum of the first-order loss function $B(k)$ above level x. Using the second form for $\beta(x)$ makes this expression simpler to compute in a spreadsheet. Using $\beta(x)$, we can express the backorder level for the (Q, r) model as

$$B(Q, r) = \frac{1}{Q} [\beta(r) - \beta(r + Q)] \tag{2.68}$$

Finally, once we have $B(Q, r)$, it is simple to compute the average inventory level in the (Q, r) model as

$$I(Q, r) = \frac{Q + 1}{2} + r - \theta + B(Q, r) \tag{2.69}$$

Normal Demand Case

If demand during replenishment lead time is approximated by a normal distribution with mean θ and standard deviation σ, then the probability density function and cumulative distribution function are given by $g(x)$ and $G(x)$, respectively, where

$$g(x) = \frac{1}{\sigma} \phi(z) \tag{2.70}$$

$$G(x) = \Phi(z) \tag{2.71}$$

and

$$z = \frac{x - \theta}{\sigma}$$

and ϕ and Φ represent the pdf and cdf, respectively, of the standard normal distribution, which are tabulated in any standard normal table and are included as standard functions in most spreadsheet programs. For example, in Excel,

$$\phi(z) = \text{NORMDIST}(z, 0, 1, \text{FALSE})$$

$$\Phi(z) = \text{NORMDIST}(z, 0, 1, \text{TRUE})$$

Here, the zero and one represent the mean and standard deviation, respectively, of the standard normal distribution and the inputs TRUE and FALSE are used to toggle between the cdf and pdf.

We can compute the fill rate for the base stock model with base stock level R as

$$S(R) = G(R) \tag{2.72}$$

Notice that this differs from the fill rate expression in the Poisson demand case. The reason is that the normal distribution is continuous. So while the fill rate is still given by $P(X < R)$, where X represents the (random) demand during the replenishment lead time, $P(X < R) = P(X \le R) = G(R)$, because there is no mass at the point $X = R$ for a continuous distribution. When demand is modeled using a discrete distribution, there *is* probability mass at $X = R$ and hence $P(X < R) = P(X \le R - 1) = G(R - 1)$.

The expected backorder level in the base stock model with base stock level R is given by

$$\begin{aligned} B(R) &= \int_R^\infty (x - R) g(x) \, dx \\ &= (\theta - R)[1 - \Phi(z)] + \sigma \phi(z) \end{aligned} \tag{2.73}$$

where $z = (R - \theta)/\sigma$. The second form is obviously better suited to use in a spreadsheet since it does not have an integral.

Using $B(R)$, we can compute the average inventory level as

$$I(R) = R - \theta + B(R) \tag{2.74}$$

Now we turn to the performance measures for the (Q, r) model under the assumption of normal demand. As we did for the Poisson case, we can compute the fill rate and backorder level by averaging these measures from the base stock case. However, since the normal distribution is continuous, we must average over the range from r to $r + Q$ (instead of from $r + 1$ to $r + Q$) and we must use an integral instead of a sum. For the fill rate, this yields

$$\begin{aligned} S(Q, r) &= \frac{1}{Q} \int_r^{r+Q} G(x) \, dx \\ &= 1 - \frac{1}{Q}[B(r) - B(r + Q)] \end{aligned} \tag{2.75}$$

Since we have a simple form for the $B(x)$ function, the second form of the above is easily computed in a spreadsheet.

We can do the same type of averaging to get an expression for the average backorder level

$$B(Q, r) = \frac{1}{Q} \int_r^{r+Q} B(x) \, dx \tag{2.76}$$

However, this is not simple to evaluate in a spreadsheet, since it involves an integral. We can simplify it by defining the continuous analog to the second-order loss function $\beta(x)$ as

$$\begin{aligned} \beta(x) &= \int_x^\infty B(y) \, dy \\ &= \frac{\sigma^2}{2} \{(z^2 + 1)[1 - \Phi(z)] - z\phi(z)\} \end{aligned} \tag{2.77}$$

where $z = (x - \theta)/\sigma$. This allows us to simplify the expression for $B(Q, r)$ to

$$B(Q, r) = \frac{1}{Q}[\beta(r) - \beta(r + Q)] \qquad (2.78)$$

Finally, we can express the average inventory level as

$$I(Q, r) = \frac{Q}{2} + r - \theta + B(Q, r) \qquad (2.79)$$

Notice that this differs from the average inventory level in the Poisson case by a quantity of one-half. The reason for this is that we are using a continuous model of demand, which views the decline of inventory as smooth rather than in unit steps. Since almost all real-world systems involve discrete inventory, it generally makes sense to use the discrete inventory formula (2.69) even when using a continuous model to compute Q and r.

We conclude by reiterating that *all* the formulas are straightforward to compute in a spreadsheet. Therefore, once we have computed R for a base stock model or Q and r for a (Q, r) model using any heuristic—either one of those suggested in this chapter or another one—we can compute the exact performance that will result by using the formulas in this appendix. While approximate objective functions can be useful for purposes of computing the decision variables, there is *no excuse* for using approximate measures in reporting the resulting performance or for checking these against desired target values.

Study Questions

1. Harris, in the original 1913 paper on the EOQ model, suggested that "most managers, indeed, have a rather hazy idea as to just what this [setup] cost amounts to."
   a. Do you think that setup cost, as defined in the EOQ model, is more easily specified today than in 1913? Why or why not?
   b. Give some examples of costs that might make up this setup cost.
   c. What might setup cost in the model actually be serving as a surrogate for in the real system?

2. Analogous to item 1c above, what might inventory carrying cost in the EOQ model serve as a surrogate for in the real system? With this in mind, comment on the suggestion (once fairly common in textbooks) that "a charge of 10 percent on stock is a fair one to cover both interest and depreciation." What is another name for this "charge"?

3. Harris wrote that "higher mathematics" is required to solve the EOQ model. What is the name of this branch of mathematics? Who invented it and when? When do most Americans study this subject in the current educational system? Was this really "higher mathematics" in 1913?

4. Consider the following situations. Label them as either A for appropriate or L for less appropriate for application of the EOQ model.
   a. Automobile manufacturer ordering screws from a vendor
   b. Automobile manufacturer deciding on how many cars to paint per batch of a particular color
   c. A job shop ordering bar stock
   d. Office ordering copier paper
   e. A steel company deciding how many slabs to move at once between the casting furnace and the rolling mill

5. A basic modeling assumption underlying the EOQ model is constant and level demand over the infinite time horizon. Of course, this is never satisfied exactly in practice. What options does one have for lot sizing in the face of nonconstant demand?

6. What is the key difference in the modeling assumptions between the EOQ and the Wagner–Whitin models?

7. Does the Wagner–Whitin property offer a fundamental insight into plant behavior? If so, what is it? What problems are there with this property as a guide for manufacturing practice?

8. Give at least three criticisms of the validity of the Wagner–Whitin model.

9. What is the key difference between the EOQ model and the (Q, r) model? Between the base stock model and the (Q, r) model?

10. Why is the statement "The reorder point r affects customer service, while the replenishment quantity Q affects replenishment frequency" true in rough terms but not precisely true?

11. Why does increasing the variability of the demand process tend to require a higher level of safety stock (i.e., a higher reorder point)?

12. Suppose you are stocking parts purchased from vendors in a warehouse. How could you use a (Q, r) model to determine whether a vendor of a part with a higher price but a shorter lead time is offering a good deal? What other factors should you consider in deciding to change vendors?

13. In a multiproduct reorder point problem subject to an aggregate service constraint, what will be the effect of increasing the cost of one of the parts on the fill rate of that part? On the fill rates of the other parts?

14. A man was discovered trying to carry a bomb onto an airplane. When he was removed, his excuse was: "Everyone knows that the probability of there being a bomb on an airplane is extremely low. Imagine how low the probability of *two* bombs on the airplane must be! I had no intention of blowing up the plane. By carrying a bomb on board, I was only trying to make it safer!"

   What do you think of the man's reasoning? (*Hint:* Use conditional probability.)

Problems

1. Perform the two-coin toss experiment discussed in Appendix 2A by flipping two coins (a penny and a nickel) 50 times and recording the outcome (H or T for each coin) for each flip.
   a. Estimate the probability of two heads given at least one head by counting the number of (H, H) outcomes and dividing by the number of outcomes that have at least one head. How does this compare to the true value of one-third computed in Appendix 2A?
   b. Estimate the probability of two heads given that the penny is a head by counting the number of (H, H) outcomes and dividing by the number of outcomes for which the penny is a head. How does this compare to the true value of one-half computed in Appendix 2A?

2. Recall the game show "Let's Make a Deal." You are a contestant and there is a fabulous prize behind door number 1, door number 2, or door number 3. You have chosen door number 1. The host of the show opens door number 3 revealing a not-so-fabulous prize, and asks you if you want to change your mind. You have watched the show for a number of years and have noticed that the host always offers contestants the option of switching doors. Moreover, you know that when the host has a choice of doors to open (e.g., the prizes behind both doors 2 and 3 are duds), he chooses randomly. Should you switch to door 2 or stick with door 1 in order to maximize your chances of winning the fabulous prize?

3. A gift shop sells Little Lentils—cuddly animal dolls stuffed with dried lentils—at a very steady pace of 10 per day, 310 days per year. The wholesale cost of the dolls is $5, and the gift shop uses an annual interest rate of 20 percent to compute holding costs.
   a. If the shop wants to place an average of 20 replenishment orders per year, what order quantity should it use?
   b. If the shop orders dolls in quantities of 100, what is the implied fixed order cost?
   c. If the shop estimates the cost of placing a purchase order to be $10, what is the optimal order quantity?

4. Quarter-inch stainless-steel bolts, one and one-half inches long are consumed in a factory at a fairly steady rate of 60 per week. The bolts cost the plant two cents each. It costs the plant $12 to initiate an order, and holding costs are based on an annual interest rate of 25 percent.
   a. Determine the optimal number of bolts for the plant to purchase and the time between placement of orders.
   b. What is the yearly holding and setup cost for this item?
   c. Suppose instead of small bolts we were talking about a bulky item, such as packaging materials. What problem might there be with our analysis?

5. Reconsider the bolt example in Problem 4. Suppose that although we have estimated demand to be 60 per week, it turns out that it is actually 120 per week (i.e., we have a 100 percent forecasting error).
   a. If we use the lot size calculated in the previous problem (i.e., using the erroneous demand estimate), what will the setup plus holding cost be under the true demand rate?
   b. What would the cost be if we had used the optimum lot size?
   c. What percentage increase in cost was caused by the 100 percent demand forecasting error? What does this tell you about the sensitivity of the EOQ model to errors in the data?

6. Consider the bolt example from Problem 4 yet again, assuming that the demand of 30 per week is correct. Now, however, suppose the minimum reorder interval is one month and all order cycles are placed on a power-of-2 multiple of months (that is, one month, two months, four months, eight months, etc. in order to permit truck sharing with orders of other parts.
   a. What is the least-cost reorder interval under this restriction?
   b. How much does this add to the total cost?
   c. How is the effectiveness of powers-of-2 order intervals related to the result of the previous problem regarding the effect of demand forecasting errors?

7. Danny Steel, Inc., fabricates various products from two basic inputs, bar stock and sheet stock. Bar stock is used at a steady rate of 1,000 units per year and costs $200 per bar. Sheet stock is used at a rate of 500 units per year and costs $300 per sheet. The company uses a 20 percent annual holding cost rate, and the fixed cost to place an order is $50, of which $10 is the cost of placing the purchase order and $40 is the fixed cost of a truck delivery. The variable (i.e., per unit charge) trucking cost is included in the unit price. The plant runs 365 days per year.
   a. Use the EOQ formula with the full fixed order cost of $50 to compute the optimal order quantities, order intervals, and annual cost for bar stock and sheet stock. What fraction of the total annual (holding plus order) cost consists of fixed trucking cost?
   b. Using a week (seven days) as the base interval, round the order intervals for bar stock and sheet stock to the nearest power of 2. If you charge the fixed trucking fee only once for deliveries that coincide, what is the annual cost now?
   c. Leave the order quantity for bar stock as in part *b*, but reduce the order interval for sheet stock to match that of bar stock. Recompute the total annual cost and compare to part *b*. Explain your result.
   d. Based on your observation in part *c*, propose an approach for computing a replenishment schedule in a multiproduct environment like this, where part of the fixed order cost corresponds to a fixed trucking fee that is only paid once per delivery regardless of how many different parts are on the truck.

8. Consider the following table resulting from lot sizing by the Wagner–Whitin algorithm:

Month	Demand	Min. Cost	Order Period
1	69	85	1
2	29	114	1
3	36	186	1
4	61	277	3
5	61	348	4
6	26	400	4
7	34	469	5
8	67	555	8
9	45	600	8
10	67	710	10
11	79	789	10
12	56	864	11

   a. Develop the "optimal" ordering schedule.
   b. What will the schedule be if your planning horizon was only six months?

9. Nozone, Inc., a manufacturer of Freon recovery units (for automotive air conditioner maintenance), experiences a strongly seasonal demand pattern, driven by the summer air conditioning season. This year Nozone has put together a six-month production plan, where the monthly demands D_t for recovery units are given in the table below. Each recovery unit is manufacured from one chassis assembly plus a variety of other parts. The chassis assemblies are produced in the machining center. Since there is a single chassis assembly per recovery unit, the demands in the table below also represent demands for chassis assemblies. The unit cost, fixed setup cost, and monthly holding cost for chassis assemblies are also given in this table. The fixed setup cost is the firm's estimate of the cost to change over the machining center to produce chassis assemblies, including labor and materials cost and the cost of disruption of other product lines.

t	1	2	3	4	5	6
D_t	1,000	1,200	500	200	800	1,000
c_t	50	50	50	50	50	50
A_t	2,000	2,000	2,000	2,000	2,000	2,000
h_t	10	10	10	10	10	10

   a. Use the Wagner–Whitin algorithm to compute an "optimal" six-month production schedule for chassis assemblies.
   b. Comment on the appropriateness of using monthly planning periods. What factors should influence the choice of a planning period?
   c. Comment on the validity of using a fixed order cost to consider the capacity constraint at the machining center.

10. YB Sporting Apparel prints up novelty T-shirts commemorating major sports events (e.g., the Super Bowl, the World Series, Northwestern University winning the NCAA Basketball Tournament). The T-shirts cost $5 to make and distribute and sell for $20. Company policy is to dispose of any excess inventory after the event by discounting the T-shirts by 80 percent, that is, sell for $4. In 1994, YB printed shirts for the World Cup soccer playoffs in Chicago. It estimated demand at 12,000 shirts, with a significant amount of uncertainty. Because of this uncertainty, YB printed only 10,000 shirts. What do you think of this decision? What quantity would you have recommended printing?

11. Slaq Computer Company manufactures notebook computers. The economic lifetime of a particular model is only four to six months, which means that Slaq has very little time to make adjustments in production capacity and supplier contracts over the production run. For a soon-to-be-introduced notebook, Slaq must negotiate a contract with a supplier of motherboards. Because supplier capacity is tight, this contract will specify the number of motherboards in advance of the start of the production run. At the time of contract negotiation, Slaq has forecasted that demand for the new notebook is normally distributed with a mean of 10,000 units and a standard deviation of 2,500 units. The net revenue from a notebook sale is $500 (note that this includes the cost of the motherboard, as well as all other material, production, and shipping costs). Motherboards cost $200 and have no salvage value (i.e., if they are not used for this particular model of notebook, they will have to be written off).

    a. Use the news vendor model to compute a purchase quantity of motherboards that balances the cost of lost sales and the cost of excess material.

    b. Comment on the appropriateness of the news vendor model for this capacity planning situation. What factors are not considered that might be important?

12. Chairish-Is-The-Word, Inc., manufactures top-end hardwood chairs that are sold through a variety of retail outlets. The most popular model sells (wholesale) for $400 per chair and costs $300 to make. Past data show that average monthly demand is 1,000 chairs with a standard deviation of 200 chairs and that the normal distribution is a reasonable fit. CITW uses a 20 percent annual interest charge to estimate inventory carrying costs, so that the cost to carry one chair in stock for one month is $300(0.20)/12 = \$5.

    a. If all orders are backlogged and the cost of lost customer goodwill from carrying a single chair on backorder is $20, what order-up-to (base stock) level should CITW use?

    b. If any order not filled from stock is lost (i.e., the customer buys it from the competition), what order-up-to level should CITW use?

    c. Explain the reason for the difference between your answers in parts *a* and *b*.

13. Jill, the office manager of a desktop publishing outfit, stocks replacement toner cartridges for laser printers. Demand for cartridges is approximately 100 per year and is quite variable (i.e., can be represented using the Poisson distribution). Cartridges cost $100 each. Jill uses a (Q, r) approach to control stock levels.

    a. If Jill wants to restrict replenishment orders to twice per year on average, what batch size Q should she use? If she wants to ensure a service level (i.e., probability of having the cartridge in stock when needed) of at least 98 percent, what reorder point r should she use? (*Hint:* Use Table 2.6.)

    b. If Jill is willing to increase the number of replenishment orders per year to six, how do Q and r change? Explain the difference in r.

    c. If the supplier of toner cartridges offers a quantity discount of $10 per cartridge for orders of 50 or more, how does this affect the relative attractiveness of ordering twice per year versus six times per year? Try to frame your answer in definite economic terms.

14. Moonbeam-Musel (MM), a manufacturer of small appliances, has a large injection molding department. Because MM's CEO, Crosscut Sal, is a stickler for keeping machinery running, the company stocks quick-change replacement modules for the two most common types of failure. Type A modules cost $150 each and have been used at a rate of about seven per month, while type B modules cost $15 and have been used at a rate of about 30 per month, and for simplicity we assume a month is 30 days. Both modules are purchased from a supplier; replenishment lead times are one month and one-half month (15 days) for modules 1 and 2, respectively.

    a. Suppose MM wishes to follow a base stock policy. Assuming that demand is Poisson-distributed, what should the base stock levels be for type A and type B modules in order to ensure a fill rate of at least 98 percent for each module? What are the expected backorder level and the expected inventory level (in dollars)?

    b. Suppose MM estimates the cost to place a replenishment order (regardless of type) to be $5 and the holding cost interest rate to be three percent per month. Use the EOQ model to compute order quantities (where the EOQ values are rounded to the nearest integer to get

Q). Using these order quantities, what should the reorder points be to achieve a 98 percent fill rate for both modules? How do these reorder points and the resulting average backorder level and inventory level compare to those in part a? Explain any difference.

c. Suppose MM estimates the cost per month per unit of backorder to be $15. Use approximation (2.49) to compute reorder points for type A and type B modules (again rounding to the nearest integer). Using the order quantities from part b along with these new reorder points, compare the average total inventory, backorder level, and fill rate with those in part b. Comment on any difference. (Note that the average fill rate is computed by $(D_1 S_1 + D_2 S_2)/(D_1 + D_2)$, where D_1, D_2 are the monthly demand rates and S_1, S_2 are the fill rates for type A and type B components, respectively.)

d. Recompute the reorder points as in part c, but this time assume that replenishment lead times are variable with standard deviations of 7 and 15 days for type A and type B modules, respectively. How much of an effect does this have on the reorder points?

15. Walled-In Books stocks the novel *War and Peace*. Demand averages 15 copies per month, but is quite variable (i.e., is well represented by a Poisson distribution). Replenishments from the publisher require a two-week lead time. The wholesale cost is $12, and Walled-In uses a weekly holding cost rate of one-half percent. It also estimates that the fixed cost of placing and receiving a replenishment order is $5.

a. Compute the approximately optimal order quantity, using the EOQ formula and rounding to the nearest integer. Using this order quantity, find the reorder point that makes the fill rate at least 90 percent. Compute the resulting average inventory (in dollars).

b. Using the order quantity computed in part a, find the reorder point that makes the type I approximation of fill rate at least 90 percent. Compute the true fill rate and inventory level resulting from this reorder point and compare to the values in part a. What does this say about the accuracy of the type I service approximation?

c. Using the order quantity computed in part a, find the reorder point that makes the type II approximation of fill rate at least 90 percent. Compute the true fill rate and inventory level resulting from this reorder point, and compare to the values in part a. What does this say about the accuracy of the type II service approximation? How does the value of Q affect the accuracy of the type II approximation?

d. Cut the order quantity from part a in half, and compute the reorder point needed to make fill rate at least 90 percent. How does the resulting inventory compare to that from part a? Does this imply that the EOQ approximation is poor? Why or why not?

3   THE MRP CRUSADE

Unlike many other approaches and techniques, material requirements planning "works," which is its best recommendation.

Joseph Orlicky, 1974

3.1   Material Requirements Planning—MRP

By the early 1960s, many companies were using digital computers to perform routine accounting functions. Given the complexity and tedium of scheduling and inventory control, it was natural to try to extend the computer to these functions as well. One of the first experimenters in this area was IBM, where Joseph Orlicky and others developed what came to be called **material requirements planning (MRP).** Although it started slowly, MRP got a tremendous boost in 1972 when the American Production and Inventory Control Society (APICS) launched its "MRP Crusade" to promote its use. Since that time, MRP has become the principal production control paradigm in the United States. By 1989, sales of MRP software and implementation support exceeded $1 billion.

Because it is so prevalent, any well-trained manufacturing manager must have some familiarity with how MRP works. Therefore, in this chapter we describe the MRP paradigm and that of its immediate successor, **manufacturing resources planning (MRP II),** as well as its current incarnation, **enterprise resources planning (ERP).** We also highlight the basic insights represented by MRP as well as some difficulties it leaves unresolved.

3.1.1   The Key Insight of MRP

As we noted in Chapter 2, before MRP, most production control systems were based on some variant of statistical reorder points. Essentially this meant that production of any part, finished product, or component was triggered by inventory for that part falling below a specified level. Orlicky and the other originators of MRP recognized that this approach is much better suited to final products than components. The reason is that demand for final products originates outside the system and is therefore subject to uncertainty. However, because components are used to produce final products, demand for components is a function of demand for final products and is therefore *known* for any given final assembly schedule. Treating the two types of demand equivalently, as is done in a statistical reorder point system, ignores the dependence of component demand on final product demand and therefore leads to inefficiencies in scheduling production.

Any demand that originates outside the system is called **independent demand.** This includes all demand for final products and possibly some demand for components (e.g., when they are sold as replacement parts). **Dependent demand** is demand for components that make up independent demand products. Using these terms, the key insight of MRP can be stated as follows:

> Dependent demand is different from independent demand. Production to meet dependent demand should be scheduled so as to explicitly recognize its linkage to production to meet independent demand.

As we will see, the basic mechanics of MRP do exactly this. By working backward from a production schedule of an independent-demand item to derive schedules for dependent-demand components, MRP adds the link between independent and dependent demand that is missing from statistical reorder point systems. MRP is therefore called a **push** system since it computes schedules of what should be started (or *pushed*) into production based on demand. This is in contrast to **pull** systems, such as Toyota's **kanban** system, that authorize production as inventory is consumed. We will discuss kanban in greater detail in Chapter 4 and provide a more complete comparison of push and pull systems in Chapter 10.

3.1.2   Overview of MRP

The basic function of MRP is revealed by its name—to plan material requirements. MRP is used to coordinate orders from within the plant and from outside. Outside orders are called **purchase orders,** while orders from within are called **jobs.** The main focus of MRP is on scheduling jobs and purchase orders to satisfy material requirements generated by external demand.

MRP deals with two basic dimensions of production control: quantities and timing. The system must determine appropriate production quantities of all types of items, from final products that are sold, to components used to build final products, to inputs purchased as raw materials. It must also determine production timing (i.e., job start times) that facilitates meeting order due dates.

In many MRP systems, time is divided into **buckets,** although some systems use continuous time. A bucket is an interval that is used to break time and demand into discrete chunks. The demand that accumulates over the time interval (bucket) is all considered due at the beginning of the bucket. Thus, if the bucket length is one week and during the third week (bucket) there is demand for 200 units on Monday, 250 on Tuesday, 100 on Wednesday, 50 on Thursday, and 350 on Friday, then demand for the third bucket is 950 units and is due on Monday morning. In the past, when data processing was more expensive, typical bucket sizes were one week or longer. Today, most modern MRP systems use daily buckets, although there are still many systems using weeks.

MRP works with both finished products, or **end items,** and their constituent parts, called **lower-level items.** The relationship between end items and lower-level items is described by the **bill of material (BOM),** as shown in Figure 3.1. Demand for end items generates dependent demand for lower-level items. As we noted above, all demand for end items is independent demand, while typically most demand for lower-level items is dependent demand. However, there can be independent demand for lower-level items in the form of spare parts, parts for research and quality tests, and so on.

FIGURE 3.1

Two bills of material

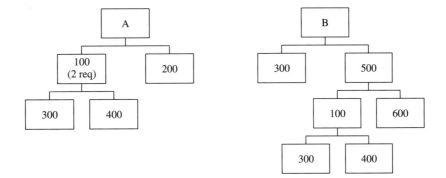

To facilitate the MRP processing, each item in the BOM is given a **low-level code (LLC).** This code indicates the lowest level in a bill of material that a particular part is ever used.[1] End items (that are not a part of any other item) have LLCs of zero. A subassembly that is used only by end items has an LLC of one. A component that is used only by subassemblies having an LLC of one will have an LLC of two, and so on. For example, in Figure 3.1 parts A and B are end items with LLCs of zero. Requirements for these parts come from independent demand. At first glance, it might appear that part 100 should have an LLC of one since it is used directly in part A. However, because it is also a component part for part 500 (whose LLC is one), it is assigned an LLC of two. Similarly, since part 300 is required to make part B with an LLC of zero, but is also required to make part 100 that has an LLC of two, it is given an LLC of three.

Most commercial MRP packages include a **BOM processor** that is used to maintain the bills of material and automatically assign low-level codes. Other functions of the BOM processor include generating "goes-into" lists (where parts are used) and BOM printing.

In addition to the BOM information, MRP requires information concerning independent demand, which comes from the **master production schedule (MPS).** The MPS contains **gross requirements,** the current inventory status known as **on-hand** inventory, and the status of outstanding orders (both purchased and manufacturing) known as **scheduled receipts.**

The basic MRP procedure is simple. We will discuss each of the steps in detail. But briefly, for each level in the bill of material, beginning with end items, MRP does the following for each part:

1. *Netting:* Determine **net requirements** by subtracting on-hand inventory and any scheduled receipts from the gross requirements. The gross requirements for level-zero items come from the MPS, while those for lower-level items are the result of previous MRP operations.

2. *Lot sizing:* Divide the netted demand into appropriate **lot sizes** to form jobs.

3. *Time phasing:* Offset the due dates of the jobs with **lead times** to determine start times.

4. *BOM explosion:* Use the start times, the lot sizes, and the BOM to generate gross requirements of any required components at the next level(s).

5. *Iterate:* Repeat these steps until all levels are processed.

---

[1]Unfortunately, *low*-level codes have the property that the *lower* a part is in the bill of material, the *higher* its *low*-level code.

FIGURE 3.2

Schematic of MRP

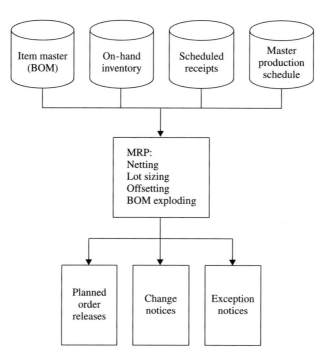

As each part in the bill of material is processed, requirements are generated for lower levels. MRP processes all parts for one level before beginning the next level. Because of the way low-level codes are defined, doing this generates all the gross demand for a lower-level part before it is processed. We will describe each of these steps in detail in Section 3.1.4. The basic outputs of an MRP system are planned order releases, change notices, and exception reports. These we will define in Section 3.1.3. Figure 3.2 presents a schematic of the overall process.

We now illustrate this procedure with a simple example. Suppose the demand for part A is given by the gross requirements from the following master production schedule:

Part A	1	2	3	4	5	6	7	8
Gross requirements	15	20	50	10	30	30	30	30

Suppose further that there are no scheduled receipts (these are a bit tricky and we will discuss them later) and there are 30 units on hand in inventory. We assume that the lot size for part A is 75 units and the lead time is one week. The MRP processing goes as follows.

Netting.   The 30 units on hand will cover all the demand in week one and 15 units left over. The remaining 15 leave five units of the demand of 20 in week two uncovered. Thus, net requirements are as follows:

Part A		1	2	3	4	5	6	7	8
Gross requirements		15	20	50	10	30	30	30	30
Projected on-hand	30	15	−5	—	—	—	—	—	—
Net requirements		0	5	50	10	30	30	30	30

Lot Sizing.   The first uncovered demand is in week two. Therefore, the first **planned order receipt** will be in week two for 75 units (the lot size). Since only five units are needed in week two, 70 units are carried over to week three, which has a demand of 50. This leaves 20 for week four, which has a demand of 10. After covering week four, the remainder is insufficient to cover the demand of 30 units in week five. Thus, we need another lot of 75 to arrive at the beginning of week five. After subtracting 30 units, we have 55 available for week six, which also has a demand of 30, leaving 25 for week seven. The 25 units are not sufficient to cover the demand of 30, and so we need another lot of 75 to arrive in week seven. This lot covers both the remaining demand in week seven (five) and the 30 needed in week eight. We show the results of these calculations in the following tableau:

Part A		1	2	3	4	5	6	7	8
Gross requirements		15	20	50	10	30	30	30	30
Projected on-hand	30	15	−5	—	—	—	—	—	—
Net requirements		0	5	50	10	30	30	30	30
Planned order receipts			75			75		75	

Time Phasing.   To determine when to release the jobs (if made in-house) or purchase orders (if bought from someone else), we simply subtract the lead time from the time of the planned order receipts to obtain the planned order releases. The result using planned lead times of one week is shown below:

Part A		1	2	3	4	5	6	7	8
Gross requirements		15	20	50	10	30	30	30	30
Projected on-hand	30	15	−5	—	—	—	—	—	—
Net requirements		0	5	50	10	30	30	30	30
Planned order receipts			75			75		75	
Planned order releases		75			75		75		

BOM Explosion.   Once we have determined start times and quantities for part A, it is a simple matter to generate demand requirements for all its components. For instance, each unit of part A requires two units of part 100. Therefore, gross requirements for part 100 to produce part A are computed by simply doubling the planned order releases for part A. The gross requirements for part 100 generated by part A must be added to those generated by other parts (e.g., part 500) in order to compute the total gross requirements for part 100. As long as we process parts in order (low to high) of their low-level code, we will have accumulated all the gross requirements for each part before processing it.

3.1.3   MRP Inputs and Outputs

The basic inputs to MRP are a forecast of demand for end items, the associated bills of material, and the current inventory status, plus any data needed to specify production policies. These data come from three sources: (1) the item master file, (2) the master production schedule, and (3) the inventory status file.

The Master Production Schedule.   The master production schedule is the source of demand for the MRP system. It gives the quantity and due dates for all parts that have independent demand. This will include demand for all end items as well as external demand for lower-level parts (e.g., demand for spare parts).

The minimum information contained in the master production schedule is a set of records containing a part number, a need quantity, and a due date for each purchase order. This information is used by MRP to obtain the gross requirements that initiate the MRP procedure. The MPS typically uses the part number to link to the item master file where other processing information is located.

The Item Master File.   The item master file is organized by part number and contains, at a minimum, a description of the part, bill-of-material information, lot-sizing information, and planning lead times.

The BOM data for a part typically list the components and quantities that are directly required to make only that part. The bill-of-material processor uses this information to display complete bills of material for any item, although such detailed information is not needed for MRP processing.

By using low-level codes, MRP accumulates all the demand of a part *before* it processes that part. To see why this is necessary, suppose it were not done. In our example, MRP might process part 100 after processing parts A and B but before processing part 500. If so, it would not have the demand for part 100 generated by part 500. If we go back and schedule more production of part 100, we may wind up with many small jobs of part 100 instead of a few large ones. Several small jobs could easily have the same due date. The result would be a failure to exploit any economies of scale from sharing setups on critical equipment. The use of low-level codes prevents this from happening.

Two other pieces of information needed to perform MRP processing are the **lot-sizing rule (LSR)** and the **planning lead time (PLT).** The LSR determines how the jobs will be sized in order to balance the competing desires of reducing inventory (by using smaller lots) and increasing capacity (by using larger lots to avoid frequent setups). EOQ and Wagner–Whitin, as discussed in Chapter 2, are possible lot-sizing rules. We discuss the use of these and other rules later in this chapter.

The PLT is used to determine job start times. In MRP, this procedure is simple: The start time is equal to the due date minus the PLT. Thus, if the lead times were always precisely equal to the PLTs, MRP would result in parts being ready exactly when needed (i.e., just in time). However, actual lead times vary and are never known in advance. Thus, deciding what planned lead times to use in an MRP system can be a difficult question and one that we will discuss further, in this chapter and in Chapter 5.

On-Hand Inventory.   On-hand inventory data are stored by part number and contain information describing the part, where it is located, and how many are currently on hand. On-hand inventory includes raw material stock, "crib" stock (i.e., inventory that has been processed since being raw material and kept within the plant), and assembly stock. On-hand inventory may also contain information about **allocation** that indicates how many parts are reserved for jobs to be released.

Scheduled Receipts.   This file contains all previously released orders, either **purchase orders** or manufacturing **jobs.** A **scheduled receipt (SR)** is a **planned order release** that has actually been released. For purchased parts, this involves executing a purchase order (PO) and sending it to a vendor. For manufactured parts, this entails gathering all necessary routing and manufacturing information, allocating the necessary inventory for the job, and releasing the job to the plant. Once the PO or job has been released, the planned order release is deleted in the database and the scheduled receipt is created. Thus, SRs are jobs and orders resulting from previous MRP runs and either are currently in process or have not yet been received from the vendor. Jobs that have not yet arrived at an inventory location are considered part of **work in process (WIP).** When the job is completed (i.e., it has finished its routing and goes into stock), the scheduled receipt is deleted from the database and the on-hand inventory is updated to reflect the amount of the part that was completed. A corresponding procedure follows the receipt of a purchased part from a vendor.

The minimum information contained for each scheduled receipt is an identifier (PO number or job number), due date, release date, unit of measure, quantity needed, and current quantity. Other information may include price or cost, routing data, vendor data, material requirements, special handling, anticipated ending quantity, anticipated completion date, etc.

Knowledge of on-hand inventory and scheduled receipts is important to determining net requirements. This procedure is often called **coverage analysis,** and it involves determining how much demand is "covered" by current inventory, purchase orders, and manufacturing jobs.

If demands never changed and jobs always finished on time, all existing scheduled receipts would correspond exactly to subsequent requirements. Unfortunately, demands do change and jobs do not always finish on time, and so scheduled receipts sometimes need to be adjusted. Such adjustments are indicated in **change notices,** described below.

MRP Outputs.   The output of an MRP system includes planned order releases, change notices, and exception reports. Planned order releases eventually become the jobs that are processed in the plant.

A **planned order release (POR)** contains at least three pieces of information: (1) the part number (there can be only one per POR), (2) the number of units required, and (3) the due date for the job. A job or a POR need not correspond to an individual customer order and, in most cases, will not. Indeed, in a situation where there are many common parts, PORs for common components will often be for many different assemblies, not to mention customers. However, if all jobs finish on their due dates, all customer orders will be filled on time. This is accomplished automatically in the MRP processing that we discuss in detail next.

Change notices indicate modifications of existing jobs, such as changes in due dates or priorities. Moving a due date earlier is called **expediting** while making a due date later is known as **deferring.**

Exception reports, as in any large management information system, are used to notify the users that there are discrepancies between what is expected and what will transpire. Such reports might indicate job count differences, inventory discrepancies, imminently tardy jobs, and the like.

3.1.4   The MRP Procedure

While the basic ideas of MRP are simple, the details can get very messy. In this section we go through the MRP procedure in enough detail to give the reader an idea of the basic workings of most commercial MRP systems. To do this, we make use of the following notation. For each part, define:

D_t = gross requirements (demand) for period t (e.g., a week)

S_t = quantity currently scheduled to complete in period t (i.e., a scheduled receipt)

I_t = projected on-hand inventory for end of period t, where current on-hand inventory is given by I_0

N_t = net requirements for period t

With these we will now describe the four basic steps of MRP: netting, lot sizing, time phasing, and BOM explosion.

Netting.   Netting, or coverage analysis, provides two important functions. (1) It adjusts scheduled receipts by expediting those that are currently scheduled to arrive too late and deferring those currently scheduled to arrive too soon, and (2) it computes net demand.

Most MRP systems assume that all SRs will be received before any newly created job can be completed. This makes sense; since SRs are already "on the way," it is unlikely that any new planned order release would be able to "pass" the SR to become available sooner. If an SR is outstanding with a vendor, it should be easier to expedite the existing order than to start a new one. Likewise an SR that is currently in the shop should finish before one that we start now. Therefore, we will assume that coverage will come first from on-hand inventory, second from SRs (regardless of their due date), and finally from new PORs. To compute when the first SR should arrive, we first determine how far into the future the on-hand inventory will cover demand. We compute

$$I_t = I_{t-1} - D_t$$

starting with $t = 1$ and with I_0 equal to current on-hand inventory. We increment t and continue to compute I_t until it becomes less than zero. The period in which this occurs is when the first scheduled receipt should arrive. If the current due date of the first SR is different from this, it should be changed. This will give rise to a change notice indicating a deferral if the SR is to be pushed back and an expedite if it is to be moved forward.[2] Once the SR is changed, the projected on-hand inventory should reflect the change; that is,

$$I_t(\text{after change in SR}) = I_t(\text{before change in SR}) + S_t$$

---

[2]Of course, this automatic changing of due dates occurs only within the database unless someone acts. The change notices are used to propagate this information to the "expediter" who is responsible for ensuring that a job finishes on its due date. This is all very easy in theory, but many times a job may be expedited to a point where it is impossible to finish on time. Such instances lead to occasions when the data in the MRP database do not reflect the true situation on the shop floor.

where S_t is the quantity of SR that is moved into period t. If I_t remains less than zero, the next SR should also be moved to period t. This is repeated until either I_t becomes nonnegative or there are no more scheduled receipts.

Once the projected on-hand inventory is made nonnegative in period t, we continue the procedure by moving forward in time, computing

$$I_t = I_{t-1} - D_t$$

until again I_t becomes less than zero. We repeat this procedure until either we exhaust the scheduled receipts or we have reached the end of the time horizon. If this happens while there are remaining scheduled receipts, a change notice should be issued to either cancel those orders/jobs or defer them to a very late date, since there is no demand for them at this time. More often we will run out of on-hand inventory and SRs before we have exhausted demand. The demands beyond what the on-hand inventory and the scheduled receipts can cover are the **net requirements.**

Once scheduled receipts have been adjusted, the net requirements are easily determined. Let t^* be the first period with a negative projected on-hand inventory *after* the SRs have been properly adjusted.[3] Then the net requirements will be zero for all the periods prior to t^*, equal to the magnitude of the first negative projected on-hand inventory for period t^*, and equal to the gross requirements for the periods beyond t^*. Using our notation,

$$N_t = \begin{cases} 0 & \text{for } t < t^* \\ -I_t & \text{for } t = t^* \\ D_t & \text{for } t > t^* \end{cases}$$

The net requirements are then used in the lot-sizing procedure.

Before we move on to lot sizing, consider an example to illustrate these coverage analysis procedures. Table 3.1 contains the gross requirements from the master production schedule for part A, three scheduled receipts, and the current on-hand inventory count.

TABLE 3.1   Input Data for Example

Part A		1	2	3	4	5	6	7	8
Gross requirements		15	20	50	10	30	30	30	30
Scheduled receipts		10	10		100				
Adjusted SRs									
Projected on-hand	20								
Net requirements									
Planned order receipts									
Planned order releases									

---

[3]Notice that if we did not adjust SRs first, this could happen in more than one time period. Then we would have an early net requirement, followed by a scheduled receipt, followed by more net requirements.

We begin by computing the projected on-hand inventory. Starting with 20 units in stock, we subtract 15 for the gross requirements in period 1, leaving five remaining on-hand. Notice we do not consider the SR of 10 in period 1 since we always use on-hand inventory before using scheduled receipts.

Moving to the second period, we see that the gross requirement of 20 exceeds the five in stock, and so we issue a change notice to defer the SR with 10 from period 1 to period 2. However, this still provides only a total of 15 units, five less than what is needed. Therefore we add the second SR to period 2, bringing the total to 25 units. Notice that since this SR is already scheduled for period 2, we do not need to generate a change notice. After adjusting the first two SRs to period 2 and subtracting the gross requirements, we have an on-hand inventory of five. Since this quantity is insufficient to cover the third demand of 50, we issue an expedite notice to change the due date of the third SR from period 4 to period 3, yielding an on-hand inventory of 55. In some systems the job could be split, expediting only that portion which is needed at the earlier date. In this example, however, we expedite the entire job. This more than covers the 10 units in period 4, leaving 45, as well as the 30 in period 5, leaving 15 units. The demand in period 6 exceeds the projected on-hand inventory, and there are no more SRs to be adjusted. Thus, the first uncovered demand occurs in period 6 and is equal to 15. Table 3.2 summarizes the coverage analysis calculations used to generate projected on-hand inventory.

The net requirements are now easily computed, as shown in Table 3.2. For periods 1 through 5 they are zero because projected on-hand inventory is greater than zero. For period 6 they are 15, simply the negative of projected on-hand inventory. For periods 7 and 8 the net requirements are equal to the gross requirements, both of which are 30.

Lot Sizing.   Once we have computed the net requirements, we must schedule production quantities to satisfy them. Because MRP assumes demands are deterministic but nonconstant over time, this is exactly the same problem we addressed in Chapter 2 and solved "optimally" using the Wagner–Whitin algorithm. We will discuss this and other lot-sizing techniques in Section 3.1.6. For clarity and to illustrate the basic MRP computations, we restrict our attention at this point to two very simple lot-sizing rules.

TABLE 3.2   Adjusted Scheduled Receipts, Projected On-Hand, and Net Requirements

Part A		1	2	3	4	5	6	7	8
Gross requirements		15	20	50	10	30	30	30	30
Scheduled receipts		10	10		100				
Adjusted SRs			20	100					
Projected on-hand	20	5	5	55	45	15	−15	—	—
Net requirements							15	30	30
Planned order receipts									
Planned order releases									

TABLE 3.3   Planned Order Receipts and Releases

Part A		1	2	3	4	5	6	7	8
Gross requirements		15	20	50	10	30	30	30	30
Scheduled receipts		10	10		100				
Adjusted SRs			20	100					
Projected on-hand	20	5	5	55	45	15	−15	—	—
Net requirements							15	30	30
Planned order receipts							45		30
Planned order releases					45		30		

The simplest lot-sizing rule, known as **lot for lot,** states that the amount to be produced in a period is equal to that period's net requirements. This policy is easier to use than the fixed quantity policy in the example in Section 3.1.2, and is consistent with just-in-time philosophy (see Chapter 4) of making only what is needed.

Another simple rule is known as **fixed order period (FOP),** also sometimes called **period order quantity.** This rule attempts to reduce the number of setups by combining the net requirements of P periods. Note that when $P = 1$, FOP is equivalent to lot-for-lot.

Returning to our example, assume that the lot-sizing rule for parts A and B is fixed order period with $P = 2$ and for all other parts we use lot-for-lot. Then, for part A, we plan on receiving 45 units in period 6 (combining net demand from periods 6 and 7) and 30 units in period 8 (we cannot combine beyond our planning horizon). The results of these lot-sizing calculations are shown in Table 3.3

Time Phasing.   Almost universally, MRP systems assume that the time to make a part is fixed, although a few systems do allow for the planned lead time to be a function of the job size. Regardless of the specifics, however, MRP treats lead times as attributes of the part and possibly the job, but *not* of the status of the shop floor. This can cause problems, as we will see later.

If we return to our example and assume that the planned lead time for part A is two periods, we are able to compute the planned order releases as shown in Table 3.3.

BOM Explosion.   Table 3.3 shows the final result of processing part A. Recall that part A is made up of two units of part 100 and one unit of part 200 (see Figure 3.1). Thus, the planned order releases generated for part A create gross requirements for parts 100 and 200. Specifically, we need 90 units of part 100 in period 4 (two are needed for each unit of A) and 60 units in period 6. Similarly, we require 45 units of part 200 in period 4 and 30 units in period 6. These demands must be added to any requirements already accumulated for these parts (e.g., if we have already processed other parts that require them as subcomponents). To illustrate this, we will pursue our example a bit further.

The next step is to process any other parts having a low-level code of zero. In this example, we would process part B next. Suppose that the master production schedule for part B is as follows:

t	1	2	3	4	5	6	7	8
Demand	10	15	10	20	20	15	15	15

Furthermore, assume the following inventory and part data for parts B, 100, 300, and 500 (for brevity, we will not treat part 200, 400, or 600).

Part Number	Current On-Hand	SRs Due	SRs Quanity	Lot-Sizing Rule	Lead Time
B	40	0		FOP, 2 weeks	2 weeks
100	40	0		Lot-for-lot	2 weeks
300	50	2	100	Lot-for-lot	1 week
500	40	0		Lot-for-lot	4 weeks

Since there are no scheduled receipts for part B, the MRP calculations for this part are simple. Table 3.4 shows the completed tableau.

We have now completed processing all parts with an LLC of zero (i.e., parts A and B). Of the remaining parts we are considering, only part 500 has an LLC of one. Therefore we treat it next.

The only source of demand for part 500 is from part B (i.e., part A does not require part 500, and there is no external demand for part 500). Because each unit of B requires one unit of part 500, the planned order releases for part B become the gross requirements for part 500. Again, there are no scheduled receipts. The MRP processing is shown in Table 3.5.

Because the lead time for part 500 is four weeks, there is not enough time to finish the first 25 units before week four. Therefore, a planned order release is scheduled

TABLE 3.4   MRP Processing for Part B

Part B		1	2	3	4	5	6	7	8
Gross requirements		10	15	10	20	20	15	15	15
Scheduled receipts									
Adjusted SRs									
Projected on-hand	40	30	15	5	−15	—	—	—	—
Net requirements					15	20	15	15	15
Planned order receipts					35		30		15
Planned order releases			35		30		15		

TABLE 3.5 MRP Calculations for Part 500

Part 500		1	2	3	4	5	6	7	8
Gross requirements			35		30		15		
Scheduled receipts									
Adjusted SRs									
Projected on-hand	40	40	5	5	−25	—	—	—	—
Net requirements					25		15		
Planned order receipts					25		15		
Planned order releases		25*	15						

*Indicates a late start

for week one (as soon as possible) with an indication on an exception report that it is expected to be late.

We now turn to level 2 and part 100. Part 100 has two sources of demand, two units for each unit of part A and one unit for each unit of part 500. There are no scheduled receipts. The MRP processing is shown in Table 3.6

The only part at level 3 we consider is part 300. It has requirements from parts B and 100. Also, there is a scheduled receipt of 100 units in week two. Since it arrives at the time of the first uncovered requirement, no adjustments are necessary. The MRP processing is shown in Table 3.7.

We have now completed the MRP processing for all the parts of interest (processing for parts 200 and 400 is entirely analogous to that done for the other parts). Table 3.8

TABLE 3.6 MRP Calculations for Part 100

Part 100		1	2	3	4	5	6
Required from A					90		60
Required from 500		25	15				
Gross requirements		25	15		90		60
Scheduled receipts							
Adjusted SRs							
Projected on-hand	40	15	0	0	−90	—	—
Net requirements					90		60
Planned order receipts					90		60
Planned order releases			90		60		

TABLE 3.7   MRP Calculations for Part 300

Part 300	1	2	3	4	5	6	7	8
Required from B		35		30		15		
Required from 100		90		60				
Gross requirements		125		90		15		
Scheduled receipts		100						
Adjusted SRs		100						
Projected on-hand	50	50	25	25	−65	—	—	—	—
Net requirements				65		15		
Planned order receipts				65		15		
Planned order releases			65		15			

TABLE 3.8   Summary of MRP Output

Transaction	Part Number	Old Due Date or Release Date	New Due Date	Quantity	Notice
Change notice	A	1	2	10	Defer
Change notice	A	4	3	100	Expedite
Planned order release	A	4	6	45	OK
Planned order release	A	6	8	30	OK
Planned order release	B	2	4	35	OK
Planned order release	B	4	6	30	OK
Planned order release	B	6	8	15	OK
Planned order release	100	2	4	90	OK
Planned order release	100	4	6	60	OK
Planned order release	300	3	4	65	OK
Planned order release	300	5	6	15	OK
Planned order release	500	1	4	25	Late
Planned order release	500	2	6	15	OK

gives a summary of the outputs that an MRP system would generate from the above calculations. For each change notice, the system reports the quantity and part number affected, old due date, new due date, and whether it is an expedite or deferral. For each new planned order release, it reports the release date, the (new) due date, the release quantity, and whether it is anticipated to be late.

3.1.5   Special Topics in MRP

Up to now, we have focused on the mechanics of MRP processing. We now consider several technical issues that affect MRP performance. In particular, we address the question of what can be done to improve performance when things do not go as planned.

Updating Frequency.   A key determinant of the effectiveness of an MRP system is the frequency of updating. If we update too frequently, the shop can be inundated with exception reports and constantly changing planned order releases.[4] If, on the other hand, we update too infrequently, we can end up with old plans that are often out of date. In designing an MRP system, one must balance the need for timeliness against the need for stability.

Firm Planned Orders.   Changing the production schedule frequently can cause it to become very unstable. This makes it difficult for managers to shift workers effectively and prepare for setups. Therefore, it is desirable to minimize schedule disruption due to changes. One way to do this is by using **firm planned orders.** A firm planned order is a planned order release that is held fixed; that is, it will be released regardless of changes in the system. Consequently, firm planned orders are treated in MRP processing as if they were scheduled receipts (i.e., they must be included in the coverage analysis). By converting all planned order releases within a specified time interval to firm planned orders, the production plans become more stable. This is particularly important in the short term for managerial control purposes. Firm planned orders are also useful for reducing system **nervousness,** which is discussed in greater detail below.

Troubleshooting in MRP.   A wise man named Murphy once said, "If something can go wrong, it will go wrong." In an MRP system, there are many things that can go wrong. Jobs can finish late, parts can be scrapped, demands can change, and so on. As a result, over the years MRP systems have acquired features to assist the planner as conditions change. Examples include the techniques of pegging and bottom-up replanning.

Pegging allows the planner to see the source of demand that results in a given planned order release. It is facilitated by providing a link from the gross requirements of an item to all its sources of demand. For example, consider the planned order release of 65 units of part 300 in week three shown in Table 3.7. Pegging would link this to the individual requirements of 60 units of part 100 and 30 units of part B in week four. These, in turn, could be linked to their demand sources, namely, part B to the master production schedule and part 100 to the 60 units needed to make part A in week six (see Table 3.6).

One of the uses of pegging is in **bottom-up replanning.** This is best illustrated with an example. Suppose we discover that the scheduled receipt of 100 units of part 300 due in week two will not be coming in (someone found the purchase order that was supposed to be sent to the vendor behind a file cabinet). Of course, the appropriate action would be to place the order immediately, call the vendor, and see if the order can be expedited. If this is not possible, we can use bottom-up replanning to investigate the impact of the late delivery.

From Table 3.7, we see that the gross requirements affected are the 125 required in week two. If the scheduled receipt will not be coming in, then we have only the 50 that are on-hand to cover demand, leaving 75 units uncovered. Of the 125 demanded, 35 are for part B, a level 0 item, and 90 are for part 100, a level 2 item. If we attempt to cover the lowest-level items first (reasoning that these have the potential for causing the greatest disruption), then we see that we can cover only 50 of the 90 units of part 100 needed in period 2. Further pegging shows that these requirements are from 90 units of demand for part A, for which we can now cover only 50 units. At this point we might

---

[4]In the past, when computer systems were small in memory and slow in processing, the cost of computer processing could also be prohibitive. However, with the dramatic increases in computer power in recent years, this is much less a factor in choosing a regeneration frequency.

want to contact the customer for the 90 units of part A and see if we can deliver 50 when requested and the other 40 later.

Alternatively, we might use the 50 units on hand to cover the demand for part B first (the idea here is to cover the items that generate revenue). If we do this, we can cover the 35 units of demand for B and are left with 15 units to cover the 90 required for part 100. Again pegging these to their original demand shows that 75 of the 90 units of part A required in period 4 would not be covered. If the demand for part B in the MPS is for an actual customer, while that for part A is only a forecast, we might want to cover B first. Of course, a different option is to split the 50 on hand to cover some of the demand for part B and some for part 100. The "correct" choice depends on the customers involved, their willingness to accept late orders, and so on.

Instead of pegging, we could have eliminated the scheduled receipt of 100 units of part 300 and made a complete regeneration of MRP. This would have resulted in a planned order release in week one with an exception notice that it is expected to be late. However, a regeneration of MRP cannot determine which customer orders will be late as a result of this delay. Bottom-up replanning and pegging provide the planner with this ability. The use of firm planned orders allows the planner to remedy a schedule by overriding standard MRP processing.

3.1.6   Lot Sizing in MRP

To demonstrate basic MRP processing, we have described two simple lot-sizing rules—fixed order period and lot-for-lot. In this section, we will discuss issues surrounding the lot-sizing problem and describe other, more complex lot-sizing rules.

The lot-sizing problem deals with the basic tradeoff between having many small jobs, which tend to increase setup costs (materials, tracking costs, labor, etc.) and/or decrease capacity, versus having a few large jobs, which tend to increase inventory.

Recall that in Chapter 2 we formulated the Wagner–Whitin (WW) approach to the lot-sizing problem by assuming infinite capacity and known setup and inventory carrying costs. Under these assumptions, we showed that the lot-sizing problem can be solved optimally using the WW algorithm. Of course, the questions with this approach are whether anyone *can* know the setup and inventory carrying costs and whether capacities will be binding. As one wag remarked about setup costs, "I have yet to write out a check to a machine." In many instances, setup "cost" is used as proxy for limited *capacity*. The idea is to design lot-sizing rules so that higher setup costs result in larger lots (e.g., the EOQ). Since larger lots require fewer setups, less capacity is consumed. Conversely, when capacity is not tight, smaller setup costs can be used to reduce lot sizes (and thereby inventory) at the expense of more setups. Thus, by adjusting setup costs, the planner can trade inventory for capacity.

Unfortunately, the so-called Wagner–Whitin property of producing only when inventory levels reach zero is *not* optimal when capacity is a constraint. Nonetheless, many of the lot-sizing rules that have been suggested possess the WW property and are typically compared to the WW algorithm when their performance is assessed. Thus, although many of the assumptions may be invalid in realistic situations, it would appear that most lot-sizing rule designers have accepted the Wagner–Whitin paradigm. Interestingly, we know of no commercial MRP package that actually uses the WW algorithm. The reasons usually given are that it is too complicated or that it is too slow. But with the advent of fast computers, speed is no longer an issue—an efficient WW algorithm runs quickly on a modern personal computer. A more likely reason may be found in the observation that "People would rather live with a problem they cannot solve than accept a solution they do not understand." Regardless of the reason, a host of alternative lot-sizing algorithms have been suggested and are offered in various forms in most commercial MRP systems. We will discuss here some of the more commonly used methods.

Lot-for-Lot.   As we have already noted, lot-for-lot (LFL) is the simplest of the lot-sizing rules—simply produce in period t the net requirements for period t. Since this leaves no inventory at the end of *any* period (given the assumptions of MRP), this method minimizes inventory (assuming that it is possible to produce the demand in each period). However, under the Wagner–Whitin paradigm, since there is a "setup" in every period with demand, this method also maximizes total setup cost. Despite this, lot-for-lot is attractive in several respects. First, it is simple. Second, it is consistent with the just-in-time philosophy (see Chapter 4) of making only what is needed when it is needed. Finally, since the procedure does not lump requirements together in some periods and produce nothing in others, it tends to generate a smoother production schedule. In situations where setup times (costs) are minimal, it is probably the best policy to use.

Fixed Order Quantity and EOQ.   A second very simple policy is to order a prede-termined quantity whenever an order is placed. We use this rule, fixed order quantity, in our first example. It is commonly used for two simple reasons.

First, when there are certain sized totes, carts, or other fixtures used to transport jobs in the shop, it makes sense to create jobs only in these sizes. In some cases, different sized totes are used at different points in the shop. For instance, fenders are usually carried in smaller quantities than spark plugs. To avoid leftovers, it makes sense to coordinate the sizes of the quantities. One way to do this is to choose power-of-2 (1, 2, 4, 8, 16, etc.) lot sizes.

Second, fixing the job size influences the number of setups. Since the basic tradeoff is between setup cost and inventory carrying cost, the problem of choosing an appropriate fixed order quantity is very similar to that of the economic order quantity problem dis-cussed in Chapter 2. The primary difference is that the EOQ model assumed a constant demand rate. In MRP, demand need not be constant. However, we can make use of the EOQ model by replacing the constant demand of that model with an estimate of the average demand $\bar{D}$. Then, using A to represent the setup cost and h to denote the in-ventory carrying cost per annum, we can use the EOQ formula we derived in Chapter 2

$$Q = \sqrt{\frac{2A\bar{D}}{h}}$$

to compute the fixed order quantity Q. As discussed previously, we may want to round this quantity to the nearest power of 2. The ratio of A/h can be adjusted to achieve a desired setup frequency. Making A/h larger will reduce the setup frequency, while reducing A/h will increase the setup frequency. After some experience, a value that is compatible with the capacity of the line can be found. Of course, since this value will depend on the actual orders, it may change frequently.

Unlike the lot-for-lot rule, the fixed order quantity method (whether or not one uses the EOQ to obtain the order size) will *not* have the Wagner–Whitin property of producing only when inventory reaches zero. This means that it can result in incurring cost to carry inventory that does not eliminate a setup—an obvious inefficiency (under the assumptions of Wagner–Whitin).[5]

---

[5]Of course, as a practical measure, we will probably not plan to run out of inventory exactly when receiving the next order. Nonetheless, we can use safety stock (discussed in the next section) to provide some cushion and then insist on the Wagner–Whitin property for the cycle stock (i.e., the stock that is intended to be used).

However, we can modify the rule slightly to consider only job sizes that are equal to the exact demand of one or more periods, and then choose the one that is closest to the desired fixed job size. This practice recovers the Wagner–Whitin property. Consider the following example. Suppose our fixed order quantity is 50 units and the net requirements are these:

Net requirements	15	15	60	65	55	15	20	10

Then, to preserve the Wagner–Whitin property, our planned order receipts would be

Planned order receipts	30		60	65	55	45		

In period 1, 30 is closer to 50 than is 15, so we ordered two periods' worth of demand instead of one. In period 3, 60 is closer than 125, so we ordered one period's worth instead of two, and so on.

Fixed Order Period.   The fixed order period (FOP) rule was used in the MRP processing example in Section 3.1.4. Its operation is simple: If you are going to produce in period t, then produce all the demand for period $t, t + 1, \ldots, t + P - 1$, where P is a parameter of the policy. If $P = 1$, the policy is lot-for-lot, since we only produce for the current period. Since each production quantity is for the exact amount required in a given set of periods, the policy has the Wagner–Whitin property.

While simple, the policy does have some subtlety. The policy *does not* state that production will occur once every P periods. If there are periods with no demand, they are skipped. Consider the following example with $P = 3$.

Period	1	2	3	4	5	6	7	8	9
Net requirements		15	45			25	15	20	15
Planned order receipts		60				60			15

We skip the first period since there is no demand. The first demand occurs in period 2 and so we accumulate the demand for periods 2, 3, and 4 (note there is no demand in period 4) and therefore order 60 units for period 2. We again skip period 5, as it has no demand, and accumulate periods 6, 7, and 8 with a planned order receipt of 60 units in period 6. Finally, we order 15 units for period 9 and look no farther out since we are at the end of our time horizon.

One way to determine an "optimal" value for P is to use the EOQ formula and the average demand in a fashion similar to that used for the fixed order quantity rule. In the preceding example, the total demand for nine periods is 135 units, so the average demand is 15 units per period. Suppose the setup cost is \$150 and the carrying cost per period is \$2. We can then compute the EOQ as

$$Q = \sqrt{\frac{2AD}{h}} = \sqrt{\frac{2 \times 150 \times 15}{2}} = 47.4$$

We can then compute the order period P as

$$P = \frac{Q}{D} = \frac{47.4}{15} = 3.16 \approx 3 \text{ periods}$$

Of course, the validity of computing P using this method has all the limitations of the EOQ method that were noted in Chapter 2.

Part-Period Balancing.   Part-period balancing (PPB) is a policy that combines the assumptions of the Wagner–Whitin paradigm with the mechanics of the EOQ. One of the properties of the EOQ solution to the lot-sizing problem is that it sets the average inventory carrying cost equal to the setup cost.

The idea of PPB is to balance (i.e., set equal) the inventory carrying cost and setup cost. To describe this, we need to define the notion of a **part-period** as the product of the number of parts in a lot times the number of periods they are carried in inventory. For instance, 1 part carried for 10 periods, 5 parts carried for 2 periods, and 10 parts carried for 1 period all represent 10 part-periods and incur the same inventory carrying cost. Part-period balancing seeks to make the carrying cost as close to the setup cost as possible. We can demonstrate this by using the data of the previous example.

By considering only those quantities that preserve the Wagner–Whitin property, we reduce our choices to a relative few. Since there are no requirements in period 1, there will be no production in period 1. The choices for period 2 are 15 (produce for period 2 only), 60 (produce for periods 2 and 3), 85 (produce for periods 2, 3, and 6), and so on. The following table shows the part-periods and the costs involved.

Quantity for Period 2	Setup Cost ($)	Part-Periods	Inventory Carrying Cost ($)
15	150	0	0
60	150	$45 \times 1 = 45$	90
85	150	$45 + 25 \times 4 = 145$	290

Since \$90 is the closest to \$150 of the options available, we elect to make 60 units in period 2. Since there are no requirements, we will make nothing in periods 3, 4, and 5. For period 6 the choices are 25, 40, 60, and 75 units. Again we present the computations in a table.

Quantity for Period 6	Setup Cost ($)	Part-Periods	Inventory Carrying Cost ($)
25	150	0	0
40	150	$15 \times 1 = 15$	30
60	150	$15 + 20 \times 2 = 55$	110
75	150	$55 + 15 \times 3 = 100$	200

The inventory carrying cost closest to $150 results from making 60 units in period 6. This covers requirements for periods 6, 7, and 8, leaving 15 for period 9. Note that this is exactly the same schedule that resulted from the FOP policy.

Other Methods.   A host of other methods for lot sizing have been proposed by researchers. Most of these attempt to provide a near-optimal solution according to the Wagner–Whitin criteria. Whether these criteria are appropriate is a matter of debate, as we have discussed. Baker (1993) gives a good review of many of the lot-sizing methods that have been suggested.

Finally, we note that although the Wagner–Whitin algorithm is optimal under certain conditions, other rules may perform better in practice. For instance, Bahl et al. (1987) report in a review of the lot-sizing literature that the fixed order quantity method, *without* modification to give it the Wagner–Whitin property, tends to work better than rules that do possess the Wagner–Whitin property in multilevel production systems with capacity limitations. They conclude that the often-imposed Wagner–Whitin property may not be practical in real settings, since "the remnants avoided by almost all (other lot-sizing rules) become an asset in terms of on-time delivery of end items." This makes sense, since these remnants become a form of safety stock, an issue that we explore in the next section.

3.1.7   Safety Stock and Safety Lead Times

Operations management researchers have long debated the role of safety stock and safety lead times in MRP systems. Orlicky felt that these had no place in the system except, possibly, for end items. Lower-level items, he believed, were more than adequately covered by the workings of the system. Since Orlicky's time, many researchers have disagreed. Because MRP is deterministic, the logic goes, something should be done to account for uncertainty and randomness.

There are several sources of uncertainty. First, in all except pure make-to-order systems, neither the demand quantity nor the timing of the demand is known exactly. Second, production timing is almost always subject to variation, due to machine breakdowns, quality problems, fluctuations in staffing, and so on. Third, production quantities are uncertain because the number of good parts that finish can be less than the quantity that start due to **yield loss** or **fallout.**

Safety stock and **safety lead time** can be used as protection against these problems. Vollmann et al. (1992) suggest that *safety stock* should be used to protect against uncertainties in production and demand *quantities,* while *safety lead time* should be used to protect against uncertainties in production and demand *timing.*

Providing safety stock (SS) in an MRP system is fairly straightforward. Suppose we wish to maintain a safety stock level of 10 units for part B (refer to Table 3.4). This time we compute the first net requirement as we did before, but we subtract an additional 10 units for the desired safety stock. The projected on-hand *minus safety stock* first becomes negative in period 3 (as opposed to period 4 before), as we see in Table 3.9.

Thus, our first planned order release is for five units needed to bring the inventory to the desired safety stock level, plus 20 units for actual demand.

Introducing safety lead time into the MRP calculations is a bit different. If the nominal lead time is two weeks and we desire a safety lead time of one week, we perform the offsetting in two stages: the first for the safety lead time regarding the planned order *receipt* date (i.e., the due date) and the second using the usual MRP method, to obtain the planned order *release* date. We demonstrate the use of a safety lead time of one week, using the same data as in the previous example in Table 3.10.

TABLE 3.9   MRP Computations for Part B with Safety Stock

Part B		1	2	3	4	5	6	7	8
Gross requirements		10	15	10	20	20	15	15	15
Scheduled receipts									
Adjusted SRs									
Projected on-hand	40	30	15	5	—	—	—	—	—
Projected on-hand—SS	30	20	5	−5	—	—	—	—	—
Net requirements				5	20	20	15	15	15
Planned order receipts				25		35		30	
Planned order releases		25		35		30			

TABLE 3.10   MRP Calculations for Part B with Safety Lead Time

Part B		1	2	3	4	5	6	7	8
Gross requirements		10	15	10	20	20	15	15	15
Scheduled receipts									
Adjusted SRs									
Projected on-hand	40	30	15	5	−15	—	—	—	—
Net requirements					15	20	15	15	15
Planned order receipts					35		30		15
Adjusted planned order receipts				35		30		15	
Planned order releases		35		30		15			

The one additional step beyond the usual MRP calculation is shown in the "Adjusted planned order receipts" line, which backs up these receipts according to the one week safety lead time. Notice that the effect on planned order releases is identical to simply inflating the planned lead times. However, the due dates on the jobs are earlier in a system using safety lead times than in one without it. The effect of safety lead times on a single part is fairly simple. Bringing parts in a week early means they will be available unless delivery is late by more than a week. However, things are more subtle when we consider multiple parts and assemblies.

For instance, suppose a plant manufactures a part that requires 10 components to come together at assembly. Suppose also that the actual manufacturing lead times can be well approximated using a normal distribution with a mean of three weeks and a standard deviation of one week. To maintain good customer service, we want assemblies to start on time at least 95 percent of the time. If s is the service level (i.e., the probability of on-time delivery) for each component, then the probability that all 10 components are available on time (assuming independent deliveries) is given by

$$\Pr \{\text{on-time start of assembly}\} = s^{10}$$

Since we want this probability to equal 0.95, we can solve for s as follows:

$$s = (0.95)^{1/10} = 0.9949$$

Since the manufacturing lead times are normally distributed, this represents approximately 2.6 standard deviations above the mean, or around 5.6 weeks—about twice the mean lead time for the planned lead time.

Of course, this analysis assumes that the 10 items are arriving to the assembly operation independently of one another, an assumption that may not be true if they are all being fabricated in the same plant. Nonetheless, the point is made—if we are to try to guarantee any level of service for an assembly, the service for the component parts must be *much* greater.

In conclusion, although safety stock and safety lead times can be useful in an MRP system, we must be cognizant of the fact that both procedures *lie* to the system. Safety stock requires the intentional production of quantities for which there is no customer need, while safety lead times set due dates earlier than are really required. Both situations will make **available-to-promise** calculations (used to quote deliveries to customers, discussed below) less accurate. Excess safety stocks and long safety lead times will result in customers being turned away due to perceived schedule infeasibility even though the schedule is actually feasible. In addition, there is always the risk that once safety stock and/or lead times are discovered by the users, an informal system of "real" quantities and due dates will appear. Such behavior can lead to a subversion of the formal system and can degrade its performance.

3.1.8   Accommodating Yield Losses

The above discussion and examples illustrate the use of hedges against uncertainties in demand and timing. However, hedging against random scrapping of parts during production—yield loss—involves an additional computation. Suppose the net demand is N_t units and the average yield fraction is y. Also suppose, for this example, that N_t is a large number, so that we do not have to worry about integer quantities. Thus, if we start $N_t(1/y)$ units, we will, on average, finish with N_t units, the net demand. However, if $N_t(1/y)$ is a large number, it is very unlikely that we will finish with exactly N_t. We will, with roughly equal probability, finish with either more or less than the net demand.

Finishing with more means that we will carry the extra parts in inventory until they are netted from future demand. If the product is highly customized, this can be a problem. On the other hand, if we finish with less, a *new* job will be required to make up the difference, and it is unlikely that the order will ship on time.

Safety stock can improve customer service and responsiveness in this case. We inflate the size of the job to $N_t(1/y)$ as before and carry safety stock to accommodate instances when production is less than the average yield. Another strategy is to carry no safety stock but to inflate the job by more than $1/y$. In this case, it is likely that the job will finish with more than the net demand and that the extra stock will be carried in inventory. The two procedures are essentially equivalent since both result in better service at the expense of additional inventory.

Lastly, we should point out that the effectiveness of any yield strategy depends on the *variability* of the yields themselves. For instance, if a job starts with 100 units, each unit having an independent probability of 0.9 of being completed, then the mean and standard deviation of the number of units finishing will be 90 and 3, respectively. Thus, by starting 120 (that is, $100/0.9 + 3 \times 3$) units, we have a probability of greater than 0.99 (three standard deviations above the mean) that we will finish with at least 100 units. However, if the yield situation is more of an all-or-nothing type, so that either all the units that start finish properly or none of them do (as in a batch process), then we need to release two separate jobs of 100 each to obtain a 0.99 probability of finishing 100 on time. In the first (independent) case, the average increase in inventory would be eight units ($120 \times 0.9 - 100$). In the second (batch) case, it would be 80 units ($200 \times 0.9 - 100$). The moral is that *average* yield rate is not enough to determine an effective yielding strategy. The mechanism and variability of the processing causing the yield fallout must also be considered.

3.1.9 Problems in MRP

Despite enthusiastic support of MRP by early proponents—Orlicky's book was subtitled *A New Way of Life*—several problems were recognized early on. Three of the most severe were (1) capacity infeasibility of MRP schedules, (2) long planned lead times, and (3) system "nervousness." These and other problems first led to new MRP procedures and spawned a new generation of MRP, called **manufacturing resources planning** or **MRP II,** which, in turn evolved into **enterprise resources planning (ERP),** as we will discuss in the next section.

Capacity Infeasibility.   The basic working model of MRP is a production line with a fixed lead time. Since this lead time does not depend on how much work is in the plant, there is an implicit assumption that the line will always have sufficient capacity regardless of the load. In other words, MRP assumes all lines have infinite capacity. This can create problems when production levels are at or near capacity.

One way to address this problem is to make sure that the master production schedule that supplies demand to the system is capacity-feasible. A check of this is provided by a procedure called rough-cut capacity planning (RCCP), as we will see later. As its name implies, RCCP is an approximation. A more detailed capacity assessment of the resulting MRP plans can be made by using a procedure known as **capacity requirements planning (CRP).** Both RCCP and CRP are modules that are often found in MRP II.

Long Planned Lead Times.   As we saw in our earlier discussion of safety lead times, there are many pressures to increase planned lead times in an MRP system. In Part II, we will see that long lead times invariably lead to large inventories. However, as long as the penalty for a late job is greater than that for excess inventory (which is typically the case, since inventory does not scream but dissatisfied customers do!), production control managers will tend toward long planned lead times.

The problems caused by long planned lead times are further exacerbated by the fact that MRP uses *constant* lead times when, in fact, actual manufacturing times vary continually. To compensate, a planner will typically choose pessimistic (long) estimates for the planned lead times. Suppose for example, the average manufacturing lead time is three weeks, with a standard deviation of one week. To maintain good customer service, the planned lead time is set to five weeks. Since the actual lead times are random, some will be less than the mean of three weeks and others will be greater. If these follow an approximately normal distribution, then the most likely lead time will be three weeks, so the most likely holding time in inventory will be two weeks. The result can be a large amount of inventory.

The longer the planned lead times, the longer parts will wait for the next operation, and so the more inventory there will be in the system. Since setting planned lead times equal to the average manufacturing time yields a service level of only 50 percent for each component (and therefore much worse service for finished assemblies), managers will virtually always choose lead times that are much longer than average manufacturing times. Such behavior results in a lack of responsiveness as well as high inventory levels.

System Nervousness.   Nervousness in an MRP system occurs when a small change in the master production schedule results in a large change in planned order releases. This can lead to strange effects. For instance, as we demonstrate with the following example, it is actually possible for a *decrease* in demand to cause a formerly feasible MRP plan to become infeasible.

The following example is taken from Vollmann et al. (1992). We consider two parts. Item A has a lead time of two weeks and uses the fixed order period (FOP) lot-sizing rule with an order period of five weeks. Each unit of A requires one unit of component B, which has a lead time of four weeks and uses the FOP rule with an order period of five weeks. Tables 3.11 and 3.12 give the MRP calculations for both parts.

TABLE 3.11   MRP Calculations for Item A before Change in Demand

Item A		1	2	3	4	5	6	7	8
Gross requirements		2	24	3	5	1	3	4	50
Scheduled receipts									
Adjusted SRs									
Projected on-hand	28	26	2	−1	—	—	—	—	—
Net requirements				1	5	1	3	4	50
Planned order receipts				14					50
Planned order releases		14					50		

TABLE 3.12   MRP Calculations for Component B before Change in Demand

Component B		1	2	3	4	5	6	7	8
Gross requirements		14					50		
Scheduled receipts		14							
Adjusted SRs		14							
Projected on-hand	2	2	2	2	2	2	−48	—	—
Net requirements							48		
Planned order receipts							48		
Planned order releases			48						

TABLE 3.13   MRP Calculations for Item A after Change in Demand

Item A		1	2	3	4	5	6	7	8
Gross requirements		2	23	3	5	1	3	4	50
Scheduled receipts									
Adjusted SRs									
Projected on-hand	28	26	3	0	−5	—	—	—	—
Net requirements					5	1	3	4	50
Planned order receipts					63				
Planned order releases			63						

We now *reduce* the demand in period 2 from 24 to 23. It would seem obvious that any schedule that is feasible for 24 parts in period 2 should also be feasible for 23 parts in the same period. But notice what happens to the calculations in Table 3.13. The aggregation of demand during lot sizing causes a drastically different set of planned order releases. In the case of component B (Table 3.14), the planned order releases are no longer even feasible.

There have been several remedies offered to reduce nervousness. One is the proper use of lot-sizing rules. Clearly, if we use lot-for-lot, the magnitude of the change to the planned order releases will be no larger than the changes to the MPS. However, lot-for-lot may result in too many setups, so we need to look for other cures.

Vollmann et al. (1992) recommend the use of different lot-sizing rules for different levels in the BOM, with fixed order quantity for end items, either fixed order quantity or lot-for-lot for intermediate levels, and fixed order period for the lowest levels. Since order sizes do not change at the higher levels, this tends to dampen nervousness due to

TABLE 3.14   MRP Calculations for Component B after Change in Demand

Component B		1	2	3	4	5	6	7	8
Gross requirements			63						
Scheduled receipts		14							
Adjusted SRs			14						
Projected on-hand	2	2	−47	—	—	—	—	—	—
Net requirements			47				48		
Planned order receipts			47						
Planned order releases		47*							

*Indicates a late start

changes in lot size. Of course, care must be taken when establishing the magnitude of the fixed lot size.

While the use of proper lot-sizing rules can reduce system nervousness, other measures can alleviate some of its effects. One obvious way is to reduce changes in the input itself. This can be done by freezing the early part of the master production schedule. This reduces the amount of change that can occur in the MPS, thereby reducing changes in planned order releases. Since early planned order releases are the ones in which change is most disruptive, a **frozen zone,** an initial number of periods in the MPS in which changes are not permitted, can dramatically reduce the problems caused by nervousness.

In some companies the first X weeks of the MPS are considered frozen. However, in most real systems, the term *frozen* may be too strong, since changes are resisted but not strictly forbidden. (Perhaps *slushy zone* would be a more accurate metaphor.) The concept of **time fences** formalizes this type of behavior. The earliest time fence, say for four weeks out, is absolutely frozen—no changes can be made. The next fence, maybe five to seven weeks out, is restricted but less rigid. Changes might be accepted in model options if the options are available, and possibly resulting in a financial penalty to the customer. The next fence, perhaps 8 to 12 weeks out, is less rigid still. In this case, changes in part number might be accepted if all components are on hand. In the final fence, 13 weeks and beyond, anything goes.

Another way to reduce the consequences of nervousness is to make use of **firm planned orders.** Unlike frozen zones or time fences, firm planned orders fix planned order releases. By converting early planned order releases to firm planned orders, we eliminate all system nervousness early in the schedule, where it is most disruptive. Consider what would happen if the first planned order release in Table 3.11 were made into a firm planned order before the change in demand. This would result in its being treated just like a scheduled receipt in the MRP processing. With this change there is no nervousness, as is shown in Tables 3.15 and 3.16.

Of course, the use of firm planned orders and time fencing means that the frozen part of the schedule will be less responsive to changes in demand. Another drawback is that the firm planned orders represent tedious manual entries that must be managed by planners.

TABLE 3.15   MRP Calculations for Item A with FPO

Item A		1	2	3	4	5	6	7	8
Gross requirements		2	23	3	5	1	3	4	50
Scheduled receipts									
Firm planned orders				14					
Projected on-hand	28	26	3	14	9	8	5	1	−49
Net requirements									49
Planned order receipts				[14]					49
Planned order releases		[14]					49		

TABLE 3.16   MRP Calculations for Component B with FPO

Component B		1	2	3	4	5	6	7	8
Gross requirements		14					49		
Scheduled receipts		14							
Adjusted SRs									
Projected on-hand	2	2	2	2	2	2	−47	—	—
Net requirements							47		
Planned order receipts							47		
Planned order releases			47						

3.2   Manufacturing Resources Planning—MRP II

Material requirements planning offered a systematic method for planning and procuring materials to support production. The ideas were relatively simple and easily implemented using a computer. However, some problems remained.

As we have mentioned, issues such as capacity infeasibility, long planned lead times, system nervousness, and others can undermine the effectiveness of an MRP system. Over time, additional procedures were developed to address some of these problems. These were incorporated into a larger construct known as **manufacturing resources planning,** or **MRP II.**

Beyond simply addressing deficiencies of MRP, MRP II also brought together other functions to make a truly integrated manufacturing management system. The additional functions subsumed by MRP II included demand management, forecasting, capacity planning, master production scheduling, rough-cut capacity planning, capacity requirements planning, dispatching, and input/output control. In this section we describe the

MRP II hierarchy into which these functions fit and discuss some of the associated modules. Our presentation is somewhat abbreviated for two reasons. First, MRP and MRP II are subjects that can occupy an entire volume themselves. We recommend Vollmann et al. (1992) as an excellent comprehensive reference. Second, we take up the issue of hierarchical production planning (in the context of pull systems) in Chapter 13. There we will address generic issues associated with any planning hierarchy such as time scales, forecasting, demand management, and so forth in greater detail.

3.2.1   The MRP II Hierarchy

Figure 3.3 depicts an instance of the MRP II hierarchy. We use the word *instance* because there are probably as many different hierarchies for MRP II as there are MRP II software vendors (and there are many such vendors, although most call themselves ERP on "enterprise" software vendors now).

3.2.2   Long-Range Planning

At the top of the hierarchy we have **long-range planning.** This involves three functions: resource planning, aggregate planning, and forecasting. The length of the time horizon for long-range planning ranges from around six months to five years. The frequency for replanning varies from once per month, to once per year, with two to four times per year being typical. The degree of detail is typically at the part family level (i.e., a grouping of end items having similar demand and production characteristics).

FIGURE 3.3

MRP II hierarchy

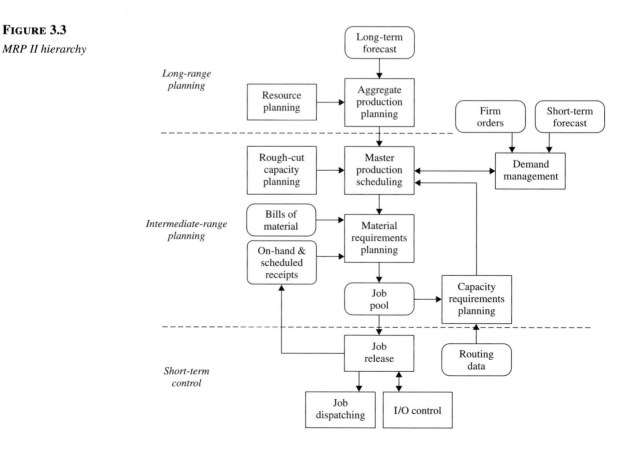

The **forecasting** function seeks to predict demands in the future. Long-range forecasting is important to determining the capacity, tooling, and personnel requirements. Short-term forecasting converts a long-range forecast of part families to short-term forecasts of individual end items. Both kinds of forecasts are input to the intermediate-level function of **demand management.** We describe specific forecasting techniques in detail in Chapter 13.

Resource planning is the process of determining capacity requirements over the long term. Decisions such as whether to build a new plant or to expand an existing one are part of the capacity planning function. An important output of resource planning is projected available capacity over the long-term planning horizon. This information is fed as a parameter to the aggregate planning function.

Aggregate planning is used to determine levels of production, staffing, inventory, overtime, and so on over the long term. The level of detail is typically by month and for part families. For instance, the aggregate planning function will determine whether we build up inventories in anticipation of increased demand (from the forecasting function), "chase" the demand by varying capacity using overtime, or do some combination of both. Optimization techniques such as linear programming are often used to assist the aggregate planning process. We discuss aggregate planning and models for supporting it in greater detail in Chapter 16.

3.2.3   Intermediate Planning

At the intermediate level, we have the bulk of the production planning functions. These include demand management, rough-cut capacity planning, master production scheduling, material requirements planning, and capacity requirements planning.

The process of converting the long-term aggregate forecast to a detailed forecast while tracking individual customer orders is the function of **demand management.** The output of the demand management module is a set of actual customer orders plus a forecast of anticipated orders. As time progresses, the anticipated orders should be "consumed" by actual orders.

This is accomplished with a technique known as **available to promise (ATP).** This feature allows the planner to know which orders on the MPS are already committed and which are available to promise to new customers. ATP combined with a capacity-feasible MPS facilitates negotiation of realistic due dates. If more orders than expected are received, so that quoted lead times become excessive, additional capacity (e.g., overtime) might be required. On the other hand, if fewer than expected orders arrive, sales might want to offer discounts or some other incentives to increase demand. In either case, the forecast and possibly the aggregate plan should be revised.

Master production scheduling takes the demand forecast along with the firm orders from the demand management module and, using aggregate capacity limits, generates an anticipated build schedule at the highest level of planning detail. These are the "demands" (i.e., part number, quantity, and due date) used by MRP. Thus, the master production schedule contains an order quantity in each time bucket for every item with independent demand, for every planning date. For most industries, these are given at the **end item** level. However, in some cases, it makes more sense to plan for groups of items or models instead of end items. An example of this is seen in the automobile industry where the exact make and specification of a car are not determined until the last minute on the assembly line. In these situations, a **final assembly schedule** determines when the exact end items are produced while the master production schedule is used to schedule models. A key input to this type of planning is the **superbill of material** that contains forecast percentages for the different options of each particular model. For a complete discussion of superbills in final assembly scheduling, the reader is referred to Vollman et al. (1992).

Rough-cut capacity planning (RCCP) is used to provide a quick capacity check of a few critical resources to ensure the feasibility of the master production schedule. Although more detailed than aggregate planning, RCCP is less detailed than capacity requirements planning (CRP), which is another tool for performing capacity checks after the MRP processing. RCCP makes use of a **bill of resources** for each end item on the MPS. The bill of resources gives the number of hours required at each critical resource to build a particular end item. These times include not only the end item itself but all the exploded requirements as well. For instance, suppose part A is made up of components A_1 and A_2. Part A requires one hour of process time in process center 21 while components A_1 and A_2 require one-half hour and one hour, respectively. Thus the bill of resource for part A would show two and one-half hours for process center 21 for each unit of A. Suppose we also have part B with no components that requires two hours in process center 21.

To continue the example, suppose we have the following information regarding the master production schedule for parts A and B:

Week	1	2	3	4	5	6	7	8
Part A	10	10	10	20	20	20	20	10
Part B	5	25	5	15	10	25	15	10

The bills of resources for parts A and B are given by

Process Center	Part A	Part B
21	2.5	2.0

Then the RCCP calculations for parts A and B at process center 21 are as follows:

Week	1	2	3	4	5	6	7	8
Part A (hour)	25	25	25	50	50	50	50	25
Part B (hour)	10	50	10	30	20	50	30	10
Total (hour)	35	75	35	80	70	100	80	35
Available	65	65	65	65	65	65	65	65
Over(+)/under(−)	30	−10	30	−15	−5	−35	−15	30

If we had considered only the sum of the eight periods in aggregate, we would have concluded that there was sufficient capacity—520 hours versus a requirement of 510 hours. However, after performing RCCP, we see that several periods have insufficent capacity while others have an excess. It is now up to the planner to determine what can be done to remedy the situation. Her options are to (1) adjust the MPS by changing due dates or (2) adjust capacity by adding or taking away resources, using overtime, or subcontracting some of the work.

Notice that RCCP does not perform any offsetting. Thus, the periods used must be long enough that the part, its subassemblies, and its components can all be completed within a single period. RCCP also assumes that the demand can be met without regard to how the work is scheduled within the process center (i.e., without any induced idle time). In this way, RCCP provides an optimistic estimate of what can be done.

On the other hand, RCCP does not perform any netting. While this may be acceptable for end items (demand for these can be netted against finished goods inventory relatively easily), it is less acceptable for subassemblies and components, particularly when there are many shared components and WIP levels are large. This aspect of RCCP tends to make it conservative.

These two effects make the behavior of RCCP difficult to gauge. Usually the first approximation tends to dominate the second, making RCCP an optimistic estimation of what can be done, but not always. Consequently, rough-cut capacity planning can be very rough indeed.

Capacity requirements planning (CRP) provides a more detailed capacity check on MRP-generated production plans than RCCP. Necessary inputs include all planned order releases, existing WIP positions, routing data, as well as capacity and lead times for all process centers. In spite of its name, capacity requirements planning does *not* generate finite capacity analysis. Instead, CRP performs what is called **infinite forward loading.** CRP predicts job completion times *for each process center,* using given *fixed* lead times, and then computes a predicted loading over time. These loadings are then compared against the available capacity, but no correction is made for an overloaded situation.[6]

To illustrate how CRP works, consider a simple example for a process center that has a three-day lead time and a capacity of 400 parts per day. At the start of the current day, 400 units have just been released into the process center, 500 units have been there for one day, and 300 have been there for two days. The planned order releases for the next five days are as follows:

Day	**1**	**2**	**3**	**4**	**5**
Planned order releases	300	350	400	350	300

Using the three-day lead time, we can compute when the parts will depart the process center. If we ever predict more than 400 units departing in a day, the process center is considered to be overloaded. The resulting **load profile** is shown in Figure 3.4. The first day shows the load to be 300 (these are the same 300 units that have been in the process center for two days and depart at the end of day one). The second day shows 500; again these are the same 500 that were in for one day at the start of the procedure. Since 500 is greater than the capacity of 400 per day, this represents an overloaded condition.

---

[6]Unlike MRP and CRP, true finite capacity analysis does not assume a fixed lead time. Instead the time to go through a manufacturing operations depends on how many other jobs are already there and their relative priority. Most finite capacity analysis packages do some sort of deterministic simulation of the flow of the jobs through the facility. As a result, finite capacity analysis is much more complex than CRP.

FIGURE 3.4

CRP load profile

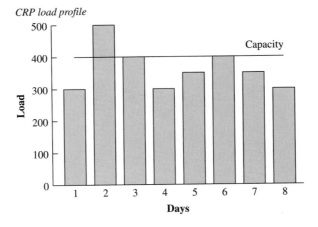

Note that even when load exceeds capacity, CRP assumes that the time to go through the process center does not change. Of course, we know that it will take longer to get through a heavily loaded process center than a lightly loaded one. Hence, all the estimates of CRP beyond such an overloaded condition will be in error. Therefore, CRP is typically not a good predictor of load conditions except in the very near term. Another problem with CRP is that it only tells the planner that there is a problem; it offers nothing about what caused the problem or what can be done to alleviate it. To determine this, the planner must first obtain a report that disaggregates the load to determine which jobs are causing the problem, and then must use pegging to track the cause back to demand on the MPS. This can be quite tedious.

A fundamental flaw with CRP is that, like MRP itself, it implicitly assumes an infinite capacity. This assumption comes from the assumption of fixed lead times that do not depend on the load of the process center. Consider the same process center having no work in it at the start and the following planned order releases, produced with a lot-sizing rule that tends to group demand to avoid setups:

Day	1	2	3	4	5
Planned order releases	1,200	0	0	1,200	0

Using CRP, the load profile will show an overloaded condition on day three and day six. If we were to perform *finite* capacity loading, we would see a very different picture. There would be no output for two days (the first release needs to work its way through), and then we would see 400 units output each day for the next six days. The second release on day four would arrive just as the last of the first release was being pulled into the process center. The basic relations between capacity, work in process, and the time to traverse a process center are the subject of Chapter 7.

Thus, in spite of its hopeful introduction and worthy goals, there are fundamental problems with CRP. First, there are enormous data requirements, and the output is voluminous and tedious. Second is the fact that it offers no remedy to an overloaded situation. Finally, since the procedure uses infinite loading and many modern systems can perform true finite capacity loading, fewer and fewer companies are seriously using CRP.

The **material requirements planning** module of all early versions of MRP II and many modern ERP systems is identical to the MRP procedure described earlier. The output of MRP is the **job pool,** consisting of planned order releases. These are released onto the shop floor by the **job release** function.

3.2.4   Short-Term Control

The plans generated in the long- and intermediate-term planning functions are implemented in the short-term control modules, of **job release, job dispatching,** and **input/output control.**

Job release converts planned order releases to scheduled receipts. One of the important functions of job release is **allocation.** When there are several high-level items that use the same lower-level part, a conflict can arise when there is an insufficient quantity on hand. By allocating parts to one job or another, the job release function can rationalize these conflicts. Suppose there are two planned order releases that require component A. Suppose further that there is enough stock on hand of component A for either job to be released but not for both. The first POR also requires component B for which there is plenty of stock, while the other POR requires component C for which there is insufficient stock. The job release function will allocate the available stock to the first POR since there is enough stock of both components A and B to start the job. A shortage notice would be generated for the second POR, which would remain in the job pool until it could be released.

Once a job or purchase order is released, some control must be maintained to make sure it is completed on time with the correct quantity and specification. If the job is for purchased components, the purchase order must be tracked. This is a straightforward practice of monitoring when orders arrive and tracking outstanding orders. If the job is for internal manufacture, this falls under the function known as **shop floor control (SFC)** or **production activity control (PAC).** Throughout this book we use the term SFC, as it is more traditional and more widely used. Within SFC are two main functions: **job dispatching** and **input/output control.**

Job Dispatching.   The basic idea behind job dispatching is simple: Develop a rule for arranging the queue in front of each workstation that will maintain due date integrity while keeping machine utilization high and manufacturing times low. Many rules have been proposed for doing this.

One of the simplest dispatching rules is known as **shortest process time,** or **SPT.** Under SPT, jobs at the process center queue are sorted with the shortest jobs first in line. Thus, the job in the queue having the shortest processing time will always be performed next. The effect is to clear out small jobs and get them through the plant quickly. Use of SPT typically decreases average manufacturing times and increases machine utilization. *Average* due date performance is also generally quite good, even though due dates are not considered in the ordering.

Problems with SPT occur whenever there are particularly *long* jobs. In such cases, jobs can sit for a long time without ever being started. Thus, while average due date performance of SPT is good, the variance of the lateness can be quite high. One way to avoid this is to use a rule known as SPT^x, where x is a parameter. By this rule, the next job to be worked will be the one with the shortest processing time *unless* a job has been waiting x time units or longer, in which case it becomes the next job. This rule seems to yield reasonably good performance in many situations.

If jobs are all approximately the same size and routings are fairly consistent, a good dispatching rule is **earliest due date,** or **EDD.** Under EDD, the job closest to its due date is worked on next. EDD exhibits reasonably good performance under the above conditions, but typically does not work better than SPT under more general conditions.

Here are three other common rules.

Least slack: The slack for a job is its due date minus the remaining process time (including setups) minus the current time. The highest priority is the job with the lowest slack value.

Least slack per remaining operation: This is similar to the least slack rule except we take the slack and divide it by the number of operations remaining on the routing. Again, the highest-priority job has the smallest value.

Critical ratio: Jobs are sorted according to an index computed by dividing the time remaining (i.e., due date minus the current time) by the number of hours of work remaining. If the index is greater than one, the job should finish early. If it is less than one, the job will be late; and if it is negative, it is already late. Again, the highest-priority job has the smallest value of the critical ratio.

There are at least 100 different dispatching rules that have been offered in the operations management literature. A good survey of many of these is found in Blackstone et al. (1982), where the authors test various rules by using a simulated factory under a broad range of conditions.

Of course, no dispatching rule can work well all the time, because, by their very nature, dispatching rules are myopic. The only consistent way to achieve good schedules is to consider the shop as a whole. The problem with doing this is that (1) the shop scheduling problem is extremely complex and can require an enormous amount of computational time and (2) the resulting schedules are often not intuitive. We will address the scheduling problem more fully in Chapter 15.

Input/Output Control.   Input/output (I/O) control was first suggested by Wight (1970) as a way to keep lead times under control. I/O control works in the following way:

1. Monitor the WIP level in each process center.
2. If the WIP goes above a certain level, then the current release rate is too high, so reduce it.
3. If it goes below a specified lower level, then the current release rate is too low, so increase it.
4. If it stays between these control levels, the release rate is correct for the current conditions.

The actions—reduce and increase—must be done by changing the MPS.

I/O control provides an easy way to check releases against available capacity. However, by waiting until WIP levels have become excessive, the system has, in many respects, already gone out of control. This may be one reason that so-called pull systems (e.g., Toyota's kanban system) may work better than push systems such as MRP, MRP II, and ERP. While these systems control releases (via the MPS) and measure WIP levels (via I/O control), kanban systems control WIP directly and measure output rates daily. Thus, kanban does not allow WIP levels to become excessive and detects problems (i.e., production shortfalls) quickly. Kanban is discussed in greater detail in Chapter 4, while the basics of push and pull are explored more fully in Chapter 10.

3.3   Beyond MRP II—Enterprise Resources Planning

In the years following the development of MRP II, a number of would-be successors were offered by vendors and consultants. MRP III never quite caught on, nor did the indigestibly acronymed BRP (business requirements planning). Finally, in spite of its gastronomically unpleasant acronym, enterprise resources planning (ERP) has emerged victorious.

This is due largely to the success of a few vendors, notably SAP, who have targeted not only manufacturing operations but *all* operations (e.g., manufacturing, distribution, accounting, financial, and personnel) of a company. Hence, the system offered is designed to control the entire *enterprise.*

SAP's R/3 software is typical of an interwoven comprehensive ERP system. The system can "act as a powerful network that can speed decision-making, slash costs, and give managers control over global empires at the click of a mouse," according to *Business Week* (Edmonson 1997). Within such "trade hype" is a kernel of truth. ERP systems are linking information together in ways that make it much easier for upper management to have a more global picture of operations in almost real time.

Advantages of this integrated approach include

1. Integrated functionality
2. Consistent user interfaces
3. Integrated database
4. Single vendor and contract
5. Unified architecture and tool set
6. Unified product support

But there are also disadvantages, including

1. Incompatibility with existing systems
2. Long and expensive implementation
3. Incompatibility with existing management practices
4. Loss of flexibility to use tactical point systems
5. Long product development and implementation cycles
6. Long payback period
7. Lack of technological innovation

In spite of any of these perceived drawbacks, ERP has enjoyed remarkable success in the marketplace, as we discuss below.

3.3.1   History and Success of ERP

The success of ERP is at least partly due to three coincident undercurrents preceding its development. The first is recognition of a field that has come to be called **supply-chain management (SCM).** In many ways, SCM extends traditional inventory control methods over a broader scope to include distribution, warehousing, and multiple production locations. Importantly, defining a function called supply-chain management has led to an appreciation of the importance of logistical issues. We see the importance of this area reflected in the growth of trade organizations such as the Council of Logistics Management, which grew from 6,256 members in 1990 to almost 14,000 in 1997.

The second trend that spurred acceptance of ERP was the **business process reengineering (BPR)** movement (see Hammer and Champy 1993). Prior to the 1990s, few companies would have been willing to radically change their management structures to support a new software package. But BPR has taught managers to think in terms of radical change. Today, many managers feel that one of the benefits of ERP implementation is the chance to reengineer their operations.

The third trend is the explosive growth in distributed processing and the power of smaller computers. An MRP run that took a weekend to run on a million-dollar computer in the 1960s can now be done on a laptop in a few seconds. Instead of a central repository for all corporate data, information is now stored where used on a personal computer or a workstation. These are linked via an intracompany network, and the data are shared by all functions. The latest offerings of ERP vendors are designed with exactly this architecture in mind (Parker 1997).

The growth of ERP sales indicates the degree of its acceptance. In 1989 total sales for MRP II at $1.2 billion accounted for just under one-third of the total software sales in the United States (*Industrial Engineering* 1991). Worldwide sales for the top 10 vendors of ERP alone were $2.8 billion in 1995, $4.2 billion in 1996, and $5.8 billion in 1997 (Michel 1997). One company, SAP, alone sold more than $3.2 billion in ERP software in 1997 (Edmonson 1997).

However, large sales of software are not the whole picture. Many companies are disenchanted at the sometimes staggering cost of implementation. In a survey of *Fortune* 1000 firms that had implemented ERP, 44 percent reported they spent at least four times as much on implementation help (e.g., consultants) as on the software itself. We are aware of several companies that have canceled projects after spending millions, not wanting to "throw good money after bad."

Nonetheless, in spite of the high cost, some companies report enormous productivity improvements. Bob Barett, vice-president at Monsanto Co., finished installing the accounting module of SAP in July 1996. He cited the software as responsible for a reduction in the planning cycle from six weeks to three, lower inventories, less working capital, an increase in its bargaining power with suppliers, all of which led to an estimated savings of $200 million per year to the company (Edmonson 1997).

3.3.2   An Example: SAP R/3

The software offering of SAP known as R/3 is a typical ERP system. R/3 utilizes client/server computer technology to provide what SAP calls a **data warehouse.** This allows common access by all applications to a single data set. It also provides the capability to share data with other software via a general interface.

Like most ERP systems, R/3 is a large, transaction-oriented software package. SAP has organized it into four application suites: financial, human resource, manufacturing and logistics, and sales and distribution. Each of these has numerous application programs. Among them, interestingly, is a simple material requirements planning module that is almost logically identical to that written by Orlicky 30 years ago.

SAP's R/3 is constantly being updated with additional modules, including modules for specific industries. A key desire is to establish what are best practices and then to incorporate these into the software. Many companies using SAP have radically changed their management procedures to conform to the software and these best practices. Indeed, this radical reduction in individuality on the part of corporate managers may ultimately prove to be the greatest social consequence of the SAP success. But since codifying practices that are several years old is hardly the best strategy for maintaining a competitive advantage, this may leave some ERP users vulnerable to more creative competitors.

3.3.3 Manufacturing Execution Systems

A **manufacturing execution system (MES)** is an automated implementation of what MRP II called *shop floor control.* Unlike SFC, however, MES tracks work in process automatically; records process, yield, and quality data; executes a schedule; releases new jobs into the system; etc. Whether the MES is part of ERP or not can generate a hot debate among consultants and software purveyors. Nonetheless, with the increasing integration of more and more business functions by software offerings like the SAP R/3, it is doubtful that MES will remain an independent entity for long.

3.3.4 Advanced Planning Systems

While ERP systems integrate company data, **advanced planning systems (APS)** are used to analyze the data up and down the organization. The capabilities of APS are as varied as the vendors supplying the software. Most APS applications are memory-based algorithms that perform functions. These include finite capacity scheduling, forecasting, available to promise, demand management, warehouse management, distribution and traffic management, etc. In many cases, ERP vendors partner with more specialized software developers to provide these functions. Interestingly, this add-on approach has frequently resembled the earlier MRP II approach to "fixing" the MRP problem of infinite backward scheduling of reworking the schedule *after* it has been generated.

3.4 Conclusions

Material requirements planning evolved from the fundamental recognition of the difference between dependent and independent demand. It was also the first major application of modern computers in production control. MRP provides a simple method for ordering materials based on needs, as established by a master production schedule and bills of material. As such, it is well suited for use in controlling the purchasing of components. However, in the control of production, there are still problems.

Manufacturing resources planning, or MRP II, was developed to address the problems of MRP and to further integrate business functions into a common framework. MRP II has provided a very general control structure that breaks the production control problem into a hierarchy based on time scale and product aggregation. Without such a hierarchical approach, it would be virtually impossible to address the huge problem of coordinating thousands of orders with hundreds of tools for thousands of end items made up of additional thousands of components. More recently, ERP has integrated this hierarchical approach into a formidable management tool that can consolidate and track enormous quantities of data.

Despite the important contributions of MRP, MRP II, and ERP to the body of manufacturing knowledge, there are fundamental problems with the basic model underlying these systems (i.e., the assumptions of infinite capacity and fixed lead times that are found even in some of the most sophisticated ERP systems). As we will discuss further in Chapter 5, a critical issue for the long term is how to resolve the basic difficulties of MRP while retaining its simplicity and broad applicability. We will address this problem in Part III, after we have taken note of the insights offered by the just-in-time (JIT) movement in Chapter 4 and have developed some basic relationships concerning factory behavior in Part II.

Study Questions

1. What is the difference between raw material inventory, work-in-process (WIP) inventory, and finished goods inventory?

2. What is the difference between independent demand and dependent demand? Give several examples of each.

3. What level is an end item in a bill of material? What is a low-level code? What is the low-level code for an end item? Draw a bill of material for which component 200 occurs on two different levels and has a low-level code of three.

4. What is the master production schedule, and what does it provide for an MRP system?

5. How do you convert gross requirements to net requirements? What is this procedure called?

6. Why are scheduled receipts adjusted before any net requirements are computed?

7. Which lot-sizing rule results in the least inventory?

8. What are the tradeoffs considered in lot sizing?

9. In what respect is the Wagner–Whitin algorithm optimal? How is it sometimes impractical (i.e., what does it ignore)?

10. Which of the following lot-sizing rules possess the so-called Wagner–Whitin property?
    a. Wagner–Whitin
    b. Lot-for-lot
    c. Fixed order quantity (e.g., all jobs have size of 50)
    d. Fixed order period
    e. Part-period balancing

11. How do planned lead times differ from actual lead times? Which is typically bigger, the planned lead time or the average actual lead time? Why?

12. What assumption in MRP makes the implicit assumption of infinite capacity? What is the impact of this assumption on planned lead times? On inventory?

13. What is the difference between a planned order receipt and a planned order release? How does a scheduled receipt differ from a planned order release?

14. What is the difference between a scheduled receipt and a firm planned order? How are they similar?

15. Why do we perform all the MRP processing for one level before going to the next-lower level? What would happen if we did not?

16. What is the bill-of-material explosion?

17. What is pegging? How does it help in bottom-up replanning?

18. What is the effect of having safety stock when computing net requirements?

19. What is the difference between having a safety lead time of one period and simply adding one period to the planned lead time? What is the same?

20. What is nervousness in an MRP system? How is it caused? Why is it bad? What are some things that can be done to prevent it?

21. What is MRP II? Why was it created?

22. Why might rough-cut capacity planning be optimistic? Why might it be pessimistic?

23. Why is capacity requirements planning not very accurate? What assumptions are made in CRP that are the same as those in MRP?

24. What is the purpose of dispatching? What are dispatching rules? Why does shortest process time seem to work pretty well? When does earliest due date work well?

25. What is the purpose of input/output control? Why is it often "too little, too late"?

Problems

1. Suppose an assembly requires five components from five different vendors. To guarantee starting the assembly on time with 90 percent confidence, what must the service level be for each of the five components? (Assume the same service level for each component.)

2. End item A has a planned lead time of two weeks. There are currently 120 units on hand and no scheduled receipts. Compute the planned order releases using lot-for-lot and the MPS shown here:

Week	1	2	3	4	5	6	7	8	9	10
Demand	41	44	84	42	84	86	7	18	49	30

3. Using the information in Problem 2, compute the planned order releases using part-period balancing where the ratio of setup cost to the holding cost is 200.

4. *(Challenge)* With the information in Problem 2, compute the planned order releases using Wagner–Whitin, where the ratio of setup cost to holding cost is 200. How much lower is the cost of the plan than in the previous case?

5. Rework Problem 2 with 50 units of safety stock. What is different from Problem 2?

6. Rework Problem 2 with a planned lead time of two periods and a safety lead time of one period. What is different from Problem 2?

7. Suppose demand for a power steering gear assembly is given by

Gear	1	2	3	4	5	6	7	8	9	10
Demand	45	65	35	40	0	0	33	0	32	25

Currently there are 150 parts on hand. Production is planned using the **fixed order period** method and two periods. The lead time is three periods. Determine the planned order release schedule.

8. Consider the previous problem, but assume that a scheduled receipt for 50 parts is scheduled to arrive in period five.
   a. What changes, if any, need to be made to the scheduled receipt?
   b. Using the same lot-sizing rule and lead time, compute the planned order release schedule.

9. Demand for a power steering gear assembly is given by

Gear	1	2	3	4	5	6	7	8	9	10
Demand	14	12	12	13	5	90	20	20	20	20

Currently there are 50 parts on hand. The lot-sizing rule is, again, **fixed order period** using two periods. Lead time is four periods.
   a. Determine the planned order release schedule for the gear.
   b. Suppose each gear assembly requires two pinions. Currently there are 175 pinions on hand, the lot-sizing rule is lot-for-lot, and the lead time is one period. Determine the gross requirements and then the planned order release schedule for pinions.

c. Suppose management decreases the demand forecast for the first period to 12. What happens to the planned order release schedule for gears? What happens to the planned order release schedule for pinions?

10. Consider an end item composed of a single component. Demand for the end item is 20 in week one, four in week two, two in week three, and zero until week eight when there is a demand of 50. Currently there are 25 units on hand and no scheduled receipts. For the component there are 10 units on hand and no scheduled receipts.

    Planned order releases for all items are computed using the Wagner–Whitin algorithm with a setup cost of $248 and a carrying cost of $1 per week. The planned lead time for the end item is one week, and for the component it is three weeks.
    a. Compute the planned order releases for the end item and the component. Are there any problems?
    b. The forecast for demand in week eight has been changed to 49. Recompute the planned order releases for the end item and the component. Are there any problems?
    c. Suppose the first two weeks' planned order releases from part *a* had been converted to *firm planned orders*. Do the computation again after changing the demand in week 8 to 49. Are there any problems? Comment on nervousness and the use of firm planned orders.

11. Generate the MRP output for items A, 200, 300, and 400 using the following information. (*Note:* End item A is the same as in Problem 3.)

    - Bills of material:

        A: Two 200 and one 400
        200:   Raw material
        300:   Raw material
        400:   One 200 and one 300

    - Master production schedule:

Week	1	2	3	4	5	6	7	8	9	10
Demand (A)	41	44	84	42	84	86	7	18	49	30

    - Item Master and Inventory Data:

Item	Amount on Hand	Amount on Order	Due	Lead Time (Weeks)	Lot Sizing Rule (Setup/Hold)
A	120	0		2	PPB (200)
200	300	200 100	3 5	2	Lot-for-lot
300	140	100 100	4 7	2	Lot-for-lot
400	200	0		3	Lot-for-lot

12. Consider a circuit-board plant that makes three kinds of boards: Trinity, Pecos, and Brazos. The bills of material are shown here:

Trinity: 1 subcomposite 111 and 1 subcomposite 112
Pecos: 1 subcomposite 211 and 1 subcomposite 212
Brazos: 1 subcomposite 311 and 1 subcomposite 312
Subcomposite 111: Core 1
Subcomposite 112: Core 2
Subcomposite 211: Core 1
Subcomposite 212: Core 1
Subcomposite 311: Core 1
Subcomposite 312: Core 2
All cores: raw material

Recently, the Lamination and the Core Circuitize operations have been bottlenecks. The unit hours (i.e., time for a single board on the bottleneck tools) in these areas are given below. These times are in hours and include inefficiencies such as operator unavailability, downtime, setups, and so forth.

Board	Trinity	Pecos	Brazos
Lam	0.020	0.022	0.020
Core Cir	0.000	0.000	0.000

Board	S111	S112	S211	S212	S311	S312
Lam	0.015	0.013	0.015	0.013	0.015	0.015
Core Cir	0.025	0.023	0.028	0.023	0.027	0.028

Board	Core 1	Core 2
Lam	0.008	0.008
Core Cir	0.000	0.000

The anticipated demand for the next six weeks is as follows:

Week	1	2	3	4	5	6
Trinity	7,474	2,984	5,276	5,516	3,818	3,048
Pecos	6,489	5,596	7,712	7,781	3,837	4,395
Brazos	3,898	3,966	3,858	6,132	5,975	6,051
Total	17,861	12,546	16,846	19,429	13,630	13,494

a. Construct bills of capacity for Trinity, Pecos, and Brazos at Lamination and Core Circuitize.

b. Use these bills to determine the load for each of the next six weeks at both Lamination and Core Circuitize. The process centers operate five days per week for three shifts per day (24 hours per day). Breaks and lunches are included in the unit hour data. There are six Lamination presses and eight expose machines (the bottleneck) in Core Circuitize. Which weeks are over- or underloaded? What should be done?

13. The Wills and Duncan parts must pass through process center 22. Wills is released to process center 22 while Duncan must first pass through process center 21 before going to process center 22. The planned lead time for going through process center 22 is three days, while the time to go through process center 21 is two days. There are 16 hours of capacity at process center 22 per day. Each Wills takes 0.04 hour while a Duncan takes 0.025 hour at process center 22. Currently there are 300 Wills units that have been in process center 22 for one day and 200 units that have been there for two days. Releases to the process center (i.e., Wills to 22 and Duncan to 21) are shown below. There are also 225 of the Duncan parts that have been in the process center for one day and 200 that have been there for two days. There are also 250 units in process center 21 that have been there for one day and 200 units that have been there for two days. The releases are as follows:

Day	Today	1	2	3	4	5
Wills	250	300	350	300	300	300
Duncan	250	150	150	150	150	150

a. Determine how many Wills parts will leave process center 22 on each day.
b. Determine how many Duncan parts will leave process center 22 on each day.
c. Compute the load profile for process center 22.

4   THE JIT REVOLUTION

I tip my hat to the new constitution
Take a bow for the new revolution
Smile and grin at the change all around
Pick up my guitar and play
Just like yesterday
Then I get on my knees and pray
WE DON'T GET FOOLED AGAIN!

The Who

4.1   The Origins of JIT

In the 1970s and 1980s, while American manufacturers were (or were not) joining the MRP crusade, something entirely different was afoot in Japan. Much like the Americans had done in the 19th century, the Japanese were evolving a distinctive style of manufacturing that would eventually spark a period of huge economic growth. The manufacturing techniques behind the phenomenal Japanese success have become collectively known as just-in-time (JIT). They represent an important chapter in the history of manufacturing management.

The roots of JIT undoubtedly extend deep into Japanese cultural, geographic, and economic history. Because of their history of living with space and resource limitations, the Japanese are inclined toward conservation. This has made tight material control policies easier to accept in Japan than in the "throw-away society" of America. Eastern culture is also more systems-oriented than Western culture with its reductionist scientific roots. Policies that cut across individual workstations, such as cross-trained floating workers and *total* quality management, are more natural in this environment. Geography has also certainly influenced Japanese practices. Policies involving delivery of materials from suppliers several times per day are simply easier in Japan, where industry is spatially concentrated, than in America with its wide-open spaces. Many other structural reasons for the Japanese success have been advanced. However, since a manufacturing firm has no control over these factors, they are of limited interest to us here.

Of greater relevance are the JIT practices themselves. The most direct source for many of the ideas represented by JIT is the work of Taiichi Ohno at Toyota Motor Company. According to Ohno, Toyota began its innovative journey in 1945 when Toyoda Kiichiro, president of Toyota, demanded that his company "catch up with America in three years. Otherwise, the automobile industry of Japan will not survive" (Ohno 1988, 3). At the time, Japan's economy was shattered by the war, labor productivity was one-ninth that of the United States, and automotive production was at minuscule levels. Obviously, Toyota did not catch up to the Americans in three years, but it set in motion an effort that would eventually achieve Toyoda's goal and would spark the most fundamental changes in manufacturing management since the scientific management movement of the 1920s.

Ohno, who moved to Toyota Motor from Toyoda Spinning and Weaving in 1943, recognized that the only way to become competitive with America would be to close the huge productivity gap between the two countries. This, he argued, could only be done through waste elimination aimed at lowering costs. But unlike the American automobile companies, Toyota could not reduce costs by exploiting economies of scale in giant mass production facilities. The market for Japanese automobiles was simply too small. Thus, the managers at Toyota decided that their manufacturing strategy had to be to produce many models in small numbers.

The principal challenge from a production control standpoint was to maintain a smooth production flow in the face of a varied product mix. Moreover, to avoid waste, this had to be accomplished without large inventories. Ohno described the system evolved at Toyota to address this challenge as resting on two pillars:

1. **Just-in-time.**
2. **Autonomation,** or automation with a human touch.

He attributed the motivation for the just-in-time idea to Toyoda Kiichiro, who used the words to describe the ideal automobile assembly process. Ohno's model for JIT was the American-style supermarket, which appeared in Japan in the mid-1950s. In a supermarket, customers get what is needed, at the time needed, and in the amount needed. In Ohno's factory analogy, a workstation is a customer that gets materials from an upstream workstation that acts as a sort of store. Of course, in a supermarket, stock is replenished from a storeroom or by means of deliveries, while in a factory, replenishment requires production by an upstream workstation. His goal was to have each workstation acquire the required materials from upstream workstations precisely as needed, or **just in time.**

Just-in-time flow requires a very smoothly operating system. If materials are not available when a workstation requires them, the entire system may be disrupted. As we discuss in the next section, this has serious implications for the production environment. One means for avoiding disruptions is Ohno's concept of **autonomation,** which refers to machines that are both *automated,* so that one worker can operate many machines, and *foolproofed,* so that they automatically detect problems. Ohno received his inspiration for the idea of autonomation from Toyoda Sakichi, inventor of the automatically activated loom used at Toyoda Spinning and Weaving. Automation was essential for achieving the productivity improvements necessary to catch up with Americans. Foolproofing, which helps operators intervene in an automated process at the right time, is primarily what Ohno meant by "automation with a human touch." He viewed the combination as necessary to avoid disruptions in a JIT environment.

Between the late 1940s and the 1970s, Toyota instituted a host of procedures and systems for implementing just-in-time and autonomation. These included the now famous kanban system, which we will discuss in detail later, as well as a variety of systems related to setup reduction, worker training, vendor relations, quality control, and many others. While not all the efforts were successful, many were, and the overall effect was to raise Toyota from an inconsequential player in the automotive market in 1950 to one of the largest automobile manufacturers in the world by the 1990s.

4.2   JIT Goals

To achieve Ohno's goal of workstations acquiring materials just in time, a pristine production environment is necessary. Perhaps as a result of the Japanese propensity to speak metaphorically,[1] or perhaps because of the difficulty of translating Japanese descriptions to English (the words translate, but the cultural context does not), this need has often been stated in terms of absolute ideals. For example, Robert Hall, one of the first American authors to describe JIT, used terms like **stockless production** and **zero inventories.** However, he did not literally mean that firms should operate without inventory. Rather, he wrote

> *Zero Inventories* connotes a level of perfection not ever attainable in a production process. However, the concept of a high level of excellence is important because it stimulates a quest for constant improvement through imaginative attention to both the overall task and to the minute details. (Hall 1983, 1)

Edwards (1983) pushed the use of absolute ideals to its limit by describing the goals of JIT in terms of the **seven zeros,** which are required to achieve **zero inventories.** These, along with the logic behind them, are summarized as follows:

1. **Zero defects.** To avoid disruption of the production process in a JIT environment where parts are acquired by workstations only as they are needed, it is essential that the parts be of good quality. Since there is no excess inventory with which to make up for the defective part, a defect will cause a delay. Thus, it is essential that every part be made correctly the first time. The only acceptable defect level is zero, and it is not possible to wait for inspection points to check quality. Quality must occur at the source.

2. **Zero (excess) lot size.** In a JIT system, the goal is to replenish stock taken by a downstream workstation as it is taken. Since the downstream workstations may take parts of many types, maximum responsiveness is maintained if each workstation is capable of replacing parts one at a time. If, instead, the workstation can only produce parts in large batches, then it may not be possible to replenish the stocks of all parts quickly enough to avoid delays. This goal is more frequently stated as a **lot size of one.**

3. **Zero setups.** The most common reason for large batch sizes in production systems is the existence of significant setup times. If it takes several hours to change a die on a machine to produce a different part type, then it only makes sense that large batches of each part will be run between setups. Small lot sizes would lead to frequent setups and thereby seriously degrade capacity. Hence, eliminating setups is a precondition for achieving lot sizes of one.

4. **Zero breakdowns.** Without excess WIP in the system to buffer machines against outages, breakdowns will quickly bring production to a halt throughout the line. Therefore, an ideal JIT environment cannot tolerate unplanned machine failures (or operator unavailability, for that matter).

5. **Zero handling.** If parts are made exactly in the quantities and at the times required, then material must not be handled more than is absolutely necessary. No extra

---

[1] Shigeo Shingo, who along with Ohno was influential in developing the Toyota system, writes such things as "the Toyota production wrings water out of towels that are already dry" (Shingo 1990, 54) and "there is nothing more important than planting 'trees of will' " (Shingo 1990, 172).

moves to and from storage can be tolerated. The ideal is to feed the material directly from workstation to workstation with no intermediate pauses. Any additional handling will move the system away from just-in-time operation, since parts will have to be produced early to accommodate the additional time spent in handling.

6. **Zero lead time.** When perfect just-in-time parts flow occurs, a downstream workstation requests parts and they are provided immediately. This requires zero lead time on the part of the upstream workstation. Of course, lot sizes of one go a long way toward reducing the effective lead time required to produce parts, but the actual processing time per part is also important, as is waiting (queuing) time. The goal of zero lead time is very close to the core of the zero inventories objective.

7. **Zero surging.** In a JIT environment where parts are produced only as needed, the flow of material through the plant will be smooth as long as the production plan is smooth. If there are sudden changes (surges) in the quantities or product mix in the production plan, then, since no excess WIP in the system can be used to level these changes, the system will be forced to respond. Unless there is substantial excess capacity in the system, this will be impossible and the result will be disruptions and delays. A level production plan and a uniform product mix are thus important inputs to a JIT system.

Obviously, the seven zeros are no more achievable in practice than is zero inventory. Zero lead time with no inventory literally means instantaneous production, which is physically impossible. The purpose of such goals, according to the JIT proponents who make use of them, is to inspire an environment of continual improvement. No matter how well a manufacturing system is running, there is always room for improvement. Gauging progress against absolute ideals provides both an incentive and a measure of success.

4.3   The Environment as a Control

The JIT ideals suggest an aspect of the Japanese production techniques that is truly revolutionary: the extent to which the Japanese have regarded the production environment as a control. Rather than simply reacting to such things as machine setup times, vendor deliveries, quality problems, production schedules, and so forth, they have worked proactively to shape the environment. By doing this, they have consciously made their manufacturing systems easier to manage.

In contrast, Americans, with their scientific management roots and reductionist tendencies, have been prone to isolating individual aspects of the production problem and working to "optimize" them separately. Americans took setup times (or costs) as fixed and tried to come up with optimal lot sizes (e.g., the EPL model). The Japanese tried to eliminate—or at least reduce—setups and thereby eliminate the lot-sizing problem. Americans took due dates as exogenously provided and attempted to optimize the production schedule (e.g., the Wagner–Whitin model). The Japanese realized that due dates are negotiated with customers and worked to integrate marketing and manufacturing to provide production schedules that do not require precise optimization or abrupt changes. Americans took infrequent, expensive deliveries from vendors as given and tried to compute optimal order sizes (e.g., the EOQ model). The Japanese worked to set up long-term agreements with a few vendors to make frequent deliveries feasible. Americans took quality defects as given and set up elaborate inspection procedures to find them. The Japanese worked to ensure that both vendors outside the plant and operators inside the plant were aware of quality requirements and equipped with the necessary tools to main-

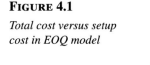

FIGURE 4.1

Total cost versus setup cost in EOQ model

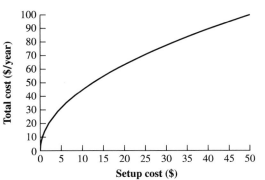

tain them. American manufacturing engineers got product specifications "thrown over the wall" from design engineers and did their best to adapt the manufacturing process to accommodate them. Japanese manufacturing and design engineers worked together to ensure designs that are practical to manufacture.

These distinctions between America and Japan are not a direct indictment of American models themselves. Indeed, as we highlighted in Chapter 2, models can offer valuable insights. For instance, the EOQ model suggests that total cost (i.e., setup plus inventory carrying cost) depends on the cost per setup according to the formula

$$\text{Annual cost} = \sqrt{2ADh}$$

where A is the setup cost (in dollars), D is the demand rate (in units per year), and h is the unit carrying cost (in dollars per unit per year). If we let $D = 100$ and $h = 1$ for purposes of illustration, then we can plot the relationship between total cost and setup cost as in Figure 4.1. This figure, and hence the model, clearly indicates that there are benefits to be gained from reducing the cost per setup. Since this cost presumably decreases with setup time, the EOQ model *does* point up the value of setup time reduction. However, while the insight is there, the sense of its strategic importance is not. Consequently, serious setup time reduction methodologies were evolved not in America, but in Japan.

In setups and many others areas, the Japanese have taken a holistic, systems view of manufacturing. Consequently, they have been able to identify policies that cut across traditional functions and to manage the interfaces between functions. Thus, while the specific techniques of JIT (which we shall discuss below) are important, the *systems approach* to transforming the manufacturing environment and the *constant attention to detail* over an extended period of time are fundamental. Ohno was urging just this with his admonition to "ask why five times," by which he meant that one should iteratively seek out and remove obstacles to the primary objective. A typical sequence of what Ohno had in mind might go as follows: A workstation becomes starved for work. Why? An upstream machine went down. Why? A pump failed. Why? It ran out of lubricant. Why? A leaky gasket was not detected. Why? And so on. This type of relentless pursuit of understanding and improvement may well be the real reason for Japan's remarkable success.

4.4 Implementing JIT

As the previous discussion makes clear, JIT is more than a system of frequent materials delivery or the use of **kanban** to control work releases. At the heart of the manufacturing systems developed by Toyota and other Japanese firms is a careful restructuring of the production environment. Ohno (1988, 3) was very clear about this:

> Kanban is a tool for realizing just-in-time. For this tool to work fairly well, the production process must be managed to flow as much as possible. This is really the basic condition. Other important conditions are leveling production as much as possible and always working in accordance with standard work methods.

Only when the environmental changes have been made can the specific JIT techniques be effective. We now turn to the key environmental issues that must be addressed in order to implement JIT.

4.4.1   Production Smoothing

As called for by the zero surging ideal, JIT requires a relatively smooth production plan. If either the volume or product mix varies greatly over time, it will be very difficult for workstations to replenish stock just in time. To return to the supermarket analogy, if all customers decided to do their shopping on Tuesday, or if all shoppers decided to buy canned tomatoes at the same time, stockouts would be very likely. However, because customers are spread over time and buy different mixes of products, the supermarket is able to replenish the shelves a little at a time and, for the most part, avoid stockouts.

In a manufacturing system, requirements are ultimately generated by customer demand. However, the sequence in which products are manufactured need not match the sequence in which they will be purchased by customers. Indeed, since customer demands are almost never completely known by the manufacturer in advance, this is not even possible. Instead, plants make use of a **master production schedule (MPS)** that specifies which products are to be produced in each time interval. As we noted in the previous chapter, MRP systems typically make use of time intervals (buckets) of a week or longer for their MPS.

A first condition for JIT, therefore, is to ensure that the MPS is reasonably level over time. As we noted in Chapter 3, many ERP systems contain MPS modules for facilitating the smoothing process. This development was stimulated in part by the Japanese JIT movement.

But even a smoothed MPS that specifies only weekly or monthly requirements could allow surges within the week or month that exceed the system's ability to meet the demands in a just-in-time fashion. Hence, the Toyota system and virtually all other JIT systems make use of a **final assembly schedule (FAS),** which specifies daily, or even hourly, requirements. Developing a level FAS from the MPS involves two steps:

1. Smoothing aggregate production requirements.
2. Sequencing final assembly.

Smoothing aggregate production is straightforward. If the MPS calls for monthly production of 10,000 units and there are 20 working days in the month, then the FAS will call for 500 units per day. If there are two shifts, this translates into 250 units per shift. If each shift is 480 minutes long, then the average time between outputs will have to be $480/250 = 1.92$ minutes per unit. In a perfect situation, this means we should produce at a rate of exactly one unit every 1.92 minutes. A system in which discrete parts are produced at a fairly steady flow rate is called a **repetitive manufacturing** environment. The kanban system developed by Toyota, which we will discuss later, is well suited only to repetitive manufacturing environments.

In reality, we are unlikely to produce exactly one unit every 1.92 minutes. Small deviations are not a problem; if the line falls behind during one hour but catches up during the next, fine. However, if the system departs from the specified rate over a period exceeding a shift or a day, corrective action (e.g., overtime) is typically required. Maintaining a steady, predictable output stream is the only means by which a JIT system can consistently meet customer due dates. Hence, JIT systems generally include measures to promote maintenance of a steady flow (e.g., incentives for making production quotas).

Once the aggregate requirements of the MPS have been translated to daily rates, we must translate the product-specific requirements to a production sequence. We do this by breaking out the daily requirements according to the product proportions from the MPS. For instance, if the 10,000 units to be produced during the month consist of 50 percent (5,000 units) product A, 25 percent (2,500 units) product B, and 25 percent (2,500 units) product C, then this means that the daily production of 500 units should consist of

$$0.5 \times 500 = 250 \text{ units of A}$$
$$0.25 \times 500 = 125 \text{ units of B}$$
$$0.25 \times 500 = 125 \text{ units of C}$$

Furthermore, the products should be sequenced on the line such that these proportions are maintained as uniformly as possible. Thus, the sequence

$$A–B–A–C–A–B–A–C–A–B–A–C–A–B–A–C \cdots$$

will maintain a 50-25-25 mix of A, B, and C over time. Obviously, this requires a line that is flexible enough to support this type of **mixed model production** (i.e., producing several products at once on the same line), which is impossible unless setups between products are very short or nonexistent. Furthermore, since the production rate is one unit every 1.92 minutes, this sequence implies that the times between outputs of product A will be $2 \times 1.92 = 3.84$ minutes. Times between outputs of products B and C will be $4 \times 1.92 = 7.68$ minutes. The assembly line, as well as the rest of the plant, must be physically capable of handling these times.

Of course, most production requirements will not lend themselves to such simple sequences. In that case, it may be reasonable to slightly adjust the demand figures (e.g., when demands are actually rough forecasts) to accommodate a simple sequence; or it may be reasonable to depart slightly from a simple sequence by spreading leftover units as evenly as possible throughout the daily schedule. The objective, however, remains as level a flow as possible. This is in sharp contrast with the traditional American practice of producing a large batch of one product before shifting to the next and emphasizing attainment of production quotas only at the end of the month.

4.4.2 Capacity Buffers

An apparent difficulty with JIT lies in coping with unexpected disruptions, such as order cancellations or machine failures. In an MRP system, when production requirements change, the schedule is simply regenerated, some jobs may be expedited, and things continue. However, in a JIT system, where great pains have been taken to ensure a constant flow, another approach is required. Similarly, if a machine failure causes production to fall behind, the netting operation in MRP will include the unmet requirements in the next pass. The JIT system with its level production quotas has no intrinsic way to keep track of such shortages.

This rigidity is certainly a problem with "ideal" JIT. But ideal JIT only works in an ideal environment—as does almost anything. (If demand is absolutely level, perfectly predictable, and within capacity capabilities, then MRP will work extremely well and will result in just-in-time production.) However, real-world JIT systems are never ideal and out of necessity contain measures for dealing with unanticipated disruptions. An approach commonly used by the Japanese is that of a capacity buffer. By scheduling the facility to less than 24 hours per day, the line can catch up if it falls behind. If production gets ahead of the desired rate, then workers are either sent home or directed to other tasks. If production falls behind the desired rate, either because of problems in the line or because of changes in the requirements, then the extra time is used. One way to allow for this is **two-shifting,** in which two shifts are scheduled per day, separated by a down period (Schonberger 1982, 137). The down period can be used for preventive maintenance or catch-up, if necessary. A popular approach is to schedule shifts "4–8–4–8," in which two eight-hour shifts are separated by four-hour down periods.

The capacity buffer offered by the availability of overtime serves as an alternative to the WIP buffers found in most MRP systems. If an unexpected occurrence, such as a machine outage, causes production to fall behind at a workstation, then WIP buffers can prevent other workstations from starving. In a JIT system where the WIP buffers are very small, a failure is very likely to cause starvation somewhere in the system. Thus, to keep the production rate constant, overtime will be needed. In effect, the Japanese have reduced WIP, so that production occurs just-in-time, but they have maintained excess capacity, just-in-case.

4.4.3   Setup Reduction

A work sequence like that suggested earlier, A–B–A–C–A–B–A–C–A–B–A–C–, is probably not workable if there are significant setup times required to switch production from one product to another. For instance, if each of the three products requires a different die that takes several hours to change over, there is no way to achieve the desired daily rate of 500 units while using a sequence that requires a die change after each part. In America these setups were traditionally regarded as given, and large lot sizes were used to keep the number of changeovers to a manageable level. In Japan, reducing the setup times to the point where changeovers no longer prevent a uniform sequence became something of an art form. Ohno reported setups at Toyota that were reduced from three hours in 1945 to three minutes in 1971 (Ohno 1988).

A number of good references provide specifics on the many clever techniques that have been used to speed machine changeovers (Hall 1983; Monden 1983; Shingo 1985), so we will not go deeply into details here. Instead, we will make note of some general principles that have been invoked to guide setup reduction efforts.

The key to a general approach to setup reduction is the distinction between an **internal setup** and an **external setup.** Internal setup operations are those tasks that take place when the machine is stopped (i.e., not producing product), while external setup operations are those tasks that can be completed while the machine is still running. For instance, removing a die is an internal task, while collecting the necessary tools to remove it is an external task. It is the internal setup that is disruptive to the production process, and hence this is the portion of the overall setup process that deserves the most intense attention. With this distinction in mind, Monden (1983) identifies four basic concepts for setup reduction:

1. *Separate the internal setup from the external setup.* The fact that current practice has the machine stopped while certain tasks are being completed does not guarantee that they are internal tasks. The setup reduction process must start by asking which tasks *must* be done with the machine stopped.

2. *Convert as much as possible of the internal setup to the external setup.* For example, if some components can be preassembled before shutting down the machine,

or if a die casting can be preheated before installing it, the internal setup time can be substantially reduced.

3. *Eliminate the adjustment process.* This frequently accounts for 50 to 70 percent of the internal setup time and is therefore critical. Jigs, fixtures, or sensors can greatly speed or even eliminate adjustments.

4. *Abolish the setup itself.* This can be done by using a uniform product design (e.g., the same bracket for all products), by producing various parts at the same time (e.g., stamping parts A and B in a single stroke and separating them later), or by maintaining parallel machines, each set up for a different product.

The references cited offer a host of techniques for implementing these concepts, ranging from quick-release bolts, to standardized tools and procedures, to parallel operations (e.g., two workers performing the setup in parallel), to color coding schemes, and so on. The real lesson from this diversity of ideas is, perhaps, the old maxim "Necessity is the mother of invention." The uniform production sequences used in JIT *demanded* quick changeovers, and the diligent efforts of Japanese engineers provided them.

4.4.4   Cross-training and Plant Layout

Ohno interpreted productivity improvement as a crucial goal for Toyota very early on. However, because of his concern with ensuring smooth material flow without excess WIP, productivity improvements could not be achieved by having workers produce large lots on individual machines. It rapidly became clear that a JIT system is much better served by multifunctional workers who can move where needed to maintain the flow. Furthermore, having workers with multiple skills adds flexibility to an inherently inflexible system, greatly increasing a JIT system's ability to cope with product mix changes and other exceptional circumstances.

To cultivate a multiskilled workforce, Toyota made use of a worker rotation system. The rotations were of two types. First, workers were rotated through the various jobs in the shop.[2] Then, once a sufficient number of workers were cross-trained, rotations on a daily basis were begun. Daily rotations served the following functions:

1. To keep multiple skills sharp.
2. To reduce boredom and fatigue on the part of the workers.
3. To foster an appreciation for the overall picture on the part of everyone.
4. To increase the potential for new idea generation, since more people would be thinking about how to do each job.

These cross-training efforts did indeed help the Japanese catch up with the Americans in terms of labor productivity. But they also fostered a great deal of flexibility, which Americans, with their rigid job classifications and history of confrontational labor relations, found difficult to match.

With cross-training and autonomation, it becomes possible for a single worker to operate several machines at once. The worker loads a part into a machine, starts it up, and moves on to another machine while the processing takes place. But remember, in a JIT system with very little WIP, it is important to keep parts flowing. Hence, it is not practical to have a worker staffing a number of machines that perform the same operation in a large, isolated process center. There simply will not be enough WIP to feed such an operation.

---

[2]It is interesting to note that managers were also rotated through the various jobs, in order to prove their abilities to the workers.

FIGURE 4.2

U-shaped manufacturing cell

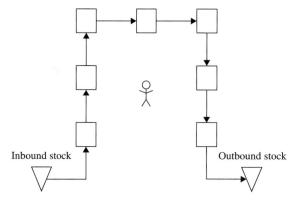

A better layout is to have machines that perform successive operations located close to one another, so that the products can flow easily from one to another. A linear arrangement of machines, traditionally common to American facilities,[3] accommodates the product flow well, but is not well suited to having workers tend multiple machines because they must walk too far from machine to machine. To facilitate material flow and reduce walking time, the Japanese have tended toward U-shaped lines, or cells, as shown in Figure 4.2.

The advantages of U-shaped cells are as follows:

1. One worker can see and attend all the machines with a minimum of walking.

2. They are flexible in the number of workers they can accommodate, allowing adjustments to respond to changes in production requirements.

3. A single worker can monitor work entering and leaving the cell to ensure that it remains constant, thereby facilitating just-in-time flow.

4. Workers can conveniently cooperate to smooth out unbalanced operations and address other problems as they surface.

The use of cellular layouts in JIT systems precipitated a trend that gathered steam in the United States during the 1980s. One now sees U-shaped manufacturing cells in a variety of production environments, to the point where cellular manufacturing has become much more prevalent than the JIT systems that spawned it.

4.4.5   Total Quality Management

Although the basic techniques of quality control were developed and espoused long ago by Americans, particularly Shewhart (1931), Feigenbaum (1961), Juran (1965), and Deming (1950a, 1950b, 1960), it was within the Japanese JIT systems that quality was lifted to new and strategic importance. Schonberger (1983, 50) offers two possible reasons for why quality control "took" in Japan so much more readily than in America:

1. The Japanese historical abhorrence for wasting scarce resources (i.e., by making bad products).

---

[3]Linear layouts were essential in colonial water-powered plants, where machines were driven by belts from a central driveshaft. By the time steam and electricity replaced water power, straight production lines had become the norm in America.

2. The Japanese innate resistance to specialists, including quality control experts, which made it more natural to ensure quality at the point of production than to check it later at a quality control station.

Beyond these cultural factors is the simple fact that JIT *requires* a high level of quality to function. Under JIT, a machine operator does not have a large batch of parts to sift through to find one suitable for use. He or she may have only one to choose from; if it is bad, the line stops. If this were to happen often enough, the consequences would be devastating. The analogy that many JIT writers have used is that of water in a stream with rocks on the bottom. The water represents WIP, the rocks are problems. As long as the water is high, the rocks are covered. However, when the water level is lowered, the rocks are exposed. Similarly, when the WIP level in a plant is reduced, problems, such as defects, become very noticeable.

Notice that JIT not only highlights the fact that there are quality problems, but also facilitates identification of their source. If WIP levels are high and quality inspections are made at separate stations, operators may get relatively little feedback about their own quality levels. Moreover, what they get will not be timely. In contrast, in a JIT environment, the parts made by an operator will be used rapidly by a downstream operator, who will have a strong incentive to notify the upstream operator of defects. This will serve to alert the operator of a potential problem while there is still time to do something about it. It also induces substantial psychological motivation to "do it right the first time." JIT advocates claim that this results in an overall increase in quality awareness and improved quality to the customer.

Analogous to the effect it had on setup reduction techniques, the pressure exerted by JIT fostered a burst of creativity in quality improvement methodologies. A huge volume of literature has detailed these over the past decade (see, e.g., DeVor 1992; Garvin 1988; Juran 1988; Shingo 1986), and so we will not go into great detail here. Instead, we will summarize seven principles identified by Schonberger (1983, 55) as essential to the quality practices of the Japanese:

1. **Process control.** The Japanese devoted a great deal of effort to enable the workers themselves to make sure their production processes were operating properly. This included use of statistical process control (SPC) charts and other statistical methods, but also involved simply giving workers responsibility for quality and the authority to make changes when needed.

2. **Easy-to-see quality.** As they were urged to do by Juran and Deming in the 1950s, the Japanese made use of extensive visual displays of quality measures. Display boards, gauges, meters, plaques, and awards were used to "put quality on display." These practices were aimed partly at providing feedback to the workforce and partly at proving that quality level is high to inspectors from customer plants.

3. **Insistence on compliance.** Japanese workers were encouraged to demand compliance with quality standards at every level in the system. If materials from a supplier did not measure up, they were sent back. If a part in the line was defective, it was not accepted. The attitude was that quality comes first and output second.

4. **Line stop.** The Japanese emphasized the "quality first" ideal to the extent that each worker had the authority to stop the line to correct quality problems. At some plants, yellow (for a problem) and red (for a line-stopping problem) lights were used to signal quality problems to the entire line. Where these techniques were used, quality really did come before throughput.

5. **Correcting one's own errors.** In contrast to the rework lines often found in American plants, the Japanese typically required the worker or work group that produced a defective item to fix it. This gave the workers full responsibility for quality.

6. **The 100 percent check.** The long-range goal was to inspect every part, not just a random sample. Simple or automated inspection techniques are desirable; foolproof (autonomous) machines that monitor quality during production are even better. However, in some situations where true 100 percent inspection was not feasible, the Japanese made use of the $N = 2$ method, in which the first and last parts of a production run are inspected. If both are good, then it is assumed that the machine was not out of adjustment and therefore that the intermediate parts are also good.

7. **Continual improvement.** In contrast to the Western notion of an acceptable defect level, the Japanese looked toward the ideal of zero defects. In this context, there is always room for further quality improvements.

Like the impact it had on cellular plant layout, JIT has engendered a revolution in quality that has grown far beyond its role in kanban and other JIT systems. The 1980s have been labeled the *quality decade* and have seen the emergence of such high-visibility initiatives as the Malcolm Baldrige Award and the ISO 9000 standards. The current heightened awareness of quality around the world is directly rooted in the JIT revolution.

4.5   Kanban

The single technique most closely associated with the JIT practices of the Japanese is the kanban system developed at Toyota. The word *kanban* is Japanese for *card,*[4] and in the Toyota kanban system, cards were used to govern the flow of materials through the plant.

To describe the Toyota kanban system, it is useful to distinguish between **push** and **pull** production control systems.[5] In a push system, such as MRP, work releases are *scheduled.* In a pull system, releases are *authorized.* The difference is that a schedule is prepared in advance, while an authorization depends on the status of the plant. Because of this, a push system directly accommodates customer due dates, but has to be forced to respond to changes in the plant (e.g., MRP must be regenerated). Similarly, a pull system directly responds to plant changes, but must be forced to accommodate customer due dates (e.g., by matching a level production plan against demand and using overtime to ensure that the production rate is maintained).

Figure 4.3 gives a schematic comparison of MRP and kanban. In the MRP system, releases into the production line are triggered by the schedule. As soon as work on a part is complete at a workstation, it is "pushed" to the next workstation. As long as machine operators have parts, they continue working under this system.

In the kanban system, production is triggered by a demand. When a part is removed from the final inventory point (which may be finished goods inventory) the last workstation in the line is given authorization to replace the part. This workstation then sends an authorization signal to the upstream workstation to replace the part it just used. Each station does the same thing, replenishing the downstream void and sending authorization to the next workstation upstream. In the kanban system, an operator requires both parts *and* an authorization signal (kanban) to work.

The kanban system developed at Toyota made use of two types of cards to authorize production and movement of product. This **two-card** system is illustrated in Figure 4.4.

---

[4]Ohno translates *kanban* as *sign board,* but we will use the more common translation of *card.*

[5]See Chapter 10 for a more detailed discussion and comparison of push and pull systems.

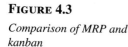

FIGURE 4.3

Comparison of MRP and kanban

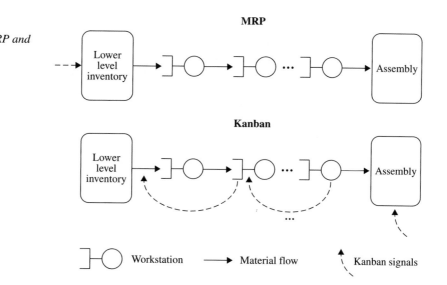

FIGURE 4.4

Toyota-style two-card kanban system

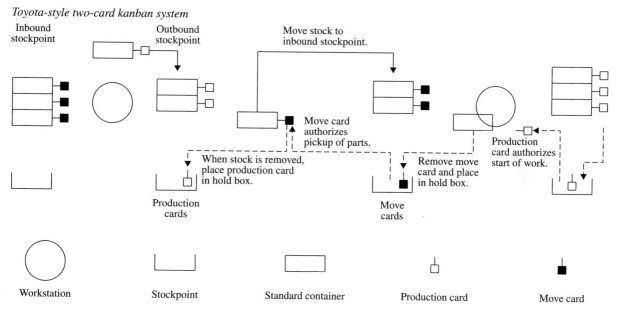

The basic mechanics are as follows. When a workstation becomes available for a new task, the operator takes the next **production card** from a box. This card tells the operator that a particular part is required at a downstream workstation. He or she looks to the inbound stockpoint for the materials required to make that part. If they are there, the operator removes the **move cards** attached to them and places them in another box. If the materials are not available, the operator chooses another production card. Whenever the operator finds both a production card and the necessary materials, he or she processes the part, attaches the production card, and places it in the outbound stockpoint.

Periodically, a **mover** will check the box containing move cards and will pick up the cards. He or she will get the materials indicated by the cards from their respective outbound stockpoints, replace their production cards with the move cards, and move them to the appropriate inbound stockpoints. The removed production cards will be deposited in the boxes of the workstations from which they came, as signals to replenish the inventory in the outbound stock points.

The rationale for the two-card system used by Toyota is that when workstations are spatially distributed, it is not feasible to achieve instantaneous movement of parts from one station to the next. Therefore, in-process inventory will have to be stored in two places, namely, an outbound stockpoint, when it has just finished processing on a machine, and an inbound stockpoint, when it has been moved to the next machine. The move cards serve as signals to the movers that material needs to be transferred from one location to another.

In a system with workstations close to one another, WIP can effectively be "handed" from one process to the next. In such settings, two inventory storage points are not necessary, and a **one-card** system, like that illustrated in Figure 4.5, can be used. In this system, an operator still requires a production card and the necessary materials to begin processing. However, instead of removing a move card from the incoming materials, the worker simply removes the production card from the upstream process and sends it back upstream. If one looks closely, it is apparent that a two-card system is identical to a one-card system in which the move operations are treated as workstations. Hence, the choice of one over the other depends on the extent to which we wish to regulate the WIP involved in move operations. If these operations are fast and predictable, it is probably unnecessary. If they are slow and irregular, regulation of move WIP may be helpful.

The key controls in a kanban system (one- or two-card) are the card counts at each station. These govern the amount of WIP in the system and, by affecting the frequency with which machines are starved for parts, determine the throughput rate. We will examine the relationship between WIP and throughput in detail in Part II. For now, it is worthwhile to note the similarity between kanban and the reorder point methods

FIGURE 4.5

One-card kanban system

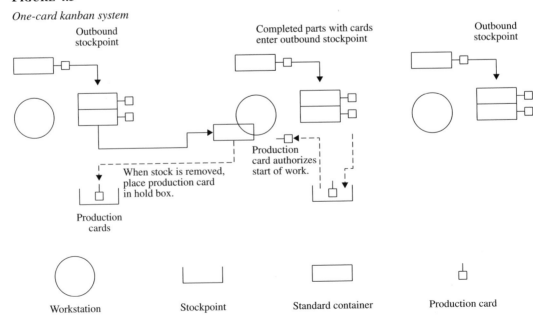

we discussed in Chapter 2. Consider the one-card kanban system with m production cards at a given station. Each time inventory in the downstream stockpoint falls below m, production cards are freed up, authorizing the station to replenish the buffer. The mechanics of this process are therefore precisely the same as those of the base stock model, with the downstream station acting as the demand and the card count m serving as the base stock level. The intuition we developed for this system in Chapter 2 carries over to the kanban system. However, the model does not directly apply because it assumes independent lead times for replenishing stock (i.e., the time to fill the nth and $(n + 1)$st orders are independent). Since the time to fill consecutive orders may well be correlated (i.e., if it takes a long time to fill the nth order, the $(n + 1)$st order may also have to wait a long time), a somewhat different model is required. We will develop models of this sort in Part II.

4.6   The Lessons of JIT

The range of issues touched on in this chapter makes it clear that JIT is not a simple procedure or technique. Nor can it be said to be a coherent, well-defined management strategy. Rather, it is an assortment of attitudes, philosophies, priorities, and methodologies that have been collectively labeled JIT. The real thread connecting them is that they have been practiced in recent times by a number of Japanese companies with notable success.

While JIT may not offer comprehensive policies for managing a manufacturing facility, its originators at Toyota and elsewhere have clearly demonstrated true genius in generating creative solutions to specific problems. Inherent in these solutions are some key insights that deserve a prominent place in the history of manufacturing management:

1. *The production environment itself is a control.* Strategies that involve reducing setups, changing product designs with manufacturing in mind, leveling production schedules, and so on, can have greater impact on the effectiveness of the production process than any decisions actually made on the factory floor.

2. *Operational details matter strategically.* Ohno and others reinforced the 100-year-old insight of Carnegie that the small details of the production process can confer a substantial competitive advantage. Like Carnegie, the JIT advocates concentrated on cost of manufacture and were willing to examine the most mundane aspects of the manufacturing process in their efforts to reduce waste.

3. *Controlling WIP is important.* The importance of the smooth and rapid flow of materials through the system was recognized by Ford in the 1910s and was echoed with emphasis by Ohno in the 1980s. Virtually all the benefits of JIT either are a direct consequence of low WIP levels (e.g., short cycle times) or are spurred by the pressure low WIP levels create (e.g., high quality levels).

4. *Flexibility is an asset.* JIT is inherently inflexible. In its essential form it calls for an absolutely steady rate and mix of production, virtually minute by minute. However, perhaps in reaction to this tendency toward inflexibility, the advocates of JIT have developed an acute appreciation for the value of flexibility in responding to a volatile marketplace. They have tempered JIT with a host of practices designed to promote flexibility, including short setup times, capacity cushions, worker cross-training, cellular plant layout, and many others.

5. *Quality can come first.* Although many of the basic quality concepts used by the Japanese in their JIT systems had long been championed by American quality experts,

Japanese firms were far more effective at putting these ideas into practice than were their American counterparts. They demonstrated to the world that a system in which quality takes precedence over throughput and is assured at the source not only works, but is profitable as well.

6. *Continual improvement is a condition for survival.* In sharp contrast to Henry Ford's belief in a perfectible product and process, the Japanese recognize that manufacturing is a continually changing game. Standards that sufficed yesterday will not be adequate tomorrow. Despite our terming JIT a "revolution," it took about 25 years (from the 1940s to the late 1960s) of constant attention for Toyota to reduce setups from three hours to three minutes. More than anything, the successful practitioners of JIT have been devoted to doing things better and better, a little bit at a time.

Discussion Point

1. Consider the following statement:

   Henry Ford practiced short-cycle manufacturing in the 1910s. The basic tools of Total Quality Management were developed and practiced at Western Electric in the 1920s. Kanban is equivalent to a base stock system, which was well known since the 1930s. Thus, just-in-time is nothing more than a repackaging of traditional American ideas, for which its Japanese proponents have been greatly overpraised.

   a. Comment on the accuracy of this statement.
   b. What aspects of JIT seem radically distinct from older techniques? Do these justify terming JIT a *revolution?*
   c. What aspects of JIT are particularly rooted in Japanese culture? What implications might this have for the transferability of JIT to America?

Study Questions

1. What are the seven zero goals of JIT? Of these, which are actually achievable? Which are completely outrageous if taken literally?
2. Discuss the fundamental difference between the zero defects goal in JIT and the acceptable quality level of former times. What does this have to do with the adage, "If you don't have time to do it right, when will you find time to do it over?"
3. Why is zero setup time desirable? Why is zero lead time?
4. Under the JIT philosophy, why is inventory often said to be evil?
5. What is meant by the common analogy of a stream, where WIP is represented by water and problems by rocks? What difficulties might arise from the perspective this analogy suggests?
6. What does Ohno mean by the "five whys"?
7. In what way does Ohno describe an American-style supermarket as an inspiration for JIT? What potential problems exist with using a supermarket as an analogy for a manufacturing system?
8. What role does total quality management (TQM) play in JIT? Does JIT depend on TQM, promote TQM, or both?
9. Describe autonomation.
10. Why is flexible labor important in a JIT system?
11. What are manufacturing cells? What role do they play in a JIT system?
12. What are the advantages of mixed model production?
13. Explain how two-card kanban works.

14. How is two-card kanban equivalent to one-card kanban? What is left out in the two-card case?

15. What is the "magic" of kanban? Is it the fact that stock is pulled from one station to the next, or is it something more fundamental?

16. Give at least two reasons that Toyota's kanban system has not been universally adopted by industry in America (or Japan).

17. Why are a relatively constant volume and relatively stable product mix essential to kanban?

18. List three ways in which the intrinsic rigidity of JIT is compensated for in practice.

19. What is the fundamental difference between a pull production system and a push production system?

20. In a serial production line, at which station (first, last, middle, etc.) would it be best to have the bottleneck in a push system? Where in a pull system? Explain your reasoning.

21. For each of the following situations, indicate whether kanban or MRP would be more effective.

    a. An auto plant producing three styles of vehicle

    b. A custom job shop

    c. A circuit board plant with 40,000 active part numbers

    d. A circuit board with 12 active part numbers

    e. A plant with one assembly line where all parts are purchased

5   WHAT WENT WRONG

Look ma, the emperor has no clothes!
Hans Christian Andersen

Our task now is not to fix the blame for the past, but to fix the course for the future.
John F. Kennedy

5.1   Introduction

By the 1980s, there were signs that not all was well with American manufacturing. Slowdowns in productivity growth (see Dertouzous, Lester, and Solow (1989) and Baumol, Blackman, and Wolff (1989) for discussions), declines in American shares of various markets, widespread perception that many goods in America were inferior in quality to their foreign counterparts, persistently large trade deficits, and many other troubling trends reminded us daily that America's once-undisputed manufacturing supremacy was no more. The "decline" of American manufacturing served as a serious wakeup call that we had entered a new globally competitive era. From that point forward, long-term success would require world-class performance and continual improvement on a variety of fronts: product development, marketing, human resource management, finance, *and* operations management. One of the main lessons of the Japanese success in the 1980s was that operations management can be (must be?) part of an effective modern manufacturing business strategy.

Conventional American operations management practices employed between World War II and 1990 can be roughly grouped into three schools of thought:

1. *Scientific management* is characterized by a rational, deductive, quantitative, modeling-oriented view of manufacturing systems. The original scientific management movement of the early 20th century spawned the quantitative methods for inventory control, scheduling, forecasting, aggregate planning, and many other manufacturing functions.

2. *Material requirements planning* is characterized by a central computerized planning approach to production control and integration. As more functions were incorporated into the system, the original MRP evolved into manufacturing resources planning (MRP II) and then into enterprise resources planning (ERP).

3. *Just-in-time* is characterized by a low-inventory, flow-oriented focus on the manufacturing environment. The original emphasis on Japanese kanban methods expanded

into the broader view of lean manufacturing. JIT was also the impetus for total quality management, which both was part of JIT and evolved into a separate movement.

While each of these certainly offers good ideas, none has been consistently successful in elevating firms to the level required to thrive in the competitive environment of the next century. In this chapter, we trace the reasons that the above approaches have failed to offer a comprehensive solution to the competitiveness problem. We will build on these negative insights, as well as on the positive ones of the previous four chapters, to develop an integrated approach to operations management in Parts II and III of this book.

5.2   Trouble with Scientific Management

In Chapter 1, we discussed two cultural tendencies we feel have had a significant influence on the way operations management has developed and been viewed in America:

1. *Faith in the scientific method,* which runs deep in the American soul, has motivated academics and practitioners to emphasize methods that are precise, quantitative, and high-technology. Taylor, with his shoveling formulas, clearly took this approach, as did the developers of the inventory control methods discussed in Chapter 2.

2. *The frontier ethic,* which glorifies wide-open spaces, rugged individualism, and sweeping adventure, is fundamentally in conflict with an attitude of careful husbanding of resources. This, coupled with the almost total lack of serious global competition in the first half of the 20th century, led many of the best and brightest in America to shun operations for more exciting careers in marketing, finance, or other fields.

It is not that either a frontier mentality or a quantitative outlook is intrinsically bad. However, the unique American combination of the two proved deadly in the 1970s and 1980s. The emphasis on marketing and finance took top management out of the loop as far as operations were concerned. This caused responsibility to devolve to middle managers, who lacked the perspective to see operations management in a strategic context. As a result, middle managers and the academic research community that supported them approached operations from an extremely narrow, reductionist perspective. Given this, our scientific bent led us to devote tremendous energy to applying increasingly sophisticated techniques to increasingly irrelevant problems.

This technical but unrealistic approach to operations management was already evident in 1913 when Harris published his original EOQ paper. Writing at the height of the early scientific management movement, Harris placed great emphasis on precision and elegance. For this reason, he made a number of simplifying assumptions about the lot-sizing problem that allowed him to derive his appealing "square root formula," but which rendered his results highly questionable for many real-world production systems. As we discussed in Chapter 2, these unrealistic assumptions included

- A fixed, known setup cost.
- Constant, deterministic demand.
- Instantaneous delivery (infinite capacity).
- A single product or no product interactions.

Because of these assumptions, EOQ makes much more sense in purchasing environments than in the production environments for which Harris intended it. In a purchasing environment, setups (i.e., purchase orders) may adequately be characterized with a constant cost. However, in manufacturing systems, setups cause all kinds of other problems (e.g., product mix implications, capacity effects, variability effects), as we will discuss in Part II. The assumptions of EOQ completely gloss over these important issues.

Even worse than the simplifying assumptions themselves was the myopic perspective toward lot sizing that the EOQ model promoted. By treating setups as exogenously specified constraints to be worked around, the EOQ model and its successors blinded operations management researchers and practitioners to the possibility of deliberately reducing the setups. It took the Japanese, approaching the problem from an entirely different perspective, to fully recognize the benefits of setup reduction.

In Chapter 2 we discussed similar aspects of unrealism in the assumptions behind the Wagner–Whitin, base stock, and (Q, r) models. In each case, the flaw of the model was not that it did not start with a real problem or a real insight. It did. As we have noted, the EOQ insight into the tradeoff between inventory and setups is fundamental to the behavior of a plant. So is the (Q, r) insight into the tradeoff between inventory (safety stock) and service. However, with our fascination for things scientific, these insights rapidly became secondary to the mathematics. Realism was sacrificed for precision and elegance. Instead of working to broaden and deepen the insights by studying the behavior of different types of real systems, we focused on faster computational procedures for solving the simplified problems. Instead of working to integrate disparate insights into a strategic framework, we concentrated on ever smaller pieces of the overall problem in order to achieve neat mathematical formulas.

Although the separation between models and reality existed right from the start of the operations management (OM) literature, it grew steadily worse. As OM became increasingly established as an academic discipline, fewer and fewer researchers drew directly on manufacturing facilities as a source of problems. Stylized standard problems became objects of volumes of research.

A classic example of this trend occurred in the field of flow shop scheduling, which was initiated by the publication of a paper by Johnson in 1954. Johnson's paper considered the problem of minimizing the total amount of time to process a fixed number of jobs (called **makespan**) on a two-machine production line. The processing times were assumed fixed and known, but not identical. The only issue, therefore, was the order in which to do the jobs on the machines. Johnson derived a simple and intuitive algorithm for computing an optimal schedule for this problem.

Unfortunately, the problem itself virtually never occurs in industry. Most manufacturing settings have jobs entering the system continually, so the issue of how to schedule a fixed number of jobs to minimize makespan is not relevant. However, the problem *is* of interest mathematically, because when the number of machines in the line is larger than three, it becomes very difficult (in a theoretical mathematical sense). Because researchers drew their inspiration from the literature and not from industry, Johnson's paper spawned an enormous number of follow-on papers addressing variations of his original problem. For the most part the variations were no more realistic than the original, and a recent survey of the flow shop scheduling research could find almost no evidence of influence on scheduling practice. Dudek, Panwalkar, and Smith (1992) summed up the history of this research area as follows:

> At this time, it appears that one research paper (that by Johnson) set a wave of research in motion that devoured scores of person-years of research time on an intractable problem of little practical consequence.

Similar stories can be told for other areas of the operations management literature, such as aggregate planning, inventory control, equipment replacement, and capacity planning. Throughout the OM field, far more was published than practiced.

The fact that most academic research had little impact on industry certainly did not help the competitiveness of American manufacturing, but it probably did not directly hurt it much either. A more insidious consequence of this research affected university teaching. By carrying the disjointed, models-oriented, unempirical approach of their research to the classroom, professors encouraged generations of engineering and business students to look at operations in a narrow, exclusively technical manner.

In engineering schools, operations management became operations research and focused almost exclusively on methodologies, such as linear programming and probability modeling. Even in courses aimed at production topics, methodologies often came first. Many scheduling classes, reflecting the scheduling literature, virtually became mathematics classes as they concentrated more on the complexity of the algorithms than on the issues involved in real scheduling situations. The contents of many "operations" texts emphasized operations research methodology over production applications.

In business schools, students were less patient and less interested in mathematics for its own sake. Therefore, as operations management courses became collections of quantitative methods applied to a host of loosely related problems (e.g., inventory control, scheduling, quality assurance, maintenance), they grew increasingly unpopular. In the 1970s and 1980s, some schools dropped OM from the curriculum! Others watered down the courses until the courses were mere compilations of anecdotal case studies.

The effect was that neither the engineering nor the business students were given much preparation for dealing with real-life operations problems. At best, this simply meant they were on their own to invent ad hoc solutions to problems as best they could. At worst, it meant they applied the mathematical methods they learned in school to situations for which the methods were ill adapted. (Our impression is that most industry practitioners have intelligently opted for the former and have largely ignored their academic training in production.)

By the late 1980s, stiff competition from the Japanese, Germans, and others made academics and practitioners alike realize that a change was necessary. Numerous distinguished voices called for a new emphasis on operations. For instance, professors from Harvard Business School stressed the strategic importance of operational details (Hayes, Wheelwright, and Clark 1988, 188):

> Even tactical decisions like the production lot size (the number of components or subassemblies produced in each batch) and department layout have a significant cumulative impact on performance characteristics. These seemingly small decisions combine to affect significantly a factory's ability to meet the key competitive priorities (cost, quality, delivery, flexibility, and innovativeness) that are established by its company's competitive strategy. Moreover, the fabric of policies, practices, and decisions that make up the manufacturing system cannot easily be acquired or copied. When well integrated with its hardware, a manufacturing system can thus become a source of sustainable competitive advantage.

Their counterparts across town at Massachusetts Institute of Technology agreed, calling for operations to play a larger role in the training of managers (Dertouzos, Lester, and Solow 1989, 161):

> For too long business schools have taken the position that a good manager could manage anything, regardless of its technological base…. Among the consequences was that courses on production or operations management became less and less central to business-school curricula. It is now clear that this view is wrong. While it is not necessary for every manager to have a science or engineering degree, every manager does need to understand how technology relates to the strategic positioning of the firm …

But while there is now increasing agreement that operations management is important, there is not yet agreement on what should be taught or how to teach it. The old approach of presenting operations solely as a series of mathematical models has been widely discredited. The pure case study approach is still in use at some business schools and may be superior because cases can provide insights into realistic production problems. However, covering hundreds of cases in a short time only serves to strengthen the notion that executive decisions can be made with little or no knowledge of the fundamental operational details. Moreover, the factory physics approach in Part II is our attempt to provide both the fundamentals and an integrating framework. In it we build upon past insights surveyed in the present section and make use of the precision of mathematics to clarify and generalize these insights. Better insight builds better intuition, and good intuition is necessary for good decision making. We are not alone in seeking a framework for building practical operations intuition via models (see Askin and Stanbridge 1993, Buzacott and Shantikumar 1993, and Suri 1998 for others). We take this as a hopeful sign that a new paradigm for operations education is emerging.

By the 1990s the mantle of scientific management had been picked up by business process reengineering (BPR). At its core, BPR was systems analysis applied to management.[1] But in keeping with the American proclivity for the big and the bold, the emphasis was heavily on *radical* change. Leading proponents of BPR defined it as "the fundamental rethinking and radical redesign of business processes to achieve dramatic improvements in critical, contemporary measures of performance, such as cost, quality, service, and speed" (Hammer and Champy 1993). Because most of the redesign efforts spawned by BPR involved eliminating jobs, it soon became synonymous with downsizing.

As a buzzword, BPR fell out of favor as quickly as it arose. By the late 1990s it had been banished from most corporate vocabularies. Still it left some lasting legacies. The layoffs of the 1990s, during bad times and good, certainly had a positive effect on labor productivity. But because the layoffs affected both labor and middle management to an unprecedented degree, they undermined worker loyalty.[2] Moreover, BPR represented an extreme backlash against the placid stability of the golden era of the 1960s; radical change was not only no longer feared, it was sought. This paved the way for more revolutions. For example, it is hard to imagine management embracing the ERP systems of the late 1990s, which required fundamental restructuring of processes to fit software as opposed to the other way around, without first having been conditioned by BPR to think in revolutionary terms. It is ironic that BPR, with its roots in the ultra-rational field of systems analysis, may actually have left American manufacturing more vulnerable to irrational buzzword fads than ever before.

The bottom line is that the scientific management school of thought contains valuable tools for addressing the problem of manufacturing competitiveness, but is not itself a comprehensive solution. The original insight of scientific management—that management is a discipline that can be studied—certainly remains valid. But Taylor's original efficiency-oriented framework of scientific management is too narrow to encompass the modern customer-oriented manufacturing environment. The quantitative models spawned by scientific management are useful for understanding and solving subproblems. But they can be dangerous if confused with the manufacturing system itself. Systems analysis is a powerful problem-solving tool that offers much promise as the basis for a balanced approach to continual improvement. But by pushing extreme solutions (radical change) or a narrow class of solutions (downsizing), it loses its balance and can

---

[1] Systems analysis is a rational means-ends approach to problem solving in which actions are evaluated in terms of a specific objective function. We discuss it in greater detail in Chapter 6.

[2] The enormous popularity of "Dilbert" cartoons, which poke liberal fun at BPR and other management fads, tapped into the growing sense of alienation of the workforce in corporate America. Ironically, some companies actually responded by banning them from office cubicles.

become the basis for personality-driven fads. The challenge, therefore, is to keep the essential components of the scientific management school, but to develop a framework in which they are applied in the right way to the problems of greatest strategic importance. This is precisely what we seek to do with factory physics in Part II.

5.3   Trouble with MRP

From at least one perspective, MRP was a stunning success. The number of MRP systems in use by American industry grew from a handful in the early 1960s, to 150 in 1971 (Orlicky 1975). The American Production and Inventory Control Society (APICS) launched its MRP Crusade to publicize and promote MRP in 1972. By 1981, claims were made that the number of MRP systems in America had risen as high as 8,000 (Wight 1981). In 1984 alone, 16 companies sold $400 million in MRP software (Zais 1986). In 1989, $1.2 billion worth of MRP software was sold to American industry, constituting just under one-third of the entire American market for computer services (*Industrial Engineering* 1989). By the late 1990s, ERP had grown to a $10 billion industry— ERP consulting was an even larger industry—and SAP, the largest ERP vendor, was the fourth-largest software company in the world (Edmondson and Reinhardt 1997). So, unlike many of the inventory models we discussed in Chapter 2, MRP *was,* and is, used in industry.

But has it worked? Were the companies who implemented MRP systems better off as a result? There is considerable evidence that suggests not.

First, from a macro perspective, American manufacturing inventory turns remained roughly constant during the 1970s and 1980s, during and after the MRP crusade (see Figure 5.1). (Note that inventory turns have increased in the 1990s, but this is almost certainly a consequence of the pressure to reduce inventory generated by the JIT movement and not directly related to MRP.) But of course many firms were not using MRP during this period. So while it appears that MRP did not revolutionize the efficiency of the entire manufacturing sector, these figures alone do not make a clear statement about MRP's effectiveness at the individual firm level.

FIGURE 5.1

U.S. manufacturing inventory turns, 1943–1994

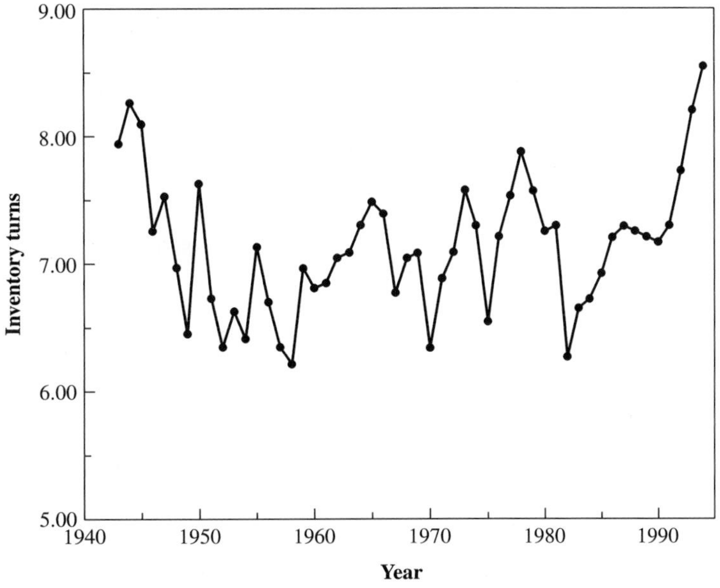

At the micro level, various surveys of MRP users did not paint a rosy picture either. Booz, Allen, and Hamilton, from a 1980 survey of more than 1,100 firms, reported that much less than 10 percent of American and European companies were able to recoup their investment in an MRP system within two years (Fox 1980). In a 1982 APICS-funded survey of 679 APICS members, only 9.5 percent regarded their companies as being class A users (Anderson et al. 1982).[3] Fully 60 percent reported their firms as being class C or class D users. To appreciate the significance of these responses, we must note that the respondents in this survey were both APICS members and materials managers—people with strong incentive to see MRP in as good a light as possible! Hence, their pessimism is most revealing. A smaller survey of 33 MRP users in South Carolina arrived at similar numbers concerning system effectiveness; it also reported that the eventual total average investment in hardware, software, personnel, and training for an MRP system was $795,000, with a standard deviation of $1,191,000 (LaForge and Sturr 1986).

Such discouraging statistics and mounting anecdotal evidence of problems led many critics of MRP to make strong disparaging statements, such as MRP is a "$100 billion mistake," "90 percent of MRP users are unhappy," and "MRP perpetuates such plant inefficiencies as high inventories" (Whiteside and Arbose 1984).

This barrage of criticism prompted the proponents of MRP to defend it. While not denying that it was far less successful than they had hoped when the MRP crusade was launched, they did not attribute this lack of success to the system itself. The APICS literature (e.g., Orlicky as quoted by Latham 1981), cited a host of reasons for most MRP system failures but never questioned the system itself. John Kanet, a former materials manager for Black & Decker who wrote a glowing account of its MRP system in 1984, but had by 1988 turned sharply critical of MRP, summarized the excuses for MRP failures as follows.

> For at least ten years now, we have been hearing more and more reasons why the MRP-based approach has not reduced inventories or improved customer service of the U.S. manufacturing sector. First we were told that the reason MRP didn't work was because our computer records were not accurate. So we fixed them; MRP still didn't work. Then we were told that our master production schedules were not "realistic." So we started making them realistic, but that did not work. Next we were told that we did not have top management involvement; so top management got involved. Finally we were told that the problem was education. So we trained everyone and spawned the golden age of MRP-based consulting.

Because these efforts still did not make MRP effective, Kanet and many others concluded that there is something more fundamental wrong with MRP. The real reason for MRP's inability to perform plant performance is that *MRP is based on a flawed model*. As we discussed in Chapter 3, the key calculation underlying MRP is performed by using fixed lead times to "back out" releases from due dates. These lead times are functions only of the part number and are not affected by the status of the plant. In particular, lead times do not consider the loading of the plant. An MRP system assumes that the time for a part to travel through the plant is the same whether the plant is empty or overflowing with work. As the following quote from Orlicky's original book shows, this separation of lead times from capacity was deliberate and basic to MRP (Orlicky 1975, 152):

---

[3]The survey used four categories proposed by Oliver Wight (1981) to classify MRP systems, classes A, B, C, and D. Roughly, Class A users represent firms with fully implemented, effective systems. Class B users have fully implemented, but less than fully effective systems. Class C users have partially implemented, modestly effective systems. And class D users have marginal systems providing little benefit to the company.

An MRP system is capacity-insensitive, and properly so, as its function is to determine what materials and components will be needed and when, in order to execute a given master production schedule. There can be only one correct answer to that, and it cannot therefore vary depending on what capacity does or does not exist.

But unless capacity is infinite, the time for a part to get through the plant *does* depend on the loading. Since all plants have finite capacity, the fixed-lead-time assumption is always an approximation of reality. Moreover, because releasing jobs too late can destroy the desired coordination of parts at assembly or cause finished products to come out too late, there is strong incentive to inflate the MRP lead times to provide a buffer against all the contingencies that a part may have to contend with (waiting behind other jobs, machine outages, etc.). But inflating lead times lets more work into the plant, increases congestion, and increases the flow time through the plant. Hence, the result is yet more pressure to increase lead times. The net effect is that MRP, touted as a tool to reduce inventories and improve customer service, can actually make them worse.

This flaw in MRP's underlying model is so simple, so obvious, that it may seem incredible that we came this far along the MRP path without noticing (or at least worrying about) it. To some extent, this is 20–20 hindsight. When viewed historically, MRP makes perfect sense and is, in some ways, the quintessential American production control system. When scientific management met the computer, MRP was the result. Unfortunately, the computer that scientific management met was a computer of the 1960s which had very limited power. Consequently, MRP is poorly suited to the environment and computers of the 1990s.

As we pointed out in Chapter 3, the original, laudable goal of MRP was to explicitly consider dependent demand, rather than to treat all demands as independent and use reorder point methods for lower-level inventories. This requires performing a bill-of-material explosion and netting demands against current inventories—both tedious data processing tasks in systems with complicated bills of material. Hence there was strong incentive to computerize.

The state of the art in computer technology in the mid-1960s, however, was an IBM 360 that used "core" memory with each *bit* represented by a magnetic doughnut about the size of the letter *o* on this page. When the IBM 370 was introduced in 1971, integrated circuits replaced the core memory. At that time a one-fourth inch square chip would typically hold less than 1,000 characters. As late as 1979, a mainframe computer with more than 1,000,000 bytes of RAM was a large machine. With such limited memory, performing all the MRP processing in RAM was out of the question. The only hope for realistically sized systems was to make MRP transaction-based. That is, individual part records would be brought in from a storage medium (probably tape), processed, and then written back to storage. As we pointed out in Chapter 3, the MRP logic is exquisitely adapted to a transaction-based system.

Thus, if one views the goal as explicitly addressing dependent demands in a transaction-based environment, MRP is not an unreasonable solution. The hope of the MRP proponents was that through careful attention to inputs, control, and special circumstances (e.g., expediting), the flaw of the underlying model could be overcome and MRP would represent a substantial improvement over older production control methods. This was exactly the intent of MRP II modules such as CRP and RCCP. Unfortunately, these were far from successful, and MRP II was roundly criticized in the 1980s while Japanese firms were strikingly successful by going back to methods that resemble the old reorder point approach. JIT advocates were quick to sound the death knell of MRP.

But MRP did not die, largely because MRP II handled important nonproduction data maintenance and transaction processing functions that were not replaced by JIT.

So MRP persisted into the 1990s, expanded in scope to include other business functions and multiple facilities, and was rechristened ERP. Simultaneously, computer technology advanced to the point where the transaction-based restriction of old MRP was no longer necessary. A host of independent companies emerged in the 1990s offering various types of finite-capacity schedulers to replace basic MRP calculations. However, because these were ad hoc and varied, many industrial users were reluctant to adopt them until they were offered as parts of comprehensive ERP packages. As a result, a host of alliances, licensing agreements, and other arrangements between ERP vendors and application software developers emerged.

There is much that is positive about the recent evolution of ERP systems. The integration and connectivity they provide make more data available to decision makers in a more timely fashion than ever before. Some finite-capacity scheduling modules are promising as replacements for old MRP logic in some environments. However, as we will discuss in Chapter 15, scheduling problems are notoriously difficult. It is not reasonable to expect a uniform solution for all environments. For this reason, ERP vendors are beginning to customize their offerings according to "best practices" in various industries. But the resulting systems are more monolithic than ever, often requiring firms to restructure their businesses to comply with the software. Although many firms, conditioned by the BPR movement to think in revolutionary terms, seem willing to do this, it may be a dangerous trend. The more firms conform to a uniform standard in the structure of their operations management, the less able they will be to use it as a strategic weapon, and the more vulnerable they will be to creative innovators in the future.

By the late 1900s, more cracks began to appear in the ERP landscape. In 1999, SAP AG, the largest ERP supplier in the world, was stung by two well-publicized implementation glitches at Whirlpool Corp., which resulted in the delay of appliance shipments to many customers and at Hershey Foods Corp., which left the shelves of candy retailers empty just before Halloween. Meanwhile, several companies decided to pull the plug on SAP installations costing between $100 and $250 million (Boudette 1999). Moreover, a survey by Meta Group of 63 companies revealed an average return on investment of a *negative* $1.5 million for an ERP installation (Stedman 1999).

Nonetheless, the original insight of MRP—that independent and dependent demands should be treated differently—remains fundamental. The hierarchical planning structure of MRP II and ERP provides coordination and a logical structure for maintaining and sharing data. However, to make effective use of the data processing power and scheduling sophistication promised by ERP systems of the future will require tailoring the operations system to a firm's business needs, not the other way around. This implies a sound understanding of core processes and the effects of specific planning and control decisions on them. *Factory Physics* provides a framework for understanding these core processes and the relationships between performance measures, as we show in Part II. In Part III, we use the insights of Part II to develop a planning hierarchy that parallels the MRP II hierarchy but incorporates advantages of pull production systems as well. We specifically focus on the scheduling problem, including approaches for working with MRP, in Chapter 15.

5.4   Trouble with JIT

As we noted in Chapter 4, the collection of ideas, priorities, and techniques that became collectively known as just-in-time (JIT) contains many creative and powerful insights.

The key question, however, is whether or not JIT represents a *system* and, if so, whether this system is transportable from Japan to America.

The early literature on JIT was somewhat contradictory on this point. On one hand, the first books on Japanese manufacturing techniques published in America suggested that these techniques are eminently transportable. In the first widely available book on JIT, Schonberger (1982, vii) said so directly: "I believe that the approaches travel easily to other countries ... Japanese production and quality management works in non-Japanese settings." Monden (1983, v) concurred, in his book describing the Toyota system: "The author firmly believes the Toyota production system can play a great role in the task for improving the constitutions of American and European companies..." Hall (1983), in his widely read JIT text, never even questioned whether JIT is a system, and proceeded to give detailed information on implementing it through such steps as flow balancing, quality improvements, and setup reduction.

In contrast to these optimistic viewpoints, other early observers of Japanese manufacturing practices were not sure that the Japanese had a system at all, let alone a transportable one. The Toyota kanban system was far from universal; in fact, it was almost exclusive to Toyota. Moreover, in a tour of six Japanese facilities, Robert Hayes (1981) did not find prevalent use of modern automation technology, quality circles, or uniform compensation systems. In short, he found "no exotic, strikingly different Japanese way of doing things."

Which view was correct? Were the Japanese practicing a well-defined JIT system that was responsible for their success, or were we simply observing a number of highly successful Japanese firms using a variety of disparate approaches?[4] The answer, perhaps, is that both views are partially correct. While Hayes did not see widely common procedures, he did observe two common effects:

1. Japanese plants were very clean and orderly.
2. Japanese plants exhibited much less work-in-process inventory than their American counterparts.[5]

Assuming that orderliness is an indicator of (or side effect from, or support to) a smoothly running system, these two effects are in fundamental agreement with the basic JIT tenets of *establishing a flow* and *eliminating waste* as described by Ohno (1988). While the Japanese firms may not have used the same methods, they do seem to have exhibited philosophical commonalities. But philosophy is trickier to transport than tools, leading Hayes (1981, 57) to be far less sanguine than Schonberger about the transferability of JIT:

> The modern Japanese factory is not, as many Americans believe, a prototype of the factory of the future. If it were, it might be, curiously, far less of a threat. We in the United States, with our technical ability and resources, ought then to be able to duplicate it. Instead, it is something much more difficult for us to copy; it is the factory of today running as it should.

Some of the controversy about its transportability was due to the fact that the term *JIT* did not mean the same thing to all people. Judging from the usage of the term in

---

[4]It is worthwhile to note here that the most successful Japanese firms are precisely the ones we saw most. Less-well-run Japanese companies simply could not survive the rigorous competitive task of marketing their goods halfway around the globe. As a result, we almost certainly overestimated the quality of overall Japanese manufacturing.

[5]A survey by Booz, Allen, and Hamilton of 1,000 firms in the United States, Europe, and Japan confirmed this observation, reporting that inventory turns were 50 percent higher in Japan than in the United States or Europe and that the gap was growing (Fox 1980).

the academic and practitioner literature, it appears that JIT represented both a *system* of beliefs and a *collection* of methods. This distinction was undoubtedly responsible for some of the considerable confusion over JIT in industry. We have heard of more frequent supplier deliveries, quality circles, smaller lot sizes, cellular layouts, material handling changes, worker participation programs, and so forth, all presented as JIT systems. The reality seems to be that whatever is systematic about them is a result of the company's own invention. Not surprisingly, some of these "systems" worked while others did not.

Zipkin (1991) aptly described the dichotomy of views on JIT expressed in the literature by separating *romantic JIT* from *pragmatic JIT*. By romantic JIT, he meant the stirring rhetoric that was built up around the idealized goals of zero inventories, zero defects, lot sizes of one, and so on, and was embodied in such lyrical slogans as "Simplify, and goods will flow like water" (Schonberger 1982). From this perspective, Zipkin (1991, 42) says, "JIT represents an aesthetic ideal, a natural state of simplicity. To implement JIT, what we need to do is to strip away needless layers of complexity." Although Schonberger generally acknowledges that working toward the JIT ideals requires grappling with myriad details, in passages of romantic fervor he has gone so far as to imply that JIT is easy, almost trivial, to implement (Schonberger 1990, 308): "Kanban is something that can be installed between any successive pair of processes in fifteen minutes, using a few containers and masking tape." While such statements are in fact true, there is a difference between *installing* kanban and *implementing* kanban.[6] Such statements invited readers, particularly those skimming through the literature rather than reading it carefully, to confuse simplicity of ideals with simplicity of implementation. As a result, many practitioners were led to believe that not only is JIT better than traditional American practices, but also it is easier.

While a senior manager who is far removed from the factory floor might be content with blithely contemplating the image of the ideal factory portrayed in romantic JIT, junior managers and operators on the shop floor who are charged with actually carrying out the revolution had no choice but to confront the other side of JIT—pragmatic JIT. Zipkin (1991, 41) describes pragmatic JIT as consisting of a host of nuts-and-bolts methods, including "engineering techniques to facilitate change-overs, cleaner plant layouts, quality-control training, scheduled maintenance, simpler product designs, and much more." The books of Hall (1983), Monden (1983), Shingo (1989), and Schonberger (1982, 1983) are replete with detailed descriptions of mechanical devices, plant layouts, and organizational structures with which to implement JIT. It is from this smorgasbord of techniques that practitioners were to achieve the environment of continuous improvement called for in romantic JIT.

Unfortunately, for all their detail, the methods described in the pragmatic JIT literature were far from being off-the-shelf technology. Indeed, they formed a much less complete system than MRP, which itself has been widely criticized as requiring an enormous amount of institutional attention to implement. To choose appropriate pragmatic JIT methods and construct a coherent set of operating policies requires a huge creative effort on the part of the practitioner.

Undoubtedly, the pioneers at Toyota *were* able to achieve a steady stream of improvements via the methods they described as JIT. But they were the creators of the methods, and they were very clever. Also, as the methods developed were specifically tailored to address *their* manufacturing environments, it is no wonder they were effective. Genius coupled with steadfast attention is a strong combination indeed.

---

[6]We will find in Part II that such "easy" installations do reduce work in process but at the expense of lost throughput and revenue.

The vast majority of companies did not have the benefit of a genius such as Ohno or Shingo (or Taylor or Ford, for that matter). Leading a revolution is a very risky and tricky business. Everyone with a vested interest in the status quo will be against the revolutionary, while those who are interested in change will offer only lukewarm support (Machiavelli 1532). Ford owned the business. He could do whatever he wanted. Ohno and Shingo were in a unique situation of "do or die" and were therefore allowed a free rein.

Although less likely to result in a complete success, imitation is a far less risky practice to the manager. If it is successful—great! If not, who can be blamed for doing what the "best in the business" were doing? Imitation, euphemistically called "benchmarking," became a standard practice for American companies in the 1980s. Unfortunately, it was based on compartmentalized descriptions of pragmatic JIT that detailed individual techniques, but could not evoke the spark of creativity required to select, develop, and balance them in a particular manufacturing setting. The lack of systematic guidance on where and how to apply the pragmatic JIT methods, coupled with the deceptively alluring visions of simplicity conjured up by romantic JIT, led too many managers to adopt specific JIT techniques with little overall coordination or prioritization.

Although some Americans may have perceived them as such, Ohno and Shingo never intended their methods as any sort of quick-fix panacea for every manufacturing environment. As we noted earlier, the dramatic setup reductions at Toyota were actually achieved by 25 years of slow, incremental work. Shingo (1989) seemed somewhat amused by the thought that Americans could rapidly adopt JIT methods successfully and quipped

> Some people imagine that Toyota has put on a smart new set of clothes, the kanban system, so they go out and purchase the same outfit and try it on. They quickly discover that they are much too fat to wear it.

Moreover, it is clear that the early JIT pioneers considered their developments a competitive advantage. Ohno admits that the Japanese used deliberately confusing terms to describe JIT. He once stated, "If the U.S. had understood what Toyota was doing, it would have been no good for us" (Myers 1990). Terms such as *JIT, zero inventories,* and *stockless production* may have served to delude Americans into thinking that JIT is far simpler than it is.

The fundamental difficulty in combining the ideals of romantic JIT with the details of pragmatic JIT into a coherent system lies in the fact that the ideals stress multiple, sometimes conflicting objectives. Throughput, quality, regularity of flow, flexibility, worker involvement, and other objectives are often cited as central to JIT. But which of these should take precedence? How is one to evaluate a policy that promotes some objectives but impedes others? The romantic JIT literature tended to oversimplify and minimize the difficulty of balancing conflicting concerns. Schonberger (1990, viii) went so far as to ban the word *tradeoff* (he calls it the *t-word*) from civilized conversation![7] But refusing to talk about them does not make tradeoffs go away. The Japanese originators of JIT *did* balance these tradeoffs—but subtly, artfully, and in the context of their specific manufacturing environments. The subtlety of the Japanese system for making tradeoffs allowed it to be easily overlooked, and consequently this aspect of JIT was lost in popular American descriptions of it.

But the failure of the American JIT literature to develop the intuition and systematic framework needed for balancing competing objectives was a serious one. The balance

---

[7]Zipkin (42) relates a story of a company that took Schonberger's overzealous advice literally and found itself inventing euphemisms for the word *tradeoff* in order to have meaningful discussions of options.

struck by Toyota and the other JIT pioneers was probably more important than any particular methodology. Ignoring it was tantamount to throwing away the banana and keeping the peel.

As a specific example, consider the fondness of the JIT literature of referring to inventory as the "root of all evil." Without a perspective on tradeoffs, this simple slogan implies that removing inventory can only benefit the system. In fact, in the often-cited JIT analogy of a stream, with WIP as water and problems as rocks, lowering the water (i.e., removing WIP) is necessary for promoting improvement. Thus, many firms in the 1980s ambitiously pursued WIP reduction programs.

Without question, many firms ultimately benefited from such efforts because inventory levels were too high. But how many went too far?[8] How many caused themselves unnecessary disruption by removing WIP before eliminating the environmental flaws that necessitated the WIP? Inman (1993) has observed that inventory is better described as the "flower" rather than the "root" of all evil, since high levels of inventory are a consequence of other problems. To pursue the stream analogy a bit further, it would be better to use sonar to locate the rocks, remove them, and then lower the water, rather than to lower the water and smash into the rocks in order to find them. Unfortunately, JIT, as described in the American literature, offered neither sonar (i.e., models that predict the effects of system changes) nor a sense of the relative economics of level reduction versus rock removal—that is, procedures for evaluating the tradeoffs between the benefits of WIP reduction and the costs of eliminating problems.

Thus, American firms implementing JIT struck, explicitly or implicitly, their own balance among competing objectives. Those that did this with a basic understanding of their fundamental processes created effective systems. The rest were probably disappointed in their JIT experiences. In any case, because putting together a coherent JIT system is a daunting task, firms across the board have frequently relied on outside consultants to help them in JIT implementation. The expense of such consulting, plus the substantial training expenses that are required, can make JIT a costly option. Indeed, Inman and Mehra (1990) reported that such expenses can put JIT beyond the reach of many small companies. So despite some well-publicized success stories and a great deal of romantic JIT hyperbole, just-in-time has proved to be neither simple nor inexpensive.

In addition to the legacy of low-inventory, flow-oriented production, the JIT movement left another important mark on the manufacturing landscape, namely, total quality management (TQM). Originally an essential component of JIT—low inventory production cannot be implemented without good quality and good quality is impossible without low inventory levels and short cycle times—TQM soon spawned a movement of its own. TQM quickly eclipsed JIT and became the preeminent manufacturing buzzword of the 1980s. By the end of the 1980s virtually all American companies had some type of TQM program, whether or not they were making use of other JIT techniques. Quality was elevated from a low-level staff function to the executive suite through the appointment of vice-presidents of quality and proclamations by CEOs of the central role of quality (e.g., Bob Galvin of Motorola stated emphatically: "No company has ever hurt profits by improving quality"). Uniform quality "standards" (e.g., ISO 9000) became part of the business landscape. The 1980s were dubbed the "decade of quality."

But by the middle 1990s quality was passé. Companies discontinued programs and renamed positions. Business students objected to TQM courses as "out of date."

---

[8]We know of a furniture manufacturer that nearly put itself out of business via inventory reduction. The reason was that rising wood prices in recent years meant that competitors who carried more inventory were able to buy it earlier at lower prices and therefore had lower costs.

Depending on the commentator, the 1990s were the decade of "speed" or "agility" or anything but quality.

This was due in part to the success of the TQM movement. The quality of American manufactured goods really did improve during the 1980s. American firms in industries threatened by higher-quality imports (e.g., the auto industry) managed to close the gap through significant investments in facilities and procedures. But gaps still exist, and most companies are still nowhere near their "parts per million" or "six-sigma" targets. Moreover, customers, conditioned by competition to demand higher standards, are still far from ecstatic about most manufactured products. So opportunties still exist to gain competitive advantage through higher quality, although it is no longer fashionable to speak in these terms.

Another reason that TQM diminished as a thrust is that quality is not always the most promising competitive lever. Ford's concentration on quality (and cost) in the 1920s and 1930s almost destroyed the firm because it neglected the diversity factor so successfully introduced by General Motors. A similar dynamic was at play in the semiconductor industry in the 1980s and 1990s, where yield losses in microprocessor wafer fabs almost never reached one in 100 let alone one in one million, before the next generation of technology was introduced. The reason, of course, was that the benefits of rapid product development outweighed the benefits of extremely high levels of quality.

These factors may explain why quality fell out of fashion as a buzzword, but they do not diminish its importance as a competitive dimension. Just as quality did not eliminate cost as a concern—indeed, one of the main challenges of TQM was to elevate quality without increasing cost—the new dimensions of speed or flexibility do not replace quality. A key to competitiveness in the future will be the ability to elevate quality even while delivering products to customers faster and in greater variety than ever before. This will require steep learning curves, which precludes reliance on trial-and-error methods. The only way to do this is to have a sufficiently sophisticated body of theory to predict performance and "do it right the first time."

We can summarize the outcome of the JIT and TQM movements by noting that both have produced a number of important and useful insights about manufacturing management. At the same time, what was described in the American JIT literature as a *system* is really a loosely coordinated *collection* of techniques infused with an inspiring stream of romantic rhetoric. The well-publicized success of the Japanese in the 1980s, appealing JIT slogans, and the apparent simplicity of JIT techniques led us to expect far more than we received from the JIT "revolution." Similarly, the elevated awareness of quality and the specific statistical tools of TQM are unquestionably essential components of modern manufacturing management. But like JIT, TQM was sold in romantic terms with near religious fervor. As a result, it failed to develop a coordinated system with which to integrate the pieces, balance quality with other business objectives, and facilitate compression of the learning curve. In both TQM and JIT, what is lacking is a fundamental paradigm that connects practices with business performance. We propose one such paradigm with factory physics in Part II and specifically examine the science of pull in Chapter 10 and the relationships between quality and logistics in Chapter 12.

5.5   Where from Here?

In Part I of the book, and particularly this chapter, we have made the following points:

1. Scientific management, particularly the quantitative methods, has reduced the manufacturing management problem to analytically tractable subproblems, often with

unrealistic modeling assumptions, to the point where they provide little useful guidance from an overall perspective. The mathematical methods and some of the original insights can certainly still be useful, but we need a better framework for applying these in the context of an overall business strategy. Business process reengineering exhorts managers to rethink their processes, but does not provide a framework and became too closely identified with exclusively radical solutions and downsizing to provide a balanced alternative.

2. MRP is fundamentally flawed, not in the details, but in the basics, because it uses an infinite-capacity, fixed-lead-time approach to control work releases. "Patches," such as MRP II and CRP, may help but cannot rectify this basic problem. The original insight of MRP, that dependent demands are distinct from independent ones, is still valid; the planning hierarchy established by MRP II is useful; and the data maintenance and sharing functions of MRP systems are essential. Finite-capacity scheduling modules in ERP systems offer the potential for a fix. However, a single scheduling approach is unlikely to be effective in all types of systems. Moreover, by building "best practices" into the systems demanded by their software, ERP could have the undesirable effect of stifling creativity and preventing firms from crafting systems well suited to their needs. To design appropriate scheduling modules and apply them effectively in an effective planning hierarchy will require careful coordination with the principles governing production system behavior.

3. JIT and TQM are collections of methods and slogans, not systems. Because of this, it is simply not possible to imitate the Japanese successes of the 1980s in cookbook fashion. The many central and creative insights of the JIT and TQM founders need to be appreciated and built upon. However, only by establishing a framework for balancing competing objectives can we develop effective manufacturing management systems.

The real lesson in all this is that there is *no easy solution*. We Americans seem to have a resolute faith in a swift and permanent resolution of the manufacturing problem. Witness the famous economist John Kenneth Galbraith who stated years ago that we had "solved the problem of production" and could move on to other things (Galbraith 1958). Even though it quickly became apparent that the production problem was far from solved, our faith in the possibility of solving it remained unshaken. Each successive approach to manufacturing management—scientific management, operations research, MRP, JIT, TQM, BPR, ERP, etc.—has been sold as *the* solution. Each one has disappointed us, but we continue to look for the elusive "technological silver bullet" to save American manufacturing.

When will we learn? Manufacturing is complex, large scale, multiobjective, rapidly changing, and highly competitive. There *cannot* be a simple, uniform solution that will work well across a spectrum of manufacturing environments. Moreover, even if a firm can come up with a system that performs extremely well today, failure to continue improving is an invitation to be overtaken by the competition. Ultimately, each firm is on its own to develop an effective manufacturing strategy, support it with appropriate policies and procedures, and continue to improve these over time. As global competition intensifies, the extent to which a firm does this will become not just a matter of profitability, but one of survival.

Discussion Points

1. Consider the following quote referring to the two-machine minimize-makespan scheduling problem:

   > At this time, it appears that one research paper (that by Johnson) set a wave of research in motion that devoured scores of person-years of research time on an intractable problem of little practical consequence. (Dudek, Panwalkar, and Smith 1992)

   a. Why would academics work on such a problem?
   b. Why would academic journals publish such research?
   c. Why didn't industry practitioners either redirect academic research or develop effective scheduling tools on their own?

2. Consider the following quotes:

   > An MRP system is capacity-insensitive, and properly so, as its function is to determine what materials and components will be needed and when, in order to execute a given master production schedule. There can be only one correct answer to that, and it cannot therefore vary depending on what capacity does or does not exist. (Orlicky 1975)

   > For at least ten years now, we have been hearing more and more reasons why the MRP-based approach has not reduced inventories or improved customer service of the U.S. manufacturing sector. First we were told that the reason MRP didn't work was because our computer records were not accurate. So we fixed them; MRP still didn't work. Then we were told that our master production schedules were not "realistic." So we started making them realistic, but that did not work. Next we were told that we did not have top management involvement; so top management got involved. Finally we were told that the problem was education. So we trained everyone and spawned the golden age of MRP-based consulting. (Kanet 1988)

   a. Who is right? Is MRP fundamentally flawed, or can its basic paradigm be made to work?
   b. What types of environment are best suited to MRP?
   c. What approaches can you think of to make an MRP system account for finite capacity?
   d. Suggest opportunities for integrating JIT concepts into an MRP system.

Study Questions

1. Why have relatively few CEOs of American manufacturing firms come from the manufacturing function, as opposed to finance or accounting, in the past half century? What factors may be changing this situation now?

2. In what way did the American faith in the scientific method contribute to the failure to develop effective OM tools?

3. What was the role of the computer in the evolution of MRP?

4. In which of the following situations would you expect MRP to work well? To work poorly?
   a. A fabrication plant operating at less than 80 percent of capacity with relatively stable demand
   b. A fabrication plant operating at less than 80 percent of capacity with extremely lumpy demand
   c. A fabrication plant operating at more than 95 percent of capacity with relatively stable demand
   d. A fabrication plant operating at more than 95 percent of capacity with extremely lumpy demand

    e. An assembly plant that uses all purchased parts and highly flexible labor (i.e., so that effective capacity can be adjusted over a wide range)

    f. An assembly plant that uses all purchased parts and fixed labor (i.e., capacity) running at more than 95 percent of capacity

5. Could a breakthrough in scheduling technology make ERP the perfect production control system and render all JIT ideas unnecessary? Why or why not?

6. What is the difference between *romantic* and *pragmatic* JIT? How may this distinction have impeded the effectiveness of JIT in America?

7. Name some JIT terms that may have served to cause confusion in America. Why might such terms be perfectly understandable to the Japanese but confusing to Americans?

8. How long did it take Toyota to reduce setups from three hours to three minutes? How frequently have you observed this kind of diligence to a low-level operational detail in an American manufacturing organization?

9. How would history have been different if Taiichi Ohno had chosen to benchmark Toyota against the American auto companies of the 1960s instead of using other sources (e.g., Toyota Spinning and Weaving Company, American supermarkets, and the ideas of Henry Ford expressed in the 1920s)? What implications does this have for the value of benchmarking in the modern environment of global competition?

II   FACTORY PHYSICS

A theory should be as simple as possible, but no simpler.
Albert Einstein

6   A SCIENCE OF MANUFACTURING

I often say that when you can measure what you are speaking about, and express it in numbers, you know something about it; but when you cannot express it in numbers, your knowledge is of a meager and unsatisfactory kind; it may be the beginning of knowledge, but have scarcely, in your thoughts, advanced to the stage of Science, whatever the matter may be.

Lord Kelvin

6.1   The Seeds of Science

These are confusing times for manufacturing managers. The barrage of books, short courses, software packages, videotapes, Web sites, and other sources pushing competing manufacturing philosophies and tools is enough to overwhelm even the most experienced professional. Moreover, as we saw in Part I, major approaches to manufacturing management (e.g., classical inventory control, MRP, and JIT) are not fully compatible with one another and suffer individually from serious flaws.

Many in manufacturing have come to view their discipline in terms of a blizzard of management buzzwords (for example, MRP, MRP II, ERP, JIT, CIM, FMS, OPT, TQM, BPR) and a succession of gurus. Micklethwait and Woolridge (1996) describe this trend in their revealingly titled book *The Witchdoctors*.

While they frequently offer kernels of truth, the very nature of buzzword approaches is to sell a single solution for all situations. Hence, they provide little balanced perspective on what works well and when. This has often led to a "management by bandwagon" mentality with unfortunate results. Employees, battered by one "revolution" after another, settle into a cynical attitude that "this too will pass." But undaunted, many managers keep to the faith, believing that someone, somewhere has a silver bullet that will solve all their operations problems. As a result, buzzword books and consultants prosper, but little real progress is made.

Certainly part of the confusion stems from the excessive hyperbole used by vendors and consultants to market their wares. Glitzy promotional materials built around vague, sweeping claims make it difficult for managers to accurately compare systems. However, we suspect the roots of the problem are deeper than this. We believe that a large measure of the confusion is a direct consequence of our lack of an underlying **science of manufacturing.**

6.1.1 Why Science?

In a field such as physics, where the objective is to understand the physical universe, the need for science is obvious. But manufacturing management is an applied field, where the objective is financial performance, not discovery of knowledge. So why does it need science?

The simplest response is that many applied fields rely on science. Medicine is based on biology, chemistry, and other sciences. Civil engineering is premised on statics, dynamics, and other branches of physics. Electrical engineering depends on the sciences of electricity and magnetism. In each case, the scientific foundation provides a powerful set of tools, but is not in itself the complete applied discipline. For example, the practice of medicine involves much more than simply applying the principles of biology.

More specifically, science offers a number of uses in the context of manufacturing management.

First, science offers precision. As the quote at the beginning of this chapter attests, "when you cannot express it in numbers, your knowledge is of a meager and unsatisfactory kind." So one reason to develop a science of manufacturing is to provide more precise characterization of how systems will work. Relations that provide predictions are the **basics** of science. For example, $F = ma$ is a basic relation of physics. Probability tools, like those we used to model demand uncertainty in inventory systems in Chapter 2, are examples of important basics of factory physics.

Science also offers **intuition.** The formula $F = ma$ is intuitive. Double the force and, for the same mass, acceleration doubles. Elementary school students are required to take science courses, not so they can calculate the outcome of an experiment, but so they can better understand the world around them. Knowing that water expands when it freezes and that expanding ice can crack an engine block convinces one of the need for antifreeze (whether or not one can compute the molality of a solution). Similarly, a manager frequently does not have time to conduct a detailed analysis of a decision. In such cases, the real value of models is to sharpen intuition. Good intuition enables managers to focus their energies on issues of maximum leverage.

Finally, science facilitates **synthesis** of disparate perspectives by providing a unified framework. For instance, for many years, electricity and magnetism and optics were thought to be different fields. James Clerk Maxwell unified them with four equations. In manufacturing, key performance measures, such as work in process and cycle time, are often treated as if they are independent. But as we will see in Chapter 7, there are well-defined and useful relationships between these measures. Moreover, manufacturing enterprises are complex systems involving people, equipment, and money. As such, they can be reasonably viewed in a variety of ways: as a collection of people with shared values, as a creative community for developing new products, as a set of interrelated physical processes, as a network of material flows, or as a set of cost centers. By providing a consistent framework, a science of manufacturing offers a means to synthesize these disparate views. Bringing the different parts of a system into an effective whole is close to the core of the management function.

To further highlight the need for a science of manufacturing, we consider two examples.

Example: A Product Design

First, suppose the marketing department of an automotive company has proposed a concept for a new car that entails

- A mass of 1,000 kilograms (about 2,200 pounds), for safety and luxury.
- Acceleration from 0 to 60 in 10 seconds (approximately 2.7 meters per second squared), for sportiness.
- An engine that generates no more than 200 newtons of force (approximately 45 lb), for fuel efficiency.

Can it be done?

When framed in such simple terms, we can give a simple answer to this question—*no way!* The elementary relationship from physics

$$F = ma$$

or, in this case,

$$200 \text{ N} \ll (1{,}000 \text{ kg})(2.7 \text{ m/s}^2) = 2{,}700 \text{ N}$$

clearly shows that the above requirements are inconsistent. Additionally, this physics analysis indicates where changes can be made to come up with a feasible design. Assuming that the acceleration requirement is fixed, we must either reduce the mass or increase the force of the engine. Hence, we need to consult more sophisticated aspects of the theory behind automotive engineering to find ways to decrease the mass while maintaining stability and safety and/or increase the force of the engine while maintaining fuel economy.

Readers with physics and engineering backgrounds will be quick to point out that this example is oversimplified—that the size of the engine should be rated in units of power and torque, not force, and that the torque generated by the engine will vary with speed. But while these considerations would complicate the analysis, they would not change the fundamental point: that there *is* a theory that enables us to determine the feasibility of a given set of requirements.

Many design decisions, for products ranging from semiconductors to bridges, are made on the basis of well-developed theoretical sciences. Although the sciences differ from one another, they all have the following features in common:

1. They offer *quantitative* relationships describing system behavior (e.g., $F = ma$).
2. They are founded on theories for *simple systems,* around which theories for more complex, real-world systems are built (e.g., the classical mechanics relationships are all stated for systems without air resistance or friction).
3. They contain *intuitive* key relationships. For example, $F = ma$ clearly indicates that doubling the mass halves the acceleration under a constant force. For a given set of observations, a much more complex formula than $F = ma$ might actually fit the data better, but would not provide the same clear insights and hence would be less powerful.

Example: A Factory

Next, suppose we are given specifications for a factory instead of for a product. Specifically, the vice president of manufacturing has demanded that a printed-circuit board (PCB) plant produce

FIGURE 6.1

Relation between cycle time and demand

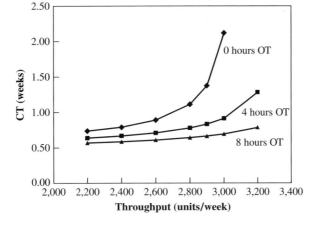

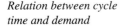

FIGURE 6.1

Relation between cycle time and demand

- 3,000 PCBs per week to meet demand.
- With an average cycle time (delay between job release and completion) of not more than one week, to maintain responsiveness,
- No overtime (workweek of 40 hours), to keep costs low.

Can it be done?

This time, the answer is not so clear. The equivalent of $F = ma$ of factory design is not widely known;[1] and the factory analogs to the more sophisticated elements of automotive engineering have not even been developed.

If it did exist, what might a theory of factory design show us? One possibility would be the relationships necessary to generate the graph in Figure 6.1 for the PCB plant. The x axis indicates the throughput with the y axis, showing the resulting average cycle time. The three curves show the relationship for the cases of no overtime, four hours of overtime, and eight hours of overtime per week.

From this graph, we see that the immediate answer to the vice president's question is no. If we insist on no more than one week of average cycle time and no overtime, the best we can do is 2,600 units per week. If we insist on an average cycle time of less than one week and 3,000 units per week, we will need an additional four hours per week of overtime. As long as the characteristics of the plant yield this set of curves, there is no way to comply with the vice president's demand. This does not mean it is impossible, only that it cannot be done with the current plant configuration. Presumably, therefore, the next thing we want from our theory is an indication of how to change the plant in a cost-effective manner so as to alter Figure 6.1 to meet the vice president's requirements.

Notice that the relationships in Figure 6.1 satisfy the previously cited properties of design sciences: they are *quantitative, simple* (we will see how they are derived for simple systems later on as we develop the results upon which these curves are based), and *intuitive*. Thus, even if they were not used to answer numerical questions, such as that posed by the vice president of manufacturing, relationships like these contain valuable management insights. They indicate that efforts to increase release rate (and

---

[1]A plausible analog to $F = ma$ for factory design does exist, as we will see in Chapter 7, but it is not sufficient by itself to answer the question posed here.

hence throughput) may result in a sharp increase in cycle time. They also show that adding capacity (in this case overtime) makes cycle time less sensitive to release rate. We will conjecture laws that govern this and other behavior in the remainder of Part II.

6.1.2 Defining a Manufacturing System

Before we can develop a science of manufacturing, we must define precisely what we mean by a manufacturing system. We use the following definition, a modification of one suggested by Deuermeyer (1994):

> A **manufacturing system** is an *objective*-oriented *network* of *processes* through which *entities flow*.

The key words of the definition are emphasized in italic. First, a manufacturing system has an **objective.** This generally relates to making money, but as we discuss below, specifying the fundamental objective requires some care. A manufacturing system contains **processes.** These may include the usual physical processes (cutting, grinding, welding, etc.), but can also include other steps that support the direct manufacturing processes (order entry, kitting, shipping, maintenance, etc.). **Entities** include not only the parts being manufactured, but also the information that is used to control the system. The **flow** of the entities through the system describes how materials and information are processed. Management of this flow is a major part of a manufacturing manager's job. Finally, it is important to recognize that a manufacturing system is a **network** of interacting parts. Managing the interactions is as important as managing individual processes and entities, if not more so.

This definition of a manufacturing system serves to highlight the roles of the different disciplines that deal with manufacturing. For instance, mechanical and electrical engineering deal principally with manufacturing processes and the design of the entities (products), while industrial engineering (and chemical engineering in continuous flow systems, such as oil refineries and chemical plants) focuses on the flows and the network. Management is concerned with ensuring compliance with the objective and measuring progress toward that goal.

6.1.3 Prescriptive and Descriptive Models

In the previous examples we used **descriptive models** to determine whether our system would meet the desired specifications. In manufacturing management teaching and research, most of the models used are **prescriptive models.** That is, they seek to *prescribe* or *optimize* design or control of a production system. Clearly prescriptive models are needed, but it is essential to understand the basic relationships governing a system *before* attempting to optimize it.

Prescriptive models are typically derived from a set of *mathematical* assumptions. As such, they differ from models used in the sciences such as physics and chemistry which are statements about nature. They are not derived from mathematics, but instead are essentially independent conjectures. For example, the overarching goal of physics is to explain the most phenomena with the fewest elementary conjectures. The resulting descriptive models provide the foundation for prescriptive models used by practitioners in applied fields such as mechanical and chemical engineering for guidance in designing and controlling complex systems (such as chemical plants).

As an example, consider the problem faced by a civil engineer in selecting a bridge design. Each available design strategy represents a prescriptive solution based on both experience and models. For instance, over a long span, a suspension bridge is often a good option. Suspension bridges are supported by cables made of steel, which can accommodate enormous *tensile* stresses but are almost worthless when faced with *compression* stresses. In contrast, a shorter span is often better served with a reinforced-concrete bridge, where the supporting members curve upward slightly, producing compression stresses in the load-bearing members. Concrete can support large compression stresses but does not work well under tension.

How do civil engineers know these things? Early in their education, before taking a course on building large structures, they take a set of engineering science courses. One of these, statics and dynamics, covers compression and tension forces. Here one learns how an arch transmits load from its top to its base. Another early course describes the strength of materials such as steel and concrete. In our parlance, these are descriptive courses. Only after these basic concepts are understood, does the prospective engineer begin to take design or prescriptive courses.

One could argue that the models traditionally taught in operations management courses represent the descriptive model foundation of manufacturing management. Like the models taught in engineering science courses, they are elementary and are used as building blocks for more complex systems. However, there is a fundamental difference. As Little (1992) pointed out, most of the mathematical models used in operations management and industrial engineering (IE) are *tautologies.* That is, given a particular set of assumptions, the system can be proved to behave in a particular manner. The emphasis is on proper derivation from the assumptions to the conclusions and not on whether the model is a realistic representation of an actual system. In essence, the *truth* of the model is self-contained. Little even demonstrated that a "law" named for himself (and one that we will explore in Chapter 7) is not a law at all but is a tautology. Since it can be shown to hold mathematically, there is no point in checking Little's law with empirical data.

Unlike mathematical tautologies, the models taught in engineering science courses *do* make conjectures about the outside world. They invite the student to check particular statements against empirical evidence (and students do exactly this in laboratory sections). The formula $F = ma$ is one such conjecture. This law is certainly not a mathematical tautology; indeed it isn't even strictly true (it is only correct for speeds that are slow compared to the speed of light). Nonetheless, it is enormously useful and is at the heart of many complex engineering models. Important results in physics, such as $F = ma$ and other Newtonian laws are also remarkable for their simplicity. However, as any sophomore engineering student can attest, the field of statics and dynamics is anything but simple, even though it is based solely on a small set of extremely simple statements about nature.

It is important to note that no scientific law can ever be proven. Derivation from first principles is not a proof since the first principles are themselves conjectured laws. Since we can never observe all possible situations (unlike mathematical induction), we can never know if our current explanation of observed phenomena is the right one or whether some other better explanation will come along. If history is any guide, it is a good bet that all the laws of physics will eventually be challenged and overthrown.

However, the practice of science is not as hopeless as it might seem. An unproved or even refuted law (such as $F = ma$) can be quite useful. The key is to understand where it does and *does not* apply. This is why it is important not to seek to verify our hypotheses but instead to try our best to refute them. The more we refute, the more we learn about the system and the better the surviving law will be (Polya 1954). We call this process **conjecture and refutation** (Popper 1963). In many ways, conjecture and refutation is to science what "ask why five times" is to JIT implementation. Both represent procedures for getting beyond the obvious and down to root causes.

While there is yet no universally accepted basic science of operations management, a number of researchers and teachers are beginning to address this gap (see Askin and Standridge 1993, Buzacott and Shantikumar 1993, and Schwarz 1996). This book represents our attempt to structure a science of manufacturing. Admittedly it is far from complete. The factory physics relationships we can offer at this time are a combination of insights from historical practices, recent developments by researchers and practitioners, equations from queueing theory, and a few results from our own research. However, factory physics is no buzzword. It is not easy nor does it pretend to offer a solution for all situations. Factory physics simply provides the basic relationships among fundamental manufacturing quantities such as inventory, cycle time, throughput, capacity, variability, customer service, and so on. It is our hope that understanding these relationships in the context of a science of manufacturing, even an incomplete one, will better equip the reader to design and control effective manufacturing enterprises.

6.2   Objectives, Measures, and Controls

Developing a science of manufacturing is not a trivial task. Just as hard is applying this science to solve manufacturing problems. A process that is helpful in both regards is the systems approach.

6.2.1   The Systems Approach

The notion of conjecture and refutation is not only a vehicle for scientific research. It is also the foundation for an extremely useful problem-solving methodology, known as the **systems approach,** or **systems analysis.** Systems analysis (SA) has been studied formally for at least 30 years (see, e.g., Ackoff 1956, Churchman 1968, and Miser and Quade 1985, 1988 for discussions), but has been part of management thought, in spirit if not name, since at least as far back as the work of Chester Barnard (1938).

Briefly, systems analysis is a structured problem-solving approach characterized by

1. *A systems view.* The problem is viewed in the context of a system of interacting subsystems (e.g., a factory is a system composed of product flows supported by various subsystems consisting of different departments, shifts, lines, etc.). The emphasis is on taking a broad, holistic view of the problem, rather than a narrow, reductionist perspective.

2. *Means-ends analysis.* The objective is always specified first, and then alternatives are sought and evaluated in terms of this objective. Note that this is in sharp contrast with the "means first" approach frequently used in the political arena, in which alternatives are posed first and objectives are only introduced as expedient to the consensus-forming process.[2] For instance, a systems analysis might use the objective "to deliver finished goods swiftly and conveniently to customers," but would *not* use the objective "to improve the efficiency of processing purchase orders." The latter is a means-first approach, which could rule out potentially attractive options (e.g., doing away with purchase orders under an entirely new procedure).

---

[2]Lindblom (1959) terms the means-first approach *disjointed incrementalism* and argues that it may be better suited to the political process than the systems approach.

3. *Creative alternative generation.* With the objective in mind, the systems approach seeks as broad a range of alternative policies as possible. Many formalized brainstorming techniques have been developed to aid in this process. Regardless of the method used, however, the intent is to find nonobvious ways to improve the system. For instance, to reduce manufacturing cycle time (the time it takes to make a product), we should go beyond simply considering how to speed up the individual steps and think about ways to eliminate entire portions of the production process.

4. *Modeling and optimization.* To compare alternatives in terms of the objective, some kind of quantification is required. The modeling/optimization step for doing this may be as simple as computing costs for each alternative and choosing the cheapest one, or it may require analysis of a sophisticated mathematical model. The appropriate level of detail will vary depending on the complexity of the system under study and the magnitude of the potential impact of the actions (e.g., it makes no sense to do $50,000 worth of analysis to save $52,000).

5. *Iteration.* In almost every systems analysis, the objective, alternatives, and model are revised repeatedly. This is not because we are dumb; it is because real-world systems are complex. Catching errors and oversights is a natural part of the conjecture and refutation process.

Figure 6.2 depicts a schematic of the systems analysis process in four basic phases: operations analysis, systems design, implementation, and evaluation. As indicated by the feedback arrows, these phases are not sequential. Iteration can and should take place both within and between phases. Furthermore, the focus of the study generally shifts

FIGURE 6.2

A systems analysis paradigm

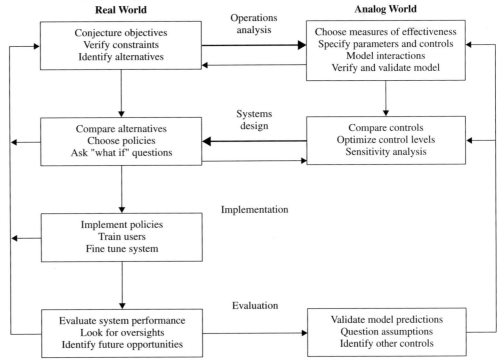

back and forth between the real world and the analog (or modeling) world as the analysis proceeds.

The process begins with the **operations analysis** phase, which focuses on the essentially *scientific* task of observing the actual system and developing an appropriate and useful model. To do this, we attempt to define objectives, constraints, and alternatives for the project. Although initially they might seem obvious, they usually prove to be more elusive than expected. Hence, we must conjecture them tentatively and then look for refutations. As the project proceeds, new objectives, constraints, and alternatives may arise, and the relative importance among them may change.

Another issue that frequently arises in reference to the iterative consideration of objectives and constraints is the choice of how to represent them. Often a particular preference may be sensibly stated as either an objective or a constraint. For example, minimizing the number of customer orders that are filled after their due dates could be an objective. Alternatively, requiring that less than two percent of orders be filled late could be a constraint that addresses the same concern. This technique of converting objectives to constraints is known as **satisficing** and is widely used in systems analysis (Majone 1985).

Iteratively considering objectives, constraints, and alternatives for the actual system is only the starting point for the operations analysis phase. In truly complex systems we cannot obtain a thorough understanding of the system and evaluate alternatives by looking at the actual system alone. There are two reasons for this. First, high-level objectives (e.g., maximize customer satisfaction) are generally not measurable. And second, real-world systems are typically too complex to allow direct description of the interaction between the various system components and the effects on the system of specific alternatives. To deepen our understanding of the system and its control alternatives, we develop an analog or model of the essential aspects of the system.

Modeling a real-world system begins with specification of low-level, quantifiable measures of effectiveness to serve as proxies for the true system objectives (e.g., fraction of jobs that are late to represent customer satisfaction). We then specify descriptive parameters, controllable variables, and their interactions to represent the system in some form of model. The art of developing a model that is sophisticated enough to capture the essential features of the system and yet simple enough to allow practical analysis is a complex task requiring a staff with skills in creative problem solving and mathematical methods. Models require both **verification** (i.e., checking the logic of the model) and **validation** (i.e., comparing the model results to reality). Model validation involves repeated iteration between the modeling and observation aspects of the analysis, and should take place throughout the study.

The **systems design** phase is the beginning of the predominantly *engineering* portion of the systems analysis paradigm. While in the operations analysis phase we work primarily from the real world to the analog world through modeling; in the systems design phase we work primarily from the analog world to the real world by translating results from the model to implementable policies. We do this by "optimizing" the model with respect to the chosen measures of effectiveness and then examining the robustness of the solution via sensitivity analysis. We then translate these mathematical or symbolic solutions to actual policies and examine the practicality of these policies in the actual environment. It is important to remember that no matter how good a mathematical model is, it is still a simplification of reality. Like developing appropriate models, interpreting the results to develop sensible courses of actions is an art that can never be fully mechanized.

A good systems analysis does not end with the presentation of the proposed policies. The **implementation** phase of the paradigm offers us the opportunity to see that they are adopted properly and to identify unanticipated problems while there is still time to deal with them effectively.

Finally, in the **evaluation** phase, we review the system after the policies have been implemented and assess the results in terms of the original objectives. This is an extremely important phase because it offers the best opportunity to validate the usefulness of the model in improving the actual system, as opposed to simply describing the behavior of the system. Since systems analyses are *applied* problem solving exercises, the degree to which the desired objectives were met must always be the bottom line of the study. However, since most real-world systems are complex and constantly changing, the end of a particular study should not mark the end of analysis. Opportunities for future improvements in both the model and the actual system should be identified as input to further cycling of the systems analysis procedure.

6.2.2 The Fundamental Objective

Since, for our purposes, we have defined a manufacturing system as an *objective*-oriented network, the obvious starting point for a science of manufacturing is the **fundamental objective.** This is a broad goal that all parties can agree on. It is generally vague since it describes a long-term aspiration that may or may not be completely quantifiable. In some companies the fundamental objective is formalized into a *mission statement.* However, this deceptively complex exercise frequently becomes an exercise in rhetoric in which scores of person-hours are wasted in "workshops" that do little more than generate a new (and often cumbersome) slogan. It is important to recognize, therefore, that systems analysis only begins with identifying the fundamental objective. By itself, a fundamental objective (or mission statement) is of little tangible value.

"Use money to make more money" is an obvious choice as a fundamental objective. However, it presents problems when we consider that there are many ways to make money, including selling off the firm's assets (possibly good in the short term, but terrible for the future) and dealing in illicit drugs (profitable, but illegal and immoral). Other popular slogans such as "Give the customers what they want" are similarly incomplete—customers would be very satisfied if we provided them with better products for free! To gain widespread support, a fundamental objective must balance the concerns of all parties involved in the organization. The following statement is vague enough to serve as the fundamental objective for almost any manufacturing firm:

> Increase the well-being of the stakeholders (stockholders, employees, and customers) over the long term.

We realize that this is a "Mom and apple pie" statement, which is too vague to yield much concrete guidance. But it does provide a point of common ground for all the stakeholders and stresses that many parties at interest may be affected by changes to a manufacturing system.

6.2.3 Hierarchical Objectives

As soon as we specify a fundamental objective, conflicts arise, since what is good for one stakeholder is not always good for another. Cost reduction through lower wages is good for profitability and hence stockholders, but is not good for employees. To strike a balance, we need to narrow our fundamental objective slightly, perhaps to something like

> Make a "good" return on investment (ROI) over the long term.

This statement will satisfy stockholders because ROI supports stock price. It will also satisfy employees in one regard since they will continue to be employed and in a position to receive better wages. Finally, customers will be satisfied, because if they are not, good returns will be impossible over the long run. Thus, this statement, while still very high level, relates to the concerns of the primary parties at interest and is directly measurable.

But we cannot simply inform the workers of the firm's high-level objectives. No amount of encouraging slogans plastered about the plant exhorting workers to achieve a good return on investment will stimulate manufacturing excellence. People have to know how *their* jobs affect the fundamental objective in order to be able to influence it in a positive fashion. For this, we need to identify measures more directly related to production.

First, note that profit and return on investment (ROI) are computed from three financial quantities—(1) **revenue,** (2) **assets,** and (3) **costs**—as follows:

$$\text{Profit} = \text{Revenue} - \text{Costs}$$

$$\text{ROI} = \frac{\text{Profit}}{\text{Assets}}$$

But even these measures are too high-level for day-to-day plant operation.

The plant-level equivalents of revenue, assets, and costs are (1) **throughput,** the amount of product *sold* per unit time (it does no good to make it and not sell it); (2) **assets,** particularly *controllable* assets such as inventory; and (3) **costs,** consisting of operating expenditures of the plant, particularly cost variances such as overtime, subcontracting, and scrap. These three basic measures provide the link between the high-level financial measures (say, ROI), and the lower-level measures (e.g., machine availability) that are more directly related to manufacturing activities.

Figure 6.3 illustrates a sample hierarchy of objectives from the fundamental objective to various supporting **subordinate objectives.** From the formula for profit, we see

FIGURE 6.3

Hierarchical objectives in a manufacturing organization

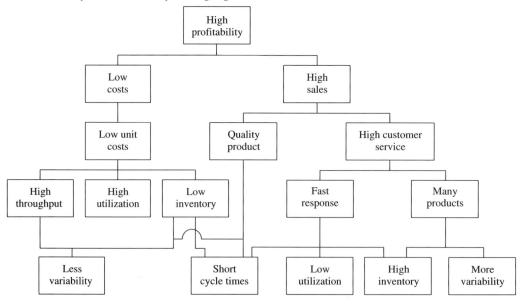

that high profitability requires low costs and high throughput (sales). Low costs imply low unit costs, which require high throughput, high utilization, and low inventory. As we will see later in Part II, less variability in production is required to achieve low inventory and high throughput. On the other side of the hierarchy, to increase sales requires a high-quality product that people want to buy, plus good customer service. High customer service requires fast response and many products (anything the customer wants). Fast response requires short cycle times, low equipment utilization, and/or high inventory levels. To keep many products available requires high inventory levels and more variability (in product). However, to obtain high quality, we need less variability (in processing) and short cycle times (to catch defects when they occur).

Note that this hierarchy contains some conflicts. For instance, we want high inventory for fast response, but low inventory for keeping total assets low so that the return on assets will be high. We want high utilization to keep unit costs down, but low utilization for good responsiveness. We want more variability for greater product variety, but less variability to keep inventory low and throughput high. Despite the reluctance of some JIT advocates to use the "t word," we have no choice but to make **tradeoffs** to resolve these conflicts.

Finally, it is useful to observe from Figure 6.3 that short cycle times support both lower costs and higher sales. This is the motivation behind the emphasis during the 1990s on speed, embodied in slogans such as **quick response manufacturing** and **time-based competition.** We will take up the important topic of cycle time reduction in Part III, after establishing basic relationships involving variability later on in Part II.

6.2.4   Control and Information Systems

Manufacturing managers face a wide array of controls with which to try to achieve their objectives. Product design, facility design, equipment maintenance, work scheduling, personnel policies, and many other areas present opportunities for controlling a manufacturing system. Despite our focus on flows in factory physics, it is important not to concentrate too narrowly on controls directly related to movement of material (e.g., scheduling). Other controls may seem less closely related to generating throughput, but may be just as important in attaining the fundamental objective of the system.

To provide a structure for thinking about the range of alternatives, Schwartz (1998) compared the practice of an operations manager to that of a financial portfolio manager. A portfolio manager mixes securities in order to get a good and stable return on investment. An operations manager has three basic assets to manage to generate return on investment: information, control, and buffers. Information involves what is known about the system (e.g., inventory status data from the ERP system). Control involves operating policies that affect system behavior (e.g., inventory stocking rules). And buffers involve protection against variability (e.g., safety stocks). The three components must be managed together to obtain effective overall performance. If any one component is lacking (e.g., imperfect information regarding demand), it must be compensated by some combination of the other two (e.g., more control by assigning due dates or more buffers by carrying safety stock).

As an example, consider a make-to-stock operation controlled by an MRP system. The information system collects data on current inventory, scheduled receipts, and capacity and forecasts demand. The control system uses MRP to translate this information to actual jobs released to the floor and then tracks them as they are completed. The control system might also involve expediting as demands change. Buffers include safety stock, safety lead time, and excess capacity in the system. These are needed because the forecast is never exactly right and because MRP is an imperfect model of the production process.

Any of the three parts of the portfolio offers opportunities for improvement, as does adjusting their mix to improve overall system performance. For instance, better (earlier) information about demand would reduce the need for inventory buffers by allowing more of the demand to be satisfied in make-to-order fashion. A completely flexible workforce (more control) would reduce the need for excess capacity (less buffer). Better prediction of the flows through the system (better information than is offered by the MRP model) would reduce the need for excessive safety lead times and WIP levels (buffers). These are the types of policies espoused in lean manufacturing, which is fundamentally about reducing the need for buffers through better use of information and control. However, we will find in Chapter 9 that no matter how perfect the information and how powerful the controls, there will always be a need for buffers.

6.3   Models and Performance Measures

A hierarchy of objectives like that in Figure 6.3 presents two practical questions. First, how do we resolve the conflicts it identifies? Second, how do we translate these high-level objectives to detailed operational policies?

The answer to the first question is the use of *models* to quantify tradeoffs. The challenge is to develop models that are accurate enough to represent these tradeoffs appropriately, but simple enough to give us good intuition. Much of the remainder of Part II is devoted to several such models that will underlie our discussion of operating procedures in Part III.

6.3.1   The Danger of Simple Models

As we discussed in Chapter 5, the use of fixed lead times in MRP leads to systems that are unresponsive to customers and bloated with inventory. The main reason is the underlying model of cycle time. MRP assumes that cycle time (CT) will be the same regardless of the work-in-process (WIP) level of the line. That is, no matter how much we load into the system, jobs will take the same amount of time to be completed. Mathematically, the MRP model is simply $CT = T_{MRP}$.

A more sophisticated model of cycle time can be constructed by separating the approximation into two cases, one for when the line is relatively empty and one for when it is saturated. When the line is relatively empty, we use an MRP-like model of cycle time $CT = T_{approx}$, where T_{approx} represents the time for a job to go through an uncongested line. When it is saturated, we assume that the line can produce at most C_{approx}, which is the capacity of the line. Hence, the time to process a quantity of WIP will be $CT = WIP/C_{approx}$. Since the cycle time cannot be less than T_{approx}, the complete model of cycle time is

$$CT = \max \left\{ T_{approx}, \frac{WIP}{C_{approx}} \right\}$$

We call this the **conveyor model** of a production line because it behaves as a conveyor. The time to go down a conveyor is constant until the conveyor is full. Once it is full, the time to get down the conveyor is computed by dividing the amount of work on and in front of the conveyor by the rate of the conveyor.

In practice, it makes sense to set T_{approx} slightly higher than the actual time to go through an empty line (to account for some congestion) and $CT = T_{approx}$ slightly below the maximum capacity of the line (to account for some inefficiency in the line).

The conveyor model is only slightly more complex than the MRP model, since it requires estimation of two parameters instead of one. However, it is much more accurate. Figure 6.4 shows a sample of the actual relationship between WIP and cycle time, along with the MRP and conveyor models. While the MRP model fits very poorly, particularly for high WIP levels, the conveyor model tracks the basic relationship much more closely.

6.3.2   Building Better Prescriptive Models

Better descriptive models provide the basis for better prescriptive models. For instance, we can use the conveyor model to solve capacitated scheduling problems, as we illustrate in the following example.

Consider a line with a capacity of 100 units per day that requires three days to process jobs (when there is no congestion). Currently there are 50 units in finished goods inventory (FGI), 95 units that have been in the line for three days (i.e., that will start coming out immediately), 95 units that have been in for two days, and 100 units that have been in one day. Since less than 100 units were started two and three days ago, output for those days is limited by available WIP. Thus, the maximum output from this point forward is 50 (immediately from FGI), 95 today, 95 tomorrow, and 100 from that point on. Demands for the next 10 days are as follows: 100, 120, 100, 0, 200, 0, 200, 120, 0, 80. The 340 units in finished goods and currently in WIP cover demand for periods 1, 2, and 3 as well as 20 units of the 200 due on day four. Thus, the netted demand is 0, 0, 0, 0, 180, 0, 200, 120, 0, 80. If we offset these by three days to find out how much should be started, we obtain starting demands of 0, 180, 0, 200, 120, 0, and 80. Our task at hand is to find a start schedule that minimizes inventory and is capacity-feasible.

We first solve the problem by using the MRP model with a lead time of four days. Since cycle time is assumed constant, releases are simply netted demand offset by four days, or 80, 0, 200, 120, 0, 80, for the next six days. But given the capacity restriction, we will finish 100 on the first day, the remaining 80 on the second day, and 100 on all subsequent days. This results in a shortage of 20 units on the eighth day. If instead we use a lead time of five days in an attempt to eliminate the shortage, the problem becomes infeasible (from an MRP standpoint because we would need to start product before the first day).

As a more sophisticated procedure based on the MRP model of cycle time, we could use the Wagner–Whitin algorithm from Chapter 2 and use the setup cost as a surrogate for a capacity constraint. The schedule becomes feasible only when we make the setup

FIGURE 6.4

Relation between cycle time and WIP

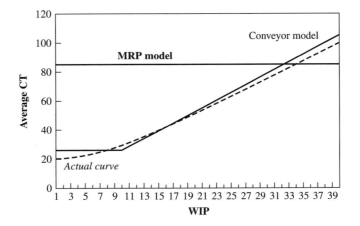

cost so high that all production is started in the first period. Specifically, the ratio of setup to holding cost must be at least 1,200. This results in an inventory carrying cost of $340h$, where h is the cost to carry one unit of inventory for one period. If we reduce the setup to holding cost ratio to 1,199, the schedule becomes infeasible with a shortage of 120 units in period 9. Further reductions of the ratio only make the infeasibility worse. Hence, Wagner–Whitin generates either high-inventory or infeasible solutions.

Alternatively, we can formulate a simple solution procedure based on the conveyor model (we develop this further in Chapter 15). To do this, let

$$D_t = \text{demand on day } t$$
$$X_t = \text{production on day } t, \text{ decision variable}$$
$$I_t = \text{ending inventory on day } t$$
$$C_t = \text{capacity available on day } t$$

We compute production quantities backward from day 10. First, we set the desired ending inventory for day 10 to zero: $I_{10} = 0$. Then for each day t the production quantity is given by

$$X_t = \min \{C_t, \ D_t + I_t\} \tag{6.1}$$

Notice that unlike in the MRP model, the conveyor model assumes that production on a day is limited by capacity. Therefore, cycle time is a function of inventory (and backlogged demand). To proceed to day $t - 1$, we compute

$$I_{t-1} = I_t + D_t - X_t \tag{6.2}$$

and continue with Equation (6.1). If the ending inventory for period 0 is greater than zero, then demands *cannot* be met using the current capacity no matter what the schedule.

In our example, we can apply this procedure to the netted demands to obtain a start schedule of 100, 100, 100, 100, 100, 0, 80. The inventory holding cost of this schedule is $260h$. Unlike the MRP schedule, this schedule meets all the demands. It also carries 24 percent less inventory than the "optimal" Wagner–Whitin algorithm. This procedure can be extended to multistage production systems, as we show in Chapter 15. The conveyor model can also be applied to throughput tracking and due date quoting, as we discuss in Chapters 13 and 15.

6.3.3   Accounting Models

The mathematical models one normally studies in a course on operations management (EOQ, MRP, forecasting models, linear programming models, etc.) are by no means the only models for measuring performance and evaluating management policies in manufacturing systems. Indeed, some of the most common models used by manufacturing managers are those related to accounting methods. Although accounting is sometimes viewed as mere bookkeeping or cost tracking, it is actually based on models and is therefore subject to the same pitfalls concerning assumptions that face any modeling exercise.

One of the key functions of cost accounting is to estimate how much individual products cost to make. Such estimates are widely used to make both long-term decisions (Should we continue to make this product in house?) and short-term decisions (What price should we quote to this customer?). But because many costs in manufacturing systems are not directly attributable to individual products, they can only be estimated by means of a model.

Direct costs, such as raw materials, are simple to assign. If castings are purchased and machined into switch housings, then the price of the castings must be included in the unit cost of the switches. Direct labor can be slightly more difficult to assign if workers produce more than one type of product. For instance, if a machinist makes two types of switch housings, then we must decide what fraction of her time she spends on each, in order to allocate the cost of her time accordingly. But this is still a relatively simple computation.

The difficulty, and hence the need for a model, arises in the allocation of **overhead** costs. Overhead (also called **fixed costs** or **burden**) refers to costs that are not directly associated with products. Mortgage payments on the factory, the salary of the chief executive officer, the cost of a research and development laboratory, and the cost of the company mail room are examples of costs that do not vary directly with the production of individual products. But since they are part of the cost of doing business, they are indirectly part of the cost of producing products. The challenge is to apportion the overhead cost among the different products in a reasonable manner.

The traditional approach (model) for allocating overhead costs was to use labor hours. That is, if a particular product used two percent of the hours spent by workers producing products, then it would be assigned two percent of the overhead cost. The rationale for this was that at the turn of the century, when "modern" accounting techniques were developed, direct labor and material typically represented up to 90 percent of the total cost of a product (see Johnson and Kaplan 1987 for an excellent history of accounting methods). Today, direct labor constitutes less than 15 percent of the cost of most products, and hence the traditional methods have been increasingly challenged as inappropriate. The title of the book by Johnson and Kaplan is *Relevance Lost*.

The leading contender to replace traditional cost accounting techniques is known as **activity-based costing (ABC).** ABC differs from traditional methods in that it seeks to link overhead costs to *activities* instead of directly to products. For instance, purchasing might be an activity that is responsible for overhead costs. By measuring the amount of purchasing activity in units of purchase orders and then allocating the purchasing overhead costs to each product on the basis of the fraction of purchase orders it generates, the ABC approach tries to accurately apportion this part of the overhead cost. Similar allocations are done for any other portions of the overhead cost that can be assigned to specific activities. Appendix 6A gives an example illustrating the mechanics of ABC and contrasting it with the traditional labor-hour approach.

Because ABC divides overhead costs into categories, it can promote better understanding, and eventually reduction, of these costs. As such, it is a positive step in the area of cost modeling. However, it is by no means a panacea. Cost-based models, however detailed, can sometimes be misleading.

First, there are cases when the *allocation* of costs is simply a poor modeling focus from a systems point of view. One of the authors worked in a chemical plant in which considerable debate and analysis were devoted to determining the price that should be exchanged for a commodity that was a by-product of one product and a raw material for another. The users of the commodity argued that the price should be zero since it would be wasted if they were not using it. The producers of the commodity argued that the users should pay what it would cost if they had to produce the product themselves. In actuality, neither of the processes would have been profitable as a stand-alone operation, but they were quite profitable when taken together. A better focus for the analysis and debate would have been on how and where to improve yields (how much product was produced) of the two processes.

Second, no matter how detailed the model, it is extremely difficult to accurately represent the value of limited resources by using a cost-based approach common to all accounting methodologies. This applies to both the **full costing** or **absorption costing** method described above and **variable costing** where overhead is not considered.

Full absorption costing is appropriate if we are building a new plant and so are concerned with all the costs of the plant. Variable costing is suited to operating an existing plant, where we should only concern ourselves with costs that can be controlled within a short time frame. For instance, in a new plant, machine and labor costs should all be considered. If one plan requires more setups and those setups take labor to perform, then that plan will truly cost more than a plan requiring fewer setups. On the other hand, in an existing plant we should completely ignore the cost of machines since they have already been purchased. It is a **sunk cost.** Managers are sometimes tempted to run more product on a more expensive machine in order to "recover its cost." But from an overall perspective this may not make sense, especially if the more expensive machine is less suited to running some products than a cheaper one is.

Most product costing (ABC included) is based on fully absorbed and not variable costs. This can lead to bad decisions. For instance, if a customer is asking for a part that requires a long time at the process center that currently has the most work, the cost is great. But if there is demand for an item that flows only through processes that currently have little work to do, the cost is essentially raw materials cost. In essence, the machines and labor are both free since they are there with little else to do. The following example illustrates the danger of using fully absorbed costs to make production decisions.

Example: Production Planning

Consider a plant consisting of three machines that make two products, A and B, as illustrated in Figure 6.5. Product A costs $50 in raw material and requires two hours on machine 1 and two hours on machine 3. Product B costs $100 in raw material and requires two and one-half hours on machine 2 and one and one-half hours on machine 3. Thus, both products require four hours of machine and four hours of labor time. Labor cost is $20 per hour (including benefits, etc.). The plant runs an average of 21 days per month with two shifts or 16 hours per day (workers relieve one another for breaks, etc.), for a total of 336 hours per month. Nonmaterial expenditures to run the plant (i.e., labor, supervision, administration, etc.) are $100,000 per month. Both products sell for $600 per unit and make use of exactly the same amount of overhead activities. Marketing estimates a demand of no more than 140 units per month for both products. Also, to maintain market position, the company needs to produce at least 75 units of product A per month. Table 6.1 summarizes the data for this example.

Suppose we cost the products by using an absorption method and then use these costs to help plan how much of each product to make. Since both products require the same number of labor hours and activities, they will receive the same overhead charge regardless of how we allocate overhead. Since these would not affect the *relative* costs of the two products, we can ignore them when choosing between products to produce. The profit per unit of A sold (neglecting overhead and labor costs) is $600 − $50 = $550, while the profit per unit of B sold is $600 − $100 = $500. Since A is more profitable, it would seem that our production plan should favor production of A.

There are 21 × 16 = 336 hours available in a month. Since each unit of B requires two hours of time on machines 1 and 3 to produce, maximum monthly production of

FIGURE 6.5

Layout of two-product plant

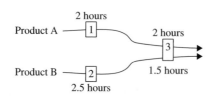

TABLE 6.1   Data for Two-Product Plant Example

Product Name	Price ($)	Raw Material Cost ($)	Total Labor Hours	Unit Cost ($)	Minimum; Maximum Demand per Month
A	600	50	4	130	75; 140
B	600	100	4	180	0; 140

either is $336/2 = 168$ units. Since potential demand is only 140, it seems reasonable to set production to maximum demand level for A (140 units per month) which, of course, meets our minimum demand requirement of 75. This uses up 280 hours per month on machine 3, leaving $336 - 280 = 56$ hours on machine 3 for the production of B. Hence, we can produce $56/1.5 = 37$ units of B per month (actually 37.33, but we round to the largest integer quantity).[3]

The monthly profit from this plan can be computed by multiplying the production quantities of A and B by their unit profits and subtracting the nonmaterial costs:

$$\text{Profit} = 140(\$550) + 37(\$500) - \$100,000 = -\$4,500$$

This plan loses money!

Instead of relying on an accounting model, we could have used an optimization model based on **linear programming.** The idea behind linear programming is to formulate a model to maximize profit subject to the demand and capacity constraints. For this example, we can state our problem as follows:

> Maximize   Profit
>
> Subject to: Time used on M1 $\leq$ 336 hours
>
> Time used on M2 $\leq$ 336 hours
>
> Time used on M3 $\leq$ 336 hours
>
> $75 \leq$ amount of A ≤ 140
>
> $0 \leq$ amount of B ≤ 140

Defining X_A and X_B to represent the monthly production quantities of products A and B, we can formalize our model as

> Maximize $550X_A + 500X_B - 100,000$
>
> Subject to: $2X_A \leq 336$
>
> $2.5X_B \leq 336$
>
> $2X_A + 1.5X_B \leq 336$
>
> $75 \leq X_A \leq 140$
>
> $0 \leq X_B \leq 140$

This model is an example of a **linear program.**[4] We will go into detail on how to formulate and solve linear programs in Chapter 16. For now, we simply note that there

---

[3]Note that we did not have to worry about machine 2, since it is only used by product B. The entire 336 hours per month are available for production of B, which is enough to produce $336/2.5 = 134$ units. Hence, it is capacity on machine 3 that determines how much B we can produce.

[4]It is *linear* because the objective function and constraints involve the decision variables X_A and X_B in linear expressions (i.e., multiplied by constants and summed). The term *program* comes from the historical fact that the technique was devised to find optimal programs (i.e., schedules) of resource use; it has nothing to do with the fact that these kinds of problems are generally solved by means of a computer program.

are very efficient techniques for solving this type of optimization model, and we report that for this example the solution results in a plan calling for 75 units of A and 124 units of B per month. Notice that this plan is completely counterintuitive when we consider the "cost" of the products; we are making more of the lower-profit product! However, the profit from this plan is

$$\text{Profit} = 75(\$550) + 124(\$500) - \$100{,}000 = \$3{,}250$$

which is profitable!

The moral of this example is that *the value of limited resources depends on how they are used.* A static cost-based model, no matter how detailed, cannot accurately assign costs to limited resources, such as machines subject to capacity constraints, and therefore may produce misleading results. Only a more sophisticated optimization model, which dynamically determines the costs of such resources as it computes the optimal plan, can be guaranteed to avoid this.

In addition to offering an alternative to the cost accounting perspective, constrained optimization models are useful in a wide variety of operations management problems. In Part III, we will specifically address problems related to scheduling, long-range production planning, and workforce planning with such models. Methods for analyzing constrained optimization models, such as linear programming, are therefore key tools for the manufacturing manager.

6.3.4   Tactical and Strategic Modeling

As useful as models are, it is important to remember that they are only tools, not reality. The appropriate formulation of a model depends on the decision it is intended to assist. Parameters that are reasonably considered constrained for the purposes of tactical decision making are often subject to control at the strategic level. Thus, while one model may be effective in planning production quantities over the intermediate term, another (possibly still a constrained optimization model) is needed for planning over the long term. Chapter 13 explains the hierarchical relationship between production planning and control models in greater detail. Here we will highlight the distinctions between tactical and strategic planning by means of the previous example.

Because the above example focused on the tactical problem of planning production over the next few months, it made perfect sense to treat capacity and demand as constrained. Over the longer strategic term, however, both capacity and demand are subject to influence. Capacity could be increased by adding a third shift or decreased by reducing the second shift. Price discounts could boost demand, while an announcement of a competing (e.g., next-generation) product could reduce demand.

Models can clarify the relationships between tactical and strategic decisions and help ensure consistency between them. For instance, by using the sensitivity analysis capabilities of linear programming (Chapter 16), we can determine that the constraint to produce at least 75 units per month of product A is detrimental to profit. In fact, if we eliminate this constraint and re-solve the model, it generates a plan to produce 68 units of A and 133 units of B, which yields a monthly profit of $3,900, an increase of $650 per month.

This suggests that we should consider the strategic reasons for the constraint to produce at least 75 units per month of A. If the reason is a firm commitment to a specific customer, it may be necessary. But if it is only an approximation of the number needed to meet our commitments, then using a lower limit of 68 might be just as reasonable, and more profitable.

Another piece of information provided by the sensitivity analysis function of linear programming is that for every additional hour of time available at machine 3 (up to seven extra hours per day), profits increase by $275. Since overtime does not cost nearly $275 per hour, we should probably consider adding some to the short-term plan. But in the longer term, the tactical decision of whether to use overtime relates to the strategic decisions of whether to increase the size of the workforce, add equipment, subcontract production, and so on. Thus, the model also suggests that these be considered as potential future options.

Effective planning calls for the use of different models for different problems and coordination between models. A tactical model, such as the constrained optimization model used earlier to generate a production plan for the next few months, can provide intuition (i.e., what variables are important), sensitivity information (i.e., where there is leverage), and data (e.g., identification of the current bottleneck resource) for use in strategic planning. Conversely, a strategic model, such as a long-term capacity planning model, can provide data (e.g., capacity constraints) and suggest alternatives (e.g., dynamic subcontracting) for use at the tactical level. We will discuss coordination in Chapter 13 and specific models for various levels throughout Part III.

6.3.5   Considering Risk

There are many sources of uncertainty in manufacturing management situations, including demand fluctuations, disruptions in materials procurement, variable yield loss, machine breakdowns, labor unrest, actions by competitors, and so on. In some cases, uncertainty should be explicitly represented in models. In other cases, as we will see in Part III, uncertainty can be safely ignored in the modeling process. But in all cases related to both modeling and management, the existence of uncertainty makes it essential to consider in some fashion what will happen if an assumption fails to hold.

As a high-level strategic example, consider the experience of a major American automobile manufacturer. In the late 1970s and early 1980s, many people in the corporation recognized a need to invest in improved product quality and proposed product and process changes to achieve this. However, funding for many of these projects was denied as not financially justified. The implicit assumption on the part of the corporate staff was that the competitive position of the company's products relative to the competition would remain unchanged. Hence, the cost of such products could not be justified by the promise of greater sales revenues. But when the competition upgraded the quality of its products at a faster pace than anticipated, the corporation experienced a disastrous loss of market share, and only in the 1990s, after a decade of huge losses and widespread plant closings, did the company return to profitability (but nowhere near its former market share).

The flaw in the firm's analysis was fundamental. The quality improvement projects were evaluated on the basis of their potential to improve profits instead of their need to avoid lost profits. Thus, management failed to consider adequately what would happen if the competition outpaced the company by offering better products. Product and process improvement should not have been viewed as an option for increased profitability but rather as a constraint to stay in business.

The procedure of evaluating the potential negative consequences in an uncertain situation is known as **risk analysis** and has been widely used in riskier industries such as petroleum exploration. Using a model, the analyst conjectures several possible scenarios and assigns a probability of each occurring.[5] Since the scenarios often involve strategic moves on the part of the competition, such analyses are generally undertaken by a senior manager working with a technical expert and a model. One approach for evaluating potential decisions is to weight the various outcomes with the probabilities and to compute an expected value of some performance measure (e.g., profit). An alternative, and sometimes more realistic, approach is to examine the various scenarios and choose a course of action that prevents really bad things from happening. This is the **minimax** (i.e., minimize the maximum damage) strategy that is often used by the military.

Had the previously mentioned automotive company employed a minimax strategy, most likely it would have approved many more product and process improvement projects than it did, as a hedge against improvements by the competition. Of course, since hindsight is 20/20, it is easy for us to say this in retrospect. The best policy is generally not obvious in advance. Indeed, the primary job of upper-level management is to chart reasonable long-term strategies in the face of considerable uncertainty about the future. These executives are highly paid in large part because their task is so difficult. (The question of whether they are smart or just lucky is moot so long as the company is successful.)

At the plant level, operations managers must perform an analogous function to that of upper management, only with a shorter time horizon and on a smaller scale. For example, consider the commonly faced operations problem of selecting machines for a new line.

Example: Risk Analysis

Suppose all the machines for a planned line, except one particular machine, the 3C 273, are capable of switching to any conceivable new product that the firm may choose to produce in the near future. A different machine, the 4C 273, could be substituted for the 3C 273 at a cost of an additional $100,000. The 4C 273 has all the same process characteristics (speed, availability, quality, etc.) as the 3C 273, but is also flexible enough to process any of the new products that might be introduced in the future. The problem is to choose between the 3C 273 and the 4C 273.

First, we articulate the possible scenarios. Either the firm will decide to produce a new product in the near future, or it will not. If it does not, then either the 3C 273 or the 4C 273 will suffice. If it does, then the 4C 273 will be required. If we install the 4C 273 now, it will cost an additional $100,000. But if we install the 3C 273 and the firm chooses to produce a new product, then we will need to replace it with the 4C 273. Suppose that this will cost $375,000 for the new machine plus $200,000 in lost revenue during the installation period and that the old machine can be sold for $50,000. Hence, the net cost incurred if we install the 3C 273 and then decide to produce a new product will be[6]

$$375 + 200 - 50 = \$525,000$$

Table 6.2 summarizes the costs of the four possible decision-scenario pairs.

---

[5]One can also perform scenario analysis without the use of probabilities for contingency planning. See, e.g., Wack (1985).

[6]Note that we are not considering the fact that this cost will actually be incurred in the future. To more accurately compare it to the $100,000 cost of installing the 4C 273 now, we should really multiply it by an appropriate discount factor to represent the time value of money. But for the sake of simplicity we will omit this.

TABLE 6.2   Costs of Decision-Scenario Pairs for Machine Installation Example

	Decision	
Scenario	**3C 273**	**4C 273**
Don't introduce new product	0	100
Introduce new product	525	100

Next we apply a decision criterion to the data. If we use the mini-max approach, we select the decision that minimizes the maximum cost. In this case, the maximum cost for the 3C 273 decision is \$525,000, while the maximum cost for the 4C 273 decision is \$100,000, so the mini-max criterion recommends installing the 4C 273.

However, the mini-max criterion may be overly conservative. If it is very unlikely that the firm will decide to produce a new product, then it may make more sense to install the 3C 273 and take our chances. To incorporate the likelihood of the different scenarios into our analysis, we might choose to use an expected value approach. Letting p represent the probability that the firm will introduce a new product, we see the expected cost of installing the 3C 273 is

$$0 \times (1 - p) + 525 \times p = \$525p$$

The expected cost of installing the 4C 273 is \$100,000 (since we incur this cost regardless of the scenario that occurs). Hence, the expected cost of the two scenarios is equal when

$$525p = 100$$
$$p = \frac{100}{525} = 0.19$$

Thus, if p is more than 0.19, the expected cost of installing the 4C 273 is smaller than that of installing the 3C 273. If p is less than 0.19, then the expected cost of the 3C 273 is smaller. To use the expected-value criterion to make a decision, therefore, we need only decide which region p lies in.

There are two important things to note about the above analysis:

1. Instead of guessing a value for p and using it to compute the expected costs of the two options, we worked backward to find the cutoff point for p that makes one option preferred to the other. The reason for this is that it is sometimes difficult to choose a value for something as intangible as the likelihood of a new product being introduced. Decision makers are generally more comfortable making the rough decision of whether a parameter lies in one range or another than trying to pin it down precisely. Since this decision does not necessarily require a highly accurate estimate of p to resolve, we set up the analysis so as not to ask for it.

2. We treated the need to meet demand for a new product as a constraint. In actuality, of course, this is a decision that will be addressed in the future. However, in order to consider the uncertainty concerning this decision when making the current equipment selection, we simply treat it as a scenario that may or may not unfold.

Modeling decision problems under uncertainty is a broad subject treated in the field of **decision analysis.** The books by Raiffa (1968), Brown (1974), and French (1986) provide good introductions to this vast discipline.

6.4   Conclusions

This chapter lays the foundation for our factory physics approach to developing the basics, intuition, and synthesis skills needed by the modern manufacturing manager. The main observations about the scientific, systems analysis, and modeling paradigm represented by this approach are as follows:

1. *Manufacturing management needs a science.* Although considerable folk wisdom exists about manufacturing, there is still only a small body of empirically verified, generalizable knowledge for supporting the design, control, and management of manufacturing facilities. If we are to move beyond fads and slogans, researchers and practitioners need to join forces to evolve a true science of manufacturing.

2. *The systems approach is a valuable manufacturing management tool.* By encouraging a holistic view of manufacturing enterprises and promoting a clear link between policies and objectives, systems analysis is the logical foundation for almost all manufacturing problem solving.

3. *Good descriptive models lead to good prescriptive models.* Trying to optimize a system we do not understand is futile. We need descriptive models to sharpen our intuition and focus our attention on the parameters with maximum leverage. Furthermore, policies based on accurate descriptions of system behavior are more likely to work with, rather than against, the system's natural tendencies. Such policies are apt to be more robust than those that try to force the system to behave unnaturally.

4. *Models are a necessary, but not complete, part of a manufacturing manager's skill set.* Because systems analysis demands that alternatives be evaluated with respect to objectives, some form of model is needed to make tradeoffs for virtually all manufacturing decision problems. Models can range from simple quantification procedures to sophisticated optimization and analysis methodologies. The *art* of modeling is in the selection of the proper model for a given situation and the coordination of the many models used to assist the decision-making process.

5. *Cost accounting typically provides poor models of manufacturing operations.* The purpose of accounting is to tell where the money went, not where to spend new money. Operations decisions require good characterization of *marginal,* not fully absorbed, costs and appropriate consideration of resource constraints.

From this base, we will now turn to developing specific models that describe the behavior of manufacturing systems.

APPENDIX 6A
ACTIVITY-BASED COSTING

There are four basic steps to ABC cost allocation (Baker 1994):

1. Determine the relevant activities.
2. Allocate overhead costs to these activities.

TABLE 6.3 Calculations for ABC Example

Category	Requisition	Engineering	Shipping	Sales	Sum
Total cost	$50,000	$65,000	$35,000	$100,000	$250,000
Units used, hot	600	2,500	6,000	400	—
Units used, mild	300	2,500	3,000	200	—
Unit cost	$55.56	$13.00	$3.89	$166.67	—
Total OH, hot	$33,336	$32,500	$23,333	$66,667	$155,836
Total OH, mild	$16,664	$32,500	$11,667	$33,333	$94,164

3. Select an allocation *base* appropriate for each activity.
4. Allocate cost to products using the base.

To illustrate the mechanics of ABC and contrast it with the traditional labor-hour approach, let us consider an example. Suppose a production facility makes two different products, hot and mild, and sells 6,000 units per month of hot and 3,000 units per month of mild. Total overhead costs are $250,000 per month. The plant runs 5,000 hours per month, of which 2,500 hours are devoted to hot and 2,500 to mild.

Traditional accounting would allocate the overhead equally among the two products because the number of labor hours devoted to each is the same. Hence, we would add $125,000 to the total cost of each product. This implies a unit charge of $125,000/6,000 = $20.83 for hot and $125,000/3,000 = $41.67 for mild. The unit cost of each product would then be computed by adding these unit overhead charges to the direct material and labor costs per unit. Notice that because fewer units of mild are produced, this procedure serves to inflate its unit cost more than that of hot.

Now reconsider the overhead allocation problem using the ABC approach. Suppose that we determine that the principal activities that account for the overhead (OH) cost are (1) requisition of material, (2) engineering support, (3) shipping, and (4) sales. Furthermore, suppose we can allocate the overhead cost to each activity as follows: $50,000 for requisition, $65,000 for engineering, $35,000 for shipping, and $100,000 for sales. The base (i.e., unit of measure) for requisition is the number of purchase orders (a total of 900); for engineering, the number of machine hours (5,000 hours); for shipping, the number of units shipped (9,000); and for sales, the number of sales calls made (600). Using these, a cost per base unit can be computed. The overhead allocation for a given product is then determined by the number of the base units used by that product times the cost per base unit. Finally, the unit overhead allocation is computed by dividing the total overhead allocation by the number of units. Table 6.3 summarizes the data and calculations for this example.

The unit overhead charge for hot is the sum of the "Total OH, Hot" entries divided by the number of units sold, that is, $155,833/6,000 = $25.97. Similarly, the unit overhead charge for mild is $94,164/3,000 = $31.38. Notice that while mild still receives a higher unit overhead charge than hot (due to its smaller volume), the difference is not as great as that resulting from the traditional labor-hour approach. The reason is that ABC recognizes that because of its higher volumes, greater effort, and hence cost, in the activities of requisition, engineering, and sales is devoted to hot. The net effect is to make mild look relatively more profitable than it would under traditional accounting methods.

Study Questions

1. What relevance does something as abstract as a "science of manufacturing" have to manufacturing management?

2. How many consistent observations does it take to prove a conjecture? How many inconsistent observations does it take to disprove a conjecture?

3. How can the concept of "conjectures and refutations" be used in a practical problem-solving environment?

4. List some dimensions along which manufacturing environments differ. How might these affect the "laws" governing their behavior? Do you think that a single science of manufacturing is possible for every manufacturing environment?

5. Indicate how each of the following might promote and impede the objective to maximize long-run profitability:
   a. Decrease average cycle time
   b. Decrease WIP
   c. Increase product diversity
   d. Improve product quality
   e. Improve machine reliability
   f. Reduce setup times
   g. Enhance worker cross-training
   h. Increase machine utilization

6. Why do you think that many writers in the JIT and TQM literature are loath to acknowledge the existence of tradeoffs? Do you think this has had positive, negative, or both impacts?

7. Why might the objective to maximize profits be difficult to use at the plant level? What advantages, or disadvantages, are there to using "minimize unit cost" instead?

8. We have suggested net profit and return on investment as firm-level measures. Do these capture the essence of a healthy firm? What characteristics are not adequately reflected in these measures? Can you suggest alternatives?

9. We have suggested

   • revenue (total quantity of good product sold per unit time)
   • operating expenses (operating budget of the plant)
   • assets (money tied up in plant, including inventories)

   as plant-level measures. How do these translate to the firm-level measures of total profit and ROI? Are there plant-level activities that are not reflected in the plant-level measures that affect the firm-level objectives? How might these be addressed?

10. Why does the distinction between objectives and constraints tend to blur in actual decision-making practice?

11. Give a specific example where "gaming behavior" (i.e., considering the other guy) is important in a manufacturing environment.

Problems

1. Consider a two-station production line in which no inventory is allowed (i.e., the stations are tightly coupled). Station 1 consists of a single machine that has potential daily production of one, two, three, four, five, or six units, each outcome being equally likely (i.e., potential production is determined by the roll of a single die). Station 2 consists of a single machine that has potential daily production of either three or four units, both of which are equally likely (i.e., it produces three units if a fair coin comes up heads and four units if it comes up tails).
   a. Compute the capacity of each station (i.e., in units per day). Is the line balanced (i.e., do both stations have the same capacity)?
   b. Compute the expected daily throughput of the line. Why does this differ from your answer to *a*?

   c. Suppose a second identical machine is added to station 1. How much does this increase average throughput? What implications might this result have concerning the desirability of a balanced line?

   d. Suppose a second identical machine is added to station 2 (but not station 1). How much does this increase average throughput? Is the impact the same from adding a machine at stations 1 and 2? Explain why or why not.

2. A manufacturer of vacuum cleaners produces three models of canister-style vacuum cleaners—the X-100, X-200, and X-300—on a production line with three stations—motor assembly, final assembly, and test. The line is highly automated and is run by three operators, one for each station. Data on production times, material cost, sales price, and bounds on demand are given in the following tables:

Product	Material Cost ($/Unit)	Price ($/Unit)	Minimum Demand (Units per Month)	Maximum Demand (Units per Month)
X-100	80	350	750	1500
X-200	150	500	0	500
X-300	160	620	0	300

Product	Motor Assembly (Minimum per Unit)	Final Assembly (Minimum per Unit)	Test (Minimum per Unit)
X-100	8	9	12
X-200	14	12	7
X-300	20	16	14

Labor costs $20 per hour (including benefits), and overhead for the line is $460,000 per month. The current production plan calls for production of X-100, X-200, and X-300 to be 625, 500, and 300 units per month, respectively.

   a. What is the monthly profit that results from the current production plan (i.e., sales revenue minus labor cost minus material cost minus overhead)?

   b. Estimate the profit per unit of each model, using direct labor hours to allocate the overhead cost per month. Which product appears most profitable? Is the current production plan consistent with these estimates? If not, propose an alternate production plan and compute its monthly profit.

   c. Suppose overhead costs are categorized into plant and equipment, management, purchasing, and sales and shipping. Plant and equipment costs use square footage as a base, where floor space dedicated to specific products (e.g., product-specific inventory sites) is assigned to individual products, while shared space is allocated equally. Management costs use labor hours as the base (i.e., as used in part *b* for all overhead costs). Purchasing uses purchase orders, where parts ordered for a specific product are counted toward that product and common parts are divided equally. Sales and shipping costs are allocated according to customer orders, where, again, orders for unique products are counted by product and orders for multiple products are split equally. The breakdown of overhead costs and the allocation of base units by product are given as follows:

Category	Plant and Equipment	Management	Purchasing	Sales and Shipping
Total cost	$250,000	$100,000	$60,000	$50,000
Base	Square feet	Labor hours	Purchase orders	Customer orders
Total units used	120,000	49,625	2,000	150
Units X-100	40,000	18,125	500	100
Units X-200	50,000	16,500	600	30
Units X-300	30,000	15,000	900	20

i. Compute the unit profit for each product, using an ABC allocation of overhead cost based on the above breakdowns. Compare these with the estimates of unit profits obtained by using a labor-hours allocation scheme.

ii. Do the ABC unit profits suggest a different production plan? If not, suggest one and compute its monthly profit and compare to that of the current plan and that suggested by the labor-hours cost allocation.

iii. What is wrong with using the approach of computing unit profits for each product and then using them to produce as much as possible of the most profitable products?

7   BASIC FACTORY DYNAMICS

I do not know what I may appear to the world; but to myself I seem to have been only like a boy playing on the seashore, and diverting myself in now and then finding a smoother pebble or a prettier shell than ordinary, whilst the great ocean of truth lay all undiscovered before me.

Isaac Newton

7.1   Introduction

In the previous chapter, we argued that manufacturing management needs a science of manufacturing. In this chapter, we begin the process of fleshing out such a science by examining some basic behavior of production lines.

To motivate the measures and mechanics on which we will focus, we begin with a realistic example. HAL, a computer company, manufactures printed-circuit boards (PCBs), which are sold to other plants, where the boards are populated with components ("stuffed") and then sent to be used in the assembly of personal computers. The basic processes used to manufacture PCBs are as follows:

1. *Lamination.* Layers of copper and prepreg (woven fiberglass cloth impregnated with epoxy) are pressed together to form cores (blank boards).
2. *Machining.* The cores are trimmed to size.
3. *Circuitize.* Through a photographic exposing and subsequent etching process, circuitry is produced in the copper layers of the blanks, giving the cores "personality" (i.e., a unique product character). They are now called *panels*.
4. *Optical test and repair.* The circuitry is scanned optically for defects, which are repaired if not too severe.
5. *Drilling.* Holes are drilled in the panels to connect circuitry on different planes of multilayer boards. Note that multilayer panels must return to lamination after being circuitized to build up the layers. Single-layer panels go through lamination only once and do not require drilling or copper plating.
6. *Copper plate.* Multilayer panels are run through a copper plating bath, which deposits copper inside the drilled holes, thereby connecting the circuits on different planes.

7. *Procoat.* A protective plastic coating is applied to the panels.

8. *Sizing.* Panels are cut to final size. In most cases, multiple PCBs are manufactured on a single panel and are cut into individual boards at the sizing step. Depending on the size of the board, there could be as few as two boards made from a panel, or as many as 20.

9. *End-of-line test.* An electrical test of each board's functionality is performed.

HAL engineers monitor the capacity and performance of the PCB line. Their best estimates of capacity are summarized in Table 7.1, which gives the average process rate (number of panels per hour) and average process time (hours) at each station. (Note that because panels are often processed in batches and because many processes have parallel machines, the rate of a process is not the inverse of the time.) These values are averages, which account for the different types of PCBs manufactured by HAL and also the different routings (e.g., some panels may visit lamination twice). They also account for "detractors," such as machine failures, setup times, and operator efficiency. As such, the process rate gives an approximation of how many panels each process could produce per hour if it had unlimited inputs. The process time represents the average time a typical panel spends being worked on at a process, which includes time waiting for detractors but *does not* include time waiting in queue to be worked on.

The main performance measures emphasized by HAL are throughput (how many PCBs are produced), cycle time (the time it takes to produce a typical PCB), work in process (inventory in the line), and customer service (fraction of orders delivered to customers on time). Over the past several months, throughput has averaged about 1,100 panels per day, or about 45.8 panels per hour (HAL works a 24-hours a day). WIP in the line has averaged about 37,000 panels, and manufacturing cycle time has been roughly 34 days, or 816 hours. Customer service has averaged about 75 percent.

The question is, how is HAL doing?

We can answer part of this question immediately. HAL management is not happy with 75 percent customer service because it has a corporate goal of 90 percent. So this aspect of performance is not good. However, perhaps the reason for this is that overzealous salespersons are promising unrealistic due dates to customers. It may not be an indication of anything wrong with the line.

The other measures—throughput, WIP and cycle time—are more difficult to deal with. We need to establish some sort of baseline against which to compare them. One

TABLE 7.1   Capacity Data for HAL Printed-Circuit Board Line

Process	Rate (parts per hour)	Time (hour)
Lamination	191.5	1.2
Machining	186.2	5.9
Circuitize	150.5	6.9
Optical test/repair	157.8	5.6
Drilling	185.9	10.0
Copper plate	136.4	1.5
Procoat	146.2	2.2
Sizing	126.5	2.4
EOL test	169.5	1.8

way to do this would be to benchmark against a competitor's operation. But even if HAL could get such data, there would still be the question of how comparable they really were. After all, every facility is unique. To be better or worse than a different type of facility does not necessarily mean much. A better baseline would be one that compares actual performance against what is theoretically possible for this facility.

In this chapter, we examine the extremes of behavior that are possible for simple idealized production lines, and we use the resulting models to develop a scale with which to rate actual facilities. We will return to the HAL example and use this scale to evaluate the performance of its PCB line. But first we must define our terms.

7.2   Definitions and Parameters

The scientific method absolutely requires precise terminology. Unfortunately, use of manufacturing terms in industry and the OM literature is far from standardized. This can make it extremely difficult for managers and engineers from different companies (and even the same company) to communicate and learn from one another. What it means for us is that the best we can do is to define our terms carefully and warn the reader that other sources will use the same terms differently or use different terms in place of ours.

7.2.1   Definitions

In Part II, we focus on the behavior of production *lines,* because these are the links between individual processes and the overall plant. Therefore, the following terms are defined in a manner that allows us to describe lines with precision. Some of these terms also have broader meanings when applied to the plant, as we note in our definitions and will occasionally adopt in Part III. However, to develop sharp intuition about production lines, we will maintain these rather narrow definitions for the remainder of Part II.

Workstation: A **workstation** is a collection of one or more machines or manual stations that perform (essentially) identical functions. Examples include a turning station made up of several vertical lathes, an inspection station made up of several benches staffed by quality inspectors, and a burn-in station consisting of a single room where components are heated for testing purposes. In **process-oriented layouts,** workstations are physically organized according to the operations they perform (e.g., all grinding machines located in the grinding department). Alternatively, in **product-oriented layouts** they are organized in lines making specific products (e.g., a single grinding machine dedicated to an individual line). The terms **station, workcenter,** and **process center** are synonymous with *workstation.*

Part: A **part** is a piece of raw material, a component, a subassembly, or an assembly that is worked on at the workstations in a plant. **Raw material** refers to parts purchased from outside the plant (e.g., bar stock). **Components** are individual pieces that are assembled into more complex products (e.g., gears). **Subassemblies** are assembled units that are further assembled into more complex products (e.g., transmissions). **Assemblies** (or final assemblies) are fully assembled products or end items (e.g., automobiles). Note that one plant's final assemblies may be another's raw material. For instance, transmissions are the final assemblies of a transmission plant, but are raw materials or purchased components to the automotive assembly plant.

End item: A part that is sold directly to a customer, whether or not it is an assembly, is called an **end item.** The relationship between end items and their constituent parts (raw materials, components, and subassemblies) is maintained in the **bill of material (BOM),** which Chapter 3 presented in detail.

Consumable: For the most part, **consumables** are materials such as bits, chemicals, gases, and lubricants that are used at workstations but do not become part of a product that is sold. More formally, we distinguish between parts and consumables in that parts are listed on the bill of material, while consumables are not. This means that some items that do become part of the product, such as solder, glue, and wire, can be considered either parts if they are recorded on the bill of material or consumables if they are not. Since different purchasing schemes are typically used for parts and consumables (e.g., parts might be ordered according to an MRP system, while consumables are purchased through a reorder point system), this choice may influence how such items are managed.

Routing: A **routing** describes the sequence of workstations passed through by a part. Routings begin at a raw material, component, or subassembly stock point and end at either an intermediate stock point or finished-goods inventory. For instance, a routing for gears may start at a stock point of raw bar stock; pass through cutting, hobbing, and deburring; and end at a stock point of finished gears. This stock of gears might in turn feed another routing that builds gear subassemblies. The bill of material and the associated routings contain the basic information needed to make an end item.

Order: A **customer order** is a request from a customer for a particular part number, in a particular quantity, to be delivered on a particular date. The paper or electronic **purchase order** sent by the customer might contain several customer orders. Henceforth, we will refer to a customer order as simply an **order.** Inside the plant, an order can also be an indication that certain inventories (e.g., safety stocks) need to be replenished. While timing may be more critical for orders originating with customers, both types of orders represent demand.

Job: A **job** refers to a set of physical materials that traverses a routing, along with the associated logical information (e.g., drawings, BOM). Although every job is triggered by either an actual customer order or the anticipation of a customer order (e.g., forecasted demand), there is frequently not a one-to-one correspondence between jobs and orders. This is because (1) jobs are measured in terms of specific parts (uniquely identified by a part number), not the collection of parts that may make up the assembly required to satisfy an order, and (2) the number of parts in a job may depend on manufacturing efficiency considerations (e.g., batch size considerations) and thus may not match the quantities ordered by customers.

Throughput (TH): The average output of a production process (machine, workstation, line, plant) per unit time (e.g., parts per hour) is defined as the system's **throughput,** or sometimes **throughput rate.** At the firm level, throughput is defined as the production per unit time that is *sold.* However, managers of production lines generally control what is made rather than what is sold. Therefore, for a plant, line, or workstation, we define throughput to be the average quantity of *good* (nondefective) parts (the manager does have control over quality) produced per unit time. In a line made up of workstations in tandem dedicated to a single family of products and where all products pass through each station exactly once, the throughput at every station will be the same (provided there is no yield loss). In a more complex plant, where workstations service multiple routings (e.g., a job shop), the throughput of an individual station will be the sum of the throughputs of the routings passing through it.

Capacity: An upper limit on the throughput of a production process is its **capacity.** In most cases, releasing work into the system at or above the capacity causes the system to become unstable (i.e., build up WIP without bound). Only very special systems can operate stably at capacity. Because this concept is subtle and important, we will investigate it more thoroughly later in this chapter, once we have introduced the appropriate notation and concepts.

Raw material inventory (RMI): As noted, the physical inputs at the start of a production process are typically called **raw material inventory.** This could represent bar stock that is cut up and then milled into gears, sheets of copper and fiberglass that are laminated together to make circuit boards, wood chips that become pulp and then paper stock, or rolls of sheet steel that are pressed into automobile fenders. Typically, the stock point at the beginning of a routing is termed raw material inventory even though the material may have already undergone some processing.

"Crib" and finished goods inventory (FGI): The stock point at the end of a routing is either a **crib inventory location** (i.e., an intermediate inventory location) or **finished goods inventory.** Crib inventories are used to gather different parts within the plant before further processing or assembly. For instance, a routing to produce gear assemblies may be fed by several crib inventories containing gears, housings, crankshafts, and so on. Finished goods inventory is where end items are held prior to shipping to the customer.

Work in process (WIP): The inventory between the start and end points of a product routing is called **work in process (WIP)**. Since routings begin and end at stock points, WIP is all the product between, but not including, the ending stock points. Although in colloquial use WIP often includes crib inventories, we make a distinction between crib inventory and WIP to help clarify the discussion.

Inventory turns: A commonly used measure of the efficiency with which inventory is used is **inventory turns,** or the **turnover ratio,** which is defined as the ratio of throughput to average inventory. Typically, throughput is stated in yearly terms, so that this ratio represents the average number of times the inventory stock is replenished or turned over. Exactly which inventory is included depends on what is being measured. For instance, in a warehouse, all inventory is FGI, so turns are given by TH/FGI. In a plant, we generally consider both WIP (inventory still in the line) and FGI (inventory waiting to ship), so turns are given by TH/(WIP + FGI). In any case, it is essential to make sure that throughput and inventory are measured in the same units. Since inventory is usually measured in cost dollars (i.e., rather than price or sales dollars), throughput should also be measured in cost dollars.

Cycle time (CT): The **cycle time** (also called variously **average cycle time, flow time, throughput time,** and **sojourn time**) of a given routing is the average time from release of a job at the beginning of the routing until it reaches an inventory point at the end of the routing (i.e., the time the part spends as WIP).[1] Although this is a precise definition of cycle time, it is also narrow, allowing us to define cycle time only for individual routings. It is common for people to refer to the cycle time of a product that is composed of many complex subassemblies (e.g., automobiles). However, it is not clear exactly what is meant by this. When does the clock start for an automobile? When the chassis starts down the assembly line? When the engine begins production? Or, as in Henry Ford's terms, when the ore is mined from the ground? We will discuss measuring cycle time for such assembled parts later, but for now we restrict our definition to single routings.

Lead time, service level, and fill rate: The **lead time** of a given routing or line is the time allotted for production of a part on that routing or line. As such, it is a management constant.[2] In contrast, cycle times are generally random. Therefore, in a line functioning

---

[1]Cycle time also has another meaning in assembly lines as the time allotted for each station to complete its task. It can also refer to the processing time of an individual machine (e.g., the time for a punch press to cycle). We will avoid these other uses of the term *cycle time* to prevent confusion.

[2]Recall that the time phasing function of MRP is critically dependent on the choice of such lead times.

in a *make-to-order* environment (i.e., it produces parts to satisfy orders with specific due dates), an important measure of line performance is **service level,** which is defined as

$$\text{Service level} = P\{\text{cycle time} \leq \text{lead time}\}$$

Notice that this definition implies that for a given distribution of cycle time, service level can be influenced by manipulating lead time (i.e., the higher the lead time, the higher the service level).

If the line is functioning in a *make-to-stock* environment (i.e., it fills a buffer from which customers or other lines expect to be able to obtain parts without delay), then a different performance measure may be more appropriate than service level. A logical choice is **fill rate,** which is defined as the fraction of orders that are filled from stock and was discussed in Chapter 2. Since fill rate and many other performance measures are often referred to as "service levels," the reader is cautioned to look for a precise definition whenever this term is encountered. We will consistently use the former definition of service level throughout Part II, but will return to the fill rate measure in Chapter 17.

Utilization: The **utilization** of a workstation is the fraction of time it is not idle for lack of parts. This includes the fraction of time the workstation is working on parts or has parts waiting and is unable to work on them due to a machine failure, setup, or other detractor. We can compute utilization as

$$\text{Utilization} = \frac{\text{Arrival rate}}{\text{Effective production rate}}$$

where the effective production rate is defined as the maximum average rate at which the workstation can process parts, considering the effects of failures, setups, and all other detractors that are relevant over the planning period of interest.

7.2.2 Parameters

Parameters are numerical descriptors of manufacturing processes and therefore vary in value from plant to plant. Two key parameters for describing an individual line (routing) are the bottleneck rate and the raw process time. We define these below, along with a third parameter, the *critical* WIP level, that can be computed from them.

Bottleneck rate (r_b): The **bottleneck rate** of the line, r_b, is the rate (parts per unit time or jobs per unit time) of the workstation having the highest long-term utilization. By long term we mean that outages due to machine failures, operator breaks, quality problems, etc., are averaged out over the time horizon under consideration. This implies that the proper treatment of outages will differ depending on the planning frequency. For example, for daily replanning, outages typically experienced during a day should be included; but unplanned long outages, such as those resulting from a major upset, should not. In contrast, for planning over a year-long horizon, time lost to major upsets should be included, if such occurrences are not unlikely over the course of a year.

In lines consisting of a single routing in which each station is visited exactly once and there is no yield loss, the arrival rate to every workstation is the same. Hence, the workstation with the highest utilization will be that with the least long-term capacity (i.e., slowest effective rate). However, in lines with more complicated routings or yield loss, the bottleneck may not be at the slowest workstation. A faster workstation that experiences a higher arrival rate may have higher utilization. For this reason, it is important to define the bottleneck in terms of utilization as we have done here.

Raw process time (T_0): The **raw process time** of the line, T_0, is the sum of the *long-term average* process times of each workstation in the line. Alternatively, we can define raw process time as the average time it takes a single job to traverse the empty line (i.e., so it does not have to wait behind other jobs). Again, we must be concerned about the length of the planning horizon when deciding what to include in the "average" process times. Over the long term, T_0 should include infrequent random and planned outages, while over a shorter term it should include only the more frequent delays.

Critical WIP (W_0): The **critical WIP** of the line, W_0, is the WIP level for which a line with given values of r_b and T_0 but having no variability achieves maximum throughput (that is, r_b) with minimum cycle time (that is, T_0). We show below that critical WIP is defined by the bottleneck rate and raw process time by the following relationship:

$$W_0 = r_b T_0$$

7.2.3 Examples

We now illustrate these definitions by means of two simple examples.

Penny Fab One.   Penny Fab One consists of a simple production line that makes giant one-cent pieces used exclusively in Fourth of July parades. The line consists of four machines in sequence that use well-known, stable processes. The first machine is a punch press that cuts penny blanks, the second stamps Lincoln's face on one side and the Memorial on the back, the third places a rim on the penny, and the fourth cleans away any burrs. Each machine takes exactly two hours to perform its operation. (We will relax this requirement that process times be deterministic later.) After each penny is processed, it is moved immediately to the next machine. The line runs 24 hours per day, with breaks, lunches, etc., covered by spare operators. For our purposes, the market for giant pennies can be assumed to be unlimited, so that all product made is sold; thus, more throughput is unambiguously better for this system.

Since this is a tandem line with no yield loss, the bottleneck is defined as the slowest workstation. However, the *capacity* of each machine is the same and equals one penny every two hours, or one-half part per hour. Hence, any of the four machines can be regarded as the bottleneck and

$$r_b = 0.5 \text{ penny per hour}$$

or 12 pennies per day. Such a line is said to be **balanced,** since all stations have equal capacity.

Next, note that the raw process time is simply the sum of the processing times at the four stations, so

$$T_0 = 8 \text{ hours}$$

The critical WIP level is given by

$$W_0 = r_b T_0 = 0.5 \times 8 = 4 \text{ pennies}$$

We will illustrate that this is indeed the level of WIP that causes the line to achieve throughput of $r_b = 0.5$ penny per hour and cycle time of $T_0 = 8$ hours. Notice that W_0 is equal to the number of machines in the line. This is *always* the case for balanced lines, since having one job per machine is just enough to keep all machines busy at all times. However, as we will see, it is not true for unbalanced lines.

Penny Fab Two.   Now consider a somewhat more complex Penny Fab Two, which represents an unbalanced line with multimachine stations. Penny Fab Two still produces giant pennies in four steps: punching, stamping, rimming, and deburring; but the workstations now have different numbers of machines and processing times, as shown in Table 7.2.

The presence of multimachine stations complicates the capacity calculations somewhat. For a single machine, the capacity is simply the reciprocal of the process time (e.g., if it takes one-half hour to do one job, the machine can do two jobs per hour). The capacity of a station consisting of several identical machines in parallel must be calculated as the individual machine capacity times the number of machines. For example, in Penny Fab Two, the capacity per machine at station 3 is

$$\frac{1}{10} \text{ penny per hour}$$

so the capacity of the station is

$$6 \times \frac{1}{10} = 0.6 \text{ penny per hour}$$

Notice that the station capacity can be computed directly by dividing the number of machines by the process time. This is done for each station in Table 7.2.

The capacity of the line with multimachine stations is still defined by the rate of the bottleneck, or slowest station in the line. In Penny Fab Two, the bottleneck is station 2, so

$$r_b = 0.4 \text{ penny per hour}$$

Notice that the bottleneck is neither the station that contains the slowest machines (station 3) nor the one with the fewest machines (station 1).

The raw process time of the line is still the sum of the process times. Notice that adding machines at a station does not decrease T_0, since a penny can be worked on by only one machine at a time. Hence, the raw process time for Penny Fab Two is

$$T_0 = 20 \text{ hours}$$

Regardless of whether the line has single- or multiple machine stations, the critical WIP level is always defined as

$$W_0 = r_b T_0 = 0.4 \times 20 = 8 \text{ pennies}$$

In Penny Fab Two, as in Penny Fab One, W_0 is a whole number. This, of course, need not be the case. If W_0 comes out to a fraction, it means that there is no constant WIP level that will achieve throughput of exactly r_b jobs per hour and cycle time of T_0 hours. Furthermore, notice that the critical WIP level in Penny Fab Two (eight pennies) is less than the number of machines (11). This is because the system is not balanced (i.e., stations have different amounts of capacity), and therefore some stations will not be fully utilized.

TABLE 7.2   Penny Fab Two: An Unbalanced Line

Station Number	Number of Machines	Process Time (hour)	Station Capacity (Jobs per Hour)
1	1	2	0.50
2	2	5	0.40
3	6	10	0.60
4	2	3	0.67

7.3   Simple Relationships

Now, in the pursuit of a science of manufacturing, we ask the fundamental question, What are the relationships among WIP, throughput, and cycle time in a single production line? Of course, the answer will depend on the assumptions we make about the line. In this section, we will give a precise (i.e., quantitative) description of the range of possible behavior. This will serve to sharpen our intuition about how lines perform and will give us a scale on which to rate (benchmark) actual systems.

A problem with characterizing the relationship between measures such as WIP and throughput is that in real systems they tend to vary simultaneously. For instance, in an MRP system, the line may be flooded with work one month (due to a heavy master production schedule) and very lightly loaded the next. Hence, both WIP and throughput are apt to be high during the first month and low during the second. For clarity of presentation, we will eliminate this problem by controlling the WIP level in the line so as to hold it constant over time. For instance, in the Penny Fabs, we will start the lines with a specified number of pennies (jobs) and then release a new penny blank into the line each time a finished penny exits the line.[3]

7.3.1   Best-Case Performance

To analyze and understand the behavior of a line under the best possible circumstances, namely, when process times are absolutely regular, we will *simulate* Penny Fab One. This is easily done by using a piece of paper and several pennies, as shown in Figure 7.1.

We begin by simulating the system when only one job is allowed in the line. The first penny spends two hours successively at stations 1, 2, 3, and 4, for a total cycle time of eight hours. Then a second penny is released into the line, and the same sequence is repeated.

FIGURE 7.1

Penny Fab One with WIP = 1

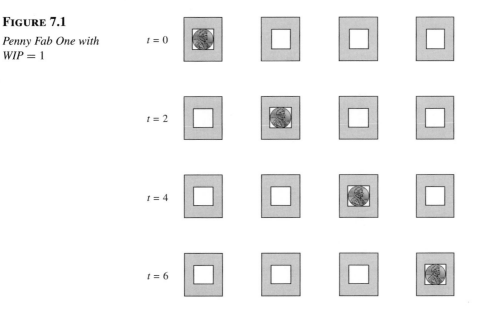

Since this results in one penny coming out of the line every eight hours, the throughput is one-eighth penny per hour. Notice that the cycle time is equal to the raw process time $T_0 = 8$, while the throughput is one-fourth of the bottleneck rate $r_b = 0.5$.

Now we add a second penny to the line (starting both at the front of the line). After two hours, the first penny completes processing at station 1 and starts on station 2. Simultaneously, the second penny starts processing on station 1. Thereafter, the second penny will follow the first, switching stations every two hours, as shown in Figure 7.2. After the initial wait experienced by the second penny, it never waits again. Hence, once the system is running in steady state, every penny released into the line still has a cycle time of exactly eight hours. Moreover, since two pennies exit the line every eight hours, the throughput increases to two-eighths penny per hour, double that when the WIP level was 1 and 50 percent of line capacity ($r_b = 0.5$).

We add a third penny. Again, after an initial transient period in which pennies wait at the first station, there is no waiting, as shown in Figure 7.3. Hence, cycle time stays at 8 h, while throughput increases to three-eighths part per hour, or 75 percent of r_b.

When we add a fourth penny, we see that all the stations stay busy all the time once steady state has been reached. Because there is no waiting at the stations, cycle time is still $T_0 = 8$ h. Since the last station is busy all the time, it outputs a penny every other hour, so throughput becomes one-half penny per hour, which equals the line capacity r_b. This very special behavior, in which cycle time T_0 (its minimum value) and throughput r_b (its maximum value) are only achieved when the WIP level is set at the critical WIP level, which we recall for Penny Fab One is

$$W_0 = r_b T_0 = 0.5 \times 8 = 4 \text{ pennies}$$

Now we add a fifth penny to the line. Because there are only four machines, a penny will wait at the first station, even after the system has settled into steady state. Since we measure cycle time as the time from when a job is released (the time it enters the queue at the first station) to when it exits the line, it now becomes 10 hours, due to the extra two hours of waiting time in front of station 1. Hence, for the first time, cycle time becomes larger than its minimal value $T_0 = 8$. However, since all stations are always busy, the throughput remains at $r_b = 0.5$ penny per hour.

FIGURE 7.2

Penny Fab One with WIP = 2

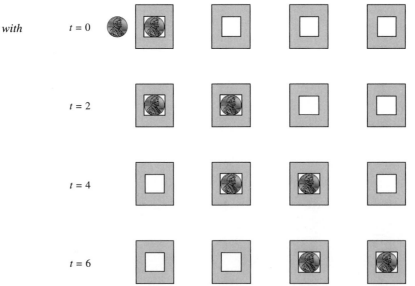

FIGURE 7.3

Penny Fab One with
WIP = 3

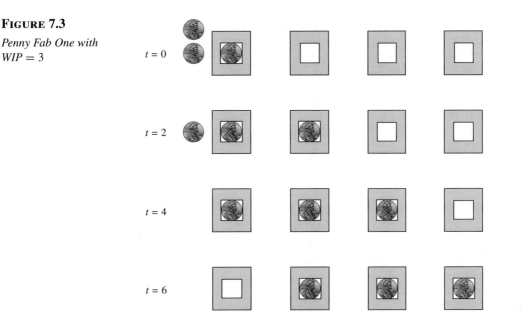

$t = 0$

$t = 2$

$t = 4$

$t = 6$

Finally, consider what happens when we allow 10 pennies in the line. In steady state, a queue of six pennies persists in front of the first station, meaning that an individual penny spends 12 hours from the time it is released to the line until it begins processing at station 1. Hence, the cycle time is 20 hours. As before, all machines remain busy all the time, so throughput is still $r_b = 0.5$ penny per hour. It should be clear at this point that each penny we add increases cycle time by two hours with no increase in throughput.

We summarize the behavior of Penny Fab One with no variability for various WIP levels in Table 7.3, and we present the results graphically in Figure 7.4. From a performance standpoint, it is clear that Penny Fab One runs best when there are four pennies in WIP. Only this WIP level results in minimum cycle time T_0 and maximum throughput r_b—any less and we lose throughput with no decrease in cycle time; any more and we increase cycle time with no increase in throughput. This special WIP level is the critical WIP (W_0) that was defined previously.

In this particular example, the critical WIP is equal to the number of machines. This is always the case when the line consists of stations with equal capacity (i.e., a balanced line). For unbalanced lines, W_0 will be less than the number of machines, but still has the property of being the WIP level that achieves maximum throughput with minimum cycle time, and is still defined by $W_0 = r_b T_0$.

It is important to note that while the critical WIP is optimal in the case with zero variability, it will *not* be optimal in other cases. Indeed, the concept of an optimal WIP level is not even well defined in the presence of variability because, in general, increasing WIP will increase both throughput (good) and cycle time (bad).

Little's Law.   Close examination of Table 7.3 reveals an interesting, and fundamental, relationship among WIP, cycle time, and throughput. At every WIP level, WIP is equal to the product of throughput and cycle time. This relation is known as *Little's law* (named for John D. C. Little, who provided the mathematical proof) and represents our first *factory physics* relationship:

Law (Little's Law):

$$WIP = TH \times CT$$

TABLE 7.3   WIP, Cycle Time, and Throughput of Penny Fab One

WIP	CT	% T_0	TH	% r_b
1	8	100	0.125	25
2	8	100	0.250	50
3	8	100	0.375	75
4	8	100	0.500	100
5	10	125	0.500	100
6	12	150	0.500	100
7	14	175	0.500	100
8	16	200	0.500	100
9	18	225	0.500	100
10	20	250	0.500	100

FIGURE 7.4

Cycle time and throughput versus WIP for Penny Fab One

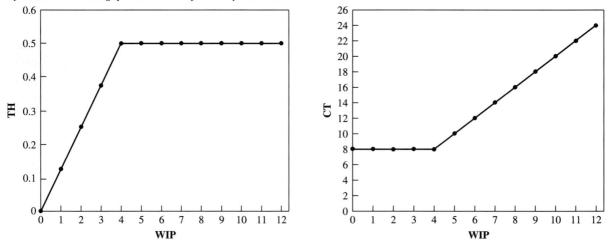

It turns out that Little's law holds for *all* production lines, not just those with zero variability. As we discussed in Chapter 6, Little's law is not a *law* at all but a *tautology*. For special cases (e.g., the case of observing the system for a time that goes to infinity), the relationship can be proved mathematically. However, it does not entirely hold in the less-than-infinite case (which, of course, involves all real cases) except for other special cases. Nonetheless, we will use it as a conjecture about the nature of manufacturing systems and use it as an approximation when it is not exact.

Little's law is quite useful in that it can be applied to a single station, a line, or an entire plant. As long as the three quantities are measured in consistent units, the above relationship will hold over the long term. This makes it immensely applicable to practical situations. Some straightforward uses of Little's law include these:

1. *Queue length calculations.* Since Little's law applies to individual stations, we can use it to calculate the expected queue length and utilization (fraction of time busy) at each station in a line. For instance, consider Penny Fab Two, which was summarized in Table 7.2, and suppose it is running at the bottleneck rate (that is, 0.4 job per hour). From Little's law, the expected WIP at the first station will be

$$\text{WIP} = \text{TH} \times \text{CT} = 0.4 \text{ job per hour} \times 2 \text{ hour} = 0.8 \text{ job}$$

Since there is only one machine at station 1, this means that it will be utilized 80 percent of the time. Similarly, at station 3, Little's law predicts an average WIP of four jobs. Since there are six machines, the average utilization will be $4/6 = 66.7$ percent. Notice that this is equal to the ratio of the rate of the bottleneck to the rate of station 3 (that is, 0.4/0.6), as we would expect.

2. *Cycle time reduction.* Since Little's law can be written as

$$\text{CT} = \frac{\text{WIP}}{\text{TH}}$$

it is clear that reducing cycle time implies reducing WIP, provided throughput remains constant. Hence, large queues are an indication of opportunities for reducing cycle time, as well as WIP. We will discuss specific measures for WIP and cycle time reduction in Chapter 17.

3. *Measure of cycle time.* Measuring cycle time directly can sometimes be difficult, since it entails registering the entry and exit times of each part in the system. Since throughput and WIP are routinely tracked, it might be easier to use the ratio WIP/TH as a perfectly reasonable indirect measure of cycle time.

4. *Planned inventory.* In many systems, jobs are scheduled to finish ahead of their due dates in order to ensure a high level of customer service. Because, in our era of inventory consciousness, customers often refuse to accept early deliveries, this type of "safety lead time" causes jobs to wait in finished goods inventory prior to shipping. If the **planned inventory** time is n days, then according to Little's law, the amount of inventory in FGI will be given by nTH (where TH is measured in units per day).

5. *Inventory turns.* Recall that inventory turns are given by the ratio of throughput to average inventory. If we have a plant in which all inventory is WIP (i.e., product is shipped directly from the line so there is no finished goods inventory), then turns are given by TH/WIP, which by Little's law is simply 1/CT. If we include finished goods, then turns are TH/(WIP + FGI). But Little's law still applies, so this ratio represents the inverse of the total average time for a job to traverse the line plus the finished goods crib. Hence, intuitively, inventory turns are one divided by the average residence time of inventory in the system.

In a sense, Little's law is the "$F = ma$" of factory physics. It is a broadly applicable equation that relates three fundamental quantities. At the same time, Little's law can be viewed as a truism about units. It merely indicates the obvious fact that we can measure WIP level in a station, line, or system in units of jobs or time. For instance, a line that produces 100 crankcases per day and has a WIP level of 500 crankcases has five days of WIP in it. Little's law is a statement that this unit's conversion is valid for average WIP, cycle time, and throughput, or

$$\text{CT} = \frac{\text{WIP}}{\text{TH}}$$

or

$$5 \text{ days} = \frac{500 \text{ crankcases}}{100 \text{ crankcases per day}}$$

We can now generalize the results shown in Table 7.3 and Figure 7.4 to achieve our original objective of giving a precise summary of the relationship between WIP and throughput for a "best-case" (i.e., zero-variability) line. We can then apply Little's law to extend this to describe the relationship between WIP and cycle time. Since these relationships were derived for perfect lines with no variability, the following expressions indicate the *maximum throughput* and *minimum cycle time* for a given WIP level for any system having parameters r_b and T_0. The resulting equations are our next *Factory Physics* law.

Law (Best-Case Performance):   *The minimum cycle time for a given WIP level w is given by*

$$\text{CT}_{\text{best}} = \begin{cases} T_0 & \text{if } w \leq W_0 \\ \dfrac{w}{r_b} & \text{otherwise} \end{cases}$$

The maximum throughput for a given WIP level w is given by

$$\text{TH}_{\text{best}} = \begin{cases} \dfrac{w}{T_0} & \text{if } w \leq W_0 \\ r_b & \text{otherwise} \end{cases}$$

One conclusion we can draw from this is that, contrary to the popular slogan, zero inventory is *not* a realistic goal. Even under perfect deterministic conditions, zero inventory yields zero throughput and therefore zero revenue. A more realistic "ideal" WIP is the critical WIP W_0.

Penny Fab One represents an ideal (zero-variability) situation, in which it is optimal to maintain a WIP level equal to the number of machines. Of course, in the real world there are not many factories that run with such low WIP levels. Indeed, in many production lines the WIP-to-machines ratio is closer to 20:1 (Bradt 1983). If this ratio were to hold for Penny Fab One, the cycle time would be almost seven days with 80 jobs in WIP. Obviously, this is much worse than a cycle time of eight hours at a WIP level of four jobs (i.e., the "optimal" level). Why, then, do actual plants operate so far from the ideal of the critical WIP level?

Unfortunately, Little's law offers little help. Since TH = WIP/CT, we can have the same throughput with large WIP levels and long cycle times, or with low WIP levels and short cycle times. The problem is that Little's law is only one relation among three quantities. We need a second relation if we are to uniquely determine two quantities, given the third (e.g., predict both WIP and cycle time from throughput). Sadly, there is no universally applicable second relationship among WIP, cycle time, and throughput. The best we can do is to characterize the behavior of a line under specific assumptions. In addition to the best case, which we considered above, we will treat two other scenarios, which we term the **worst case** and the **practical worst case.**

7.3.2   Worst-Case Performance

Instead of imagining the best possible behavior of a line, we consider the worst. Specifically, we seek the *maximum cycle time* and *minimum throughput* possible for a line with bottleneck rate r_b and raw process time T_0. This will enable us to bracket the behavior and gauge the performance of real lines. If a line is closer to the worst case than to the best case, then there are some real problems (or opportunities, depending on your perspective).

To facilitate our discussion of the worst case, recall that we are assuming a constant amount of work is maintained in the line at all times. Whenever a job finishes, another is started. One way that this could be achieved in practice would be to transport jobs through the line on *pallets*. Whenever a job is finished, it is removed from its pallet and the pallet immediately returns to the front of the line to carry a new job. The WIP level, therefore, is equal to the (fixed) number of pallets.

Now, imagine yourself sitting on a pallet riding around and around a best-case line with WIP equal to the critical WIP (e.g., Penny Fab One with four jobs). Each time you arrive at a station, a machine is available to begin work on the job immediately. It is precisely because there is no waiting (queueing) that this line achieves the minimum possible cycle time of T_0.

To get the longest possible cycle times for this system, we must somehow increase the waiting time without changing the *average* processing times (otherwise we would change r_b and T_0). The very worst we could possibly make waiting time would be that every time our pallet reached a station, we found ourselves waiting behind *every* other job in the line. How could this possibly occur?

Consider the following. Suppose that you are riding on pallet number 4 in a modified Penny Fab One with four pallets. However, instead of all jobs requiring exactly two hours at each station, suppose that jobs on pallet 1 require eight hours, while jobs on pallets 2, 3, and 4 require zero hours. The average processing time at each station is

$$\frac{8+0+0+0}{4} = 2 \text{ hours}$$

as before, and hence we still have $r_b = 0.5$ job per hour and $T_0 = 8$ hours. However, every time your pallet reaches a station, you find pallets 1, 2, and 3 ahead of you (see Figure 7.5). The slow job on pallet 1 causes all the other jobs to pile up behind it at all times. This is the absolute maximum amount of waiting time it is possible to introduce, and hence this represents the worst case.

The cycle time for this system is

$$8 + 8 + 8 + 8 = 32 \text{ hours}$$

or $4T_0$, and since four jobs are output each time pallet 1 finishes on station 4, the throughput is

$$\tfrac{4}{32} = \tfrac{1}{8} \text{ job per hour}$$

or $1/T_0$ jobs per hour. Notice that the product of throughput and cycle time is $\frac{1}{8} \times 32 = 4$, which is the WIP level, so, as always, Little's law holds.

Let us summarize these results for a general line as our next factory physics law.

Law (Worst-Case Performance): *The worst-case cycle time for a given WIP level w is given by*

$$\text{CT}_{\text{worst}} = wT_0$$

The worst-case throughput for a given WIP level w is given by

$$\text{TH}_{\text{worst}} = \frac{1}{T_0}$$

It is interesting to note that both the best-case and worst-case performances occur in systems with no randomness. There is *variability* in the worst-case system, since jobs have different process times; but there is no *randomness,* since all process times are completely predictable. The literature on quality management stresses the need

FIGURE 7.5

*Evolution of worst-case
line*

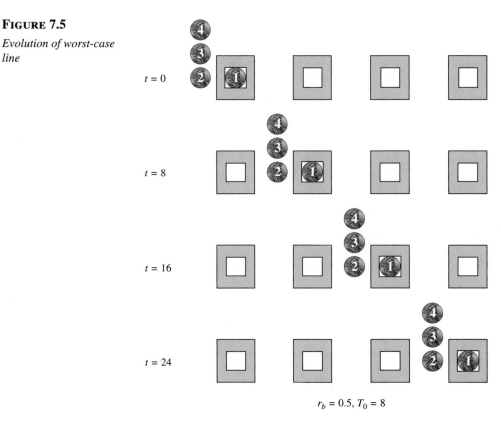

$r_b = 0.5$, $T_0 = 8$

for variability reduction, but sometimes implies that variability and randomness are synonymous. The above *Factory Physics* results show that this is not the case; variability can be the result of randomness or *bad control* (or both). We will examine this distinction in greater depth after we have developed the tools for treating variability in Chapters 8 and 9.

Finally, the reader may be justifiably skeptical about the realism of the worst case. After all, we arrived at this case by forcing the maximum amount of waiting time (in order to make cycle times as long as possible) by making the processing times as variable as possible. To do this, we assumed jobs on one of the pallets had long processing times, while all the others had zero processing times. Surely this could never happen in real life.

But it can and (at least to some extent) does happen. To see how, suppose that the four pallets used to carry jobs in Penny Fab One (when WIP equals four jobs) are themselves moved between stations with a forklift. Further, suppose that because the forklift has other obligations, it cannot afford to make the number of trips necessary to move each pallet individually. Instead, it waits until all four jobs are finished on a station and then moves them as a group to the next station. Similarly, it waits until all four pallets are empty at the end of the line to bring them back to the front to receive new jobs. Assuming that processing times of each job at each station are two hours (as in the original Penny Fab One), and that move times on the forklift are sufficiently short as to be reasonably treated as zero, the progress of the system will be *exactly* the same as that shown in Figure 7.5. Hence, worst-case behavior can result from **batch moves.**

Of course, it is rare to find real plants in which batch moves are so extreme as to cause every job in the line to travel together. More commonly, the WIP in a line will be transported in several batches, possibly of varying size. While this kind of more modest batching will not produce worst-case behavior, it is one factor that can push the performance of a line closer to that of the worst case than the best case. Consequently, batching is a genuine problem (opportunity) in many production systems.

7.3.3   Practical Worst-Case Performance

Virtually no real-world line behaves literally according to either the best case or the worst case. Therefore, to better understand the behavior between these two extreme cases, it is instructive to consider an intermediate case. We do this by means of a case that, unlike the previous two, involves randomness. In fact, in a sense, it represents the "maximum randomness" case. We term this the **practical worst case** to express our belief that virtually any system with worse behavior is a target for improvement.

To describe the practical worst case and show why it can be regarded as the maximum randomness case, we must first define the concept of a system *state*. The state of the system is a complete description of the jobs at all the stations: how many there are and how long they have been in process. Under special conditions, which we assume here and describe below, the only information needed is the number of jobs at each station. Hence, we can give a concise summary of a state by using a vector with as many elements as there are stations in the line.

For instance, in a line with four stations and three jobs, the vector (3, 0, 0, 0) represents the state in which all three jobs are at the first station, while the vector (1, 1, 1, 0) represents the state in which there is one job each at stations 1, 2, and 3. There are 20 possible states for a system consisting of four machines and three jobs, which are enumerated in Table 7.4.

Depending on the specific assumptions about the line, not all states will necessarily occur. For instance, if all processing times in the four-station, three-job system are one hour and it behaves according to the best case, then only four states—(1, 1, 1, 0), (0, 1, 1, 1), (1, 0, 1, 1), and (1, 1, 0, 1)—will be repeated as illustrated in Figure 7.6. Similarly, if it behaves according to the worst case, then four different states—(3, 0, 0, 0), (0, 3, 0, 0), (0, 0, 3, 0), and (0, 0, 0, 3)—will be repeated, as illustrated in Figure 7.7. Because both of these systems have no randomness, other states are never reached.

TABLE 7.4   Possible States for a System with Four Machines and Three Jobs

State	Vector	State	Vector
1	(3, 0, 0, 0)	11	(1, 0, 2, 0)
2	(0, 3, 0, 0)	12	(0, 1, 2, 0)
3	(0, 0, 3, 0)	13	(0, 0, 2, 1)
4	(0, 0, 0, 3)	14	(1, 0, 0, 2)
5	(2, 1, 0, 0)	15	(0, 1, 0, 2)
6	(2, 0, 1, 0)	16	(0, 0, 1, 2)
7	(2, 0, 0, 1)	17	(1, 1, 1, 0)
8	(1, 2, 0, 0)	18	(1, 1, 0, 1)
9	(0, 2, 1, 0)	19	(1, 0, 1, 1)
10	(0, 2, 0, 1)	20	(0, 1, 1, 1)

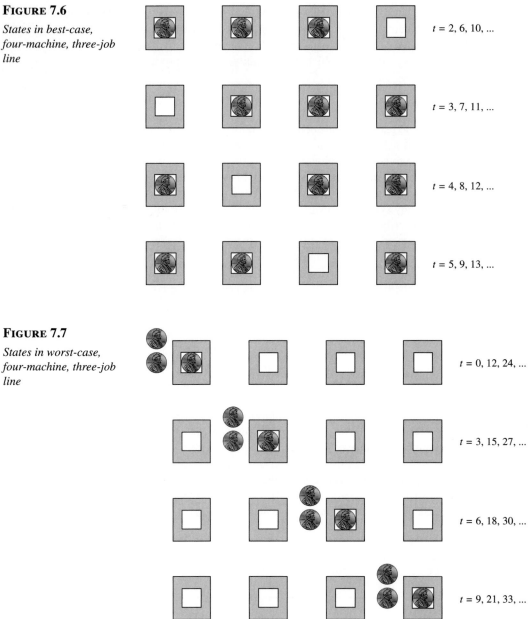

FIGURE 7.6

States in best-case, four-machine, three-job line

t = 2, 6, 10, ...

t = 3, 7, 11, ...

t = 4, 8, 12, ...

t = 5, 9, 13, ...

FIGURE 7.7

States in worst-case, four-machine, three-job line

t = 0, 12, 24, ...

t = 3, 15, 27, ...

t = 6, 18, 30, ...

t = 9, 21, 33, ...

When randomness is introduced into a line, more states become possible. For instance, suppose the processing times are deterministic, but every once in a while a machine may break down for several hours. Then most of the time we will observe "spread out" states, like those in Figure 7.6, but occasionally we will see "clumped up" states, like those in Figure 7.7. If there is only a little randomness (e.g., machine failures are very rare), then the frequency of the spread-out states will be very high, whereas if there is a lot of randomness (e.g., machines are failing right and left), then *all* the states shown in Table 7.4 may occur quite often. Hence, we define the *maximum randomness* scenario to be that which causes every possible state to occur with equal frequency.

In order for all states to be equally likely, three special conditions are required:

1. The line must be balanced (i.e., all stations must have the same average process times).

2. All stations must consist of single machines. (This assumption also allows us to avoid the complexities of parallel processing and jobs passing one another.)

3. Process times must be random and occur according to a specific probability distribution known as the **exponential distribution.** The exponential is the *only* continuous distribution that has a special property known as the **memoryless property** (see Appendix 2A). What this means is that if the processing time on a machine is exponentially distributed, then knowledge of how long a part has been in process offers no information about when it will be finished. For instance, if process times on a machine are exponential with mean one hour and the current job has been in process for five seconds, then the expected remaining process time is one hour. If the current job has been in process for one hour, the remaining process time is one hour. If the current job has been in process for 942 hours, the expected remaining process time is one hour.[4] It is as if the machine forgets its past work when predicting the future—hence the term *memoryless*. Thus, if process times are exponentially distributed, there is no need to know about how long a job has been in process to completely define the system state.

To understand how the practical worst case (PWC) works, return to the thought experiment in which you envisioned yourself riding around on a pallet that cycles through the line again and again. Suppose there are N (single machine) stations, each with average processing times of t, and a constant level of w jobs in the line. Thus, the raw process time is $T_0 = Nt$, and the bottleneck rate is $r_b = 1/t$ for this line.

Since the above three conditions guarantee that all states are equally likely, then, from your vantage point on a pallet, you would expect to see on average the $w - 1$ other jobs equally distributed among the N stations each time you arrive at a station. So the expected number of jobs ahead of you upon arrival is $(w - 1)/N$. Since the average time you spend at the station will be the time for the other jobs to complete processing plus the time for your job to be processed, we can write

$$\text{Average time at a station} = \text{Time for other jobs} + \text{Time for your job}$$

$$= \frac{w - 1}{N} t + t$$

$$= \left(1 + \frac{w - 1}{N}\right) t$$

By assuming that the $(w-1)/N$ jobs ahead of you require an average of $[(w-1)/N]t$ time to complete, we are ignoring the fact that the job in process at the station was partially finished when you arrived. It is the memoryless property of the exponential distribution that enables us to do this.

Finally, since all stations are assumed identical, we can compute the average cycle time by simply multiplying the average time at each station by the number of stations N, to get

---

[4]Although it may be a stretch to imagine processing times behaving in this way, there certainly seem to be examples of this type of behavior in daily life, for instance, times until departure of delayed flights, times until the arrival of trains on certain railways, times until some contractors finish home improvement jobs, etc.

$$CT = N \left(1 + \frac{w-1}{N}\right) t$$

$$= Nt + (w-1)t$$

$$= T_0 + \frac{w-1}{r_b}$$

To get the corresponding throughput, we simply apply Little's law:

$$TH = \frac{WIP}{CT}$$

$$= \frac{w}{T_0 + (w-1)/r_b}$$

$$= \frac{w}{W_0/r_b + (w-1)/r_b}$$

$$= \frac{w}{W_0 + w - 1} r_b$$

This provides our definition of practical worst-case performance.

Definition (Practical Worst-Case Performance):   *The practical worst-case (PWC) cycle time for a given WIP level w is given by*

$$CT_{PWC} = T_0 + \frac{w-1}{r_b}$$

The PWC throughput for a given WIP level w is given by

$$TH_{PWC} = \frac{w}{W_0 + w - 1} r_b$$

Notice that the behavior of this case is reasonable for both extremely low and extremely high WIP levels. At one extreme, when there is only one job in the system ($w = 1$), cycle time becomes raw process time T_0, as we would expect. At the other extreme, as the WIP level grows very large (that is, $w \to \infty$), throughput approaches capacity r_b, while cycle time increases without bound. The intuition behind this latter result is that achieving throughput close to capacity in systems with high variability requires high WIP levels, in order to ensure high utilization of machines. But this also ensures a great deal of waiting and hence high cycle times.

The throughput and cycle time of the practical worst case are always between those of the best case and the worst case. As such, the PWC provides a useful midpoint that approximates the behavior of many real systems. By collecting data on average WIP, throughput, and cycle time (actually, because of Little's law, any two of these will suffice) for a real production line, we can determine whether it lies in the region between the best and practical worst cases, or between the practical worst and worst cases. Systems with better performance than the PWC (i.e., that have larger throughput and smaller cycle time for a given WIP level) are "good," and systems with worse performance are "bad." It makes sense to focus our improvement efforts on the bad lines because they are the ones with room for improvement. Thus, our three cases offer a sort of **internal benchmarking** methodology (i.e., as opposed to **external benchmarking** in which comparisons are made against outside systems).

For further guidance on *how* to improve a bad line, we can look to the three assumptions under which the PWC was derived:

1. Balanced line.
2. Single machine stations.
3. Exponential (memoryless) processing times.

Since these three conditions were chosen to maximize randomness in the line, improving any of them will tend to improve the performance of the line.

First, we could unbalance the line by adding capacity at a station. This could be accomplished by adding physical equipment, reducing downtime due to worker breaks or equipment failures, speeding up the process through more efficient work methods, and so on. Obviously, if we increase capacity at all stations, throughput will increase. But even if we increase capacity at only some stations, so that r_b does not change, this serves to reduce randomness (i.e., the states in Table 7.4 are no longer equally likely) and therefore causes the throughput-versus-WIP curve to increase more rapidly (i.e., less WIP in the system achieves the same throughput). We realize that line *unbalancing* is somewhat counter to the traditional industrial engineering emphasis on line balancing. However, as we will see in Chapter 18, line balancing is primarily applicable to *paced* assembly lines, not a line of independent workstations like those we are considering here.

Second, we could make use of parallel machines in place of single machines at workstations. If this is accomplished by adding extra machines, then it serves to increase capacity and therefore has essentially the same effects as those discussed above. But even replacing single machines with parallel ones with the same capacity can improve performance in some cases. For instance, reconsider Penny Fab One under the assumption that process times are exponential instead of deterministic with *average* process times still two hours at each station. Suppose stations 3 and 4 (rimming and deburring) are collapsed into a single station with two parallel machines, where the machines perform both rimming and deburring in a single step and take twice as long as before (i.e., an average of four hours per penny). Since the capacity of the station is one-half penny per hour, the bottleneck rate of the line is still $r_b = 0.5$. Also, the raw process time remains $T_0 = 8$ hours. But in the former arrangement, two pennies could have wound up at either rimming or deburring, with the consequence that one has to wait. In the revised line, anytime there are two pennies in rimming or deburring, we are guaranteed that both are being worked on. The result will be less waiting, and hence shorter cycle times, for a given WIP level in the revised system with parallel machines.

Finally, we could reduce the variability of the processing times to less than that implied by the exponential distribution. Reducing the likelihood of jobs clumping up behind stations, and hence waiting, will improve throughput and cycle time for a given WIP level. We will examine what is meant by variability reduction relative to the exponential in Chapter 8, and we will discuss practical methods for achieving it in Part III.

Figures 7.8 and 7.9 illustrate some of these concepts by plotting cycle time and throughput as a function of WIP level for Penny Fab Two under the assumption of exponentially distributed process times at all stations. For comparison, we have also plotted the best, worst, and practical worst cases for the same bottleneck rate and raw process time (i.e., for $r_b = 0.4$ and $T_0 = 20$). Even though processing times are exponential, because Penny Fab Two has an unbalanced line and parallel machine stations, it outperforms the practical worst case. If we were to reduce the variability of the processing times, this would improve it even more.

7.3.4 Bottleneck Rates and Cycle Time

Since the 1980s, a great deal of attention has been focused on the importance of bottlenecks in production systems (see, e.g., Goldratt and Cox 1984). Our discussion here

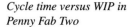

FIGURE 7.8

Cycle time versus WIP in Penny Fab Two

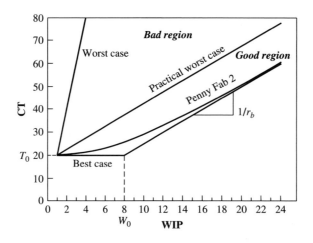

FIGURE 7.9

Throughput time versus WIP in Penny Fab Two

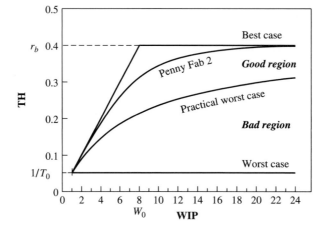

certainly concurs that the bottleneck rate r_b is important, since it establishes the capacity of the line. But the factory physics laws also give us insights into the role of bottlenecks beyond this obvious conclusion.

First, if we are operating a "good" line (i.e., throughput greater than the practical worst case for any WIP level), then at typical WIP levels (e.g., between 5 and 10 times W_0) the cycle time will be very close to w/r_b, where w is the WIP level. (This can be observed in Figures 7.8 and 7.9.) Hence, increasing the bottleneck rate r_b will reduce cycle time for any given WIP level.

Unfortunately, there are times when it is physically or economically impractical to speed up the bottleneck. For example, suppose the copper plater is the bottleneck in the HAL plant we described at the beginning of the chapter. The rate at which this machine runs is governed by the chemistry of the process. Therefore, if it is already running for the maximum number of hours per day (i.e., it does not suffer from staffing or maintenance problems that could be resolved to increase the effective capacity), then the only way to increase capacity is to add another plater. This is an extremely expensive option that would probably be overkill, since it would result in a 100 percent increase in capacity. In a situation like this, it may make economic sense to consider increasing capacity of nonbottleneck resources.

To see this, consider a system with four single machine stations. Each station takes 10 minutes to perform a job except the last station (the bottleneck) which takes 15 minutes. Thus, the bottleneck rate is four jobs per hour.

Now, suppose we speed up the bottleneck to 10 minutes per job (6 jobs per hour), thereby balancing the line. Figure 7.10 illustrates the impact on the throughput versus WIP curve for the line. Notice that the improved line has a higher limiting production rate (a new r_b), but the throughput curve stays further from it than the original system. The reason is that a balanced line tends to starve its bottleneck more frequently than an unbalanced line, and hence requires more WIP for throughput to approach capacity. Nonetheless, speeding up the bottleneck causes throughput to increase for any WIP level.

Alternatively, suppose we speed up all of the nonbottleneck processes so that they require only five minutes, but keep bottleneck time at 15 minutes. Figure 7.11 shows that this also increases throughput for any WIP level. Indeed, for small WIP levels, the increase in throughput is actually greater than that achieved by speeding up the bottleneck. However, for large WIP levels (six or above), increasing the bottleneck rate achieves a greater increase in throughput than does the increase in nonbottleneck rates. Also we note that we made a bigger change to the nonbottleneck stations than we did to the bottleneck station (i.e., we cut the process time in half at three machines as opposed to reducing the time at a single machine by 33 percent). If we had the freedom to reduce any process time by five minutes, the best place to do it would be the bottleneck, *always!* But since this is not always possible (economical), it is good to know that performance gains can be achieved by improving nonbottleneck resources.

7.3.5   Internal Benchmarking

We now have the tools to reconsider the HAL example from the beginning of the chapter. We can evaluate the PCB line by comparing actual performance to the best, worst, and practical worst cases. To do this, we must estimate the bottleneck rate r_b and raw process

FIGURE 7.10

Change in throughput curve due to increase in bottleneck rate

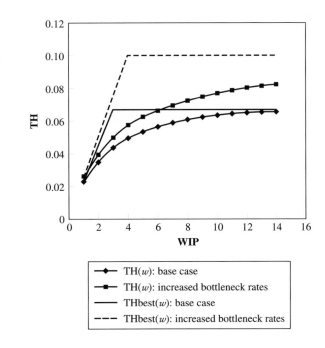

- ◆ TH(w): base case
- ■ TH(w): increased bottleneck rates
- — THbest(w): base case
- --- THbest(w): increased bottleneck rates

FIGURE 7.11

Change in throughput curve due to increase in rate of nonbottlenecks

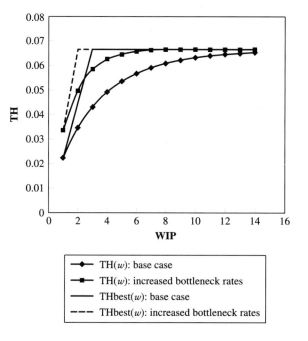

time T_0. If we ignore multiple visits to some workstations (e.g., lamination), since this was considered in the rate and time data, the bottleneck is simply the process with the smallest capacity. This is sizing with $r_b = 126.5$ panels per hour. The raw process time is simply the sum of the process times in Table 7.1, which is $T_0 = 33.1$ hours. Hence, the critical WIP for the line is

$$W_0 = r_b \times T_0 = 126.5 \times 33.1 = 4,187 \text{ panels}$$

Recalling that the actual throughput was 45.8 panels per hour, actual cycle time was 816 hours, and actual WIP level was 37,000 panels, we can make some quick observations. First we make a quick Little's law check of the data:

$$\text{TH} \times \text{CT} = 1,100 \text{ panels/day} \times 34 \text{ days} = 37,400 \text{ panels} \approx 37,000 \text{ panels}$$

Since Little's law applies precisely only to long-term averages, we would not expect it to hold exactly. However, this is certainly well within the precision of the data and hence suggests no problems.

Second, we place these actual measures in context by noting that throughput is $45.8/126.5 = 36$ percent of the bottleneck rate, cycle time is $816/33.1 = 24.6$ times the raw process time, and WIP is $37,000/4,187 = 8.8$ times critical WIP. None of these look very good. However, we must be careful about drawing conclusions from any single measure. For instance, simply knowing that the WIP level is 8.8 times critical WIP does not by itself mean that the line is performing poorly. Even a very good line will require high WIP to attain a throughput level close to the bottleneck rate. But when WIP is high *and* throughput is low, this is a bad sign. Just how bad can be determined by comparing to the practical worst case.

There are two ways we can compare actual performance to the PWC. One way is to compute the throughput level that would be achieved by a PWC line with the same r_b, T_0, and WIP level as the HAL line and to compare to actual throughput. Using the formula from the PWC definition, we get

$$\text{TH}_{\text{PWC}} = \frac{w}{W_0 + w - 1} r_b = \frac{37,400}{4,187 + 37,400 - 1}(126.5) = 113.8 \text{ panels per hours}$$

Actual throughput of 45.8 panels per hour is less than one-half this level, indicating performance that is much worse than that in the practical worst case.

Alternatively, we can compute what WIP level would be required in a PWC line with the same r_b and T_0 as the HAL line, to achieve the observed level of throughput. That is,

$$\text{TH}_{\text{PWC}} = \frac{w}{W_0 + w - 1} r_b = 45.8 = 0.36 r_b$$

which yields

$$\frac{w}{W_0 + w - 1} = 0.36$$

or
$$w = \frac{0.36}{0.64}(W_0 - 1) = 2,354 \text{ panels}$$

Actual WIP is more than 15 times this level, again indicating that the HAL line is far less efficient at converting WIP to throughput than the PWC.

We can put these calculations in graphical terms by plotting the best, worst, and practical worst throughput versus WIP curves and plotting the actual performance. This results in the graph in Figure 7.12. From this we can see dramatically that the WIP/throughput pair of $(37,400, 45.8)$ is well into the "bad" region between the worst and practical worst cases. Clearly, lines that exhibit such behavior offer much more opportunity for improvement than lines in the "good" region between the practical worst and best cases.

This example shows that the models presented in this chapter can help diagnose a production line and determine whether it is operating efficiently or not. But they do not tell us why a line is operating poorly and therefore do not help us determine how to improve it. For this, we require a deeper investigation of what causes some lines to be

FIGURE 7.12

Throughput versus WIP in HAL example

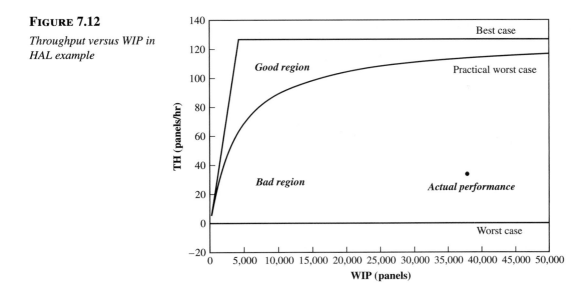

very efficient at converting WIP to throughput and others to be very inefficient. This is the subject of the next two chapters.

7.4   Labor-Constrained Systems

Throughout this chapter, we have focused on lines in which machines are the primary constraint. We have implicitly assumed that if there are human operators, they are assigned to machines and can therefore be viewed as part of the workstations. However, in some systems, workers perform multiple tasks or tend more than one workstation. These types of systems exhibit more complex behavior than the simple lines considered so far, since the flow of work is affected by the number and characteristics of both machines and operators.

Although the subject of flexible labor is much too broad for us to treat comprehensively here, we can make some observations about how labor-constrained lines relate to the simple lines presented earlier. We do this by considering three situations below.

7.4.1   Ample Capacity Case

We begin with the case in which labor is the only constraint on output. That is, we assume sufficient equipment at each workstation to ensure that a worker is never blocked for lack of a machine. While one might think that such a situation would never arise in practice, there are realistic situations that approximate this behavior. An example the authors encountered was that of a prepress graphical production facility of catalogs and other marketing materials. This firm received content (text, photos, etc.) from its clients and converted these to electronic engraving data via a series of steps (e.g., scanning, color correction, page finishing), which it then sent to a printer to be made into paper products. Most of the prepress steps required a computer along with some peripheral equipment. Because computer equipment was inexpensive relative to the cost of delays, the firm installed enough duplicates of each station to ensure that technicians virtually never had to wait for equipment to perform the various tasks. The result was many more machines than people, which meant that labor was the key constraint in the system.

A primary reason the graphics company installed ample capacity at its stations was to facilitate its flexible labor policy. Instead of having specialists for each operation, the company had cross-trained the workforce so that almost everyone could do almost every operation. This allowed the company to assign workers to jobs instead of stations. A worker would follow a job through the system, performing each operation on the appropriate workstation, as shown in Figure 7.13. The extra computers made it very

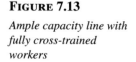

FIGURE 7.13

Ample capacity line with fully cross-trained workers

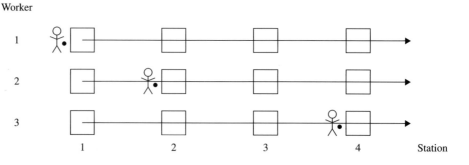

unlikely that someone would ever have to wait for equipment at a station. Having workers stay with a job all the way through the system meant that customers had a single person to contact and also made one person clearly responsible for quality.

In a system like this, capacity is defined by labor rather than equipment. To characterize capacity, we will continue to let T_0 represent the average time for one job to traverse the system, which we assume is independent of which worker is assigned to the job. Furthermore, we suppose that once a worker starts a job, he or she continues with it until it is done. Stopping work midway through a job cannot improve throughput and will only increase cycle time, so unless some customers have higher priority than others, there is no reason to do this. Under these assumptions, jobs are released into the system only when a worker becomes available, and since there is no blocking due to equipment, cycle time is always T_0. If there are n workers in the line, all working at the same rate, then each puts out a job every T_0 time units, which means that throughput is n/T_0.

Since the ample capacity case is an ideal situation, any changes to our assumptions can only decrease throughput. Examples of such changes include less-than-ample equipment so that blocking occurs, intermittent arrival of work that may cause starving, partial cross-training so that jobs may have to wait for a "specialist" at some stations, or any other change that prevents workers from being completely busy. Hence, we can state the following factory physics law.

Law (Labor Capacity): *The maximum capacity of a line staffed by n cross-trained operators with identical work rates is*

$$\text{TH}_{\max} = \frac{n}{T_0}$$

This law provides a way to introduce labor into the capacity calculations. For instance, in a line that has more stations than workers, the bottleneck rate of the equipment r_b may be a poor estimate of the capacity of the line. Where throughput is constrained by labor, n/T_0 may be a more realistic and useful upper bound on capacity. This bound is applicable to a wide range of systems, including those with fully or partially cross-trained workers.

One class of systems to which it does not apply, however, is that in which a worker can process more than one job simultaneously. For instance, a manufacturing cell where a single operator can tend several automated machines at the same time may have throughput exceed n/T_0. Such systems are often appropriately viewed as equipment-constrained, where operator unavailability acts as a capacity detractor and variability inflator. We will examine detractors in Chapter 8.

7.4.2   Full Flexibility Case

To deepen our insight into how both equipment and labor affect capacity, we next consider the case in which workers are completely cross-trained (i.e., can operate every station in the line). Furthermore, we begin by assuming that workers are tied to jobs as in the ample capacity case. However, unlike in the ample capacity case, equipment is limited so workers may become blocked, as shown in Figure 7.14. Once a worker finishes a job at the end of the line, she goes back to the beginning and starts a new one.

If the workers in Figure 7.14 have identical work rates, then this line is logically identical to the CONWIP lines we considered previously, except that the WIP level is now the number of workers. Hence, the behavior of the line will lie somewhere between the best and worst cases, with the practical worst case defining the division between

FIGURE 7.14

*Line with fully
cross-trained workers tied
to jobs*

good and bad lines. Furthermore, all the improvement strategies we listed earlier—increasing capacity, reducing line balance, using parallel machine stations, and reducing variability—still apply to this case.

The assumption of fully cross-trained workers who walk jobs all the way through the line may not be realistic in many situations. For instance, if the workstations require very different skills, it may make sense to have workers pass jobs from one to another. One mechanism is the *bucket brigade* (see Bartholdi and Eisenstein 1996). In this system, whenever the worker farthest downstream in the line completes a job, he or she moves up the line and takes the job from the next worker upstream. That worker in turn moves upstream and takes the job from the next worker. And so on, until the worker farthest upstream takes a new job. If all workers work at the same speed and there is no delay due to the handing off of the jobs, then there is no logical difference in this system from the one depicted in Figure 7.14. The line still operates as a CONWIP line with the WIP level set by the number of workers. Only the identities of the workers assigned to each job are changed.

While the bucket brigade system may not differ logically from the system with workers tied to jobs, it does differ practically. Each worker will tend to operate machines in a zone. Indeed, in the case where all process times are perfectly deterministic (i.e., the best case), the line will settle into a repetitive cycle where each worker processes jobs through the same sequence of stations. The cross-training and job transfers allow the line to balance itself so that each worker spends the same amount of time with a job. This type of system has been used effectively in automobile seat construction (see Chapter 10 for a discussion of this system at Toyota), warehouse picking, and fast-food sandwich construction (Subway).

Notice that blocking is still possible in the bucket brigades. Whenever an upstream worker catches up with the next worker downstream, she or he will be blocked unless the station has extra equipment. Hence, it makes sense to organize the workers so as to minimize the frequency with which this happens, by placing the fastest workers downstream and the slowest workers upstream. Bartholdi and Eisenstein (1996) show that this arrangement from slowest to fastest can significantly improve throughput and observed that this tends to be the practice in industry where such systems are used.

7.4.3   CONWIP Lines with Flexible Labor

If workers stay tied to jobs (or hand off jobs directly to one another as in the bucket brigade system), then the number of jobs in the system always equals the number of workers and the system behaves logistically as a CONWIP line. But in many, if not most, systems, the number of jobs will typically exceed the number of workers. If workers can rove through the system and work at different stations, then the performance of the system will depend on how effectively labor is allocated to promote flow through the system. This can get complex, since there are countless ways that labor can be dynamically allocated in the system.

One approach, which is a natural extension of the bucket brigade system to the case with more jobs than workers, is to have any worker who becomes free take the next job upstream, either from the upstream worker or from a buffer (see Figure 7.15 for an illustration of the mechanics). Whenever a worker becomes blocked because a downstream station is busy, the worker drops the job in the buffer in front of the station and moves upstream to get another job. This continues as long as the total number of jobs in the system does not exceed some preset limit (without such a limit, a fast worker at the front of the line would flood the line with WIP).

If all stations consist of single machines, so that no passing is possible, then at any time worker n (the last worker in the line) will be working on the job farthest downstream. Worker $n - 1$ will be working on the next-farthest job downstream that is not blocked by worker n. And so on. If passing on multimachine stations is possible, then the workers can get out of order. But the basic intent is still to keep workers working whenever possible on the jobs farthest downstream. Keeping workers busy tends to maximize throughput; working on downstream jobs tends to minimize cycle times. Hence, we would expect this policy to work reasonably well.

Of course, other flexible labor policies are possible. Which is appropriate depends on a variety of factors, including the degree of worker cross-training, the relative speed of the workers at the different stations, and the efficiency with which jobs can be passed from one worker to another. If there is no difference in the speed of workers, then the throughput of the system depends entirely on how often unblocked jobs are idle for lack of a worker. If this never happens, then the system will operate like a regular CONWIP line. If it happens so frequently that the workers might just as well be tied to one job each, then the system will operate as a CONWIP line with only as many jobs as workers. Hence, we can bound the throughput of a CONWIP line with flexible workers as in the following factory physics law.

Law (CONWIP with Flexible Labor): *In a CONWIP line with n identical workers and w jobs, where w ≥ n, any policy that never idles workers when unblocked jobs are available will achieve a throughput level* TH(w) *bounded by*

$$\text{TH}_{\text{CW}}(n) \leq \text{TH}(w) \leq \text{TH}_{\text{CW}}(w)$$

where $\text{TH}_{\text{CW}}(x)$ *represents the throughput of a CONWIP line with all machines staffed by workers and x jobs in the system.*

This law can give us some insight into the value of cross-training in a system. For instance, in a line with fixed workers, where the number of workers is at least equal to critical WIP and performance is close to the best case, there is clearly little benefit to cross-training. The reason is that the throughput of a CONWIP line with any WIP above critical WIP will be close to the bottleneck rate, so $\text{TH}_{\text{CW}}(n)$ will be approximately equal to $\text{TH}_{\text{CW}}(w)$. The reason is that because there is little variability in the system, there will not be many occasions in which the capability of moving workers between stations will be of value.

On the other hand, if a line has significant variability, then the potential improvement from cross-training can be substantial. To see this, consider a practical worst-case line

FIGURE 7.15

CONWIP line using bucket brigade with job dropping

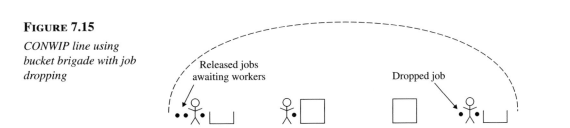

FIGURE 7.16

*Performance improvement
in a CONWIP line through
use of flexible labor*

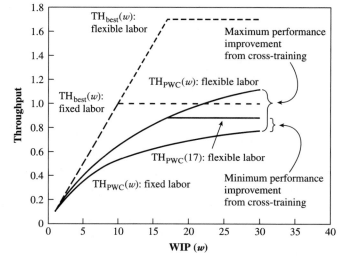

with bottleneck rate of $r_b = 1$ per hour and raw process time $t_0 = 10$ hours, which is staffed by $n = 17$ workers. Currently, the line is behaving as the practical worst case (the TH-versus-WIP curve in Figure 7.16 is labeled "$TH_{PWC}(w)$: Fixed Labor"). But suppose that we were to cross-train the workers so that they could staff any task. This would enable workers to shift to stations where they are needed when fluctuations in work require it. From the labor capacity law, we know that this could increase the effective capacity up to as much as $n/T_0 = 17/10 = 1.7$ jobs per hour. If it does and the line still behaves as in the practical worst case, then the throughput curve will shift upward accordingly. By the CONWIP with flexible labor law, the actual throughput will lie between $TH_{PWC}(n)$ and $TH_{PWC}(w)$. So while we cannot say exactly how large the performance improvement will be, it is clear that it is significant. The conclusion is that by dynamically balancing the line, the cross-trained workers are able to increase its effective capacity and thereby achieve increased output.

From our previous analyses, we know that systems with high process variability and a high degree of balance will tend to look more like the practical worst case than low-variability/low-balance systems. Hence, these conditions will tend to make cross-training attractive. The reason is that balanced, high-variability systems tend to starve workers who are tied to stations. Therefore, allowing workers to follow jobs prevents some of this starvation and hence increases throughput.

We should note, however, that while single-machine stations also tend to make systems behave as in the practical worst case, they do not generally make cross-training more attractive. Parallel machine stations actually facilitate flexible work policies by reducing the frequency with which workers are blocked for lack of a machine. In the extreme case, where there is sufficient parallel capacity to prevent blocking, the system can approach the behavior of the ample capacity case where labor becomes the only constraint.

7.5   Conclusions

In this chapter we examined the fundamental behavior of a single production line by studying the relationships among cycle time, WIP, throughput, and capacity. We observed the following:

1. A single line can be reasonably summarized by two independent parameters: the bottleneck rate r_b and the raw process time T_0. However, as we observed, a wide range of behavior is possible for lines with the same r_b and T_0. We will investigate the causes of this disparity in the next two chapters.

2. Little's law (WIP = TH $\times$ CT) provides a fundamental relationship between three long-term average measures of the performance of *any* production station, line, or system.

3. The best case defines the maximum throughput and minimum cycle time for a given WIP level for any line with specified values of r_b and T_0. The worst case defines the minimum throughput and maximum cycle time for any line with specified values of r_b and T_0. The practical worst case provides an intermediate scenario that serves as a useful demarcation between "good" and "bad" systems.

4. The critical WIP level, defined as $W_0 = r_b T_0$, represents a realistic ideal WIP level (as opposed to the unrealistic ideal of zero inventory, which would result in zero throughput). At W_0, a best-case (i.e., zero-variability) line achieves both maximum throughput (i.e., r_b) and minimum cycle time (i.e., T_0).

5. Both the best case and the worst case occur in systems with zero randomness. The worst case results from high variability caused by bad control rather than randomness. The practical worst case represents the maximum randomness situation.

6. When WIP levels are high, reducing raw process time T_0 has little effect on cycle times, while increasing r_b can have a great impact.

7. Other things being equal (that is, r_b and T_0 are the same), unbalanced lines exhibit less congestion than balanced lines.

8. Production lines can be constrained by a combination of equipment and labor. Equipment capacity is bounded by the bottleneck rate r_b, while labor capacity is bounded by n/T_0, where n is the number of workers in the line.

9. Systems with high process variability and balanced stations are most amenable to cross-training and flexible labor policies. In addition, parallel machine stations help facilitate flexible work policies.

A thread that has emerged from this analysis of basic factory dynamics is that a line can achieve the same thoughput at a lower WIP level by either increasing capacity or improving the efficiency of the line. As we hinted in our treatment of the practical worst case, a primary way of increasing line efficiency is by reducing variability at individual stations. To be able to evaluate the relative effectiveness of capacity increases versus variability reduction, we must further develop the science of factory physics to describe the behavior of production systems involving randomness. We do this next in Chapters 8 and 9.

Study Questions

1. Suppose throughput TH is near capacity r_b. Using Little's law, relate
   a. WIP and cycle time in a production line
   b. Finished goods inventory and time spent in finished goods inventory
   c. The number of cars waiting at a toll booth and the average wait time

2. Is it possible for a line to have the same throughput with both high WIP with high cycle time and low WIP with low cycle time? Which would you rather have? Why?

3. For a given set of production line characteristics (i.e., raw process time T_0 and bottleneck rate r_b) and a given WIP level w, what is the best cycle time that can be achieved? What is the "worst"? What is the corresponding throughput for these two cases?

4. What are the conditions for the practical worst-case throughput? What types of behavior can lead to performance worse than that in this case? What would this do to throughput? To cycle times?

5. Can the critical WIP level W_0 ever exceed the number of machines in the line?

Problems

1. Consider a four-station line in which all stations consist of single machines. Station 2 has average processing times of two hours per job, while the remaining stations have average processing times of one hour per job. Answer the following, under the assumption that the line behaves according to the best case.
   a. What are r_b and T_0 for this line?
   b. How do r_b and T_0 change if a second identical machine is added to station 2? What effects will this have on performance?
   c. How do r_b and T_0 change if the machine at station 2 is speeded up to have average processing times of one hour? What effects will this have on performance?
   d. How do r_b and T_0 change if a second identical machine is added to station 1? What effects will this have on performance?
   e. How do r_b and T_0 change if the machine at station 1 is speeded up to have average processing times of one-half hour? What effects will this have on performance? Do your results agree or disagree with the statement "An hour saved at a nonbottleneck is a mirage (i.e., of no value)"?

2. Repeat Problem 1 under the assumption that the line behaves according to the worst case.

3. Repeat Problem 1 under the assumption that the line behaves according to the practical worst case.

4. Consider the following three-station production line with a single product that must visit stations 1, 2, and 3 in sequence:

   • Station 1 has 5 identical machines with average processing times of 15 minutes per job.
   • Station 2 has 12 identical machines with average processing times of 30 minutes per job.
   • Station 3 has 1 machine with average processing times of 3 minutes per job.

   a. What are the bottleneck rate r_b, the raw process time T_0, and the critical WIP w_0?
   b. Compute the average cycle time when the WIP level is set at 20 jobs, under the assumptions of
      i. the best case
      ii. the worst case
      iii. the practical worst case
   c. We desire the throughput to be 90 percent of the bottleneck rate. Find the WIP level required to achieve this under the assumptions of
      i. the best case
      ii. the worst case
      iii. the practical worst case
   d. If the cycle time at the critical WIP is 100 minutes, where does performance fall relative to the three cases? Is there much room for improvement?

5. Positively Rivet Inc. is a small machine shop that produces sheet metal products. It had one line dedicated to the manufacture of light-duty vent hood shells, but because of strong demand it recently added a second line. The new line makes use of higher-capacity automated equipment but consists of the same basic four processes as the old line. In addition, the new line makes use of one machine per workstation, while the old line has parallel machines at the workstations. The processes, along with their machine rates, number of machines per station, and average times for a lone job to go through a station (i.e., not including queue time), are given for each line in the following table:

Process	Old Line			New Line		
	Rate per Machine (parts/hour)	**Number Machines per Station**	**Time (minute)**	**Rate per Machine (parts/hour)**	**Number Machines per Station**	**Time (minute)**
Punching	15	4	4.0	120	1	0.50
Braking	12	4	5.0	120	1	0.50
Assembly	20	2	3.0	125	1	0.48
Finishing	50	1	1.2	125	1	0.48

Over the past three months, the old line has averaged 350 parts per day, where one day consists of one eight-hour shift, and has had an average WIP level of 400 parts. The new line has averaged 680 parts per eight-hour day with an average WIP level of 350 parts. Management has been dissatisfied with the performance of the old line because it is achieving lower throughput with higher WIP than the new line. Your job is to evaluate these two lines to the extent possible with the above data and identify potentially attractive improvement paths for each line by addressing the following questions.

a. Compute r_b, T_0, and W_0 for both lines. Which line has the larger critical WIP? Explain why.

b. Compare the performance of the two lines to the practical worst case. What can you conclude about the relative performance of the two lines compared to their underlying capabilities? Is management correct in criticizing the old line for inefficiency?

c. If you were the manager in charge of these lines, what option would you consider first to improve throughput of the old line? Of the new line?

6. Floor-On, Ltd., operates a line that produces self-adhesive tiles. This line consists of single-machine stations and is almost balanced (i.e., station rates are nearly equal). A manufacturing engineer has estimated the bottleneck rate of the line to be 2,000 cases per 16-hour day and the raw process time to be 30 minutes. The line has averaged 1,700 cases per day, and cycle time has averaged 3.5 hours.

a. What would you estimate average WIP level to be?

b. How does this performance compare to the practical worst case?

c. What would happen to the throughput of the line if we increased capacity at a nonbottleneck station and held WIP constant at its current level?

d. What would happen to the throughput of the line if we replaced a single-machine station with four machines whose capacity equaled that of the single machine and held the WIP constant at its current level?

e. What would happen to the throughput of the line if we began moving cases of tiles between stations in large batches instead of one at a time?

7. T&D Electric manufactures high-voltage switches and other equipment for electric utilities. One line that is staffed by three workers assembles a particular type of switch. Currently the three workers have fixed assignments; each worker fastens a specific set of components onto the switch and passes it downstream on a rolling conveyor. The conveyor has capacity to allow a queue to build up in front of each worker. The bottleneck is the middle station with a rate of 11 switches per hour. The raw process time is 15 minutes. To improve the efficiency of the line, management is considering cross-training the workers and implementing some sort of flexible labor system.

a. If current throughput is 10.5 switches per hour with an average WIP level of five jobs, how much potential do you think there is for a flexible work system?

b. If current throughput is eight switches per hour with an average WIP level of seven jobs, how much potential do you think there is for a flexible work system?

c. If all three workers were fully cross-trained and equipped to assemble the entire switch in parallel (i.e., no passing of jobs to one another) and were able to maintain the current work pace of each operation, what would the capacity of the system be? What real-world problems might make such a policy unattractive?

   d. Suggest a flexible work system that could improve the efficiency of a line like this with less than full cross-training (i.e., with workers trained and equipped to assemble only certain components).

8. Consider a balanced line consisting of five single-machine stations with exponential process times. Suppose the utilization is 75 percent and the line runs under the CONWIP protocol (i.e., a new job is started each time a job is completed).

   a. What is the WIP level in the line?

   b. What is the cycle time as a percentage of T_0?

   c. What happens to WIP, CT, and TH relative to the original system if you make each of the following changes (one at a time)?

      i. Increase the WIP level

     ii. Decrease the variability of one station

    iii. Decrease the capacity at one station

    iv. Increase the capacity of all stations

Intuition-Building Exercises

1. Simulate Penny Fab Two by taking a piece of paper and drawing a schematic of the line (see Figure 7.17). Draw the squares large enough to contain a penny. To the right of each square, write the time of the completion of the job occupying that square (as the simulation progresses, you will cross out the old time and replace it with the next time). The simulation progresses by setting the current "simulated time" to be the earliest completion time and moving the pennies accordingly.

   a. Run your simulation for several simulated hours with seven pennies. Note how the second station sometimes starves.

FIGURE 7.17

Penny Fab Two with
$w = 9$, 22 hours into the
simulation

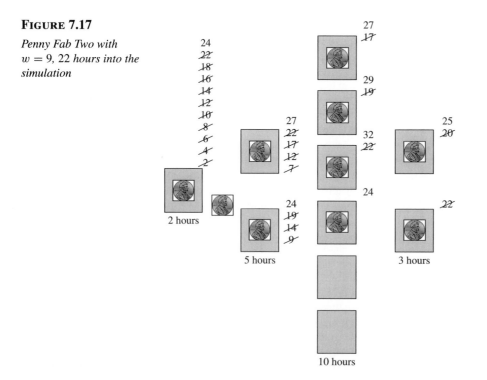

    b. Run your simulation for several simulated hours with eight pennies. Observe that station 2 never starves and there is never any queueing once the initial transient queue is dissipated in front of the first station.

    c. Run your simulation for several simulated hours with nine pennies (Figure 7.17 illustrates this scenario after 22 simulated hours). Note that after the initial transient, there is always a queue in front of second station.

2. Simulate Penny Fab Two for 25 hours starting with an empty line and eight pennies in front. Record the cycle time of each penny that finishes during this time (i.e., record its start time and finish time and compute cycle time as the difference).

    a. What is the average cycle time CT?

    b. How many jobs finish during the 25 hours?

    c. What is the average throughput TH over 25 hours? Does average WIP equal CT times TH? Why or why not? (*Hint:* Did Little's law hold for the first two hours of our simulation of Penny Fab One?) What does this tell you about the use of Little's law over short time intervals?

8   VARIABILITY BASICS

God does not play dice with the universe.
Albert Einstein

Stop telling God what to do.
Niels Bohr

8.1   Introduction

Little's law (TH = CT/WIP) implies that it is possible to achieve the same throughput with long cycle time and large WIP or short cycle time and small WIP. Of course, the short-cycle-time, low-WIP system is preferable. But what causes the difference? The answer, in a great many instances, is *variability*.

Penny Fab One from Chapter 7 achieves full throughput (one-half job per hour) at a WIP level of $W_0 = 4$ jobs (the critical WIP) if it behaves like the best case. But if it behaves like the practical worst case, it requires a WIP level of 27 jobs to achieve 90 percent of capacity (57 jobs to achieve 95 percent of capacity). If it behaves like the worst case, 90 percent of capacity is not even feasible. Why the big difference? *Variability!*

Briar Patch Manufacturing has two very similar workstations as part of its plant. Both are composed of a single machine that runs at a rate of four jobs per hour (when it is not down). Both are subject to the same pattern of demand with an average work load of 69 jobs per day (2.875 jobs per hour). And both are subject to periodic unpredictable outages. However, for one workstation, consisting of a Hare X19 machine, outages are rather infrequent but tend to be quite long when they occur. For the other station, consisting of a Tortoise 2000 machine, outages are much more frequent and correspondingly shorter. Both machines have an *availability* (i.e., the long-term fraction of the time that the machine is not down for repair) of 75 percent. Thus, the capacity of both stations is $4(0.75) = 3$ jobs per hour. Since the two stations have the same capacity and are subject to the same demand, they should have the same performance—cycle time, WIP, lead time, and customer service—right? Wrong! It turns out that the Hare X19 is substantially worse on all measures than the Tortoise 2000. Why? Again, the answer is *variability!*

Variability exists in all production systems and can have an enormous impact on performance. For this reason, the ability to measure, understand, and manage variability is critical to effective manufacturing management. In this chapter we will develop basic tools and intuition for characterizing variability in production systems. In the next chapter, we probe more deeply into the manner in which variability degrades system performance and how it can be managed.

8.2   Variability and Randomness

What, exactly, is variability? A formal definition is the *quality of nonuniformity of a class of entities.* For example, a group of individuals who all weigh exactly the same have no variability in weight, while a group with vastly different weights is highly variable in this regard. In manufacturing systems, there are many attributes in which variability is of interest. Physical dimensions, process times, machine failure/repair times, quality measures, temperatures, material hardness, setup times, and so on are examples of characteristics that are prone to nonuniformity.

Variability is closely associated with (but not identical to) **randomness.** Therefore, to understand the causes and effects of variability, one must understand the concept of randomness and the related subject of **probability.** In this chapter we develop the necessary ideas in as loose and intuitive a manner as possible. However, for precision, there are points at which we must invoke the formal language of probability. In particular, the concept of a **random variable** and its characterization via its **mean and standard deviation** are essential. The reader for whom this terminology is new or rusty should refer to the review of basic probability in Appendix 2A before proceeding with this chapter.

As mentioned above, both the worst and practical worst cases represent systems whose performance is degraded by variability. However, the variability in the worst case is completely predictable—a consequence of *bad control*—while the variability in the practical worst case is due to unpredictable randomness. To understand the difference, we must distinguish between controllable variation and random variation.

Controllable variation occurs as a direct result of decisions. For instance, if several products are produced in a plant, there will be variability in the product descriptors (e.g., their physical dimensions, time to manufacture, etc.). Likewise, if material is moved in batches from one process to the next, the first part to finish will have to wait longer to move than the last part, and so waiting times will be more variable than if moved one at a time.

In contrast, **random variation** is a consequence of events beyond our immediate control. For example, the times between customer demands are not generally under our control. Thus, we should expect the load at any particular workstation to fluctuate. Likewise, we do not know when a machine might fail. Such downtime adds to the effective process time of a job, since the job must wait for the machine to be repaired before completing processing. Since such contingencies cannot be predicted or controlled (at least in the immediate term), machine outages increase the variability of effective process times in a random fashion.

Although both types of variation can be disruptive to a plant, the effects of random variation are more subtle and require more sophisticated tools to describe. For this reason, we will focus mainly on random variation in this chapter.

8.2.1   The Roots of Randomness

Unfortunately, the very notion of randomness gives most people (including philosophers) trouble. How can something occur that is independent of its initial conditions? Does this not violate the notion of cause and effect? While it is beyond our scope to discuss this philosophical dilemma thoroughly, it is interesting to make some basic observations about the nature of randomness.

One interpretation of randomness is that because we have imperfect (or incomplete) information, systems *appear* to behave randomly. The underlying premise of this view is that if we knew all the laws of physics and had a complete description of the universe at some time, then, in theory, we could predict every detail of its evolution from then on with certainty.

A second interpretation is that the universe actually *behaves* randomly. In other words, having a complete description of the universe and the laws of physics is not enough to predict the future. At best, these can provide only *statistical* estimates of what will happen. Furthermore, identical starting conditions may not yield identical futures. Because of the apparent violation of the principle of cause and effect, this viewpoint has been roundly criticized in philosophy circles. However, its proponents have pointed out that the cause-and-effect principle can be recovered by defining other, more fundamental quantities that are not affected by randomness.[1]

The debate between these two schools of thought became quite heated within the physics community during the early part of the 20th century. Einstein sided with the first view (incomplete knowledge) and stated emphatically that "God does not play dice." Bohr and others believed in the second (random universe) view and suggested that Einstein "not tell God what to do" (see Planck 1936 for a discussion of this controversy). In recent years, experimental evidence has tended to side with the random universe view, much to the distaste of some philosophers.

Regardless of whether randomness is elemental or due to a lack of knowledge, the effects are the same—many facets of life, including manufacturing management, are inherently unpredictable. This means that the results of management actions can never be guaranteed. In fact, starting with the same conditions and using the same control policy on different days may well lead to different outcomes.

This does not mean that we should give up on managing the factory, only that we need to be concerned with finding *robust* policies. A robust policy is one that works well *most of the time*. This differs from an *optimal* policy, which is the best policy for a specific set of conditions. A robust policy is almost never optimal but is usually "pretty good." In contrast, an optimal policy may work extremely well for the set of conditions for which it was designed, but perform very poorly for many others. The most powerful tool a manager can have for identifying effective and robust policies in the face of randomness is *good probabilistic intuition*. Unfortunately, such intuition appears to be rare. A major goal of this chapter is to develop this critical skill.

8.2.2   Probabilistic Intuition

Intuition plays an important part in many aspects of our everyday lives. Most decisions we make are based upon some form of intuition. For instance, we slow down when making turns in an automobile because of our intuition developed after driving for some time, rather than our detailed understanding of automotive physics. We decide whether to refinance our house by appealing to our intuition about the economy, rather than a formal economic analysis. We time our request for a raise according to our intuitive sense of the boss's mood, rather than deep theory about his or her psychological profile.

In many situations, our intuition is quite good with respect to "first-order" effects. For example, if we speed up the bottleneck (busiest workstation) in a production line,

---

[1]Quantities known as *quantum numbers* are well-defined and determine the probability distributions of random observables, such as location and velocity, instead of actual outcomes.

without changing anything else, we expect to get out more product. This type of intuition typically comes from acting as though the world were **deterministic,** that is, without randomness. In the language of probability and statistics, such reasoning is based on the **first moment** or the **mean** (average) of the random variables involved. As long as the change in the mean quantity (e.g., increase in average speed of a machine) is large relative to the randomness involved, first-order intuition usually works well.

Our intuition tends to be much less developed for second moments (i.e., for quantities involving the variance of random variables). For instance, which is more variable, the time to process an individual part or the time to process a batch of parts? Which are more disruptive, short, frequent machine failures or long, infrequent ones? Which will result in a greater improvement in line performance, reducing the variability of process times at stations near the front of the line or near the back? These and other variability-related questions concerning plant behavior require much more subtle intuition than that required to see that speeding up the bottleneck will improve throughput.

Because people frequently lack well-developed intuition regarding second moments, they often misinterpret random phenomena. A typical example occurs in the classroom when students who made low grades on a first examination show relative improvement on the second examination, while students who made high scores on the first examination do worse on the second. This is an example of the phenomenon known as **regression to the mean.** An extreme score (high or low) on the first examination is likely to be at least partially due to randomness (e.g., lucky or unlucky guesses, a headache on test day, etc.). Since the random effects for a given student are unlikely to be extreme twice in a row, the student with an extreme score on the first examination is likely to have a more moderate score on the second. Unfortunately, many teachers interpret these results as a sign that they have finally reached the slower students and are beginning to lose the better ones. In reality, simple randomness may well account for the effect.

Misinterpretation of the general tendency for regression to the mean also occurs among manufacturing managers. After a particularly slow period of output, a manager may react with harsh appraisals and disciplinary action. Sure enough, production goes up. Similarly, after outstanding performance and much praise, production declines—clear evidence that the workers have grown complacent. Of course, the same behavior—better following bad and worse following good—is likely to happen *even when there has been no change,* whenever randomness is present.

In addition to the first two moments (mean and variance), random phenomena are influenced by the third (skewness), the fourth (kurtosis), and higher moments. The effects of these higher moments are generally much less pronounced than those associated with the first two, so we will focus on only the mean and the variance. Furthermore, as noted above, since effects associated with the mean are fairly intuitive, while effects associated with the variance are much more subtle, we will devote particular attention to understanding variance.

8.3   Process Time Variability

The random variable of primary interest in factory physics is the **effective process time** of a job at a workstation. We use the label *effective* because we are referring to the total time "seen" by a job at a station. We do this because from a logistical point of view, if machine B is idle because it is waiting for a job to finish on machine A, it does not matter whether the job is actually being processed or is being held up because machine A is being repaired, undergoing a setup, reworking the part due to a quality problem, or waiting for its operator to return from a break. To machine B, the effects are the same. For this reason, we will combine these and other effects into one aggregate measure of variability.

8.3.1   Measures and Classes of Variability

To effectively analyze variability, we must be able to quantify it. We do this by using standard *measures* from statistics to define a set of factory physics variability *classes*.

Variance, commonly denoted by σ^2 (sigma squared), is a measure of *absolute* variability, as is the **standard deviation** σ, defined as the square root of the variance. Often, however, absolute variability is less important than *relative* variability. For instance, a standard deviation of 10 micrometers (μm) would indicate extremely low variability in the length of bolts with a nominal length of two inches, but would represent a very high level of variation for line widths on a chip whose mean width is five micrometers. A reasonable relative measure of the variability of a random variable is the standard deviation divided by the mean, which is called the **coefficient of variation (CV).** If we let t denote the mean (we use t because the primary random variables we are considering here are times) and σ denote the variance, the coefficient of variation c can be written

$$c = \frac{\sigma}{t}$$

In many cases, it turns out to be more convenient to use the **squared coefficient of variation (SCV)**

$$c^2 = \frac{\sigma^2}{t^2}$$

We will make extensive use of the CV and the SCV for representing and analyzing variability in production systems. We will say that a random variable has **low variability (LV)** if its CV is less than 0.75, that it has **moderate variability (MV)** if its CV is between 0.75 and 1.33, and that it has **high variability (HV)** if the CV is greater than 1.33. Table 8.1 presents these cases and provides examples.

8.3.2   Low and Moderate Variability

When we think of process times, we tend to think of the actual time that a machine or an operator spends on the job (i.e., not including failures or setups). Such times tend to have probability distributions that look like the classic bell-shaped curve. Figure 8.1 shows the probability distribution for process times with a mean of 20 minutes and a standard deviation of 6.3 minutes. Notice how most of the area under the curve is symmetrically distributed around 20. The CV for this case is around 0.32, so it is in the low variability (LV) range. It is a characteristic of most LV process times to have a bell-shaped probability density.

TABLE 8.1   Classes of Variability

Variability Class	Coefficient of Variation	Typical Situation
Low (LV)	$c < 0.75$	Process times without outages
Moderate (MV)	$0.75 \leq c < 1.33$	Process times with short adjustments (e.g., setups)
High (HV)	$c \geq 1.33$	Process times with long outages (e.g., failures)

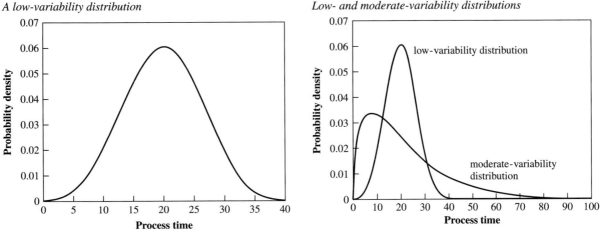

Figure 8.1

A low-variability distribution

Figure 8.2

Low- and moderate-variability distributions

Now consider a situation with a mean process time of 20 minutes but for which the CV is around 0.75, the beginning of the moderate-variability case. An example might be process times of a manual operation in which most of the time the operation is easy but occasionally difficulties occur. Figure 8.2 compares the two distributions. Notice that the LV case has most of its probability concentrated near the mean of 20. In the moderate-variability (MV) case, the most likely times are actually lower than the mean, around nine minutes. However, while the LV plot tails off around 40, the MV plot does not do so until around 80. Thus the means are the same, but the variances are much different. As we will see, this difference is critical to the operational performance of a workstation.

To get a sense of the operational effects of variability, suppose the LV process is feeding the MV process. For a while, the MV process will be able to keep up easily. However, once a long process time occurs, a queue of work begins to build in front of the second process. Offhand we might think that the long process times will be offset by the short process times, but this does not happen. A string of short process times at the second station might deplete the queue, causing the second station to become idle. When this occurs, capacity is lost and cannot be "saved up" for the next period of longer process times.[2]

Another way to look at this is to note that when one process feeds another, what comes in must go out; that is, there is **conservation of material.** Unless we turn off the stream of work from the first process whenever the second process is full (a procedure called *blocking* and one which we will discuss later), the amount of work in front of the second process can grow freely. Since there are times when the second station runs much faster than the first and since the *average* rate out must equal the average rate in, there will tend to be a queue of work.

We will discuss this more fully in Section 8.6. For now, we note that the greater the variability in effective process times, the greater the average queue. Given Little's law, this also implies that the greater the variability, the longer the cycle time.

---

[2]In the moderate-variability process shown in Figure 8.2, 20 percent of the process times are nine minutes or less, and another 20 percent are 31 minutes or more. For the mean to remain at 20, both have to occur.

8.3.3   Highly Variable Process Times

It may be hard to imagine process times whose CV is greater than 1.33. However, it is easy to construct *effective* process times with this much variability. Suppose a machine has an average process time of 15 minutes with a CV of 0.225 when there are no outages. This would be less variable than the previous low-variability case. But now suppose the machine has outages that average 248 minutes and occur, on average, after 744 minutes of production. We can show (details are given later) that this results in an effective mean process time of 20 minutes (as before) and an effective CV of a whopping 2.5! Figure 8.3 compares this high-variability (HV) distribution with the previous LV distribution. Because the HV distribution is taller and thinner, at first glance, it might appear less variable than the LV distribution. This is because we cannot see what is happening farther out in time. Once past 40 minutes or so, the picture changes. Figure 8.4 compares the distributions on a different scale for time greater than 40 minutes. Here we see the LV distribution immediately drops to almost no probability while the HV distribution appears almost uniform. It is going down very slowly indeed. This implies that there is a small probability that the process times will be extremely long. It is also the reason that the distribution for the highly variable process times appears to have a lower mean on the other plot. Most of the time, it takes around 15 minutes. However, about 1 out of every 50 jobs takes around 17 times as long. This inflates the mean to around 20 and drives the CV up to 2.5.

The effect of this level of variability on the production line can be severe. For instance, suppose the throughput is one job every 22 minutes. There should be no problem from a capacity perspective since the average process time *including* outages is 20 minutes. However, an outage of 250 minutes will build up a queue of almost 12 jobs. When the machine comes back up, the rate at which this queue is depleted is $\frac{1}{15} - \frac{1}{22} \simeq \frac{1}{47}$. Thus, the time to clear the queue formed would be around 536 minutes, *assuming no more outages occur!* If an outage occurs during this time, it adds to the queue. Under conditions commonly found with complex equipment (i.e., times to failure that are exponentially distributed), the probability of such an outage is $1 - e^{-536/744} = 0.51$. This means that more than 50 percent of the time an outage occurs before the queue would be cleared. Thus the average queue will be greater than 12 jobs and is, in fact, around 20 (as we will see in Section 8.6).

FIGURE 8.3

Comparison of high- and low-variability distributions

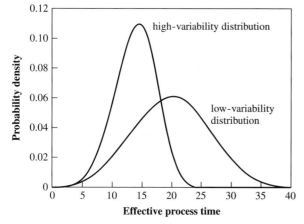

FIGURE 8.4

Comparison of high- and low-variability distributions above 40 minutes

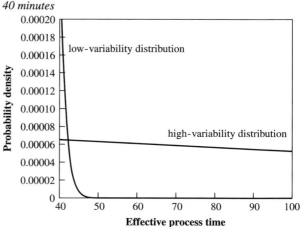

8.4   Causes of Variability

To identify strategies for managing production systems in the face of variability, it is important to first understand the causes of variability. The most prevalent sources of variability in manufacturing environments are:

- "Natural" variability, which includes minor fluctuations in process time due to differences in operators, machines, and material.
- Random outages.
- Setups.
- Operator availability.
- Recycle.

We discuss each of these separately below.

8.4.1   Natural Variability

Natural variability is the variability inherent in **natural process time,** which excludes random downtimes, setups, or any other external influences. In a sense, this is a catch-all category, since it accounts for variability from sources that have not been explicitly called out (e.g., a piece of dust in the operator's eye). Because many of these unidentified sources of variability are operator-related, there is typically more natural variability in a manual process than in an automated one. But even in the most tightly controlled processes, there is always some natural variability. For instance, in fully automated machining operations, the composition of the material might differ, causing processing speed to vary slightly.

We let t_0 and σ_0 denote the mean and standard deviation, respectively, of natural process time. Thus, we can express the coefficient of variation of natural process time as

$$c_0 = \frac{\sigma_0}{t_0}$$

In most systems, natural process times are LV and so $c_0 < 0.75$.

Natural process times are only the starting point for evaluating effective process times. In any real production system, workstations are subject to various **detractors,** including machine downtime, setups, operator unavailability, and so on. As discussed earlier, these detractors serve to inflate both the mean *and* the standard deviation of effective process time. We now provide a way to quantify this effect.

8.4.2   Variability from Preemptive Outages (Breakdowns)

In the high-variability example discussed earlier, we saw that unscheduled downtimes can greatly inflate both the mean and the CV of effective process times. Indeed, in many systems, this is the single largest cause of variability. Fortunately, there are often practical ways to reduce its effects. Since this is a common problem, we will discuss it in detail.

We refer to breakdowns as **preemptive outages** because they occur whether we want them to or not (e.g., they can occur right in the middle of a job). Power outages, operators being called away on emergencies, and running out of consumables (e.g., cutting oil) are other possible sources of preemptive outages. Since these have similar effects on the behavior of production lines, it makes sense to combine them and treat them all as machine breakdowns in the fashion discussed (i.e., include outages due to these other sources, as well as true machine breakdowns, when computing MTTF and MTTR). We discuss **nonpreemptive outages** (i.e., stoppages that occur between, rather than during, jobs) in the next section.

To see how machine outages cause variability, let us return to the Briar Patch Manufacturing example and provide some numerical detail. Both the Hare X19 and the Tortoise 2000 have a natural process time mean of $t_0 = 15$ minutes and a *natural* standard deviation of $\sigma_0 = 3.35$ minutes. Thus, both stations have a natural CV of $c_0 = \sigma_0/t_0 = 3.35/15.0 = 0.05$. Both machines are subject to failures and have the same long-term availability (i.e., fraction of uptime) of 75 percent. However, the Hare X19 has long but infrequent outages, while the Tortoise 2000 has short, frequent ones. Specifically, the Hare X19 has a mean time to failure (MTTF), denoted by m_f, of 12.4 hours, or 744 minutes, and a mean time to repair (MTTR), denoted by m_r, of 4.133 hours, or 248 minutes. The Tortoise 2000 has an MTTF of $m_f = 1.90$ hours, or 114.0 minutes, and MTTR of $m_r = 0.633$ hours, or 38.0 minutes. Note that the times to failure and times to repair are both three times greater for the Hare X19 than for the Tortoise 2000. Finally, we suppose that repair times are variable and have CV $= 1.0$ (moderate variability) for both machines.

Most capacity planning tools used in industry account for random outages when computing *average* capacity. This is done by computing the **availability,** which is given in terms of m_f and m_r by

$$A = \frac{m_f}{m_f + m_r} \tag{8.1}$$

Hence, for both machines, the availability A is

$$A = \frac{744}{744 + 248} = \frac{114}{114 + 38} = 0.75$$

Adjusting the natural process time t_0 to account for the fraction of time the machine is unavailable results in an **effective mean process time** t_e of

$$t_e = \frac{t_0}{A} \tag{8.2}$$

So in both cases, $t_e = 20$ minutes. Recall that in Chapter 7 we derived the capacity of a workstation to be the number of machines m divided by the effective mean process time. So if r_0 is the natural capacity (rate), then the **effective capacity** (rate) r_e is

$$r_e = \frac{m}{t_e} = A\frac{m}{t_0} = Ar_0 = 0.75(4 \text{ jobs/hour}) = 3 \text{ jobs/hour} \tag{8.3}$$

So the effective capacity of the Hare X19 and the Tortoise 2000 is the same. Since almost all maintenance systems used in industry to analyze breakdowns only consider the effects on availability and capacity, the two workstations would generally be regarded as equivalent.

However, when we include variability effects, the workstations are very different. To see why, consider how they will behave as part of a production line. If the Hare X19 fails for 12.4 hours (its average failure duration), it will need 12.4 hours of WIP to keep from starving. On the other hand, the Tortoise 2000 needs less than one-sixth as much WIP to be covered for an average-length failure. Since failures are, by their very nature, random, the WIP in the downstream buffer must be maintained at all times to provide protection against throughput loss. Clearly, a line with the Tortoise 2000 will be able to achieve the same level of protection, and hence the same level of throughput, with less

WIP, than same line with the Hare X19.[3] The net effect is that the line with the Hare X19 will be less efficient (i.e., will achieve lower throughput for a given WIP level or will have higher WIP and cycle time for the same throughput) than the line with the Tortoise 2000.

Earlier, we stated that the CV for the Hare X19 was 2.5. We obtained this by using a mathematical model, which we now describe. We assume the times to failures are exponentially distributed (i.e., they are MV).[4] However, we make no particular assumptions about the repair times other than that they are from some probability distribution. We define σ_r to be the standard deviation of these repair times and $c_r = \sigma_r/m_r$ to be the CV. In our example c_r is 1.0 (i.e., we assume repair times have moderate variability).

Under these assumptions we can calculate the mean, variance, and squared coefficient of variation (SCV) of the effective process time (t_e, σ_e^2, and c_e^2, respectively) as

$$t_e = \frac{t_0}{A} \tag{8.4}$$

$$\sigma_e^2 = \left(\frac{\sigma_0}{A}\right)^2 + \frac{(m_r^2 + \sigma_r^2)(1 - A)t_0}{Am_r} \tag{8.5}$$

$$c_e^2 = \frac{\sigma_e^2}{t_e^2} = c_0^2 + (1 + c_r^2)A(1 - A)\frac{m_r}{t_0} \tag{8.6}$$

The CV of effective process time c_e can be computed by taking the square root of c_e^2.

Notice that the mean effective process time, given by Equation (8.4), depends only on the mean natural process time and the availability and is hence the same for both stations:

$$t_e = \frac{t_0}{A} = \frac{15}{0.75} = 20.0 \text{ minutes}$$

However, the SCV of effective process time in Equation (8.6) depends on more than the mean process time and availability. To understand the effects involved, we can rewrite (8.6) as

$$c_e^2 = c_0^2 + A(1 - A)\frac{m_r}{t_0} + c_r^2 A(1 - A)\frac{m_r}{t_0}$$

The first term is due to the natural (unaccounted for) variability in the process. The second term is due to the fact that there are random outages. Note that this term would be there even if the outages themselves (i.e., the repair times) were constant (i.e., even if $c_r = 0$). For instance, a periodic adjustment that always takes the same time to complete would have $c_r^2 = 0$. Thus eliminating variability in repair time will do nothing to reduce this term. However, the last term is due explicitly to the variability of the repair times and would vanish if this variability were eliminated. Notice that both of the second two terms are increasing in m_r for a fixed availability. Hence, all other things being equal, long repair times induce more variability than short ones.

---

[3]Actually, the line with the Hare X19 will require more than 12.4 hours of WIP, and the line with Tortoise 2000 will require more than 4.133 hours of WIP, because these are only *average* downtimes. But the point remains the same: The line with the Hare X19 requires substantially more WIP to achieve the same throughput as the line with the Tortoise 2000.

[4]This is frequently a good assumption in practice, particularly for complex equipment since such machines tend to be combinations of old and new components. Thus, the memoryless property of the exponential tends to hold for the time between *any* outage, which could be caused by failure of an old component or a new one.

Substituting numbers into these equations yields

$$c_e^2 = 0.05 + (1+1)0.75(1-0.75)\frac{248}{15} = 6.25$$

or $c_e = 2.5$, which shows that the Hare X19 is well up in the HV range. However, the Tortoise 2000 has

$$c_e^2 = 0.05 + (1+1)0.75(1-0.75)\frac{38}{15} = 1.0$$

and so $c_e = 1$, which shows that it is in the MV range.

Hence a line with the Hare X19 will exhibit much more variability than one with the Tortoise 2000. How this affects WIP and cycle time will be explored more fully in Section 8.6.

This analysis leads to the conclusion that a machine with frequent but short outages is preferable to one with infrequent but long outages, provided that the availabilities are the same. This may be somewhat contrary to our nonprobabilistic intuition, which might suggest that we would be better off with a major headache once per month than a minor throb every day. But logistically speaking, the daily throb is easier to manage.

This is a potentially valuable insight, since in practice we may be able to convert long, infrequent failures to shorter, more frequent ones (e.g., through preventive maintenance procedures). However, lest the reader become complacent—no failures at all are even better than short, frequent ones. Nothing here should be construed to deflect attention from efforts to improve overall reliability.

8.4.3   Variability from Nonpreemptive Outages

Nonpreemptive outages represent downtimes that will inevitably occur but for which we have some control as to exactly when. In contrast, a preemptive outage, which might be caused by catastrophic failure of a machine or when the machine becomes radically out of adjustment, forces a stoppage whether or not the current job is completed. An example of a nonpreemptive outage occurs when a tool starts to become dull and needs to be replaced or when the mask used to expose a circuit board begins to wear out. In situations like these we can wait until the current piece or job is finished before stopping production.

Process changeovers (setups) can be regarded as nonpreemptive outages when they occur due to changes in the production process (such as changing a mask) as opposed to changes in the product. Changeovers due to changes in product (e.g., setting up for a new part) are more under our control (we decide how many to make before changing to a new part) and are the subject of Chapters 9 and 15. Other nonpreemptive outages include preventive maintenance, breaks, operator meetings, and (we hope) shift changes. These typically occur *between* jobs, rather than *during* them. Nonpreemptive outages require somewhat different treatment than preemptive outages. Since the most common source of nonpreemptive outages is machine setups, we will frame our discussion in these terms. However, the approach is applicable to any form of nonpreemptive outage, just as our analysis of breakdowns is applicable to any form of preemptive outage.

As with preemptive outages, ordinary capacity calculations do not fully analyze the impacts of nonpreemptive setups. Average capacity analysis only tells us that short setups are better than long ones. It cannot evaluate the differences between a slow machine with short setups and a fast one with long setups that have the same effective capacity.

For example, consider the decision of whether to replace a relatively fast machine requiring periodic setups with a slower flexible machine that does not require setups. Machine 1, the fast one, can do an average of one part per hour, but requires a two-hour setup every four parts on average. Machine 2, the flexible one, requires no setups but is slower, requiring an average of 1.5 hours per part. The effective capacity r_e for machine 1 is

$$r_e = \frac{4 \text{ parts}}{6 \text{ hours}} = \frac{2}{3} \text{ parts/hour}$$

Since this is a single-machine workstation, the effective process time is simply the reciprocal of the effective capacity, so $t_e = 1.5$ hours. Thus, machines 1 and 2 have the same effective capacity.

Traditional capacity analysis, which considers only mean capacity, would consider the two machines equivalent and hence would offer no support for replacing machine 1 with machine 2. However, our previous factory physics treatment of machine breakdowns showed that considering variability can be important in evaluating machines with breakdowns. All other things being equal, machine 2 will have less variable effective process times than machine 1 (i.e., because every fourth job at machine 1 will have a long setup time included in its effective process time). Thus, replacing machine 1 with machine 2 will serve to reduce the process time CV and therefore will make the line more efficient. This *variability reduction* effect provides further support for the JIT preference for short setups and is a clear motivation for *flexible manufacturing* technology.

However, the evaluation of the benefits of flexibility can be subtle. The above condition of "all other things being equal" requires that the natural variability of both machines 1 and 2 be the same (i.e., so that the setups for machine 1 will unambiguously increase the CV of effective process times). But what if the flexible machine also has more natural variability? In this case, we must compute and compare the CV of effective process times for both machines.

To compute the CV of effective process times for a machine with setups, we first require data on the natural process times, namely, the mean t_0 and variance σ_0^2. (Equivalently, we could use the mean t_0 and the CV c_0, since $\sigma_0^2 = c_0^2 t_0^2$.) Next we must describe the setups, which we do by assuming that the machine processes an average of N_s parts (or jobs) between setups, where the setup times have a mean duration of t_s and a CV of c_s. We also assume that the probability of doing a setup after any part is equal.[5] That is, if an average of 10 parts are processed between setups, there will be a 1-in-10 chance that a setup will be performed after the current part, regardless of how many have been done since the last setup.

Under these assumptions, the equations for the mean, variance, and SCV of effective process time are, respectively,

$$t_e = t_0 + \frac{t_s}{N_s} \tag{8.7}$$

$$\sigma_e^2 = \sigma_0^2 + \frac{\sigma_s^2}{N_s} + \frac{N_s - 1}{N_s^2} t_s^2 \tag{8.8}$$

$$c_e^2 = \frac{\sigma_e^2}{t_e^2} \tag{8.9}$$

To illustrate the usefulness of these equations, consider another example that compares two machines. Machine 1 is a flexible machine, with no setups, but has somewhat

---

[5]This assumption implies that the number of parts processed between setups is moderately variable (i.e., the mean and standard deviation are equal). Similar analysis can be done for other assumptions regarding the variability of the time between setups.

variable process times. Specifically, the natural process time has a mean of $t_0 = 1.2$ hours and a CV of $c_0 = 0.5$. Machine 2 performs an average of $N_s = 10$ parts between setups and has natural process times with a mean of $t_0 = 1.0$ hours and a CV of $c_0 = 0.25$. The average setup time is $t_s = 2$ hours with a CV of $c_s = 0.25$. Which machine is better?

First, consider the effective capacity. Machine 1 has

$$r_e = \frac{1}{t_0} = \frac{1}{1.2} = 0.833$$

while machine 2 has

$$r_e = \frac{1}{t_e} = \frac{1}{1 + \frac{2}{10}} = 0.833$$

so the two machines are equivalent in this regard. Therefore, the question of which is better becomes, Which machine has less variability?

Using Equation (8.9), we can compute $c_e^2 = 0.31$ for machine 2, as compared to $c_e^2 = c_0^2 = 0.25$ for machine 1. Thus, machine 1, the more variable machine without setups, has less overall variability than machine 2, the less variable machine with setups.

Of course, this conclusion was a consequence of the specific numbers in the example. Flexible machines do not always have less variability. For instance, consider what happens if machine 2 has a shorter setup ($t_s = 1$ hour) after an average of $N_s = 5$ parts. The effective capacity remains unchanged. However, the effective variability for machine 2 is significantly less, with $c_e^2 = 0.16$. In this case, machine 2 with setups would be the better choice.

8.4.4   Variability from Recycle

Another major source of variability in manufacturing systems is quality problems. The simplest quality case to analyze is that of rework on a single workstation. This happens when a workstation performs a task and then checks to see whether the task was done correctly. If it was not, the task is repeated. If we think of the additional processing time spent "getting the job right" as an outage, it is easy to see that this situation is equivalent to the nonpreemptive outage case. Hence, rework has analogous effects to those of setups, namely, that it both robs capacity and contributes greatly to the variability of the effective process times.

As with breakdowns and setups, the traditional reason for reducing rework is to prevent a loss of effective capacity (i.e., reduce waste). Of course, as with traditional analyses of breakdowns and setups, this perspective would regard two machines with the same effective capacity but different rework fractions as equivalent. However, an analysis like that done above for setups shows that the CV of effective process times increases as the fraction of rework increases. Hence, more rework implies more variability. More variability causes more congestion, WIP, and cycle time. Hence, these variability impacts, coupled with the loss of capacity, make rework a disruptive problem indeed. We will return to this important interface between quality and operations in greater detail in Chapter 12.

8.4.5   Summary of Variability Formulas

The computations for t_e, σ_e^2, and c_e^2 for both the preemptive and the nonpreemptive cases are summarized in Table 8.2. Note that if we have a situation involving both preemptive and nonpreemptive outages (e.g., both breakdowns and setups), then these formulas must be applied consecutively. For instance, we begin with the natural process time

TABLE 8.2 **Summary of Formulas for Computing Effective Process Time Parameters**

Situation	Natural	Preemptive	Nonpreemptive
Examples	Reliable Machine	Random Failures	Setups; Rework
Parameters	t_0, c_0^2 (basic)	Basic plus m_f, m_r, c_r^2	Basic plus N_s, t_s, c_s^2
t_e	t_0	$\dfrac{t_0}{A},\ A = \dfrac{m_f}{m_f + m_r}$	$t_0 + \dfrac{t_s}{N_s}$
σ_e^2	$t_0^2 c_0^2$	$\dfrac{\sigma_0^2}{A^2} + \dfrac{(m_r^2 + \sigma_r^2)(1 - A)t_0}{Am_r}$	$\sigma_0^2 + \dfrac{\sigma_s^2}{N_s} + \dfrac{N_s - 1}{N_s^2}t_s^2$
c_e^2	c_0^2	$c_0^2 + (1 + c_r^2)A(1 - A)\dfrac{m_r}{t_0}$	$\dfrac{\sigma_e^2}{t_e^2}$

parameters t_0 and c_0^2. Then we incorporate the effects of failures by computing t_e, σ_e, and c_e^2 for the effective process times, using the preemptive outage formulas. Finally, we incorporate the effects of setups by using these values of t_e, σ_e, and c_e^2 in place of t_0, σ_e, and c_0^2 in the nonpreemptive outage formulas. The final mean t_e, standard deviation σ_e, and SCV c_e^2 will thus be "inflated" to reflect both types of outage.

8.5   Flow Variability

All the above discussion focused solely on process time variability at individual workstations. But variability at one station can affect the behavior of other stations in a line by means of another type of variability, which we call **flow variability.** Flows refer to the transfer of jobs or parts from one station to another. Clearly if an upstream workstation has highly variable process times, the flows it feeds to downstream workstations will also be highly variable. Therefore, to analyze the effect of variability on the line, we must characterize the variability in flows.

8.5.1   Characterizing Variability in Flows

The starting point for studying flows is the arrival of jobs to a single workstation. The departures from this workstation will in turn be arrivals to other workstations. Therefore, once we have described the variability of arrivals to one workstation and determined how this affects the variability of departures from that workstation (and hence arrivals to other workstations), we will have characterized the flow variability for the entire line.

The first descriptor of arrivals to a workstation is the **arrival rate,** measured in jobs per unit time. For consistency, the units of arrival rate must be the same as those of capacity. For instance, if we state capacities of workstations in units of jobs per hour, then arrival rates must also be stated in jobs per hour. Then just as we can characterize capacity by either the mean process time t_e or the average rate of the station r_e, we can characterize the arrival rate to the station by either the **mean time between arrivals,** which we denote by t_a, or the average arrival rate, denoted by r_a. These two measures are simply the inverse of each other

$$r_a = \frac{1}{t_a}$$

and so are entirely equivalent as information.

In order for the workstation to be able to keep up with arrivals, it is essential that capacity exceed the arrival rate, that is,

$$r_e > r_a$$

In virtually all realistic cases (i.e., those with variability present), the capacity must be *strictly* greater than the arrival rate to keep the station from becoming overloaded. We will examine why more precisely below.

Just as there is variability in process times, there is also variability in interarrival times. A reasonable variability measure for interarrival times can be defined in exactly the same way as for process times. If σ_a is the standard deviation of the time between arrivals, then the coefficient of variation of the interarrival times c_a is

$$c_a = \frac{\sigma_a}{t_a}$$

We refer to this as the **arrival CV,** to distinguish it from the **process time CV,** denoted by c_e. Intuitively, a low arrival CV indicates regular, or evenly spaced, arrivals, while a high arrival CV indicates uneven, or "bursty" arrivals. The difference is illustrated in Figure 8.5. The arrival CV c_a, along with the mean interarrival time t_a, summarizes the essential aspects of the arrival process to a workstation.

The next step is to characterize the departures from a workstation. We can use measures analogous to those used to describe arrivals, namely, the **mean time between departures** t_d, the **departure rate** $r_d = 1/t_d$, and the **departure CV** c_d. In a serial production line, where all the output from workstation i becomes input to workstation $i + 1$, the departure rate from i must equal the arrival rate to $i + 1$, so

$$t_a(i + 1) = t_d(i)$$

Indeed, in a serial production line without yield loss or rework, the arrival rate to *every* workstation is equal to the throughput TH. Also, in a serial line where departures from i become arrivals to $i + 1$, the departure CV of workstation i is the same as the arrival CV of workstation $i + 1$

$$c_a(i + 1) = c_d(i)$$

These relationships are depicted graphically in Figure 8.6.

FIGURE 8.5

Arrival processes with low and high CVs

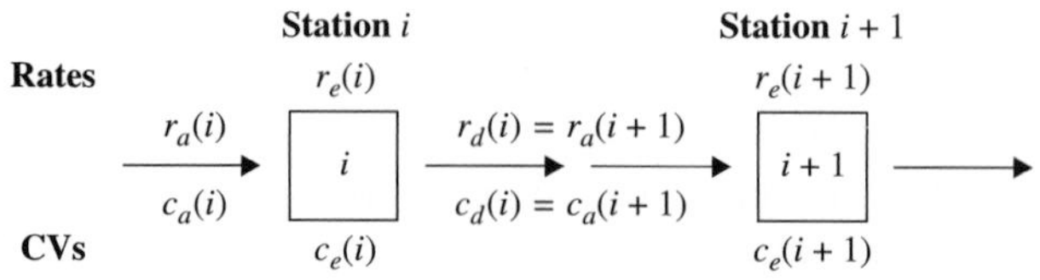

Low CV arrivals

High CV arrivals

FIGURE 8.6

Propagation of variability between workstations in series

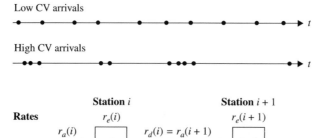

The one remaining issue to resolve concerning flow variability is how to characterize the variability of departures from a station in terms of information about the variability of arrivals and process times. Variability in departures from a station are the result of both variability in arrivals to the station and variability in the process times. The relative contribution of these two factors depends on the **utilization** of the workstation. Recall that the utilization of a workstation, denoted by u, is the fraction of time it is busy over the long run and is defined formally for a workstation consisting of m identical machines as

$$u = \frac{r_a t_e}{m}$$

Notice that u increases with both the arrival rate and the mean effective process time. An obvious upper limit on the utilization is one (that is, 100 percent), which implies that the effective process times must satisfy

$$t_e < \frac{m}{r_a}$$

If u is close to one, then the station is almost always busy. Therefore, under these conditions, the interdeparture times from the station will be essentially identical to the process times. Thus, we would expect the departure CV to be the same as the process time CV (that is, $c_d = c_e$).

At the other extreme, when u is close to zero, the station is very lightly loaded. Virtually every time a job is finished, the station has to wait a long time for another arrival to work on. Because process time is a small fraction of the time between departures, interdeparture times will be almost identical to interarrival times. Thus, under these conditions we would expect the arrival and departure CVs to be the same (that is, $c_d = c_a$).

A good, simple method for interpolating between these two extremes is to use the square of the utilization as follows:[6]

$$c_d^2 = u^2 c_e^2 + (1 - u^2) c_a^2 \tag{8.10}$$

If the workstation is always busy, so that $u = 1$, then $c_d^2 = c_e^2$. Similarly, if the machine is (almost) always idle, so that $u = 0$, then $c_d^2 = c_a^2$. For intermediate utilization levels, $0 < u < 1$, the departure SCV c_d^2 is a combination of the arrival SCV c_a^2 and the process time SCV c_e^2.

When there is more than one machine at a station (that is, $m > 1$), the following is a reasonable way to estimate c_d^2 (although there are others; see Buzacott and Shanthikumar 1993):

$$c_d^2 = 1 + (1 - u^2)(c_a^2 - 1) + \frac{u^2}{\sqrt{m}}(c_e^2 - 1) \tag{8.11}$$

Note that this reduces to Equation (8.10) when $m = 1$.

The net result is that flow variability, like process time variability, can vary widely in practical situations. Using the same classification scheme we used for process time variability, we can classify arrivals according to the arrival CV c_a as follows:

Low variability (LV)	$c_a \leq 0.75$
Moderate variability (MV)	$0.75 < c_a \leq 1.33$
High variability (HV)	$c_a > 1.33$

Departures can be classified in the same manner according to the departure CV c_d.

For example, departures from a heavily loaded LV workstation will tend to be LV, while departures from a heavily loaded HV workstation will tend to be HV. MV

---

[6]Notice that once again an equation involving CVs is written in terms of their SCVs.

workstations fed by MV arrivals will produce MV departures. All these departures in turn become arrivals to other stations, so all types of arrivals can occur in practice.

Another way that MV arrivals can arise in practice is when a workstation is fed by many sources. For instance, a heat-treating operation may receive jobs from many different lines. When this is the case, the time since the last arrival does not provide much information about when the next arrival is likely to occur (because it could come from many places). Thus, the interarrival times will tend to be *memoryless* (i.e., exponential), and therefore c_a will be close to one. Even when the arrivals from any given source are quite regular (i.e., LV), the *superposition* of all the arrivals tends to look MV.

8.5.2   Batch Arrivals and Departures

One important cause of flow variability is **batch arrivals.** These happen whenever jobs are batched together for delivery to a station. For example, suppose a forklift brings 16 jobs once per shift (eight hours) to a workstation. Since arrivals always occur in this way with no randomness whatever, one might reasonably interpret the variability and the CV to be zero.

However, a very different picture results from looking at the interarrival times of the jobs in the batch from the perspective of the individual jobs. The interarrival time (i.e., time since the previous arrival) for the first job in the batch is eight hours. For the next 15 jobs it is zero. Therefore, the mean time between arrivals t_a is one-half hour (eight hours divided by 16 jobs), and the variance of these times is given by

$$\sigma_a^2 = [\tfrac{1}{16}(8^2) + \tfrac{15}{16}(0^2)] - t_a^2 = \tfrac{1}{16}(8^2) - 0.5^2 = 3.75$$

The arrival SCV is therefore

$$c_a^2 = \frac{3.75}{(0.5)^2} = 15$$

In general, if we have a batch size k, this analysis will yield $c_a^2 = k - 1$.

So which is correct, $c_a^2 = 15$ or $c_a^2 = 0$? The answer is that the system will behave "somewhere in between." The reason is that batching confounds two different effects. The first effect is due to the batching itself. This is not really a randomness issue, but rather one of *bad control,* like that we discussed for the worst case in Chapter 7. The second is the variability in the batch arrivals themselves (i.e., as characterized by the arrival CV for the batches). We will examine the relationship between batching and variability more carefully in Chapter 9.

8.6   Variability Interactions—Queueing

The above results for process time variability and flow variability are building blocks for characterizing the effects of variability in the overall production line. We now turn to the problem of evaluating the impact of these types of variability on the key performance measures for a line, namely, WIP, cycle time, and throughput.

To do this, we first observe that actual process time (including setups, downtime, etc.) typically represents only a small fraction (5 to 10 percent) of the total cycle time in a plant. This has been documented in numerous published surveys (e.g., Bradt 1983). The majority of the extra time is spent *waiting* for various resources (e.g., workstations, transport devices, machine operators, etc.). Hence, a fundamental issue in factory physics is to understand the underlying causes of all this waiting.

The science of waiting is called **queueing theory.** In Great Britain, people do not stand in line, they stand in a **queue.** So, queueing theory is the theory of standing in lines.[7] Since jobs "stand in line" while waiting to be processed, waiting to move, waiting for parts, and so on, queueing theory is a powerful tool for analyzing manufacturing systems.

A **queueing system** combines the components that have been considered so far: an arrival process, a service (i.e., production) process, and a queue. Arrivals can consist of individual jobs or batches. Jobs can be identical or have different characteristics. Interarrival times can be constant or random. The workstation can have a single machine or several machines in parallel, which can have constant or random process times. The queueing discipline can be first-come first-served (FCFS), last-come first-served (LCFS), earliest due date (EDD), shortest process time (SPT), or any of a host of priority schemes. The queue space can be unlimited or finite. The variety of queueing systems is almost endless.

Regardless of the queueing system under consideration, the job of queueing theory is to characterize performance measures in terms of descriptive parameters. We do this below for a few queueing systems that are most applicable to manufacturing settings.

8.6.1   Queueing Notation and Measures

To use queueing theory to describe the performance of a single workstation, we will assume we know the following parameters:

r_a = rate of arrivals in jobs per unit time to station. In a serial line without yield loss or rework, r_a = TH at every workstation.

t_a = $1/r_a$ = average time between arrivals

c_a = arrival CV

m = number of parallel machines at station

b = buffer size (i.e., maximum number of jobs allowed in system)

t_e = mean effective process time. The rate (capacity) of the workstation is given by $r_e = m/t_e$.

c_e = CV of effective process time

The performance measures we will focus on are

p_n = probability there are n jobs at station

CT_q = expected waiting time spent in queue

CT = expected time spent at station (i.e., queue time plus process time)

WIP = average WIP level (in jobs) at station

WIP_q = expected WIP (in jobs) in queue

In addition to the above parameters, a queueing system is characterized by a host of specific assumptions, including the type of arrival and process time distributions, dispatching rules, balking protocols, batch arrivals or processing, whether it consists of a network of queueing stations, whether it has single or multiple job classes, and many others. A partial classification of single-station, single-job-class queueing systems is given by *Kendall's notation,* which characterizes a queueing station by means of four parameters:

$$A/B/m/b$$

---

[7]**Queueing** is also the only word we can think of with five vowels in a row, which could be useful if one is a contestant on a game show.

where A describes the distribution of interarrival times, B describes the distribution of process times, m is the number of machines at the station, and b is the maximum number of jobs that can be in the system. Typical values for A and B, along with their interpretations, are

D: constant (deterministic) distribution

M: exponential (Markovian) distribution

G: completely general distribution (e.g., normal, uniform)

In many situations, queue size is not explicitly restricted (e.g., the buffer is very large). We indicate this case as $A/B/m/\infty$ or simply as $A/B/m$.

For example, the $M/G/3$ queueing system refers to a three-machine station with exponentially distributed interarrival times and generally distributed process times and an infinite buffer.

We will focus initially on the $M/M/1$ and $M/M/m$ queueing systems because they yield important intuition and serve as building blocks for more general systems. We will then consider the $G/G/1$ and $G/G/m$ queueing systems because they are directly useful for modeling manufacturing workstations. Finally, we discuss what happens when we limit the buffer in the $M/M/1/b$ and the $G/G/1/b$ cases.

For simplicity, we will restrict our consideration to systems with a single job class (i.e., a single product). Of course, most manufacturing systems have multiple products. But we can develop the key insights into the role of variability in production systems with single-job-class models. Moreover, these models can sometimes be used to approximate the behavior of multiple-job-class systems. Details on how to do this and the development of more sophisticated multiple-job-class models are given in Buzacott and Shanthikumar (1993).

8.6.2   Fundamental Relations

Before considering specific queueing systems, we note that some important relationships hold for all single-station systems (i.e., regardless of the assumptions about arrival and process time distributions, number of machines, etc.). First is the expression for **utilization,** which is the probability that the station is busy, and is given by

$$u = \frac{r_a}{r_e} = \frac{r_a t_e}{m} \tag{8.12}$$

Second is the relation between mean total time spent at the station CT and mean time spent in queue CT_q. Since means are additive,

$$\text{CT} = \text{CT}_q + t_e \tag{8.13}$$

Third, applying Little's law to the station yields a relation among WIP, CT, and the arrival rate:

$$\text{WIP} = \text{TH} \times \text{CT} \tag{8.14}$$

And fourth, applying Little's law to the queue alone yields a relation among WIP_q, CT_q, and the arrival rate:

$$\text{WIP}_q = r_a \times \text{CT}_q \tag{8.15}$$

Using the above relations and knowledge of any one of the four performance measures (CT, CT_q, WIP, or WIP_q), we can compute the other three.

8.6.3   The *M/M/*1 Queue

One of the simplest queueing systems to analyze is the $M/M/1$. This model assumes exponential interarrival times, a single machine with exponential process times, a first-come first-served protocol, and unlimited space for jobs waiting in queue. While not an accurate representation of most manufacturing workstations, the $M/M/1$ queue is tractable and offers valuable insight into more complex and realistic systems.

The key to analyzing the $M/M/1$ queue is the *memoryless property* of the exponential distribution. To see why, consider what information is needed to characterize the future (probabilistic) evolution of the system. That is, what do we need to know about the current status of the system in order to answer such questions as How likely is it that the system will be empty by a certain time? or How likely is it that a job will wait less than a specified amount of time before being served? The issue is not *how* to compute the answers to such questions, but simply *what information* about the system would be needed to do so.

To begin, we require information about the interarrival and process times. Since both are assumed to be exponential, all we need to know are the means (i.e., because the standard deviation is equal to the mean for the exponential distribution). The mean time between arrivals is t_a, so that the arrival rate is $r_a = 1/t_a$. The mean process time is t_e, so the process rate is $r_e = 1/t_e$.

Beyond these, the *only* other information we need is how many jobs are currently in the system. Because the interarrival and process time distributions are memoryless, the time since the last arrival and the time the current job has been in process are irrelevant to the future behavior of the system. Because of this, the **state** of the system can be expressed as a single number n, representing the number of jobs currently in the system. By computing the long-run probability of being in each state, we can characterize all the long-term (steady state) performance measures, including CT, WIP, CT_q, and WIP_q. We do this for the $M/M/1$ queue in the following Technical Note.

Technical Note

Define p_n to be the long-run probability of finding the system in state n (i.e., with a total of n jobs in process and in queue).[8] Since jobs arrive one at a time and the machine works on only one job at a time, the system state can only change by one unit at a time. For instance, if there are currently n jobs at the station, then the only possible state changes are an increase to $n + 1$ (an arrival) or a decrease to $n - 1$ (a departure). The rate the system moves from state n to state $n + 1$, **given it is currently in state** n, is r_a, the arrival rate. Likewise, the conditional rate to move from n to $n - 1$, **given the system is currently in state** n, is r_e, the process rate. The dynamics of the system are graphically illustrated in Figure 8.7.

It follows that the unconditional (i.e., steady-state) **rate** at which the system moves from state $n - 1$ to state n is given by $p_{n-1}r_a$, that is, the probability of being in state $n - 1$ times the rate from $n - 1$ to n, given the system is in state n. Similarly, the rate at which the system moves from state n to state $n - 1$ is $p_n r_e$. In order for the system to be stable, these two rates must be equal (i.e., otherwise the probability of being in any given state would "drift" over time). Hence,

$$p_{n-1}r_a = p_n r_e$$

---

[8]These probabilities are only meaningful in **steady state** (i.e., after the system has been running so long that the current state does not depend on the starting conditions). This means that we can only compute long-term measures from the p_n values. Fortunately, our key measures CT, WIP, CT_q, and WIP_q are long-term measures. Analysis of the **transient** (i.e., short-term) behavior of queueing systems is difficult and will not be discussed here.

FIGURE 8.7

State transition diagram for M/M/1 queue

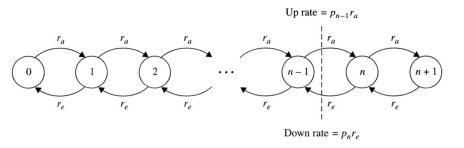

or
$$p_n = \frac{r_a}{r_e} p_{n-1} = u p_{n-1} \qquad (8.16)$$

where $u = r_a t_e = r_a/r_e$ is the utilization which, if there is no blocking, will be the long-run fraction of time the machine is busy.

By the definition of utilization, it follows that the probability (long-run fraction of time) that the station is not busy is $1 - u$. Since the machine is only idle when there are no jobs in the system, this implies that $p_0 = 1 - u$. This gives us one of the p_n values. To get the rest, we write out Equation (8.16) for $n = 1, 2, 3, \ldots$, which yields

$$p_1 = u p_0 = u(1 - u)$$
$$p_2 = u p_1 = u \cdot u(1 - u) = u^2(1 - u)$$
$$p_3 = u p_2 = u \cdot u^2(1 - u) = u^3(1 - u)$$
$$\vdots$$

Continuing in this manner shows that for any state

$$p_n = u^n(1 - u) \qquad n = 0, 1, 2, \ldots \qquad (8.17)$$

These p_n values are probabilities and therefore must sum to 1, so

$$p_0 + p_1 + p_2 + \cdots = (1 + u + u^2 + \cdots)p_0 = 1$$

or
$$p_0 = 1 - u \qquad (8.18)$$

However, if $u \geq 1$, then the sum in the parentheses will be infinite, which violates the properties of probabilities. Therefore, in order for the station to have stable long-run behavior (i.e., not have a queue that "blows up"), we must have $u < 1$ (i.e., utilization *strictly* less than 100 percent).[9]

The most straightforward performance measure to compute is WIP (i.e., expected number in the system). For the $M/M/1$ case

$$\text{WIP} = \sum_{n=0}^{\infty} n p_n$$

$$= (1 - u) \sum_{n=0}^{\infty} n u^n$$

$$= u(1 - u) \sum_{n=1}^{\infty} n u^{n-1} \qquad (8.19)$$

---

[9]If $u < 1$, then by noting that $1 + u + u^2 + \cdots = 1 + u(1 + u + u^2 + \cdots)$ and letting $x = 1 + u + u^2 + \cdots$, we see that $x = 1 + ux$. Solving for x yields $1 - ux = 1$, or $x = (1 - u)^{-1}$. Since $xp_0 = 1$, this shows that $p_0 = 1 - u$, as we showed above by considering utilization.

It is easy to show that $\sum_{n=1}^{\infty} n u^{n-1} = (1-u)^{-2}$, so Equation (8.19) yields a concise expression for WIP.[10]

8.6.4 Performance Measures

The various steady-state performance measures can be computed from the results derived in the Technical Note. The expression for expected WIP follows from Equation (8.19) and is given by

$$\text{WIP}(M/M/1) = \frac{u}{1-u} \tag{8.20}$$

Using this and Little's law yields a relation for average cycle time

$$\text{CT}(M/M/1) = \frac{\text{WIP}(M/M/1)}{r_a} = \frac{t_e}{1-u} \tag{8.21}$$

Then from Equation (8.13) we can compute the average time in queue

$$\text{CT}_q(M/M/1) = \text{CT}(M/M/1) - t_e = \frac{u}{1-u} t_e \tag{8.22}$$

Finally, for the WIP in queue, Little's law again yields

$$\text{WIP}_q(M/M/1) = r_a \times \text{CT}_q(M/M/1) = \frac{u^2}{1-u} \tag{8.23}$$

Observe that WIP, CT, CT_q, and WIP_q are all increasing in u. Not surprisingly, busy systems exhibit more congestion than lightly loaded systems. Also, for a fixed u, CT and CT_q are increasing in t_e. Hence, for a given level of utilization, slower machines cause more waiting time. Finally, notice that since these expressions have the term $1 - u$ in the denominator, all the congestion measures "explode" as u gets close to one. What this means is that WIP levels and cycle times increase very rapidly (i.e., nonlinearly) as utilization approaches 100 percent. We will discuss the implications of this in greater detail in Chapter 9.

Example:
Recall that in the Briar Patch Manufacturing example, the arrival rate to the Tortoise 2000 was 2.875 jobs per hour ($r_a = 2.875$). Assume now that times between arrival are exponentially distributed (not a bad assumption if jobs are arriving from many different locations). Also, recall that the production rate is three jobs per hour (or $t_e = \frac{1}{3}$) and that $c_e = 1.0$. Since the effective process times have a CV of one, just as the exponential distribution does, it is reasonable to use the $M/M/1$ model to represent the Tortoise 2000.[11] The utilization is computed as $u = 2.875/3 = 0.9583$, and the performance measures are given below:

$$\text{WIP} = \frac{u}{1-u} = \frac{0.9583}{1 - 0.9583} = 23 \text{ jobs}$$

$$\text{CT} = \frac{\text{WIP}}{\text{TH}} = \frac{23}{2.875} = 8 \text{ hours}$$

---

[10]This is because $\sum_{n=1}^{\infty} n u^n$ is the derivative of $\sum_{n=0}^{\infty} u^n$, which we saw is equal to $1/(1 - u)$. Since the derivative of the sum is the sum of the derivatives, $\sum_{n=1}^{\infty} n u^n$ is equal to the derivative of $1/(1 - u)$, which is $1/(1 - u)^2$. Notice that this is only valid as long as $u < 1$, which was already required for the queue to be stable.

[11]The process times are not actually exponential, however, since $c_e = 1$ was the result of failures superimposed on low-variability natural process times. So the $M/M/1$ queue is not exact, but will be a reasonable approximation.

$$CT_q = CT - t_e = 8 - 0.3333 = 7.6667 \text{ hours}$$

$$WIP_q = TH \times CT_q = 2.875 \times 7.6667 = 22.0417 \text{ jobs}$$

We see that WIP and CT are much smaller than those for the Hare X19 under the same demand conditions. However, to model the nonexponential Hare X19, we need a more general model than the $M/M/1$.

8.6.5   Systems with General Process and Interarrival Times

Most real-world manufacturing systems do not satisfy the assumptions of the $M/M/1$ queueing model. Process times are seldom exponential. When workstations are fed by upstream stations whose process times are not exponential, interarrival times are also unlikely to be exponential. To address systems with nonexponential interarrival and process time distributions, we must turn to the $G/G/1$ queue.

Unfortunately, without the memoryless property of the exponential to facilitate analysis, we cannot compute exact performance measures for the $G/G/1$ queue. But we can estimate them by means of a "two-moment" approximation, which makes use of only the mean and standard deviation (or CV) of the interarrival and process time distributions. Although cases can be constructed for which this approximation works poorly, it is reasonably accurate in typical manufacturing systems (i.e., for most cases except those with c_e and c_a much larger than one, or u larger than 0.95 or smaller than 0.1). Because it works well, this approximation is the basis of several commercially available manufacturing queueing analysis packages.

As we did for the $M/M/1$ case, we will proceed by first developing an expression for the waiting time in queue CT_q and then computing the other performance measures. The approximation for CT_q, which was first investigated by Kingman (1961) (see Medhi 1991 for a derivation), is given by

$$CT_q(G/G/1) = \left(\frac{c_a^2 + c_e^2}{2} \right) \left(\frac{u}{1-u} \right) t_e \tag{8.24}$$

This approximation has several nice properties. First, it is exact for the $M/M/1$ queue.[12] It also happens to be exact for the $G/G/1$ queue, although this is not evident from our discussion here. Finally, it neatly separates into three terms: a dimensionless **variability term** V, a **utilization term** U, and a **time term** T, as

$$CT_q(G/G/1) = \underbrace{\left(\frac{c_a^2 + c_e^2}{2} \right)}_{V} \underbrace{\left(\frac{u}{1-u} \right)}_{U} \underbrace{t_e}_{T}$$

or
$$CT_q = VUT \tag{8.25}$$

We refer to this as **Kingman's equation** or as the *VUT* **equation.** From it, we see that if the V factor is less than one, then the queue time, and hence other congestion measures, for the $G/G/1$ queue will be smaller than those for the $M/M/1$ queue. Conversely, if V is greater than one, congestion will be greater than in the $M/M/1$ queue. Thus, the VUT equation shows that the $M/M/1$ case represents an intermediate case for single stations analogous to that represented by the practical worst case for lines.

---

[12]When c_a and c_e are both equal to one, the first fraction becomes one and the other term is the waiting time in queue for the $M/M/1$ queue $CT_q(M/M/1)$.

Example:

Let us return to the Briar Patch Manufacturing example and consider the Hare X19. Recall that this machine has high variability ($c_e^2 = 6.25$). Again, assume the time between job arrivals is exponential (that is, $c_a^2 = 1$). Utilization of the Hare X19 is $u = 0.9583$. Hence, we can use the VUT equation to compute the expected queue time as

$$CT_q = \left(\frac{c_a^2 + c_e^2}{2}\right)\left(\frac{u}{1-u}\right)t_e$$

$$= \left(\frac{1 + 6.25}{2}\right)\left(\frac{0.9583}{1 - 0.9583}\right)20$$

$$= 1{,}667.5 \text{ minutes} = 27.79 \text{ hours}$$

which is what we reported in the introduction to the chapter.

Now suppose that the Hare X19 feeds the Tortoise 2000. There is no yield loss, so the rate into the Tortoise 2000 is the same as that into the Hare X19; and since the two machines have the same effective rate, they will have the same utilization $u = 0.9583$. However, to use the VUT equation, we must find the arrival CV c_a to the Tortoise 2000. We do this by first finding the departure CV from the Hare c_d by using linking Equation (8.10)

$$c_d^2 = c_e^2 u^2 + c_a^2(1 - u^2)$$

$$= 6.25(0.9583^2) + 1.0(1 - 0.9583^2)$$

$$= 5.8216$$

Since the Hare X19 feeds the Tortoise 2000, c_a^2 for the Tortoise 2000 is equal to c_d^2 for the Hare X19. Hence, the expected queue time at the Tortoise 2000 will be

$$CT_q = \left(\frac{c_a^2 + c_e^2}{2}\right)\left(\frac{u}{1-u}\right)t_e$$

$$= \left(\frac{5.82 + 1.0}{2}\right)\left(\frac{0.9583}{1 - 0.9583}\right)20$$

$$= 1{,}568.97 \text{ minutes} = 26.15 \text{ hours}$$

which again is what we reported in the introduction.

Notice that the queue time at the Tortoise 2000 is almost as large as that for the Hare X19, even though the Hare X19 has much higher process variability. The reason for this is the high variability of arrivals to the Tortoise 2000 ($c_a = \sqrt{5.8216} = 2.41$). If the Tortoise 2000 were fed by moderately variable arrivals (with $c_a = 1.0$), then its performance would be represented by the $M/M/1$ queue, which predicts average queue time of 7.67 hours. The excess time (and congestion) is a consequence of the propagation of variability from the upstream Hare X19.

8.6.6   Parallel Machines

The VUT equation gives us a tool for analyzing workstations consisting of single machines. However, in real-world systems, workstations often consist of multiple machines in parallel. The reason, of course, is that often more than a single machine is required to achieve the desired workstation capacity. To analyze and understand the behavior of parallel machine stations, we need a more general model.

The simplest type of parallel machine station is the case in which interarrival times are exponential ($c_a = 1$) and process times are exponential ($c_e = 1$). This corresponds to the $M/M/m$ queueing system. In this model, all jobs wait in a single queue for the next available machine (unlike in most grocery stores where each server has a separate queue, but like in most banks where there is a single queue for all the servers). Although the steady-state probabilities for the $M/M/m$ queue *can* be computed exactly, they are messy and provide little additional intuition. More useful is the following closed-form approximation for the waiting time in queue proposed by Sakasegawa (1977) that both offers intuition and is quite accurate (see Whitt (1993) for a discussion of its merits and uses):

$$\text{CT}_q(M/M/m) = \frac{u^{\sqrt{2(m+1)}-1}}{m(1-u)} t_e \tag{8.26}$$

Note that when $m = 1$, this expression reduces to Equation (8.22), which is the exact expression for queue time in the $M/M/1$ queue. Using this expression, along with universal relations (8.13) to (8.15), we can obtain expressions for $\text{CT}(M/M/m)$, $\text{WIP}(M/M/m)$, and $\text{WIP}_q(M/M/m)$.

Example:
Consider the Briar Patch Manufacturing example again. Recall that the Tortoise 2000 had process times with $c_e = 1$ and hence is well approximated by an exponential model. Suppose now, however, that arrivals to the Tortoise 2000 occur at a rate of 207 jobs per day and have exponential interarrival times ($c_a = 1$). Since this is beyond the capacity of a single Tortoise 2000, we now assume that Briar Patch Manufacturing has three machines.

First, consider what would happen if each of the three machines had its own arrival stream. That is, each machine sees one-third of the total demand, or 69 jobs per day (2.875 jobs per hour). Since process times are one-third hour, the utilization of each machine is $u = 2.875(\frac{1}{3}) = 0.958$. Hence, the situation for each machine is precisely that which we modeled in Section 8.6.4, where we computed the average time in queue to be 7.67 hours.

Now suppose that the three Tortoise 2000s are combined into a single station so that the entire demand of 207 jobs per day, or 8.625 jobs per hour, arrives to a single queue that is serviced by the three machines in parallel. Utilization is the same, since

$$u = \frac{r_a t_e}{m} = \frac{(8.625)(\frac{1}{3})}{3} = 0.958$$

However, average time in queue is now

$$\begin{aligned}
\text{CT}_q &= \frac{u^{\sqrt{2(m+1)}-1}}{m(1-u)} t_e \\
&= \frac{(0.958)^{\sqrt{2(3+1)}-1}}{3(1-0.958)} \left(\frac{1}{3}\right) = 2.467 \text{ hours}
\end{aligned}$$

which is significantly lower than the case where the three machines had separate queues. We conclude that when variability and utilization are the same, a station with parallel machines will outperform one with dedicated machines. The reason, as anyone who has ever chosen the wrong line at the grocery store knows, is that a long process time will delay everyone waiting in the queue at a dedicated machine. When the queue is combined, as at the bank, the machine experiencing a long process time gets bypassed and therefore does not have such a damaging effect on average queue time. This is an example of the more general property of **variability pooling,** which we discuss in Section 8.8.

8.6.7   Parallel Machines and General Times

A parallel machine station with general (nonexponential) process and interarrival times is represented by a $G/G/m$ queue. To develop an approximation for this situation, note that approximation (8.24) can be rewritten as

$$\mathrm{CT}_q(G/G/1) = \left(\frac{c_a^2 + c_e^2}{2}\right)\mathrm{CT}_q(M/M/1)$$

where $\mathrm{CT}_q(M/M/1) = [u/(1-u)]t_e$ is the waiting time in queue for the $M/M/1$ queue. This suggests the following approximation for the $G/G/m$ queue (see Whitt 1983 for a discussion)

$$\mathrm{CT}_q(G/G/m) = \left(\frac{c_a^2 + c_e^2}{2}\right)\mathrm{CT}_q(M/M/m) \tag{8.27}$$

Using Equation (8.26) to approximate $\mathrm{CT}_q(M/M/m)$ in Equation (8.27) yields the following closed-form expression for the waiting time in the $G/G/m$ queue:

$$\mathrm{CT}_q(G/G/m) = \left(\frac{c_a^2 + c_e^2}{2}\right)\left(\frac{u^{\sqrt{2(m+1)}-1}}{m(1-u)}\right)t_e \tag{8.28}$$

Expression (8.28) is the parallel machine version of the VUT equation. The V and T terms are identical to the single-machine version given in expression (8.25), but the U term is different. Although it may appear complicated, it does not require any type of iterative algorithm to solve and is therefore easily implementable in a spreadsheet program. This makes it possible to couple the single-station approximation (8.28) with the multimachine "linking equation" (8.11) to create a spreadsheet tool for analyzing the performance of a line.

8.7   Effects of Blocking

Thus far, we have considered only systems in which there is no limit to how large the queue can grow. Indeed, in every system we have examined, the average queue (and cycle time) grows to infinity as utilization approaches 100 percent. But in the real world, queues never become infinite. They are bounded by limitations of space, time, or operating policy. Therefore, an important topic in the science of factory physics is the behavior of systems with finite queueing space.

8.7.1   The $M/M/1/b$ Queue

Consider the case where process and interarrival times are exponential, as they are in the $M/M/1$ queue, but where there is only enough space for b units in the system (in queue and in process). In Kendall's notation this corresponds to the $M/M/1/b$ queue. This system behaves in much the same way as the $M/M/1$ queue except now whenever the system becomes full, the arrival process is stopped. When this happens, the machine is said to be **blocked.** This model represents a very common situation in manufacturing applications.

For instance, consider a manufacturing cell consisting of two stations with a finite buffer in between. The first machine processes raw material and delivers it to the buffer of the second machine. If we can assume that raw material is always available (e.g., raw material is bar stock or sheet metal, which is in ample supply), then the $M/M/1/b$ model can be a good approximation of the behavior of the second machine. Indeed, if both machines have exponential process times, the model will be exact. This type of configuration is not uncommon. In fact, by their very nature all kanban systems exhibit blocking behavior.

In a queueing model with blocking, like the $M/M/1/b$, the arrival rate r_a takes on a different meaning than it does in models with unbounded queues. Here it represents the rate of *potential* arrivals, assuming that the system is not full. Thus, $u = r_a t_e$, is no longer the long-run probability that the machine is busy, but instead represents what the utilization would be if no arrivals were turned away. Consequently, u can equal or exceed one. We compute the probabilities and measures for the $M/M/1/b$ queue in the next Technical Note.

Technical Note

As in the $M/M/1$ queue, we define the state of the $M/M/1/b$ queue to be the number of jobs in the system. However, unlike in the $M/M/1$ case, the $M/M/1/b$ queue has a finite number of states $n = 0, 1, 2, \ldots, b$. Proceeding as we did for the $M/M/1$ queue, we can show that the long-run probability of being in state n is

$$p_n = u^n p_0$$

for the $M/M/1/b$ queue. A little algebra shows that in order to have $p_0 + \cdots + p_b = 1$, we must have

$$p_0 = \frac{1-u}{1-u^{b+1}} \tag{8.29}$$

Thus,
$$p_n = \frac{u^n(1-u)}{1-u^{b+1}} \tag{8.30}$$

Note that Equations (8.29) and (8.30) reduce to those for the $M/M/1$ queue as b goes to infinity (because $u^{b+1} \to 0$ as $b \to \infty$).

Equation (8.30) is valid as long as $u \neq 1$. For the special case where $u = 1$, all states of the system are equally likely and have the same probability, so

$$p_n = \frac{1}{b+1} \quad \text{for } n = 0, 1, \ldots, b \tag{8.31}$$

We can compute the average WIP level from

$$\text{WIP} = \sum_{n=0}^{b} n p_n \tag{8.32}$$

Since the system accepts arrivals whenever it is not full and the rate in equals the rate out, we can compute throughput from

$$\text{TH} = (1 - p_b)r_a \tag{8.33}$$

For the case where $u \neq 1$, the average WIP and throughput are

$$\text{WIP}(M/M/1/b) = \frac{u}{1-u} - \frac{(b+1)u^{b+1}}{1-u^{b+1}} \tag{8.34}$$

$$\text{TH}(M/M/1/b) = \frac{1-u^b}{1-u^{b+1}}r_a \tag{8.35}$$

For the case where $u = 1$, WIP and throughput simplify to

$$\text{WIP}(M/M/1/b) = \frac{b}{2} \tag{8.36}$$

$$\text{TH}(M/M/1/b) = \frac{b}{b+1}r_a = \frac{b}{b+1}r_e \tag{8.37}$$

For either case, we can use Little's law to compute the cycle time, queue time, and queue length as

$$\text{CT}(M/M/1/b) = \frac{\text{WIP}(M/M/1/b)}{\text{TH}(M/M/1/b)} \tag{8.38}$$

$$\text{CT}_q(M/M/1/b) = \text{CT}(M/M/1/b) - t_e \tag{8.39}$$

$$\text{WIP}_q(M/M/1/b) = \text{TH}(M/M/1/b) \times \text{CT}_q(M/M/1/b) \tag{8.40}$$

We can gain some useful insights from these formulas by interpreting the $M/M/1/b$ model as a system of two machines in series. The first machine is assumed to have enough raw material so that it never starves. Similarly, the second machine can always move its product out (i.e., it is never blocked). However, the buffer between the two machines is finite and is equal to B. If both machines have exponential process times, the model for the behavior of the second machine and the buffer is given by the $M/M/1/b$ queue, where $b = B + 2$. The two extra buffer spaces are the two machines themselves.

Notice that the WIP for the $M/M/1/b$ queue will *always* be less than that for the $M/M/1$ system. This is because the second machine has blocking, which prevents the WIP level from growing beyond b. If b is small, the effect can be dramatic. Indeed, kanban, which acts just like a finite buffer, is specifically intended to prevent WIP buildup.

However, WIP has a price—lost throughput. Recall that in the $M/M/1$ case the arrival rate is equal to the output rate. This is because, in steady state, whatever comes in must go out. This is not so in the case with blocking since the input rate is equal to the output rate (throughput) plus the balking rate (rate at which arrivals are rejected). Using Equations (8.35) and (8.37), we see that

$$\text{TH} = \frac{1 - u^b}{1 - u^{b+1}}ur_e < ur_e$$

if $u \neq 1$, and

$$\text{TH} = \frac{b}{b+1}r_e < r_e$$

if $u = 1$. These last expressions show that the throughput in a system with blocking will always be less than that in a system without blocking. Furthermore, the smaller the buffer size b, the greater the reduction in throughput.

Example:
Consider a line consisting of two machines in series. The first machine takes, on average, $t_e(1) = 21$ minutes to complete a job. The second machine takes $t_e(2) = 20$ minutes. Both machines have exponential process times ($c_e(1) = c_e(2) = 1$). Between the two machines there is enough room for two jobs, so $b = 4$ (two in the buffer and two at the machines themselves).

First consider what would happen if there were an infinite buffer. Since the first machine runs constantly, the arrival rate to the second machine is simply the rate of the first machine. Hence, utilization of the second machine is $u = r_a/r_e = \frac{1}{21}/\frac{1}{20} = 0.9524$.

The other performance measures for the second machine can be computed by using the $M/M/1$ formulas to be

$$\text{WIP} = \frac{u}{1 - u} = \frac{0.9524}{1 - 0.9524} = 20 \text{ jobs}$$

$$\text{TH} = r_a = \tfrac{1}{21} \text{ minute} = 0.0476 \text{ job/minute}$$

$$\text{CT} = \frac{\text{WIP}}{\text{TH}} = 420.18 \text{ minutes}$$

Now, consider the finite buffer case. We first compute TH, using the $M/M/1/b$ queueing model.

$$\text{TH} = \frac{1 - u^b}{1 - u^{b+1}} r_a$$

$$= \frac{1 - 0.9524^4}{1 - 0.9524^5} \left(\frac{1}{21} \right)$$

$$= 0.039 \text{ job/minute}$$

We can now compute the *partial WIP* (denoted by WIPP) in the system represented by the $M/M/1/b$ model, namely, the second machine, the two-job buffer, and the buffer involving the first machine. We note that WIP at the first machine is only included in WIPP if it is in queue (i.e., when the first machine is blocked). WIP that is being processed at the first machine is not included, since it is viewed as "on its way" to the system represented by the $M/M/1/b$ model. From Equation (8.34), the partial WIP is

$$\text{WIPP} = \frac{u}{1 - u} - \frac{(b + 1)u^{b+1}}{1 - u^{b+1}}$$

$$= 20 - \frac{5(0.9524^5)}{1 - 0.9524^5} = 20 - 18.106 = 1.894 \text{ jobs}$$

The cycle time for the line is the time spent in partial WIP at the second machine plus the time in process at the first machine. Note that we do not consider any queue time at the first machine since it would be infinite due to the assumption of unlimited raw materials.

$$\text{CT} = \frac{\text{WIPP}}{\text{TH}} + t_e(1) = \frac{1.894}{0.039} + 21 = 69.57 \text{ minutes}$$

A second application of Little's law shows that the WIP in the system line is

$$\text{WIP} = \text{TH} \times \text{CT} = 0.039 \text{ job/minute} \times 69.57 \text{ minutes} = 2.71 \text{ jobs}$$

Comparison of the buffered and unbuffered cases is revealing. Limiting the interstation queue greatly reduces WIP and CT (by more than 83 percent) but also reduces TH (but by only 18 percent). However, a decline in throughput of 18 percent could more than offset the savings in inventory costs. This highlights why kanban cannot be implemented simply by reducing buffer sizes. The loss in throughput is typically too great. The only way to reduce WIP and CT without sacrificing too much throughput is to also reduce variability (i.e., we have to remove the rocks, not just lower the water). Unfortunately, we cannot examine variability reduction with the $M/M/1/b$ model because it assumes exponential process times. We discuss nonexponential models in the next section.

A second observation we can make using the $M/M/1/b$ model is that finite buffers force stability regardless of r_a and r_e. The reason is that WIP, and consequently CT, cannot "blow up" in a system with a finite buffer. For instance, suppose the speeds of the two machines above were reversed with the faster one feeding the slower one. If the buffer were infinite, WIP would go to infinity (in the long run), as would CT. But in the finite buffer case $u = 21/20 = 1.05$, so

$$\text{TH} = \frac{1 - u^b}{1 - u^{b+1}} r_a = \frac{1 - 1.05^4}{1 - 1.05^5} \left(\frac{1}{20} \right) = 0.0390 \text{ job/minute}$$

The partial WIP is

$$\text{WIPP} = \frac{u}{1 - u} - \frac{(b + 1)u^{b+1}}{1 - u^{b+1}}$$

$$= \frac{1.05}{1 - 1.05} - \frac{5(1.05^5)}{1 - 1.05^5}$$

$$= 2.097 \text{ jobs}$$

and cycle time is

$$\text{CT} = \frac{\text{WIP}}{\text{TH}} + t_e(1) = \frac{2.097}{0.0390} + 20 = 73.78 \text{ minutes}$$

Finally, WIP in the line is

$$\text{WIP} = \text{TH} \times \text{CT} = 0.0390 \times 73.78 = 2.88 \text{ jobs}$$

which is somewhat larger than in the case with the faster machine in second position, because the rate of arrival to the system is greater. However, throughput is unaffected by the order of the machines. This latter result is known as *reversibility* and holds for lines with more than two machines and general process times (see Muth 1979 for a proof). It is a fascinating theoretical result, but since firms seldom get the opportunity to run their lines backward, it does not often come up in practice.

8.7.2   General Blocking Models

To analyze variability effects, we need to extend the $M/M/1/b$ model to more general process and interarrival time distributions. In general, this is very difficult. We refer the interested reader to Buzacott and Shanthikumar (1993, Chapter 4) for a more complete treatment. However, we can make some useful approximations by modifying the $M/M/1/b$ queue in a manner analogous to the way we modified the $M/M/1$ queue to model the $G/G/1$ queue.

We consider three cases: (1) when the arrival rate is less than the production rate ($u < 1$), (2) when the arrival rate exceeds the production rate ($u > 1$), and (3) when the arrival and production rates are the same ($u = 1$).

Arrival Rate Less than Production Rate.   First we compute the expected WIP in the system without any blocking, denoted by WIP_{nb}, by using Kingman's equation and Little's law.

$$\text{WIP}_{\text{nb}} \approx r_a \left(\frac{c_a^2 + c_e^2}{2} \right) \left(\frac{u}{1 - u} \right) t_e + t_e$$

$$= \left(\frac{c_a^2 + c_e^2}{2} \right) \left(\frac{u^2}{1 - u} \right) + u \tag{8.41}$$

Now recall that for the $M/M/1$ queue, $\text{WIP} = u/(1 - u)$, so that

$$u = \frac{\text{WIP} - u}{\text{WIP}}$$

We can use WIP_{nb} in analogous fashion to compute a "corrected" utilization ρ

$$\rho = \frac{\text{WIP}_{nb} - u}{\text{WIP}_{nb}} \tag{8.42}$$

Then we substitute ρ for (almost) all the u terms in the $M/M/1/b$ expression for TH to obtain

$$\text{TH} \approx \frac{1 - u\rho^{b-1}}{1 - u^2\rho^{b-1}} r_a \tag{8.43}$$

By combining Kingman's equation (to compute ρ) with the $M/M/1/b$ model, we incorporate the effects of both variability and blocking. Although this expression is significantly more complex than that for the $M/M/1/b$ queue, it is straightforward to evaluate by using a spreadsheet. Furthermore, because we can easily show that $\rho = u$ if $c_a = c_e = 1$, Equation (8.43) reduces to the exact expression (8.35) for the case in which interarrival and process times are exponential.

Unfortunately, the expressions for expected WIP and CT become much more messy. However, for small buffers, WIP will be close to (but always less than) the size of the buffer (that is, $b - 1$). For large buffers, WIP will approach (but always be less than) that for the $G/G/1$ queue. Thus,

$$\text{WIP} < \min\{\text{WIP}_{nb}, b - 1\} \tag{8.44}$$

From Little's law, we obtain an approximate bound on CT

$$\text{CT} > \frac{\min\{\text{WIP}_{nb}, b - 1\}}{\text{TH}} \tag{8.45}$$

with TH computed as above. It is only an approximate bound because the expression for TH is an approximation.

Arrival Rate Greater than Production Rate.   In the earlier example for the $M/M/1/b$ queue, we saw that the average WIP level was different, but not too different, when the order of the machines was reversed. This motivates us to approximate the WIP in the case in which the arrival rate is greater than the production rate by the WIP that results from having the machines in reverse order. When we switch the order of the machines, the production process becomes the arrival process and vice versa, so that utilization is $1/u$ (which will be less than 1 since $u > 1$). The average WIP level of the reversed line is approximated by

$$\text{WIP}_{nb} \approx \left(\frac{c_a^2 + c_e^2}{2}\right)\left(\frac{1/u^2}{1 - 1/u}\right) + \frac{1}{u} \tag{8.46}$$

We can compute a "corrected" utilization ρ_R for the reversed line in the same fashion as we did for the case where $u < 1$, which yields

$$\rho_R = \frac{\text{WIP}_{nb} - 1/u}{\text{WIP}_{nb}}$$

We then define $\rho = 1/\rho_R$ and compute TH as before. Once we have an approximation for TH, we can use inequalities (8.44) and (8.45) for bounds on WIP and CT, respectively.

Arrival Rate Equal to Production Rate.   Finally, the following is a good approximation of TH for the case in which $u = 1$ (Buzacott and Shanthikumar 1993):

$$\text{TH} \approx \frac{c_a^2 + c_e^2 + 2(b - 1)}{2(c_a^2 + c_e^2 + b - 1)} \tag{8.47}$$

Again, with this approximation of TH, we can use inequalities (8.44) and (8.45) for bounds on WIP and CT.

Example:
Let us return to the example of Section 8.7.1, in which the first machine (with 21-minute process times) fed the second machine (with 20-minute process times) and there is an interstation buffer with room for two jobs (so that $b = 4$). Previously, we assumed that the process times were exponential and saw that limiting the buffer resulted in an 18 percent reduction in throughput. One way to offset the throughput drop resulting from limiting WIP is to reduce variability. So let us reconsider this example with reduced process variability, such that the effective coefficients of variation (CVs) for both machines are equal to 0.25.

Utilization is still $u = r_a/r_e = \frac{1}{21}/\frac{1}{20} = 0.9524$, so we can compute the WIP without blocking to be

$$\text{WIP}_{nb} = \left(\frac{c_a^2 + c_e^2}{2}\right)\left(\frac{u^2}{1-u}\right) + u$$

$$= \left(\frac{0.25^2 + 0.25^2}{2}\right)\left(\frac{0.9524^2}{1 - 0.9524}\right) + 0.9524$$

$$= 2.143$$

The corrected utilization is

$$\rho = \frac{\text{WIP}_{nb} - u}{\text{WIP}_{nb}} = \frac{2.143 - 0.9524}{2.143} = 0.556$$

Finally, we compute the throughput as

$$\text{TH} = \frac{1 - u\rho^{b-1}}{1 - u^2\rho^{b-1}}r_a$$

$$= \frac{1 - 0.9524(0.556^3)}{1 - 0.9524^2(0.556^3)}\frac{1}{21}$$

$$= 0.0473$$

Hence, the percentage reduction in throughput relative to the unbuffered rate ($\frac{1}{21} = 0.0476$) is now less than one percent. Reducing process variability in the two machines made it possible to reduce the WIP by limiting the interstation buffer without a significant loss in throughput. This highlights why variability reduction is such an important component of JIT implementation.

8.8   Variability Pooling

In this chapter we have identified a number of causes of variability (failures, setups, etc.) and have observed how they cause congestion in a manufacturing system. Clearly, as we will discuss more fully in Chapter 9, one way to reduce this congestion is to reduce variability by addressing its causes. But another, and more subtle, way to deal with congestion effects is by combining multiple sources of variability. This is known as **variability pooling,** and it has a number of manufacturing applications.

An everyday example of the use of variability pooling is financial planning. Virtually all financial advisers recommend investing in a diversified portfolio of financial instruments. The reason, of course, is to hedge against risk. It is highly unlikely that a wide spectrum of investments will perform extremely poorly at the same time. At the same time, it is also unlikely that they will perform extremely well at the same time. Hence, we expect less variable returns from a diversified portfolio than from any single asset.

Variability pooling plays an important role in a number of manufacturing situations. Here we discuss how it affects batch processing, safety stock aggregation, and queue sharing.

8.8.1   Batch Processing

To illustrate the basic idea behind variability pooling, we consider the question, Which is more variable, the process time of an individual part or the process time of a batch of parts? To answer this question, we must define what we mean by *variable*. In this chapter we have argued that the coefficient of variation is a reasonable way to characterize variability. So we will frame our analysis in terms of the CV.

First, consider a single part whose process time is described by a random variable with mean t_0 and standard deviation σ_0. Then the process time CV is

$$c_0 = \frac{\sigma_0}{t_0}$$

Now consider a batch of n parts, each of which has a process time with mean t_0 and standard deviation σ_0. Then the mean time to process the batch is simply the sum of the individual process times

$$t_0(\text{batch}) = nt_0$$

and the variance of the time to process the batch is the sum of the individual variances

$$\sigma_0^2(\text{batch}) = n\sigma_0^2$$

Hence, the CV of the time to process the batch is

$$c_0(\text{batch}) = \frac{\sigma_0(\text{batch})}{t_0(\text{batch})} = \frac{\sqrt{n}\sigma_0}{nt_0} = \frac{\sigma_0}{\sqrt{n}t_0} = \frac{c_0}{\sqrt{n}}$$

Thus, the CV of the time to process decreases by one over the square root of the batch size. We can conclude that process times of batches are less variable than process times of individual parts (provided that all process times are independent and identically distributed). The reason is analogous to that for the financial portfolio. Having extremely long or short process times for all n parts is highly unlikely. So the batch tends to "average out" the variability of individual parts.

Does this mean that we should process parts in batches to reduce variability? Not necessarily. As we will see in Chapter 9, batching has other negative consequences that may offset any benefits from lower variability. But there are times when the variability reduction effect of batching is very important, for instance, in sampling for quality control. Taking a quality measurement on a batch of parts reduces the variability in the estimate and hence is a standard practice in the construction of statistical control charts (see Chapter 12).

8.8.2   Safety Stock Aggregation

Variability pooling is also of enormous importance in inventory management. To see why, consider a computer manufacturer that sells systems with three different choices each of processor, hard drive, CD ROM, removable media storage device, RAM configurations, and keyboard. This makes a total of $3^6 = 729$ different computer configurations. To make the example simple, we suppose that all components cost \$150, so that the cost of finished goods for any computer configuration is $6 \times \$150 = \$900$. Furthermore, we assume that demand for each configuration is Poisson with an average rate of 100 units per year and that replenishment lead time for any configuration is three months.

First suppose that the manufacturer stocks finished goods inventory of all configurations and sets the stock levels according to a base stock model. Using the techniques of Chapter 2, we can show that to maintain a customer service level (fill rate) of 99 percent requires a base stock level of 38 units and results in an average inventory level of $\$11,712.425$ for each configuration. Therefore, the total investment in inventory is $729 \times \$11,712.425 = \$8,538.358$.

Now suppose that instead of stocking finished computers, the manufacturer stocks only the components and then assembles to order. We assume that this is feasible from a customer lead time standpoint, because the vast majority of the three-month replenishment lead time is presumably due to component acquisition. Furthermore, since there are only 18 different components, as opposed to 729 different computer configurations, there are fewer things to stock. However, because we are assembling the components, each must have a fill rate of $0.99^{1/6} = 0.9983$ in order to ensure a customer service level of 99 percent.[13] Assuming a three-month replenishment lead time for each component, achieving a fill rate of 0.9983 requires a base stock level of 6,306 and results in an average inventory level of $\$34,655.447$ for each component. Thus, total inventory investment is now $18 \times \$34,655.447 = \$623,798$, a 93 percent reduction!

This effect is not limited to the base stock model. It also occurs in systems using the (Q, r) or other stocking rules. The key is to hold *generic* inventory, so that it can be used to satisfy demand from multiple sources. This exploits the variability pooling property to greatly reduce the safety stock required. We will examine additional assemble-to-order types of systems in Chapter 10 in the context of push and pull production.

8.8.3   Queue Sharing

We mentioned earlier that grocery stores typically have individual queues for checkout lanes, while banks often have a single queue for all tellers. The reason banks do this is to reduce congestion by pooling variability in process times. If one teller gets bogged down serving a person who insists that an account is *not* overdrawn, the queue keeps moving to the other tellers. In contrast, if a cashier is held up waiting for a price check, everyone in that line is stuck (or starts lane hopping, which makes the system behave more like the combined-queue case, but with less efficiency and equity of waiting time).

In a factory, queue sharing can be used to reduce the chance that WIP piles up in front of a machine that is experiencing a long process time. For instance, in Section 8.6.6 we gave an example in which cycle time was 7.67 hours if three machines had individual queues, but only 2.467 hours, (a 67 percent reduction) if the three machines shared a single queue.

Consider another instance. Suppose the arrival rate of jobs is 13.5 jobs per hour (with $c_a = 1$) to a workstation consisting of five machines. Each machine nominally takes 0.3 hours per job with a natural CV of 0.5 (that is, $c_0^2 = 0.25$). The mean time to failure for any machine is 36 hours, and repair times are assumed exponential with a mean time to repair of four hours. Using Equation (8.6), we can compute the effective SCV to be 2.65, so that $c_e = \sqrt{2.65} = 1.63$.

---

[13]Note that if component costs were different we would want to set different fill rates. To reduce total inventory cost, it makes sense to set the fill rate higher for cheaper components and lower for more expensive ones. We ignore this since we are focusing on the efficiency improvement possible through pooling. Chapter 17 presents tools for optimizing stocking rules in multipart inventory systems.

Using the model in Section 8.6.6, we can model both the case with dedicated queues and the case with a single combined queue. In the dedicated queue case, average cycle time is 5.8 hours, while in the combined-queue case it is 1.27 hours, a 78 percent reduction (see Problem 6). Here the reason for the big difference is clear. The combined queue protects jobs against long failures. It is unlikely that all the machines will be down simultaneously, so if the machines are fed by a shared queue, jobs can avoid a failed machine by going to the other machines. This can be a powerful way to mitigate variability in processes with shared machines.

However, if the separate queues are actually different job types and combining them entails a time-consuming setup to switch the machines from one job type to another, then the situation is more complex. The capacity savings by avoiding setups through the use of dedicated queues might offset the variability savings possible by combining the queues. We will examine the tradeoffs involved in setups and batching in systems with variability in Chapter 9.

8.9   Conclusions

This chapter has traversed the complex and subtle topic of variability all the way from the fundamental nature of randomness to the propagation and effects of variability in a production line. Points that are fundamental from a factory physics perspective include the following:

1. *Variability is a fact of life.* Indeed, the field of physics is increasingly indicating that randomness may be an inescapable aspect of existence itself. From a management point of view, it is clear that the ability to deal effectively with variability and uncertainty will be an important skill for the foreseeable future.

2. *There are many sources of variability in manufacturing systems.* Process variability is created by things as simple as work procedure variations and by more complex effects such as setups, random outages, and quality problems. Flow variability is created by the way work is released to the system or moved between stations. As a result, the variability present in a system is the consequence of a host of process selection, system design, quality control, and management decisions.

3. *The coefficient of variation is a key measure of item variability.* Using this unitless ratio of the standard deviation to the mean, we can make consistent comparisons of the level of variability in both process times and flows. At the workstation level, the CV of *effective* process time is inflated by machine failures, setups, recycle, and many other factors. Disruptions that cause long, infrequent outages tend to inflate CV more than disruptions that cause short, frequent outages, given constant availability.

4. *Variability propagates.* Highly variable outputs from one workstation become highly variable inputs to another. At low utilization levels, the flow variability of the output process from a station is determined largely by the variability of the arrival process to that station. However, as utilization increases, flow variability becomes determined by the variability of process times at the station.

5. *Waiting time is frequently the largest component of cycle time.* Two factors contribute to long waiting times: high utilization levels and high levels of variability. The queueing models discussed in this chapter clearly illustrate that both increasing effective capacity (i.e., to bring down utilization levels) and decreasing variability (i.e., to decrease congestion) are useful for reducing cycle time.

6. *Limiting buffers reduces cycle time at the cost of decreasing throughput.* Since limiting interstation buffers is logically equivalent to installing kanban, this property is

the key reason that variability reduction (via production smoothing, improved layout and flow control, total preventive maintenance, and enhanced quality assurance) is critical in just-in-time systems. It also points up the manner in which capacity, WIP buffering, and variability reduction can act as substitutes for one another in achieving desired throughput and cycle time performance. Understanding the tradeoffs among these is fundamental to designing an operating system that supports strategic business goals.

7. *Variability pooling reduces the effects of variability.* Pooling variability tends to dampen the overall variability by making it less likely that a single occurrence will dominate performance. This effect has a variety of factory physics applications. For instance, safety stocks can be reduced by holding stock at a generic level and assembling to order. Also, cycle times at multiple-machine process centers can be reduced by sharing a single queue.

In the next chapter, we will use these insights, along with the concepts and formulas developed, to examine how variability degrades the performance of a manufacturing plant and to provide ways to protect against it.

Study Questions

1. What is the rationale for using the coefficient of variation c instead of the standard deviation σ as a measure of variability?
2. For the following random variables, indicate whether you would expect each to be LV, MV or HV.
   a. Time to complete this set of study questions
   b. Time for a mechanic to replace a muffler on an automobile
   c. Number of rolls of a pair of dice between rolls of seven
   d. Time until failure of a recently repaired machine by a good maintenance technician
   e. Time until failure of a recently repaired machine by a not-so-good technician
   f. Number of words between typographical errors in the book *Factory Physics*
   g. Time between customer arrivals to an automatic teller machine
3. What type of manufacturing workstation does the $M/G/2$ queue represent?
4. Why must utilization be *strictly* less than 100 percent for the $M/M/1$ queueing system to be stable?
5. What is meant by *steady state?* Why is this concept important in the analysis of queueing models?
6. Why is the number of customers at the station an adequate state for summarizing current status in the $M/M/1$ queue but not the $G/G/1$ queue?
7. What happens to CT, WIP, CT_q, and WIP_q as the arrival rate r_a approaches the process rate r_e?

Problems

1. Consider the following sets of interoutput times from a machine. Compute the coefficient of variation for each sample, and suggest a situation under which such behavior might occur.
   a. 5, 5, 5, 5, 5, 5, 5, 5, 5, 5
   b. 5.1, 4.9, 5.0, 5.0, 5.2, 5.1, 4.8, 4.9, 5.0, 5.0
   c. 5, 5, 5, 35, 5, 5, 5, 5, 5, 42
   d. 10, 0, 0, 0, 0, 10, 0, 0, 0, 0
2. Suppose jobs arrive at a single-machine workstation at a rate of 20 per hour and the average process time is two and one-half minutes.
   a. What is the utilization of the machine?

   b. Suppose that interarrival and process times are exponential,
      i. What is the average time a job spends at the station (i.e., waiting plus process time)?
      ii. What is the average number of jobs at the station?
      iii. What is the long-run probability of finding more than three jobs at the station?
   c. Process times are not exponential, but instead have a mean of two and one-half minutes and a standard deviation of five minutes
      i. What is the average time a job spends at the station?
      ii. What is the average number of jobs at the station?
      iii. What is the average number of jobs in the queue?

3. The mean time to expose a single panel in a circuit-board plant is two minutes with a standard deviation of one-half minutes.
   a. What is the natural coefficient of variation?
   b. If the times remain independent, what will be the mean and variance of a job of 60 panels? What will be the coefficient of variation of the job of 60?
   c. Now suppose times to failure on the expose machine are exponentially distributed with a mean of 60 hours and the repair time is also exponentially distributed with a mean of two hours. What are the *effective* mean and CV of the process time for a job of 60 panels?

4. Reconsider the expose machine of Problem 3 with mean time to expose a single panel of two minutes with a standard deviation of one and one-half minutes and jobs of 60 panels. As before, failures occur after about 60 hours of run time, but now happen only between jobs (i.e., these failures do not *preempt* the job). Repair times are the same as before. Compute the effective mean and CV of the process times for the 60 panel jobs. How do these compare with the results in Problem 3?

5. Consider two different machines A and B that could be used at a station. Machine A has a mean effective process time t_e of 1.0 hours and an SCV c_e^2 of 0.25. Machine B has a mean effective process time of 0.85 hour and an SCV of four. (*Hint:* You may find a simple spreadsheet helpful in making the calculations required to answer the following questions.)
   a. For an arrival rate of 0.92 job per hour with $c_a^2 = 1$, which machine will have a shorter average cycle time?
   b. Now put two machines of type A at the station and double the arrival rate (i.e., double the capacity and the throughput). What happens to cycle time? Do the same for machine B. Which type of machine produces shorter average cycle time?
   c. With only one machine at each station, let the arrival rate be 0.95 job per hour with $c_a^2 = 1$. Recompute the average time spent at the stations using both machines A and B. Compare with *a*.
   d. Consider the station with one machine of type A.
      i. Let the arrival rate be one-half. What is the average time spent at the station? What happens to the average time spent at the station if the arrival rate is increased by one percent (i.e., to 0.505)? What percentage increase in wait time does this represent?
      ii. Let the arrival rate be 0.95. What is the average time spent at the station? What happens to the average time spent at the station if the arrival rate is increased by one percent (i.e., to 0.9595)? What percentage increase in wait time does this represent?

6. Consider the example in Section 8.8. The arrival rate of jobs is 13.5 jobs per hour (with $c_a^2 = 1$) to a workstation consisting of five machines. Each machine nominally takes 0.3 hour per job with a natural CV of $\frac{1}{2}$ (that is, $c_0^2 = 0.25$). The mean time to failure for any machine is 36 hours, and repair times are exponential with a mean time to repair of four hours.
   a. Show that the SCV of effective process times is 2.65.
   b. What is the utilization of a single machine when it is allocated one-fifth of the demand (that is, 2.7 jobs per hour) assuming c_a is still equal to one?
   c. What is the utilization of the battery with an arrival rate of 13.5 jobs per hour?
   d. Compute the mean cycle time at a single machine when allocated one-fifth of the demand.
   e. Compute the mean cycle time at the station serving 13.5 jobs per hour.

7. A car company sells 50 different basic models (additional options are added at the dealership after purchases are made). Customers are of two basic types: (1) those who are willing to order the configuration they desire from the factory and wait several weeks for delivery and (2) those who want the car quickly and therefore buy off the lot. The traditional mode of handling customers of the second type is for the dealerships to hold stock of models they think will sell. A newer strategy is to hold stock in regional distribution centers, which can ship cars to dealerships within 24 hours. Under this strategy, dealerships only hold show inventory and a sufficient variety of stock to facilitate test drives.

   Consider a region in which total demand for each of the 50 models is Poisson with a rate of 1,000 cars per month. Replenishment lead time from the factory (to either a dealership or the regional distribution center) is one month.

   a. First consider the case in which inventory is held at the dealerships. Assume that there are 200 dealerships in the region, each of which experiences demand of $1,000/200 = 5$ cars of each of the 50 model types per month (and demand is still Poisson). The dealerships monitor their inventory levels in continuous time and order replenishments in lots of one (i.e., they make use of a base stock model). How many vehicles must each dealership stock to guarantee a fill rate of 99 percent?

   b. Now suppose that all inventory is held at the regional distribution center, which also uses a base stock model to set inventory levels. How much inventory is required to guarantee a 99 percent fill rate?

8. Frequently, natural process times are made up of several distinct stages. For instance, a manual task can be thought of as being comprised of individual motions (or "therbligs" as Gilbreth termed them).

   Suppose a manual task takes a single operator an average of one hour to perform. Alternatively, the task could be separated into 10 distinct six-minute subtasks performed by separate operators. Suppose that the subtask times are independent (i.e., uncorrelated), and assume that the coefficient of variation is 0.75 for both the single large task and the small subtasks. Such an assumption will be valid if the relative shapes of the process time distributions for both large and small tasks are the same. (Recall that the variances of independent random variables are additive.)

   a. What is the coefficient of variation for the 10 subtasks taken together?

   b. Write an expression relating the SCV of the original tasks to the SCV of the combined task.

   c. What are the issues that must be considered before dividing a task into smaller subtasks? Why not divide it into as many as possible? Give several pros and cons.

   d. One of the principles of JIT is to standardize production. How does this explain some of the success of JIT in terms of variability reduction?

9. Consider a workstation with 11 machines (in parallel), each requiring one hour of process time per job with $c_e^2 = 5$. Each machine costs $10,000. Orders for jobs arrive at a rate of 10 per hour with $c_a^2 = 1$ and must be filled. Management has specified a maximum allowable average response time (i.e., time a job spends at the station) of two hours. Currently it is just over three hours (check it).

   Analyze the following options for reducing average response time.

   a. Perform more preventive maintenance so that m_r and m_f are reduced, but m_r/m_f remains the same. This costs $8,000 and does not improve the average process time but does reduce c_e^2 to one.

   b. Add another machine to the workstation at a cost of $10,000. The new machine is identical to existing machines, so $t_e = 1$ and $c_e^2 = 5$.

   c. Modify the existing machines to make them faster without changing the SCV, at a cost of $8,500. The modified machines would have $t_e = 0.96$ and $c_e^2 = 5$.

   What is the best option?

10. (This problem is fairly involved and could be considered a small project.) Consider a simple two-station line as shown in Figure 8.8. Both machines take 20 minutes per job and have

FIGURE 8.8

Two-station line with a finite buffer

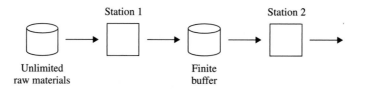

SCV = 1. The first machine can always pull in material, and the second machine can always push material to finished goods. Between the two machines is a buffer that can hold only 10 jobs (see Sections 8.7.1 and 8.7.2).

a. Model the system using an $M/M/1/b$ queue. (Note that $b = 12$ considering the two machines.)

   i. What is the throughput?
   ii. What is the partial WIP (i.e., WIP waiting at the first machine or at the second machine, but not in process at the first machine)?
   iii. What is the total cycle time for the line (not including time in raw material)? (*Hint:* Use Little's law with the partial WIP and the throughput and then add the process time at the first machine.)
   iv. What is the total WIP in the line? (*Hint:* Use Little's law with the total cycle time and the throughput.)

b. Reduce the buffer to one (so that $b = 3$) and recompute the above measures. What happens to throughput, cycle time, and WIP? Comment on this as a strategy.

c. Set the buffer to one and make the process time at the second machine equal to 10 minutes. Recompute the above measures. What happens to throughput, cycle time, and WIP? Comment on this as a strategy.

d. Keep the buffer at one, make the process times for both stations equal to 20 minutes (as in the original case), but set the process CVs to 0.25 (SCV = 0.0625).

   i. What is the throughput?
   ii. Compute an upper bound on the partial WIP at the second machine and waiting at the first machine.
   iii. Compute an (approximate) upper bound on the total cycle time. (*Hint:* Use Little's law with the partial WIP upper bound and the throughput, and then add the time at the first machine.) Is even this upper bound an acceptable cycle time?
   iv. Compute an (approximate) upper bound on the total WIP in the line. (*Hint:* Use Little's law with the upper bound on the total cycle time and the throughput estimate.) Is this upper bound an acceptable WIP level?
   v. Comment on reducing variability as a strategy.

9   THE CORRUPTING INFLUENCE OF VARIABILITY

When luck is on your side, you can do without brains.
Giordano Bruno, burned at the stake in 1600

The more you know the luckier you get.
J. R. Ewing of *Dallas*

9.1   Introduction

The previous chapter developed tools for characterizing and evaluating variability in process times and flows. In this chapter, we use these tools to describe fundamental behavior of manufacturing systems involving variability.

As we did in Chapter 7, we state our main conclusions as laws of factory physics. Some of these "laws" are *always* true (e.g., the Conservation of Material Law), while others hold *most of the time*. On the surface this may appear unscientific. However, we point out that physics laws, such as Newton's second law $F = ma$ and the law of the conservation of energy, hold only approximately. But even though they have been replaced by deeper results of quantum mechanics and relativity, these laws are still very useful. So are the laws of factory physics.

9.1.1   Can Variability Be Good?

The discussions of Chapters 7 and 8 (and the title of this chapter) may give the impression that variability is evil. Using the jargon of lean manufacturing (Womack and Jones 1996), one might be tempted to equate variability with *muda* (waste) and conclude that it should always be eliminated.[1]

But we must be careful not to lose sight of the fundamental objective of the firm. As we observed in Chapter 1, Henry Ford was something of a fanatic about reducing variability. A customer could have any color desired as long as it was *black*. Car models

---

[1] *Muda* is the Japanese word for "waste" and is defined as "any human activity that absorbs resources but creates no *value*." Ohno gave seven examples of *muda:* defects in products, overproduction of goods, inventories of goods awaiting further processing or consumption, unnecessary processing, unnecessary movement, unnecessary transport, and waiting.

were changed infrequently with little variety within models. By stabilizing products and keeping operations simple and efficient, Ford created a major revolution by making automobiles affordable to the masses. However, when General Motors under Alfred P. Sloan offered greater product variety in the 1930s and 1940s, Ford Motor Company lost much of its market share and nearly went under. Of course, greater product variety meant greater variability in GM's production system. Greater variability meant GM's system could not run as efficiently as Ford's. Nonetheless, GM did better than Ford. Why?

The answer is simple. Neither GM nor Ford were in business to reduce variability or even to reduce *muda*. They were in business to *make a good return on investment over the long term*. If adding product variety increases variability and hence *muda* but increases revenues by an amount that more than offsets the additional cost, then it can be a sound business strategy.

9.1.2   Examples of Good and Bad Variability

To highlight the manner in which variability can be good (a necessary implication of a business strategy) or bad (an undesired side effect of a poor operating policy), we consider a few examples.

Table 9.1 lists several causes of undesirable variability. For instance, as we saw in Chapter 8, unplanned outages, such as machine breakdowns, can introduce an enormous amount of variability into a system. While such variability may be unavoidable, it is not something we would deliberately introduce into the system.

In contrast, Table 9.2 gives some cases in which effective corporate strategies consciously introduced variability into the system. As we noted above, at GM in the 1930s and 1940s the variability was a consequence of greater product variety. At Intel in the 1980s and 1990s, the variability was a consequence of rapid product introduction in an environment of changing technology. By aggressively pushing out the next generation of microprocessor before processes for the last generation had stabilized, Intel stimulated demand for new computers and provided a powerful barrier to entry by competitors. At Jiffy Lube, where offering while-you-wait oil changes is the core of the firm's business strategy, demand variability is an unavoidable result. Jiffy Lube could reduce this variability by scheduling oil changes as in traditional auto shops, but doing so would forfeit the company's competitive edge.

Regardless of whether variability is good or bad in business strategy terms, it causes operating problems and therefore must be managed. The specific strategy for dealing with variability will depend on the structure of the system and the firm's strategic goals.

TABLE 9.1   Examples of Bad Variability

Cause	Example
Planned outages	Setups
Unplanned outages	Machine failures
Quality problems	Yield loss and rework
Operator variation	Skill differences
Inadequate design	Engineering changes

TABLE 9.2   Examples of (Potentially) Good Variability

Cause	Example
Product variety	GM in the 1930s and 1940s
Technological change	INTEL in the 1980s and 1990s
Demand variability	Jiffy Lube

In this chapter, we present laws governing the manner in which variability affects the behavior of manufacturing systems. These define key tradeoffs that must be faced in developing effective operations.

9.2   Performance and Variability

In the systems analysis terminology of Chapter 6, management of any system begins with an **objective.** The decision maker manipulates **controls** in an attempt to achieve this objective and evaluates performance in terms of **measures.** For example, the objective of an airplane trip is to take passengers from point *A* to point *B* in a safe and timely manner. To do this, the pilot makes use of many controls while monitoring numerous measures of the plane's performance. The links between controls and measures are well known through the science of aeronautical engineering. Analogously, the objective of a plant manager is to contribute to the firm's long-term profitability by efficiently converting raw materials to goods that will be sold. Like the pilot, the plant manager has many controls and measures to consider. Understanding the relationships between the controls and measures available to a manufacturing manager is the primary goal of factory physics.

A concept at the core of how controls affect measures in production systems is variability. As we saw in Chapter 7, best-case behavior occurs in a line with no variability, while worst-case behavior occurs in a line with maximum variability. In Chapter 8 we observed that several important measures of station performance, such as cycle time and work in process (WIP), are increasing functions of variability.

To understand how variability impacts performance in more general production systems than the idealized lines of Chapter 7 or the single stations of Chapter 8, we need to be more precise about how we define performance. We do this by first discussing *perfect* performance in a production system. Then, by observing the dimensions along which this performance can degrade, we define a set of measures. Finally, we discuss the manner in which the relative weights of these measures depend on both the manufacturing environment and the firm's business strategy.

9.2.1   Measures of Manufacturing Performance

Anyone who has ever peeked into a cockpit knows that the performance of an airplane is not evaluated by a single measure. The impressive array of gauges, dials, meters, LED readouts, etc., is proof that even though the objective is simple (travel from point *A* to point *B*), measuring performance is not. Altitude, direction, thrust, airspeed, groundspeed, elevator settings, engine temperature, etc., must be monitored carefully in order to attain the fundamental objective.

In the same fashion, a manufacturing enterprise has a relatively simple fundamental objective (make money) but a wide array of potential performance measures, such as throughput, inventory, customer service, and quality (see Figure 9.1). Appropriate numerical definitions of performance measures depend on the environment. For example, a styrene plant might measure throughput in straightforward units of pounds per day. A manufacturer of seed planters (devices pulled behind tractors to plant and fertilize as few as 4 or as many as 30 rows at once) might not want to measure throughput in the obvious units of planters per day. The reason is that there is wide variability in size among planters. Measuring throughput in row units per day might be a better measure of aggregate output. Indeed in some systems with many products and complex flows,

FIGURE 9.1

*The manufacturing
control panel*

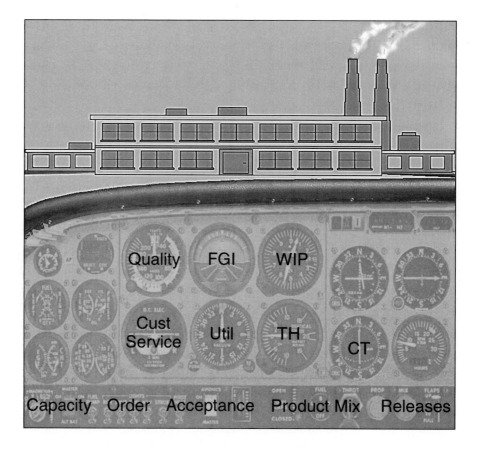

throughput is measured in dollars per day in order to aggregate output into a single number.

The relative importance of performance measures also depends on the specific system and its business strategy. For example, Federal Express, whose competitive advantage is delivery speed and traceability, places a great deal of weight on measures of responsiveness (lead time) and customer service (on-time delivery). The U.S. Postal Service, in contrast, competes largely on price and therefore emphasizes cost-related measures, such as equipment utilization and amount of material handling. Even though both organizations are in the package delivery industry, they have different business strategies targeted at different segments of the market and therefore require different measures of performance.

Given the broad range of production environments and business strategies, it is not possible to define a single set of performance measures for all manufacturing systems. However, to get a sense of what types of measures are possible and to see how these relate to variability, it is useful to consider performance of a simple single-product production line. In principle, measures for more complex multiproduct lines can be developed as extensions of the single-product line measures, and measures for systems made up of many lines can be constructed as weighted combinations of the line measures.

Chapter 7 used throughput, cycle time, and WIP to characterize performance of a simple serial production line. Clearly these are important measures, but they are not comprehensive. Because cost matters, we must also consider equipment utilization. Since the line is fed by a procurement process, another measure of interest is raw material inventory. When we consider customers, lead time, service and finished goods inventory become relevant measures. Finally, since yield loss and rework are often realities, quality is a key performance measure. A perfect single-product line would have throughput exactly equal to demand, full utilization of all equipment, average cycle and lead times as short as possible, perfect customer service (no late or backordered jobs), perfect quality (no scrap or rework), zero raw material or finished goods inventory, and minimum (critical) WIP.

We can characterize each of these measures more precisely in terms of a quantitative efficiency value. For each efficiency, a value of one indicates perfect performance, while zero represents the worst possible performance. To do this, we make use of the following notation, where for specificity we will measure inventories in units of parts and time in days:

$r_e(i)$ = effective rate of station i including detractors such as downtime, setups, and operator efficiency (parts/day)

$r^*(i)$ = ideal rate of station i not including detractors (parts/day)

r_b = bottleneck rate of line including detractors (parts/day)

r_b^* = bottleneck rate of line not including detractors (parts/day)

T_0 = raw process time including detractors (days)

T_0^* = raw process time not including detractors (days)

$W_0 = r_b T_0$ = critical WIP including detractors (parts)

$W_0^* = r_b^* T_0^*$ = critical WIP not including detractors (parts)

D = average demand rate (parts/day)

WIP = average work in process level in line (parts)

FGI = average finished goods inventory level (parts)

RMI = average raw material inventory level (parts)

CT = average cycle time from release to stock point, which is either finished goods or an interline buffer (days)

LT = average lead time quoted to customer; in systems where lead time is fixed, LT is constant; where lead times are quoted individually to customers, it represents an average (days)

TH = average throughput given by ouput rate from line (parts/day)

TH(i) = average throughput (output rate) at station i, which could include multiple visits by some parts due to routing or rework considerations (parts/day)

Notice that the starred parameters, $r^*(i), r_b^*, T_0^*$, and W_0^* are ideal versions of $r_e(i), r_b, T_0$, and W_0. The reason we need them is that a line running at the bottleneck rate and raw process time may actually *not* be exhibiting perfect performance because r_b and T_0 can include many inefficiencies. Perfect performance, therefore, involves two levels. First, the line must attain the best possible performance given its parameters; this is what the best case of Chapter 7 represents. Second, its parameters must be as good as they can be. Thus, perfect performance represents the *best of the best*.

Using the above parameters, we can define seven efficiencies that measure the performance of a single-product line.

Throughput is defined as the rate of parts produced by the line that are *used*. Ideally, this should exactly match demand. Too little production, and we lose sales; too much, and we build up unnecessary finished goods inventory (FGI). Since we will have another measure to penalize excess inventory, we define **throughput efficiency** in terms of whether output is adequate to satisfy demand, so that

$$E_{\text{TH}} = \frac{\min\{\text{TH}, D\}}{D}$$

If throughput is greater than or equal to demand, then throughput efficiency is equal to one. Any shortage will degrade this measure.

Utilization of a station is the fraction of time it is busy. Since unused capacity implies excess cost, an ideal line will have all workstations 100 percent utilized.[2] Furthermore, since a perfect line will not be plagued by detractors, utilization will be 100 percent relative to the best possible (no detractors) rate. Thus, for a line with n stations, we define **utilization efficiency** as

$$E_u = \frac{1}{n} \sum_{i=1}^{n} \frac{\text{TH}(i)}{r^*(i)}$$

Inventory includes RMI, FGI, and WIP. A perfect line would have no raw material inventory (suppliers would deliver literally just-in-time), no finished goods inventory (deliveries to customers would also be made just-in-time), and only the minimum WIP needed for the given throughput, which by Little's Law is $\sum_i \text{TH}(i)/r^*(i)$. Thus a measure of **inventory efficiency** is,

$$E_{\text{inv}} = \frac{\sum_i \text{TH}(i)/r^*(i)}{\text{RMI} + \text{WIP} + \text{FGI}}$$

Cycle time is important to both costs and revenue. Shorter cycle time means less WIP, better quality, better forecasting, and less scrap—all of which reduce costs. It also means better responsiveness, which improves sales revenue. By Little's Law, average cycle time is fully determined by throughput and WIP. Hence, a line with perfect throughput efficiency and inventory efficiency is guaranteed to have perfect cycle time efficiency. However, for imperfect lines WIP is not completely characterized by inventory efficiency (since it involves RMI and FGI), and hence cycle time becomes an independent measure. We define **cycle time efficiency** as the ratio of the best-possible cycle time (raw process time with no detractors) to actual cycle time:

$$E_{\text{CT}} = \frac{T_0^*}{\text{CT}}$$

Lead time is the time quoted to the customer, which should be as short as possible for competitive reasons. Indeed, in make-to-stock systems, lead time is zero, which is clearly as short as possible. However, zero is not a reasonable target for a make-to-order system. Therefore, we define **lead time efficiency** as the ratio of the ideal raw process time to the actual lead time, provided lead time (LT) is at least as large as the ideal raw process time. If lead time is less than this, then we define the lead time efficiency to be one. We can write this as follows:

$$E_{\text{LT}} = \frac{T_0^*}{\max\{\text{LT}, T_0^*\}}$$

Notice that in a make-to-order system we could quote unreasonably short lead times (less than T_0^*) and ensure that this measure is one. But if the line is not capable of delivering product this quickly, the measure of customer service will suffer.

---

[2]Note that 100 percent utilization is only possible in *perfect* lines. In realistic lines containing variability, pushing utilization close to one will seriously degrade other measures. It is critical to remember that system performance is measured by *all* the efficiencies, not by any single number.

Customer service is the fraction of demands that are satisfied on time. In a make-to-stock situation, this is the fill rate (fraction of demands filled from stock, rather than backordered). In a make-to-order system, customer service is the fraction of orders that are filled within their lead times (i.e., cycle time is less than or equal to lead time). Hence, we define **service efficiency** as the customer service itself:

$$E_S = \begin{cases} \text{fraction of demand filled from stock in make-to-stock system} \\ \text{fraction of orders filled within lead time in make-to-order system} \end{cases}$$

Quality is a complex characteristic of the product, process, and customer (see Chapter 12 for a discussion). For operational purposes, the essential aspect of quality is captured by the fraction of parts that are made correctly the first time through the line. Any scrap or rework decreases this value. Hence, we measure **quality efficiency** as

$$E_Q = \text{fraction of jobs that go through line with no defects on first pass}$$

These efficiencies are stated specifically for a single-product line. However, one could extend these measures to a multiproduct line by aggregating the flows and inventories (e.g., in dollars) and measuring cycle time, lead time, and service individually by product (see Problem 1).

A perfect single-product line will have all seven of the above efficiencies equal to one. For example, Penny Fab One of Chapter 7 has no detractors, so $r_b = r_b^*$ and $T_0 = T_0^*$. If raw materials are delivered just in time (one penny blank every two hours), customer orders are promised (and shipped) every two hours, and the CONWIP level is set at WIP $= W_0^*$, then inventory, lead time, and service efficiencies will all be one. Finally, since there are no quality problems, quality efficiency is also one. Obviously we would not expect to see such perfect performance in the real world. All realistic production systems will have some efficiencies less than one.

In less than perfect lines, performance is a composite of these efficiencies (or similar ones suited to the specific environment of the line). In theory, we could construct a single-number measure of efficiency as a weighted average of these efficiencies. As we noted, however, the individual weights would be highly dependent on the nature of the line and its business. For instance, a commodity producer with expensive capital equipment would stress utilization and service efficiency much more than inventory efficiency, while a specialty job shop would stress lead time efficiency at the expense of utilization efficiency.

Consider the example shown in Figure 9.2, which represents a card stuffing operation line feeding an assembly operation in a "box plant" making personal computers. In this case, finished goods inventory is really intermediate stock for the final assembly operation controlled by a kanban system. The five percent rework through the last station represents cards that must be touched up. Cards that are reworked never need to be reworked again.

FIGURE 9.2

Operational efficiency example

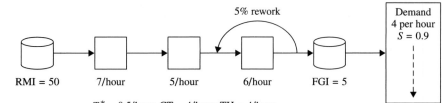

RMI = 50   7/hour   5/hour   6/hour   FGI = 5

5% rework

Demand
4 per hour
S = 0.9

$T_0^* = 0.5$/hour, CT = 4/hour, TH = 4/hour

Since TH is equal to demand, throughput efficiency E_{TH} is equal to one. Cycle time efficiency is given by $E_{CT} = T_0^*/CT = 0.5/4 = 0.125$. Utilization efficiency is the average of the individual station utilizations. To get this, we must first compute the throughput at each station. Because there is five percent rework at station 3,

$$TH(3) = TH + 0.05TH = 1.05(4) = 4.2$$

Since there is no rework at stations 1 and 2, $TH(1) = TH(2) = 4$. Thus, utilization efficiency is

$$E_u = \frac{1}{3} \sum_{i=1}^{3} \frac{TH(i)}{r^*(i)} = \frac{\frac{4}{7} + \frac{4}{5} + \frac{4.2}{6}}{3} = 0.6905$$

According to the problem data, service efficiency E_S is 0.9. Since production is controlled by a kanban system, lead time is zero so that $E_{LT} = 1.0$. Quality efficiency E_Q is also given as part of the data and is 0.95. To compute inventory efficiency, we must first compute WIP from Little's Law: WIP = TH × CT = (4 cards per hour)(4 hours) = 16 cards; and the ideal WIP is given by $\sum_i TH(i)/r^*(i) = \frac{4}{7} + \frac{4}{5} + \frac{4.2}{6} = 2.071$. Then we compute

$$E_{inv} = \frac{\sum_i TH(i)/r^*(i)}{RMI + WIP + FGI} = \frac{2.071}{50 + 16 + 5} = 0.0292$$

Now suppose we increase the kanban level so that, on average, there are 15 cards in FGI; and suppose that this change causes the service level to increase to 0.999. While the other efficiencies stay the same, E_S becomes 0.999 and E_{inv} goes down to 0.0256. Table 9.3 compares the two systems.

Which system is better? It depends on whether the firm's business strategy deems it more important to have high customer service or low inventory. Most likely in this environment the modified system is better, since the stuffing line's customer is the assembly line and shutting it down 10 percent of the time would probably result in unacceptable service to the ultimate customer.

9.2.2   Variability Laws

Now that we have defined performance in reasonably concrete terms, we can characterize the effect of variability on performance. Variability can affect supplier deliveries, manufacturing process times, or customer demand. If we examine these carefully, we see that increasing any source of variability will degrade at least one of the above efficiency measures. For instance, if we increase the variability of process times while holding throughput constant, we know from the *VUT* equation of Chapter 8 that WIP will

TABLE 9.3   System Efficiency Comparison

Measure	Card Stuffing System	Modified Card Stuffing System
Cycle time	0.1250	0.1250
Utilization	0.6905	0.6905
Service	0.9000	0.9990
Quality	0.9500	0.9500
Inventory	0.0292	0.0256

increase, thereby degrading inventory efficiency. If we place a restriction on WIP (via kanban or CONWIP), then by our analysis of queueing systems with blocking we know that, in general, throughput will decline (because the bottleneck will starve), thereby degrading throughput efficiency.

These observations are specific instances of the following fundamental law of factory physics.

Law (Variability): *Increasing variability always degrades the performance of a production system.*

This is an extremely powerful concept, since it implies that higher variability of any sort must harm some measure of performance. Consequently, variability reduction is central to improving performance, regardless of the specific weights a firm attaches to the individual performance measures. Indeed, much of the success of JIT methods was a consequence of recognizing the power of variability reduction and developing methods for achieving it (e.g., production smoothing, setup reduction, total quality management, and total preventive maintenance).

We can deepen the insight of the Variability Law by observing that increasing variability impacts the system along three general dimensions: inventory, capacity, and time. Clearly, inventory efficiency measures the inventory impact. Production and utilization efficiency are measures of the capacity impact. Cycle time and lead time efficiency measure the time impact, as does service efficiency, since the customer must wait for parts that are not ready. Finally, quality efficiency impacts the system in all three dimensions: Scrap or rework requires additional capacity, redoing an operation requires additional time, and parts being (or waiting to be) repaired or redone add inventory to the system.

Another way to view these three impacts is as **buffers** with which we control the system. Worse performance corresponds to more buffering. We can summarize this as the following factory physics law.

Law (Variability Buffering): *Variability in a production system* will *be buffered by some combination of*

1. *Inventory*
2. *Capacity*
3. *Time*

This law is an enormously important extension of the Variability Law because it enumerates the ways in which variability can impact a system. While there is no question that variability will degrade performance, we have a choice of *how* it will do so. Different strategies for coping with variability make sense in different business environments. For instance, in the earlier board-stuffing example, the modified system used a larger inventory buffer to enable a smaller time (service) buffer, a change that made good business sense in that environment. We offer some additional examples of the different ways to buffer variability.

9.2.3   Buffering Examples

The following examples illustrate (1) that variability must be buffered and (2) how the appropriate buffering strategy depends on the production environment and business strategy. We deliberately include some nonmanufacturing examples to emphasize that the variability laws apply to production systems for services as well as for goods.

Ballpoint pens. Suppose a retailer sells inexpensive ballpoint pens. Demand is unpredictable (variable). But since customers will go elsewhere if they do not find the item in stock (who is going to backorder a cheap ballpoint pen?), the retailer cannot buffer this variability with time. Likewise, because the instant-delivery requirement of the customer rules out a make-to-order environment, capacity cannot be used as a buffer. This leaves only inventory. And indeed, this is precisely what the retailer creates by holding a stock of pens.

Emergency service. Demand for fire or ambulance service is necessarily variable, since we obviously cannot get people to schedule their emergencies. We cannot buffer this variability with inventory (an inventory of trips to the hospital?). We cannot buffer with time, since response time is *the* key performance measure for this system. Hence, the only available buffer is capacity. And indeed, utilization of fire engines and ambulances is very low. The "excess" capacity is necessary to cover peaks in demand.

Organ transplants. Demand for organ transplants is variable, as is supply, since we cannot schedule either. Since the supply rate is fixed by donor deaths, we cannot (ethically) increase capacity. Since organs have a very short usable life after the donor dies, we cannot use inventory as a buffer. This leaves only time. And indeed, the waiting time for most organ transplants is very long. Even medical production systems must obey the laws of factory physics.

The Toyota Production System. The Toyota production system was the birthplace of JIT and remains the paragon of lean manufacturing. On the basis of its success, Toyota rose from relative obscurity to become one of the world's leading auto manufacturers. How did they do it?

First, Toyota reduced variability at every opportunity. In particular:

1. *Demand variability.* Toyota's product design and marketing were so successful that demand for its cars consistently exceeded supply (the Big Three in America also did their part by building particularly shoddy cars in the late 1970s). This helped in several ways. First, Toyota was able to limit the number of options of cars produced. A maroon Toyota would always have maroon interior. Many options, such as chrome packages and radios, were dealer installed. Second, Toyota could establish a production schedule months in advance. This virtually eliminated all demand variability seen by the manufacturing facility.

2. *Manufacturing variability.* By focusing on setup reduction, standardizing work practices, total quality management, error proofing, total preventive maintenance, and other flow-smoothing techniques, Toyota did much to eliminate variability inside its factories.

3. *Supplier variability.* The Toyota-supplier relationship in the early 1980s hinted of feudalism. Because Toyota was such a large portion of its suppliers' demand, it had enormous leverage. Indeed, Toyota executives often sat as directors on the boards of its suppliers. This ensured that (1) Toyota got the supplies it needed when it needed them, (2) suppliers adopted variability reduction techniques "suggested" to them by Toyota, and (3) the suppliers carried any necessary buffer inventory.

Second, Toyota made use of capacity buffers against remaining manufacturing variability. It did this by scheduling plants for less than three shifts per day and making use of preventive maintenance periods at the end of shifts to make up any

FIGURE 9.3

"Pay me now or pay me later" scenario

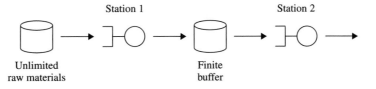

Station 1   Station 2

Unlimited
raw materials

Finite
buffer

shortfalls relative to production quotas. The result was a very predictable daily production rate.

Third, despite the propensity of American JIT writers to speak in terms of "zero inventories" and "evil inventory," Toyota did carry WIP and finished goods inventories in its system. But because of its vigorous variability reduction efforts and willingness to buffer with capacity, the amount of inventory required was far smaller than was typical of auto manufacturers in the 1980s.

9.2.4   Pay Me Now or Pay Me Later

The Buffering Law could also be called the "law of pay me now or pay me later" because if you do not pay to reduce variability, you *will* pay in one or more of the following ways:

- Lost throughput.
- Wasted capacity.
- Inflated cycle times.
- Larger inventory levels.
- Long lead times and/or poor customer service.

To examine the implications of the Buffering Law in more concrete manufacturing terms, we consider the simple two-station line shown in Figure 9.3. Station 1 pulls in jobs, which contain 50 pieces, from an unlimited supply of raw materials, processes them, and sends them to a buffer in front of station 2. Station 2 pulls jobs from the buffer, processes them, and sends them downstream. Throughout this example, we assume station 1 requires 20 minutes to process a job and is the bottleneck. This means that the theoretical capacity is 3,600 pieces per day (24 hours/day $\times$ 60 minutes/hour $\times$ 1 job/20 minutes $\times$ 50 pieces/job).[3]

To start with, we assume that station 2 also has average processing times of 20 minutes, so that the line is balanced. Thus, the theoretical minimum cycle time is 40 minutes, and the minimum WIP level is 100 pieces (one job per station). However, because of variability, the system cannot achieve this ideal performance. Below we discuss the results of a computer simulation model of this system under various conditions, to illustrate the impacts of changes in capacity, variability, and buffer space. These results are summarized in Table 9.4.

Balanced, Moderate Variability, Large Buffer.   As our starting point, we consider the balanced line where both machines have mean process times of 20 minutes per job and are moderately variable (i.e., have process CVs equal to one, so $c_e(1) = c_e(2) = 1$)

---

[3]This is the same system that was considered in Problem 10 of Chapter 8.

TABLE 9.4   Summary of Pay-Me-Now-or-Pay-Me-Later Simulation Results

Case	Buffer (Jobs)	$t_e(2)$ (Minutes)	CV	TH (per Day)		CT (Minutes)	WIP (Pieces)
				E_{TH}	E_u	E_{CT}	E_{inv}
1	10	20	1	3,321		150	347
				0.9225	0.9225	0.2667	0.2659
2	1	20	1	2,712		60	113
				0.7533	0.7533	0.6667	0.6667
3	1	10	1	3,367		36	83
				0.9353	0.7015	0.8333	0.8451
4	1	20	0.25	3,443		51	123
				0.9564	0.9564	0.7843	0.7776

and the interstation buffer holds 10 jobs (500 pieces).[4]   A simulation of this system for 1,000,000 minutes (694 days running 24 hours/day) estimates throughput of 3,321 pieces/day, an average cycle time of 150 minutes, and an average WIP of 347 pieces. We can check Little's Law (WIP = TH × CT) by noting that throughput can be expressed as 3,321 pieces/day ÷ 1,440 minutes/day = 2.3 pieces/minute, so

$$347 \text{ pieces} \approx 2.3 \text{ pieces/minute} \times 150 \text{ minutes} = 345 \text{ pieces}$$

Because we are simulating a system involving variability, the estimates of TH, CT, and WIP are necessarily subject to error. However, because we used a long simulation run, the system was allowed to stabilize and therefore very nearly complies with Little's Law.

Notice that while this configuration achieves reasonable throughput (i.e., only 7.7 percent below the theoretical maximum of 3,600 pieces per day), it does so at the cost of high WIP and long cycle times. The reason is that fluctuations in the speeds of the two stations causes the interstation buffer to fill up regularly, which inflates both WIP and cycle time. Hence, the system is using WIP as the primary buffer against variability.

Balanced, Moderate Variability, Small Buffer.   One way to reduce the high WIP and cycle time of the above case is by fiat. That is, simply reduce the size of the buffer. This is effectively what implementing a low-WIP kanban system without any other structural changes would do. To give a stark illustration of the impacts of this approach, we reduce buffer size from 10 jobs to 1 job. If the first machine finishes when the second has one job in queue, it will wait in a nonproductive *blocked* state until the second machine is finished.

---

[4]Note that because the line is balanced and has an unlimited supply of work at the front, utilization at both machines would be 100 percent if the interstation buffer were infinitely large. But this would result in an unstable system in which the WIP would grow to infinity. A finite buffer will occasionally become full and block station 1, choking off releases and preventing WIP from growing indefinitely. This serves to stabilize the system and makes it more representative of a real production system, in which WIP levels would never be allowed to become infinite.

Our simulation model confirms that the small buffer reduces cycle time and WIP as expected, with cycle time dropping to around 60 minutes and WIP dropping to around 113 pieces. However, throughput also drops to around 2,712 pieces per day (an 18 percent decrease relative to the first case). Without the high WIP level in the buffer to protect station 2 against fluctuations in the speed of station 1, station 2 frequently becomes starved for jobs to work on. Hence, throughput and revenue seriously decline. Because utilization of station 2 has fallen, the system is now using capacity as the primary buffer against variability. However, in most environments, this would not be an acceptable price to pay for reducing cycle time and WIP.

Unbalanced, Moderate Variability, Small Buffer.   Part of the reason that stations 1 and 2 are prone to blocking and starving each other in the above case is that their capacities are identical. If a job is in the buffer and station 1 completes its job before station 2 is finished, station 1 becomes *blocked;* if the buffer is empty and station 2 completes its job before station 1 is finished, station 2 becomes *starved.* Since both situations occur often, neither station is able to run at anything close to its capacity.

One way to resolve this is to unbalance the line. If either machine were significantly faster than the other, it would almost always finish its job first, thereby allowing the other station to operate at close to its capacity. To illustrate this, we suppose that the machine at station 2 is replaced with one that runs twice as fast (i.e., has mean process times of $t_e(2) = 10$ minutes per job), but still has the same CV (that is, $c_e(2) = 1$). We keep the buffer size at one job.

Our simulation model predicts a dramatic increase in throughput to 3,367 pieces per day, while cycle time and WIP level remain low at 36 minutes and 83 pieces, respectively. Of course, the price for this improved performance is wasted capacity—the utilization of station 2 is less than 50 percent—so the system is again using capacity as a buffer against variability. If the faster machine is inexpensive, this might be attractive. However, if it is costly, this option is almost certainly unacceptable.

Balanced, Low Variability, Small Buffer.   Finally, to achieve high throughput with low cycle time and WIP *without* resorting to wasted capacity, we consider the option of reducing variability. In this case, we return to a balanced line, with both stations having mean process times of 20 minutes per job. However, we assume the process CVs have been reduced from 1.0 to 0.25 (i.e., from the moderate-variability category to the low-variability category).

Under these conditions, our simulation model shows that throughput is high, at 3,443 pieces per day; cycle time is low, at 51 minutes; and WIP level is low, at 123 pieces. Hence, if this variability reduction is feasible and affordable, it offers the best of all possible worlds. As we noted in Chapter 8, there are many options for reducing process variability, including improving machine reliability, speeding up equipment repairs, shortening setups, and minimizing operator outages, among others.

Comparison.   As we can see from the summary in Table 9.4, the above four cases are a direct illustration of the pay-me-now-or-pay-me-later interpretation of the Variability Buffering Law. In the first case, we "pay" for throughput by means of long cycle times and high WIP levels. In the second case, we pay for short cycle times and low WIP levels with lost throughput. In the third case we pay for them with wasted capacity. In the fourth case, we pay for high throughput, short cycle time, and low WIP through variability reduction. While the Variability Buffering Law cannot specify which form of payment is best, it does serve warning that some kind of payment *will* be made.

9.2.5   Flexibility

Although variability always requires some kind of buffer, the effects can be mitigated somewhat with **flexibility.** A flexible buffer is one that can be used in more than one way. Since a flexible buffer is more likely to be available when and where it is needed than a fixed buffer is, we can state the following corollary to the buffering law.

Corollary (Buffer Flexibility):   *Flexibility reduces the amount of variability buffering required in a production system.*

An example of flexible capacity is a cross-trained workforce. By floating to operations that need the capacity, flexible workers can cover the same workload with less total capacity than would be required if workers were fixed to specific tasks.

An example of flexible inventory is generic WIP held in a system with late product customization. For instance, Hewlett-Packard produced generic printers for the European market by leaving off the country-specific power connections. These generic printers could be assembled to order to fill demand from any country in Europe. The result was that significantly less generic (flexible) inventory was required to ensure customer service than would have been required if fixed (country-specific) inventory had been used.

An example of flexible time is the practice of quoting variable lead times to customers depending on the current work backlog (i.e., the larger the backlog, the longer the quote). A given level of customer service can be achieved with shorter average lead time if variable lead times are quoted individually to customers than if a uniform fixed lead time is quoted in advance. We present a model for lead time quoting in Chapter 15.

There are many ways that flexibility can be built into production systems, through product design, facility design, process equipment, labor policies, vendor management, etc. Finding creative new ways to make resources more flexible is the central challenge of the mass customization approach to making a diverse set of products at mass production costs.

9.2.6   Organizational Learning

The pay-me-now-or-pay-me-later example suggests that adding capacity and reducing variability are, in some sense, interchangeable options. Both can be used to reduce cycle times for a given throughput level or to increase throughput for a given cycle time. However, there are certain intangibles to consider. First is the ease of implementation. Increasing capacity is often an easy solution—just buy some more machines—while decreasing variability is generally more difficult (and risky), requiring identification of the source of excess variability and execution of a custom-designed policy to eliminate it. From this standpoint, it would seem that if the costs and impacts to the line of capacity expansion and variability reduction are the same, capacity increases are the more attractive option.

But there is a second important intangible to consider—*learning.* A successful variability reduction program can generate capabilities that are transferable to other parts of the business. The experience of conducting systems analysis studies (discussed in Chapter 6), the resulting improvements in specific processes (e.g., reduced setup times or rework), and the heightened awareness of the consequences of variability by the workforce are examples of benefits from a variability reduction program whose impact can spread well beyond that of the original program. The mind-set of variability reduction promotes an environment of continual process capability improvement. This can be a source of significant competitive advantage—anyone can buy more machinery, but not everyone can constantly upgrade the ability to use it. For this reason, we believe that variability reduction is frequently the preferred improvement option, which should be considered seriously before resorting to capacity increases.

9.3   Flow Laws

Variability impacts the way material flows through the system and how much capacity can be actually utilized. In this section we describe laws concerning material flow, capacity, utilization, and variability propagation.

9.3.1   Product Flows

We start with an important law that comes directly from (natural) physics, namely *Conservation of Material.* In manufacturing terms, we can state it as follows:

Law (Conservation of Material):   *In a stable system, over the long run, the rate out of a system will equal the rate in, less any yield loss, plus any parts production within the system.*

The phrase *in a stable system* requires that the input to the system not exceed (or even be equal to) its capacity. The next phrase, *over the long run,* implies that the system is observed over a significantly long time. The law can obviously be violated over shorter intervals. For instance, more material may come out of a plant than went into it—for awhile. Of course, when this happens, WIP in the plant will fall and eventually will become zero, causing output to stop. Thus, the law cannot be violated indefinitely. The last phrases, *less any yield loss* and *plus any parts production* are important caveats to the simpler statement, *input must equal output.* Yield losses occur when the number of parts in a system is reduced by some means other than output (e.g., scrap or damage). Parts production occurs whenever one part becomes multiple parts. For instance, one piece of sheet metal may be cut into several smaller pieces by a shearing operation.

This law links the utilization of the individual stations in a line with the throughput. For instance, in a serial line with no yield loss operating under an MRP (push) protocol, throughput at any station i, TH(i), plus the line throughput itself, TH, equals the release rate r_a into the line. The reason, of course, is that what goes in must come out (provided that the release rate is less than the capacity of the line, so that it is stable). Then the utilization at each station is given by the ratio of the throughput to the station capacity (for example, $u(i) = \text{TH}(i)/r_e(i) = r_a/r_e(i)$ at station i).

Finally, this law is behind our choice to define the bottleneck as the *busiest* station, not necessarily the *slowest* station. For example, if a line has yield loss, then a slower station later in the line may have a lower utilization than a faster station earlier in the line (i.e., because the earlier station processes parts that are later scrapped). Since the earlier station will serve to constrain the performance of the line, it is rightly deemed the bottleneck.

9.3.2   Capacity

The Conservation of Material Law implies that the capacity of a line must be at least as large as the arrival rate to the system. Otherwise, the WIP levels would continue to grow and never stabilize. However, when one considers variability, this condition is not strong enough. To see why, recall that the queueing models presented in Chapter 8 indicated that both WIP and cycle time go to infinity as utilization approaches one if there is no limit on how much WIP can be in the system. Therefore, to be stable, all workstations in the system must have a processing rate that is *strictly greater* than the arrival rate to that station. It turns out that this behavior is not some sort of mathematical oddity, but is, in fact, a fundamental principle of factory physics.

To see this, note that if a production system contains variability (and all real systems do), then regardless of the WIP level, we can always find a possible sequence of events that causes the system bottleneck to *starve* (run out of WIP). The only way to ensure that the bottleneck station does not starve is to *always* have WIP in the queue. However, no matter how much WIP we begin with, there exists a set of process and interarrival times that will eventually exhaust it. The only way to *always* have WIP is to start with an *infinite* amount of it. Thus, for r_a (arrival rate) to be equal to r_e (process rate), there must be an infinite amount of WIP in the queue. But by Little's Law this implies that cycle time will be infinite as well.

There is one exception to this behavior. When both c_a^2 and c_e^2 are equal to zero, then the system is completely deterministic. For this case, we have *absolutely no* randomness in either interarrival or process time, and the arrival rate is *exactly* equal to the service rate. However, since modern physics ("natural," not "factory") tells us that there is always some randomness present, this case will never arise in practice.

At this point, the reader with a practical bent may be skeptical, thinking something like, "Wait a minute. I've been in a lot of plants, many of which do their best to set work releases equal to capacity, and I've yet to see a single one with an *infinite* amount of WIP." This is a valid point, which brings up the important concept of **steady state.**

Steady state is related to the notion of a "stable system" and "long-run" performance, discussed in the conservation of material law. For a system to be in steady state, the parameters of the system must *never* change and the system must have been operating long enough that initial conditions no longer matter.[5] Since our formulas were derived under the assumption of steady state, the discrepancy between our analysis (which is correct) and what we see in real life (which is also correct) must lie in our view of the steady state of a manufacturing system.

The Overtime Vicious Cycle.   What really happens in steady state is that a plant runs through a series of "cycles," in which system parameters are changed over time. A common type of behavior is the "overtime vicious cycle," which goes as follows:

1. Plant capacity is computed by taking into consideration detractors such as random outages, recycle, setups, operator unavailability, breaks, and lunches.
2. The master production schedule is filled according to this effective capacity. Release rates are now essentially the same as capacity.[6]
3. Sooner or later, due to randomness in job arrivals, in process times, or in both, the bottleneck process starves.

---

[5]Recall in the Penny Fab examples of Chapter 7 that the line had to run for awhile to work out of a transient condition caused by starting up with all pennies at the first station. There, steady state was reached when the line began to cycle through the same behavior over and over. In lines with variability, the actual behavior will not repeat, but the probability of finding the system in a given state will stabilize.

[6]Notice that if there has been some wishful thinking in computing capacity, release rates may well be *greater* than capacity.

4. More work has gone in than has gone out, so WIP increases.

5. Since the system is at capacity, throughput remains relatively constant. From Little's Law, the increase in WIP is reflected by a nearly proportional increase in cycle times.

6. Jobs become late.

7. Customers begin to complain.

8. After WIP and cycle times have increased enough and customer complaints grow loud enough, management decides to take action.

9. A "one-time" authorization of overtime, adding a shift, subcontracting, rejection of new orders, etc., is allowed.

10. As a consequence of step 9, effective capacity is now significantly greater than the release rate. For instance, if a third shift was added, utilization dropped from 100 percent to around 67 percent.

11. WIP level decreases, cycle times go down, and customer service improves. Everyone breaths a sigh of relief, wonders aloud how things got so out of hand, and promises to never let it happen again.

12. *Go to step 1!*

The moral of the overtime vicious cycle is that although management may *intend* to release work at the rate of the bottleneck, in steady state, it *cannot.* Whenever overtime, or adding a shift, or working on a weekend, or subcontracting, etc., is authorized, plant capacity suddenly jumps to a level significantly greater than the release rate. (Likewise, order rejection causes release rate to suddenly fall below capacity.) Thus, over the long run, *average* release rate is *always* less than *average* capacity. We can sum up this fact of manufacturing life with the following law of factory physics.

Law (Capacity): *In steady state, all plants* will *release work at an average rate that is strictly less than the average capacity.*

This law has profound implications. Since it is impossible to achieve true 100 percent utilization of plant resources, the real management decision concerns whether measures such as excess capacity, overtime, or subcontracting will be part of a planned strategy or will be used in response to conditions that are spinning out of control. Unfortunately, because many manufacturing managers fail to appreciate this law of factory physics, they unconsciously choose to run their factories in constant "fire-fighting" mode.

9.3.3   Utilization

The Buffering Law and the *VUT* equation suggest that there are two drivers of queue time: utilization and variability. Of these, utilization has the most dramatic effect. The reason is that the *VUT* equation (for single- or multiple-machine stations) has a $1 - u$ term in the denominator. Hence as utilization u approaches one, cycle time approaches infinity. We can state this as the following law.

Law (Utilization): *If a station increases utilization without making any other changes, average WIP and cycle time will increase in a highly nonlinear fashion.*

In practice, it is the phrase *in a highly nonlinear fashion* that generally presents the real problem. To illustrate why, suppose utilization is $u = 97$ percent, cycle time is two days, and the CVs of both process times c_e and interarrival times c_a are equal to one. If we increase utilization by one percent to $u = 0.9797$, cycle time becomes 2.96 days, a 48 percent increase. Clearly, cycle time is very sensitive to utilization. Moreover, this effect becomes even more pronounced as u gets closer to one, as we can see in Figure 9.4. This graph shows the relationship between cycle time and utilization for $V = 1.0$ and $V = 0.25$, where $V = (c_a^2 + c_e^2)/2$. Notice that both curves "blow up" as u gets close to 1.0, but the curve corresponding to the system with higher variability ($V = 1.0$) blows up faster. From Little's Law, we can conclude that WIP similarly blows up as u approaches one.

A couple of technical caveats are in order. First, if $V = 0$, then cycle time remains constant for all utilization levels up to 100 percent and then becomes infinite (infeasible) when utilization becomes greater than 100 percent. In analogous fashion to the best-case line we studied in Chapter 7, a station with absolutely no variability can operate at 100 percent utilization without building a queue. But since all real stations contain some variability, this never occurs in practice.

Second, no real-world station has space to build an infinite queue. Space, time, or policy will serve to cap WIP at some finite level. As we saw in the blocking models of Chapter 8, putting a limit on WIP without any other changes causes throughput (and hence utilization) to decrease. Thus, the qualitative relationship in Figure 9.4 still holds, but the limit on queue size will make it impossible to reach the high utilization/high cycle time parts of the curve.

The extreme sensitivity of system performance to utilization makes it very difficult to choose a release rate that achieves both high station efficiency and short cycle times. Any errors, particularly those on the high side (which are likely to occur as a result of optimism about the system's capacity, coupled with the desire to maximize output), can result in large increases in average cycle time. We will discuss structural changes for addressing this issue in Chapter 10 in the context of push and pull production systems.

9.3.4   Variability and Flow

The Variability Law states that variability degrades performance of all production systems. But how much it degrades performance can depend on *where* in the line the variability is created. In lines without WIP control, increasing process variability at any station will (1) increase the cycle time at that station *and* (2) propagate more variability to downstream stations, thereby increasing cycle time at them as well. This observation

FIGURE 9.4

Relation between cycle time and utilization

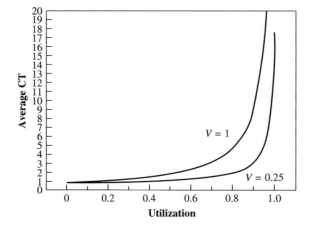

motivates the following corollary of the Variability Law and the propagation property of Chapter 8.

Corollary (Variability Placement): *In a line where releases are independent of completions, variability early in a routing increases cycle time more than equivalent variability later in the routing.*

The implication of this corollary is that efforts to reduce variability should be directed at the front of the line first, because that is where they are likely to have the greatest impact (see Problem 12 for an illustration).

Note that this corollary applies *only* where releases are independent of completions. In a CONWIP line, where releases are directly tied to completions, the flow at the first station is affected by flow at the last station just as strongly as the flow at station $i + 1$ is affected by the flow at station i. Hence, there is little distinction between the front and back of the line and little incentive to reduce variability early as opposed to late in the line. The variability placement corollary, therefore, is applicable to push rather than pull systems.

9.4 Batching Laws

A particularly dramatic cause of variability is batching. As we saw in the worst-case performance in Chapter 7, maximum variability can occur when moving product in large batches even when process times themselves are constant. The reason in that example was that the effective interarrival times were large for the first part in a batch and zero for all others (because they arrived simultaneously). The result was that each station "saw" highly variable arrivals, hence the average cycle time was as bad as it could possibly be for a given bottleneck rate and raw process time. Because batching can have such a large effect on variability, and hence performance, setting batch sizes in a manufacturing system is a very important control. However, before we try to compute "optimal" batch sizes (which we will save for Chapter 15 as part of our treatment of scheduling), we need to understand the effects of batching on the system.

9.4.1 Types of Batches

An issue that sometimes clouds discussions of batching is that there are actually two kinds of batches. Consider a dedicated assembly line that makes only one type of product. After each unit is made, it is moved to a painting operation. What is the batch size?

On one hand, you might say it is *one* because after each item is complete, it can be moved to the painting operation. On the other hand, you could argue that the batch size is *infinity* since you never perform a changeover (i.e., the number of parts between changeovers is infinite). Since one is not equal to infinity, which is correct?

The answer is that both are correct. But there are two different kinds of batches: **process batches** and **transfer batches.**

> **Process batch.** There are two types of process batches. The **serial batch** size is the number of jobs of a common family processed before the workstation is changed over to another family. We call these *serial* batches because the parts are produced serially (one at a time) on the workstation. **Parallel batch** size is the number of parts produced simultaneously in a true batch workstation, such as a furnace or heat treat operation. Although serial and parallel batches are very different physically, they have similar operational impacts, as we will see.

The size of a serial process batch is related to the length of a changeover or setup. The longer the setup, the more parts must be produced between setups to achieve a given capacity. The size of a parallel process batch depends on the demand placed on the station. To minimize utilization, such machines should be run with a full batch. However, if the machine is not a bottleneck, then minimizing utilization may not be critical, so running less than a full load may be the right thing to do to reduce cycle times.

Transfer batch. This is the number of parts that accumulate before being transferred to the next station. The smaller the transfer batch, the shorter the cycle time since there is less time waiting for the batch to form. However, smaller transfer batches also result in more material handling, so there is a tradeoff. For instance, a forklift might be needed only once per shift to move material between adjacent stations in a line if moves are made in batches of 3,000 units. However, the operator would have to make 30 trips per shift to move material between the stations in batches of 100 units.

Strictly speaking, if one considers the material handling operation between stations to be a process, a transfer batch is simply a parallel process batch. The forklift can transfer 10 parts as quickly as one, just as a furnace can bake 10 parts as quickly as one. Nonetheless, since it is intuitive to think of material handling as distinct from processing, we will consider transfer and process batching separately.

The distinction between process and transfer batches is sometimes overlooked. Indeed, from the time Ford Harris first derived the EOQ in 1913 until recently, most production planners simply assumed that these two batches should be equal. But this need not be so. In a system where setups are long but processes are close together, it might make good sense to keep process batches large and transfer batches small. This practice is called **lot splitting** and can significantly reduce the cycle time (we discuss this in greater detail in Section 9.5.3).

9.4.2   Process Batching

Recall from Chapter 4 that JIT advocates are fond of calling for batch sizes of one. The reason is that if processing is done one part at a time, no time is spent waiting for the batch to form and less time is spent waiting in a queue of large batches. However, in most real-world systems, setting batch sizes equal to one is not so simple. The reason is that batch size can affect *capacity*. It may well be the case that processing in batches of one will cause a workstation to become overutilized (due to excessive setup time or excessive parallel batch process time). The challenge, therefore, is to balance these capacity considerations with the delays that batching introduces (see Karmarkar (1987) for a more complete discussion). We can summarize the key dynamics of serial and parallel process batching in the following factory physics law.

Law (Process Batching): *In stations with batch operations or significant changeover times:*

1. *The minimum process batch size that yields a stable system may be greater than one.*

2. *As process batch size becomes large, cycle time grows proportionally with batch size.*

3. *Cycle time at the station will be minimized for some process batch size, which may be greater than one.*

We can illustrate the relationship between capacity and process batching described in this law with the following examples.

Example: Serial Process Batching

Consider a machining station that processes several part families. The parts arrive in batches where all parts within batches are of like family, but the batches are of different families. The arrival rate of batches is set so that parts arrive at a rate of 0.4 part per hour. Each part requires one hour of processing regardless of family type. However, the machine requires a five-hour setup between batches (because it is assumed to be switching to a different family). Hence, the choice of batch size will affect both the number of setups required (and hence utilization) and the time spent waiting in a partial batch. Furthermore, the cycle time will be affected by whether parts exit the station in a batch when the whole batch is complete or one at a time if lot splitting is used.

Notice that if we were to use a batch size of one, we could only process one part every six hours (five hours for the setup plus one hour for processing), which does not keep up with arrivals. The smallest batch size we can consider is four parts, which will enable a capacity of four parts every nine hours (five hours for setup plus four hours to process the parts), or a rate of 0.44 part per hour.

Figure 9.5 graphs the cycle time at the station for a range of batch sizes with and without lot splitting. Notice that minimum feasible batch size yields an average cycle time of approximately 70 hours without lot splitting and 68 hours with lot splitting. Without lot splitting, the minimum cycle time is about 31 hours and is achieved at a batch size of eight parts. With lot splitting, it is about 22 hours and is achieved at a batch size of nine parts. Above these minimal levels, cycle time grows in an almost straight-line fashion, with the lot splitting case outperforming (achieving smaller cycle times than) the nonsplitting case by an increasing margin.

The Process Batching Law implies that it may be necessary, even desirable, to use large process batches in order to keep utilization, and hence cycle time and WIP, under control. But one should be careful about accepting this conclusion without question. The need for large serial batch sizes is caused by long setup times. Therefore, the first priority should be to try to reduce setup times as much as economically practical. For instance, Figure 9.5 shows the behavior of the machining station example, but with average setup times of two and one-half hours instead of five hours. Notice that with shorter setup times, minimal cycle times are roughly 50 percent smaller (around 16 hours without lot splitting and 11 hours with lot splitting) and are attained at smaller batch sizes (four parts without lot splitting and five parts with lot splitting). So the full implication of the above law is that batching *and* setup time reduction must be used in concert to achieve high throughput and efficient WIP and cycle time levels.

FIGURE 9.5

Cycle time versus serial batch size at a station with five-hour and two-and-one-half-hour setup times

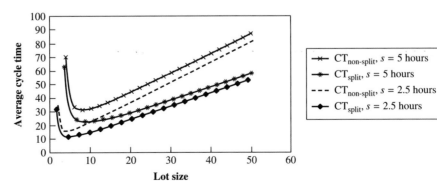

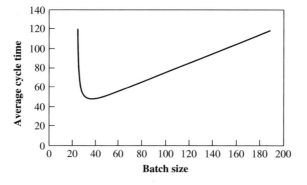

Example: Parallel Process Batching

Consider the burn-in operation of a facility that produces medical diagnostic units. The operation involves running a batch of units through multiple power-on and diagnostic cycles inside a temperature-controlled room, and it requires 24 hours regardless of how many units are being burned in. The burn-in room is large enough to hold 100 units at a time. Suppose units arrive to burn in at a rate of one per hour (24 per day). Clearly, if we were to burn in one unit at a time, we would only have capacity of $\frac{1}{24}$ per hour, which is far below the arrival rate. Indeed, if we burn in units in batches of 24, then we will have capacity of one per hour, which would make utilization equal to 100 percent. Since utilization must be less than 100 percent to achieve stability, the smallest feasible batch size is 25.

Figure 9.6 plots the cycle time as a function of batch size. It turns out that cycle time is minimized at a batch size of 32, which achieves a cycle time of 43 hours. Since 24 hours of this is process time, the rest is queue time and wait-to-batch time. We will develop the formulas for computing these quantities later.

Serial Batching.   We can give a deeper interpretation of the batching–cycle time interactions underlying the process batching law by examining the models behind the above examples. We begin with the serial batching case of Figure 9.5 in the following technical note.

Technical Note—Serial Batching Interactions

To model serial batching, in which batches of parts arrive at a single machine and are processed with a setup between each batch, we make use of the following notation:

$$k = \text{serial batch size}$$
$$r_a = \text{arival rate (parts per hour)}$$
$$t = \text{time to process a single part (hour)}$$
$$s = \text{time to perform a setup (hour)}$$
$$c_e^2 = \text{effective SCV for processing time of a batch, including both process time and}$$
$$\text{setup time}$$

Furthermore, we make these simplifying assumptions: (1) The SCV c_e^2 of the effective process time of a batch is equal to 0.5 regardless of batch size[7] and (2) the arrival SCV (of batches) is always one.

---

[7]We could fix the CV for processing individual jobs and compute the CV for a batch as a function of batch size. However, the model assuming a constant arrival CV for batches exhibits the same principal behavior—a sharp increase in cycle time for small batches and the linear increase for large batches—and is much easier to analyze.

Since r_a is the arrival rate of *parts,* the arrival rate of batches is r_a/k. The effective process time for a batch is given by the time to process the k parts in the batch plus the setup time

$$t_e = kt + s \tag{9.1}$$

so machine utilization is

$$u = \frac{r_a}{k}(kt + s) = r_a\left(t + \frac{s}{k}\right) \tag{9.2}$$

Notice that for stability we must have $u < 1$, which requires

$$k > \frac{sr_a}{1 - tr_a}$$

The average time in queue CT_q is given by the *VUT* equation

$$CT_q = \left(\frac{1 + c_e^2}{2}\right)\left(\frac{u}{1 - u}\right)t_e \tag{9.3}$$

where t_e and u are given by Equations (9.1) and (9.2).

The total average cycle time at the station consists of queue time plus setup time plus wait-in-batch time (WIBT) plus process time. WIBT depends on whether lots are split for purposes of moving parts downstream. If they are not (i.e., the entire batch must be completed before any of the parts are moved downstream), then all parts wait for the other $k - 1$ parts in the batch, so

$$WIBT_{nonsplit} = (k - 1)t$$

and total cycle time is

$$\begin{aligned}
CT_{nonsplit} &= CT_q + s + WIBT_{nonsplit} + t \\
&= CT_q + s + (k - 1)t + t \\
&= CT_q + s + kt \tag{9.4}
\end{aligned}$$

If lots are split (i.e., individual parts are sent downstream as soon as they have been processed, so that transfer batches of one are used), then wait-in-batch time depends on the position of the part in the batch. The first part spends no time waiting, since it departs immediately after it is processed. The second part waits behind the first part and hence spends t waiting in batch. The third part spends $2t$ waiting in batch, and so on. The average time for the k jobs to wait in batch is therefore

$$WIBT_{split} = \frac{k - 1}{2}t$$

so that

$$\begin{aligned}
CT_{split} &= CT_q + s + WIBT_{split} + t \\
&= CT_q + s + \frac{k - 1}{2}t + t \\
&= CT_q + s + \frac{k + 1}{2}t \tag{9.5}
\end{aligned}$$

Equations (9.4) and (9.5) are the basis for Figure 9.5. We can give a specific illustration of their use by using the data from the Figure 9.5 example ($r_a = 0.4$, $c_a^2 = 1$, $t = 1$, $c_e^2 = 0.5$, $s = 5$) for $k = 10$, so that

$$t_e = s + kt = 5 + 10 \times 1 = 15 \text{ hours}$$

Machine utilization is

$$u = \frac{r_a t_e}{k} = \frac{(0.4 \text{ part/hour})(15 \text{ hours})}{10} = 0.6$$

The expected time in queue for a batch is

$$CT_q = \left(\frac{1 + 0.5}{2}\right)\left(\frac{0.6}{1 - 0.6}\right)15 = 16.875 \text{ hours}$$

So if we do not use lot splitting, average cycle time is

$$\mathrm{CT_{nonsplit}} = \mathrm{CT}_q + s + kt = 16.875 + 5 + 10(1) = 31.875 \text{ hours}$$

If we do split process batches into transfer batches of size one, average cycle time is

$$\mathrm{CT_{split}} = \mathrm{CT}_q + s + \frac{k+1}{2}t = 16.875 + 5 + \frac{10+1}{2}(1) = 27.375 \text{ hours}$$

which is smaller, as expected.

The main conclusion of this analysis of serial batching is that if setup times can be made sufficiently short, then using serial process batch sizes of one is an effective way to reduce cycle times. However, if short setup times are not possible (at least in the near term), then cycle time can be sensitive to the choice of process batch size and the "best" batch size may be significantly greater than one.

Parallel Process Batching.   Depending on the control policy, a serial batching operation can start on a batch before the entire batch is present at the station and can release jobs in the batch before the entire batch has been processed. (We will examine the manner in which this causes cycle time to "overlap" at stations in the next section.) But in a parallel batching operation, such as a heat treat furnace, a bake oven, or a burn-in room, the entire batch is processed at once and therefore must begin and end processing at the same time. This makes analysis of parallel process batching slightly different from analysis of serial process batching.

Total cycle time at a parallel batching station includes wait-to-batch time (the time to accumulate a full batch), queue time (the time full batches wait in queue), and processing time. We develop formulas for these in the following technical note.

Technical Note—Parallel Batching Interactions

We assume that parts arrive one at a time to the parallel batch operation. They wait to form a batch, may wait in a queue of batches, and then are processed as a batch. We make use of the following notation, which is similar to that used for the serial batching case.

k = parallel batch size
r_a = arrival rate (parts per hour)
c_a = CV of interarrival times
t = time to process batch (hour)
c_e = effective CV for processing time of batch
B = maximum batch size (number of parts that can fit into process)

To calculate the average wait-to-batch time (WTBT), note that the average time between arrivals is $1/r_a$. The first part in a batch waits for $k-1$ other parts to arrive and hence waits $(k-1)/r_a$ hour. The last part in a batch does not wait at all to form a batch. Hence, the average time a part waits to form a batch is the average of these two extremes, or

$$\mathrm{WTBT} = \frac{k-1}{2r_a}$$

Once k arrivals have occurred, we have a full batch to move either into the queue or into the process. Hence, the interarrival times of batches are equal to the sum of k interarrival times of parts. As we saw in Chapter 8, adding k independent, identically distributed random variables with SCVs of c^2 results in a random variable with an SCV of c^2/k. Therefore, the arrival SCV of batches is given by

$$c_a^2(\text{batch}) = \frac{c_a^2}{k}$$

The capacity of the process with batch size k is k/t, so the maximum capacity is B/t. To keep utilization below 100 percent, effective capacity must be greater than demand, so we require

$$u = \frac{r_a}{k/t} < 1$$

or
$$k > r_a t$$

If B is less than or just equal to $r_a t$, then there is insufficient capacity to meet demand.

Once a batch is formed, it goes to the batch process. If utilization is high and there is variability, there is likely to be a queue. The queue time can be computed by using the *VUT* equation to be

$$CT_q = \left(\frac{c_a^2/k + c_e^2}{2} \right) \left(\frac{u}{1-u} \right) t$$

Consequently, total cycle time is

$$CT = WTBT + CT_q + t$$

$$= \frac{k-1}{2r_a} + \left(\frac{c_a^2/k + c_e^2}{2} \right) \left(\frac{u}{1-u} \right) t + t$$

$$= \frac{k-1}{2ku} t + \left(\frac{c_a^2/k + c_e^2}{2} \right) \left(\frac{u}{1-u} \right) t + t \tag{9.6}$$

where the last equality follows from the fact that $u = r_a/(k/t)$ so $r_a = uk/t$.

Notice that Equation (9.6) implies that cycle time becomes large when u approaches zero, as well as when it approaches one. The reason is that when utilization is low, arrivals are slow relative to process times and hence the time to form a batch becomes long.

As we saw in Figure 9.6, the cycle time at a parallel batch operation is significantly impacted by the batch size. Depending on the capacity of the operation, it may be optimal to run less-than-full batches. To find the optimal batch size, we could implement the expressions from the above technical note in a spreadsheet and use trial and error. Alternatively, we could use an analytical approach, like that presented in Chapter 15.

9.4.3   Move Batching

On a tour of an assembly plant, our guide proudly displayed one of his recent accomplishments—a manufacturing cell. Castings arrived at this cell from the foundry and, in less than an hour, were drilled, machined, ground, and polished. From the cell, they went to a subassembly operation. Our guide indicated that by placing the various processes in close proximity to one another and focusing on streamlining flow within the cell, cycle times for this portion of the routing had been reduced from several days to one hour. We were impressed—until we discovered that castings were delivered to the cell and completed parts were moved to assembly by forklift in totes containing approximately 10,000 parts! The result was that the first part required only one hour to go through the cell, but had to wait for 9,999 other parts before it could move on to assembly. Since the capacity of the cell was about 100 parts per hour, the tote sat waiting to be filled for 100 hours. Thus, although the cell had been designed to reduce WIP and cycle time, the actual performance was the closest we have ever seen to the worst case of Chapter 7.

The reason the plant had chosen to move parts in batches of 10,000 was the mistaken (but common) assumption that transfer batches should equal process batches. However, in most production environments, there is no compelling need for this to be the case. As we noted above, splitting of batches or lots can reduce cycle time tremendously. Of course, smaller lots also imply more material handling. For instance, if parts in the above cell were moved in lots of 1,000 (instead of 10,000), then a tote would need to be moved every 10 hours (instead of every 100 hours). Although the assembly plant was large and interprocess moves were lengthy, this additional material handling was clearly manageable and would have reduced WIP and cycle time in this portion of the line by a factor of 10.

The behavior underlying this example is summarized in the following law of factory physics.

Law (Move Batching): *Cycle times over a segment of a routing are roughly proportional to the transfer batch sizes used over that segment, provided there is no waiting for the conveyance device.*

This law suggests one of the easiest ways to reduce cycle times in some manufacturing systems—reduce transfer batches. In fact, it is sometimes so easy that management may overlook it. But because reducing transfer batches can be simple and inexpensive, it deserves consideration before moving on to more complex cycle time reduction strategies. Of course, smaller transfer batches will require *more* material handling, hence the caveat *provided there is no waiting for the conveyance device.* If the more often we move parts between stations, the longer they wait for the material handling device, then this additional queue time might cancel out the reduction in wait-to-batch time. Thus, the Move Batching Law describes the cycle time reduction that is possible through move batch reduction, *provided* there is sufficient material handling capacity to carry out the moves without delay.

To appreciate the relationship between cycle time and move batch size, note that the dynamics are identical to those of a parallel batch process in which the material handling device is the parallel batch operation. If batches are too small, utilization will grow and cause the queue waiting for the material handler to become excessive. We illustrate these mechanics more precisely by means of a mathematical model in the following technical note.

Technical Note—Transfer Batches

Consider the effects of batching in the simple two-station serial line shown in Figure 9.7. The first station receives single parts and processes them one at a time. Parts are then collected into transfer batches of size k before they are moved to the second station, where they are processed as a batch and sent downstream as single parts. For simplicity, we assume that the time to move between the stations is zero.

Letting r_a denote the arrival rate to the line and $t(1)$ and $c_e(1)$ represent the mean and CV, respectively, of processing time at the first station, we can compute the utilization as $u(1) = r_a t(1)$ and the expected waiting time in queue by using the *VUT* equation.

$$\text{CT}_q(1) = \left(\frac{c_a^2(1) + c_e^2(1)}{2} \right) \left(\frac{u(1)}{1 - u(1)} \right) t \tag{9.7}$$

The total time spent at the first station includes this queue time, the process time itself, and the time spent forming a batch. The average batching time is computed by observing

Figure 9.7

A batching and unbatching example

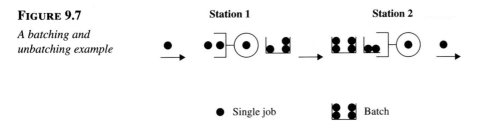

Station 1   Station 2

● Single job   Batch

that the first part must wait for $k - 1$ other parts, while the last part does not wait at all. Since parts arrive to the batching process at the same rate as they arrive to the station itself r_a (remember conservation of flow), the average time spent forming a batch is the average between $(k - 1)(1/r_a)$ and 0, which is $(k - 1)/(2r_a)$. Since $u(1) = r_a t(1)$, we have

$$\text{average wait-to-batch-time} = \frac{k - 1}{2r_a} = \frac{k - 1}{2u(1)} t(1)$$

As we would expect, this quantity becomes zero if the batch size k is equal to one. We can now express the total time spent by a part at the first station $\text{CT}(1)$ as

$$\text{CT}(1) = \text{CT}_q(1) + t(1) + \frac{k - 1}{2u(1)} t(1) \tag{9.8}$$

To compute average cycle time at the second station, we can view it as a queue of whole batches, a queue of single parts (i.e., partial batch), and a server. We can compute the waiting time in the queue of whole batches $\text{CT}_q(2)$ by using Equation (9.7) with the values of $u(2)$, $c_a^2(2)$, $c_e^2(2)$, and $t(2)$ adjusted to represent batches. We do this by noting that interdeparture times for batches are equal to the sum of k interdeparture times for parts. Hence, because, as we saw in Chapter 8, adding k independent, identically distributed random variables with SCVs of c^2 results in a random variable with an SCV of c^2/k, the arrival SCV of batches to the second station is given by $c_d^2(1)/k = c_a^2(2)/k$. Similarly, since we must process k separate parts to process a batch, the SCV for the batch process times at the second station is $c_e^2(2)/k$, where $c_e^2(2)$ is the process SCV for individual parts at the second station. The effective average time to process a batch is $kt(2)$ and the average arrival rate of batches is r_a/k. Thus, as we would expect, utilization is

$$u(2) = \frac{r_a}{k} kt(2) = r_a t(2)$$

Hence, by the *VUT* equation, average cycle time at the second station is

$$\text{CT}_q(2) = \left(\frac{c_a^2(2)/k + (c_e^2(2)/k)}{2} \right) \left(\frac{u(2)}{1 - u(2)} \right) kt(2)$$

$$= \left(\frac{c_a^2(2) + c_e^2(2)}{2} \right) \left(\frac{u(2)}{1 - u(2)} \right) t(2)$$

Interestingly, the waiting time in the queue of whole batches is the same as the waiting time we would have computed for single parts (because the k's cancel, leaving us with the usual *VUT* equation).

In addition to the queue of full batches, we must consider the queue of partial batches. We can compute this by considering how long a part spends in this partial queue. The first piece arriving in a batch to an idle machine does not have to wait at all, while the last piece in the batch has to wait for $k - 1$ other pieces to finish processing. Thus, the average time that parts in the batch have to wait is $(k - 1)t(2)/2$.

The total cycle time of a part at the second station is the sum of the wait time in the queue of batches, the wait time in a partial batch, and the actual process time of the part:

$$\text{CT}(2) = \text{CT}_q(2) + \frac{k - 1}{2} t(2) + t(2) \tag{9.9}$$

We can now express the total cycle time for the two-station system with batch size k as

$$
\begin{aligned}
\text{CT}_{\text{batch}} &= \text{CT}(1) + \text{CT}(2) \\
&= \text{CT}_q(1) + t(1) + \frac{k-1}{2u(1)}t(1) + \text{CT}_q(2) + \frac{k-1}{2}t(2) + t(2) \\
&= \text{CT}_{\text{single}} + \frac{k-1}{2u(1)}t(1) + \frac{k-1}{2}t(2)
\end{aligned}
\tag{9.10}
$$

where $\text{CT}_{\text{single}}$ represents the cycle time of the system without batching (i.e., with $k = 1$).

Expression (9.10) quantitatively illustrates the Move Batching Law—cycle times increase proportionally with batch size. Notice, however, that the increase in cycle time that occurs when batch size k is increased has nothing to do with process or arrival variability (i.e., the terms in Equation (9.10) that involve k do not include any coefficients of variability). There *is* variability—some parts wait a long time due to batching while others do not wait at all—but it is variability caused by *bad control* or *bad design* (similar to the worst case in Chapter 7), rather than by process or flow uncertainty.

Finally, we note that the impact of transfer batching is largest when the utilization of the first station is low, because this causes the $(k-1)t(1)/[2u(1)]$ term in Equation (9.10) to become large. The reason for this is that when arrival rate is low relative to processing rate, it takes a long time to fill up a transfer batch. Hence, parts spend a great deal of time waiting in partial batches. This is very similar to what happens in parallel process batches (see Equation (9.6)). The only difference between Equations (9.6) and (9.10) is that in the former we did not model the move process as having limited capacity. If we had, the two situations would have been identical.

Cellular Manufacturing.   The fundamental implication of the Move Batching Law is that large transfer batches directly inflate cycle times. Hence, reducing them can be a useful cycle time reduction strategy. One way to keep transfer batches small is through **cellular manufacturing,** which we discussed in the context of JIT in Chapter 4.

In theory, a cell positions all workstations needed to produce a family of parts in close physical proximity. Since material handling is minimized, it is feasible to move parts between stations in small batches, ideally in batches of one. If the cell truly processes only one family of parts, so there are no setups, the process batch can be one, infinity, or any number in between (essentially controlled by demand).

If the cell handles multiple families, so that there are significant setups, we know from our previous discussions that serial process batching is very important to the capacity and cycle time of the cell. Indeed, as we will see in Chapter 15, it may make sense to set the process batch size differently for different families and even vary these over time. Regardless of how process batching is done, however, it is an independent decision from move batching. Even if large process batches are required because of setups, we can use lot splitting to move material in small transfer batches and take advantage of the physical compactness of a cell.

9.5 Cycle Time

Having considered issues of utilization, variability, and batching, we now move to the more complicated performance measure, cycle time. First we consider the cycle time at a single station. Later we will describe how these station cycle times combine to form the cycle time for a line.

9.5.1   Cycle Time at a Single Station

We begin by breaking down cycle time at a single station into its components.

Definition (Station Cycle Time): *The average cycle time at a station is made up of the following components:*

$$\text{Cycle time} = \text{move time} + \text{queue time} + \text{setup time} + \text{process time}$$
$$+ \text{ wait-to-batch time} + \text{wait-in-batch time}$$
$$+ \text{ wait-to-match time} \tag{9.11}$$

Move time is the time jobs spend being moved from the previous workstation. **Queue time** is the time jobs spend waiting for processing at the station or to be moved to the next station. **Setup time** is the time a job spends waiting for the station to be set up. Note that this could actually be less than the station setup time if the setup is partially completed while the job is still being moved to the station. **Process time** is the time jobs are actually being worked on at the station. As we discussed in the context of batching, **wait-to-batch time** is the time jobs spend waiting to form a batch for either (parallel) processing or moving, and **wait-in-batch time** is the average time a part spends in a (process) batch waiting its turn on a machine. Finally, **wait-to-match time** occurs at assembly stations when components wait for their mates to allow the assembly operation to occur.

Notice that of these, only process time actually contributes to the manufacture of products. Move time could be viewed as a necessary evil, since no matter how close stations are to one another, some amount of move time will be necessary. But all the other terms are sheer inefficiency. Indeed, these times are often referred to as non-value-add time, waste, or *muda.* They are also commonly lumped together as delay time or queue time. But as we will see, these times are the consequence of very different causes and are therefore amenable to different cures. Since they frequently constitute the vast majority of cycle time, it is useful to distinguish between them in order to identify specific improvement policies.

We have already discussed the batching times, so now we deal with wait-to-match time before moving on to cycle times in a line.

9.5.2   Assembly Operations

Most manufacturing systems involve some kind of assembly. Electronic components are inserted into circuit boards. Body parts, engines, and other components are assembled into automobiles. Chemicals are combined in reactions to produce other chemicals. Any process that uses two or more inputs to produce its output is an assembly operation.

Assemblies complicate flows in production systems because they involve **matching.** In a matching operation, processing cannot start until all the necessary components are present. If an assembly operation is being fed by several fabrication lines that make the components, shortage of any one of the components can disrupt the assembly operation and thereby all the other fabrication lines as well. Because they are so influential to system performance, it is common to subordinate the scheduling and control of the fabrication lines to the assembly operations. This is done by specifying a **final assembly schedule** and working backward to schedule fabrication lines. We will discuss assembly operations from a quality standpoint in Chapter 12, from a shop floor control standpoint in Chapter 14, and from a scheduling standpoint in Chapter 15.

For now, we summarize the basic dynamics underlying the behavior of assembly operations in the following factory physics law.

Law (Assembly Operations):   *The performance of an assembly station is degraded by increasing any of the following:*

1. *Number of components being assembled.*
2. *Variability of component arrivals.*
3. *Lack of coordination between component arrivals.*

Note that each of these could be considered an increase in variability. Thus, the Assembly Operations Law is a specific instance of the more general Variability Law. The reasoning and implications of this law are fairly intuitive. To put them in concrete terms, consider an operation that places components on a circuit board. All components are purchased according to an MRP schedule. If any component is out of stock, then the assembly cannot take place and the schedule is disrupted.

To appreciate the impact of the number of components on cycle time, suppose that a change is made in the bill of material that requires one more component in the final product. All other things being equal, the extra component can only inflate the cycle time, by being out of stock from time to time.

To understand the effect of variability of component arrivals, suppose the firm changes suppliers for one of the components and finds that the new supplier is much more variable than the old supplier. In the same fashion that arrival variability causes queueing at regular nonassembly stations, the added arrival variability will inflate the cycle time of the assembly station by causing the operation to wait for late deliveries.

Finally, to appreciate the impact of lack of coordination between component arrivals, suppose the firm currently purchases two components from the same supplier, who always delivers them at the same time. If the firm switches to a policy in which the two components are purchased from separate suppliers, then the components may not be delivered at the same time any longer. Even if the two suppliers have the same level of variability as before, the fact that deliveries are uncoordinated will lead to more delays. Of course, this neglects all other complicating factors, such as the fact that having two components to deliver may cause a supplier to be less reliable, or that certain suppliers may be better at delivering specific components. But all other things being equal, having the components arrive in synchronized fashion will reduce delays. We will discuss methods for synchronizing fabrication lines to assembly operations in Chapter 14.

9.5.3   Line Cycle Time

In the Penny Fab examples in Chapter 7, where all jobs were processed in batches of one and moves were instantaneous, cycle times were simply the sum of process times and queue times. But when batching and moving are considered, we cannot always compute the cycle time of the line as the sum of the cycle times at the stations. Since a batch may be processed at more than one station at a time (i.e., if lot splitting is used), we must account for overlapping time at stations. Thus, we define the cycle time in a line as follows.

Definition (Line Cycle Time):   *The average cycle time in a line is equal to the sum of the cycle times at the individual stations less any time that overlaps two or more stations.*

To illustrate the impact of overlapping cycle times, we consider the two lines in Table 9.5. Lines 1 and 2 are both three-station lines with no process variability that experience (deterministic) arrivals of batches of $k = 6$ jobs every 35 hours. A setup is done for each batch, after which jobs are processed one at a time and are sent to the next station. The only difference is that the process and setup times are different in the two lines (line 2 is the reverse of line 1). Hence, in line 1 the utilizations of the stations are increasing, with station 1 at 49 percent, station 2 at 75 percent, and station 3 at 100 percent utilization. In line 2 these are reversed. For modeling purposes we use $t(i)$ and $s(i)$ to represent the unit process time and setup time, respectively, at station i.

Consider line 1. Since we are processing jobs in series on stations with setups and letting them go as they are finished, we can apply Equation (9.5) to compute the cycle time at each station. At station 1, this yields

$$CT(1) = CTq + s(1) + \frac{k+1}{2}t(1) = 0.0 + 5 + \frac{6+1}{2}(2) = 12$$

where the queue time is zero because there is no variability in the system.

For stations 2 and 3, we can do the same thing to get

$$CT(2) = CTq + s(2) + \frac{k+1}{2}t(2) = 0.0 + 8 + \frac{6+1}{2}(3) = 18.5$$

$$CT(3) = CTq + s(3) + \frac{k+1}{2}t(3) = 0.0 + 11 + \frac{6+1}{2}(4) = 25$$

which yields a total cycle time of

$$CT = CT(1) + CT(2) + CT(3) = 12 + 18.5 + 25 = 55.5$$

But this is not right. The first job in a batch at station 2 or 3 is already in process while the last job in the batch is still at the previous station. Therefore, the wait-in-batch time component of Equation (9.5) overestimates the total delay at stations 2 and 3 due to batching.

For this deterministic example, we can compute the cycle time by following the jobs in a batch one at a time through the station. As shown in Figure 9.8, the first job to arrive at station 2 has a cycle time of $s(2)+t(2)$. The second finishes at $s(2)+2t(2)$ but arrived $t(1)$ hour later than the first job, so its cycle time at station 2 is $s(2) + 2t(2) - t(1)$. Likewise, the third has a cycle time of $s(2) + 3t(2) - 2t(1)$. This continues until the kth (last) job in the batch, which starts at $(k - 1)t(1)$ and completes at $s(2) + kt(2)$ for a

---

TABLE 9.5 Examples Illustrating Cycle Time Overlap

	Station 1	Station 2	Station 3
	Line 1		
Setup time (hour)	5	8	11
Unit process time (hour)	2	3	4
	Line 2		
Setup time (hour)	11	8	5
Unit process time (hour)	4	3	2

FIGURE 9.8

Lot splitting: faster to slower

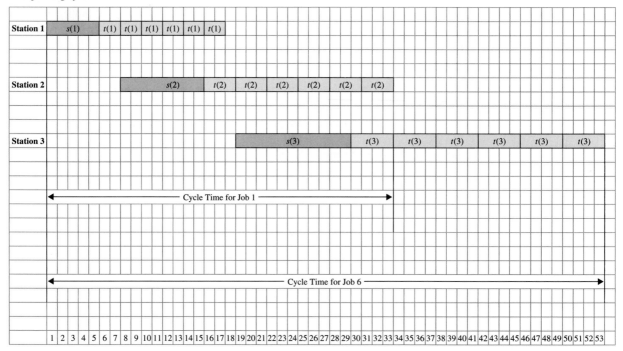

cycle time of $s(2) + kt(2) - (k-1)t(1)$. The average cycle time at station 2 is therefore

$$CT(2) = \frac{1}{k}[ks(2) + (1 + 2 + \cdots + k)t(1) - (1 + 2 + \cdots + k - 1)t(1)]$$

$$= s(2) + \frac{k+1}{2}t(2) - \frac{k-1}{2}t(1)$$

$$= 8 + 3.5(3) - 2.5(2) = 13.5$$

The term $[(k-1)/2]t(1) = 5$ hours represents the batch *overlap* time.

The situation at station 3 is similar to that at station 2 and leads to a cycle time at station 3 of

$$CT(3) = s(3) + \frac{k+1}{2}t(3) - \frac{k-1}{2}t(2)$$

$$= 11 + 3.5(4) - 2.5(3) = 17.5$$

Thus, the correct total time through line 1 is computed by adding the corrected versions of CT(1), CT(2) and CT(3), which yields

$$CT(line) = s(1) + s(2) + s(3) + t(1) + t(2) + \frac{k+1}{2}t(3) = 43 \text{ hours}$$

This is illustrated in Figure 9.8, which shows that the cycle time of the first job in the batch is 33 hours, while the cycle time of the sixth job is 53 hours, so the average cycle time is $(33 + 53)/2 = 434$ hours. Note that this is considerably less than the 55.5 hours arrived at by summing the cycle times at the stations.

If we were to compute the cycle time for line 2, using Equation (9.5) at each station, and add the results, we would get the same answer as for line 1, or 55.5 hours. The

FIGURE 9.9

Lot splitting: slower to faster

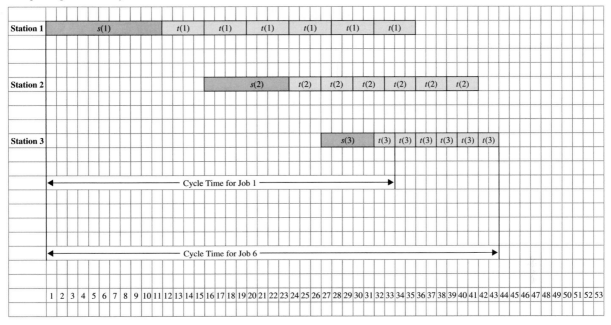

reason is that without variability the equation is unaffected by the order of the line. However, now if we work through the mechanics of the line directly, we find that the true average cycle time is 38 hours (see Figure 9.9, which shows that the cycle times of the first and sixth jobs are 33 hours and 43 hours respectively, so the average cycle time is $(33 + 43)/2 = 38$ hours). Again, this is considerably less than our initial estimate. It is also much less than the first case (there is more overlapping when slower processes are first). The point is that not only are overlapping cycle times important to determining the cycle time of a line, but also the mechanics are such that the order of the stations matters.

Although the behavior of lines with batching is complex, we can gain insight into the line cycle time by following a single job through the line. As in the above example, we assume that

1. Jobs arrive in batches.[8]
2. The first job in each batch sees a full setup at each station (i.e., we are not allowed to start setups before the first job in the batch arrives, although we do allow the case where all setup times at a station are zero).
3. Jobs are moved one at a time between stations.

Under these conditions, we develop upper and lower bounds on the cycle time of a line in the following technical note.

Technical Note—Cycle Time Bounds

We refer to nonqueueing (i.e., time in batch, setup time, and process time) time as *total in-process time.* We can bound the total in-process time by considering a line with no variability

---

[8]Since a full batch is committed to enter the line once the first job is released to the line, for the purposes of computing cycle time it is reasonable to assume that the entire batch arrives to the line simultaneously.

(and therefore no queueing) and examining the time it takes for the first job T_1 and time for the last job T_k of a batch to go through the line.[9] For a k-station line with $s(i)$ and $t(i)$ being the setup and process times, respectively, at station i, the first job will require a setup and a single process time at each station

$$T_1 = \sum_{i=1}^{K} s(i) + t(i)$$

The last job will require this time plus the time spent waiting behind the other jobs in the batch. The longest time this could possibly be occurs if the last job encountered all the $k - 1$ other jobs at the process with the longest process time (see Figure 9.8). Thus,

$$T_k \leq T_1 + (k - 1)t_b$$

where $t_b = \max_i\{t(i)\}$. An upper bound for the average total in-process time is the average of T_1 and T_k, which yields

$$\text{total in-process time} \leq \sum_{i=1}^{K}[s(i) + t(i)] + \frac{k - 1}{2}t_b \qquad (9.12)$$

Because all jobs arrive to the first station at one time, the last job will *always* finish after the other $k - 1$ jobs at the last station. The smallest delay that can occur is seen if the last station has the *fastest* process time *and* there is no idle time at the last station (see Figure 9.9). So a lower bound on the average total in-process time can be computed by using $t_f = \min_i\{t(i)\}$ in place of t_b

$$T_k \geq T_1 + (k - 1)t_f$$

and so

$$\text{total in-process time} \geq \sum_{i=1}^{K}[s(i) + t(i)] + \frac{k - 1}{2}t_f \qquad (9.13)$$

To get bounds on cycle time, we must consider queue time in addition to total in-process time. To do this, recall our discussion of batch moves. There, the total queue time did not depend on the batch size (remember how the k's "canceled out"). If we can assume that this is approximately true for the serial batching case, then a good approximation of the queue time can be made by using the *VUT* equation to compute the average time that *full batches* wait in queue at each station. At the first station, since arrivals occur in batches, this approximation is as accurate as the *VUT* equation itself. At other stations, where arrivals occur one at a time, more error is introduced by not really knowing c_a^2. Of course, this problem exists in systems without batching as well. Experience with a limited number of examples shows that the accuracy is no worse than the accuracy of the equations developed for single jobs (in Chapter 8).

Letting $\text{CT}_q^b(i)$ represent the average time that full batches wait at station i (which is computed by using the *VUT* equation in the usual way), we can express approximate upper and lower bounds on total cycle time in a line with serial batching as

$$\sum_{i=1}^{n}[\text{CT}_q^b(i) + s(i) + t(i)] + \frac{k - 1}{2}t_f \leq \text{CT}$$
$$\leq \sum_{i=1}^{n}[\text{CT}_q^b(i) + s(i) + t(i)] + \frac{k - 1}{2}t_b \qquad (9.14)$$

where $t_f = \min_i\{t(i)\}$, and $t_b = \max_i\{t(i)\}$.

---

[9]The authors would like to express their gratitude to Dr. Greg Diehl at Network Dynamics, Inc., for his assistance in the development of these equations.

Example: Bounding Cycle Time

Reconsider the two lines in Table 9.5. If there is no process or arrival variability, then the sum of the queue times is zero and the sum of the setup and process times is 33. Hence the cycle time bounds are

$$33 + \frac{6-1}{2}(2) \leq CT \leq 33 + \frac{6-1}{2}(4)$$

$$38 \leq CT \leq 43$$

For line 1, the upper bound is tight. For line 2, the lower bound is tight. However, if we switch things around so that the slowest station is at the front and the fastest station is in the middle, then it turns out that CT = 40.5, which is between the bounds. Likewise, if we place the slowest station in the middle and the fastest station at the end, CT = 39.5, which is also between the bounds. In these examples, no idle time occurs within batches (i.e., no machine goes idle between jobs of the same batch). However, this can occur and indeed does occur in this system if the slowest station is first and the fastest is second (see Problem 15).

The cycle time bounds in Equation (9.14) will be very close to one another for lines in which process times are similar (i.e., so that $t_f \approx t_b$). But for lines where the fastest machine is much faster than the slowest one (e.g., because it also has a very long setup time), these bounds can be quite far apart. Tighter bounds require more complex calculations (see Benjaafar and Sheikhzadeh 1997).

9.5.4   Cycle Time, Lead Time, and Service

In a manufacturing system with infinite capacity and absolutely no variability, the relation between cycle time and customer lead time is simple—they are the same. The lucky manager of such a system could simply quote a lead time to customers equal to the cycle time required to make the product and be assured of 100 percent service. Unfortunately, all real systems contain variability, and so perfect service is not possible and there is frequently confusion regarding the distinction between lead time, cycle time, and their relation to service level. Although we touched on these issues briefly in Chapters 3 and 7, we now define them more precisely and offer a law of factory physics that relates variability to lead time, cycle time, and service.

Definitions.   Throughout this book we have used the terms *cycle time* and *average cycle time* interchangeably to denote the average time it takes a job to go through a line. To talk about lead times, however, we need to be a bit more precise in our terminology. Therefore, for the purposes of this section, we will define **cycle time** as a *random variable* that gives the time an individual job takes to traverse a routing. Specifically, we define T to be a random variable representing cycle time, with a mean of CT and a standard deviation of σ_{CT}.

Unlike cycle time, **lead time** is a *management constant* used to indicate the anticipated or maximum allowable cycle time for a job. There are two types of lead time: customer lead time and manufacturing lead time. **Customer lead time** is the amount of time allowed to fill a customer order from start to finish (i.e., multiple routings), while the **manufacturing lead time** is the time allowed on a particular routing.

In a **make-to-stock** environment, the customer lead time is zero. When the customer arrives, the product either is available or is not. If it is not, the service level (usually called **fill rate** in such cases) suffers. In a **make-to-order** environment, the customer lead time is the time the customer allows the firm to produce and deliver an item. For this case, when variability is present, the lead time must generally be greater than the average cycle time in order to have acceptable service (defined as the percentage of on-time deliveries).

One way to reduce customer lead times is to build lower-level components to stock. Since customers only see the cycle time of the remaining operations, lead times can be significantly shorter. We discuss this type of **assemble-to-order** system in the context of push and pull production in Chapter 10.

Relations.   With complex bills of material, computing suitable customer lead times can be difficult. One way to approach this problem is to use the manufacturing lead time that specifies the anticipated or maximum allowable cycle time for a job on a specific routing. We denote the manufacturing lead time for a specific routing with cycle time T as ℓ. Manufacturing lead time is often used to plan releases (e.g., in an MRP system) and to track service.

Service s can now be defined for routings operating in make-to-order mode as the probability that the cycle time is less than or equal to the specified lead time, so that

$$s = \Pr\{T \leq \ell\} \tag{9.15}$$

If T has distribution function F, then Equation (9.15) can be used to set ℓ as

$$s = F(\ell) \tag{9.16}$$

If cycle times are normally distributed, then for a service level of s

$$\ell = \mathrm{CT} + z_s \sigma_{\mathrm{CT}} \tag{9.17}$$

where z_s is the value in the standard normal table for which $\Phi(z_s) = s$. For instance, if cycle time on a given routing has a mean of eight days and a standard deviation of three days, the value for z_s for 95 percent is 1.645, so the required lead time is

$$\ell = 8 + 1.645(3) = 12.94 \approx 13 \text{ days}$$

Figure 9.10 shows both the distribution function F and its associated density function f for cycle time. The additional five days above the mean is called the **safety lead time.**

By specifying a high enough service level (to guarantee that jobs generally finish on time), we can compute customer lead times by simply adding the longest manufacturing lead times (when several routings come together in an assembly) for each level in the bill of material. For example, Figure 9.11 illustrates a system with two fabrication lines feeding an assembly operation followed by several more operations. The manufacturing lead time for assembly and the subsequent operations is four days for a service level of 95 percent. Since assembly represents level 0 in the bill of material (recall low-level codes from Chapter 3), we have that the level 0 lead time is four days. Similarly, the 95 percent manufacturing lead time is four days for the top fabrication line and six days for the bottom one, so that the lead time for level 1 is six days. Thus, total customer lead time is 10 days.

Unfortunately, the overall service level using a customer lead time of 10 days will be something less than 95 percent. This is because we did not consider the possibility of **wait-to-match time** in front of assembly. As we noted in the assembly operation law, wait-to-match time results when variability causes the fabrication lines to deliver product to assembly in an unsynchronized fashion. Because of this, whenever we have assembly operations, we must add some safety lead time.

We can now summarize the fundamental principle relating variability in cycle time to required lead times in the following law of factory physics.

FIGURE 9.10

Distribution function for cycle time and required lead time

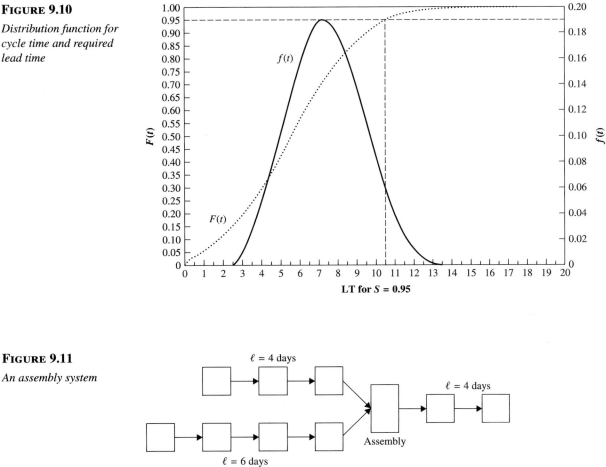

FIGURE 9.11

An assembly system

Law (Lead Time): *The manufacturing lead time for a routing that yields a given service level is an increasing function of both the mean and standard deviation of the cycle time of the routing.*

Intuitively, this law suggests that we view manufacturing lead times as given by the cycle time plus a "fudge factor" that depends on the cycle time standard deviation. The larger the cycle time standard deviation, the larger the fudge factor must be to achieve a given service level. In a make-to-order environment, where we want manufacturing lead times short in order to keep customer lead times short, we need to keep both the mean and the standard deviation of cycle time low.

The factors that inflate *mean* cycle time are generally the same as those that inflate the *standard deviation* of process time, as we noted in Chapter 8. These include operator variability, random outages, setups, rework, and the like. However, from a cycle time perspective, *rework* is particularly disruptive. Whenever there is a chance that a job will be required to go back through a portion of the line, the variability of cycle time increases dramatically. We will return to this and other issues related to cycle time variability when we discuss the impact of quality on logistics in Chapter 12.

9.6   Diagnostics and Improvements

The factory physics laws discussed describe fundamental aspects of the behavior of manufacturing systems and highlight key tradeoffs. However, by themselves they do not yield design and management policies. The reason is that the "optimal" operational structure depends on environmental constraints and strategic goals. A firm that competes on customer service needs to focus on swift and responsive deliveries, while a firm that competes on price needs to focus on equipment utilization and cost. Fortunately, the laws of factory physics can help identify areas of leverage and opportunities for improvement, regardless of the system specifics.

The following examples illustrate the use of the principles developed in this chapter to improve an existing system with regard to three key performance measures: throughput, cycle time, and customer service.

9.6.1   Increasing Throughput

Throughput of a line is given by

$$\text{TH} = \text{bottleneck utilization} \times \text{bottleneck rate}$$

Therefore, the two ways to increase throughput are to increase utilization of the bottleneck or increase its rate. It may sound blasphemous to talk of increasing utilization, since we know that increasing utilization increases cycle time. But different objectives call for different policies. In a system without restrictions on WIP, high utilization causes queueing and hence increases cycle time. But, as we saw in the pay-me-now-or-pay-me-later examples, in systems with constraints on WIP (finite buffers or logical limitations such as those imposed by kanban), blocking and starving will limit utilization of the bottleneck and hence degrade throughput.

A basic checklist of policies for increasing throughput is as follows.

1. **Increase bottleneck rate** by increasing the effective rate of the bottleneck. This can be done through equipment additions, staff additions or training, covering stations through breaks or lunches, use of flexible labor, quality improvements, product design changes to reduce time at the bottleneck, and so forth.

2. **Increase bottleneck utilization** by reducing blocking and starving of the bottleneck. There are two basic ways to do this:

   - *Buffer bottleneck with WIP.* This can be done by increasing the size of the buffers (or equivalently, the number of kanban cards) in the system. Most effective are buffer spaces immediately in front of the bottleneck (where allowing a queue to grow helps prevent starvation) and immediately after the bottleneck (where building a queue helps prevent blocking). Buffer space farther away from the bottleneck can still help, but will have a smaller effect than space close to it.

   - *Buffer bottleneck with capacity.* This can be done by increasing the effective rates of nonbottleneck stations. Faster stations upstream from the bottleneck make starving less frequent, while faster stations downstream make blocking less frequent. Adding capacity to the highest-utilization nonbottleneck stations will generally have the largest impact, since these are the stations most likely to cause blocking/starving. These can be made through the usual capacity enhancement policies, such as those listed above for increasing capacity of the bottleneck station.

Example: Throughput Enhancement

HAL Computer has a printed-circuit board plant that contains a line with two stations. The first station (resist apply) applies a photoresist material to circuit boards. The second station (expose) exposes the boards to ultra-violet light to produce a circuit pattern that is later etched onto the boards. Because the expose operation must take place in a clean room, space for WIP between the two processes is limited to 10 jobs. Capacity calculations show the bottleneck to be expose, which requires an average of 22 minutes to process a job, with an SCV of one. Resist apply requires 19 minutes per job, with an SCV of 0.25. In addition (and not included in the above process times), expose has a mean time to failure (MTTF) of $3\frac{1}{3}$ hours and a mean time to repair (MTTR) of 10 minutes, while resist apply has an MTTF of 48 hours and an MTTR of 8 hours. Jobs arrive to resist apply with a fair amount of variability, so we assume an arrival SCV c_a^2 of one. The desired throughput rate is 2.4 jobs per hour.

From past experience, HAL knows the line to be incapable of achieving the target throughput. To remedy this situation, the responsible engineers are in favor of installing a second expose machine. However, in addition to being expensive, a second machine would require expanding the clean room, which would add significantly to the cost and would result in substantial lost production during construction. The challenge, therefore, is to use factory physics to find a better solution.

The two principal tools at our disposal are the *VUT* equation for computing queue time

$$\mathrm{CT}_q = \left(\frac{c_a^2 + c_e^2}{2}\right)\left(\frac{u}{1-u}\right)t \tag{9.18}$$

and the linking equation

$$c_d^2 = u^2 c_e^2 + (1 - u^2)c_a^2 \tag{9.19}$$

Using these in conjunction with the formulas presented in Chapter 8 for the effective squared coefficient of variation, we can analyze the reasons why the line is failing to meet its throughput target.

Formulas (9.18) and (9.19) (along with additional calculations to compute the average process times $t_e(1)$ and $t_e(2)$, and the process SCVs $c_e^2(1)$ and $c_e^2(2)$, which we will come back to later), we estimate the waiting time in queue station to be 645 minutes at resist apply and 887 minutes at expose, when the arrival rate is set at 2.4 jobs per hour. The average WIP levels are 25.8 and 35.5 jobs at stations 1 and 2, respectively.

This reveals why the system cannot make 2.4 jobs per hour, even though the utilization of the bottleneck (expose) is only 92.4 percent. Namely, the clean room can hold only 20 jobs, while the model predicts an average number in queue of 35.5 jobs. Since the real system cannot allow WIP in front of expose to reach this level, resist apply will occasionally become *blocked* (i.e., idled due to a lack of space in the downstream buffer to which to send completed parts). The resulting lost production at resist apply eventually causes expose to become *starved* (i.e., idled due to a lack of parts to work on). The result is that neither station can maintain the utilization necessary to produce 2.4 parts per hour.[10]

Thus, we conclude that the problem is rooted in the long queue at expose. By Little's Law, reducing average queue length is equivalent to reducing average queue time. So

---

[10]Note that we could also have analyzed this situation by using the blocking model of Section 8.7.2. The reader is invited to try Problem 13 to see how this more sophisticated tool can be used to obtain the same qualitative result, albeit with greater quantitative precision.

we now consider the queue time at expose more closely:

$$CT_q(2) = \left(\frac{c_a^2(2) + c_e^2(2)}{2}\right)\left(\frac{u(2)}{1 - u(2)}\right)t_e(2)$$
$$= (3.16)(12.15)(23.1 \text{ minutes})$$
$$= 887 \text{ minutes}$$

The third term $t_e(2)$ is the effective process time at expose, which is simply raw process time divided by availability

$$t_e(2) = \frac{t(2)}{A(2)} = \frac{t(2)}{m_f(2)/(m_f(2) + m_r(2))}$$
$$= 22\left(\frac{31/3 + 1/6}{31/3}\right)$$
$$= 23.1 \text{ minutes}$$

Since this only slightly larger than the raw process time of 22 minutes, there is little room for improvement by increasing availability.

The second term in the expression for $CT_q(2)$ is the utilization term $u(2)/(1-u(2))$. Although at first glance a value of 12.15 may appear large, it corresponds to a utilization of 92.4 percent, which is large but not excessive. Although increasing the capacity of this station would certainly reduce the queue time (and queue size), we have already noted that this is an expensive option.

So we look to the first term, the variability inflation factor $(c_a^2(2) + c_e^2(2))/2$. Recall that moderate variability in arrivals (that is, $c_a^2(2) = 1$) and moderate variability in process times (that is, $c_e^2(2) = 1$) result in a value of one for this term. Therefore, a value of 3.16 is unambiguously large in any system. To investigate why this occurs, we break it down into its constituent parts, which reveals

$$c_e^2(2) = 1.04$$
$$c_a^2(2) = 5.27$$

Obviously, the arrival process is the dominant source of variability. This points to the problem lying upstream in the resist apply process. So we now investigate the cause of the large $c_a^2(2)$. Recall that $c_a^2(2) = c_d^2(1)$, which from Equation (9.19) is given by

$$c_d^2(1) = u^2(1)c_e^2(1) + [1 - u^2(1)]c_a^2(1)$$
$$= (0.887^2)(6.437) + (1 - 0.887^2)(1.0)$$
$$= 5.05 + 0.22$$
$$= 5.27$$

The component that makes $c_d^2(1)$ large is $c_e^2(1)$, the effective SCV of the resist apply machine. This coefficient is in turn made up of two components: a natural SCV, $c_0^2(1)$ and an inflation term due to machine failures. Using formulas from Chapter 8, we can break down $c_e^2(1)$ as follows:

$$A(1) = \frac{m_f(1)}{m_f(1) + m_r(1)} = \frac{48}{48 + 8} = 0.8571$$

$$t_e(1) = \frac{t(1)}{A(1)} = \frac{19}{0.8571} = 22.17 \text{ minutes}$$

$$c_e^2(1) = c_0^2(1) + \frac{2m_r(1)A(1)[1 - A(1)]}{t(1)}$$

$$= 0.25 + \frac{2(480)(0.8571)(0.1429)}{19} = 6.44$$

The lion's share of $c_e^2(1)$ is a result of the random outages. This suggests that an alternative to increasing capacity at expose is to improve the breakdown situation at resist apply. It is important to note that resist apply is the problem even though expose is the bottleneck. Because variability propagates through a line, a congestion problem at one station may actually be the result of a variability problem at an upstream station.

Various practical options might be available for mitigating the outage problem at resist apply. For instance, HAL could attempt to reduce the mean time to repair by holding "field-ready" spares for parts subject to failures. If such a policy could halve the MTTR, the resulting increase in effective capacity and reduction in departure SCV from resist apply would cause queue time to fall to 146 minutes at resist apply (less than one-fourth of the original) and 385 minutes at expose (less than one-half of the original).

Alternatively, HAL could perform more frequent preventive maintenance. Suppose we could avoid the long (eight-hour) failures by shutting down the machine every 30 minutes to perform a five-minute adjustment. The capacity will be the same as in the original case (i.e., because availability is unchanged), but because outages are more regular, queue time is reduced to 114 minutes at resist apply and 211 minutes at expose. Using Little's Law, this translates to an average of 8.4 jobs at expose, which is well within the space limit.

With either of the above improvements in place, it turns out to be feasible to run at (actually slightly above) the desired rate of 2.4 jobs per hour. Any other policy that would serve to reduce the variability of inter output times from resist apply would have a similar effect. Because improving the repair profile of resist apply is likely to be less expensive and disruptive than adding an expose machine, these alternatives deserve serious consideration.

9.6.2 Reducing Cycle Time

Combining the definitions of station and line cycle time, we can break down cycle times in a production system into the following:

1. Move time.
2. Queue time.
3. Setup time.
4. Process time.
5. Process batch time (wait-to-batch and wait-in-batch time).
6. Move batch time (wait-to-batch and wait-in-batch time).
7. Wait-to-match time.
8. *Minus* station overlap time.

In most production systems, actual process and move times are a small fraction (5 to 10 percent) of total cycle time (Bradt 1983). Indeed, lines for which these terms dominate are probably already very efficient with little opportunity for improvement. For inefficient lines, the major leverage lies in the other terms. The following is a brief checklist of generic policies for reducing each of these terms.

Queue time is caused by utilization and variability. Hence, the two categories of improvement policies are as follows:

1. *Reduce utilization* by increasing the effective rate at the bottleneck. This can be done by either increasing the bottleneck rate (by adding equipment, reducing setup times, decreasing time to repair, making process improvements, spelling operators through breaks and lunches, cross-training workers to take advantage of flexible capacity, etc.) or reducing flow into the bottleneck (by scheduling changes to route flow to nonbottlenecks, improving yield, or reducing rework).

2. *Reduce variability* in either process times or arrivals at any station, but particularly at high-utilization stations. Process variability can be reduced by reducing repair times, reducing setup times, improving quality to reduce rework or yield loss, reducing operator variability through better training, etc. Arrival variability can be reduced by decreasing process variability at upstream stations, by using better scheduling and shop floor control to smooth material flow, eliminating batch releases (i.e., releases of more than one job at a time), and installing a pull system (see Chapter 10).

Process batch time is driven by process batch size. The two basic means for reducing (serial or parallel) process batch size are as follows:

1. *Batching optimization* to better balance batch time with queue time due to high utilization. We gave some insight into this tradeoff earlier in this chapter. We pursue more detailed optimization in Chapter 15.

2. *Setup reduction* to allow smaller batch sizes without increasing utilization. Well-defined techniques exist for analyzing and reducing setups (Shingo 1985).

Wait-to-match time is caused by lack of synchronization of component arrivals to an assembly station. The main alternatives for improving synchronization are as follows:

1. *Fabrication variability reduction* to reduce the volatility of arrivals to the assembly. This can be accomplished by the same variability reduction techniques used to reduce queue time.

2. *Release synchronization* by using the shop floor control and/or scheduling systems to coordinate releases in the line to completions at assembly. We discuss shop floor mechanisms in Chapter 14 and scheduling procedures in Chapter 15.

Station overlap time. Unlike the other "times," we would like to *increase* station overlap time because it is subtracted from the total cycle time. It can be increased by the use of lot splitting where feasible. Streamlined material handling (e.g., through the use of cells) makes the use of smaller transfer batches possible and hence enhances the cycle time benefits of lot splitting.

Example: Cycle Time Reduction

SteadyEye, a maker of commercial camera mounts, sells its products in make-to-order fashion to the motion picture industry. Lately the company has become concerned that customer lead times are no longer competitive. SteadyEye offers 10-week lead times, quoted from the end of two-week order buckets. (For instance, if an order is received anywhere in the two-week interval between September 5, 1999, and September 18, 1999, it is quoted a delivery date 10 weeks from September 18, 1999.) However, their major competitor is offering five-week lead times from the date of the order. Worse yet, SteadyEye's inventory levels are at record levels, average cycle time (currently nine weeks) is as long as it has ever been, and customer service (fraction of orders delivered on-time) is poor (less than 70 percent) and declining.

SteadyEye's process begins with the entry of customer orders, which is done by a clerk daily. Much to the clerk's frustration, it seems that most of the orders seem to come at the end of the two-week interval, which forces her to fall behind even though she puts in significant overtime every other weekend. Using the most recent customer orders, an ERP system generates a daily set of purchase orders and dispatch lists. These lists are sent to each process center but are especially important at the assembly area because that is where parts are matched to fill orders. Unfortunately, it is common for lists to be ignored because the requisite parts are not available.

SteadyEye manufactures legs, booms, and other structural components of its camera mounts, as well as gears and gearboxes that go into the control assembly. It purchases all motors and electronics from outside suppliers. Raw materials and subassemblies are received at the receiving dock. Bar stock is sawed to the correct lengths for the various gears and is then sent to the milling operation on a pallet carried by a forklift. Because of long changeover times at the mills, process batches are very large. Other operations include drilling, grinding, and polishing. The polisher is very fast, and so there is only one. Unfortunately, it is also difficult to adjust, and so downtimes are very long and generate a lot of parts that need to be scrapped. The heat treat operation takes three hours and involves a very large oven that can hold nearly 1,000 parts. Since most process batches are larger than those required by a single order, parts are returned to a crib inventory location after each operation.

The root of SteadyEye's problem is excessive cycle time, which from factory physics is a consequence of variability (arrival and process) and utilization. Thus, improvement policies must focus on these.

To begin, the arrival variability is being unnecessarily magnified by the order processing system. By establishing a two-week window within which all orders are quoted the same due date, the system encourages procrastination on the part of the customers and sales engineers. (Why get an order in before the end of the time window if it won't be shipped any earlier?) The resulting last-minute behavior creates a burst of arrivals to the system, thereby greatly increasing the effective c_a^2. Fortunately, this problem can be remedied by simply eliminating the order window. A better policy would have orders received on day t promised delivery on day $t + \ell$ (where ℓ is a lead time, which we hope to get down to five weeks or less). Orders can still be batched within the system by pulling in orders later on the master production schedule, but this can be transparent to customers.

Next, variability analysis of the effective process times shows that the polisher has an enormous c_e^2 of around seven. This is further aggravated by the fact that utilization of the polisher, after considering the various detractors, is greater than 90 percent. An attractive improvement policy, therefore, is to analyze the parameters affecting the polisher to find ways to reduce the time needed for adjustment. This will also reduce scrap and the need to expedite small jobs of parts to replace those that were scrapped. The net effect will be to reduce c_e^2 and u at a bottleneck operation, which will significantly reduce queueing, and hence average cycle time. Since these measures will also reduce cycle time variability, they will enable reduction of customer lead time by even more than the reduction in average cycle time.

Another large source of variability and cycle time in this system is batching, so we turn to it next. Batching is driven by both material handling and processing considerations. Move batches are large (typically a full pallet) because processes are far apart so that forklift capacity does not permit frequent transfers. An appealing policy therefore would be to organize processes into cells near the assembly lines. With this and some investment in material handling devices (e.g., conveyors) it may be practical to reduce move sizes to one. Process batches are large because of long setups. Hence, the logical improvement step is to implement a rigorous setup reduction program (e.g., using single minute exchange of die (SMED) techniques, see Shingo 1985). Since cutting setup times by a factor of four or more is not uncommon, such steps could enable SteadyEye to reduce process batch sizes by 75 percent or more.

In addition to these improvements in the processes themselves, there are some system changes that could further reduce cycle times. One would be to restrict use of the ERP system to providing purchase orders for outside parts and to generating "planned orders" but *not* for converting these to actual jobs. A separate module is needed to combine orders into jobs such that like orders of like families will be processed together (to share a setup at milling where setups are still significant) while still meeting due dates. The mechanics for such a module are given in Chapter 15.

Additionally, it may make sense to convert some commonly used components from make-to-order to make-to-stock parts. The crib that is now storing remnants of large batches of many parts would be converted to storage of stocks of these parts. Because batch sizes will be much smaller, all other parts will never enter the crib, but instead will be used as produced. Thus, even though stock levels of selected parts (common parts for which elimination of cycle time would appreciably reduce customer lead time) will increase, the overall stock level in the crib should be significantly less.

The net result of this battery of changes will be to substantially reduce cycle times. To go from an average cycle time of 10 weeks to less than two weeks is not an unreasonable expectation. If the company can pull it off, SteadyEye will transform its manufacturing operation from a competitive millstone to a strategic advantage.

For a more detailed example of cycle time reduction, the reader is referred to Chapter 19.

9.6.3   Improving Customer Service

In operational terms, satisfying customer needs is primarily about lead time (quick response) and service (on-time delivery). As we noted earlier, one way to radically reduce lead time is to move from a make-to-order system to a make-to-stock system, or to do this partially by making generic components to stock and assembling to order. We discuss this approach more fully in Chapter 10.

For the segment of the system that is make to order, the Lead Time Law implies

$$\text{lead time} = \text{average cycle time} + \text{safety lead time}$$

$$= \text{average cycle time} + z_s \times \text{standard deviation of cycle time}$$

where z_s is a safety factor that increases in the desired level of service. Therefore, reducing lead time for a fixed service level (or improving service for a fixed lead time) requires reducing average cycle time and/or reducing standard deviation of cycle time. Policies for reducing average cycle time were noted above. Fortunately, these same policies are effective for reducing cycle time standard deviation. However, as we noted, some policies, such as reducing long rework loops, are particularly effective at reducing cycle time variability.

Example: Customer Service Enhancement

The focus of the SteadyEye example was on reducing mean cycle time. The underlying reason for this, of course, was the firm's concern about responsiveness to customers. But it makes no sense to address lead time without simultaneously considering service. Promising short lead times and then failing to meet them is hardly the way to improve customer service. Fortunately, the improvements we suggested can enable the system to both reduce lead time and improve service.

For example, recall that one proposed policy was to reduce scrap at the polisher, which in turn will reduce the need to expedite small jobs of parts to catch up with the rest of the batch at final assembly. Doing this will significantly reduce the *standard deviation* of cycle time, as well as mean cycle time. Therefore, even if we increase service (i.e., raise the safety factor z_s), total customer lead time can still be reduced. The other variability reduction measures will have similar impacts.

To illustrate this, suppose that the original mean cycle time was nine weeks with a standard deviation of three weeks. A lead time of 10 weeks allows for only about one-third of a standard deviation for safety lead time. Since $z_{0.33} = 0.63$, this results in service of only around 63 percent, which is consistent with what is being observed.

Suppose that after all the cycle time reduction steps have been implemented, average cycle time is reduced to seven shop days (1.4 weeks) and the standard deviation is reduced to one-half week. In this case, a two-week lead time represents a safety lead time of 0.6 week, or 1.2 standard deviations, which would result in 88 percent service. A (probably more reasonable) three-week lead time represents a safety lead time of 3.2 standard deviations, which would result in more than 99.9 percent service. The combination of significantly shorter lead times than the competition *and* reliable delivery would be a very strong competitive weapon for SteadyEye.

Finally, we point out that the benefits of variability and cycle time reduction are not limited to make-to-order systems. Recall that one of the improvement suggestions for cycle time reduction was to shift some parts to make-to-stock control. For instance, suppose SteadyEye stocks a common gear for which there is average demand of 500 per week with a standard deviation of 100. The cycle time to make the part is nine weeks with a standard deviation of three weeks. Thus, the mean demand during the replenishment time is 4,500, and the standard deviation is 1,530. If we produce $Q = 500$ at a time, then we can use the (Q, r) model of Chapter 2 to compute that a reorder point of $r = 7,800$ will be needed to ensure a 99 percent fill rate. This policy will result in an average on-hand inventory of 3,555 units. However, if the variability reduction measures suggested above reduced the cycle time to 1.4 weeks with a standard deviation of 0.4 week, the reorder point would fall to $r = 1,080$ and the average on-hand inventory would decrease to 631 units, a 92 percent reduction. This makes moving to the more responsive make-to-stock control for common parts an economically viable option.

9.7   Conclusions

The primary focus of this chapter is the effect of variability on the performance of production lines. The main points can be summarized as follows:

1. *Variability degrades performance.* If variability of any kind—process, flow, or batching—is increased, something has to give. Inventory will build up, throughput will decline, lead times will grow, or some other performance measure will get worse. As a result, almost all effective improvement campaigns involve at least some amount of variability reduction.

2. *Variability buffering is a fact of manufacturing life.* All systems buffer variability with inventory, capacity, and time. Hence, if you cannot reduce variability, you will have to live with one or more of the following:

   a. Long cycle times and high inventory levels

   b. Wasted capacity

   c. Lost throughput

   d. Long lead times and/or poor customer service

3. *Flexible buffers are more effective than fixed buffers.* Having capacity, inventory, or time that can be used in more than one way reduces the total amount of buffering required in a given system. This principle is behind much of the flexibility or agility emphasis in modern manufacturing practice.

4. *Material is conserved.* What flows into a workstation will flow out as either good product or scrap.

5. *Releases are* always *less than capacity in the long run.* The intent may be to run a process at 100 percent of capacity, but when true capacity, including overtime, outsourcing, etc., is considered, this can never occur. It is better to *plan* to reduce release rates before the system "blows up" and rates have to be reduced anyway.

6. *Variability early in a line is more disruptive than variability late in a line.* High process variability toward the front of a push line propagates downstream and causes queueing at later stations, while high process variability toward the end of the line affects only those stations. Therefore, there tends to be greater leverage from variability reduction applied to the front end of a line than to the back end.

7. *Cycle time increases nonlinearly in utilization.* As utilization approaches one, long-term WIP and cycle time approach infinity. This means that system performance is very sensitive to release rates at high utilization levels.

8. *Process batch sizes affect capacity.* The interaction between process batch size and setup time is subtle. Increasing batch sizes increases capacity and thereby reduces queueing. However, increasing batch sizes also increases wait-to-batch and wait-in-batch times. Therefore, the first focus in serial batching situations should be on setup time reduction, which will enable use of small, efficient batch sizes. If setup times cannot be reduced, cycle time may well be minimized at a batch size greater than one. Likewise, depending on the capacity and demand, the most efficient batch size in a parallel process may be in between one and the maximum number that will fit into the process.

9. *Cycle times increase proportionally with transfer batch size.* Waiting to batch and unbatch can be a large source of cycle time. Hence, reducing transfer batches is one of the simplest cycle time reduction measures available in many production environments.

10. *Matching can be an important source of delay in assembly systems.* Lack of synchronization, caused by variability, poor scheduling, or poor shop floor control, can cause significant buildup of WIP, and hence delay, wherever components are assembled.

11. *Diagnosis is an important role for factory physics.* The laws and concepts of factory physics are useful to trace the sources of performance problems in a manufacturing system. While the analytical formulas are certainly valuable in this regard, it is the intuition behind the formulas that is most critical in the diagnostic process.

Because variability is not well understood in manufacturing, the ideas in this chapter are among the most useful factory physics concepts presented in this book. We will rely heavily on them in Part III to address specific manufacturing management problems.

Study Questions

1. Under what conditions is it possible for a workstation to operate at 100 percent capacity over the long term and not be unstable (i.e., not have WIP grow to infinity)? Can this occur in practice?

2. In a line with large transfer batches, why is wait-for-batch time larger when utilization is low than when it is high? What assumption about releases is behind this, and why might it not be the case in practice?

3. In what way are variability reduction and capacity expansion analogous improvement options? What important differences are there between them?

4. Consider two adjacent stations in a line, labeled A and B. A worker at station A performs a set of tasks on a job and passes the job to station B, where a second worker performs another set of tasks. There is a finite amount of space for inventory between the two stations. Currently, A and B simply do their own tasks. When the buffer is full, A is blocked. When the buffer is empty, B is starved. However, a new policy has been proposed. The new policy designates a set of tasks, some from A's original set and others from B's set, as "shared tasks." When the buffer is more than half full, A does the shared tasks before putting jobs into the buffer. When the buffer is less than half full, A leaves the shared tasks for B to do. Assuming that the shared tasks can be done equally quickly by either A or B, comment on the effect that this policy will have on overall variability in the line. Do you think this policy might have merit?

5. The JIT literature is fond of the maxim "Variability is the root of all evil." The Variability Law of factory physics states that "variability degrades performance." However, in Chapter 7, we showed that the worst possible behavior for a line with a given r_b (bottleneck rate) and T_0 (raw process time) occurs when the system is completely deterministic (i.e., there is no variation in arrivals or process times). How can these be consistent?

6. Consider a one-station plant that consists of four machines in parallel. The machines have moderately variable random process times. Note that if the WIP level is fixed at four jobs, the plant will be able to maintain 100 percent utilization, minimum cycle time, and maximum throughput whether or not the process times are random. How do you explain this apparent "perfect" performance in light of the variability that is present? (*Hint:* Consider *all* the performance measures, including those for FGI and demand, when there is no variability at all. What happens to these measures when process times are made variable and demand is still constant?)

Intuition-Building Exercises

The purpose of these exercises is to build your intuition. They are in no way intended to be realistic problems.

1. You need to make 35 units of a product in one day. If you make more than 35 units, you must pay a carrying cost of $1 per unit extra. If you make less than 35 units, you must pay a penalty cost of $10 per unit.

   You can make the product in one of two workstations (you cannot use both). The first workstation (W1) contains a single machine capable of making 35 units per day, on average. The second workstation (W2) contains 10 machines, each capable of making 3.5 units per day, on average. Which workstation should you use?

Exercise: Simulate the output of W1 by rolling a single die and multiplying the number of spots by 10. Simulate the output of W2 by rolling the die 10 times and adding the total number of spots.

Perform five replications of the experiment. Compute the amount of penalty and carrying cost you would incur for each time. Which is the better workstation to use? What implications might this have for replacing of a group of old machines with a single "flexible manufacturing system"?

2. You market 20 different products and have a choice of two different processes. In process one (P1) you stock each of the 20, maintaining a stock of five for each of the products for a total of 100 units. In process two (P2) you stock only the basic component and then give each order "personality" when the order is received. The time to do this is, essentially, no greater than that for processing the order. For this process you stock 80 of the basic components.

Every day you receive demand for each of the products. The demand is between one and six items with each level equally likely. Stock is refilled at the end of each day.

Exercise: Which process do you think would have the better fill rate (i.e., probability of having stock for an order), P1 with 100 parts in inventory or P2 with only 80? Simulate each, using a roll of a die to represent the demand for each of the 20 products, and keep track of total demand and the total number of stockouts. Repeat the simulation at least five times, and compute the average fill rate.

3. Consider a line composed of five workstations in series. Each workstation has the *potential* to produce anywhere between one and six parts on any given day, with each outcome equally likely (note that this implies the average potential production of each station is 3.5 units per day). However, a workstation in the middle of the line cannot produce more on a day than the amount of WIP it starts the day with.

Exercise 1: Perform an experiment using a separate roll of a die for the daily potential production at each station. Use matchsticks, toothpicks, poker chips, whatever, to represent WIP. Each time you roll the die, actual production at the station will be the lesser of the die roll and the available WIP.

Since you start out empty, it will take five days to fill up the line. So begin recording the output at the sixth period. Plot the cumulative output and total WIP in the line versus time up to day 25.

Exercise 2: Now reduce the WIP by employing a kanban mechanism. To do this, do not allow WIP to exceed four units at any buffer (after all, the production rate is 3.5 so we should be able to live with four). Do this by reducing the actual production at a station if it will ever cause WIP at the next station to exceed four. Repeat the above exercise under these conditions. What happens to throughput? What about WIP?

Exercise 3: Now reduce variability. To do this, change the interpretation of roll. If a roll is three or less, potential production is three units. If it is four or more, potential production is four units. Note that the average is the same as before. Now repeat both the first exercise (without the kanban mechanism) and the second exercise (with kanban). Compare your results with those of the previous cases.

Exercise 4: Finally, consider the situation where there are two types of machine in the line, one that is highly variable and another that is less variable. Should we have the more variable ones feed the less variable ones, or the other way round? Repeat the first exercise for a line where the first two machines are extremely variable (i.e., potential production is given by the number of spots on the die) and the last three are less variable (i.e., potential production is three if the roll is three or less and four if it is four or more). Repeat with a line where the last two machines are extremely variable and the first three are less variable. Compare the throughput and WIP for the two lines, and explain your results.

Problems

1. Consider a line that makes two different astronomical digital cameras. The TS-7 costs $2,000 while the TS-8, which uses a much larger chip, costs $7,000. Most of the cost of the cameras is due to the cost of the chip.

   In manufacturing, both go through the same three steps but take different amounts of time. The capacities for the TS-7 are seven, five, and six per day at workstations 1, 2, and 3, respectively (that is, if we run exclusively TS-7 product). Similarly, capacity for the TS-8 is six per day at all stations (again, assuming we run only TS-8). Five percent of TS-8 units must be reworked, which requires them to go back through all three stations a second time (process times are the same as those for the first pass). Reworked jobs never make a third pass through the line. There is no rework for the TS-7.

   Demand is three per day for the TS-7 and one per day for the TS-8. The average inventory level of chips is 20 for the TS-7 and five for the TS-8. Cycle time for both cameras is four days, while the raw process time with no detractors is one-half a day. Cameras are made to stock and sold from finished goods inventory. Average finished goods inventory is four units of the TS-7 and one unit of the TS-8, while the fill rate is 0.85 for both cameras.

   a. Compute throughput TH(i) for each station for each product.
   b. Compute utilization $u(i)$ at each station.
   c. Using dollars as the aggregate measure, compute RMI, WIP, and FGI.
   d. Compute the efficiencies E_{TH}, E_u, E_{inv}, E_{CT}, E_{LT}, E_s, and E_Q.
   e. Suppose the machine at workstation 1 costs $1 million and the machines at the second and third workstations cost $10,000 each. Suggest a different measure for E_u than that given in the text. Compute it and compare with the previous value.

2. Describe the types of buffer(s) (i.e., inventory, time, or capacity) you would expect to find in the following situations.
   a. A maker of custom cabinets
   b. A producer of automotive spare parts
   c. An emergency room
   d. Wal-Mart
   e. Amazon.com
   f. A government contractor that builds submarines
   g. A bulk producer of chemical intermediates such as acetic acid
   h. A maker of lawn mowers for K-Mart, Sam's Club, and Target
   i. A freeway
   j. The space shuttle (i.e., as a delivery system for advanced experiments)
   k. A business school

3. Compute the capacity (jobs per day) for the following situations.
   a. A single machine with a mean process time of two and one-half hours and an SCV of 1.0. There are eight work hours per day.
   b. A single machine with a mean process time of two and one-half hours and an SCV of 0.5. There are eight work hours per day.
   c. A workstation consisting of 10 machines in parallel, each having a mean process time of two and one-half hours. There are two eight-hour shifts. Lunch and breaks take one and one-fourth hours per shift.
   d. A workstation with 10 machines in parallel, each having a mean process time of two and one-half hours. There are two eight-hour shifts. Lunch and breaks take one and one-fourth hours per shift. The machines have a mean time to failure of 100 hours with a mean time to repair of four hours.
   e. A workstation with 10 machines in parallel, each having a mean process time of two and one-half hours. There are two eight-hour shifts. Lunch and breaks take one and one-fourth hours per shift. The machines have a mean time to failure of 100 hours with a mean time to repair of four hours. The machines are set up every 10 jobs, and the mean setup time is three hours.

f. A workstation with 10 machines in parallel, each having a mean process time of two and one-half hours. There are two eight-hour shifts. Lunch and breaks take one and one-fourth hours per shift. The machines have a mean time to failure of 100 hours with a mean time to repair of four hours. The machines are set up every 10 jobs, and the mean setup time is three hours. Because the operators have to attend training meetings and the like, we cannot plan more than 85 percent utilization of the workers operating the machines.

4. Jobs arrive to a two-station serial line at a rate of two jobs per hour with deterministic interarrival times. Station 1 has one machine which requires exactly 29 minutes to process a job. Station 2 has one machine which requires exactly 26 minutes to process a job, provided it is up, but is subject to failures where the mean time to failure is 10 hours and the mean time to repair is one hour.

   a. What is the SCV c_a^2 of arrivals to station 1?
   b. What is the effective SCV $c_e^2(1)$ of process times at station 1?
   c. What is the utilization of station 1?
   d. What is the cycle time in queue at station 1?
   e. What is the total cycle time at station 1?
   f. What is the SCV of arrivals to station 2?
   g. What is the utilization of station 2?
   h. What is the effective SCV $c_e^2(2)$ of process times at station 2?
   i. What is the cycle time in queue at station 2?
   j. What is the total cycle time at station 2?

5. A punch press takes in coils of sheet metal and can make five different electrical breaker boxes, denoted by B1, B2, B3, B4, and B5. Each box takes exactly one minute to produce. To switch the process from one type of box to another takes four hours. There is demand of 1,800, 1,000, 600, 350, and 200 units per month for boxes B1, B2, B3, B4, and B5, respectively. The plant works one shift, five days per week. After lunch, breaks, etc., there is seven hours available per shift. Assume 52 weeks per year.

   a. What is r_a in boxes per hour?
   b. What would utilization be if there were no setups? (Note that utilization will approach this as batch sizes approach infinity.)
   c. Suppose the SCV of the press is 0.2 no matter what the batch sizes are. What is the average cycle time for the optimal batch sizes (assume $c_a^2 = 1$)?
   d. Use trial and error to find a set of batch sizes that minimizes cycle time.
   e. On average, how many times per month do we make each type of box if we use the batch sizes computed in part *d*?

6. A heat treat operation takes six hours to process a batch of parts with a standard deviation of three hours. The maximum that the oven can hold is 125 parts. Currently there is demand for 160 parts per day (16-hour day). These arrive to the heat treat operation one at a time according to a Poisson stream (i.e., with $c_a = 1$).

   a. What is the maximum capacity (parts per day) of the heat treat operation?
   b. If we were to use the maximum batch size, what would be the average cycle time through the operation?
   c. What is the minimum batch size that will meet demand?
   d. If we were to use the minimum feasible batch size, what would be the average cycle time through the operation?
   e. Find the batch size that minimizes cycle time. What is the resulting average cycle time?

7. Consider a balanced line, having five identical stations in series, each consisting of a single machine with low-variability process times and an infinite buffer. Suppose the arrival rate is r_a, utilization of all machines is 85 percent, and the arrival SCV is $c_a^2 = 1$. What happens to WIP, CT, and TH if we do the following?

   a. Decrease the arrival rate.
   b. Increase the variability of one station.
   c. Increase the capacity at one station.
   d. Decrease the capacity of all stations.

8. Consider a two-station line. The first station pulls from an infinite supply of raw materials. Between the two stations there is a buffer with room for five jobs. The second station can always push to finished goods inventory. However, if the buffer is full when the first station finishes, it must wait until there is room in the buffer before it can start another job. Both stations take 10 minutes per job and have exponential process times ($c_e = 1$).

   a. What are TH, CT, and WIP for the line?

   b. What are TH, CT, and WIP if we increase the buffer to seven jobs?

   c. What are TH, CT, and WIP if we slow down the second machine to take 12 minutes per job?

   d. What are TH, CT, and WIP if we slow down the first machine to take 12 minutes per job?

   e. What happens to TH if we decrease the variability of the second machine so that the effective SCV is a $\frac{1}{4}$?

9. Consider a single station that processes two items, A and B. Item A arrives at a rate of 30 per hour. Setup times are five hours, and the time it takes to process one part is one minute. Item B arrives at a rate of 20 per hour. The setup time is four hours, and the unit process time is two minutes. Arrival and process variability is moderate (that is, $c_a = c_e = 1$) regardless of the batch size (just assume they are).

   a. What is the minimum lot size for A for which the system is stable (assume B has an infinite lot size)?

   b. Make a spreadsheet and find the lot sizes for A and B that minimize average cycle time.

10. Consider a balanced and stable line with moderate variability and large buffers between stations. The line uses a push protocol, so that releases to the line are independent of line status. The capacity of the line is r_b, and the utilization is fairly high. What happens to throughput and cycle time when we do the following?

   a. Reduce the buffer sizes and allow blocking at all stations except the first where jobs balk if the buffer is full (i.e., they go away if there is no room).

   b. Reduce the variability in all process times.

   c. Unbalance the line, but do not change r_b.

   d. Increase the variability in the process times.

   e. Decrease the arrival rate.

   f. Decrease the variability in the process times *and* reduce the buffer sizes as in *a*. Compare to the situation in *a*.

11. A particular workstation has a capacity of 1,000 units per day and variability is moderate, such that $V = (c_a^2 + c_e^2)/2 = 1$. Demand is currently 900 units per day. Suppose management has decided that cycle times should be no longer than one and one-half times raw process time.

   a. What is the current cycle time in multiples of the raw process time?

   b. If variability is not changed, what would the capacity have to be in order to meet the cycle time and demand requirements? What percentage increase does this represent?

   c. If capacity is not changed, what value would be needed for V in order to meet the cycle time and demand requirements? What percentage decrease does this represent (compare CVs, not SCVs)?

   d. Discuss a realistic strategy for achieving management's goal.

12. Consider two stations in series. Each is composed of a single machine that requires a rather lengthy setup. Large batches are used to maintain capacity. The result is an effective process time of one hour per job and an effective CV of 3 (that is, $t_e = 1.0$ and $c_e^2 = 9.0$). Jobs arrive in a steady stream at a rate of 0.9 job per hour, and they come from all over the plant, so $c_a = 1.0$ is a reasonable assumption (see the discussion in Chapter 8).

   Now, suppose a flexible machine is available with the same capacity but less effective variability (that is, $t_e = 1.0$ and $c_e^2 = 0.25$) and can be used to replace the machine at either station. At which station should we replace the existing machine with the new one to get the largest reduction in cycle time? (*Hint:* Use the equation $c_d^2 = u^2 c_e^2 + (1 - u^2)c_a^2$ along with the cycle time equations.)

13. Recall the throughput enhancement example in Section 9.6.1. Assuming there is an unlimited amount of raw material for the coater, answer the following.

   a. Compute t_e and c_e^2, using the data given in Section 9.6.1 for both the coater and the expose operation.
   b. Use the general blocking model of Section 8.7.2 to compute the throughput for the line, assuming there is room for 10 jobs in between the two stations (that is, $b = 12$). Will the resulting throughput meet demand?
   c. Reduce the MTTR from eight to four hours, and recompute throughput. Now does the throughput meet demand?

14. Table 9.6 gives the speed (in parts per hour), the CV, and the cost for a set of tools for a circuit-board line. Jobs go through the line in totes that hold 50 parts each (this cannot be changed). The CVs represent the *effective* process times and thus include the effects of downtime, setups, etc.

   The desired average cycle time through this line is 1.0 day. The maximum demand is 1,000 parts per day.
   a. What is the least-cost configuration that meets demand requirements?
   b. How many possible configurations are there?
   c. Find a good configuration.

15. Consider line 1 in Table 9.5. Assume batches of six jobs arrive every 35 hours with no variability in the arrivals, the setup times, or the process times. Construct a Gantt chart (i.e., time line) like that in Figure 9.8 for the system when the stations are permuted from the original order (1, 2, 3) as follows:
   a. 1, 3, 2
   b. 2, 1, 3
   c. 2, 3, 1
   d. 3, 1, 2
   Check to see if the cycle times fall within the bounds given in Section 9.5.3.

16. Suppose parts arrive in batches of 12 every 396 minutes to a three-station line having no variability. The first station has a setup time of 15 minutes and a unit process time of seven minutes, the second sets up in eight minutes and processes one part every three minutes, the third requires 12 and four minutes for setup and unit processing, respectively.
   a. What is the utilization of each station? Which is the bottleneck?
   b. What is the cycle time if parts are moved 12 at a time?
   c. What is the cycle time for the first part if parts are moved one at a time?
   d. What is the range of cycle time for the 12th part if parts are moved one at a time?
   e. What is the range of average cycle times if parts are moved one at a time?
   f. Perform a Penny Fab–like experiment to determine the average cycle time. Let 12 parts arrive each 396 minutes, and then move them one at a time.
   g. Double the arrival rate (i.e., batches of 12 arrive every 198 minutes). What happens to cycle time if parts are moved 12 at a time? What happens to cycle time if parts are moved one at a time?
   h. Now let the arrivals be Poisson with the same average time between arrivals (396 minutes). What is the added queue time at each station?
   i. Now double the Poisson arrival rate. What happens to cycle time?

TABLE 9.6   Possible Machines to Purchase for Each Work Center

Station	Possible Machines (speed [parts/hour], CV, cost [$000])			
	Type 1	**Type 2**	**Type 3**	**Type 4**
MMOD	42, 2.0, 50	42, 1.0, 85	50, 2.0, 65	10, 2.0, 110.5
SIP	42, 2.0, 50	42, 1.0, 85	50, 2.0, 65	10, 2.0, 110.5
ROBOT	25, 1.0, 100	25, 0.7, 120	—	—
HDBLD	5, 0.75, 20	5.5, 0.75, 22	6, 0.75, 24	—

10   PUSH AND PULL PRODUCTION SYSTEMS

You say yes.
I say no.
You say stop,
And I say go, go, go!

John Lennon, Paul McCartney

10.1   Introduction

Virtually all descriptions of just-in-time make use of the terms *push* and *pull* production systems. However, these terms are not always precisely defined and, as a result, may have contributed to some confusion surrounding JIT in America.

In this chapter, we offer a formal definition of push and pull at the conceptual level. By separating the concepts of push and pull from their specific implementations, we observe that most real-world systems are actually hybrids or mixtures of push and pull. Furthermore, by contrasting the extremes of "pure push" and "pure pull" production systems, we gain insight into the factors that make pull systems effective. This insight suggests that there are many different ways to achieve the benefits of pull. Which is best depends on a variety of environmental considerations, as we discuss in this chapter and pursue further in Part III.

10.2   Definitions

The father of JIT, Taiichi Ohno, used the term *pull* only in a very general sense (Ohno 1988, xiv):

> Manufacturers and workplaces can no longer base production on desktop planning alone and then distribute, or *push,* them onto the market. It has become a matter of course for customers, or users, each with a different value system, to stand in the frontline of the marketplace and, so to speak, *pull* the goods they need, in the amount and at the time they need them.

Hall (1983, 39), in one of the most prominent American texts on JIT, was more specific, defining pull systems by the fact that "material is drawn or sent for by the users of the material as needed." Although he acknowledged that different types of pull systems are possible, the only one he described in detail was the Toyota kanban system, which we discussed in Chapter 4.[1] Schonberger (1982), in the other major American JIT book, referred to pull systems strictly in the context of the Toyota-style kanban system. Hence, it is hardly surprising that the term *pull* is frequently viewed as synonymous with *kanban*.

However, we do not feel that such a narrow interpretation was Ohno's intent. In our view, limiting *pull* to mean *kanban* is downright counterproductive: It obscures the essence of pull by assigning it too much specificity. It mixes a concept (pull) with its implementation (kanban). In order for us to discuss the concept of pull from a factory physics perspective, it is important to give a general, but simple, definition of push and pull systems.

10.2.1   The Key Difference between Push and Pull

What distinguishes push from pull is the mechanism that triggers the movement of work in the system. Fundamentally, the trigger for work releases comes from *outside* a push system but from *inside* a pull system. More formally we define push and pull systems as follows:

Definition: A **push** system schedules *the release of work based on demand, while a* **pull** system authorizes *the release of work based on system status.*

The contrast between push and pull systems is depicted schematically in Figure 10.1. Strictly speaking, a push system releases a job into a production process (factory, line, or workstation) precisely when called to do so by an exogenous schedule, and the release time is not modified according to what is happening in the process itself. In contrast, a pull system only allows a job onto the floor when a signal generated by a change in line status calls for it. Typically, as in the Toyota kanban system, these authorization signals are the result of the completion of work at some point in the line. Notice that this definition has nothing to do with who actually moves the job. If an operator from a downstream process comes and gets work from an upstream process, but does it according to an exogenous schedule, then the process is push. If an upstream operator delivers work to the downstream process, but does so in response to status changes in the downstream process, then this is pull.

Another useful way to think about the distinction between push and pull systems is that push systems are inherently *make-to-order* while pull systems are *make-to-stock.* That is, the schedule that drives a push system is driven by orders (or forecasts), but not by system status. The signals that authorize releases in a pull system are voids in a stock level somewhere in the system. Viewed in this way, the base stock model, which triggers orders when stock drops below a specified level, is a pull system. An MRP system, which releases order into the system according to a schedule based on customer orders, is a push system.

Of course, most real-world systems have aspects of both push and pull. For instance, if a job is scheduled to be released by MRP, but is held out because the line is

---

[1]Hall also presented as a pull system the *broadcast* system, in which the final assembly schedule (FAS) is broadcast to all starting points in the line in order to trigger work releases. However, he noted that because the FAS is generated externally, this system does not place a restriction on the total inventory in the system. He distinguished the control in a broadcast system from that in a kanban system by referring to the FAS signals as *loose pull signals.* Because of its failure to limit WIP, we are not convinced that this system should be termed a pull system at all.

FIGURE 10.1

FIGURE 10.1

*Release triggers in push
and pull production
systems*

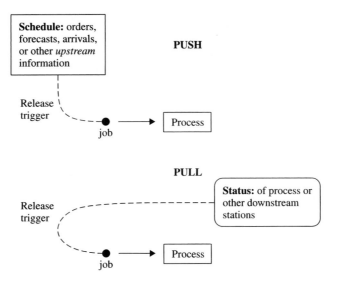

considered too congested, then the effect is a hybrid push-pull system. Conversely, if a kanban system generates a card authorizing production but the actual work release is delayed because of anticipated lack of demand for the part (i.e., it is not called for in the master production schedule), then this, too, is a hybrid system. There have been various attempts to formally combine push and pull into hybrid systems (e.g., see Wight 1970, Deelersnyder et al. 1988, and Suri 1998). We will discuss the virtues of hybrid systems and present an approach in Part III.

Our purpose in setting up a sharp distinction between push and pull is *not* to suggest that users must rigidly choose one or the other. Rather, in the spirit of factory physics, we use our definition to isolate the benefits of pull systems and trace their root causes. In a sense, we are taking a similar approach to that of (nonfactory) physics in which mechanical systems are frequently considered in frictionless environments. It is not that frictionless environments are common, but rather that the concepts of gravitation, acceleration, velocity, and so forth are clearer in this pristine framework. Just as the frictionless insights of classical mechanics underlie analysis of realistic physical systems, our observations about pure push and pure pull systems provide a foundation for analysis of realistic production systems.

10.2.2   The Push-Pull Interface

The questions of whether and how to use pull are only part of the picture; *where* to use pull is also important. Even in an individual production system, it is possible to run only part of it as a pull system. A useful concept for thinking about placement of pull mechanisms is the *push-pull interface,* which divides a production process into push and pull segments.[2] Choosing the location of this interface wisely can enable a system to take strategic advantage of the benefits of pull, while still retaining the customer-driven character of push.

---

[2]We are indebted to Corey Billington of Hewlett-Packard (HP) for the term *push-pull interface*, which was coined to help describe practices developed as part of their "design for supply chain management" efforts. See Lee and Billington (1995) for an overview of HP supply chain initiatives.

To understand the concept of the push-pull interface, it is convenient to think in terms of push being defined as make to order and pull being defined as make to stock. To see how similar lines can be divided differently into push and pull segments, consider the two systems depicted in Figure 10.2. In the front part of the QuickTaco line, tacos are produced to stock, to maintain specified inventory levels at the warming table, which makes this portion of the line behave as a pull system. The back of the line moves product (tacos) only when triggered by customer orders, and hence it acts as a push system. The push-pull interface lies at the warming table. In contrast, the movement of tacos in the TacoUltimo line is triggered solely by customer orders, so it is entirely a push system. The push-pull interface lies at the refrigerator, where raw materials are stocked according to inventory targets.

By contrasting the relative advantages of the QuickTaco and TacoUltimo lines, we can gain insight into the tradeoffs involved in positioning the push-pull interface. The TacoUltimo line, because it is entirely order-driven and holds inventory almost exclusively in the form of raw materials, has the advantage of being very flexible (i.e., it can produce virtually any taco a customer wants). The QuickTaco line, because it holds finished tacos in stock, has the advantage of being responsive (i.e., it offers shorter lead times to the customer). Hence, the tradeoff is between speed and flexibility. By moving the push-pull interface closer to the customer, we can reduce lead times, but only at the expense of reducing flexibility.

FIGURE 10.2

Illustration of push-pull interface placement

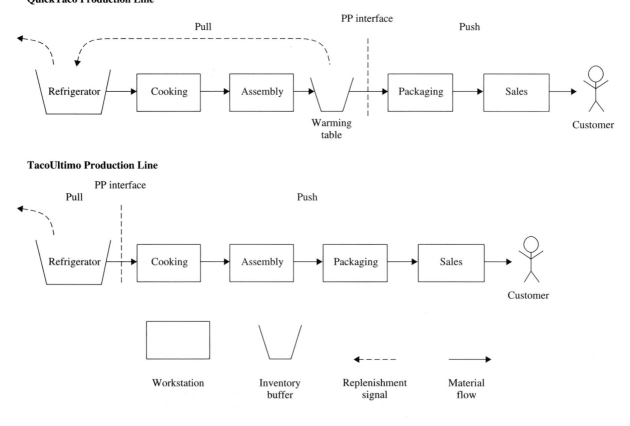

So how does one choose the location of the push-pull interface for a given system? Since it depends on both customer preferences and the physical details of the production process, this is not a simple question. But we can offer some observations and some real-world examples.

First, note that the primary reason for moving the push-pull interface closer to the customer is speed. So it only makes sense to do it when the additional speed will produce a noticeable improvement in service from the perspective of the customer. For instance, in a production system with two-hour cycle times within the line but which makes end-of-day shipments, customers might not see any difference in lead times by shortening cycle time in the line through a push-pull interface shift. Even in the fast-food industry, where speed is clearly critical, there are restaurants that make use of a TacoUltimo type of line. They do this by making sure that the cycle time of the entire line is sufficiently short to enable the system to meet customer expectations. However, during rush hour, when the pressure for speed is especially great, many TacoUltimo-type fast-food restaurants shift to the QuickTaco mode.

Second, observe that the options for positioning the push-pull interface are strongly affected by the process itself. For instance, in the taco line, we could propose a push-pull interface somewhere in the middle of assembly. That is, cook the tortilla shell and fill it with meat, but leave it open, waiting for toppings. However, this would present storage and quality problems (e.g., partially assembled tacos falling apart) and hence is probably infeasible.

Third, notice that the economics of push-pull interface placement are affected by how the product is customized as it progresses through the system. In a system with very few end items (e.g., a plywood mill that takes a few raw materials like logs and glue and produces a few different thicknesses of plywood), it may be perfectly sensible to set the push-pull interface at finished goods. However in a system with many end items (e.g., a PC assembly plant, where components can be combined into a wide range of finished computers), holding inventory at the finished goods level would be very expensive (see the safety stock aggregation example in Section 8.8.2). For example, in the taco system, locating the push-pull interface after packaging is probably a bad idea, since it would require stocking bags of tacos in all needed sizes and combinations.

Finally, note that the issue of customization is closely related to the issue of variability pooling, which we introduced in Chapter 8. In a system in which the product becomes increasingly customized as it progresses down the line, moving the push-pull interface upstream can reduce the amount of safety stock that needs to be carried as protection against demand variability. For example, Benetton made use of a system in which undyed sweaters were produced to stock and then "dyed to order." That is, they moved the push-pull interface from behind the dying process to in front of it. In doing so, they were able to pool the safety stock for the various colors of sweaters and thereby reduce inventory costs of achieving a given level of customer service.

Some other real-world examples in which the push-pull interface was relocated to improve overall system performance include these:

1. *IBM* had a printed-circuit board plant that produced more than 150 different boards from fiberglass and a few thicknesses of copper. The front part of the line produced *core blanks*—laminates of copper and fiberglass from which all circuit boards are made. There were only about eight different core blanks, which were produced in an inherently batch lamination process that was difficult to match to customer orders. Management elected to stock core blanks (i.e., move the push-pull interface from raw materials to a stock point beyond the lamination process). The result was the elimination

of a day or two of cycle time from the lead time perceived by customers at the cost of very little additional inventory.

2. *General Motors* introduced a new vehicle delivery system, starting with Cadillac in Florida, in which popular configurations were stocked at regional distribution centers (*Wall Street Journal,* October 21, 1996, A1). The goal was to provide 24-hour delivery to buyers of these "pop cons" from any dealership. Lead times for other configurations would remain at the traditional level of several weeks. So, unlike in a traditional system, in which the push-pull interface is located at the assembly plant (for build-to-order vehicles) and at the dealerships (for build-to-stock vehicles), this new system places the push-pull interface at the regional distribution centers. The hope is that by pooling inventory across dealerships, General Motors will be able to provide quick delivery for a high percentage of sales with lower total inventory costs. Note that this example illustrates that it is possible, even desirable, to have different locations for the push-pull interface for different products in the same system.

3. *Hewlett-Packard* produced a variety of printers for the European market. However, because of varying voltage and plug conventions, printers required different power supplies for different countries. By modifying the production process to leave off the power supplies, Hewlett-Packard was able to ship generic printers to Europe. There, in the distribution centers, power supplies were installed to customize the printers for particular countries (see Lee, Billington, and Carter 1993 for a discussion of this system). By locating the push-pull interface at the Europe-based distribution center instead of at the American-based factory, the entire shipping cycle time was eliminated from the customer lead time. At the same time, by delaying customization of the printers in terms of power supply, Hewlett-Packard was able to pool inventory across countries. This is an example of **postponement,** in which the product and production process are designed to allow late customization. Postponement can be used to facilitate rapid customer response in a highly customized manufacturing environment, a technique sometimes referred to as **mass customization** (Feitzinger and Lee 1997).

10.3   The Magic of Pull

What makes Japanese manufacturing systems so good? We hope that the reader gathered from Chapter 4 that there is no simple answer to this question. The success of several high-profile Japanese companies in the 1980s was the result of a variety of practices, ranging from setup reduction to quality control to rapid product introduction. Moreover, these companies operated in a cultural, geographic, and economic environment very different from that in America. If we are to understand the essence of the success of JIT, we must narrow our focus.

At a macro level, the Japanese success was premised on an ability to bring quality products to market in a timely fashion at a competitive cost and in a responsive mix. At a micro level, this was achieved via an effective production control system, which facilitated low-cost manufacture by promoting high throughput, low inventory, and little rework. It fostered high external quality by engendering high internal quality. It enabled good customer service by maintaining a steady, predictable output stream. And it allowed responsiveness to a changing demand profile by being flexible enough to accommodate product mix changes (as long as they were not too rapid or pronounced).

What is the key to all these desirable features that made the Japanese production control system such an attractive basis for a business strategy? The American JIT literature seems to suggest that the act of pulling is fundamental. Hall (1983, 39) cited a General Motors foreman who described the essence of pull as "You don't never make nothin' and *send* it no place. Somebody has to come get it."

We disagree. Our view, which we will expand upon in this chapter, is that the pulling of parts into workstations is merely a means to an end. The true underlying cause of the key benefits of a pull system is that *there is a limit on the maximum amount of inventory in the system.* In a (one-card) kanban system, the number of containers is bounded by the number of production cards. No matter what happens on the plant floor, the WIP level *cannot* exceed a prespecified limit. But this effect is not limited to kanban systems. Because a pull system authorizes releases on the basis of voids in stock levels, or equivalently is a make-to-stock system, any true pull system will establish an upper bound on the WIP level. As we discuss in the following subsections, the major benefits of JIT can be attributed to the existence of this **WIP cap,** no matter how it is achieved. The magic was in the WIP cap, not the pulling process.

10.3.1   Reducing Manufacturing Costs

If WIP is capped, then disruptions in the line (e.g., machine failures, shutdowns due to quality problems, slowdowns due to product mix changes) do not cause WIP to grow beyond a predetermined level. Note that in a pure push system, no such limit exists. If an MRP-generated schedule is followed literally (i.e., without adjustment for plant conditions), then the schedule could get arbitrarily far ahead of production and thereby bury the plant in WIP, causing a **WIP explosion.**

Of course, we never observe real-world plants with infinite amounts of WIP. Eventually, when things get bad enough, management does something. It schedules overtime. It hires temporary workers to increase capacity. It pushes out due dates and limits releases to the plant—in other words, management stops using a pure push system. And eventually things return to normal...until the next WIP explosion (see Chapter 9 for a discussion of the overtime vicious cycle). The key point here is that in a push environment, corrective action is not taken until *after* there is a problem and WIP has already spiraled out of control.

In a pull system that establishes a WIP cap, releases are choked off *before* the system has become overloaded. Output will fall off, to be sure, but this would happen regardless of whether the WIP level were allowed to soar. For example, if a key machine is down, then all the WIP in the world in front of it cannot make it produce more. But by holding WIP out of the system, the WIP cap retains a degree of flexibility that would be lost if it were released to the floor. As long as jobs exist only as orders on paper, they can accommodate engineering or scheduling priority changes relatively easily. But once the jobs are on the floor, and given "personality" (e.g., a printed-circuit board receives its circuitry), changes in scheduling priority require costly and disruptive expediting, and engineering changes may be almost impossible. Thus, a WIP cap reduces manufacturing costs by reducing costs due to expediting and engineering changes.

In addition to improving flexibility, a pull system promotes better timing of work releases. To see this, observe that a pure push system periodically allows too much work into the system (i.e., at times when congestion will prevent the new work from being processed quickly). This serves to inflate the average WIP level without improving throughput. A WIP cap, regardless of the type of pull mechanism used to achieve it, will reduce the average WIP level required to achieve a given level of throughput. This will directly reduce the manufacturing costs associated with holding inventory.

10.3.2   Reducing Variability

The key to keeping customer service high is a predictable flow through the line. In particular, we need low **cycle time variability.** If cycle time variability is low, then we know with a high degree of precision how long it will take a job to get through the plant. This allows us to quote accurate due dates to customers, and meet them. Low cycle time variability also helps us quote shorter lead times to customers. If cycle time is 10 days plus or minus 6 days, then we will have to quote a 16-day lead time to ensure a high service level. On the other hand, if cycle time is 10 days plus or minus 1 day, then a quote of 11 days will suffice.

Kanban achieves less variable cycle times than does a pure push system. Since cycle time increases with WIP level (by Little's law), and kanban prevents WIP explosions, it also prevents cycle time explosions. However, note that the reason for this, again, is the WIP cap—not the pulling at each station. Hence, any system that caps WIP will prevent the wild gyrations in WIP, and hence cycle time, that can occur in a pure push system.

Kanban is also often credited with reducing variability directly at workstations. This is the JIT "reduce the water level to expose the rocks" analogy. Essentially, kanban limits the WIP in the system, making it much more vulnerable to variability and thereby putting pressure on management to continually improve.

We illustrate the intuition behind this analogy by means of the simple example shown in Figure 10.3. The system consists of two machines, and machine 1 feeds machine 2. Machine 1 is extremely fast, producing parts at a rate of one per second, while machine 2 is slow, producing at a rate of one per hour. Suppose a (one-card) kanban system is in use, which limits the WIP between machines to five jobs. Because machine 1 is so fast, this buffer will virtually always be full whenever machine 1 is running.

However, suppose that machine 1 is subject to periodic failures. If a failure lasts longer than five hours, then machine 2, the bottleneck, will starve. Thus, depending on the frequency and duration of failures of machine 1, machine 2 could be starved a significant fraction of time, despite the tremendous speed of machine 1.

Clearly, if the buffer size (number of kanban cards) were increased, the level of starvation of machine 2 would decrease. For instance, if the buffer were increased to 10 jobs, only failures in excess of 10 hours would cause starvation. In effect, the extra WIP insulates the system from the disruptive effects of failures. But as we noted previously, a pure push system requires higher average WIP levels to attain a given throughput level. A push system will tend to mask the effects of machine 1 failures in precisely this way. The

FIGURE 10.3

Workstations connected by a finite buffer

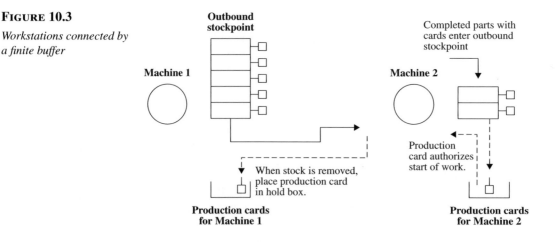

push system will have higher WIP levels throughout the system, and therefore failures will be less disruptive. As long as management is willing to live with high WIP levels, there is little pressure to improve the reliability of machine 1.

As the JIT literature correctly points out, if one wants to maintain high levels of throughput with *low* WIP levels (and short cycle times), one must reduce these disruptive sources of variability (failures, setups, recycle, etc.). We note that, again, the source of this pressure is the limited WIP level, not the mechanism of pulling at each station. To be sure, pulling at each station controls the WIP level at every point in the process, which would not necessarily be the case with a general WIP cap. However, reducing overall WIP level via a WIP cap *will* reduce the WIP between various workstations on average and thereby will apply the pressure that promotes continual improvement. Whether or not a general WIP cap will distribute WIP properly in the line is a question we will take up later.

10.3.3   Improving Quality

Quality is generally considered to be both a precondition for JIT and a benefit of JIT. As such, JIT promotes higher levels of quality out of sheer necessity and also establishes conditions under which high quality is easier to achieve.

As Chapter 4 observed, quality is a basic component of the JIT philosophy. The reason is that if WIP levels are low, then a workstation will effectively be starved for parts whenever the parts in its inbound buffer (stockpoint) do not meet quality standards. From a logistics standpoint, the effect of this is very similar to that of machine failures; once WIP levels become sufficiently low, the percentage of good parts in the system *must* be high in order to maintain reasonable throughput levels. To ensure this, kanban systems are usually accompanied by statistical process control (SPC), quality-oriented worker training, quality-at-the-source procedures, and other techniques for monitoring and improving quality levels throughout the system. Since the higher the quality, the lower the WIP levels can be, continual efforts at WIP reduction practiced in a JIT system will demand continual quality improvement.

Beyond this simple pressure for better quality, JIT can also directly facilitate improved quality because inspection is more effective in a low-WIP environment. If WIP levels are high and queues are long, a quality assurance (QA) inspection may not identify a process problem until a large batch of defective parts has already been produced. If WIP levels are low, so that the queue in front of QA is short, then defects can be detected in time to correct a process before it produces many bad parts. This, of course, is the goal of SPC, which monitors the quality of a process in real time. However, where immediate inspection is not possible, say, in a circuit-board plant where boards must be optically or electronically tested to determine quality, then low WIP levels can significantly amplify the power of a quality control program.

Notice that, once again, the benefits we are ascribing to kanban or JIT are really the consequence of WIP reduction. Hence, a simple WIP cap will serve to provide the same pressure for quality improvement and the same queue reduction for facilitating QA provided by kanban.

However, there is one further quality-related benefit that is often attributed directly to the pulling activity of kanban. The basic argument is that if workers from downstream workstations must go to an upstream workstation to get parts, then they will be able to inspect them. If the parts are not of acceptable quality, the worker can reject them immediately. The result will be quicker detection of quality problems and less likelihood of moving and working on bad parts.

This argument is not very convincing when the material handling is carried out by a separate worker, say, a forklift driver. Whether forklift drivers are "pushing" parts to the next station because they are finished or "pulling" them from the previous station because they are authorized to do so by a kanban makes little difference to their ability to conduct a quality inspection.

The argument is more persuasive when parts are small and workstations close, so that operators can move their own parts. Then, presumably, if the downstream operators go and get the parts, they will be more likely to check them for quality than if the upstream operator simply drops them off. But this reasoning unnecessarily combines two separate issues.

The first issue is whether the downstream operators inspect all parts that they receive (pushed or pulled). We have seen implementations in industry, not necessarily pull systems, in which operators had to approve material transfers by signing a routing form. Implicit in this approval was an inspection for quality.

It is a second and wholly separate issue whether to limit the WIP between two adjacent workstations. We will take up this issue later in this chapter. For now, we simply point out that the quality assurance benefits of pulling at each station can be attained via inspection transactions independently of the mechanism used for achieving the needed limit on WIP.

10.3.4   Maintaining Flexibility

A pure push system can release work to a very congested line, only to have the work get stuck somewhere in the middle. The result will be a loss of flexibility in several ways. First, parts that have been partially completed cannot easily incorporate engineering (e.g., design) changes. Second, high WIP levels impede priority or scheduling changes, as parts may have to be moved out of the line to make way for a high-priority part. And finally, if WIP levels are high, parts must be released to the plant floor well in advance of their due dates. Because customer orders become less certain as the planning horizon is increased, the system may have to rely on forecasts of future demand to determine releases. And since forecasts are never as accurate as one would like, this reliance serves to further degrade performance of the system.

A pull system that establishes a WIP cap can prevent these negative effects and thereby enhance the overall flexibility of the system. By preventing release of parts when the factory is overly congested, the pull system will keep orders on paper as long as possible. This will facilitate engineering and priority/scheduling changes. Also, releasing work as late as possible will ensure that releases are based on firm customer orders to the greatest extent possible. The net effect will be an increased ability to provide responsive customer service.

The analogy we like to use to illustrate the flexibility benefits of pull systems is that of air traffic control. When we fly from Austin, Texas, to Chicago, Illinois, we frequently wind up waiting on the ground in Austin past our scheduled departure due to what the airlines call *flow control*. What they mean is that O'Hare Airport in Chicago is overloaded (or will be by the time we get there). Even if we left Austin on time, we would only wind up circling over Lake Michigan, waiting for an opportunity to land. Therefore, air traffic control wisely (albeit maddeningly) keeps the plane on the ground in Austin until the congestion at O'Hare has cleared (or will clear by the time we get there). The net result is that we land at exactly the same time (late, that is!) as if we had left on schedule, but we use less fuel and reduce the risk of an accident. Importantly, we also keep other options open, such as that of canceling the flight if the weather becomes too dangerous.

10.3.5   Facilitating Work Ahead

The preceding discussion implies that pull systems maintain flexibility by coordinating releases with the current situation in the line (i.e., by not releasing when the line is too congested). The benefits of coordination can also extend to the situation in which plant status is favorable. If we strictly follow a pull mechanism and release work into the system whenever WIP falls below the WIP cap, then we may "work ahead" of schedule when things go well. For instance, if we experience an interval of no machine failures, staffing problems, materials shortages, and so on, we may be able to produce more than we had anticipated. A pure push system cannot exploit this stretch of good luck because releases are made according to a schedule without regard to plant status.

Of course, in practice there is generally a limit to how far we should work ahead in a pull system. If we begin working on jobs whose due dates are so far into the future that they represent speculative forecasts, then completing them now may be risky. Changes in demand or engineering changes could well negate the value of early completion. Therefore, once we have given ourselves a comfortable cushion relative to demand, it makes sense to reduce the work pace. We will discuss mechanisms for doing this in Part III.

10.4   CONWIP

The simplest way we can think of to establish a WIP cap is to *just do it!* That is, for a given production line, establish a limit on the WIP in the line and simply do not allow releases into the line whenever the WIP is at or above the limit. We call the protocol under which a new job is introduced to the line each time a job departs **CONWIP** (*con*stant *w*ork *in p*rocess) because it results in a WIP level that is very nearly constant.

Recall that in Chapter 7 we made use of the CONWIP protocol to control WIP so that we could determine the relationships among WIP, cycle time, and throughput. We now offer it as the basis of a practical WIP cap mechanism. First we describe it qualitatively, and then we give a quantitative model for analyzing the performance of a CONWIP line.

10.4.1   Basic Mechanics

We can envision a CONWIP line operating as depicted in Figure 10.4, in which departing jobs send production cards back to the beginning of the line to authorize release of new jobs. Note that this way of describing CONWIP implicitly assumes two things:

1. The production line consists of a single routing, along which all parts flow.
2. Jobs are identical, so that WIP can be reasonably measured in units (i.e., number of jobs or parts in the line).

If the facility contains multiple routings that share workstations, or if different jobs require substantially different amounts of processing on the machines, then things are not so simple. There are, however, ways to address these complicating factors. For instance, we could establish CONWIP levels along different routings. We could also state the CONWIP levels in units of "standardized jobs," which are adjusted according to the amount of processing they require on critical resources. We address these types

FIGURE 10.4

A CONWIP production line

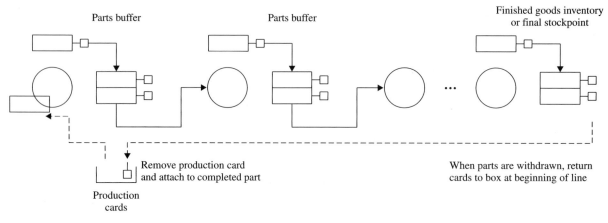

of implementation issues in Part III. For now, we focus on the single-product, single-routing production line in order to examine the essential differences between CONWIP, kanban, and MRP systems.

From a modeling perspective, a CONWIP system looks like a **closed queueing network,** in which customers (jobs) never leave the system, but instead circulate around the network indefinitely, as shown in Figure 10.5. Of course, in reality, the entering jobs are different from the departing jobs. But for modeling purposes, this makes no difference, because of the assumption that all jobs are identical.

In contrast, a pure push, or MRP, system behaves as an **open queueing network,** in which jobs enter the line and depart after one pass (also shown in Figure 10.5). Releases into the line are triggered by the material requirements plan without regard to the number of jobs in the line. Therefore, unlike in a closed queueing network, the number of jobs can vary over time.

Finally, Figure 10.5 depicts a (one-card) kanban system as a **closed queueing network with blocking.** As in the closed queueing network model of a CONWIP system, jobs circulate around the network indefinitely. However, unlike the CONWIP system, the kanban system limits the number of jobs that can be at each station, since the number of production cards at a station establishes a maximum WIP level for that station. Each production card acts exactly like a space in a finite buffer in front of the workstation. If this buffer gets full, the upstream workstation becomes blocked.

10.4.2   Mean-Value Analysis Model

To analyze CONWIP lines and make comparisons with push systems, it is useful to have a quantitative model of closed (CONWIP) systems, similar to Kingman's equation model we developed for open (push) systems in Chapter 8. For the case in which all stations consist of single machines, we can do this by using a technique known as **mean-value analysis (MVA).**[3] This approach, which we used without specifically identifying it in

---

[3]Unfortunately, MVA is not valid for the multimachine case. We can approximate a station with parallel machines with a single fast machine (i.e., so the capacity is the same). But as we know from Chapter 7, parallel machines tend to outperform single machines, given the same capacity. Therefore, we would expect this approximation to underestimate the performance of a CONWIP line with parallel machine stations.

FIGURE 10.5

CONWIP, pure push, and kanban systems

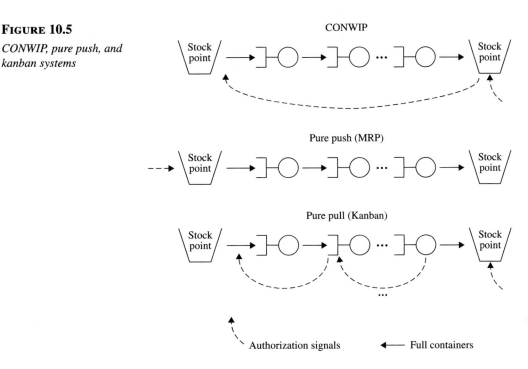

CONWIP

Pure push (MRP)

Pure pull (Kanban)

Authorization signals   ←—— Full containers

Chapter 7 to develop the throughput and cycle time curves for the practical worst case, is an iterative procedure that develops the measures of the line with WIP level w in terms of those for WIP level $w - 1$. The basic idea is that a job arriving to a station in a system with w jobs in it sees the other $w - 1$ jobs distributed according to the average behavior of a system with $w - 1$ jobs in it. This is exactly true for the case in which process times are exponential ($c_e = 1$). For general process times, it is only approximately true. As such, it gives us an approximate model, much like Kingman's model of open systems.

Using the following notation to describe an n-station CONWIP line

$$u_j(w) = \text{utilization of station } j \text{ in CONWIP line with WIP level } w$$
$$\text{CT}_j(w) = \text{cycle time at station } j \text{ in CONWIP line with WIP level } w$$
$$\text{CT}(w) = \sum_{j=1}^{n} \text{CT}_j(w) = \text{cycle time of CONWIP line with WIP level } w$$
$$\text{TH}(w) = \text{throughput of CONWIP line with WIP level } w$$
$$\text{WIP}_j(w) = \text{average WIP level at station } j \text{ in CONWIP line with WIP level } w$$

we develop an MVA model for computing each of the above quantities as functions of the WIP level w. We give the details in the following technical note.

Technical Note

As was the case with Kingman's model of open systems, the basic modeling challenge in developing the MVA model of a closed system is to compute the average cycle time at a single station. We do this by treating stations as if they behaved as $M/G/1$ queues—that is, were single-machine stations with Poisson arrivals and general (random) processing times. Three key results for the $M/G/1$ queue are as follows:

1. The long-run average probability that the server is busy is

$$P(\text{busy}) = u$$

where u is the utilization of the station.

2. The average number of jobs in service (i.e., being processed, not waiting in the queue) as seen by a randomly arriving job is

$$E[\text{no. jobs in service}] = P(\text{busy}) \times 1 + [1 - P(\text{busy})](0) = u$$

3. The average remaining process time of a job in service (which is zero if there is no job in service) as seen by a randomly arriving job (see Kleinrock 1975 for details) is

$$E[\text{remaining process time}] = P(\text{busy})E[\text{remaining process time}|\text{busy}]$$

$$\approx u\frac{t_e(c_e^2 + 1)}{2}$$

Note that if $c_e = 1$ (i.e., process times are exponential), then the expected remaining process time, given the station is busy, is simply t_e (the average processing time of a job that has just begun processing), which is an illustration of the memoryless property of the exponential distribution. When $c_e > 1$, the expected remaining process time is greater than t_e, because randomly arriving jobs are more likely to encounter long jobs in high-variability systems. Conversely, if $c_e < 1$, then the average remaining process time is less than t_e.

With these three properties, we can estimate the average time a job spends at station j in a system with w jobs as the remaining process time of the job currently in service plus the time to process the jobs in queue ahead of the arriving job plus the process time of the job itself. Since the number of jobs in queue is the number of jobs at the station minus the one (if any) in service, we can write this as

$$CT_j(w) = E[\text{remaining process time}] + (E[\text{no. jobs at station}]$$

$$-E[\text{no. jobs in service}])t_e(j) + t_e(j)$$

Now, supposing that an arriving job in a line with w jobs sees the other jobs distributed according to the average behavior of a line with $w - 1$ jobs and using the above expression for remaining process time, we can write this as

$$CT_j(w) = u_j(w - 1)\frac{t_e(j)[c_e^2(j) + 1]}{2} + [\text{WIP}_j(w - 1) - u_j(w - 1)]t_e(j) + t_e(j)$$

$$= \text{TH}(w - 1)t_e(j)\frac{t_e(j)[c_e^2(j) + 1]}{2} + [\text{WIP}_j(w - 1) - \text{TH}(w - 1)t_e(j) + 1]t_e(j)$$

$$= \frac{t_e^2(j)}{2}[c_e^2(j) - 1]\text{TH}(w - 1) + [\text{WIP}_j(w - 1) + 1]t_e(j)$$

Note that we have substituted the expression for utilization $u_j(w) = \text{TH}(w)t_e(j)$. With this formula for the cycle time at station j, we can easily compute the cycle time for the line (i.e., it is just the sum of the station cycle times). Knowing the cycle time allows us to compute the throughput by using Little's law (since the WIP level in a CONWIP line is fixed at w). And finally, by using this throughput and the cycle time for each station in Little's law, we can compute the WIP level at each station.

Letting $\text{WIP}_j(0) = 0$ and $\text{TH}(0) = 0$, the MVA algorithm computes the cycle time, throughput, and station-by-station WIP levels as a function of the number of jobs in the CONWIP line in iterative fashion by using the following:

$$CT_j(w) = \frac{t_e^2(j)}{2}[c_e^2(j) - 1]\text{TH}(w - 1) + [\text{WIP}_j(w - 1) + 1]t_e(j) \quad (10.1)$$

$$CT(w) = \sum_{j=1}^{n} CT_j(w) \quad (10.2)$$

$$TH(w) = \frac{w}{CT(w)} \tag{10.3}$$

$$WIP_j(w) = TH(w)CT_j(w) \tag{10.4}$$

These formulas are easily implemented in a spreadsheet and can be used to generate curves of $TH(w)$ and $CT(w)$ for CONWIP lines other than the best, worst, and practical worst cases. Buzacott and Shanthikumar (1993) have tested them against simulation for various sets of system parameters and found that the approximation is reasonably accurate for systems with $c_e^2(j)$ values between 0.5 and 2.

To illustrate the use of Equations (10.1) through (10.4), let us return to the Penny Fab example of Chapter 7. Recall that the Penny Fab had four stations, each with average process time $t_e = 2$ hours. Using the formulas of Chapter 7, we were able to plot $TH(w)$ and $CT(w)$ for the particular situations represented by the best, worst, and practical worst cases. Suppose, however, we are interested in considering the effect of speeding up one of the stations (i.e., to create an unbalanced line) or reducing variability relative to the practical worst case (PWC). Since the practical-worst-case formulas only consider the balanced case with $c_e = 1$ at all stations, we cannot do this with the Chapter 7 formulas. However, we can do it with the MVA algorithm above.

Consider the Penny Fab with reduced variability (relative to the PWC) so that $c_e(j) = 0.5$ for $j = 1, \ldots, 4$. Starting with $WIP_j(0) = 0$ and $TH(0) = 0$, we can compute

$$CT_j(1) = \frac{t_e^2(j)}{2}[c_e^2(j) - 1]TH(0) + [WIP_j(0) + 1]t_e(j) = t_e(j) = 2$$

for $j = 1, \ldots, 4$. Since all stations are identical, $CT(w) = 4CT_j(w)$, and therefore $CT(1) = 8$ hours. Throughput is

$$TH(1) = \frac{1}{CT(1)} = \frac{1}{8}$$

and average WIP at each station is

$$WIP_j(1) = TH(1)CT_j(1) = (\tfrac{1}{8})(2) = \tfrac{1}{4}$$

Having computed these for $w = 1$, we next move to $w = 2$ and compute the cycle time at each station as

$$CT_j(2) = \frac{t_e^2(j)}{2}[c_e^2(j) - 1]TH(1) + [WIP_j(1) + 1]t_e(j)$$

$$= \frac{2^2}{2}(0.5^2 - 1)\left(\frac{1}{8}\right) + \left(\frac{1}{4} + 1\right)2 = 2.313$$

So $CT(2) = 4CT_j(2) = 9.250$ and $TH(2) = 2/CT(2) = 0.216$. Continuing in this fashion, we can generate the numbers shown in Table 10.1.

Using the same procedure, we could also generate $TH(w)$ and $CT(w)$ for the case in which we increase capacity, for instance, by reducing the average process time at stations 1 and 2 from two hours to one hour. We have done this and plotted the results for both the reduced variability case from Table 10.1 and the increased capacity case, along with the best, worst, and practical worst cases, in Figure 10.6. Notice that both cases represent improvements over the practical worst case, since they enable the line to generate greater throughput for a given WIP level. In this example, speeding up two of the stations produced a greater improvement than reducing variability on all stations. Of course, in practice the outcome will depend on the specifics of the system. The MVA model presented here is a simple, rough-cut analysis tool for examining the effects of capacity and variability changes on a CONWIP line.

TABLE 10.1   MVA Calculations for Penny Fab with
$$c_e(j) = 0.5$$

w	$TH(w)$	$CT(w)$	$CT_j(w)$	$WIP_j(w)$
1	0.125	8.000	2.000	0.250
2	0.216	9.250	2.313	0.500
3	0.280	10.703	2.676	0.750
4	0.325	12.318	3.080	1.000
5	0.356	14.052	3.513	1.250
6	0.378	15.865	3.966	1.500
7	0.395	17.731	4.433	1.750
8	0.408	19.631	4.908	2.000
9	0.418	21.555	5.389	2.250
10	0.426	23.495	5.874	2.500
11	0.432	25.446	6.362	2.750
12	0.438	27.406	6.852	3.000

FIGURE 10.6

Effects of reducing variability and increasing capacity on Penny Fab performance curves

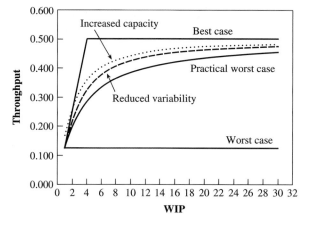

Now that we have models of both push and pull systems, we can make some comparisons to deepen our understanding of the potential benefits of pull systems. We begin by comparing CONWIP with MRP and then contrast CONWIP with kanban.

10.5   Comparisons of CONWIP with MRP

A fundamental distinction between push and pull systems is this:

> Push systems control throughput and observe WIP. Pull systems control WIP and observe throughput.

For example, in MRP, we establish a master production schedule, which determines planned order releases. These, in turn, determine what is released into the system. Depending on what happens in the line, however, the WIP level may float up and down over time. In a pull system, the WIP level is directly controlled by setting the card counts. However, depending on what happens in the line, the output rate may vary over time. Which approach is better? While this is not a simple question, we can make some observations.

10.5.1   Observability

First, and fundamentally, we note that WIP is directly observable, while throughput is not. Hence, setting WIP as the control in a pull system is comparatively simple. We can physically count jobs on the shop floor and maintain compliance with a WIP cap. In contrast, setting the release rate in a push system must be done with respect to *capacity*. If the rate chosen is too high, the system will be choked with WIP; too low, and revenue will be lost due to insufficient throughput. But estimating capacity is not simple. A host of detractors, ranging from machine outages to operator unavailability, are difficult to estimate with precision. This fact makes a push system intrinsically more difficult to optimize than a pull system.

What we are talking about here is a general principle from the field of control theory. In general, it is preferable to control the robust parameter (so that errors are less damaging) and observe the sensitive parameter (so that feedback is responsive), rather than the other way around. Since WIP is robust and observable, while throughput is sensitive and can only be controlled relative to the unobservable parameter of capacity, this is a very powerful argument in favor of pull production systems.

10.5.2   Efficiency

A second argument in favor of pull systems is that they are more efficient than push systems. By more efficient we mean that the WIP level required to achieve a given throughput is lower in a pull system than in a push system. To illustrate why this is the case, we consider a CONWIP system like that shown in Figure 10.4 with a fixed WIP level w, and we observe the throughput $\widetilde{\text{TH}}(w)$. Then we consider the (pure push) MRP system, like that shown in Figure 10.5, made up of the same machines as the CONWIP line but with releases fed into the line at rate $\widetilde{\text{TH}}(w)$. By conservation of material, the output rate of the MRP system will be the same as the input rate, namely, $\widetilde{\text{TH}}(w)$. So the CONWIP and MRP systems are equivalent in terms of throughput, and the question of efficiency hinges on which achieves this throughput with less WIP.

Let us consider a specific example in which there are five single-machine stations in tandem, each station processes jobs at a rate of one per hour, and processing times are exponentially distributed. For this simple system, the throughput of the CONWIP system as a function of the WIP level is given by the formula for the practical worst case from Chapter 7, which reduces to

$$\widetilde{\text{TH}}(w) = \frac{w}{w + W_0 - 1} r_b = \frac{w}{w + 4} \tag{10.5}$$

If we fix the release rate into the push system to be TH, where times between releases are exponential, then each station behaves as an independent $M/M/1$ queue, so the overall WIP level is given by five times the average WIP level of an $M/M/1$ queue, which we know from Chapter 8 is $u/(1 - u)$, where u is the utilization level. Since, in this case, the process time is equal to one and the arrival rate is equal to TH, $u = \text{TH}$. Therefore, the average WIP for the system is

$$\tilde{w}(\text{TH}) = 5\left(\frac{u}{1 - u}\right) = 5\left(\frac{\text{TH}}{1 - \text{TH}}\right) \tag{10.6}$$

Now suppose we choose $w = 6$ in the CONWIP system. By Equation (10.5), the throughput is $\widetilde{\text{TH}}(6) = 0.6$ job per hour. If we then fix TH $= 0.6$ in Equation (10.6), we see that WIP in the MRP system is $\tilde{w}(0.6) = 7.5$. Hence, the push system has more WIP for the same throughput level.

Notice that the WIP level in the push system will be greater than w regardless of the choice of w. To see this, we set TH $= w/(w + 4)$ in Equation (10.6):

$$\tilde{w}\left(\frac{w}{w + 4}\right) = \frac{5[w/(w + 4)]}{1 - w/(w + 4)} = \frac{5w}{4}$$

So, in this example, for *any* throughput level the average WIP level in the push system will be 25 percent higher than that in the CONWIP system.

Although the magnitude of the increase in WIP of the push system over the CONWIP system obviously depends on the specific parameters of the line, this qualitative effect is general, as we state in the following law:

Law (CONWIP Efficiency):   *For a given level of throughput, a push system will have more WIP on average than an equivalent CONWIP system.*

This law has an immediate corollary. When throughput is the same in the CONWIP and MRP systems, then Little's law and the fact that average WIP is greater in the MRP system imply the following:

Corollary: *For a given level of throughput, a push system will have longer average cycle times than an equivalent CONWIP system.*

10.5.3   Variability

We can show that MRP systems also have more variable cycle times than equivalent CONWIP systems. The reason for this is as follows. By definition, the WIP level in a CONWIP system is fixed at some level w. This fact introduces *negative correlation* between the WIP levels at different stations. For instance, if we know that there are w jobs at station 1, then we are absolutely certain that there are no jobs at any other station. In this case, knowledge of the WIP level at station 1 gives us perfect information about the WIP levels at the other stations. However, even if we knew only that there were $w/2$ jobs at station 1 (in a 10-station line, say), then we would still gain some information about the other stations. For instance, it is quite unlikely that any other station has all $w/2$ of the other jobs. This negative correlation between WIP levels tends to dampen fluctuations in cycle time.

In contrast, WIP levels at the individual stations are *independent* of one another in a push system;[4] a large WIP level at station 1 tells us nothing about the WIP levels at the other stations. Hence, it is possible for the WIP levels to be high (or low) at several stations simultaneously. Since cycle times are directly related to WIP, this means that extreme (high or low) cycle times are possible. The result is that cycle times are more variable in a push system than in an equivalent pull system.

Increased cycle time variability means that we must quote longer lead times in order to achieve the same level of customer service. This is because to achieve a given level of service, we must quote the mean cycle time plus some multiple of the standard deviation of cycle time (where the multiple depends on the desired service level). For example, Figure 10.7 illustrates two systems with a mean cycle time of 10 days. However, system

---

[4]This observation is only strictly true if processing times are exponential, but is still much closer to being true in the push system than in the pull system, even when processing times are not exponential.

FIGURE 10.7

Effect of cycle time variability on customer lead time

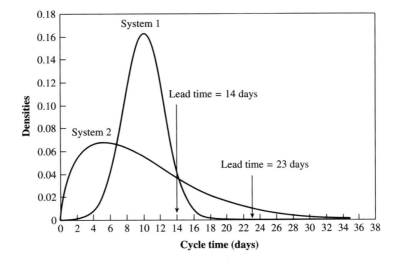

2 has a substantially higher standard deviation of cycle time than does system 1. To achieve 90 percent service, system 1 must quote a lead time of 14 days, while system 2 must quote 23 days. The increased variability of the push system gives rise to a larger standard deviation of cycle time. Notice that this is on top of the fact that, for a given throughput, the *average* cycle time of the push system is longer than that in an equivalent pull system. Thus, for the same throughput and customer service level, lead times will be longer in the push system for two reasons: longer mean cycle time and larger standard deviation of cycle time.

10.5.4 Robustness

The most important advantage of a CONWIP system over a pure push system is neither the reduction in WIP (and average cycle time) nor the reduction in cycle time variance, important as these are. Instead, the key advantage of pull systems is their robustness, which we can state as follows:

Law (CONWIP Robustness): *A CONWIP system is more robust to errors in WIP level than a pure push system is to errors in release rate.*

To make the meaning of this law clear, we suppose the existence of a very simple profit function of the form

$$\text{Profit} = p\text{TH} - hw \tag{10.7}$$

where p is the marginal profit per job, TH is the throughput rate, h is a cost for each unit of WIP (this includes costs for increased cycle time, decreased quality, etc.), and w is the average WIP level. In the CONWIP system, throughput will be a function of WIP, that is, $\widetilde{\text{TH}}(w)$, and we will choose the value of w to maximize profit. In the push system, average WIP is a function of release rate $\tilde{w}(\text{TH})$, and we will choose the value of TH that maximizes profit.

It should be clear from our earlier law that the optimal profit will be higher in the CONWIP system than in the push system (since the CONWIP system will have a lower WIP for any chosen throughput level). However, the CONWIP robustness law is concerned with what happens if w is chosen at a suboptimal level in the CONWIP system

FIGURE 10.8

*Relative robustness of
CONWIP and pure push
systems*

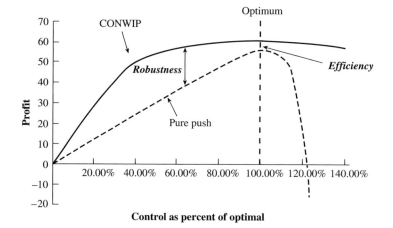

or TH is chosen at a suboptimal level in the push system. Since WIP and throughput are measured in different units, we measure suboptimality in terms of percentage error. We do this for our previous example of five machines with exponential processing times of one hour and cost coefficients $p = 100$ and $h = 1$ in Figure 10.8.

We find that the best WIP level for the CONWIP system is 16 jobs, resulting in a profit of $63.30 per hour. In the push system, the best TH turns out to be 0.776 job per hour, yielding a profit of $60.30 per hour. Thus, as expected, the optimal profit level for the CONWIP system is slightly greater (around 5 percent) than the optimal level in the push system. More important, however, is the fact that the profit function for the CONWIP system is very flat between WIP levels as low as 40 percent and as high as 160 percent of the optimal level. In contrast, the profit function for the push system declines steadily when the release rate is chosen at a level below the optimum and falls off sharply when the release rate is set even slightly above the optimum level. In fact, profit becomes negative when the release rate reaches 120 percent of the optimum level, while profit in the CONWIP system remains positive until the WIP level reaches 600 percent of the optimum level.[5]

These observations are particularly important in light of the observability issue we raised earlier. As we noted, the optimal release rate in a push system must be set relative to the real capacity of the system, which is not directly observable. Natural human optimism, combined with an understandable desire to maximize revenue by getting as much throughput out of the system as possible, provides strong incentive to set the release rate too high. As Figure 10.8 shows, this is precisely the kind of error that is most costly.

The CONWIP system, on the other hand, is controlled by setting the easily observable parameter of WIP level. This, combined with the flatness of the profit curve in the vicinity of the optimum, means that achieving a profit close to the optimum level will be much easier than in the push system. The practical consequence of all this is that the difference in performance between a CONWIP and a pure push system is likely to be substantially larger than indicated by a "fair comparison" of the type we made by using Equations (10.5) and (10.6). Hence, increased robustness is probably the most compelling reason to use a pull system, such as CONWIP, instead of a push system.

---

[5]Although we have offered only one example, this robustness result is quite general and does not depend on the assumptions made here. See Spearman and Zazanis (1992) for details.

10.6   Comparisons of CONWIP with Kanban

As shown in Figure 10.5, CONWIP and kanban are both pull systems in the sense that releases into the line are triggered by external demands. Because both systems establish a WIP cap, they exhibit similar performance advantages relative to MRP. Specifically, both CONWIP and kanban will achieve a target throughput level with less WIP than a pure push system and will exhibit less cycle time variability. Moreover, since both are controlled by setting WIP, and we know that WIP is a more robust control than release rate, they will be easier to manage than a pure push system. However, there are important differences between CONWIP and kanban.

10.6.1   Card Count Issues

The most obvious difference is that kanban requires setting more parameters than does CONWIP. In a one-card kanban system, the user must establish a card count for every station. (In a two-card system, there are twice as many card counts to set.) In contrast, in a CONWIP system there is only a single card count to set. Since coming up with appropriate card counts requires a combination of analysis and continual adjustment, this fact means that CONWIP is intrinsically easier to control. For this reason, we view CONWIP as the standard by which other systems should be evaluated. If one is to use a more complex pull system than CONWIP, such as kanban, then that system's performance should justify the added complexity. In Part III we will examine situations in which more complex systems do indeed seem worthwhile. However, for this chapter we will continue to restrict our scope to simple production lines with a series of workstations in tandem, to enable us to make basic comparisons between CONWIP and kanban.

A second important difference between CONWIP and kanban systems, not obvious from Figure 10.5, is that cards are typically *part number–specific* in a kanban system, but *line-specific* in a CONWIP system. That is, cards in a kanban system identify the part for which they are authorizing production. This is necessary in a multiproduct environment, since a workstation must know which type of stock to replenish in its outbound stock point. In a CONWIP system, on the other hand, cards do not identify any specific part number. Instead, they come to the front of the line and are matched against a **backlog,** which gives the sequence of parts to be introduced to the line. This backlog, or sequence, must be generated by a module outside the CONWIP loop, in a manner analogous to master production scheduling in an MRP system.[6] Thus, depending on the backlog, each time a particular card returns to the front of a CONWIP line, it may authorize a different part type to be released into the line.

The significance of this difference is manifested not in the mechanics of the work release process, but in what it implies for the two systems. In its pure form, a kanban system *must* include standard containers of WIP for every active part number in the line. If it did not, a downstream workstation could generate a demand on an upstream workstation that it could not meet. If, as we have seen in practice, the line produces 40,000 different part numbers, a Toyota-style kanban system would be swamped with WIP. The problem is that most of the 40,000 part numbers, while active, are produced only occasionally, in "onesies and twosies." Hence, the kanban system unnecessarily

---

[6]The primary difference between developing a backlog and an MPS is that a backlog is a *sequence* without times associated with jobs, while an MPS is a *schedule* that does indicate times for requirements. We will discuss the distinction, as well as the relative advantages, of using a sequence versus a schedule in Chapter 15.

maintains WIP on the floor for many parts that will not be produced for months. But if these low-demand parts were not stocked on the floor, then a demand at the end of the line would generate unfilled demands at each station all the way back to the beginning of the line. The time to start a job from the beginning of the line and run it all the way through the line would be much longer than the normal response to demands at the end of the line, and the just-in-time protocol would break down.

A CONWIP system, because of its use of line-specific cards and a work backlog, does not have this problem. If the card count in a CONWIP line is w, then at most w jobs can be in the line, where w will virtually always be much smaller than 40,000. If a part is not required for 6 months, then it will not show up on the backlog and therefore will not be released into the line. When a demand for a low-volume part does show up, the backlog will release it into the line with an appropriate lead time to accommodate the production time on the line. Hence, "just-in-time" performance can be maintained, even for onesies and twosies.

However, we should point out that a fundamental difference between kanban and CONWIP in that the lead time in a pure kanban system is *zero* while under CONWIP it is *small*. This is the price that CONWIP pays to maintain flexibility. Kanban is a pure make-to-stock system in which the part is supposed to be in the outbound stock point when requested. CONWIP, on the other hand, keeps cycle times short by keeping WIP levels low. If cycle times are short enough, there will be no need to change the sequence of parts, and so the added flexibility is worth the added cycle time.

10.6.2   Product Mix Issues

The experts on kanban were clearly aware that it would not work in all production environments. Hall (1983) pointed out that kanban is applicable only in **repetitive manufacturing** environments. By repetitive manufacturing, he meant systems in which material flows along fixed paths at steady rates. Large variations in either volume or product mix destroy this flow, at least when parts are viewed individually, and hence seriously undermine kanban. CONWIP, while still requiring a relatively steady volume (i.e., a level MPS), is much more robust to swings in product mix, due to the planning capability introduced by the process of generating a work backlog.

A changing product mix may have more subtle consequences than simply elevating the total WIP required in a kanban system. If the complexity of the different parts varies (i.e., the parts require different amounts of processing on the machines), the bottleneck of the line may change depending on product mix. For instance, consider the five-station line shown in Figure 10.9. Product A requires one hour of processing on all machines except machines 2 and 3, where it requires three and two and one-half hours, respectively. Product B requires one hour of processing on all machines except machines 3 and 4, where it requires 2.5 and 3 hours, respectively. Thus, if we are running product A, machine 2 is the bottleneck. If we are running product B, machine 4 is the bottleneck. However, for mixes containing between 25 and 75 percent of product A, machine 3 becomes the overall bottleneck.

FIGURE 10.9

System with a floating bottleneck

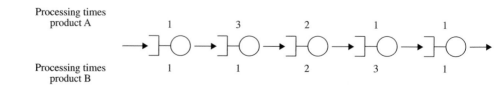

To see this, consider a 50–50 mix of products A and B. The average processing times on machines 2, 3, and 4 are

$$\text{Average time on machine 2} = 0.5(3) + 0.5(1) = 2 \text{ hours}$$

$$\text{Average time on machine 3} = 0.5(2.5) + 0.5(2.5) = 2.5 \text{ hours}$$

$$\text{Average time on machine 4} = 0.5(1) + 0.5(3) = 2 \text{ hours}$$

Only when the percentage of A exceeds 75 percent does the average time on machine 2 exceed two and one-half hours. Likewise, only when the percentage of B exceeds 75 percent (i.e., the percentage of A is less than 25 percent) does the average time on machine 4 exceed two and one-half hours.

In an ideal kanban environment, we would set the sequence of A and B to achieve a steady mix; for example, for a 50–50 mix we would use a sequence of A-B-A-B-A-B-.... In a nonideal environment, where the mix requirements are not steady (e.g., demand is seasonal or forecasts are volatile), a uniform sequence may not be practical. However, if we let the mix vary to track demand, this may cause problems with our card counts in the kanban system. We generally want to put more production cards before and after the bottleneck station, in order to protect it against starvation and blocking. But which is the bottleneck—machine 2, machine 3, or machine 4? The answer, of course, depends on the mix we are running. This means that the optimal card count allocation is a function of mix. Hence, to achieve high throughput with low WIP, we may need to dynamically vary the card counts over time. Since we have already argued that setting card counts in a kanban system is not trivial, this could be a difficult task indeed.

CONWIP, however, has only a single card count. Therefore, as long as the desired rate remains relatively steady, there is no need to alter the card count as the product mix changes. Moreover, the WIP will *naturally* accumulate in front of the bottleneck, right where we need it.[7] In our example, when we are running a mix heavy in product A, machine 2 will be the slowest and therefore will accumulate the largest queue. When the mix becomes heavy in product B, the largest queue will shift to machine 4. Happily, this all happens without our intervention, due to the natural forces governing the behavior of bottlenecks. Again, we can see that the CONWIP system is fundamentally simpler to manage than a kanban system.

10.6.3 People Issues

Finally, we complete our comparison of CONWIP and kanban with two people-oriented observations. First, the fact that kanban systems pull at every station introduces a certain amount of stress into the system. Operators in a kanban system who have raw materials but no production card cannot begin work. When the production card arrives, they must replenish the void in the system as quickly as possible, in order to prevent starvation somewhere in the line. As Klein (1989) has pointed out, this type of pressured pacing can serve as a significant source of operator stress.

In contrast, a CONWIP system acts as a push system at every station except the first one. When operators of midstream machines receive raw materials, they are authorized to work on them. Hence, the operators can work ahead to the maximum extent permitted by material availability and therefore will be subject to less pacing stress. Of course, at the first station of a CONWIP line, the operators are only able to work when authorized by a production card, so they have virtually identical working conditions to the operator of the first station in a kanban line. This is unavoidable if we are to establish a WIP cap.

---

[7]Note that blocking is not a problem in a CONWIP system, since there are no interstation card counts to restrict the transfer of completed jobs to the next station.

Thus, the CONWIP line may still introduce a certain amount of pacing stress, but less than a kanban line.

Our second people-oriented observation is that the act of pulling at each station in a kanban line may foster a closer relationship between operators of adjacent workstations. Since operators must pull needed parts in a kanban system, they will communicate with the operators of upstream machines. This provides an opportunity to check parts for quality problems and to identify and discuss any problems with adhering to the production rate. We have frequently heard this benefit cited as motivation for using a pure kanban system.

While we acknowledge that the communication and learning benefits of having operators of adjacent workstations interact can be significant, we question whether the kanban pull discipline is necessary to achieve this. Whether or not a kanban mechanism is being used between two stations, a transfer of parts from the upstream station to the downstream station must occur. To prevent transfer of bad parts, a "buy-sell" protocol, in which the downstream operator refuses to accept the parts if they do not meet quality specifications, can be used with or without kanban. To motivate workers to cooperate in solving flow-related problems, one must foster a linewide perspective among the operators. Instead of the kanban focus on keeping outbound stock points full, a CONWIP system needs a focus on adhering to the desired production rate. If operators need to float among workstations to promote this, fine. There are a host of ways work assignments might be structured to achieve the overall goal of a steady output rate. Our point is merely that while the kanban pull mechanism may be one way to promote cooperation among operators, it is not the only one. Given the logistics and simplicity considerations favoring CONWIP, it may be worthwhile to pursue these other learning motivators, rather than implementing a rigid kanban protocol.

10.7   Conclusions

In this chapter, we have made the following basic points:

1. Push systems *schedule* the release of work, while pull systems *authorize* the release of work on the basis of system status.

2. The "magic" of pull systems is that they establish a WIP cap, which prevents producing unnecessary WIP that does not significantly improve throughput. Pulling is just a means to an end. The result is that pull systems reduce average WIP and cycle times, reduce variability of cycle times, create pressure for quality improvements and (by decreasing WIP) promote more effective defect detection, and increase flexibility for accommodating change.

3. The simplest mechanism for establishing a WIP cap is CONWIP (*con*stant *w*ork *in* *p*rocess), in which the WIP level in a line is held constant by synchronizing releases to departures.

4. CONWIP exhibits the following advantages over a pure push system:

   - The WIP level is directly observable, while the release rate in a push system must be set with respect to (unobservable) capacity.
   - It requires less WIP on average to attain the same throughput.
   - It is more robust to errors in control parameters.
   - It facilitates working ahead of a schedule when favorable circumstances permit it.

5. CONWIP exhibits the following advantages over a pure kanban system:

- It is simpler in the sense that it requires setting only a single card count instead of a card count for each workstation.
- It can accommodate a changing part mix, due to its use of line-specific cards and a work backlog.
- It can accommodate a floating (mix-dependent) bottleneck, due to the natural tendency of WIP to accumulate in front of the slowest machine.
- It introduces less operator stress due to a more flexible pacing protocol.

While these observations are based on highly simplified versions of pure push, pure kanban, and pure CONWIP, they contain essential factory physics insights. We will turn to the problem of putting these insights into practice in messy, real-world environments in Part III.

Study Questions

1. Is MRP/MRP II as practiced in industry a pure push system under the definition used here? Why or why not?
2. Why is WIP more easily observable than throughput?
3. When controlling a system subject to randomness, why does it make sense to control the robust parameter and observe the sensitive one, rather than the other way around?
4. Why are pull systems more robust than push systems? What practical consequences does this have for manufacturing plants?
5. Suggest as many mechanisms as you can by which a firm could establish a WIP cap for a production line.
6. A potential benefit of "pulling everywhere" in a kanban system is that it promotes communication between stages of the line. How important is the pull mechanism to this communication? Can you suggest other procedures for improving communication?
7. How can piecework incentive systems be counterproductive in a pull environment? What other forms of compensation or incentive systems may be more suitable?

Problems

1. Consider a production line with three single-machine stations in series. Each has processing times with mean two hours and standard deviation of two hours. (Note that this makes it identical to the line represented in the practical worst case of Chapter 7.)
   a. Suppose we run this line as a push system and release jobs into it at a rate of 0.45 per hour with arrival variability given by $c_a = 1$. What is the average WIP in the line?
   b. Compute the throughput of this line if it is run as a CONWIP line with a WIP level equal to your answer in (a). Is the throughput higher or lower than 0.45? Explain this result.
2. Consider the same production line as in Problem 1. Suppose the marginal profit is $50 per piece and the cost of WIP is $0.25 per piece per hour.
   a. What is the profit from the push system if we set TH = 0.4?
   b. What is the profit from the pull system if we set WIP = 12? How does this compare to the answer of (a) and what does it imply about the relative profitability of push and pull systems?
   c. Increase TH in (a) by 20 percent to 0.48, and compute the profit for the push system. Increase WIP in (b) by 25 percent to 15, and compute the profit for the pull system.

Compare the difference to the difference computed in (*b*). What does it imply about the relative robustness of push and pull systems?

3. Consider the same production system and profit function as in Problem 2.
   a. Compute the optimal throughput level operating as a push system and the optimal WIP level operating as a CONWIP system. What is the difference in the resulting profit levels?
   b. Suppose the process times actually have a mean and standard deviation of 2.2 hours, but the throughput used for the push system and the WIP level used for the pull system are computed as if the process times had a mean and standard deviation of 2 hours (i.e., were equal to the levels computed in (*a*)). Now what is the profit level in the push and pull systems, and how do they compare? Repeat this calculation for a system in which processing times have a mean and standard deviation of 2.4 hours. What happens to the gap between the profit in the push and pull systems?

4. In the practical worst case, it is assumed that the line is balanced (that is, $t_e(j) = t$ for all j) and that processing times are exponential (that is, $c_e(j) = 1$ for all j). Show that under these conditions, the MVA formulas for $CT(w)$ and $TH(w)$ reduce to the corresponding formulas for the practical worst case

$$CT(w) = T_0 + \frac{w-1}{r_b}$$

$$TH(w) = \frac{w}{W_0 + w - 1} r_b$$

Hint: Note that because the line is balanced, $T_0 = nt$ and $r_b = 1/t$, where n is the number of stations in the line.

5. Implement MVA formulas (10.1) to (10.4) in a spreadsheet for the Penny Fab example with $t_e(j) = 2$ hours and $c_e(j) = 0.5$ for $j = 1, \ldots, 4$. (You can validate your model by checking against Table 10.1.) Now change the coefficients of variation, so that $c_e(j) = 1$ for $j = 1, \ldots, 4$ and compare your results to the practical worst case. Are they the same? (If not, you have a bug in your model.) Use the case with $c_e(j) = 1$ as your base case for the questions below.
   a. Make the following changes one at a time and observe the effects on $TH(w)$:
      i. $t_e(1) = 2.5$
      ii. $t_e(3) = 2.5$
      Is there a difference between having the bottleneck at station 1 or station 3? Explain why or why not.
   b. Leaving $t_e(3) = 2.5$ and all other $t_e(j) = 2$, make the following changes one at a time, and observe the effects on $TH(w)$.
      i. $c_e(1) = 0.5$
      ii. $c_e(3) = 0.5$
      Is it more effective to reduce variability at the bottleneck (station 3) or at a nonbottleneck? Explain.
   c. Again leaving $t_e(3) = 2.5$ and all other $t_e(j) = 2$, suppose that for the same amount of money you can speed up station 2, so that $t_e(2) = 1.5$; or you can reduce variability at all nonbottleneck machines, so that $c_e(j) = 0.5$ for $j = 1, 2, 4$. Which would be the better investment and why?

11 THE HUMAN ELEMENT IN OPERATIONS MANAGEMENT

For as laws are necessary that good manners may be preserved, so there is a need of good manners that laws may be maintained.

Machiavelli

We hold these truths to be self-evident.
Thomas Jefferson

11.1 Introduction

We begin by noting what this chapter is not. Clearly, on the basis of its short length alone, this chapter cannot provide any kind of comprehensive treatment of human issues in manufacturing management. We are not attempting to survey organizational behavior, human factors, industrial psychology, organization theory, applied behavioral science, or any of the other fields in which human issues are studied. Important as they are, this is a book on operations management, and we must adhere to our focus on operations.

Even in an operations book, however, we would be remiss if we were to leave the impression that factory management is only a matter of clever mathematical modeling or keen logistical insight. People are a critical element of any factory. Even in modern "lights out" plants with highly automated machinery, people play a fundamental role in machine maintenance, material flow coordination, quality control, capacity planning, and so on. No matter how sophisticated a physical plant, if the humans in it do not work effectively, it will not function well. In contrast, some plants with very primitive hardware and software are enormously effective, in a business strategy context, precisely because of the people in them.

What we offer here is a factory physics perspective on the role of humans in manufacturing systems. Recall that the fundamental premise of factory physics is that there are natural laws or tendencies that govern the behavior of plants. Understanding these laws and working with them facilitate better management policies. Analogous to these physical laws, we feel that there are natural tendencies of human behavior, or "personnel laws," that significantly influence the operation of a factory. In this chapter, we observe some of the most basic aspects of human behavior as they relate to operations management. It is our hope that this cursory treatment will inspire the reader to make deeper connections between the subject material of this book and that of the behavioral disciplines.

11.2   Basic Human Laws

Part of the reason why we feel the brief treatment we are about to give the human element of the factory can be useful is that poor operations decisions are generally not misguided because of a lack of appreciation of subtle psychological details; they are frequently wrongheaded because of a wholesale inattention to fundamental aspects of human nature. We offer examples later. For now we start with some basics.

11.2.1   The Foundation of Self-interest

Because the study of human behavior is well-trod territory, we could offer a host of historical perspectives on what is elemental. For instance, we could start with something like

> *Self-preservation is the first of laws.*
>
> John Dryden, 1681

Indeed, a variation on this, with a little more institutional relevance, is our first personnel law:

Law (Self-Interest):   *People, not organizations, are self-optimizing.*

By this statement we merely mean that individuals make their choices in accordance with their preferences or goals, while organizations do not. Of course, an individual's preferences may be complex and implicit, making it virtually impossible for us to trace each action to a well-defined motive. But this is beside the point, which is that organizations made up of people will not necessarily act according to organizational goals. The reason is that the sum of the actions that improve the well-being of the constituent individuals is by no means guaranteed to improve the well-being of the organization.

The self-interest law may appear entirely obvious. Indeed, examples of behavior that is self-optimizing from an individual standpoint but suboptimal from a company perspective are prevalent throughout industry. A product designer may design a difficult-to-manufacture product because her goals are to optimize the design of the product with respect to performance. A salesperson may push a product that requires capacity that is already overloaded because his goal is to maximize sales. A manufacturing manager may make long product runs before changing over the line because her goal is to maximize throughput. A repairman may stock excessively large amounts of inventory because his goal is to effect repairs as rapidly as possible. Undoubtedly anyone with experience in a plant has observed many more examples of counterproductive behavior that is perfectly logical from an individual perspective.

In spite of such readily available examples, we frequently act as though the above law were not true. As a result, we carry an implicit model of the factory too far. The specific model of the factory to which we refer is that of a constrained optimization problem. A **constrained optimization** problem is a mathematical model that can be expressed as

Optimize   objective
Subject to   constraints

In any operations management book, including this one (see, e.g., Chapters 15, 16, and 17), one will find several examples of constrained optimization problems. For instance, in an inventory management situation, we might want to minimize inventory investment subject to achieving a minimum level of customer service. Or, in a capacity-planning problem, we might want to maximize throughput subject to a constraint on budget and product demand. There are many other operations problems that can be usefully characterized as constrained optimization models.

Perhaps precisely because constrained optimization models are so common in operations, in this field one often sees the manufacturing enterprise itself expressed as a constrained optimization problem (see, e.g., Goldratt 1986). The objective is to maximize profit, and the constraints are on physical capacity, demand, raw material availability, and so forth. Although it may sometimes be useful to think of a plant in this manner, the analogy can be dangerous if one forgets about the above personnel law.

In a mathematical optimization model, eliminating or relaxing a constraint can only improve the solution. This property follows from the fact that the constraints in a mathematical model geometrically define a feasible region. Figure 11.1 represents a special case of a constrained optimization model, called a **linear program,** in which the objective and constraints are linear functions of the decision variables (see Appendix 16A for an overview of linear programming). The shaded area represents the set of points that satisfy all the constraints, that is, the **feasible region.** The "best" point in the feasible region (point A in Figure 11.1) is the optimal solution to the problem. If we relax a constraint, the feasible region grows, adding the shaded area in the figure. Thus, the original optimum is still available, as are additional points, so things cannot get worse. In this case, the objective is improved by moving to point B.

This behavior is a powerful underpinning of an elaborate theory of sensitivity analysis of constrained optimization models. Indeed, in many models it is possible to go so far as to characterize how much the objective function will improve if a constraint is relaxed slightly. However, it is important to note that this behavior *only* holds because we are assuming that we find the *best* solution in the feasible region. For instance, in Figure 11.1, we assumed that we had found the best point, point A, before we relaxed a constraint, and therefore we could still find point A if we wanted to, after the constraint was relaxed. Moreover, we would only forsake point A if we could find a better

FIGURE 11.1

An example of a constrained optimization model

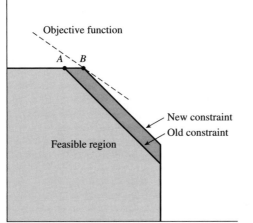

point, namely, point *B*. If we were not guaranteed to find the optimal point in the feasible region, then removing or relaxing a constraint might well lead us to an even more suboptimal solution.

In mathematics, of course, it is a given that we will find the optimum point subject to the constraints. But the significance of the preceding law is that organizations, including manufacturing systems, *do not* naturally seek the optimum within the feasible region. There is no guarantee whatsoever that the product mix of a plant is optimal in any precise sense. Nor are many other attributes, including throughput, WIP, quality level, product design, work schedule, marketing strategy, and capacity plan, likely to be optimized from a profit standpoint. Therefore, there is no guarantee that relaxing a constraint will improve the system.

Perhaps, as a complex system involving people, a factory is better likened to society than to an optimization model. Society has many constraints, in the form of laws and other behavioral restrictions. And while one might reasonably debate the appropriateness of any given law, virtually no one would argue that society would be better off with no laws at all. We clearly need some constraints to keep us from extremely bad solutions.

The same is true for manufacturing systems. There are many cases in which additional constraints actually improve the behavior of the system. In production control, a CONWIP system places constraints on the movement of material through the plant and, as we discussed in Chapter 10, works better than a pure push system without these constraints. In product design, restricting engineers to use certain standardized holes, bolts, and brackets specified by the computer-aided design (CAD) system can force them to design parts that are easier and less costly to manufacture. In sales, forcing representatives to coordinate their offerings with plant status may reduce their individual sales but increase the profitability of the plant. All manufacturing plants make use of a wide range of perfectly reasonable constraints on the system.

The point of all this is that, despite the claims of some popular manufacturing gurus, improving a manufacturing system is not simply a matter of removing constraints. Certainly some improvements can be characterized in this way. For instance, if we are seeking to improve throughput, relaxing the constraint imposed by the bottleneck machine by adding capacity may be a reasonable option. However, improving throughput may also be achieved by working on the right parts at the right time, behavior that may require *adding* constraints to achieve.[1] Realistically, the manufacturing system will not be "optimized" to start with, nor will it be "optimized" after improvements are made. The best we can do is to keep our minds open to a broad range of improvement options and select in a coherent manner. Ultimately, good management is more a matter of choosing appropriate incentives and restrictions than one of removing constraints. Certainly, narrowing our vision by using an overly restrictive view of the factory is one constraint we can do without.

11.2.2   The Fact of Diversity

All of us, as human beings, have so much in common that it is tempting to generalize. Countless philosophers, novelists, songwriters, and social scientists down through the centuries have made a living doing just that. However, before we follow suit and succumb to the urge to treat humans as just another element in our mathematical representations of the factory, we pause to point out the obvious:

---

[1]We suppose that one might characterize this type of improved coordination as removing an information constraint, but this seems overly pedantic to us.

Law (Individuality):   *People are different.*

Besides making life interesting, this personnel law has a host of ramifications in the factory. Operators work at different rates; managers interact differently with workers; employees are motivated by different things. While we all know this, it is important not to forget it when drawing conclusions from simplified models or evaluating staffing requirements in terms of standardized job descriptions.

The most apparent difference between people in the workplace lies in their level of ability. Some people simply do a job better than others. We have observed huge differences between the work pace of different workers on the same manual task. Differences in experience, manual dexterity, or just sheer discipline may have accounted for this. But regardless of the cause, such differences exist and should not be ignored.

As noted in Chapter 1, Taylor acknowledged the inherent differences between workers. His response was to have managers train workers in the proper way to do their tasks. Those who responded by achieving the standard work rate set by management would be termed "first-class men," and everyone else would be fired. In addition to the threat of termination, Taylor and his industrial engineering descendants made use of a variety of incentive schemes to motivate workers to achieve the desired work pace. More recent *participative management* styles have promoted less reliance on incentive systems and more on teamwork and the use of skilled workers to train their colleagues. But regardless of how the selection, compensation, and training of workers are done, differences persist and sometimes they matter.

A specific example of a manufacturing system in which worker differences have a significant impact on logistics decisions is the **bucket brigade system** (Bartholdi and Eisenstein 1996). This system was motivated by the Toyota Sewn Products Management System, which was commercialized by Seiki Co., a subsidiary of Toyota, for the production of many types of sewn products. Variants of it have been used in a wide range of environments including warehouse picking and sandwich assembly (at Subway). The basic system, depicted in Figure 11.2, works as follows. Workers stay with a job, carrying it from one machine to the next, until they are preempted by a downstream worker. For instance, whenever the last worker in the line (worker 3) completes a job, she walks up the line to the next worker (worker 2) and takes over his job. She then takes this job through each stage in the line from where she got it to the end of the line. The preempted worker (worker 2) similarly goes upstream to the next worker (worker 1) and takes over his job. He will then continue with that job until preempted by the downstream worker (worker 3). At the beginning of the line, worker 1 starts a new job and carries it downstream as far as he can before being preempted by worker 2.

Because each step in the line involves similar skills (i.e., use of a sewing machine in a Sewn Products System, picking parts in a warehouse, or assembling a sandwich at Subway), a worker who is adept at one stage is likely to be adept at all of them. Thus, workers can be rank-ordered according to their work speed. Bartholdi and Eisenstein (1996) have shown that arranging workers from slowest to fastest (i.e., worker 1 is the slowest and worker 3 is the fastest) naturally balances the production line and is guaranteed to be close to optimal (in the sense of maximizing throughput). Their empirical studies of companies using the bucket brigade type of systems support the conclusion that a slowest-to-fastest assignment of workers is an effective policy. This work is an excellent example of how mathematical models can be productively used to help manage effectively a system involving differing skill levels.

Other differences in human ability levels beyond simple variations in work pace can also have important consequences for operations management decisions. For instance, a manager with a remarkable memory and a "gift" for manipulating a schedule may make

FIGURE 11.2

The bucket brigade system

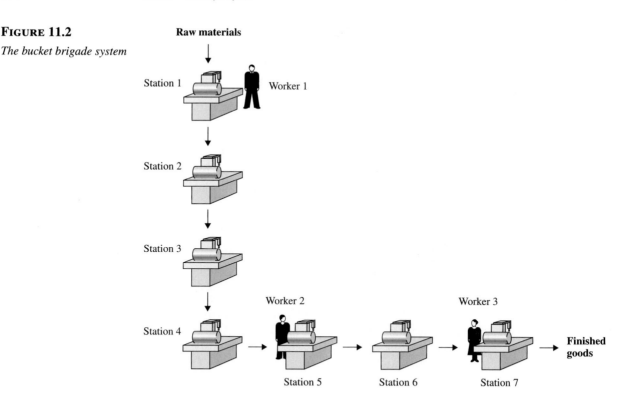

a scheduling system appear effective. But when another manager takes over, things may deteriorate drastically. While the new manager is likely to be blamed for not matching the performance of the genius predecessor, the real fault may well lie in the scheduling system.

A few years ago, we observed an example along these lines of a disaster waiting to happen in a small plant that manufactured institutional cabinetry from sheet metal. The plant used a computerized production control system that generated "cutting orders" for the presses, detailing the shape of each sheet-metal component required to build the end products. These components were cut and sent in unordered stacks on carts to the assembly area. At assembly, the computer system provided only a list of the finished product requirements, with no guidance regarding which components were needed for each product. The only bill-of-material information available was contained in the head of a man named John. John looked at the list of products and then put together "kits" of components for each. Having worked in the plant for decades, he knew the requirements for the entire list of products made by the plant. No one else in the entire organization had John's expertise. When John was sick, productivity dropped dramatically as others floundered around to find components. Although management seemed satisfied with the system, our guess was that because John was in his middle 60s, the satisfaction would not last long.

Beyond variations in skill or experience, people also differ with respect to their basic outlook on life. The basic American axiom that "all men are created equal" does not imply that all people want the same thing. For better or worse, we have observed a fundamental distinction between peoples' attitudes toward their job. Some want responsibility, challenge, and variety in their jobs; others prefer stability, predictability, and the ability to leave their work behind at the end of the day. The military has explicitly recognized this distinction with its definition of the respective roles of officers and enlisted personnel. Officers have great authority, but are also ultimately responsible for anything done by those under their command. Enlisted personnel, however, are given little authority and are held responsible only for following rules and orders.

Some writers seem to feel that everyone should belong to the first category, and that only the lack of a supportive environment condemns them to the second category. For instance, Douglas McGregor (1960) proposed the **theory Y** approach to management on the assumption that workers are better motivated by responsibility and challenge than by the fear and financial incentives of traditional management, which he terms **theory X.** While it may be true that theory Y management practices can induce more workers to take an officer's view of work, it is our opinion that there will always be workers—including very good ones—who will adhere to the enlisted's view.

In the factory, the distinction between people who identify with the officer's view and people who identify with the enlisted's view implies that pushing responsibility for decision making down to the level of the worker will have varying success, depending on the workers. Many of the Japanese manufacturing techniques that give machine operators responsibility for quality control, problem identification, or stopping the line when problems occur are based on the assumption that workers want this responsibility. In our experience this is often the case, but not always. Some individuals blossom when given added responsibility and authority; others chafe and wither under the strain. A key individual who is not inclined to accept additional responsibility can seriously undermine techniques that rely on worker empowerment. This is a consideration that must be given attention when new operating policies are implemented. New procedures may require retraining or rotation of workers. There is certainly a place for "enlisted" and "officer" workers in a plant, but having an enlisted worker in an officer's job, or vice versa, can make even good operating policies go bad.

As a final observation on human differences, we note that the fact that individuals differ in their perspectives toward life and work implies that they also differ in their response to various forms of motivation. As we noted in Chapter 1, Taylor's view that workers are motivated almost solely by money has been largely discredited. The work of Hugo Munsterberg (1913), Lillian Gilbreth (1914), Elton Mayo (1933, 1945), and Mary Parker Follett (1942) provided convincing evidence that workers are motivated by social aspects of work, in addition to financial gain. Clearly, the relative weights that individuals attach to monetary and social considerations differ. But the important point from an operations standpoint is that there are nonfinancial ways to motivate workers to participate in new systems. Awards, ceremonies, increased job flexibility, recognition in company newsletters, and many other creative options can be effective, provided that they are used in an atmosphere of genuine respect for the worker. As industry moves toward the use of pull systems—in which the uncoordinated production of parts promoted by piecework incentive systems can be particularly destructive—such nonmonetary motivational techniques will become increasingly important.

11.2.3   The Power of Zealotry

As we alluded to in Part I of this book, recent years have seen a crush of activity in the factory. From the MRP crusade of the 1970s, to the JIT and TQM revolutions of the 1980s, to the TBC (time-based competition) and BPR (business process reengineering) movements of the 1990s, manufacturing managers have been under constant pressure to change the way they do things. As a result, firms have altered the responsibilities of various positions, established new positions, and set up transition teams to carry out the desired changes. Under these conditions, the role of the person in charge of the change is enormously important. In fact, we go so far as to state the following personnel law:

Law (Advocacy):   *For almost any program, there exists a champion who can make it work—at least for awhile.*

Obviously, many programs of change fail in spite of the existence of a champion. This may or may not mean that the program itself is bad. But it tautologically means that the champion was not sufficiently gifted to make the program into a success in spite of itself. The above law implies that champions can be very powerful agents of change, but that there is both an upside and a downside to our reliance on them.

The upside of champions is that they can have a tremendous influence on the success of a system. Consider the roles of Taichi Ohno and Shigeo Shingo at Toyota. These remarkable men developed, sold, and implemented the many features of the Toyota just-in-time system in a way that turned it into the backbone of an enormously successful firm. It is important to note, however, that Ohno and Shingo were far more than mere salesmen. They were thinkers and creators as well. An effective champion must be able to develop and adapt the system to fit the needs of the target application. Besides being brilliant, Ohno and Shingo had another advantage as champions: They worked full-time on site at Toyota for many years. To be truly effective, champions must be intimately involved with the systems they are trying to change.

The importance of a local champion was brought home to us by the experience of a consultant close to us. Our friend had just finished a stirring exposition before a group of managers concerning why their plant should adopt a particular production control system. As he sat down, he was confident that he had made his point, and the satisfied looks around the table confirmed this. The plant manager, while clearly impressed by the performance, responded by deliberately turning his back on our friend and asking his managers to explain *in their own words* why he should adopt the new system. When the managers were unable to even come close to the exhilarating rhetoric and confident logic of our friend, the plant manager realized he lacked an in-house champion. He dropped the program and sent our friend packing.

The downside of champions is that in today's business environment, almost every manager is being (or is trying to be) groomed for a new position. The sheer speed with which managers are rotated means that the originator of a program is very likely to leave it before it has become thoroughly institutionalized. We have seen systems that worked well enough while their originator was still in charge rapidly collapse once she is gone. A wag once observed that the definition of a rising star is "someone who keeps one step ahead of the disasters they cause." While there may be some truth to this characterization, the phenomenon it refers to may also be a result of the natural tendency for systems to degrade once their original champions leave.

The implication of these observations on the role of champions as agents of change is that we should look at the ability to survive the loss of the originator as an important measure of the quality of a new system. This is slowly occurring in academia, where repeated educational experiments that started out as promising dissolved into mechanical imitations of their original form as soon as the second set of instructors took over. Now, a routine question asked of a professor who suggests a new course or curricular innovation is, What will happen when it is turned over to someone else? The result is that some highly innovative plans may be blocked or altered; but the changes that do get through are much more likely to have a sustained impact. Similarly, asking about the future beyond the first champion may lead manufacturing firms to abandon some plans or downgrade others to proportions manageable by nonzealots.

It is probably wise to remember that the "JIT revolution" was not a revolution in Japan. Rather, it was the result of a long series of incremental improvements over a period of decades. Each successive improvement was integrated into the system gradually, allowing time for the workforce to become acclimated to the change. Consequently, the Japanese experienced a much less sweeping program of reform with JIT than did their American counterparts. Because of this stability, the Japanese were less reliant on champions for success (despite the fact that in Ohno and Shingo they had truly superior champions) than were many American companies trying to implement JIT. The lesson: While champions can be highly influential in promoting change, we should probably strive for an environment in which they are helpful, but not all-important.

11.2.4   The Reality of Burnout

The rapid pace of revolutions in manufacturing, with the associated coming and going of champions, has had another very serious negative effect as a consequence of the following personnel law:

Law (Burnout): *People get burned out.*

In virtually every plant we visit, we hear of a long line of innovations that were announced with great fanfare, championed by a true zealot, implemented with enthusiasm, and then practiced only partly, gradually forgotten, and ultimately dropped. Perhaps on the first go-around—with MRP in the 1970s—workers were true believers in the change. But it is our view that many workers, and managers, have become increasingly cynical with each additional failure. Many take the attitude that a new program is merely the "revolution of the month"; if they ignore it, it will go away. Unfortunately, it usually does.

As we noted in Chapter 3, the MRP advocates set the tone for viewing changes in operating policies in revolutionary terms by describing MRP as no less than "a new way of life" (Orlicky 1975) and proclaiming the "MRP *crusade*." In Chapter 4, we pointed out that the JIT advocates only intensified this tendency by describing just-in-time with a fervor that borders on religious. By now, the pattern has been established, and anyone with a new manufacturing idea is almost required to use revolutionary rhetoric to attract any attention at all. The danger in this is that it encourages managers to forsake small incremental changes at the local level in favor of sweeping systemwide reforms. While revolutions are occasionally necessary, declaring too many of them risks a distinct burnout problem.

We now find ourselves in the position of needing to make changes to systems with a cynical, burned-out workforce. Clearly this is not easy, since the success of any new operating system is intimately dependent on the people who use it. But there are some things we can do:

1. *Use revolutions sparingly.* Not every improvement in a plant needs to be presented as a new way of life. For instance, instead of going headlong into a full-fledged kanban system, it may make sense to adopt some limited WIP-capping procedures. As pointed out in Chapter 10, a WIP cap provides many of the logistical benefits of kanban and is far more transparent to the workers.

2. *Do not skimp on training.* If a major system change is deemed necessary, make sure that *all* workers are trained at an appropriate level. It is our view that even machine operators need to know *why* a new system is being adopted, not just *how* to use it. Basic training in statistics may be a prerequisite for a quality control program. Basic factory

physics training may be a prerequisite for a pull production or cycle time reduction program.

3. *Use pilot programs.* Rather than try to implement a program plantwide, it may make sense to target a particular line or portion of the plant. One might test a new scheduling tool on a single-process center or adopt a pull mechanism on a segment of the line. The nature of the pilot study should be given thought early on in the planning and development stages, because the system may need adaptation to be able to perform in the pilot setting. For example, if a scheduling tool is applied to only part of the plant, it must be able to function with other portions of the plant not being scheduled in the same manner. By attacking a manageable portion of the problem, the system has a higher probability of success and therefore a better chance of overcoming cynicism and garnering supporters among the workforce. By a similar token, the best place to try a pilot effort is often in a new plant, line, or product, rather than an existing one, since the newness helps overcome people's tendency to resent change and to cling to traditional methods. Once new procedures have been demonstrated in a pilot program, it is far easier to expand them to existing parts of the system.

11.3   Planning versus Motivating

There are a number of operations management arenas in which the human element can cause the distinction between planning and motivating to blur. For instance, for the sake of accuracy, a scheduling tool should probably make use of historical capacity data. However, if historical performance is deemed poor, then using it to schedule the future may be seen as accepting substandard results. We have encountered several managers who, to avoid this perception, deliberately made use of unrealistic capacity data in their scheduling procedures. As they put it to us, "If you don't set the bar high enough, workers won't deliver their best efforts."

Given our previous discussion of how individuals are motivated by different things, it is certainly reasonable to suppose that some workers perform well under the pressure of an impossible schedule. However, we have also talked with operators and line managers who were being measured against production targets that had never even been approached in the history of the plant. Some were genuinely discouraged; others were openly cynical. It is our view that most often, unrealistic capacity numbers not only fail to motivate, but also serve to undermine morale.

There can be even more serious consequences from overestimating capacity, when these figures are used to quote due dates to customers. We observed a case in which the plant manager, by fiat, raised the capacity of the plant by 50 percent virtually overnight. Almost no physical changes were made; his intent was entirely to apply pressure to increase output. However, because no one in the plant dared defy the plant manager, the new capacity figures were immediately put into use in all the plant's systems, including those used to make commitments to customers. When output failed to go up by an amount even remotely close to 50 percent, the plant quickly found itself awash in late orders.

The behavior of this plant manager is a variation of that encouraged by the popular JIT analogy of a plant as a river with WIP as water and problems as rocks. To find the problems (rocks), one must lower the WIP (water). Of course, this implies that one finds the problems by slamming headlong into them. Our plant manager found his capacity limitations in a similarly direct manner. In Part I, we suggested that perhaps sonar (in the form of appropriate models) might be a valuable addition to this analogy. With it, one might identify and remove the problems before lowering the WIP, thereby avoiding much pain and suffering. Our manager could have saved his staff, and customers, a good deal of anguish if he had made sure that the new mandated capacity figures were *not* used for determining customer requirements until or unless they had been proved feasible.

In general, any modeling, analysis, or control system will rely on various performance parameters such as throughput, yield, machine rates, quality measures, and rework. Since we naturally wish to improve these performance measures, there is a temptation to make them better than history justifies, either out of optimism or for motivational reasons. We feel that it is important to make a distinction between systems that are used for *prediction* and those used for *motivation.* Predictive systems, such as scheduling tools, due date quoting systems, and capacity planning procedures, should use the most accurate data available, including actual historical data where appropriate. Motivational systems, such as incentive mechanisms, merit evaluations, and disciplinary procedures, may rely on speculative targets, although one must still be careful not to discourage workers by using overly lofty targets.

11.4   Responsibility and Authority

The observation that evaluating people against unrealistic targets can be demoralizing is really a specific case of a broader problem. In general, people should not be punished for things that are beyond their control. Clearly, we recognize this principle in our legal system, in which minors are treated differently from adults and a plea of insanity is allowed. But we frequently ignore it in factory management, when we set targets that cannot be achieved or when we evaluate workers against measures they cannot control. We feel that this violates a management principle so basic as to be a personnel law:

Law (Responsibility): *Responsibility without commensurate authority is demoralizing and counterproductive.*

Deming (1986) gave an illustration of management practice that is inconsistent with this principle with his well-known "red beads" experiment. When he demonstrated this experiment in his short courses, he would choose a group of people from the audience to come up on stage. After some preliminaries designed to simulate the hiring and training process, he had each person dip a paddle with 50 holes in it into a container filled with red and white marbles (see Figure 11.3). Each white marble was interpreted as good quality, while each red marble was defective. The "employee" with the lowest defect rate was rewarded by being named "employee of the month," while those with high defect rates were fired or put on probation. Then the dipping process was repeated. Invariably, the "employee of the month" did worse on the second try, while most of those on probation improved. With tongue in cheek, Deming concluded that the top employee was slacking off after being rewarded, while the bottom employees were responding to his disciplinary methods. He would go through a few more iterations of promotions, demotions, firings, and disciplining to drive home his point.

Of course, the defect rate in Deming's well-known red bead experiment is entirely outside the control of the employees. The tendency of the best workers to get worse and the worst workers to improve is nothing more than an example of **regression to the mean,** which we discussed in Chapter 8. Deming's management activities, as he well knew, were responses to statistical noise. The conclusion is that in a manufacturing system with randomness (i.e., *all* manufacturing systems), some variations in performance will be due to pure chance. Effective management practices must be able to distinguish between

FIGURE 11.3

*Deming's red beads
experiment*

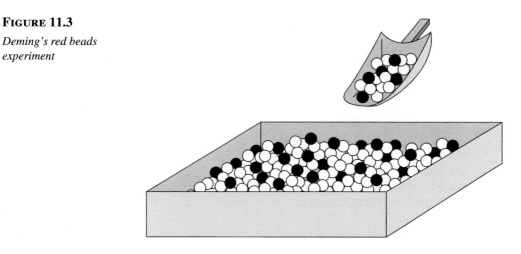

real differences and noise. If they do not, then we wind up putting workers in a position where they will be evaluated, at least partly, according to measures outside their control.

While Deming's experiment is extreme—seldom are differences in employee performance *completely* due to chance—there are partial real-world analogies. For instance, many factories still use piecework incentive systems in which worker pay is tied to the number of parts produced. If, for whatever reason, a worker does not receive sufficient raw materials from upstream, she will lose pay through no fault of her own. Similarly, if a worker gets stuck with a part for which the incentive rate is not very lucrative, he may be penalized financially even though his productivity did not decline.[2] And if a worker gets compensated only for good parts and quality defects are generated upstream, she will pay the price. If she acts in accordance with the law of self-interest, she will have an incentive to ignore quality defects. If the system forces her to inspect parts, she will be penalized by the resulting slowdown in work.

These examples illustrate some of the reasons that incentive systems have been hotly debated since the time of Taylor, and why many traditional systems have fallen into disfavor in recent years. A hundred years of tinkering have not produced a generally effective piecework incentive system, which leads us to doubt whether such a thing is even possible.

Incentive systems are not the only operations management practice that frequently give rise to a mismatch between responsibility and authority. Another is the procedure for setting and using manufacturing due dates. In general, customer due dates are established outside manufacturing, by sales, production control, or a published set of lead times (e.g., a guarantee of x-week delivery). If, as is often the case, manufacturing is held responsible for meeting customer due dates, it will wind up being punished whenever demand exceeds capacity. But since demand is not under the control of manufacturing, this violates the implication of the responsibility law that responsibility should be commensurate with authority. For this reason, we feel that it is appropriate to set separate *manufacturing due dates,* which are consistent with capacity estimates agreed to by manufacturing, but may not be identical to customer due dates. If sales overcommits relative to capacity, that department should be held responsible; if manufacturing fails to achieve output it

---

[2]Because piecework systems can make some parts more profitable to work on than others, workers tend to "cherry-pick" the most desirable parts, regardless of the overall needs of the plant. This is a natural example of the law of self-interest in action.

promised, it should be held responsible. Of course, we must not be too rigid about this separation, since it is clearly desirable to encourage manufacturing to be flexible enough to accommodate legitimate changes from sales. Chapter 15 will probe this problem in greater detail and will give specifics on how to quote customer due dates sensibly and derive a set of manufacturing due dates from them.

The disparity between responsibility and authority can extend beyond the workers into management and can be the result of subtle factors. We witnessed an example of a particular manager who had responsibility for the operational aspects of his production line, including throughput, quality, and cycle time. Moreover, he had full authority, budgetary and otherwise, to take the necessary steps to achieve his performance targets. However, he was unable to do so because of a lack of *time* to spend on operational issues; he was also responsible for personnel issues for the workforce on the line, and the majority of his time was taken up with these concerns. As a result, he was taking a great deal of heat for the poor operational performance of his line. Our impression is that this is not at all an unusual situation.

To avoid placing managers in a position in which they are unable to deal effectively with logistical concerns, we suggest using policies to explicitly *make time for operations*. One approach is to designate a manager as the "operating manager" for a specific period (e.g., a shift or day). During this time, the manager is temporarily exempted from personnel duties and is expected to concentrate exclusively on running the line. The effect will be to force the manager to appreciate the problems at an intimate level and provide time for generating solutions. This concept is analogous to the "officer of the deck" (OOD) policy used in navies around the world. When the OOD "has the con," he is ultimately responsible for the operation of the ship and is temporarily absolved from all duties not directly related to this responsibility. On a ship, having a clearly defined ultimate authority at all times is essential to making critical decisions on a split-second basis. As manufacturing practice moves toward low-WIP, short-cycle-time techniques, having a manager with the time and focus to make real-time judgments on operating issues will become increasingly important in factories as well.

11.5   Summary

We realize that this chapter is only a quick glance at the complex and multifaceted manner in which human beings function in manufacturing systems. We hope we have offered enough to convince the reader that operations management is more than just models. Even strongly technical topics, such as scheduling, capacity planning, quality control, and machine maintenance, involve people in a fundamental way. It is important to remember that a manufacturing system consists of equipment, logic, *and* people. Well-designed systems make effective use of all three components.

Beyond this fundamental observation, our main points in this chapter were these:

1. *People act according to their self-interest.* Certainly altruism exists and sometimes motives are subtle, but overwhelmingly, peoples' actions are a consequence of their real and perceived personal incentives. If these incentives induce behavior that is counterproductive to the system, they must be changed. While we cannot give here any kind of comprehensive treatment of the topic of motivation, we have tried to demonstrate that simple financial incentive systems are unlikely to be sufficient.

2. *People differ.* Because individuals differ with regard to their talents, interests, and desires, different systems are likely to work with different workforces. It makes no

sense to force-fit a control system to an environment in which the workers' abilities are ill suited to it.

3. *Champions can have powerful positive and negative influences.* We seem to be in an age when each new manufacturing management idea must be supported by a guru of godlike stature. While such people can be powerful agents for change, they can also make unsound ideas seem attractive. We would all probably be better off with a little less hype and a little more plodding, incremental improvement in manufacturing.

4. *People can burn out.* This is a real problem for the post-1990 era. We have jumped on so many bandwagons that workers and managers alike are tired of the "revolution of the month." In the future, promoting real change in manufacturing plants is likely to require less reliance on rhetoric and more on logic and hard work.

5. *There is a difference between planning and motivating.* Using optimistic capacity, yield, or reliability data for motivational purposes may be appropriate, provided it is not carried to extremes. But using historically unproven numbers for predictive purposes is downright dangerous.

6. *Responsibility should be commensurate with authority.* This well-known and obvious management principle is still frequently violated in manufacturing practice. In particular, as we move toward more rapid, low-WIP manufacturing styles, it will be increasingly important to provide managers with *time* for operations as part of their authority for meeting their manufacturing responsibilities.

We hope that these simple observations will inspire the reader to think more carefully about the human element in operations management systems. We have tried to maintain a human perspective in Part III of this book, in which we discuss putting the factory physics concepts into practice, and we encourage the reader to do the same.

Discussion Points

1. Comment on the following paraphrase of a statement by an hourly worker overheard in a plant lunchroom:

   > Management expects us to bust our butts getting more efficient and reengineering the plant. If we don't, they'll be all over us. But if we do, we'll just downsize ourselves out of jobs. So the best thing to do is make it look like we're working real hard at it, but be sure that no really big changes happen.

   a. What does this statement imply about the relationship between management and labor at that plant?
   b. Does the worker have a point?
   c. How might such concerns on the part of workers be addressed as part of a program of change?

2. Consider the following paraphrase of a statement by the owner of a small manufacturing business:

   > Twenty years ago our machinists were craftsmen and knew these processes inside and out. Today, we're lucky if they show up on a regular basis. We need to develop an automated system to control the process settings on our machines, not so much to enhance quality or keep up with the competition, but because the workers are no longer capable of doing it manually.

   a. What does this statement imply about the relationship between management and labor at that plant?

       b. Does the owner have a point?

       c. What kinds of policies might management pursue to improve the effectiveness of operators?

  3. Consider the following statement:

> JIT worked for Toyota and other Japanese companies because they had the champions who originated it. American firms were far less successful with it because they had less effective champions to sell the change.

       a. Do you think there is a grain of truth in this statement?

       b. What important differences about JIT in Japan and America does it ignore?

Study Questions

1. The popular literature on manufacturing has sometimes portrayed continual improvement as a matter of "removing constraints." Why are constraints sometimes a good thing in manufacturing systems? How could removing constraints actually make things worse?

2. When dealing with a manufacturing system that is burned out by "revolutions," what measures can a manager use to inspire needed change?

3. Many manufacturing managers are reluctant to use historical capacity data for future planning because they regard it as tantamount to accepting previous substandard performance. Comment on the dilemma between using historical capacity data for planning versus using rated capacity for motivation. What measures can a manager take to separate planning from motivation?

4. In Deming's red beads example, employees have no control over their performance. What does this experiment have to do with a situation in the real world, where employees' performance is a function both of their ability/effort and random factors? What managerial insights can one obtain from this example?

5. Contrast MRP, Kanban, and CONWIP from a human issues standpoint. What implications do each of these systems have for the working environment of the employees on the factory floor? The staff engineers responsible for generating and propagating the schedule? The managers responsible for supervising direct labor? To what extent are the human factors benefits of a particular production control system specific to that system, and therefore not to be obtained by modifying one of the other production control methods?

12   TOTAL QUALITY MANUFACTURING

Saw it on the tube
Bought it on the phone
Now you're home alone
It's a piece of crap.

I tried to plug it in
I tried to turn it on
When I got it home
It was a piece of crap.

Neil Young

12.1   Introduction

A fundamental factory physics insight is that variability plays an important role in determining the performance of a manufacturing system. As we observed in Chapters 8 and 9, variability can come from a variety of sources: machine failures, setups, operator behavior, fluctuations in product mix, and many others. A particularly important source of variability, which can radically alter the performance of a system, is quality. Quality problems almost always become variability problems. By the same token, variability reduction is frequently a vehicle for quality improvement. Since quality and variability are intimately linked, we conclude Part II with an overview of this critical issue.[1]

12.1.1   The Decade of Quality

The 1980s were the *decade of quality* in America. Scores of books were published on the subject, thousands of employees went through short courses and other training programs, and "quality-speak" entered the standard language of corporate America. In 1987, the International Standards Organization established the ISO 9000 Series of quality

---

[1] We have deliberately used the title *Total Quality Manufacturing* in place of the more conventional *Total Quality Management (TQM)* in recognition that we are covering only the subset of TQM that relates to operations management.

standards. In the same year, the Malcolm Baldrige National Quality Award was created by an act of the U.S. Congress.[2]

The concept of quality and the methods for its control, assurance, and management were not new in the 1980s. Quality control as a discipline dates back at least to 1924 when Walter A. Shewhart of Western Electric's Bell Telephone Laboratories first introduced process control charts. Shewhart published the first important text on quality in 1931. Armand Feigenbaum coined the term *total quality control* in a 1956 paper and used this as the title of a 1961 revision of his 1951 book, *Quality Control.*

But while the terms and tools of quality have been around for a long time, it was not until the 1980s that American industry really took notice of the strategic potential of quality. Undoubtedly, this interest was stimulated in large part by the dramatic increase in the quality of Japanese products during the 1970s and 1980s, much in the same way that American interest in inventory reduction was prompted by Japanese JIT success stories.

Has all the talk about quality led to improvements? Probably so, although it is difficult to measure them since, as we will discuss in this chapter, quality is a broad term that can be interpreted in many ways. Nevertheless, some surveys have suggested that consumers viewed the overall quality of American products as *declining* during the 1980s (Garvin 1988). The American Customer Satisfaction Index (ACSI), an overall gauge of customer perceptions of quality that has been tracked quarterly since 1994, also showed declining satisfaction in the 1990s. Whether these declines are due to rising customer expectations, ongoing management problems, or both, it seems clear that quality remains a significant challenge for the future.

12.1.2   A Quality Anecdote

To set the stage, we introduce the quality issue from a personal perspective. In 1991, one of the authors purchased a kitchen range that managed to present an astonishing array of quality problems. First, for styling purposes, the stove came with light-colored porcelain-coated steel cooktop grates. After only a few days of use, the porcelain cracked and chipped off, leaving a rough, unattractive appearance. When the author called the service department (and friends with similar stoves), he found that *every single* stove of this model suffered from the same defect—a 100 percent failure rate! So much for inspection and quality assurance!

The customer service department was reasonably polite and sent replacement grates, but these lasted no longer than the originals, so the author continued to complain. After three or four replacements (including one in which the service department sent two sets with the recommendation that we use one set and save the other to put on the stove when entertaining guests!), the manufacturer changed suppliers and sent dark-colored, more durable grates. So much for quality design and styling!

As the grate story was evolving, the stove suffered from a succession of other problems. For instance, the pilotless ignition feature would not shut off after the burners lit, causing a loud clicking noise whenever the stove was in use. Repair people came to fix this problem no fewer than *eight* times during the first year of use (i.e., the warranty period). During one of these visits, the repairman admitted that he really had no idea of how to adjust the stove because he had never received specifications for this model from the manufacturer and was therefore just replacing parts and hoping for the best. So much for service after the sale and for doing things right the first time!

---

[2]Tellingly, the Japanese Union of Scientists and Engineers (JUSE) had already established its major quality award, the Deming Prize, in honor of American W. Edwards Deming, in 1951.

At the end of the first year, the service department called to sell the author an extended warranty and actually said that because the stove was so unreliable (they used a much less polite term than *unreliable*) the extended warranty would be a good deal for us. So much for standing behind your product, and for customer-driven quality!

(By the way, as this book was being written, the oven door fell off. WE DID NOT MAKE ANY OF THIS UP!)

12.1.3   The Status of Quality

We do not mean to imply that this story sums up the quality level of American manufacturing. But it is fascinating (and depressing) that a company in the 1990s could be in such glaring violation of virtually every principle of good quality management. Furthermore, we suspect that this is not an isolated example. (We have more personal experiences, but will not subject the reader to them.) An exercise we are fond of in our executive courses is to challenge the participants to think about quality not from the viewpoint of a manufacturing manager, educator, or professional, but from that of a *consumer*. A dishearteningly high fraction report that they have rarely had their expectations for a product or service exceeded, but have frequently been disappointed.

Evidently, there is still a considerable gap between the rhetoric and the reality of quality. Thus, while it is convenient to speak, as we did at the beginning of the book, as if *cost* was the dimension of competition in the 1970s, *quality* was the dimension of competition in the 1980s, and *speed* is the dimension of competition in the 1990s, one should not take this apothegm literally. Quality (and cost, for that matter) will remain an important determinant of competitiveness well beyond the 1990s.

What can an individual firm do? The answer is, plenty. There is not a plant in the world that could not improve its products, processes, or systems; get closer to its customers; or better understand the influence of quality on its business. Furthermore, there is a vast literature to consult for ideas. Although the quality literature, like the JIT literature, contains an overabundance of imprecise romantic rhetoric, it offers much useful guidance as well. The literature on quality can be divided into two categories, **total quality management (TQM),** which focuses on quality in qualitative management terms (e.g., fostering an overall environment supportive of quality improvement), and **statistical quality control (SQC),** which focuses on quality in quantitative engineering terms (e.g., measuring quality and assuring compliance with specifications). Both views are needed to formulate an effective quality improvement program. All TQM with no SQC produces talk without substance, while all SQC with no TQM produces numbers without purpose.

A strong representative from the TQM literature is the work of Garvin (1988), on which some of the following discussion is based. Garvin's book offers an insightful perspective of what quality is and how it affects the firm. Other widely read TQM books include those by Crosby (1979, 1984), Deming (1986), and Juran (1989, 1992). In the SQC field there are many solid works, most of which contain a brief introductory section on TQM; these include those by Banks (1989); DeVor, Chang, and Sutherland (1992); Gitlow et al. (1989); Montgomery (1991); and Thompson and Koronacki (1993); among others. Some books, notably Juran's *Quality Control Handbook* (1988), address both the TQM and SQC perspectives.

We cannot hope to provide the depth and breadth of these references in this brief chapter. What we can do is to focus on how quality fits into the overall picture of plant operations management. The framework of factory physics allows us to synthesize the perspectives of quality and operations into elements of the same picture. We leave the reader to consult references like those mentioned, to flesh out the specifics of quality management procedures.

12.2 Views of Quality

12.2.1 General Definitions

What is quality? This question is a logical place to start our discussion. Garvin (1988) offers five definitions of quality, which we summarize as follows:

1. *Transcendent.* Quality refers to an "innate excellence," which is not a specific attribute of either the product or the customer, but is a third entity altogether. This boils down to the "I can't define it, but I know it when I see it" view of quality.

2. *Product-based.* Quality is a function of the attributes of the product (the quality of a rug is determined by the number of knots per square inch, or the quality of an automobile bumper is determined by the dollars of damage caused by a five-mile-per hour crash). This is something of a "more is better" view of quality (more knots, more crashworthiness, etc.).

3. *User-based.* Quality is determined by how well customer preferences are satisfied; thus, it is a function of whatever the customer values (features, durability, aesthetic appeal, and so on). In essence, this is the "beauty is in the eye of the beholder" view of quality.

4. *Manufacturing-based.* Quality is equated with conformance to specifications (e.g., is within dimensional tolerances, or achieves stated performance standards). Because this definition of quality directly refers to the processes for making products, it is closely related to the "do it right the first time" view of quality.

5. *Value-based.* Quality is jointly determined by the performance or conformance of the product *and* the price (e.g., a $1,000 compact disk is not high quality, regardless of performance, because few would find it worth the price). This is a "getting your money's worth" or "affordable excellence" view of quality.

These definitions bring up two points. First, quality is a multifaceted concept that does not easily reduce to simple numerical measures. We need a framework within which to evaluate quality policies, just as we needed one (i.e., factory physics) for evaluating operations management policies. Indeed, as we will discuss, the two frameworks are closely related, perhaps as two facets of the larger science of manufacturing to which we referred in Chapter 6.

Second, the definitions are heavily **product-oriented.** This is the case with most of the TQM literature and is a function of the principle that quality must ultimately be "customer-driven." Since what the customer sees is the product, quality must be measured in product terms. However, the quality of the product as seen by the customer is ultimately determined by a number of **process-oriented** factors, such as design of the product, control of the manufacturing operations, involvement of labor and management in overseeing the process, customer service after the sale, and so on.

12.2.2 Internal versus External Quality

To better understand the relationship between product-oriented and process-oriented quality, we find it useful to draw the following distinction between internal quality and external quality:

1. **Internal quality** refers to conformance with quality specifications inside the plant and is closely related to the manufacturing-based definition of quality. It is typically monitored through direct product measures such as scrap and rework rates and indirect process measures such as pressure (in an injection molding machine) and temperature (in a plating bath).

2. **External quality** refers to how the customer views the product and may be interpreted by using the transcendent, product-based, user-based, or value-based definition, or a combination of them. It can be monitored via direct measures of customer satisfaction, such as return rate, and indirect indications of customer satisfaction derived from sampling, inspection, field service data, customer surveys, and so on.

To achieve high external quality, one must translate customer concerns to measures and controls for internal quality. Thus, from the perspective of a manufacturing manager, the links between internal and external quality are key to the development of a strategically effective quality program. The following are some of the more important ways in which quality inside the plant is linked to the quality that results in customer satisfaction.

1. *Error prevention.* If fewer errors are made in the plant, fewer defects are likely to slip through the inspection process and reach the customer. Therefore, to the extent that quality as perceived by the customer is determined by freedom from defects, high "quality at the source" in the plant will engender high customer-driven quality.

2. *Inspection improvement.* If fewer defects are produced during the manufacturing process, then quality assurance will require inspection to detect and reject or correct fewer items. This tends to reduce pressure on quality personnel to "let things slide"— in other words, relax quality standards in the name of getting product out the door.[3] Furthermore, the less time spent reworking or replacing defective parts, the more time people have for tracing quality problems to the root causes. Ideally, the net effect will be an upward quality spiral, in which error prevention and error detection both improve over time.

3. *Environment enhancement.* Even if quality problems in the field cannot be traced directly to plant-level defects, high internal quality and external quality may still be linked.[4] Both types of quality are promoted by the same environmental factors (e.g., supportive management attitudes, tangible rewards for improvements, sophisticated tracking and control systems, and effective training). An organization that has fostered the right attitudes and tools inside the plant is likely to be able to do the same outside the plant.

In short, understanding quality means looking to the customer. Delivering it entails looking to manufacturing.[5] For the purposes of this chapter we will assume that the concerns of the customer have been understood and translated to quality specifications for use by the plant. Our focus will be on the relationship between quality and operations, and particularly how the two can work together as parts of a continual improvement process for the plant.

---

[3]Crosby (1979, 41) relates a story in which manufacturing viewed inspection in an adversarial mode, protesting each rejected part as if quality inspectors were personally trying to sabotage the plant.

[4]Garvin (1988, 129) offers the example of the compressor on an air conditioner failing due to corrosion caused by excess moisture seeping into the unit. Such a problem would not show up in any reasonable "burn in" period and therefore would most likely be undetected as a defect at the plant level.

[5]Here we are referring to "big M" manufacturing, including product design, production, and field service.

12.3   Statistical Quality Control

Statistical quality control (SQC) generally focuses on *manufacturing* quality, as measured by conformance to specifications. The ultimate objective of SQC is the systematic reduction of variability in key quality measures. For instance, size, weight, smoothness, strength, color, and speed (e.g., of delivery) are all measurable attributes that can be used to characterize the quality of manufacturing processes. By working to assure that these measures are tightly controlled within desired bounds, SQC functions directly at the interface between operations and quality.

12.3.1   SQC Approaches

There are three major classes of tools used in SQC to ensure quality:

1. *Acceptance sampling.* Products are inspected to determine whether they conform to quality specifications. In some situations, 100 percent inspection is used, while in others some form of statistical sampling is substituted. Sampling may be an option chosen for cost reasons or an absolute necessity (e.g., when inspection is destructive).

2. *Process control.* Processes are continuously monitored with respect to both mean and variability of performance to determine when special problems occur or when the process has gone out of control.

3. *Design of experiments.* Causes of quality problems are traced through specifically targeted experiments. The basic idea is to systematically vary controllable variables to determine their effect on quality measures. A host of statistical tools (e.g., block designs, factorial designs, nested designs, response surface analysis, and Taguchi methods) have been developed for efficiently correlating controls with outputs and optimizing processes.

Typically, as an organization matures, it relies less on after-the-fact acceptance sampling and more on at-the-source process control and continual-improvement-oriented design of experiments.

Obviously, entire books have been written on each of these subjects, so detailed coverage of them is beyond the scope of this chapter. However, because process control deals so specifically with the interface between quality and variability, we offer an overview of the basic concepts here.

12.3.2   Statistical Process Control

Statistical process control (SPC) begins with a measurable quality attribute—for example, the diameter of a hole in a cast steel part. Regardless of how tightly controlled the casting process is, there will always be a certain amount of variability in this diameter. If it is relatively small and due to essentially uncontrollable sources, then we call it **natural variability.** A process that is operating stably within its natural variation is said to be **in statistical control.** Larger sources of variability that can potentially be traced to their causes are called **assignable-cause variation.** A process subject to assignable-cause variation is said to be **out of control.** The fundamental challenge of SPC is to separate assignable-cause variation from natural variation. Because we generally observe directly only the quality attribute itself, but not the causes of variation, we need statistics to accomplish this.

FIGURE 12.1

Process control chart for average hole size in steel castings

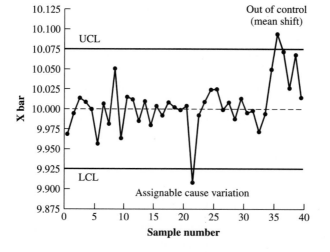

To illustrate the basic principles behind SPC, let us consider the example of controlling the diameter of a hole in a steel part made using a sand casting process. Suppose that the desired nominal diameter is 10 millimeters and we observe a casting with a diameter of 10.1 millimeters. Can we conclude that the casting process is out of control? The answer is, of course, "It depends." It may be that a deviation of 0.1 millimeter is well within natural variation levels. If this were the case and we were to adjust the process (e.g., by altering the sand, steel, or mold) in an attempt to correct the deviation, in all likelihood we would make it worse. The reason is that adjusting a process in response to random noise increases its variability (see Deming 1982, 327, for discussion of a funnel experiment that illustrates this point). Hence, to ensure that adjustments are made only in response to assignable-cause variation, we must characterize the natural variation.

In our example, suppose we have measured a number of castings and have determined that the mean diameter can be controlled to be $\mu = 10$ millimeters and the standard deviation of the diameter is $\sigma = 0.025$ millimeter. Further suppose that every two hours we take a random sample of five castings, measure their hole diameters, compute the average (which we call $\bar{x}$), and plot it on a chart like that shown in Figure 12.1. From basic statistics, we know that $\bar{x}$ is itself a random variable which has standard deviation

$$\sigma_{\bar{x}} = \frac{\sigma}{\sqrt{n}} \tag{12.1}$$

where n is the number in the sample; $n = 5$ in this example.[6]

The basic idea behind control charts is very similar to hypothesis testing. Our null hypothesis is that the process is in control; that is, the samples are coming from a process with mean μ and standard deviation σ. To avoid concluding that the process is out of control when it is not (i.e., type I error), we set a stringent standard for designating deviations as "assignable cause." Standard convention is to flag points that lie more than three standard deviations above or below the mean. We do this by specifying lower and upper control limits as follows:

$$\text{LCL} = \mu - 3\sigma_{\bar{x}} \tag{12.2}$$

$$\text{UCL} = \mu + 3\sigma_{\bar{x}} \tag{12.3}$$

---

[6]Note that this is another example of variability pooling. Choosing $n > 1$ tightens our estimate of $\bar{x}$ and therefore reduces our chances of reacting to random noise in the system.

If we observe a sample mean outside the range between LCL and UCL, then this observation is designated as assignable-cause variation. In the casting example charted in Figure 12.1, such a deviation occurred at sample 22. This might have been caused by defective inputs (e.g., steel or sand), machine problems (e.g., in the mold, the packing process, the pouring process), or operator error. SPC does not tell us why the deviation occurred—only that it is sufficiently unusual to warrant further investigation.

Other criteria besides points outside the control limits are sometimes used to signal out-of-control conditions. For instance, the occurrence of several points in a row above (or below) the target mean is frequently used to spot a potential shift in the process mean. In Figure 12.1, sample 37 is out of control. But unlike the out-of-control point at sample 22, this point is accompanied by an unusual run of above-average observations in samples 35 to 40. This is strong evidence that the cause of the problem is not unique to sample 37, but instead is due to something in the casting process itself that has caused the mean diameter to increase. Other criteria based on multiple samples, such as rules that look for trends (e.g., high followed by low followed by high again), are also used with control charts to spot assignable-cause variation.

It is important to note that because a process is in statistical control does *not* necessarily mean that it is **capable** (i.e., able to meet process specifications with regularity). For instance, suppose in our casting example that for reasons of functionality we require the hole diameter to be between a **lower specification level (LSL)** and an **upper specification level (USL).** Whether or not the process is capable of achieving these levels depends on how they compare with the **lower and upper natural tolerance limits,** which are defined as

$$\text{LNTL} = \mu - 3\sigma \qquad\qquad (12.4)$$

$$\text{UNTL} = \mu + 3\sigma \qquad\qquad (12.5)$$

Note that LNTL and UNTL are limits on the diameter of individual holes, while the LCL and UCL are limits on the average diameter of samples. Moreover, note that LNTL and UNTL are internally determined by the process itself, while LSL and USL are externally determined by performance requirements.

Let us consider some illustrative cases. The natural tolerance limits are given by $\text{LNTL} = \mu - 3\sigma = 10 - 3(0.025) = 9.925$ and $\text{LNTL} = \mu + 3\sigma = 10 + 3(0.025) = 10.075$. Suppose that the specification levels are given by $\text{LSL} = \text{LSL1} = 9.975$ and $\text{USL} = \text{USL1} = 10.025$. It is apparent from Figure 12.2 that the casting process will produce a large fraction of nonconforming parts. To be precise, if hole diameters are normally distributed, then

$$P(9.975 \leq X \leq 10.025) = P\left(\frac{9.975 - 10}{0.025} \leq Z \leq \frac{10.025 - 10}{0.025}\right)$$
$$= P(-1 \leq Z \leq 1) = \Phi(-1) + 1 - \Phi(1)$$
$$= 0.1587 + 1 - 0.8413$$
$$= 0.3174$$

This means that almost 32 percent will fail to meet specification levels.

Suppose instead that the specification levels are given by $\text{LSL} = \text{LSL2} = 9.875$ and $\text{USL} = \text{USL2} = 10.125$. Since the natural tolerance limits lie well within this range, we would expect very few nonconforming castings. Indeed, repeating the calculation above for these limits shows that the fraction of nonconforming parts will be 0.00000057.

Figure 12.2

Process capability comparing specification limits to natural tolerance limits

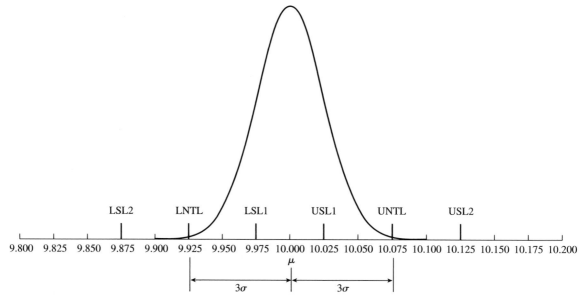

A measure of capability is the **process capability index,** which is defined as

$$C_{pk} = \frac{Z_{min}}{3} \tag{12.6}$$

where

$$Z_{min} = \min\{-Z_{LSL}, Z_{USL}\} \tag{12.7}$$

and

$$Z_{LSL} = \frac{LSL - \mu}{\sigma} \tag{12.8}$$

$$Z_{USL} = \frac{USL - \mu}{\sigma} \tag{12.9}$$

The minimum acceptable value of C_{pk} is generally considered to be one. Note that in the above examples, $C_{pk} = 1/3$ for (LSL1, USL1), but $C_{pk} = 5/3$ for (LSL2, USL2). Note that C_{pk} is sensitive to both variability (σ) and asymmetry (i.e., a process mean that is not centered between USL and LSL). Hence, it gives us a simple quantitative measure of how capable a process is of meeting its performance specifications.

Of course, a host of details needs to be addressed to implement an effective SPC chart. We have glossed over the original estimates of μ and σ; in practice, there are a variety of ways to collect these from observable data. We also need to select the sample size n to be large enough to prevent reacting to random fluctuations but not so large that it masks assignable-cause variation. The frequency with which we sample must be chosen to balance the cost of sampling with the sensitivity of the monitoring.

12.3.3   SPC Extensions

The $\bar{x}$ chart discussed is only one type of SPC chart. Many variations have been proposed to meet the needs of a wide variety of quality assurance situations. A few that are particularly useful in manufacturing management include these:

1. *Range (R charts).* An $\bar{x}$ chart requires process variability (that is, σ) to be in control in order for the control limits to be valid. Therefore, it is common to monitor this variability by charting the range of the samples. If $x_1, x_2, \ldots, x_n$ are the measurements (e.g., hole diameters) in a sample of size n, then the range is the difference between the largest and smallest observations

$$R = x_{max} - x_{min} \tag{12.10}$$

Each sample yields a range, which can be plotted on a chart. Using past data to estimate the mean and standard deviation of R, denoted by $\bar{R}$ and σ_R, we can set the control limits for the R chart as

$$LCL = \bar{R} - 3\sigma_R \tag{12.11}$$

$$UCL = \bar{R} + 3\sigma_R \tag{12.12}$$

If the R chart does not indicate out-of-control situations, then this is a sign that the variability in the process is sufficiently stable to apply an $\bar{x}$ chart. Often, $\bar{x}$ and R charts are tracked simultaneously to watch for changes in either the mean or the variance of the underlying process.

2. *Fraction nonconforming (p charts).* An alternative to charting a physical measure, as we do in an $\bar{x}$ chart, is to track the fraction of items in periodic samples that fail to meet quality standards. Note that these standards could be quantitative (e.g., a hole diameter is within specified bounds) or qualitative (e.g., a wine is approved by a taster). If each item independently has probability p of being defective, then the variance of the fraction of nonconforming items in a sample of size n is given by $p(1-p)/n$. Therefore, if we estimate the fraction of nonconforming items from past data, we can express the control limits for the p chart as

$$LCL = p - 3\sqrt{\frac{p(1-p)}{n}} \tag{12.13}$$

$$UCL = p + 3\sqrt{\frac{p(1-p)}{n}} \tag{12.14}$$

3. *Nonquality applications.* The basic control chart procedure can be used to track almost any process subject to variability. For example, we describe a procedure for statistical throughput control in Chapter 14, which monitors the output from a process in order to determine whether it is on track to attain a specified production quota. Another nonquality application of control charts is in due date quoting, which we discuss in Chapter 15. The basic idea is to attach a safety lead time to the estimated cycle time and then track customer service (e.g., as percentage delivered on time). If the system goes out of control, then this is a signal to adjust the safety lead time.

The power and flexibility of control charts make them extremely useful in monitoring all sorts of processes where variability is present. Since, as we have stressed repeatedly in this book, virtually all manufacturing processes involve variability, SPC techniques are a fundamental part of the tool kit of the modern manufacturing manager.

12.4   Quality and Operations

Closely related to variability as a link between quality and operations is cost. However, there is some disagreement about just how this link works. Here are two distinct views:

1. *Cost increases with quality.* This is the traditional industrial engineering view, which holds that achieving higher external quality requires more intense inspection, more rejects, and more expensive materials and processes. Since customers' willingness to pay for additional quality diminishes with the level of quality, this view leads to the "optimal defect level" arguments common to industrial engineering textbooks in the past.

2. *Cost decreases with quality.* This is the more recent TQM view, espoused using phrases such as *quality is free* (Crosby 1979) or *the hidden factory;* it holds that the material and labor savings from doing things right the first time more than offset the cost of the quality improvements. This view supports the zero-defects and continual-improvement goals of JIT.

Neither view is universally correct. If improving quality of a particular product means replacing a copper component with a gold one, then cost does increase with quality. Where this is the case, it makes sense to ask whether the market is willing to pay for, or will even notice, the improvement. On the other hand, if quality improvement is a matter of shifting some responsibility for inspection from end-of-line testing to individual machine operators, it is entirely possible that the reduction in rework, scrap, and inspection costs will more than offset the implementation cost. Ultimately, what matters is which view is appropriate for assessing the costs and consequences of a specific quality improvement. This is crucial for deciding which policies should be pursued while making continual improvements, and which should be tempered by the market.

In the next discussion and examples, we rely on the factory physics framework to evaluate the impacts of quality on operations and the impacts of operations on quality. Our intent is not so much to provide specific numerical estimates of the cost of quality—the range of situations that arise in industry is too varied to permit comprehensive treatment of this nature—but rather to broaden and extend the intuition we developed for the behavior of manufacturing systems in Part II to incorporate quality considerations.

12.4.1 Quality Supports Operations

In Chapter 9 we presented two manufacturing laws that are central to understanding the impact of quality on plant operations, the variability law and the utilization law. These can be paraphrased as follows:

1. Variability causes congestion.
2. Congestion increases nonlinearly with utilization.

In practice, quality problems are one of the largest and most common causes of variability. Additionally, by causing work to be done over (either as rework or as replacements for scrapped parts), quality problems often end up increasing the utilization of workstations. By affecting both variability and capacity, quality problems can have extreme operational consequences.

The Effect of Rework on a Single Machine.   To get a feel for how quality affects utilization and variability, let us consider a simple single-machine example. The machine receives parts at a rate of one every three minutes. Processing times have a mean and standard deviation of t_0 and σ_0 minute, respectively, so that the CV of the natural process time is $c_0 = \sigma_0/t_0$. However, with probability p, a given part is defective. We assume that the quality check is integral to the processing, and therefore whether the part is defective is immediately known upon its completion. If it is defective, it must be reworked, which requires another processing time with mean t_0 and standard deviation σ_0 and again has probability p of failing to produce a good part. The machine continues reworking the part until a good one is produced. We define the total time it takes to produce a good part to be the **effective processing time.**

Letting T_e represent the (random) effective processing time of a part, we can compute the mean t_e, variance σ_e^2, and squared coefficient of variation (SCV) c_e^2 of this time, as well as the utilization of the machine u, as follows:

$$t_e = E[T_e] = \frac{t_0}{1 - p} \tag{12.15}$$

$$\sigma_e^2 = \text{Var}(T_e) = \frac{\sigma_0^2}{1 - p} + \frac{p t_0^2}{(1 - p)^2} \tag{12.16}$$

$$c_e^2 = \frac{\sigma_e^2}{t_e^2} = \frac{(1 - p)\sigma_0^2 + p t_0^2}{t_0^2} = c_0^2 + p(1 - c_0^2) \tag{12.17}$$

$$u = \frac{1}{3} t_e = \frac{t_0}{3(1 - p)} \tag{12.18}$$

We can draw the following conclusions from this example:

1. *Utilization increases nonlinearly with rework rate.* This occurs because the mean time to process a job increases with the expected number of passes, while the arrival rate of new jobs remains constant. At some point, the added workload due to rework will overwhelm the station. In this example, Equation (12.18) shows that for $p > 1 - t_0/3$, utilization exceeds one, indicating that the system does not have enough capacity to keep up with both new arrivals and rework jobs over the long run.

2. *Variance of process time, given by σ_e^2, increases with rework rate.* The reason, of course, is that the more likely a job is to make multiple passes through the machine, the more unpredictable its completion time becomes.

3. *Variability of process time, as measured by the SCV, may increase or decrease with rework rate, depending on the natural variability of the process.* Although both the variance and the mean of the effective process time always increase with the rework rate, the variance does not always increase faster than the mean. Hence the SCV, which is the ratio of variance to mean, can increase or decrease. We can see from Equation (12.17) that c_e^2 increases in p if $c_0^2 < 1$, decreases in p if $c_0^2 > 1$, and is constant in p if $c_0^2 = 1$. The intuition behind this is that the effects of variability pooling (which happens when we sum the process times of repeated passes) become large enough when $c_0^2 > 1$ to cause the SCV of effective process times to decrease in p.

We can use these specific results for a single machine with rework to motivate some general observations about the effect of rework on the cycle time and lead time of a process. Since both the mean and the variance of effective process time increase with rework rate, we can invoke the lead time law of Chapter 9 to conclude that the lead time required to achieve a given service level also increases the rework rate.

The effect of rework on cycle time is not so obvious, however. The fact that the SCV of effective process time can go down when rework increases, may give the impression that rework might actually reduce cycle time. But this is not the case. The reason is that increasing rework increases utilization, which is a first-order effect on cycle time that outweighs the second-order effect from a possible reduction in variability. Hence, even in processes with high natural variability, increasing rework will inflate the mean cycle time. Moreover, because it also increases the variance of total processing time per job and the variance of the time to wait in queue, increasing rework also inflates the standard deviation of cycle time. These cycle time effects represent general observations about the impact of rework, as we summarize in the following manufacturing law.

Law (Rework):   *For a given throughput level, rework increases both the mean and standard deviation of the cycle time of a process.*

To give an illustration of this law, suppose the previously mentioned station is fed by a moderately variable arrival process (that is, $c_a = 1$) but has deterministic processing times such that $t_0 = 1$ and $c_0 = 0$. Then, using Kingman's model of a workstation introduced in Chapter 8, the cycle time at the station can be expressed as a function of p as

$$CT = \frac{c_a^2 + c_e^2}{2} \frac{u}{1 - u} t_e + t_e$$

$$= \frac{1 + p}{2} \frac{1/(3(1 - p))}{1 - 1/(3(1 - p))} \frac{1}{1 - p} + \frac{1}{1 - p}$$

Figure 12.3 plots cycle time versus rework rate. This plot shows that cycle time grows nonlinearly toward infinity as p approaches 2/3, the point at which rework reduces the effective capacity of the system below the arrival rate.

Effect of Rework on a CONWIP Line.   Of course, station level measures such as utilization, variability, and cycle time are only indirect measures; what we really care about is the throughput, WIP, and cycle time of a line. To illustrate the rework law in a line, consider the CONWIP line depicted in Figure 12.4. Processing times are two-thirds hour for machines 1, 2, and 4 and one hour for machine 3 (the bottleneck). All processing times are deterministic (that is, $c_e^2 = 0$). However, machine 2 is subject to rework. As in the previous example, we assume that each job that is processed must be reprocessed with probability p. Hence, as in the previous example, the mean effective processing time on machine 2 is given by

$$t_e(2) = \frac{2/3}{1 - p}$$

We assume that the line has unlimited raw materials, so the only source of variability is rework.

FIGURE 12.3

Cycle time as a function of rework rate

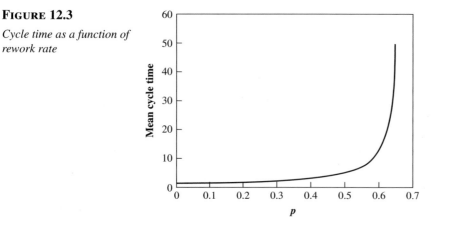

Because even this simple line is too complex to permit convenient analysis (the single-machine example was messy enough!), we turn to computer simulation to estimate the performance measures for various values of p and different WIP levels. Figures 12.5 and 12.6 summarize our simulation results.

When $p = 0$ (no rework), the system behaves as the best case we studied in Chapter 7. Thus, we can apply the formulas derived there to characterize the throughput-versus-WIP and cycle-time-versus-WIP curves. Note that without rework, the bottleneck rate r_b is one job per hour, and the raw process time T_0 is $r_b T_0 = 3$ hours. Hence, the critical WIP level is 3 jobs. At this WIP level, maximum throughput (1 job per hour) and minimum cycle time (three hours) are attained.

When $p = 1/3$, the mean effective process time on machine 2 is $t_e(2) = 1$, the bottleneck rate. Thus, r_b is not changed, but T_0 increases to 3.33 hours. This means that as WIP approaches infinity, full throughput of one job per hour will be attained. Our simulation indicates that virtually full throughput is attained at a WIP level of about 10 jobs—more than three times the WIP level required in the no-rework case. At a WIP level of 10 jobs, the average cycle time is roughly 10 hours—also three times the ideal level of the

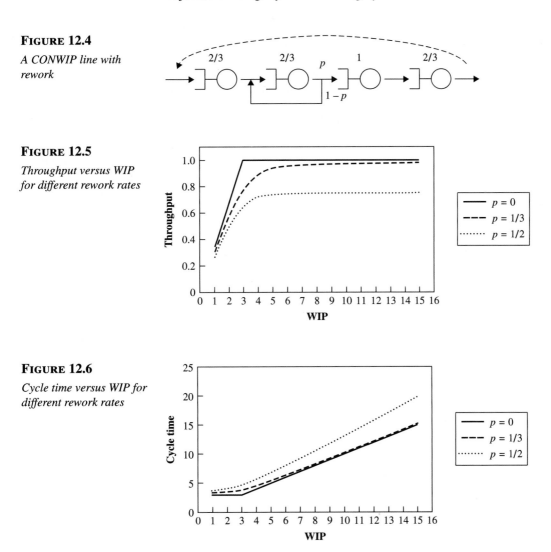

FIGURE 12.4

A CONWIP line with rework

FIGURE 12.5

Throughput versus WIP for different rework rates

FIGURE 12.6

Cycle time versus WIP for different rework rates

no-rework case. The implication here is that the primary effect of rework when $p = 1/3$ is to transform a line that behaved as the best case to one approaching the practical worst case. This illustrates the rework law in action with regard to the mean cycle time.

When $p = 1/2$, the mean effective process time on machine 2 is $t_e(2) = 4/3$, which makes it the bottleneck. Thus, even with infinite WIP, we cannot achieve throughput above $r_b = 3/4$ job per hour. As expected, Figure 12.5 shows substantially reduced throughput at all WIP levels. Figure 12.6 shows that cycle times are longer, as a consequence of the reduced capacity at machine 2, at all WIP levels. Moreover, because the bottleneck rate has been decreased, the cycle time curve increases with WIP at a faster rate than in the previous two cases.

The simulation model enables us to keep track of other line statistics. Of particular interest is the standard deviation of cycle time. Recall that the lead time law implies that if we quote customer lead times to achieve a specified service level (probability of on-time delivery), then lead times are an increasing function of both average cycle time and the standard deviation of cycle time. Larger standard deviation of cycle time means we will have to quote longer lead times, and consequently must hold items in finished goods inventory longer, to compensate for the variable production rate. As Figure 12.7 shows, the standard deviation of cycle time increases in the rework rate. Moreover, it increases in the WIP level (as there is more WIP in the line to cause random queueing delays at the stations). Since, as we noted, rework requires additional WIP in the line to achieve a given throughput level, this effect tends to aggravate further the cycle time variability problem. This is an illustration of the rework law with regard to variance of cycle time.

The results of Figures 12.5, 12.6, and 12.7 imply the following about the operations and cost impacts of quality problems.

1. *Throughput effects.* If the rework is high enough to cause a resource to become a bottleneck (or, even worse, the rework problem is *on* the bottleneck resource), it can substantially alter the capacity of the line. Where this is the case, a quality improvement can facilitate an increase in throughput. The increased revenue from such an improvement can *vastly* exceed the cost of improving quality in the line.

2. *WIP effects.* Rework on a nonbottleneck resource, even one that has plenty of spare capacity, increases variability in the line, thereby requiring higher WIP (and cycle time) to attain a given level of throughput. Thus, reductions in rework can facilitate reductions in WIP. Although the cost savings from such a change are not likely to be as large as the revenue enhancement from a capacity increase, they can be significant relative to the cost of achieving the improvement.

FIGURE 12.7

Standard deviation of cycle time versus WIP for different rework rates

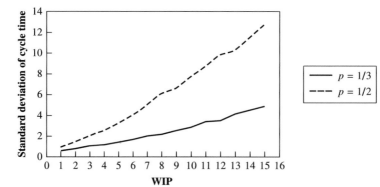

3. *Lead time effects.* By decreasing capacity and increasing variability, rework problems necessitate additional WIP in the line and hence lead to longer average cycle times. These problems also increase the variability of cycle times and hence lead to either longer quoted lead times or poorer service to the customer. The competitive advantage of shorter lead times and more reliable delivery, achieved via a reduction in rework, is difficult to quantify precisely but can be of substantial strategic importance.

Further Observations.   We conclude our discussion on the operations impacts of quality problems with some observations that go beyond the preceding examples.

To begin, we note that *the longer the rework loop, the more pronounced the consequences.* In the two examples above, we represented rework as a second pass through a single machine. In practice, rework is frequently much more involved than this. A defective part may have to loop back through several stations in the line in order to be corrected. When this is the case, rework affects the capacity and variability of effective processing time on several stations. Additionally, because each pass through the rework loop adds even more time than in the single-machine rework loop case, the effect on the standard deviation of cycle time tends to be larger. As a result, the consequences of the rework law become even more pronounced as the length of the rework loop grows.

Because rework has such a disruptive effect on a production line, manufacturing managers are frequently tempted to set up separate rework lines. Such an approach does prevent defective parts from sapping capacity and inflating variability in the main line. However, it does this by installing extra capacity somewhere else, which costs money, takes up space, and does little to eliminate the inflation of the mean and standard deviation of cycle time caused by rework. Even worse, such an approach can serve to sweep quality problems under the rug. Shunting defective parts to a separate line makes them someone else's responsibility. Making a line responsible for correcting its own problems fosters greater awareness of the causes and effects of quality problems. If such awareness can lead to quicker detection of problems, it can shorten the rework loop and mitigate the consequences. If it can lead to ways to avoid the defects in the first place, then truly major improvements can be achieved. Consequently, despite the short-term appeal of separate rework lines, it is probably better in the long run to avoid them and strive for more fundamental quality improvements.

In many manufacturing environments, internal quality problems lead to **scrap**—that is, yield loss—rather than rework, either because the defect cannot be corrected or because it is not economical to do so. Thus, it is important to point out that *scrap has similar effects to rework.* From an operations standpoint, scrapped parts are essentially identical to reworked parts that must be processed again from the beginning of the line. In this sense, scrap is the most extreme form of rework and therefore has the same effects we observed for rework, only more so.

A difference between scrap and rework, however, lies in the method used to compensate. While separate lines can be used for rework, they make no sense as a remedy for scrap. Instead, most manufacturing systems perform some form of job size inflation as protection against yield loss. (We first discussed this approach in Chapter 3 in the context of MRP but will review it again here in the context of quality and operations.) The most obvious approach is to divide the desired quantity by the expected yield rate. For example, if we have an order for 90 parts and the yield rate is 90 percent (i.e., a 10 percent scrap rate), then we could release

$$\frac{90}{0.9} = 100$$

units. Then if 10 percent are lost to scrap, we will have 90 good parts to ship to the customer.

This approach would be fine if the scrap rate were truly a deterministic constant (i.e., we *always* lose 10 percent). But in virtually all real situations, the scrap rate for a given job is a random quantity; it might range from 0 to 100 percent. When this is the case, it is not at all clear that inflating by the expected yield rate is the best approach. For instance, in the previous example, suppose the *expected* scrap rate is 90 percent, but what really happens is that 90 percent of the time the yield for a given job is 100 percent (no yield loss) and the other 10 percent of the time it is 0 percent (catastrophic yield loss). If we inflate by dividing the amount demanded by the customer by 0.9, then 90 percent of the time we will wind up with excess and the other 10 percent of the time we will be short. In this extreme case, job inflation does not improve customer service at all!

When too little good product finishes to fill an order, we must start additional parts and wait for them to finish before we can ship the entire amount to the customer. That is, it is similar to a rework loop that encompasses the entire line. Unless we have built in substantial lead time to the customer, this is likely to result in a late delivery. The costs to the firm are the (hard to quantify) cost of lost customer goodwill and the cost of disrupting the line to rush the makeup order through the line.

On the other hand, when low yield loss results in more good product finishing than required to fill an order, the excess will go into finished goods inventory (FGI) and be used to fill future orders. The cost to the firm is that incurred to hold the extra inventory in FGI. Of course, if all products are customized and cannot be used against future demand, the extra inventory will amount to scrap.

At any rate, there is no reason to expect the cost of being short on an order by n units to be equal to that of being over it by n units. In most cases, the cost of being short exceeds that of being over. Hence, from a cost minimization standpoint, it might make sense to inflate by *more* than the expected yield loss. For instance, in a situation where yield varies between 80 and 100 percent, we might divide the amount demanded by 0.85 instead of 0.9, so that we release 106 parts instead of 100 to cover an order of 90. This would allow us to ship on time as long as the yield loss was not greater than 15 percent.

But in cases where yield loss is frequently all or nothing (e.g., we get either 100 good parts or none from a release quantity of 100), inflating job size is generally futile. (We would have to start an entire second job of 100 parts to make up for the catastrophic failure of the first batch.) A more practical alternative is to carry safety stock in finished goods inventory; for example, we try to carry n jobs' worth of FGI, where n is the number of scrapped jobs we want to be able to cover. In a system with many products, this can require considerable (expensive) inventory.

The unavoidable conclusion is that scrap loss caused by variable yields is costly and disruptive. The more variable the yields, the more difficult it is to mitigate the effect with inflated job sizes or safety stocks. Thus, in the long term, the best option is to strive to minimize or eliminate scrap and rework.

12.4.2   Operations Supports Quality

The previous subsection stressed that better quality promotes better operations. Happily, the reverse is also frequently true. As pointed out frequently in the JIT literature, to the extent that tighter operations management leads to less WIP (i.e., shorter queues), it aids in the detection of quality problems and facilitates tracing them to their source.

Specifically, suppose that there tends to be a great deal of WIP between a point in a production line that causes defects and the point where these defects are detected. The defects might be caused by a machine early in the line because it has imperceptibly gone "out of control" but not be detected until an end-of-line (EOL) test. By the time a defect is detected at the EOL test, it is likely that all the parts that have been produced by the upstream machine are similarly defective. If the line has a high WIP level in it, scrap loss could be large. If the line has little WIP, scrap loss is likely to be much less.

Of course, in the real world, causes and detection of defects are considerably more complex and varied than this. There are likely to be many sources of potential defects, some of which have never been encountered before—or at least, for which there is no institutional memory. Detection of defects can occur at many places in the line, both at formal inspection points and as a result of informal observations elsewhere. While these realities serve to make understanding and managing quality a challenge, they do not alter the main point: High WIP levels tend to aggravate scrap loss by increasing the time, and hence number of items produced, between the cause and the detection of a defect.

Example: A Defect Detection

Consider again the CONWIP line depicted in Figure 12.4, only this time suppose that the rework rate at machine 2 is zero. Instead, suppose that each time a job is processed on machine 1, there is a probability q that this machine goes out of control and produces bad parts until it is fixed. However, the out-of-control status of machine 1 can only be inferred by detecting the bad parts, which does not occur until after the parts have been processed at machine 4. Each time a defective part is detected, we assume that machine 1 is corrected instantly. But all the parts that have been produced on machine 1 between the time it went out of control and the time the defect was detected at machine 4 will be defective and must be scrapped at the end of the line.

Figure 12.8 illustrates the curve of throughput (of good parts only) versus WIP for four cases of this example. First, when $q = 0$ (no quality problems) and all processing times are deterministic, we get the familiar best-case curve. Second, for comparison, we plot throughput versus WIP when $q = 0$ but processing times are exponential (i.e., they have CVs of 1). Here, throughput increases with WIP, reaching nearly maximum output at around 15 jobs. Note that this curve is somewhat better than (i.e., lies above) the practical worst case due to the imbalance in the line.

However, when $q = 0.05$ and processing times are deterministic, throughput increases and then declines with WIP. The reason, of course, is that for high WIP levels, the increased scrap loss outweighs the higher production rate it promotes. The maximum throughput occurs at a WIP level of three jobs, the critical WIP level. When $q = 0.05$

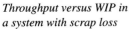

FIGURE 12.8

Throughput versus WIP in a system with scrap loss

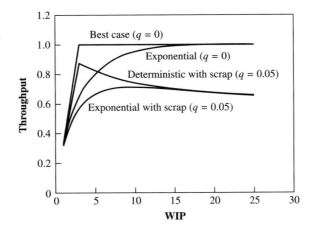

and processing times are exponential, throughput again increases and then decreases, with maximum throughput being achieved at a WIP level of nine jobs. Notice that while we can make up for the variability induced by random processing times by maintaining a high WIP level (for example, 15 jobs), the variability due to scrap loss is only aggravated by more WIP. So instead of putting more WIP in the system to compensate, we must *reduce* the WIP level to mitigate this second form of variability and thereby maximize throughput. Metaphorically speaking, this is like lowering the water to cover the rocks. Obviously, metaphors have their limits.

It is our guess that in real life, throughput-versus-WIP curves frequently do exhibit this increasing-then-decreasing type of behavior, not only because of poor quality detection but also because high WIP levels make it harder to keep track of jobs, so that more time is wasted locating jobs and finding places to put them between processes. Moreover, more WIP leads to more chances for damage. In general, we can conclude that better operations (i.e., tighter WIP control) leads to better quality (less scrap loss) and hence higher throughput (better operations again). This is a simple illustration of the fact that quality and operations are mutually supportive and therefore can be jointly exploited to promote a cycle of continual improvement.

12.5   Quality and the Supply Chain

Total quality management refers to quality outside, as well as inside, the walls of the plant. Under the topic of vendor certification (e.g., ISO 9000), the TQM literature frequently mentions the **supply chain:** the network of plants and vendors that supply raw material, components, and services to one another. Almost all plants today rely on outside suppliers for at least some of the inputs to their manufacturing process. Indeed, the tendency in recent years has been toward *vertical deintegration* through outsourcing of an increasing percentage of manufactured components.

When significant portions of a finished product come from outside sources, it is clear that internal, and perhaps external, quality at the plant can depend critically on these inputs. As computer programmers say, "garbage in, garbage out." (Or as farmers say, "you can't make a silk purse out of a sow's ear.") Whatever the metaphor, the point is that a TQM program must address the issue of purchased parts if it is to be effective. Vendor certification, working with fewer vendors, using more than price to choose between vendors, and establishing quality assurance procedures as close to the front of the line as possible—all are options for improving purchased part quality. The choice and character of these policies obviously depend on the setting. We refer the reader to the previously cited TQM references for more in-depth discussion.

Just as internal scrap and rework problems can have significant operations consequences, quality problems from outside suppliers can have strong impacts on plant performance. First, any defects in purchased parts that find their way into the production process to cause scrap or rework problems will affect operations in the fashion we have discussed. However, even if defective purchased parts are screened out before they reach the line, either at the supplier plant or at the receiving dock, these quality problems can still have negative operational effects. The reason is that they serve to inflate the *variability of delivery time.* If scrap or rework problems at the supplier plant cause some orders to be delivered late, or if some orders must be sent back because quality problems were detected upon receipt, the effective delivery time (i.e., the time between submission of a purchase order and receipt of acceptable parts) will not be regular and predictable.

12.5.1   A Safety Lead Time Example

To appreciate the effects of variable delivery times for purchased parts, consider the following example. A plant has decided to purchase a particular part from one of two suppliers on a lot-for-lot basis. That is, the company will not buy the part in bulk and stock it at the plant, but instead will bring in just the quantities needed to satisfy the production schedule. If the part is late, the schedule will be disrupted and customer deliveries may be delayed. Therefore, management chooses to build a certain amount of **safety lead time** into the purchasing lead time. The result is that, on average, parts will arrive somewhat early and wait in raw materials inventory until they are needed at the line. The key question is, How much safety lead time is required?

Figure 12.9 depicts the probability density functions (pdf's) for the delivery time from the two candidate suppliers. Both suppliers have mean delivery times of 10 days. However, deliveries from supplier 2 are much more variable than those from supplier 1 (perhaps because supplier 2 does not have sound OM and TQM systems in place). As a result, to be 95 percent certain that an order will arrive on time (i.e., when required by the production schedule), parts must be ordered with a lead time of 14 days from supplier 1 or a lead time of 23 days from supplier 2 (see Figure 12.10). The additional lead time is required for supplier 2 to make up for the variability in delivery time. Notice that this implies that an average order from supplier 1 will wait in raw materials inventory for $14 - 10 = 4$ days, while an average order from supplier 2 will wait in raw materials inventory for $23 - 10 = 13$ days—an increase of 225 percent. From Little's law, we know that raw materials inventory will also be 225 percent larger if we purchase from supplier 2 rather than from supplier 1.

12.5.2   Purchased Parts in an Assembly System

The effects of delivery time variability become even more pronounced when assemblies are considered. In many manufacturing environments, a number of components are purchased from different suppliers for assembly into a final product. To avoid a schedule disruption, *all* the components must be available on time. Because of this, the amount of safety lead time needed to achieve the same probability of being able to start on time is larger than it would be if there were only a single purchased component.

To see how this works, consider an example in which a product is assembled from 10 components, all of which are purchased from separate vendors and have the same

FIGURE 12.9

Effect of delivery time variability on purchasing lead times

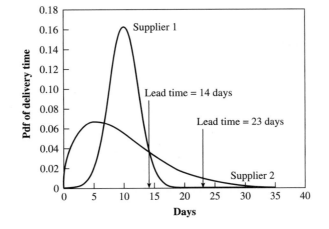

FIGURE 12.10

Setting safety lead times in multiple-component systems

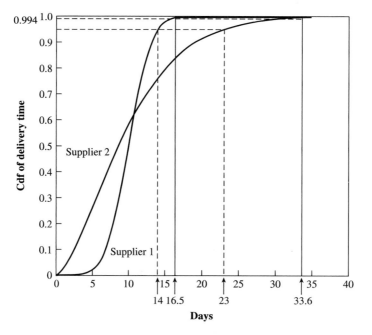

distribution (i.e., mean and variance) of delivery time. Since the parts are identical with regard to their delivery characteristics, it is sensible to choose the same purchasing lead time for all. Suppose this is done as in the previous single-component example so that each component has a 95 percent chance of being received on time. Assuming delivery times of the different components to be independent, the probability that all are on time is given by the *product* of the individual on-time probabilities

$$\text{Prob\{all 10 components arrive on time\}} = (0.95)^{10} = 0.5987$$

Assembly will be able to start on time less than 60 percent of the time!

Obviously, the plant needs longer lead times and higher individual on-time probabilities to achieve the desired 95 percent likelihood of having all components in when required by the schedule. Specifically, if we let p represent the on-time percentage for a single part, we want

$$p^{10} = 0.95$$

or $$p = 0.95^{1/10} = 0.9949$$

To ensure that the entire set of parts is available 95 percent of the time, each individual part must be available 99.49 percent of the time.

To see the operations effects of this, consider Figure 12.10, which shows the cumulative distribution function (cdf) of the delivery times from supplier 1.[7] This curve gives the probability that the delivery time is less than or equal to t for all values of t. For a single component to be available 95 percent of the time, a purchasing lead time of 14 days (i.e., a safety lead time of four days) is sufficient. However, for a single component to be available 99.49 percent of the time, in order to support the 10-component assembly system, a purchasing lead time of 16.3 days (i.e., a safety lead time of 6.3 days) is needed. Thus, purchased parts will reside in raw materials inventory for an additional 2.3 days

[7]The cdf is simply the area under the pdf shown in Figure 12.9 from 0 to t.

on average in the multicomponent assembly system, and therefore the raw materials inventories will be increased by a corresponding amount.

Since multiple-component systems require high individual on-time probabilities, the tails of the delivery time distributions are critical. For instance, the purchasing lead time required for supplier 2 in Figure 12.10 to achieve a 99.49 percent probability of on-time delivery is 33.6 days. Recall that in the single-component case, there was a difference of nine days between the required lead times for suppliers 1 and 2 (that is, 14 days for supplier 1 and 23 days for supplier 2). In the 10-component case, there is a difference of $33.6 - 16.3 = 17.3$ days. The conclusion is that reliable suppliers are *extremely* important to efficient operation of an assembly system that involves multiple purchased parts.

12.5.3   Vendor Selection and Management

The preceding discussion has something (though far from everything) to say about the problem of supplier selection. To see what, suppose components are purchased from two separate suppliers. Each has a probability p of delivering on time, so that the probability of receiving both parts on time is p^2. Now, further suppose that both parts could be purchased from a single vendor. If that vendor could provide better on-time performance than p^2 for the combined shipments, then, all other things being equal, it would be better to switch to the single vendor. Even if the purchasing cost is higher when using the single vendor, the savings in inventory and schedule disruption costs may justify the switch. Having fewer vendors providing multiple parts might produce better on-time performance than having many vendors providing single parts, for these reasons:

1. Purchases become a larger percentage, and therefore a higher-priority piece, of the supplier's business.
2. The purchasing department can keep better track of suppliers (by knowing about special circumstances that would alter the usual purchasing lead times, by being able to place "reminder" phone calls, etc.) if there are fewer of them.

The insights from these simplified examples extend to more realistic systems. Obviously, in the real world, suppliers do not have identical delivery time distributions, nor are the costs of the different components necessarily similar. For these reasons, it may make sense to set the on-time delivery probabilities differently for different components. An inexpensive component (e.g., a resistor) should probably have a very high on-time probability because the inventory cost of achieving it is low.[8] An expensive component (e.g., a cathode-ray tube display) should have a relatively lower on-time probability, in order to reduce its safety lead time and hence average inventory level. The general idea is that if a schedule disruption is going to occur, it ought to be due to a $500 cathode-ray tube, not a 2-cent resistor.

Formal algorithms exist for computing appropriate safety lead times in assembly systems with multiple nonidentical purchased components (see Hopp and Spearman 1993). But whether we use algorithms or less rigorous methods to establish safety lead times for the individual components, the result will be to set an on-time probability for each component. As our previous discussion of Figure 12.9 illustrated, for a fixed

---

[8]Actually, for really inexpensive items that are used with some regularity, it makes sense to simply order them in bulk and stock them on site to ensure that they are virtually never out of stock. However, this advice does not apply to bulky materials (e.g., packaging) for which the cost of storage space and handling makes large on-site stocks uneconomical.

on-time probability, safety lead time and raw materials inventory are both increasing in the variance of supplier delivery time. Moreover, as we observed in Figure 12.10, the more independent suppliers we order from, the higher the individual on-time probabilities required to support a given probability of maintaining schedule.

This discussion can be thought of as a quick factory physics interpretation of the JIT view on vendoring. The JIT literature routinely suggests certifying a smaller number of vendors, precisely because low delivery time variance is needed to support just-in-time deliveries. Indeed, Toyota has evolved a very extensive system of working with its suppliers that goes well beyond simple certification—to the point of sending in advisers to set up the "Toyota system," which addresses both quality and operations, in the supplier's plant. The goal is to nurture suppliers that effectively support Toyota's operation *and* are efficient enough to remain economically viable partners over the long term.

12.6   Conclusions

Quality is a broad and varied subject, which ranges from definitions of customer needs to analytical measurement and maintenance tools. In this chapter, we have tried to give a sense of this range and have suggested references for the interested reader to consult for additional depth. In keeping with the factory physics framework of this book, we have concentrated primarily on the relationship between quality and operations and have shown that the two are intimately related in a variety of ways. Specifically, we have argued the following:

1. *Good quality supports good operations.* Reducing recycle and/or scrap serves to increase capacity and decrease congestion. Thus, better quality control—through tighter control of inputs, mistake prevention, and earlier detection—facilitates increased throughput and reduced WIP, cycle time, and customer lead time.

2. *Good operations supports quality improvement.* Reducing WIP—via better scheduling, pull mechanisms for shop floor control, or (although it is hardly an imaginative option) capacity increases—serves to reduce the amount of product generated between the cause of a defect and its detection. This has the potential to reduce the scrap and rework rate and to help identify the root causes of quality problems.

3. *Good quality at the supplier level promotes good operations and quality at the plant level.* A supplier plant with fewer scrap, rework, and external quality problems will make more reliable deliveries. This enables a customer plant to use shorter purchasing lead times for these parts (e.g., just-in-time becomes a possibility), to carry smaller raw materials inventories, and to avoid frequent schedule disruption.

Based on these discussions, we conclude that both quality and operations are integral parts of a sound manufacturing management strategy. One cannot reasonably consider one without the other. Hence, perhaps we should really view total quality management more in terms of *quality of management* than *management of quality*.

Study Questions

1. Why is quality so difficult to define? Provide your own definition for a specific operation of your choosing.
2. Give three major ways that good internal quality can promote good external quality.

3. Using the following definition of the cost of quality

> *Quality costs are defined as any expenditures on manufacturing or service in excess of those that would have been incurred if the product had been built or the service had been performed exactly right the first time.*

Garvin (1988, 78)

identify the costs associated with each of the following types of quality problems:

a. A flow line with a single-product family where defects detected at any station are scrapped.

b. A flow line with a single-product family where defects detected at any station are reworked through a portion of the line.

c. A cutting machine where bit breakage destroys the part in production and brings the machine down for repair.

d. Steel burners for a kitchen range that are coated with a porcelain that cracks off after a small amount of use in the field.

e. A minivan whose springs for holding open the hatchback are prone to failure.

f. A cheap battery in new cars and light trucks that fails after about 18 months when the warranty period is 12 months.

4. For each of the following examples, would you expect cost to increase or decrease with quality? Explain your reasoning.

a. An automobile manufacturer increases expected battery life by installing more expensive batteries in new cars.

b. A publisher reduces the number of errors in newly published books by assigning extra proofreaders.

c. A steel rolling mill improves the consistency of its galvanizing process through installation of a more sophisticated monitoring system (i.e., that measures temperature, pH, etc., at various points in the chemical bath).

d. A manufacturer of high-voltage switches eliminates quality inspection of metal castings after certifying the supplier from which they are purchased.

e. An automobile manufacturer repairs an obvious defect (e.g., a defective paint job) after the warranty period has expired.

5. What quality implications could setup time reduction have in a manufacturing line?

6. How might improved internal quality make scheduling a production system easier?

7. Why do the operational consequences of rework become more severe as the length of the rework loop increases?

8. How are the operational consequences of rework similar to those of scrap? How are they different?

9. Why is it important to detect quality problems as early in the line as possible?

Problems

1. Manov Steel, Inc., has a rolling mill that produces sheet steel with a nominal thickness of 0.125 inch. Suppose that the specification limits are given by LSL $= 0.120$ and USL $= 0.130$ inch. Based on historical data, the actual thickness of a random sheet produced by the mill is normally distributed with mean and standard deviation of $\mu = 0.125$ and $\sigma = 0.0025$.

a. What are the lower and upper natural tolerance limits (LNTL and UNTL) for individual sheets of steel?

b. What are the lower and upper control limits (LSL and USL) if we use a control chart that plots the average thickness of samples of size $n = 4$?

c. What will be the percentage nonconforming, given the above values for (LNTL, UNTL) and (LSL, USL)? What is the process capability index C_{pk}? Do you consider this process capable of meeting its performance specifications?

d. Suppose that the process mean suddenly shifts from 0.125 to 0.1275. What happens to the process capability index C_{pk} and the percentage nonconforming?

e. Under the conditions of *d*, what is the probability that the $\bar{x}$ chart specified in *b* will detect an out-of-control signal on the first sample after the change in process mean?

2. A purchasing agent has requested quotes for valve gaskets with diameters of 3.0 ± 0.018 in. SPC studies of three suppliers have indicated that their processes are in statistical control and produce measurements that are normally distributed with the following statistics:

$$\text{Supplier 1: } \mu = 3 \text{ inches} \qquad \sigma = 0.009 \text{ inch}$$
$$\text{Supplier 2: } \mu = 3 \text{ inches} \qquad \sigma = 0.0044 \text{ inch}$$
$$\text{Supplier 3: } \mu = 2.99 \text{ inches} \qquad \sigma = 0.003 \text{ inch}$$

Assuming that all suppliers offer the same price and delivery reliability/flexibility, which supplier should the agent purchase from? Explain your reasoning.

3. Consider a single machine that requires one hour to process parts. With probability p, a given part must be reworked, which requires a second one-hour pass through the machine. However, all parts are guaranteed to be good after a second pass, so none go through more than twice.

a. Compute the mean and variance of the effective processing time on this machine as a function of p.

b. Use your answer from *a* to compute the squared coefficient of variation (SCV) of the effective processing times. Is it an increasing function of p? Explain.

4. Suppose the machine in Problem 1 is part of a two-station line, in which it feeds a second machine that has processing times with a mean of 1.2 hours and SCV of 1. Jobs arrive to the line at a rate of 0.8 job per hour with an arrival SCV of 1.

a. Compute the expected cycle time in the line when $p = 0.1$.

b. Compute the expected cycle time in the line when $p = 0.2$.

c. What effects does rework have on cycle time, and how do these differ in *a* and *b*?

5. Suppose a cellular telephone plant purchases electronic components from various suppliers. For one particular component, the plant has a choice between two suppliers: Supplier 1 has delivery lead times with a mean of 15 days and a standard deviation of 1 day, while supplier 2 has delivery lead times with a mean of 15 days and a standard deviation of 5 days. Both suppliers can be assumed to have normally distributed lead times.

a. Assuming that the cellular plant purchases the component on a lot-for-lot basis and wants to be 99 percent certain that the component is in stock when needed by the production schedule, how many days of lead time are needed if supplier 1 is used? Supplier 2?

b. How many days will a typical component purchased from supplier 1 wait in inventory before being used? From supplier 2? How might this information be used to justify using supplier 1 even if it charges a higher price?

c. Suppose that the cellular plant purchases (on a lot-for-lot basis) 100 parts from different suppliers, all of which have delivery times like those of supplier 1. Assuming all components are assigned the same lead time, what lead times are required to ensure that *all* components are in stock when required by the schedule? How does your answer change if all suppliers have lead times like those of supplier 2?

d. How would your answer to *a* be affected if, instead of ordering lot for lot, the cellular plant ordered the particular component in batches corresponding to five days' worth of production?

6. Consider a workstation that machines castings into switch housings. The castings are purchased from a vendor and are prone to material defects. If all goes well, machining (including load and unload time) requires 15 minutes, and the SCV of natural processing time (due to variability in the time it takes the operator to load and start the machine) is 0.1. However, two types of defect in the castings can disrupt the process.

One type of defect (a flaw) causes the casting to crack during machining. When this happens, the casting is scrapped at the end of the operation and another casting is machined. About 15 percent of castings have this first type of defect.

A second type of defect (a hard spot) causes the cutting bit to break. When this happens, the machine must be shut down, must wait for a repair technician to arrive, must be examined for damage, and must have its bit replaced. The whole process takes an average of two hours, but is quite variable (i.e., the standard deviation of the repair time is also two hours). Furthermore, since the casting must be scrapped, another one must be machined to replace it once the repair is complete. About five percent of castings have this second type of defect.

a. Compute the mean and SCV of effective process time (i.e., the time it takes to machine a good housing). (*Hint:* Use Equations (12.15) and (12.17) to consider the effects of the first type of defect, and consult Table 8.1 for formulas to address the second type of defect. *Question:* Should stoppages due to the second type of defect be modeled as preempt or nonpreempt outages?)

b. How does your answer to *a* change if the defect percentages are reversed (that is, five percent of castings have the first type of defect, while 15 percent have the second type)? What does this say about the relative disruptiveness of the two types of defects?

c. Suppose that by feeding the castings through the cutting tool more slowly, we could ensure that the second type of defect does not cause bit breakage. Under this policy, castings with the second type of defect will be scrapped, but will not cause any machine downtime (i.e., they become identical to the first type of defect). However, this increases the average time to machine a casting without defects from 15 minutes to t minutes. What is the maximum value of t for which the slower feed speed achieves at least as much capacity as the original situation in *a*?

d. Which workstation would you rather manage, that in *a* (i.e., fast feeds and bit breakages) or that in *c* (i.e., slow speeds, resulting in machining times equal to your answer to *c*, and no bit breakages)? (*Hint:* How do the effective SCVs of the two cases compare?)

III PRINCIPLES IN PRACTICE

In matters of style, swim with the current;
In matters of principle, stand like a rock.
Thomas Jefferson

13   A PULL PLANNING FRAMEWORK

We think in generalities, we live in detail.
Alfred North Whitehead

13.1   Introduction

Recall that we began this book by stating that the three critical elements of an operations management education are

1. Basics
2. Intuition
3. Synthesis

We spent almost all Parts I and II on the first two items. For instance, the tools and terminology introduced in Part I (for example, EOQ, (Q, r), BOM, MPS) and the measures of variability (e.g., coefficient of variation) and elementary queueing concepts presented in Part II are *basics* of fundamental importance to the manufacturing manager. The insights from traditional inventory models, MRP, and JIT we observed in Part I and the factory physics relationships among throughput, WIP, cycle time, and variability we developed in Part II are key components of sound *intuition* for making good operating decisions.

But, with the exception of a bit of integration of the contrasting perspectives of operations and behavioral science in Chapter 11 and the pervasive aspects of quality presented in Chapter 12, we have devoted almost no time to the third item, **synthesis.** We are now ready to fill in this important gap by establishing a framework for applying the principles from Parts I and II to real manufacturing problems.

Our approach is based on two premises:

1. Problems at different levels of the organization require different levels of detail, modeling assumptions, and planning frequency.
2. Planning and analysis tools must be consistent across levels.

The first premise motivates us to use separate tools for separate problems. Unfortunately, using different tools and procedures throughout the system can easily bring us into conflict with the second premise. Because of the potential for inconsistency, it is not uncommon to find planning tools in industry that have been extended across applications for which they are ill suited. For instance, we once worked in a plant that used a scheduling tool that calculated detailed, *minute-by-minute* production on each machine in the plant to generate *two-year* aggregate production plans. Although this tool may have been reasonable for short-term planning (e.g., a day or a week), it was far too cumbersome to run for long-term purposes (the data input and debugging alone took an entire week!). Moreover, it was so inaccurate beyond a few weeks into the future that the schedule, so painfully obtained, was virtually ignored on the plant floor.

To develop methods that are both well suited to their specific application *and* mutually consistent across applications, we recommend the following steps in developing a planning framework:

1. **Divide the overall system appropriately.** Different planning methods for different portions of the process, different product categories, different planning horizons, different shifts, etc., can be used. The key is to find a set of divisions that make each piece manageable, but still allow integration.

2. **Identify links between the divisions.** For instance, if production plans for two products with a shared process center are made separately, they should be linked via the capacity of the shared process. If we use different tools to plan production requirements over different time horizons, we should make sure that the plans are consistent with regard to their assumptions about capacity, product mix, staffing, etc.

3. **Use feedback to enforce consistency.** All analysis, planning, and control tools make use of estimated parameters (e.g., capacity, machine speeds, yields, failure and repair rates, demand rates, and many others). As the system runs, we should continually update our knowledge of these values. Rather than allow the inputs to the various tools to be estimated in an ad hoc, uncoordinated fashion, we should explicitly make use of our updated knowledge to force tools to make use of timely, consistent information.

In the remainder of this chapter, we preview a planning framework that is consistent with these principles, as well as the factory physics principles presented earlier. We do not pretend that this framework is the only one that is consistent with these principles. Rather, we offer it as one approach and try to present the issues involved at the various levels from a sufficiently broad perspective as to allow room for customization to specific manufacturing environments. Subsequent chapters in Part III will flesh out the major components of this framework in greater detail.

13.2   Disaggregation

The first step in developing a planning structure is to break down the various decision problems into manageable subproblems. This can be done explicitly, through the development of a formal planning hierarchy, as we will discuss. Or it can be done implicitly by addressing the various decisions piecemeal with different models and assumptions. Regardless of the level of foresight, some form of disaggregation *will* be done, since all real-world production systems are too complex to address with a single model.

13.2.1   Time Scales in Production Planning

One of the most important dimensions along which manufacturing systems are typically broken down is that of *time*. The primary reason for this is that manufacturing decisions differ greatly with regard to the length of time over which their consequences persist.

For example, the construction of a new plant will affect a firm's position for years or even decades, while the effects of selecting a particular part to work on at a particular workstation may evaporate within hours or even minutes. This makes it essential to use different **planning horizons** in the decision-making process. Since the decision to construct a new plant will influence operations for years, we must forecast these effects years into the future in order to make a reasonable decision. Hence, the planning horizon should be long for this problem. Clearly, we do not need to look nearly so far into the future to evaluate the decision of what to work on at a workstation, so this problem will have a short planning horizon.

The appropriate length of the planning horizon also varies across industries and levels of the organization. Some industries, oil and long-distance telephone, for example, routinely make use of horizons as long as several decades because the consequences of their business decisions persist this long. Within a given company, longer time horizons are generally used at the corporate office, which is responsible for long-range business planning, than at the plant where day-to-day execution decisions are made.

In this book we focus primarily on decisions relevant to running a plant, and we divide planning horizons in this context into **long**, **intermediate**, and **short**. At the plant level, a long planning horizon can range from one to five years with two years being typical. An intermediate planning horizon can range from a week to a year, with a month being typical. A short time horizon can range from an hour to a week, with a day being typical.

Table 13.1 lists various manufacturing decisions that are made over long, intermediate, and short planning horizons. Notice that in general, long-range decisions address **strategy,** by considering such questions as what to make, how to make it, how to finance

TABLE 13.1   Strategy, Tactics, and Control Decisions

Time Horizon	Length	Representative Decisions
Long term (strategy)	Year to decades	Financial decisions Marketing strategies Product designs Process technology decisions Capacity decisions Facility locations Supplier contracts Personnel development programs Plant control policies Quality assurance policies
Intermediate term (tactics)	Week to year	Work scheduling Staffing assignments Preventive maintenance Sales promotions Purchasing decisions
Short term (control)	Hour to week	Material flow control Worker assignments Machine setup decisions Process control Quality compliance decisions Emergency equipment repairs

it, how to sell it, where to make it, and where to get materials and general principles for operating the system. Intermediate-range decisions address **tactics,** by determining what to work on, who will work on it, what actions will be taken to maintain the equipment, what products will be pushed by sales, and so on. These tactical decisions must be made within the physical and logical constraints established by the strategic long-range decisions. Finally, short-range decisions address **control,** by moving material and workers, adjusting processes and equipment, and taking whatever actions are required to ensure that the system continues to function toward its goal. Both the long-term strategic and intermediate-range tactical decisions establish the constraints within which these control decisions must be made.

Different planning horizons imply different **regeneration frequencies.** A long-range decision that is based on information extending years into the future does not need to be reconsidered very often, because the estimates about what will happen this far into the future do not change very fast. For instance, while it is a good thing for a plant to reevaluate what products it should be making, this is not a decision that should be reconsidered every week. Typically, long-range problems are considered on a quarterly to annual basis, with very long-range issues (e.g., what business should we be in?) being considered even less frequently. Intermediate-range problems are reconsidered on roughly a weekly to monthly basis. Short-range problems are reconsidered on a real-time to daily basis. Of course, these are merely typical values, and considerable variation occurs across firms and decision problems.

In addition to differing with respect to regeneration frequency, problems with different planning horizons differ with respect to the required *level of detail*. In general, the shorter the planning horizon, the greater the amount of detail required in modeling and data collection. For instance, if we are making a long-term strategic capacity decision about what size plant to build, we do not need to know very much about the routings that parts will take. It may be enough to have a rough estimate of how much time each part will require of each process, in order to estimate capacity requirements. However, at the intermediate tactical level, we need more information about these routings, for instance, which specific machines will be visited, in order to determine whether a given schedule is actually feasible with respect to customer requirements. Finally, at the short-term control level, we may need to know a great deal about part routings, including whether or not a given part requires rework or other special attention, in order to guide parts through the system.

A good analogy for this strategy/tactics/control distinction is mapmaking. Long-term problems are like long-distance travel. We require a map that covers a large amount of distance, but not in great detail. A map that shows only major highways may be adequate for our needs. Likewise, a long-term decision problem requires a tool that covers a large amount of time (i.e., long planning horizon), but not in great detail. In contrast, short-term problems are like short-distance travel. We require a map that does not cover much distance, but gives lots of details about what it does cover. A map showing city streets, or even individual buildings, may be appropriate. Analogously, for a short-term decision problem, we require a tool that does not cover much time (i.e., short planning horizon), but gives considerable detail about what it does cover.

13.2.2   Other Dimensions of Disaggregation

In addition to time, there are several other dimensions along which the production planning and control problem is typically broken down. Because modern factories are large and complex, it is frequently impossible to consider the plant as a whole when one is making specific decisions. The following are three dimensions that can be used to break the plant into more manageable pieces for analysis and management:

1. **Processes.** Traditionally, many plants were organized according to physical manufacturing processes. Operations such as casting, milling, grinding, drilling, and heat treat were performed in separate departments in distinct locations and under different management. While such process organization has become less popular in the wake of the JIT revolution, with its flow-oriented cellular layouts, process divisions still exist. For instance, casting is operationally very different, and sometimes physically distant, from rolling in a steel mill. Likewise, mass lamination of copper and fiberglass cores in large presses is distinct—physically, operationally, and logistically—from the circuitizing process in which circuitry is etched into the copper in a photo-optical/chemical flow line process. In such situations, it frequently makes sense to assign separate managers to the different processes. It may also be reasonable to use different planning, scheduling, and control procedures.

2. **Products.** Although plants dedicated to a single product exist (e.g., a polystyrene plant), most plants today make multiple products. Indeed, the pressure to compete via variety and customization has probably served to increase the average number of different products produced by an average plant. For instance, it is not uncommon to find a plant with 20,000 distinct part numbers (i.e., counting finished products and subcomponents). Because it is difficult, under these conditions, to consider part numbers individually, many manufacturing plants aggregate part numbers into coarser categories for planning and management purposes.

One form of aggregation is to lump parts with identical routings together. Typically, there are many fewer routings through the plant than there are part numbers. For instance, a printed-circuit board plant, which produces several thousand different circuit boards, may have only two **basic routings** (e.g., for small and large boards). Frequently, however, the actual number of routings can be substantially larger than the number of basic routings if one counts minor variations (e.g., extra test steps, vendoring of individual operations, and gold plating of contact surfaces) in the basic routing. For planning, it is generally desirable to keep the number of "official" routings to a minimum by ignoring minor variations.

In systems with significant setup times, aggregation by routing may be going too far. For instance, a particular routing in a circuit board line may produce 1,000 different circuit boards. However, there may be only four different thicknesses of copper. Since the speed of the conveyor must be changed with thickness (to ensure proper etching), a setup involving lost capacity must be made whenever the line switches thicknesses. In addition, the 1,000 boards may require three different dies for punching rectangular holes in the boards. Whenever the line switches between boards requiring different dies, a setup is incurred. If all possible combinations of copper thickness and die requirement are represented in the 1,000 boards, then there are $4 \times 3 = 12$ distinct **product families** within the routing. This definition of *family* ensures that there are no significant setups *within* families but there may be setups *between* families. As we will discuss in Chapter 15, setups have important ramifications for scheduling. For this reason, aggregation of products by family can often simplify the planning process without oversimplifying it.

3. **People.** There are a host of ways that a factory's workforce can be broken down: labor versus management, union versus nonunion, factory floor versus staff support, permanent versus temporary, departments (e.g., manufacturing, production control, engineering, personnel), shifts, and so on. In a large plant, the personnel organization scheme can be almost as complex as the machinery. While a detailed discussion of workforce organization is largely beyond the scope of this book—we touched on some

of the issues involved in Chapter 11—we feel it is important to point out the logistical implications of such organizations. For instance, having separate managers for different processes or shifts can lead to a lack of coordination. Relying on temporary workers to facilitate a varying workforce can decrease the institutional memory, and possibly the skill level, of the organization. Rigidly adhering to job descriptions can preclude opportunities for cross-training and flexibility within the system. As we stressed in Chapter 11, the effectiveness of a manufacturing system is very much a function of its workforce. While it will always be necessary to classify workers into different categories for purposes of training, compensation, and communication, it is important to remember that we are not necessarily constrained to follow the procedures of the past. By taking a perspective that is sensitive to logistics and people, a good manager will seek effective personnel policies that support both.

13.2.3   Coordination

There is nothing revolutionary about the previous discussion about separating decision problems along the dimensions of time, process, product, or people. For instance, virtually every manufacturing operation in the world does some sort of long-, inter-mediate-, and short-range decision making. What distinguishes a good system from a bad one is not whether it makes such a breakdown, but how well the resulting subproblems are solved and, especially, how well they are coordinated with one another. We will examine the subproblems in some detail in the remaining chapters of Part III. For now, we begin addressing the issue of coordination by means of an illustration.

The problem of what parts to make at what times is addressed at the long-, interme-diate-, and short-term levels. Over the long term, we must worry about rough volumes and product mix in order to be able to plan for capacity and staffing. Over the intermediate term, we must develop a somewhat more detailed production plan, in order to procure materials, line up vendors, and rationally negotiate customer contracts. Over the short term, we must establish and execute a detailed work schedule that controls what happens at each process center. The basic essence of all three problems is the same; only the time frame is different. Hence, it seems obvious that the decisions made at the three different levels should be consistent, at least in expectation, with one another. As one might expect, this is easier to say than to do.

When we generate a long-range production plan, giving the quantity of each part to produce in various time buckets (typically months or quarters), we cannot possibly con-sider the production process in enough detail to determine the exact number of machine setups that will be required. However, when we develop an intermediate-range produc-tion schedule, we must compute the required number of setups, because otherwise we cannot determine whether the schedule is feasible with respect to capacity. Therefore, for the long-range plan to be consistent with the intermediate-range plan, we should make sure that the long-range planning tool subtracts an amount from the capacity of each process center that corresponds to an anticipated average number of setups. To ensure this over time, we should track the actual number of setups and adjust the long-range planning accordingly.

A similar link is needed between the intermediate- and short-term plans. When we generate an intermediate-range production schedule, we cannot anticipate all the variations in material flow that will occur in the actual production process. Machines may fail, operators may call in sick, process or quality problems may arise—none of which can be foreseen. However, at the short-range level, when we are planning minute by minute what to work on, we must consider what machines are down, what workers are absent, and many other factors affecting the current status of the plant. The result will be that actual production activities will never match planned ones exactly. Therefore, for the short-range activities to be able to generate outputs that are consistent, at least on average, with planned requirements, the intermediate-range planning tool must contain some form of buffer capacity or buffer lead time to accommodate randomness. Buffer capacity might be provided in the form of the "two-shifting" we discussed in Chapter 4 on JIT. Buffer lead times are simply additions to the times we quote to customers to allow for unanticipated delays in the factory.

Next we will discuss other links between planning levels in the context of specific problems. However, since the reader is certain to encounter planning tools and procedures other than those discussed in this book, we have raised the issue of establishing links as a general principle. The main point is that the various levels can and should be addressed with different tools and assumptions, but linked via simple mechanisms such as those discussed previously.

13.3   Forecasting

The starting point of virtually all production planning systems is forecasting. This is because the consequences of manufacturing planning decisions almost always depend on the future. A decision that looks good now may turn out later to be terrible. But since no one has a crystal ball with which to predict the future, the best we can do is to make use of whatever information is available in the present to choose the policies that we predict will be successful in the future.

Obviously, dependence on the future is not unique to manufacturing. The success or failure of government policies is heavily influenced by future parameters, such as interest rates, economic growth, inflation, and unemployment. Profitability of insurance companies depends on future liabilities, which are in turn a function of such unpredictable things as natural disasters. Cash flow in oil companies is governed by future success in drilling ventures. In cases like these, where the effectiveness of current decisions depends on uncertain outcomes in the future, decision makers generally rely on some type of **forecasting** to generate expectations of the future in order to evaluate alternate policies.

Because there are many approaches one can use to predict the future, forecasting is a large and varied field. One basic distinction is between methods of

1. Qualitative forecasting
2. Quantitative forecasting

Qualitative forecasting methods attempt to develop likely future scenarios by using the expertise of people, rather than precise mathematical models. One structured method for eliciting forecasts from experts is **Delphi.** In Delphi, experts are queried about some future subject, for instance, the likely introduction date of a new technology. This is usually done in written form, but can be done orally. The responses are tabulated and returned to the panel of experts, who reconsider and respond again, to the original and possibly some new questions as well. The process can be repeated several times, until consensus is reached or the respondents have stabilized in their answers. Delphi and techniques like it are useful for long-term forecasting where the future depends on the past in very complex ways. Technological forecasts, where predicting highly uncertain breakthroughs is at the core of the exercise, frequently use this type of approach. Martino (1983) summarizes a variety of qualitative forecasting methods in this context.

Quantitative forecasting methods are based on the assumption that the future can be predicted by using numerical measures of the past in some kind of mathematical model. There are two basic classes of quantitative forecasting models:

1. **Causal models** predict a future parameter (e.g., demand for a product) as a function of other parameters (e.g., interest rates, growth in GNP, housing starts).

2. **Time series models** predict a future parameter (e.g. demand for a product) as a function of past values of that parameter (e.g., historical demand).

Because we cannot hope to provide a comprehensive overview of forecasting, we will restrict our attention to those techniques that have the greatest relevance to operations management (OM). Specifically, because operational decisions are primarily concerned with problems having planning horizons of less than two years, the long-term techniques of qualitative forecasting are not widely used in OM situations. Therefore, we will focus on quantitative methods. Furthermore, because time series models are simple to use and have direct applicability (in a nonforecasting context) to the production tracking module, we will devote most of our attention to these.

Before we cover specific techniques, we note the following well-known laws of forecasting:

First law of forecasting: *Forecasts are always wrong!*

Second law of forecasting: *Detailed forecasts are worse than aggregate forecasts!*

Third law of forecasting: *The further into the future, the less reliable the forecast will be!*

No matter how qualified the expert or how sophisticated the model, perfect prediction of the future is simply not possible; hence the first law. Furthermore, by the concept of variability pooling, an aggregate forecast (e.g., of a product family) will exhibit less variability than a detailed forecast (e.g., of an individual product); hence the second law. Finally, the further out one goes, the greater the potential for qualitative changes (e.g., the competition introduces an important new product) that completely invalidate whatever forecasting approach we use; hence the third law.

We do not mean by these laws to disparage the idea of forecasting altogether. On the contrary, the whole notion of a planning hierarchy is premised on forecasting. There is simply no way to sensibly make decisions of how much capacity to install, how large a workforce to maintain, or how much inventory to stock without some estimate of future demand. *But* since our estimate is likely to be approximate at best, we should strive to make these decisions as robust as possible with respect to errors in the forecast. For instance, using equipment and plant layouts that enable accommodation of new products, changes in volume, and shifts in product mix, sometimes referred to as **agile manufacturing,** can greatly reduce the consequences of forecasting errors. Similarly, cross-training of workers and adaptable workforce scheduling policies can substantially increase flexibility. Finally, as we noted in Part II, shortening manufacturing cycle times can reduce dependence on forecasts.

13.3.1   Causal Forecasting

In a causal forecast, we attempt to explain the behavior of an uncertain future parameter in terms of other, observable or at least more predictable, parameters. For instance, if we are trying to evaluate the economics of opening a new fast-food outlet at a given location, we need a forecast of demand. Possible predictors of demand include population and number of competitor fast-food restaurants within some distance of the location. By collecting data on demand, population, and competition for existing comparable restaurants, we can use statistics to estimate constants in a model.

The most commonly used model is the simple linear model, of the form

$$Y = b_0 + b_1 X_1 + b_2 X_2 + \cdots + b_m X_m \tag{13.1}$$

where Y represents the parameter to be predicted (demand) and the X_i variables are the predictive parameters (population and competition). The b_i values are constants that must be statistically estimated from data.

This technique for fitting a function to data is called **regression analysis;** many computer packages, including all major spreadsheet programs, are available for performing the necessary computations. The following example briefly illustrates how regression analysis can be used as a tool for causal forecasting.

Example: Mr. Forest's Cookies

An emerging cookie store franchise was in the process of evaluating sites for future outlets. Top management conjectured that the success of a store is strongly influenced by the number of people who live within five miles of it. Analysts collected this population data and annual sales data for 12 existing franchises, as summarized in Table 13.2.

To develop a model for predicting the sales of a new franchise from its five-mile-radius population, the analysts made use of regression analysis, which is a tool for finding the "best-fit" straight line through the data. They did this by choosing the **Regression** function in Excel, which produced the output shown in Figure 13.1. The three key numbers, marked in boldface, are as follows:

1. **Intercept coefficient,** which is the estimate of b_0 in Equation (13.1), or 50.30 (rounded to two decimals) for this problem. This coefficient represents the Y intercept of the straight line being fit through the data.

2. **X_1 coefficient,** or the estimate of b_1 in Equation (13.1), which is 4.17 for this problem. This coefficient represents the slope of the straight line being fit through the data. It is indicated as "Population (000)" in Figure 13.1.

3. **R square,** which represents the fraction of variation in the data that is explained by the regression line. If the data fit the regression line perfectly, R square will

TABLE 13.2   Mr. Forest's Cookies Franchise Data

Franchise	Population (000)	Sales ($000)
1	50	200
2	25	50
3	14	210
4	76	240
5	88	400
6	35	200
7	85	410
8	110	500
9	95	610
10	21	120
11	30	190
12	44	180

FIGURE 13.1

Excel regression analysis output

SUMMARY OUTPUT

Regression Statistics						
Multiple R	0.880008188					
R Square	**0.774414411**					
Adjusted R Square	0.751855852					
Standard Error	77.79635826					
Observations	12					
ANOVA						
	df	*SS*	*MS*	*F*	*Significance F*	
Regression	1	207768.9331	207768.9331	34.32907286	0.000159631	
Residual	10	60522.73358	6052.273358			
Total	11	268291.6667				
	Coefficients	*Standard Error*	*t Stat*	*P-value*	*Lower 95%*	*Upper 95%*
Intercept	**50.30456039**	45.79857723	1.098386968	0.297777155	−51.74104657	152.3501673
Population (000)	**4.169903827**	0.711696781	5.859101711	0.000159631	2.584144304	5.755663349

FIGURE 13.2

Fit of regression line to Mr. Forest's data

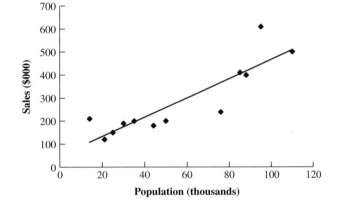

be one. The smaller R square is, the poorer the fit of the data to the regression line. In this case, R square is 0.77441441, which means that the fit is reasonably good, but hardly perfect. Excel also generates a plot of the data and the regression line, as shown in Figure 13.2, which allows us to visually examine how well the model fits the data.

Thus, the predictive model is given by

$$\text{Sales} = 50.30 + 4.17 \times \text{Population} \tag{13.2}$$

where sales are measured in thousands of dollars ($000) and population represents the five-mile-radius population in thousands. So a new franchise with a five-mile-radius population of 60 thousand would have predicted annual sales of

$$50.30 + 4.17(60) = \$300.5$$

The above equation is in thousands.

Judging from the results in Figures 13.1 and 13.2, the model appears reasonable for making rough predictions, provided that the population for the new franchise is between 15,000 and 110,000. Since the initial data set does not include populations outside this range, we have no basis for making predictions for populations smaller than 15,000 or larger than 110,000.

If the analysts for Mr. Forest want to develop a more refined model, they might consider adding other predictive variables, such as the average income of the five-mile-radius population, number of other cookie stores within a specified distance of the proposed location, and number of other retail establishments within walking distance of the proposed location. The general model of Equation (13.1), known as a **multiple regression model** (as opposed to a **simple regression model** that includes only a single predictive variable), allows such multiple predictors, as do the computer packages for performing the computations.

Packages such as Excel make the mechanics of regression simple. But full interpretation of the results requires knowledge of statistics. Given that statistics and regression are widely used throughout business—for marketing analysis, product design, personnel evaluation, forecasting, quality control, and process control—they are essential basics of a modern manager's skill set. Any good business statistics text can provide the necessary background in these important topics.

Although frequently useful, a causal model by itself cannot always enable us to make predictions about the future. For instance, if *next* month's demand for roofing materials, as seen by the manufacturer, depends on *last* month's housing starts (because of the time lag between the housing start and the replenishment purchase order placed on the manufacturer by the supplier), then the model requires only observable inputs and we can make a forecast directly. In contrast, if *next* month's demand for air conditioners depends on *next* month's average daily temperature, then we must forecast next month's temperature before we can predict demand. (Given the quality of long-term weather forecasts, it is not clear that such a causal model would be of much help, however.)

13.3.2   Time Series Forecasting

To predict a numerical parameter for which past results are a good indicator of future behavior, but where a strong cause-and-effect relationship is not available for constructing a causal model, a **time series model** is frequently used. Demand for a product often falls into this category, and therefore demand forecasting is one of the most common applications of this technique. The reason is that demand is a function of such factors as customer appeal, marketing effectiveness, and competition. Although these factors are difficult to model explicitly, they do tend to persist over time, so past demand is often a good predictor of future demand. What time series models do is to try to capture past trends and extrapolate them into the future.

Although there are many different time series models, the basic procedure is the same for all. We treat time in periods (e.g., months), labeled $i = 1, 2, \ldots, t$, where period t is the most recent data observation to be used in the forecast. We denote the actual observations by $A(i)$ and let the forecasts for periods $t + \tau$, $\tau = 1, 2, \ldots$, be represented by $f(t + \tau)$. As shown in Figure 13.3, a time series model takes as input the past observations $A(i), i = 1, \ldots, t$ (for example, $A(i)$ could represent demand in month i, where t represents the most recent month for which data are available) and generates predictions for the future values $f(t + \tau), \tau = 1, 2, \ldots$ (for example, $f(t + \tau)$ represents the forecasted demand for month $t + \tau$, which is τ months into the future).

FIGURE 13.3

Basic structure of time series models

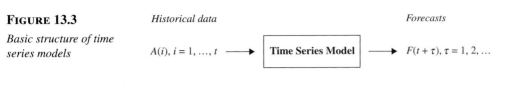

Toward this end, some models, including those discussed here, compute a **smoothed estimate** $F(t)$, which represents an estimate of the current position of the process under consideration, and a **smoothed trend** $T(t)$, which represents an estimate of the current trend of the process.

There are many different models that can perform this basic forecasting function; which is most appropriate depends on the specific application. Here we present four of the simplest and most common approaches. The **moving-average** model computes the forecast for the next period (and thereafter) as the average of the last m observations (where the user chooses the value of m). **Exponential smoothing** computes a smoothed estimate as a weighted average (where the user chooses the weights) of the most recent observation and the previous smoothed estimate. Like the moving-average model, simple exponential smoothing assumes no trend (i.e., upward or downward) in the data and therefore uses the smoothed estimate as the forecast for all future periods. **Exponential smoothing with a linear trend** estimates the smoothed estimate in a manner similar to exponential smoothing, but also computes a smoothed trend, or slope, in the data. Finally, **Winter's method** adds seasonal multipliers to the exponential smoothing with a linear trend model, in order to represent situations where demand exhibits seasonal behavior.

Moving Average.   The simplest way to convert actual observations to forecasts is to simply average them. In doing this, we are implicitly assuming that there is no trend, so that $T(t) = 0$ for all t. We then compute the smoothed estimate as the simple average and use this average for all future forecasts, so that

$$F(t) = \frac{\sum_{i=1}^{t} A(i)}{t}$$

$$f(t + \tau) = F(t) \quad \tau = 1, 2, \ldots$$

A potential problem with this approach is that it gives all past data equal weight regardless of their age. But demand data from three years ago may no longer be representative of future expectations. To capture the tendency for more recent data to be better correlated with future outcomes than old data are, virtually all time series models contain a mechanism for discounting old data. The simplest procedure for doing this is to throw data away beyond some point in the past. The time series model that does this is called the **moving-average** model, and it works in the same way as the simple average except that only the most recent m data points (where m is a parameter chosen by the user) are used in the average. Again, the trend is assumed to be zero, so $T(t) = 0$, and all future forecasts beyond the present are assumed to be equal to the current smoothed estimate:

$$F(t) = \frac{\sum_{i=t-m+1}^{t} A(i)}{m} \tag{13.3}$$

$$f(t + \tau) = F(t) \quad \tau = 1, 2, \ldots \tag{13.4}$$

Notice that the choice of m will make a difference in how the moving-average method performs. A way to find an appropriate value for a particular situation is to try various values and see how well they predict already known data. For instance, suppose we have 20 months of past demand for a particular product, as shown in Table 13.3. At any time, we can pretend that we only have data up to that point and use our moving average to generate a forecast. If we set $m = 3$, then in period $t = 3$ we can compute the smoothed estimate as the average of the first three points, or

$$F(3) = \frac{10 + 12 + 12}{3} = 11.33$$

At time $t = 3$, our forecast for demand in period 4 (and beyond, since there is no trend) is $f(4) = F(3) = 11.33$. However, once we actually get to period 4 and make another observation of actual demand, our estimate becomes the average of the second, third, and fourth points, or

$$F(4) = \frac{12 + 12 + 11}{3} = 11.67$$

Now our forecast for period 5 (and beyond) is $f(5) = F(4) = 11.67$. Continuing in this manner, we can compute what our forecast would have been for $t = 4, \ldots, 20$, as shown in Figure 13.3. We cannot make forecasts in periods 1, 2, and 3 because we need three data points before we can compute a three-period moving average.

If we change the number of periods in our moving average to $m = 5$, we can compute the smoothed estimate, and therefore the forecast, for periods $6, \ldots, 20$, as shown in Table 13.3.

Which is better, $m = 3$ or $m = 5$? It is rather difficult to tell from Table 13.3. However, if we plot $A(t)$ and $f(t)$, we can see which model's forecast came closer to the

TABLE 13.3   Moving Averages with $m = 3$ and $m = 5$

Month t	Demand $A(t)$	Forecast $f(t)$ $m = 3$	Forecast $f(t)$ $m = 5$
1	10	—	—
2	12	—	—
3	12	—	—
4	11	11.33	—
5	15	11.67	—
6	14	12.67	12.0
7	18	13.33	12.8
8	22	15.67	14.0
9	18	18.00	16.0
10	28	19.33	17.4
11	33	22.67	20.0
12	31	26.33	23.8
13	31	30.67	26.4
14	37	31.67	28.2
15	40	33.00	32.0
16	33	36.00	34.4
17	50	36.67	34.4
18	45	41.00	38.2
19	55	42.67	41.0
20	60	50.00	44.6

FIGURE 13.4

Moving average with
m = 3, 5

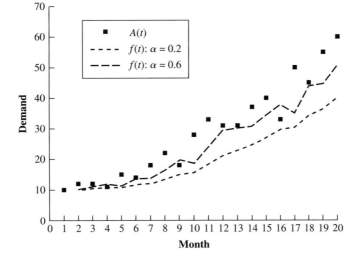

actual observed values. As we see in Figure 13.4, both models tended to underestimate demand, with the $m = 5$ model performing worse. The reason for this underestimation is that the moving-average model assumes no upward or downward trend in the data. But we can see from the plots that these data clearly have an upward trend. Therefore, the moving average of past demand tends to be less than future demand. Since the model with $m = 5$ is even more heavily tied to past demand (because it includes more, and therefore older, points), it suffers from this tendency to a greater extent.

This example illustrates the following general conclusions about the moving-average model:

1. Higher values of m will make the model more stable, but less responsive, to changes in the process being forecast.
2. The model will tend to underestimate parameters with an increasing trend, and overestimate parameters with a decreasing trend.

We can address the problem of tracking a trend in the context of the moving-average model. For those familiar with regression analysis, the way this works is to estimate a slope for the last m data points via linear regression and then make the forecast equal to the smoothed estimate plus an extrapolation of this linear trend. However, there is another, easier way to introduce a linear trend into a different time series model. Next, we will pursue this approach after presenting another trendless model below.

Exponential Smoothing.   Observe that the moving-average approach gives equal weight to each of the m most recent observations and no weight to observations older than these. Another way to discount old data points is to average the current smoothed estimate with the most recent data point. The result will be that the older the data point, the smaller the weight it receives in determining the forecast. We call this method **exponential smoothing,** and it works as follows. First, we assume, for now, that the trend is always zero, so $T(t) = 0$. Then we compute the smoothed estimate and forecast at time t as

$$F(t) = \alpha A(t) + (1 - \alpha)F(t - 1) \tag{13.5}$$

$$f(t + \tau) = F(t) \qquad \tau = 1, 2, \ldots \tag{13.6}$$

where α is a smoothing constant between 0 and 1 chosen by the user. The best value will depend on the particular data.

Table 13.4 illustrates the exponential method, using the same data we used for the moving average. Unless we start with a historical value for $F(0)$, we cannot make a forecast for period 1. Although there are various ways to initialize the model (e.g., by averaging past observations over some interval), the choice of $F(0)$ will dissipate as time goes on. Therefore, we choose to use the simplest possible initialization method and set $F(1) = A(1) = 10$ and start the process. At time $t = 1$, our forecast for period 2 (and beyond) is $f(2) = F(1) = 10$. When we reach period 2 and observe that $A(2) = 12$, we update our smoothed estimate as follows:

$$F(2) = \alpha A(2) + (1 - \alpha)F(1) = (0.2)(12) + (1 - 0.2)(10) = 10.40$$

Our forecast for period 3 and beyond is now $f(3) = F(2) = 10.40$. We can continue in this manner to generate the remaining $f(t)$ values in Table 13.4.

Notice in Table 13.4 that when we use $\alpha = 0.6$ instead of $\alpha = 0.2$, the forecasts are much more sensitive to each new data point. For instance, in period 2, when demand increased from 10 to 12, the forecast using $\alpha = 0.2$ only increased to 10.40, while the forecast using $\alpha = 0.6$ increased to 11.20. This increased sensitivity may be good, if the model is tracking a real trend in the data, or bad, if it is overreacting to an unusual observation. Hence, analogous to our observations about the moving-average method, we can make the following points about single exponential smoothing:

1. Lower values of α will make the model more stable, but less responsive, to changes in the process being forecast.
2. The model will tend to underestimate parameters with an increasing trend, and overestimate parameters with a decreasing trend.

TABLE 13.4   Exponential Smoothing with $\alpha = 0.2$ and $\alpha = 0.6$

Month t	Demand $A(t)$	Forecast $f(t)$ $\alpha = 0.2$	Forecast $f(t)$ $\alpha = 0.6$
1	10	—	—
2	12	10.00	10.00
3	12	10.40	11.20
4	11	10.72	11.68
5	15	10.78	11.27
6	14	11.62	13.51
7	18	12.10	13.80
8	22	13.28	16.32
9	18	15.02	19.73
10	28	15.62	18.69
11	33	18.09	24.28
12	31	21.08	29.51
13	31	23.06	30.40
14	37	24.65	30.76
15	40	27.12	34.50
16	33	29.69	37.80
17	50	30.36	34.92
18	45	34.28	43.97
19	55	36.43	44.59
20	60	40.14	50.83

Choosing the appropriate smoothing constant α for exponential smoothing, like choosing the appropriate value of m for the moving-average method, requires a bit of trial and error. Typically, the best we can do is to try various values of α and see which one generates forecasts that match the historical data best. For instance, Figure 13.5 plots exponential smoothing forecasts $f(t)$, using $\alpha = 0.2$ and 0.6, along with actual values $A(t)$. This plot clearly shows that the values generated using $\alpha = 0.6$ are closer to the actual data points than those generated using $\alpha = 0.2$. The increased sensitivity caused by using a high α value enabled the model to track the obvious upward trend of the data. However, because the single exponential smoothing model does not explicitly assume the existence of a trend, both sets of forecasts tended to lag behind the actual data.

Exponential Smoothing with a Linear Trend.   We now turn to a model that is specifically designed to track data with upward or downward trends. For simplicity, the model assumes the trend is linear. That is, our forecasts from the present out into the future will follow a straight line. Of course, each time we receive a new observation, we will update the slope of this line, so the method can track data that change in a nonlinear fashion, although less accurately than data with a trend that is generally linear.

The basic method updates a smoothed estimate $F(t)$ and a smoothed trend $T(t)$ each time a new observation becomes available. Using these, the forecast for τ periods into the future, denoted by $f(t + \tau)$, is computed as the smoothed estimate plus τ times the smoothed trend. The equations for doing this are as follows:

$$F(t) = \alpha A(t) + (1 - \alpha)[F(t - 1) + T(t - 1)] \qquad (13.7)$$

$$T(t) = \beta[F(t) - F(t - 1)] + (1 - \beta)T(t - 1) \qquad (13.8)$$

$$f(t + \tau) = F(t) + \tau T(t) \qquad (13.9)$$

where α and β are smoothing constants between 0 and 1 to be chosen by the user.

Notice that the equation for computing $F(t)$ is slightly different from that for exponential smoothing without a linear trend. The reason is that at period $t - 1$ the forecast for period t is given by $F(t-1) + T(t-1)$ (i.e., we need to add the trend for one period). Therefore, when we compute the weighted average of $A(t)$ and the current forecast, we must use $F(t - 1) + T(t - 1)$ as the current forecast.

FIGURE 13.5

Exponential smoothing with $\alpha = 0.2, 0.6$

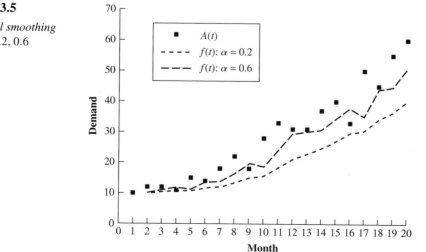

We update the trend in Equation (13.8) by computing a weighted average between the last smoothed trend $T(t-1)$ and the most recent estimate of the trend, which is computed as the difference between the two most recent smoothed estimates, or $F(t) - F(t-1)$. The $F(t) - F(t-1)$ term is like a slope. By giving this slope a weight of β (less than one), we smooth our estimate of the trend to avoid overreacting to sudden changes in the data.

As in simple exponential smoothing, we must initialize the model before we can begin. We could do this by using historical data to estimate $F(0)$ and $T(0)$. However, the simplest initialization method is to set $F(1) = A(1)$ and $T(1) = 0$. We illustrate the exponential smoothing with linear trend method using this initialization procedure, the demand data from Table 13.4, and smoothing constants $\alpha = 0.2$ and $\beta = 0.2$. For instance,

$$F(2) = \alpha A(2) + (1 - \alpha)[F(1) + T(1)] = 0.2(12) + (1 - 0.2)(10 + 0) = 10.4$$

$$T(2) = \beta[F(2) - F(1)] + (1 - \beta)T(1) = 0.2(10.4 - 10) + (1 - 0.2)(0) = 0.08$$

The remainder of the calculations are given in Table 13.5.

Figure 13.6 plots the forecast values $f(t)$ and the actual values $A(t)$ from Table 13.5 and plots the forecast that results from using $\alpha = 0.3$ and $\beta = 0.5$. Notice that these forecasts track these data much better than either the moving average or exponential smoothing without a linear trend. The linear trend enables this method to track the upward trend in these data quite effectively. Additionally, it appears that using smoothing coefficients $\alpha = 0.3$ and $\beta = 0.5$ results in better forecasts than using $\alpha = 0.2$ and $\beta = 0.2$. Next, we will discuss how to choose smoothing constants later in this section.

TABLE 13.5   Exponential Smoothing with a Linear Trend, $\alpha = 0.2$ and $\beta = 0.2$

Month t	Demand $A(t)$	Smoothed Estimate $F(t)$	Smoothed Trend $T(t)$	Forecast $f(t)$
1	10	10.00	0.00	—
2	12	10.40	0.08	10.00
3	12	10.78	0.14	10.48
4	11	10.94	0.14	10.92
5	15	11.87	0.30	11.08
6	14	12.53	0.37	12.17
7	18	13.93	0.58	12.91
8	22	16.00	0.88	14.50
9	18	17.10	0.92	16.88
10	28	20.02	1.32	18.03
11	33	23.67	1.79	21.34
12	31	26.57	2.01	25.46
13	31	29.06	2.11	28.58
14	37	32.33	2.34	31.17
15	40	35.74	2.55	34.67
16	33	37.23	2.34	38.29
17	50	41.66	2.76	39.57
18	45	44.53	2.78	44.42
19	55	48.85	3.09	47.31
20	60	53.55	3.41	51.94

FIGURE 13.6

Exponential smoothing with linear trend

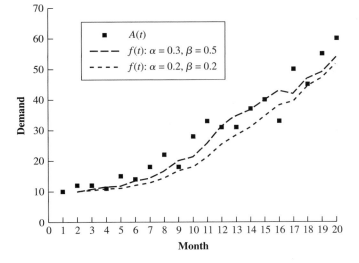

The Winters Method for Seasonality.   Many products exhibit seasonal demand. For instance, lawn mowers, ice cream, and air conditioners have peaks associated with summer, while snow blowers, weather stripping, and furnaces have winter peaks. When this is the case, the above forecasting models will not work well because they will interpret seasonal rises in demand as permanent rises and thereby will overshoot actual demand when it declines in the off season. Likewise, they will interpret the low off-season demand as permanent and will undershoot actual demand during the peak season.

A natural way to build seasonality into a forecasting model was suggested by Winters (1960). The basic idea is to estimate a multiplicative seasonality factor $c(t), t = 1, 2, \ldots$, where $c(t)$ represents the ratio of demand during period t to the average demand during the season. Therefore, if there are N periods in the season (for example, $N = 12$ if periods are months and the season is 1 year), then the sum of the $c(t)$ factors over the season will always be equal to N. The seasonally adjusted forecast is computed by multiplying the forecast from the exponential smoothing with linear trend model (that is, $F(t) + \tau T(t)$) by the appropriate seasonality factor. The equations for doing this are as follows:

$$F(t) = \alpha \frac{A(t)}{c(t - N)} + (1 - \alpha)[F(t - 1) + T(t - 1)] \qquad (13.10)$$

$$T(t) = \beta[F(t) - F(t - 1)] + (1 - \beta)T(t - 1) \qquad (13.11)$$

$$c(t) = \gamma \frac{A(t)}{F(t)} + (1 - \gamma)c(t - N) \qquad (13.12)$$

$$f(t + \tau) = [F(t) + \tau T(t)]c(t) \qquad (13.13)$$

where α, β, and γ are smoothing constants between 0 and 1 to be chosen by the user. Notice that Equations (13.10) and (13.11) are identical to Equations (13.7) and (13.8) for computing the smoothed estimate and smoothed trend in the exponential smoothing with linear trend model, except that the actual observation $A(t)$ is scaled by dividing by the seasonality factor $c(t - N)$. This normalizes all the observations relative to the average and hence places the smoothed estimate and trend in units of average (nonseasonal) demand. Equation (13.12) uses exponential smoothing to update the seasonality factor $c(t)$ as a weighted average of this season's ratio of actual demand to smoothed estimate

TABLE 13.6 **The Winters Method for Forecasting with Seasonality**

Year	Month	Time Period t	Actual Demand $A(t)$	Smoothed Estimate $F(t)$	Smoothed Trend $T(t)$	Seasonal Factor $c(t)$	Forecast $f(t)$
1997	Jan	1	4	—	—	0.480	
	Feb	2	2	–	—	0.240	
	Mar	3	5	—	—	0.600	
	Apr	4	8	—	—	0.960	
	May	5	11	—	—	1.320	
	Jun	6	13	—	—	1.560	
	Jul	7	18	—	—	2.160	
	Aug	8	15	—	—	1.800	
	Sep	9	9	—	—	1.080	
	Oct	10	6	—	—	0.720	
	Nov	11	5	—	—	0.600	
	Dec	12	4	8.33	0.00	0.480	
1998	Jan	13	5	8.54	0.02	0.491	4.00
	Feb	14	4	9.37	0.10	0.259	2.06
	Mar	15	7	9.69	0.12	0.612	5.68
	Apr	16	7	9.57	0.10	0.937	9.43
	May	17	15	9.83	0.12	1.341	12.76
	Jun	18	17	10.04	0.13	1.573	15.52
	Jul	19	24	10.26	0.13	2.178	21.97
	Aug	20	18	10.36	0.13	1.794	18.72
	Sep	21	12	10.55	0.14	1.086	11.33
	Oct	22	7	10.59	0.13	0.714	7.69
	Nov	23	8	10.98	0.15	0.613	6.43
	Dec	24	6	11.27	0.17	0.485	5.34

$A(t)/F(t)$ and last season's factor $c(t - N)$. To make the forecast in seasonal units, we multiply the nonseasonal forecast $F(t) + \tau T(t)$ by the seasonality factor $c(t)$.

We illustrate the Winters method with the example in Table 13.6. To initialize the procedure, we require a full season of seasonality factors plus an initial smoothed estimate and smoothed trend. The simplest way to do this is to use the first season of data to compute these initial parameters and then use the above equations to update them with additional seasons of data. Specifically, we simply set the smoothed estimate to be the average of the first seasons of data

$$F(N) = \frac{\sum_{t=1}^{N} A(t)}{N} \tag{13.14}$$

So, in our example, we can compute the smoothed estimate as of December 1998 to be

$$F(12) = \frac{\sum_{t=1}^{12} A(t)}{12} = \frac{4 + 2 + \cdots + 4}{12} = 8.33$$

Since we are starting with only a single season of data, we have no basis for estimating a trend, so we will assume initially that the trend is zero, so that $T(N) = T(12) = 0$. The model will quickly update the trend as seasons are added.[1] Finally, we compute

---

[1] Alternatively, one could use multiple seasons of data to initialize the model and estimate the trend from these (see Silver, Pyke, and Peterson 1998 for a method).

initial seasonality factors as the ratio of actual demand to average demand during the first season:

$$c(i) = \frac{A(i)}{\sum_{t=1}^{N} A(t)/N} = \frac{A(i)}{F(N)} \tag{13.15}$$

For instance, in our example, the initial seasonality factor for January is

$$c(1) = \frac{A(1)}{F(12)} = \frac{4}{8.33} = 0.480$$

Once we have computed values for $F(N)$, $T(N)$, and $c(1), \ldots, c(N)$, we can begin the smoothing procedure. The smoothed estimate for January 1998 is computed as

$$F(13) = \alpha \frac{A(13)}{c(13-12)} + (1-\alpha)[F(12) + T(12)]$$

$$= 0.1 \left(\frac{5}{0.480} \right) + (1 - 0.1)(8.33 + 0) = 8.54$$

The smoothed trend is

$$T(13) = \beta[F(13) - F(12)] + (1-\beta)T(12) = 0.1(8.54 - 8.33) + (1 - 0.1)(0) = 0.02$$

The updated seasonality factor for January is

$$c(13) = \gamma \frac{A(13)}{F(13)} + (1-\gamma)c(1) = 0.1 \left(\frac{5}{8.54} \right) + (1 - 0.1)(0.48) = 0.491$$

The computations continue in this manner, resulting in the numbers shown in Table 13.6. We plot the actual and forecasted demand in Figure 13.7. In this example, the Winters method works very well. The primary reason is that the seasonal spike in 1998 had a similar shape to that in 1997. That is, the proportion of total annual demand that occurred in a given month, such as July, is fairly constant across years. Hence, the seasonality factors provide a good fit to the seasonal behavior. The fact that total annual demand is growing, which is accounted for by the positive trend in the model, results in an appropriately amplified seasonal spike in the second year. In general, the Winters method gives reasonable performance for seasonal forecasting where the shape of the seasonality does not vary too much from season to season.

FIGURE 13.7

The Winters method,
$\alpha = 0.1$, $\beta = 0.1$,
$\gamma = 0.1$

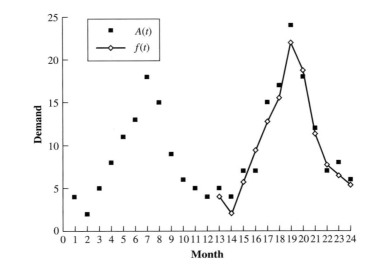

Adjusting Forecasting Parameters.   All the time series models discussed involve adjustable coefficients (for example, m in the moving-average model and α in the exponential smoothing model), which must be "tuned" to the data to yield a suitable forecasting model. Indeed, we saw in Figure 13.6 that adjusting the smoothing coefficients can substantially affect the accuracy of a forecasting model. We now turn to the question of how to find good coefficients for a given forecasting situation.

The first step in developing a forecasting model is to plot the data. This will help us decide whether the data appear predictable at all, whether a trend seems to be present, or whether seasonality seems to be a factor. Once we have chosen a model, we can plot the forecast versus actual past data for various sets of parameters to see how the model behaves. However, to find a good set of coefficients, it is helpful to be more precise about measuring model accuracy.

The three most common quantitative measures for evaluating forecasting models are the *mean absolute deviation* (MAD), *mean square deviation* (MSD), and *bias* (BIAS). Each of these takes the differences between the forecast and actual values, $f(t) - A(t)$, and computes a numerical score. The specific formulas for these are as follows:

$$\text{MAD} = \frac{\sum_{t=1}^{n} |f(t) - A(t)|}{n} \tag{13.16}$$

$$\text{MSD} = \frac{\sum_{t=1}^{n} [f(t) - A(t)]^2}{n} \tag{13.17}$$

$$\text{BIAS} = \frac{\sum_{t=1}^{n} f(t) - A(t)}{n} \tag{13.18}$$

Both MAD and MSD can only be positive, so the objective is to find model coefficients that make them as small as possible. BIAS can be positive, indicating that the forecast tends to overestimate the actual data, or negative, indicating that the forecast tends to underestimate the actual data. The objective, then, is to find coefficients that make BIAS close to zero. However, note that zero BIAS does not mean that the forecast is accurate, only that the errors tend to be balanced high and low. Hence, one would never use BIAS alone to evaluate a forecasting model.

To illustrate how these measures might be used to select model coefficients, let us return to the exponential smoothing with linear trend model as applied to the demand data in Table 13.5. Table 13.7 reports the values of MAD, MSD, and BIAS for various combinations of α and β. From this table, it appears that the combination $\alpha = 0.3$, $\beta = 0.5$ works well with regard to minimizing MAD and MSD, but that $\alpha = 0.6$, $\beta = 0.6$ is better with regard to minimizing BIAS. In general, it is unlikely that any set of coefficients will be best with regard to all three measures of effectiveness. In this specific case, as can be seen in Figure 13.6, the actual data not only have an upward trend, but also tend to increase according to a nonlinear curve (i.e., the curve has a sort of parabolic shape). This nonlinear shape causes the model with a linear trend to lag slightly behind the data, resulting in a negative BIAS. Higher values of α and β give the new observations more weight and thereby cause the model to track this upward swing more rapidly. This reduces BIAS. However, they also cause it to overshoot the occasional downward dip in the data, increasing MAD and MSD.

Table 13.7 shows that the model with $\alpha = 0.3$, $\beta = 0.5$ has significantly smaller MSD than the model with our original choice of $\alpha = 0.2$, $\beta = 0.2$. This means that it fits the past data more closely, as illustrated in Figure 13.6. Since our basic assumption in using a time series forecasting model is that future data will behave similarly to past data, we should set the coefficients to provide a good fit to past data and then use these for future forecasting purposes.

TABLE 13.7   Exponential Smoothing with Linear Trend for Various α and β

α	β	MAD	MSD	BIAS	α	β	MAD	MSD	BIAS
0.1	0.1	10.23	146.94	-10.23	0.4	0.1	4.30	30.14	-3.45
0.1	0.2	8.27	95.31	-8.27	0.4	0.2	3.89	23.78	-2.34
0.1	0.3	6.83	64.91	-6.69	0.4	0.3	3.77	22.25	-1.77
0.1	0.4	5.83	47.17	-5.43	0.4	0.4	3.75	22.11	-1.46
0.1	0.5	5.16	36.88	-4.42	0.4	0.5	3.76	22.36	-1.29
0.1	0.6	4.69	30.91	-3.62	0.4	0.6	3.79	22.67	-1.18
0.2	0.1	6.48	60.55	-6.29	0.5	0.1	4.13	27.40	-2.84
0.2	0.2	5.04	37.04	-4.49	0.5	0.2	3.91	23.61	-1.94
0.2	0.3	4.26	27.56	-3.29	0.5	0.3	3.88	23.02	-1.49
0.2	0.4	3.90	23.75	-2.51	0.5	0.4	3.90	23.26	-1.25
0.2	0.5	3.73	22.32	-2.02	0.5	0.5	3.94	23.73	-1.10
0.2	0.6	3.65	21.94	-1.71	0.5	0.6	3.97	24.27	-1.00
0.3	0.1	4.98	37.81	-4.45	0.6	0.1	4.12	26.85	-2.42
0.3	0.2	4.11	26.30	-3.03	0.6	0.2	4.03	24.63	-1.66
0.3	0.3	3.82	22.74	-2.23	0.6	0.3	4.04	24.69	-1.29
0.3	0.4	3.66	21.81	-1.77	0.6	0.4	4.09	25.35	-1.08
0.3	0.5	3.65	21.78	-1.52	0.6	0.5	4.14	26.25	-0.95
0.3	0.6	3.68	22.06	-1.38	0.6	0.6	4.21	27.29	-0.84

The enumeration offered in Table 13.7 is given here to illustrate the impact of changing smoothing coefficients. However, in practice we do not have to use a trial-and-error approach to search for a good set of smoothing coefficients. Instead, we can use the internal optimization tool, Solver, that is included in Excel to do the search for us (see Chapter 16 for details on Solver). If we set up Solver to search for the values of α and β that (1) are between zero and one and (2) minimize MSD in the previous example, we obtain the solution $\alpha = 0.284$, $\beta = 0.467$, which attains an MSD value of 21.73. This is slightly better than the $\alpha = 0.3$, $\beta = 0.5$ solution we obtained by brute-force searching, and much faster to obtain.

Notice that in our discussion of choosing smoothing coefficients we have compared the forecast for one period into the future (i.e., the lag-1 forecast) with the actual value. However, in practice, we frequently need to forecast further into the future. For instance, if we are using a demand forecast to determine how much raw material to procure, we may need to forecast several months into the future (e.g., we may require the lag-τ forecast). When this is the case, we should use the formulas to compute the forecast for τ periods from now $f(t + \tau)$ and compare this to the actual value $A(t + \tau)$ when it occurs. The model parameters should be therefore chosen with the goal of minimizing the deviations between $f(t + \tau)$ and $A(t + \tau)$, and MAD, MSD, and BIAS should be defined accordingly.

13.3.3   The Art of Forecasting

The regression model for causal forecasting and the four time series models are representative of the vast number of quantitative tools available to assist the forecasting function. Many others exist (see Box and Jenkins (1970) for an overview of more sophisticated time series models). Clearly, forecasting is an area in which quantitative models can be of great value.

However, forecasting is more than a matter of selecting a model and tinkering with its parameters to make it as effective as possible. No model can incorporate all factors that could be relevant in anticipating the future. Therefore, in any forecasting environment, situations will arise in which the forecaster must override the quantitative model with qualitative information. For instance, if there is reason to expect an impending jump in demand (e.g., because a competitor's plant is scheduled to shut down), the forecaster may need to augment the quantitative model with this information. Although there is no substitute for experience and insight, it is a good idea to occasionally look back at past forecasting experience to see what information could have been used to improve the forecast. We won't be able to predict the future precisely, but we may be able to avoid some future blunders.

13.4   Planning for Pull

A logical and customary way to break the **production planning and control (PPC)** problem into manageable pieces is to construct a hierarchical planning framework. We illustrated a typical MRP II hierarchy in Figure 3.2. However, that framework was based on the basic MRP *push* job release mechanism. As we saw in our discussion of JIT in Chapter 4 and our comparison of push and pull in Chapter 10, pull systems offer many potential benefits over push systems. Briefly, pull systems are

1. **More efficient,** in that they can attain the same throughput as a push system with less average WIP.
2. **Easier to control,** since they rely on setting (easily observable) WIP levels, rather than release rates as do push systems.
3. **More robust,** since the performance of a pull system is degraded much less by an error in WIP level than is a push system by a comparable percentage of error in release rate.
4. **More supportive of improving quality,** since low WIP levels in pull systems both require high quality (to prevent disruptions) and facilitate it (by shortening queues and quickening detection of defects).

These benefits urge us to incorporate aspects of pull into our manufacturing control systems. Unfortunately, from a planning perspective, there is a drawback to pull. Pull systems are inherently *rate-driven*, in that we fix the level of WIP and let them run. Capacity buffers (e.g., preventive maintenance periods available to be used for overtime between shifts) are used to facilitate a very steady pace, which in turn requires highly stable demand. To achieve this, the JIT literature places considerable emphasis on production smoothing.

While a rate-driven system is logistically appealing, it is not necessarily well suited to planning. There is no natural link to customer due dates in a pull system. Customers "pull" what they need, and signals (cards or whatever) trigger replenishments. But until the demands actually occur, the system offers us no information about them. Hence, a pull system provides no inherent mechanism for planning raw material procurement, staffing, opportunities for machine maintenance, etc.

In contrast, as we noted in Chapter 5, push systems can be logistical nightmares, but are extremely well suited to planning. There is a simple and direct link between customer due dates and order releases in a push system. For instance, in a lot-for-lot MRP system, the planned order releases *are* the customer requirements (only time-phased according

FIGURE 13.8

The conveyor model of a production line

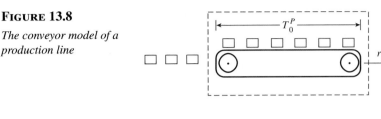

to production lead times). If only the infinite-capacity assumption of MRP did not make these lead times largely fictional, we could use them to drive all sorts of planning modules. Indeed, this is precisely what MRP II systems do.

The question then is, Can we obtain the logistical benefits of pull and still develop a coherent planning structure? We think the answer is yes. But the mechanism for linking a rate-based pull system with due dates is necessarily more complex than the simple time phasing of MRP. The simplest link we know of is the **conveyor model** of a pull production line or facility, depicted in Figure 13.8 and upon which we will rely extensively in subsequent chapters.

The conveyor model is based on the observation that a pull system maintains a fairly steady WIP level, so the speed of the line and the time to traverse it are relatively constant over time. This allows us to characterize a production line with two parameters: the **practical production rate** r_b^P and the **minimum practical lead time** T_0^P. These serve the same functions as, but are somewhat different from, the **bottleneck rate** r_b and the **raw process time** T_0 of the line as defined in Chapter 7, and their ideal realizations r_b^* and T_0^* introduced in Chapter 9. Unlike the bottleneck rate, the practical production rate is the *anticipated* throughput of the line. This rate can also be standardized according to part complexity (e.g., we could count parts in units of hours of work at a bottleneck process). Thus, since r_b is the capacity of the line, we expect $r_b^P < r_b$ with utilization $u = r_b^P / r_b$. Likewise, T_0^P is the practical minimum (i.e., no queueing) *practical* time to traverse the line. This will include detractors for short-term disruptions, such as setups and routine machine failures along with routine waiting to move and any other delays that do not involve queueing. Consequently, $T_0^P > T_0$.

Using Little's law, we see that the CONWIP level W must be

$$W = r_b^P \times T_0^P$$

We can now use the conveyor model to predict when jobs will be completed by a line or process center. For instance, suppose we release a job into the line when there are already n jobs waiting in queue to be admitted into the CONWIP line (i.e., waiting for a space on the conveyor). The time until the job will be completed ℓ is given by

$$\ell = \frac{n}{r_b^P} + T_0^P = \frac{n + W}{r_b^P} \tag{13.19}$$

For example, suppose the conveyor depicted in Figure 13.8 represents a circuit board assembly line. The line runs at an average rate of $r_b^P = 2$ jobs per hour, where a job consists of a standard-size container of circuit boards. Once started, a job takes an average of $T_0^P = 8$ hours to finish. A new job that finds $n = 3$ jobs waiting to released into the line (i.e., waiting for CONWIP authorization signals) will be completed in

$$\ell = \frac{n}{r_P} + T_0^P = \frac{3}{2} + 8 = 9.5 \text{ hours}$$

on average. We revisit this problem, adding variability to the production rate in Chapter 15 where we further refine the conveyor model.

Being able to estimate output times of specific jobs allows us to address a host of planning problems:

1. If sales personnel have a means of keeping track of factory loading, they could use the conveyor model to predict how long new orders will require to fill and therefore will be able to quote reasonable due dates to customers.

2. If we keep track of how the system will evolve (i.e., what jobs will be in the line and what jobs will be waiting in queue) over time, we can "simulate" the performance of a line. This would provide the basis for a "what if" tool for analyzing the effects of different priority rules or capacity decisions on outputs. As we noted in Chapter 3, capacity requirements planning (CRP) attempts such an analysis. However, as we pointed out there, CRP uses an infinite-capacity model that invalidates predictions beyond any point where a resource becomes fully loaded. More sophisticated, finite-capacity models for making such predictions have begun to appear on the market. While more accurate than CRP, finite models frequently have massive data needs and complex computations akin to those used in discrete event simulation models. The conveyor model can simplify both data requirements and computation, as we will discuss in various contexts throughout Part III.

3. We can use the conveyor model to determine whether completions will satisfy customer due dates to develop an optimization model for setting job release times. We will do this in Chapter 15 to generate a finite-capacity scheduling tool.

By addressing these and other problems, the conveyor model can provide the linch-pin of a planning framework for pull production systems. Where lines are simple enough to invoke it directly, it can be a powerful integrating tool. We give an outline of a framework that can exploit this integration. We will fill in the details and discuss generalizations to situations in which the conveyor model is overly simplistic in the remainder of Part III.

13.5   Hierarchical Production Planning

With the conveyor model to predict job completions, we can develop a hierarchical **production planning and control (PPC)** framework for pull production systems. Figure 13.9 illustrates such a hierarchy, spanning from long-term strategic issues at the top levels to short-term control issues at the bottom levels.

Each rectangular box in Figure 13.9 represents a separate decision problem and hence a **planning module**.[2] The rounded rectangular boxes represent *outputs* from modules, many of which are used as *inputs* in other modules. The oval boxes represent inputs to modules that are generated outside this planning hierarchy (e.g., by marketing or engineering design). Finally, the arrows indicate the *interdependence* of the modules.

The PPC hierarchy is divided into three basic levels, corresponding to long-term (strategy), intermediate-term (tactics), and short-term (control) planning. Of course, from a corporate perspective, there are levels above those shown in Figure 13.9, such as product development and business planning. Certainly these are important business

---

[2]We use the term *module* to represent the combination of analytic models, computer tools, and human judgment used to address the individual planning problems. As such, they are never fully automated, nor should they be.

FIGURE 13.9

*A production planning
and control hierarchy for
pull systems*

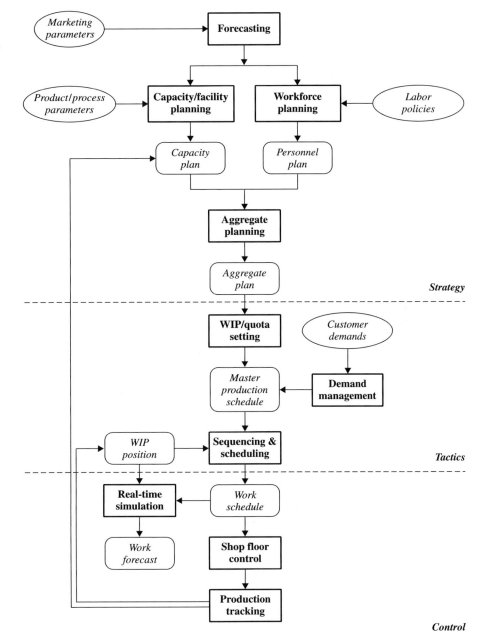

strategy decisions, and their interaction with the manufacturing function deserves serious consideration. Indeed, it is our hope that readers whose careers take them outside of manufacturing will actively pursue opportunities for greater integration of manufacturing issues into these areas. However, we will adhere to our focus on operations and assume that business strategy decisions, such as what business to be in and the nature of the product designs, have already been made. Therefore, when we speak of strategy, we are referring to *plant strategy,* which is only part of an overall business strategy.

The basic function of the long-term strategic planning tools shown in Figure 13.9 is to establish a production environment capable of meeting the plant's overall goals.

At the plant level, this begins with a **forecasting** module that takes marketing information and generates a forecast of future demand, possibly using a quantitative model like those we discussed previously. A **capacity/facility planning** module uses these demand forecasts, along with descriptions of process requirements for making the various products, to determine the needs for physical equipment. Analogously, a **workforce planning** module uses demand forecasts to generate a personnel plan for hiring, firing, training, etc., in accordance with company labor policies. Using the demand forecast, the capacity/facilities plan, and the labor plan, along with various economic parameters (material costs, wages, vendoring costs, etc.), the **aggregate planning** module makes rough predictions about future production mix and volume. The aggregate plan can also address other related issues, such as which parts to make in-house and which to contract out to external suppliers, and whether adjustments are needed in the personnel plan.

The intermediate tactical tools in Figure 13.9 take the long-range plans from the strategic level, along with information about customer orders, to generate a general plan of action that will help the plant prepare for upcoming production (by procuring materials, lining up subcontractors, etc.). A **WIP/quota-setting** module works to translate the aggregate plan into card counts and periodic production quotas required by a pull system. The production quotas form part of the **master production schedule** (MPS), which is based on the forecast demands as processed by the aggregate planning module. The MPS also contains firm customer orders, which are suitably smoothed for use in a pull production system by the *demand management* module. The **sequencing and scheduling** module translates the MPS into a work schedule that dictates what is to be worked on in the near term, for example, the next week, day, or shift.

The low-level tools in Figure 13.9 directly control the plant. The **shop floor control** module controls the real-time flow of material through the plant in accordance with this schedule, while the **production tracking** module measures actual progress against the schedule. In Figure 13.9, the production tracking module is also shown as serving a second useful function, that of feeding back information (e.g., capacity data) for use by other planning modules. Finally, the PPC hierarchy includes a **real-time simulation** module, which allows examination of what-if scenarios, such as what will happen if certain jobs are made "hot."

In the following sections, we discuss in overview fashion the issues involved at each level and the integrative philosophy for this PPC hierarchy. In this discussion, we will proceed top-down, since this helps highlight the interactions between levels. In subsequent chapters, we will provide details of how to construct the individual modules. There we will proceed bottom-up, in order to emphasize the relationship of each planning problem to the actual production process.

13.5.1   Capacity/Facility Planning

Once we have a forecast of future demand, and have made the strategic decision to attempt to fill it, we must ensure that we have adequate physical capacity. This is the function of the **capacity/facility planning** module depicted in Figure 13.9. The basic decisions to be made regarding capacity concern how much and what kind of equipment to purchase. Naturally, this includes the actual machines used to make components and final products. But it also extends to other facility issues related to the support of these machines, such as factory floor space, power supplies, air/water/chemical supplies, spare-parts inventories, material handling systems, WIP and FGI storage, and staffing levels.

Issues that can be considered in the capacity/facility planning process include the following:

1. **Product lifetimes.** The decisions of what type and how much capacity to install depend on how long we anticipate making the product. In recent years, product lifetimes have become significantly shorter, to the point where they are frequently shorter than the physical life of the equipment. This means that the equipment must either pay for itself during the product lifetime or be sufficiently flexible to be used to manufacture other future products. Because it is often difficult to predict with any degree of confidence what future products will be, quantifying the benefits of flexibility is not easy. But it can be one of the most important aspects of facility planning, since a flexible plant that can be swiftly "tooled up" to produce new products can be a potent strategic weapon.

2. **Vendoring options.** Before the characterization of the nature of the equipment to install, a "make or buy" decision must be made, for the finished product and its subcomponents. While this is a complex issue that we cannot hope to cover comprehensively here, we offer some observations.

   a. This make-or-buy decision should not be made on cost alone. Outsourcing a product because it appears that the unit cost of the vendor is lower than the (fully loaded) unit cost of making it in-house can be risky. Because unit costs depend strongly on the manner in which overhead allocation is done, a decision that seems locally rational may be globally disastrous. For example, a product that is outsourced because its unit cost is higher than the price offered by an outside supplier may not eliminate many of the overhead costs that were factored into its unit cost. Hence, these costs must be spread over the remaining products manufactured in-house, causing their unit costs to increase and making them more attractive candidates for outsourcing. There are examples of firms that have fallen into a virtual "death spiral" of repeated rounds of outsourcing on the basis of unit cost comparisons. In addition to the economic issues associated with outsourcing, there are other benefits to in-house production, such as learning effects, the ability to control one's own destiny, and tighter control over the scheduling process, that are not captured by a simple cost comparison.

   b. Consideration should be given to the long term in make-or-buy decisions. We have seen companies evolve from manufacturing into distribution/service through a sequence of outsourcing decisions. While this is not necessarily a bad transition, it is certainly one that should not be made without a full awareness of the consequences and careful consideration of the viability of the firm in the marketplace as a nonmanufacturing entity.

   c. When the make-or-buy decision concerns whether or not to make the product at all, then it is clearly a capacity planning decision. However, many manufacturing managers find it attractive to vendor a portion of the volume of certain products they have the capability to make in-house. Such vendoring can augment capacity and smooth the load on the plant. Since the decision of which products and how much volume to vendor depends on capacity *and* planned production, this is a question that spills over into the aggregate planning module, in which long-term production planning is done. We will discuss this problem in greater detail later and in Chapter 16. From a high-level strategic perspective, it is important to remember that giving business to outside vendors enables them to breed capabilities that may make them into competitors some day. We offer the example of IBM using Microsoft to supply the operating system for its personal computers as one example of what can happen.

3. **Pricing.** We have tried to ignore pricing as much as possible in this book, since it is a factor over which plants generally have little influence. However, in capacity decisions, a valid economic analysis simply cannot be done without some sort of forecast

of prices. We need to know how much revenue will be generated by sales in order to determine whether a particular equipment configuration is economically justified. Because prices are frequently subject to great uncertainty, this is an area in which sensitivity analysis is critical.

4. **Time value of money.** Typically, capacity increases and equipment improvements are made as capital requisitions and then depreciated over time. Interest rate and depreciation schedule, therefore, can have a significant impact on the choice of equipment.

5. **Reliability and maintainability.** As we discussed in Part II, reliability (e.g., mean time to failure (MTTF)) and maintainability (e.g., mean time to repair (MTTR)) are important determinants of capacity. Recall that availability A (the fraction of time a machine is working) is given by

$$A = \frac{\text{MTTF}}{\text{MTTF} + \text{MTTR}}$$

Obviously, all things being equal, we want MTTF to be big and MTTR to be small. But all things are never equal, as we point out in the next two observations.

6. **Bottleneck effects.** As should be clear from the discussions in Part II, capacity increases at bottleneck resources typically have a much larger effect on throughput than increases at nonbottleneck resources. Thus, it would seem that paying extra for high-speed or high-availability machines is likely to be most attractive at a bottleneck resource. However, aside from the fact that a stable, distinct bottleneck may not exist, there are problems with this overly simple reasoning, as we point out in the next observation.

7. **Congestion effects.** The single most neglected factor in capacity analysis, as it is practiced in American industry today, is variability. As we saw again and again in Part II, *variability degrades performance*. The variability of machines, which is substantially affected by failures, is an important determinant of throughput. When variability is considered, reliability and maintainability can become important factors at nonbottleneck resources as well as at the bottleneck.

We will discuss the capacity/facility analysis problem in greater detail in Chapter 18. For now, we point out that it should be done with an eye toward long-term strategic concerns and should explicitly consider variability at some level. In terms of our hierarchical planning structure, the output of a capacity planning exercise is a forecast of the physical capacity of the plant over a horizon at least long enough for the purposes of aggregate planning—typically on the order of two years.

13.5.2   Workforce Planning

As the capacity/facility planning module in Figure 13.9 determines what equipment is needed, the **workforce planning** module analogously determines what workforce is needed to support production. Both planning problems involve long-term issues, since neither the physical plant nor the workforce can be radically adjusted in the near term. So both planning modules work with long-range forecasts of demand and try to construct an environment that can achieve the system's goals. Of course, the actual sequence of events never matches the plan exactly, so both long-term capacity/facility and workforce plans are subject to short-term modification over time.

The basic workforce issues to be addressed over the long term concern how much and what kind of labor to make available. These questions must be answered within the constraints imposed by corporate labor policies. For instance, in plants with unionized labor, labor contracts may restrict who can be hired or laid off, what tasks different labor classifications can be assigned, and what hours people can work. Usually, management spends far more time hammering out the details of such agreements with labor than with determining what labor is required to support a long-term production plan. Although careful use of the workforce planning module cannot undo years of management-labor conflict, it can help both sides focus on issues that are of strategic importance to the firm.

At the root of most long-term workforce planning is a set of estimates of the **standard hours** of labor required by the products made by the plant. For example, a commercial vent hood might require 20 minutes (one-third hour) of a welder's time to assemble. If a welder is available 36 hours per week, then one welder has the capacity to produce $36 \times 3 = 108$ vent hoods per week. Thus, a production plan that calls for 540 vent hoods per week requires five welders.

Simple standard labor hour conversions can be a useful starting point for a workforce planning module. However, they fall far short of a complete representation of the issues involved in workforce planning. These issues include the following:

1. **Worker availability.** Estimates of standard labor hours must be sophisticated enough to account for breaks, vacations, training, and other factors that reduce worker availability. Many firms set "inflation factors" for converting the number of workers directly needed to the number of "onboard" workers. For instance, a multiplier of 1.4 would mean that 14 workers must be employed in order to have the equivalent of 10 directly on the jobs at all times during a given shift.

2. **Workforce stability.** Although production requirements may move up and down suddenly, it is generally neither possible nor desirable to rapidly increase and decrease the size of the workforce. A firm's ability to recruit qualified people, as well as its overall workplace attitude, can be strongly affected by changes in the size of the workforce. Some of these "softer issues" are difficult to incorporate into models but are absolutely critical to maintenance of a productive workforce.

3. **Employee training.** Training new recruits costs money *and* takes the time of current employees. In addition, inexperienced workers require time to reach full productivity. These considerations argue against sudden large increases in the workforce. However, when growth requires rapid expansion of the workforce, concerted efforts are needed to maintain the corporate culture (i.e., whatever it was that made growth occur in the first place).

4. **Short-term flexibility.** A workforce is described by more than head count. The degree of cross-training among workers is an important determinant of a plant's flexibility (its ability to respond to short-term changes in product mix and volume). Thus, workforce planning needs to look beyond the production plan to consider the unplanned contingencies (emergency customer orders, runaway success of a new product) with which the system should be able to cope.

5. **Long-term agility.** The standard labor hours approach views labor as simply another input to products, along with material and capital equipment. But workers represent more than this. In the current era, where products and processes are constantly changing, the workforce is a key source of agility (the plant's ability to rapidly reconfigure a manufacturing system for efficient production of new products as they are introduced). So-called **agile manufacturing** is largely dependent on its people, both managers and workers, to learn and evolve with change.

6. **Quality improvement.** As we noted in Chapter 12, quality, both internal and external, is the result of a number of factors, many under the direct control of workers. Educating machine operators in quality control methods, cross-training workers so that

they develop a systemwide appreciation of the quality implications of their actions, and moderating the influx of new employees so that a corporate consciousness of quality is not undermined—all these are critical parts of a plan to continuously improve quality. Although such factors are difficult to incorporate explicitly into manpower planning models, it is important that they be recognized in the overall workforce planning module.

Workforce planning is a deep and far-reaching subject that occupies a position close to the core of manufacturing management. As such, it goes well beyond operations management or factory physics. In Chapter 16 we will revisit this topic from an analytical perspective and will examine the relationship between workforce planning and aggregate planning. While this is a useful starting place for workforce planning, we remind the reader that it is only that. A well-balanced manpower plan must consider issues such as those listed previously and will require input from virtually all segments of the manufacturing organization.

13.5.3 Aggregate Planning

Once we have estimated future demand and have determined what equipment and labor will be available, we can generate an **aggregate plan** that specifies how much of each product to produce over time. This is the role of the **aggregate planning** module depicted in Figure 13.9. Because different facilities have different priorities and operating characteristics, aggregate plans will differ from plant to plant. In some facilities the dominant issue will be product mix, so aggregate planning will consist primarily of determining how much of each product to produce in each period, subject to constraints on demand, capacity, and raw material availability. In other facilities, the crucial issue will be the timing of production, so the aggregate planning module will seek to balance the costs of production (e.g., overtime and changes in the workforce size) with the costs of carrying inventory while still meeting demand targets. In still others, the focus will be on the timing of staff additions or reductions. In all these, we may also include the possibility of augmenting capacity through the use of outside vendors.

Regardless of the specific formulation of the aggregate planning problem, it is valuable to be able to identify which constraints are binding. For instance, if the aggregate planning module tells us that a particular process center is heavily utilized on average over the next year, then we know that this is a resource that will have to be carefully managed. We may want to institute special operating policies, such as using floating labor, to make sure this process keeps working during breaks and lunches. If the problem is serious enough, it may even make sense to go back and revise the capacity and manpower plans and requisition additional machinery and/or labor if possible.

The decisions that are addressed by the aggregate planning module require a fair amount of advance planning. For instance, if we are seeking to build up inventory for a period of peak demand during the summer, clearly we must consider the production plan for several months prior to the summer. If we want to consider staffing changes to accommodate the production plan, we may require even more advance warning. This generally means that the planning horizon for aggregate planning must be relatively long, typically a year or more. Of course, we should regenerate our aggregate plan more frequently than this, since a year-long plan will be highly unreliable toward the end. It often makes sense to update the aggregate plan quarterly or biannually.

We give specific formulations of representative aggregate planning modules in Chapter 16. Because we can often state the problem in terms of minimizing cost subject to meeting demand, we frequently use the tool of linear programming to help solve the aggregate planning problem. Linear programming has the advantages that

1. It is very fast, enabling us to solve large problems quickly. This is extremely important for using the aggregate planning module in what-if mode.

2. It provides powerful sensitivity analysis capability, for instance, calculating how much additional capacity would affect total cost. This enables us to identify critical resources and quickly gauge the effectiveness of various changes.

As we will see in Chapter 16, linear programming also offers us a great deal of flexibility for representing different aggregate planning situations.

13.5.4   WIP and Quota Setting

The **WIP/quota-setting** module, depicted in Figure 13.9 as working in close conjunction with the aggregate planning module, is needed to translate the aggregate plan to control parameters for a pull system. Recall that the key controls in a pull system are the WIP levels, or card counts, in the production lines. Also, to link the pull system to customer due dates, we need to set an additional control, namely, the production quota. By establishing a quota, and then using buffer capacity to ensure that the quota is met with regularity, we make the system behavior approximate that of the "conveyor model" discussed. The predictability of the conveyor model allows us to coordinate system outputs with customer due dates.

Card Counts.   We include **WIP setting,** or card count setting, at the intermediate level in the PPC hierarchy in Figure 13.9, instead of at the bottom level, to remind the reader that WIP levels should not be adjusted too frequently. As we noted in Chapter 10, WIP is a fairly insensitive control. Altering card counts in an effort to cause throughput to track demand is not likely to work well because the system will not respond rapidly enough. Therefore, like other decisions at this level in the hierarchy, WIP levels should be reevaluated on a fairly infrequent basis, say, quarterly.

Fortunately, the fact that WIP is an insensitive control also makes it relatively easy to set. As long as WIP levels are adequate to attain the desired throughput and are not grossly high, the system will function well. Therefore, it does not make sense to develop highly sophisticated tools for computing WIP levels. In systems that are moving from push to pull, it probably makes sense to set the initial WIP levels in the pull system equal to the average levels that were experienced under push. Then, once the system is operating stably, make incremental reductions. If a kanban-type system is used, so that WIP levels are set at different points in the line, remove cards from those stations with long queues that never or rarely empty out. If a CONWIP system is used, then the overall WIP level can be reduced incrementally. Once workable WIP levels have been established, they should be adjusted infrequently, to ensure that changes are made in response to long-term trends rather than short-term fluctuations.

If we must set the WIP level along a *new* (or reconfigured) routing that is to be run as a CONWIP line, we cannot rely on historical performance to gauge the appropriate WIP level. In this situation, the following is a reasonable rule of thumb. First, establish a desired and *feasible* cycle time for the routing CT and identify the practical production rate r_b^P (e.g., a feasible fraction of the bottleneck rate r_b). Then use Little's law to solve for the WIP level as

$$\text{WIP} = r_b^P \times \text{CT}$$

If r_b^P and CT are realistic, this method will yield a reasonable starting point for WIP, which can be adjusted over time. In general, care must be taken not to underestimate the feasible cycle time or practical production rate, since this will result in too little WIP, with the consequence that throughput will be too low.

Production Quotas.   In addition to WIP levels, the other key parameter for controlling a pull system is the production quota. Hence, **quota setting** is included with the WIP setting module in the PPC hierarchy in Figure 13.9.

The basic idea of a production quota is that we establish a periodic quantity of work that we will (almost) always complete during the quota period. The period under question might be a shift, a day, or a week. In its strictest form, a production quota means that

1. Production during the period stops when quota is reached.
2. Overtime is used at the end of the period to make up any shortage that occurred during regular time.

This allows us to count on a steady output and therefore facilitates planning and due date quoting. Of course, in practice, few quota systems adhere rigidly to this protocol. Indeed, one of the benefits of CONWIP that we cited in Chapter 10 is that it allows working ahead of the schedule when circumstances permit. However, for the purposes of planning a reasonable periodic production quota, it makes sense to model the system as if we stop when quota is reached.

Establishing an economic production quota requires consideration of both cost and capacity data. Relevant costs are those related to lost throughput and overtime. Important capacity parameters include both the mean *and* the standard deviation of production during a specified time interval (e.g., a week or a day). Standard deviation is needed because variability of output has an impact on our ability to make a given production quota. In general, the more variable the production process, the more likely we are to miss the quota.

To see this, consider Figure 13.10. Suppose we have set the production quota for regular time production (e.g., Monday through Friday) to be Q units of work.[3] If we do not make Q units during regular time, then we must run overtime (e.g., Saturday and Sunday) to make up the shortage. Because of the usual contingencies (machine failures, worker absenteeism, yield loss, etc.), the actual amount of work completed during regular time will vary from period to period. Figure 13.10 represents two possible distributions of regular time production that have the same mean μ but different standard deviations σ. The probability of missing the quota is represented by the area under each curve to the left of the value Q. Since the area under curve A, with the smaller standard deviation, is less than that under B, the probability of missing the quota is less. What this means is that if we define a probability of missing the quota that we are willing to live with—a "service level" of sorts—then we will be able to set a higher quota for curve A than for curve B. We can aim closer to capacity because the greater predictability of curve A gives us more confidence in our ability to achieve our goal with regularity.

This analysis suggests that if we knew the mean μ and standard deviation σ of regular time production,[4] a very simple way to set a production quota would be to calculate the quota we can achieve S percent of the time, where S is chosen by the user. If regular time production X can be reasonably approximated by the normal distribution, then we

---

[3] In a simple, single-product model, units of work are equal to physical units. In a more complex, multiproduct situation, units must be adjusted for capacity, for instance, by measuring them in hours required at a critical resource.

[4] We will discuss a mechanism for obtaining estimates of μ and σ from actual operating experience in Chapter 14.

FIGURE 13.10

Probability of missing quota under different production distributions

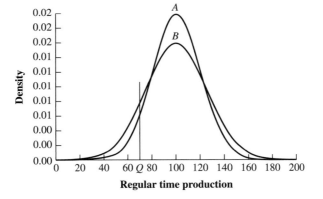

can compute the appropriate quota by finding the value Q that satisfies

$$\Phi\left(\frac{Q - \mu}{\sigma}\right) = 1 - S$$

where $\Phi(\cdot)$ represents the cdf of the standard normal distribution.

For example, suppose that $\mu = 100$, $\sigma = 10$, and we have selected $S = 85$ percent as our service level. Then the quota Q is the value for which

$$\Phi\left(\frac{Q - 100}{10}\right) = 1 - 0.85 = 0.15$$

From a standard normal table, we find that $\Phi(-1.04) = 0.15$. Therefore, we can find Q from

$$\frac{Q - 100}{10} = -1.04$$

$$Q = 89.6$$

A problem with this simple method is that it considers only capacity, not costs. Therefore it offers no guidance as to whether the chosen service level is appropriate. A lower service level will result in a higher quota, which will increase throughput but will also increase overtime costs. A higher service level will result in a lower quota, which will reduce throughput and overtime costs. We offer a model for balancing the cost of lost throughput with the cost of overtime in Appendix 13A and more complex variations on this model in Hopp et al. (1993).

13.5.5   Demand Management

The effectiveness of any production control system is greatly determined by the environment in which it operates. A simple flow line can function well with very simple planning tools, while a complex job shop can be a management nightmare even with very sophisticated tools. This is just a fact of life; some plants are easier to manage than others. But it is also a good reason to remember one of our "lessons of JIT," namely, that *the environment is a control*. For example, if managers can make a job shop look like a flow shop by dedicating machines to "cells" for making particular groups of products, they can greatly simplify the planning and control process.

One key area in which we can shape the environment "seen" by the modules in the lowest levels of the planning hierarchy is in managing customer demands. The **demand management** module shown in Figure 13.9 does this by filtering and possibly adjusting customer orders into a form that produces a manageable master production schedule. As we noted in Chapter 4, leveling demand or "production smoothing" is an essential feature of JIT. Without a stable production volume and product mix, the rate-driven, mixed-model production approach described by Ohno (1988) and the other JIT advocates cannot work. This implies that customer orders cannot be released to the factory in the random order in which they are received. Rather, they must be collected and grouped in a way that maintains a fairly constant loading on the factory. Balancing the concern for factory stability with the desire for dependable customer service and short competitive due date quotes is the challenge of the demand management module.

There are many approaches one could use to quote due dates and establish a near-term MPS within the demand management module. As we discussed, if we establish periodic production quotas, then we can use the conveyor model for predicting flow through the plant. Under these conditions, we can think of customer due date quoting as "loading the conveyor." If we do not have to worry about machine setups and have a capacity cushion, we can quote due dates in the order they are received, using the conveyor model described by Equation (13.19). However, when there is variability and little or no capacity cushion, we must quote due dates using a different procedure (see Chapter 15). Likewise, if batching products according to family (i.e., parts that share important machine setups) is important to throughput, we may want to use some of the sequencing techniques discussed in Chapter 15.

While there are many methods, the important point is not *which* method but that *some* method be employed. Almost anything that achieves consistency with the scheduling procedure will be better than the all-too-common approach of quoting due dates in near isolation from the manufacturing process.

13.5.6   Sequencing and Scheduling

The MPS is still a production *plan,* which must be translated to a work schedule in order to guide what actually happens on the factory floor. In the MRP II hierarchy, shown in Figure 3.2, this figure is carried out by MRP.[5] In the production planning and control hierarchy for pull systems shown in Figure 13.9, we include a **sequencing/scheduling module** that is the pull analog of MRP. As in MRP, the objective of this sequencing/scheduling module is to provide a schedule that governs release times of work orders and materials and then facilitates their movement through the factory.

To paraphrase Einstein, we should strive to make the work schedule as simple as possible, but no simpler. The goal should be to provide people on the floor with enough information to enable them to make reasonable control choices, but not so much as to overly restrict their options or make the schedule unwieldy. What this means in practice is that different plants will require different scheduling approaches. In a simple flow line with no significant setup times, a simple sequence of orders, possibly arranged according to earliest due date (EDD), may be sufficient. Maintaining a first-in-system-first-out (FISFO) ordering of jobs at the other stations will yield a highly predictable and easily manageable output stream for this situation.

However, in a highly complex job shop, with many routings, machine setups, and assemblies of subcomponents, a simple sequence is not even well defined, let alone useful. In the more complex situations, it will not be clear that the MPS is feasible.

---

[5]Recall from Chapter 3 that MRP ("little mrp") refers to *material requirements planning,* the tool for generating planned order releases, while MRP II ("big MRP") refers to *manufacturing resources planning,* the overarching planning system incorporating MRP. *Enterprise resource planning (ERP)* extends the MRP II hierarchy to multiple-facility systems.

Consequently, iteration between the MPS module and the sequencing/scheduling module will be necessary. A procedure for detecting schedule infeasibility and suggesting remedies (e.g., adding capacity, pushing out due dates) is called **capacitated material requirements planning,** or **MRP-C,** and is described in Chapter 15. This procedure integrates the demand management, MPS, and sequencing/scheduling functions into one. In complex situations such as this, we may need to provide a fairly detailed schedule, with specific release times for jobs and materials and predicted arrival times of jobs at workstations. Of course, the data requirements and maintenance overhead of the system required to generate such a schedule may be substantial, but this is the price we pay for complexity.

13.5.7   Shop Floor Control

Regardless of how accurate and sophisticated the scheduling tool is, the actual work sequence never follows the schedule exactly. The **shop floor control (SFC)** module shown in Figure 13.9 uses the work schedule as a source of general guidance, adhering to it whenever possible, but also making adjustments when necessary. For instance, if a machine failure delays the arrival of parts required in an assembly operation, the SFC module must determine how the work sequence should be changed. In theory, this can be an enormously complex problem, since the number of options is immense—we could wait for the delayed part, we could jump another job ahead in the sequence, we could scramble the entire schedule, and so on. But, in practice, we must make decisions quickly, in real time, and therefore cannot hope to consider every possibility. Therefore, the SFC module must restrict attention to a reasonable class of actions and help the user make effective and robust choices.

To take advantage of the pull benefits we discussed in Chapter 10, we favor an SFC module based on a pull mechanism. The CONWIP protocol is perhaps the simplest approach and therefore deserves at least initial consideration. To use CONWIP in conjunction with the sequencing/scheduling module, we establish a WIP cap and do not allow releases into the line when the WIP exceeds the maximum level. This will serve to delay releases when the plant is behind schedule and further releases cannot help. CONWIP also provides a mechanism for working ahead of the schedule when things are going well. If the WIP level falls below the WIP cap before the next job is scheduled to be released, we may want to allow the job to start anyway. As long as we do not work too far ahead of the schedule and cause a loss of flexibility by giving parts "personality" too early, this type of work-ahead protocol can be very effective.

Chapter 14 is devoted to the SFC problem; there we will discuss implementation of CONWIP-type SFC modules and will identify situations in which more complicated SFC approaches may be necessary.

13.5.8   Real-Time Simulation

In a manufacturing management book such as this, one is tempted to make sweeping admonitions of the form "Never have hot jobs," and "Always follow the published schedule." Certainly, the factory would be easier to run if such rigid rules could be followed. But the ultimate purpose of a manufacturing plant is not to make the lives of its managers easy; it is to make money by satisfying customers. Since customers change their minds, ask for favors, etc., the reality of almost every manufacturing environment is that sometimes emergencies occur and therefore some jobs must be given special treatment. One would hope that this doesn't occur all the time (although it all too frequently does, as in a plant we once visited where *every* job shown on the MRP system had been designated "rush"). But, given that it will happen, it makes sense to design the planning system to survive these eventualities, and even provide assistance with them. This is the job of the **real-time simulation** module shown in Figure 13.9.

We have found simulation to be useful in dealing with emergency situations, such as hot jobs. By simulation, however, we do not mean full-blown Monte Carlo simulation with random number generators and statistical output analysis. Instead, we are referring to a very simple deterministic model that can mimic the behavior of the factory for short periods of time. One option for doing this is to make use of the previously described conveyor model to represent the behavior of process centers and take the current position of WIP in the system, a list of anticipated releases, and a set of capacity data (including staffing), to generate a set of job output times. Such a model can be reasonably accurate in the near term (e.g., over the next week), but because it cannot incorporate unforeseen events such as machine failures, it can become very inaccurate over the longer term. Thus, as long as we restrict the use of such a model to answering short-term what-if questions—What will happen to due date performance of various other jobs if we expedite job n?—this type of tool can be very useful. Knowing the likely consequences in advance of taking emergency actions can prevent causing serious disruption of the factory for little gain.

13.5.9   Production Tracking

In the real world there will always be contingencies that require human intervention by managers. While this may seem discouraging to the designers of production planning systems, it is one of the key reasons for the existence of manufacturing managers. A good manager should strive for a system that functions smoothly most of the time, but also be ready to take corrective action when things do not function smoothly. To detect problems in a timely fashion and formulate responses, a manager must have key data at her fingertips. These data might include the location of parts in the factory, status of equipment (e.g., up, down, under repair), and progress toward meeting schedule. The **production tracking** module depicted in Figure 13.9 is responsible for tabulating and displaying this type of data in a usable format.

Many of the planning modules in Figure 13.9 rely on estimated data. In particular, capacity data are essential to several planning decisions. A widely used practice for estimating capacity of currently installed equipment is to start with the rated capacity (e.g., in parts per hour) and reduce this number according to various detractors (machine downtime, operator unavailability, setups, etc.). Since each detractor is subject to speculation, such estimates can be seriously in error. For this reason, it makes sense to use the production tracking module to collect and update capacity data used by other planning modules. As we will see in Chapter 14, we can use the technique of exponential smoothing from forecasting to generate a smoothed estimate of capacity and to monitor trends over time.

13.6   Conclusions

In this chapter, we have offered an overview of a production planning and control hierarchy that is consistent with the pull production systems we discussed in Chapters 4 and 10. This overview was necessarily general, since there are many ways a planning system could be constructed and different environments are likely to require different systems. We will fill in specifics in subsequent chapters on the individual planning modules. For now, we close with a summary of the main points of this chapter pertaining to the overall structure of a planning hierarchy:

1. *Planning should be done hierarchically.* It makes no sense to try to use a precise, detailed model to make general, long-term decisions on the basis of rough, speculative data. In general, the shorter the planning horizon, the more details are required. For this reason, it is useful to separate planning problems into long-term (strategic), intermediate-term (tactical), and short-term (control) problems. Similarly, the level of detail about products increases with nearness in time, for instance, planning for total volume in the very long term, part families in the intermediate term, and specific part numbers in the very short term.

2. *Consistency is critical.* Good individual modules can be undermined by a lack of coordination. It is important that common capacity assumptions, consistent staffing assumptions, and coordinated data inputs be used in the different planning modules.

3. *Feedback forces consistency and learning.* Some manufacturing managers continue to use poor-quality data without checking their accuracy or setting up a system for collecting better data from actual plant performance. Regardless of how it is done (e.g., manually or in automated fashion), it is important to provide some kind of feedback for updating critical parameters. Furthermore, by providing a mechanism for observing and tracking progress, feedback promotes an environment of continual improvement.

4. *Different plants have different needs.* The above principles are general; the details of implementing them must be specific to the environment. Small, simple plants can get away with uncomplicated manual procedures for many of the planning steps. Large, complex plants may require sophisticated automated systems. Although we will be as specific as possible in the remainder of Part III, the reader is cautioned against taking details too literally; they are presented for the purposes of illustration and inspiration and cannot replace the thoughtful application of basics, intuition, and synthesis.

APPENDIX 13A
A QUOTA-SETTING MODEL

The key economic tradeoff to consider in the quota-setting module is that between the cost of lost throughput and the cost of overtime. High production quotas tend to increase throughput, but run the risk of requiring more frequent overtime. Low quotas will reduce overtime, but will also reduce throughput.

To develop a specific quota-setting model, let us consider regular time consisting of Monday through Friday (three shifts per day) with Saturday available for preventive maintenance (PM) and catch-up. If catch-up time is needed, we assume a full shift is worked (e.g., union regulations or company policy requires it). Consequently the cost of overtime is essentially fixed, and we will represent it by C_{OT}. If we let the net profit per standardized unit of production be p and the total expected profit (net revenue minus expected overtime cost) be denoted by Z, the quota-setting problem can be formally stated as

$$\max_{Q} \ Z = pQ - C_{OT}\mathrm{P} \text{ (overtime is needed)} \tag{13.20}$$

Notice that, as expected, decreasing Q affects the objective by lost sales, while increasing Q will affect it by increasing the probability that overtime will be needed. The optimization problem is to find the value of Q that strikes the right balance.

Where shifts are long compared to the time to produce one part, it may be reasonable to assume that production during regular time is normally distributed with mean μ and standard deviation σ. This assumption allows us to express the weekly quota as $Q = \mu - k\sigma$. Now the question becomes,

How many standard deviations below mean production should we set the quota to be? In other words, our decision variable is now k. Under this assumption, we can rewrite Equation (13.20) as

$$\max_{k} \ Z = p(\mu - k\sigma) - C_{OT}[1 - \Phi(k)] \tag{13.21}$$

where $\Phi(k)$ represents the cumulative distribution function of the standard normal distribution.

It not difficult to show (although we will not burden the reader with the details) that the unique solution to Equation (13.21) is

$$k^* = \sqrt{2 \ln \frac{C_{OT}}{\sqrt{2\pi} \, p\sigma}} \tag{13.22}$$

We can then express the optimal quota directly in units of work, instead of units of standard deviations, as follows:

$$Q^* = \mu - k^*\sigma \tag{13.23}$$

Notice that since k^* will never be negative, Equation (13.23) implies that the optimal quota will always be less than mean regular time production. As long as overtime costs are sufficiently high to make using overtime on a routine basis unattractive, this result will be reasonable. If we were to use a quota *equal* to the mean regular time production, then we would expect to miss it, and require overtime, approximately 50 percent of the time. Hence, if overtime is sufficiently expensive, less frequent use of it will be economical; therefore we should choose a quota less than the mean regular time production, and this model is plausible.

However, it is quite possible that the profitability of additional sales outweighs the cost of overtime. In this situation, our intuition tells us that a high quota (i.e., to force additional production) may be attractive, even if it results in missing the quota more than 50 percent of the time. For instance, consider an example with the following costs and production parameters:

$$p = \$100 \qquad \mu = 5{,}000$$
$$C_{OT} = \$10{,}000 \qquad \sigma = 500$$

Notice that we can "pay" for overtime with the profits of just 100 units, which is only 2 percent of the mean regular time production. This means that there is strong incentive to use the overtime period for extra production. Using our model to analyze this issue by substituting the above numbers into expression (13.22), we get

$$k^* = \sqrt{-5.06}$$

which is mathematically ridiculous. Clearly, the model runs into trouble whenever

$$\frac{C_{OT}}{\sqrt{2\pi} \, p\sigma} < 1 \tag{13.24}$$

because the natural logarithm term in Equation (13.22) becomes negative. In economic terms, this means that the fixed cost of overtime is not large enough to discourage the use of overtime for routine production. In practical terms, it means either of the following:

1. The fixed overtime cost should be reexamined, and perhaps altered. It may also make sense to include a variable (i.e., per unit) overtime cost. Development of such a model is given in Hopp, Spearman, and Duenyas (1993).

2. It may really be economically attractive to use overtime for routine production. If this is the case, it may make sense to run continuously, without capacity cushions. To set a target quota for the purposes of quoting due dates to customers, we need to balance the cost of running at less than maximum capacity with the cost of failing to meet a promised due date. A model for this case is also described in Hopp, Spearman, and Duenyas (1993).

The above simple model can be used to give a rough measure of the economics of capacity parameters. Clearly, Equations (13.21) and (13.22) indicate that both the mean and the standard deviation of regular time production are important. By using these equations, we can compute the effect on the weekly profit of changes in various parameters. In particular, we can examine the

FIGURE 13.11

Weekly profit as a function of σ when $\mu = 100$

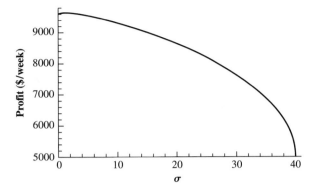

effect of changes in the mean of regular time production μ and standard deviation of regular time production σ.

To see this, consider a simple example in which $p = \$100$, $C_{OT} = \$10,000$, and μ and σ are varied to determine their impact. From Equation (13.21) it is obvious that profit will increase linearly in mean regular time capacity μ. If σ is fixed, k^* does not change when μ is varied. Therefore, each increase in μ by 1 unit increases Z by p. Obviously, we are able to make more and therefore sell more.[6]

The situation is a little more complex when μ is fixed but σ is varied. This is because (from (13.22)) k^* will change as σ is altered. Furthermore, we must be careful that the term inside the square root of Equation (13.22) does not become negative. Condition (13.24) implies that we must have

$$\sigma > \frac{C_{OT}}{\sqrt{2\pi}\,p} = \frac{10,000}{\sqrt{2\pi}\,100p} = 39.9$$

for k^* to be well defined. Figure 13.11 plots the optimal weekly profit when μ is fixed at 100 units and σ is varied from 0 to 39.9. This figure illustrates the general result that profits increase when variability is reduced. The reason for this is that when regular time production is less variable, we can set quota closer to capacity without risking frequent overtime. Thus, we can achieve greater sales revenues without incurring greater overtime costs.

Study Questions

1. Why does it make sense to address the problems of planning and control in a manufacturing system with a hierarchical system? What would a nonhierarchical system look like?

2. Is it reasonable to specify rules regarding the frequency of regeneration of particular planning functions (e.g., "aggregate planning should be done quarterly")? Why or why not?

3. Give some possible reasons why MRP has spawned elaborate hierarchical planning structures while JIT has not.

4. Why is it important for the various modules in a hierarchical planning system to achieve consistency? Why is such consistency not always maintained in practice?

5. What is the difference between *causal forecasting* and *time series forecasting?*

6. Why might an exponential smoothing model exhibit negative bias? An exponential smoothing model with a linear trend?

7. In this era of rapid change and short product lifetimes, it is common for process technology to be used to produce several generations of a product or even completely new products. How might this fact enter into the decisions related to capacity/facility planning?

---

[6]Note that this is only true because of our assumption that capacity is the constraint on sales. If demand becomes the constraint, then this is clearly no longer true, since it makes no sense to set the quota beyond what can be sold.

8. In what ways are capacity/facility planning and workforce planning analogous? How do they differ?

9. How must the capacity/facility planning and aggregate planning be coordinated? What can happen if they are not?

10. One of the functions of sequencing and scheduling is to make effective use of capacity by balancing setups and due dates. This implies that actual capacity is not known until a schedule is developed. But both the capacity/facility planning and aggregate planning functions rely on capacity data. How can they do this in the absence of a schedule (i.e., how can they be done at a higher level in the hierarchy than sequencing or scheduling)?

11. How is demand management practiced in MRP? In JIT?

12. If a plant generates a detailed schedule at the beginning of every week, does it need a shop floor control module? If so, what functions might an SFC module serve in such a system?

13. What purpose does feedback serve in a hierarchical production planning system?

Problems

1. Suppose the monthly sales for a particular product for the past 20 months have been as follows:

Month	1	2	3	4	5	6	7	8	9	10
Sales	22	21	24	30	25	25	33	40	36	39
Month	11	12	13	14	15	16	17	18	19	20
Sales	50	55	44	48	55	47	61	58	55	60

   a. Use a five-period moving average to compute forecasts of sales for months 6 to 20 and a seven-period moving average to compute forecasts for months 8 to 20. Which fits the data better for months 8 to 20? Explain.

   b. Use an exponential smoothing approach with smoothing constant $\alpha = 0.2$ to forecast sales for months 2 to 20. Change α to 0.1. Does this make the fit better or worse? Explain.

   c. Using exponential smoothing, find the value of α that minimizes the mean squared deviation (MSD) over months 2 to 20. Find the value of α that minimizes BIAS. Are they the same? Explain.

   d. Use an exponential smoothing with a linear trend and smoothing constants $\alpha = 0.4$ and $\beta = 0.2$ to predict output for months 2 to 20. Does this fit better or worse than your answers to *b*? Explain.

2. The following data give closing values of the Dow Jones Industrial Average for the 30 weeks, months, and years prior to August 1, 1999.

   a. Use exponential smoothing with a linear trend and smoothing coefficients of $\alpha = \beta = 0.1$ on each set of data to generate forecasts for the Dow Jones Industrial Average on August 1, 2000. Which data set do you think yields the best forecast?

   b. What weight does a one-year-old data point get when we use smoothing constant $\alpha = 0.1$ on the weekly data? On the monthly data? On the annual data? What smoothing constant for the monthly model that gives the same weight to one-year-old data is given by the annual model with $\alpha = 0.1$?

   c. Does using the adjusted smoothing constant computed in part *b* (for α and β) in the monthly model make it predict the closing price for August 1, 2000? If not, why not?

   d. How much value do you think time series models have for forecasting stock prices? What features of the stock market make it difficult to predict, particularly in the short term?

Weekly Data		Monthly Data		Annual Data	
Date	Close	Date	Close	Date	Close
1/4/99	9,643.3	2/1/97	6,877.7	8/1/69	836.7
1/11/99	9,340.6	3/1/97	6,583.5	8/1/70	764.6
1/18/99	9,120.7	4/1/97	7,009.0	8/1/71	898.1
1/25/99	9,358.8	5/1/97	7,331.0	8/1/72	963.7
2/1/99	9,304.2	6/1/97	7,672.8	8/1/73	887.6
2/8/99	9,274.9	7/1/97	8,222.6	8/1/74	678.6
2/15/99	9,340.0	8/1/97	7,622.4	8/1/75	835.3
2/22/99	9,306.6	9/1/97	7,945.3	8/1/76	973.7
3/1/99	9,736.1	10/1/97	7,442.1	8/1/77	861.5
3/8/99	9,876.4	11/1/97	7,823.1	8/1/78	876.8
3/15/99	9,903.6	12/1/97	7,908.3	8/1/79	887.6
3/22/99	9,822.2	1/1/98	7,906.5	8/1/80	932.6
3/29/99	9,832.5	2/1/98	8,545.7	8/1/81	881.5
4/5/99	10,173.8	3/1/98	8,799.8	8/1/82	901.3
4/12/99	10,493.9	4/1/98	9,063.4	8/1/83	1,216.2
4/19/99	10,689.7	5/1/98	8,900.0	8/1/84	1,224.4
4/26/99	10,789.0	6/1/98	8,952.0	8/1/85	1,334.0
5/3/99	11,031.6	7/1/98	8,883.3	8/1/86	1,898.3
5/10/99	10,913.3	8/1/98	7,539.1	8/1/87	2,663.0
5/17/99	10,829.3	9/1/98	7,842.6	8/1/88	2,031.7
5/24/99	10,559.7	10/1/98	8,592.1	8/1/89	2,737.3
5/31/99	10,799.8	11/1/98	9,116.6	8/1/90	2,614.4
6/7/99	10,490.5	12/1/98	9,181.4	8/1/91	3,043.6
6/14/99	10,855.6	1/1/99	9,358.8	8/1/92	3,257.4
6/21/99	10,552.6	2/1/99	9,306.6	8/1/93	3,651.3
6/28/99	11,139.2	3/1/99	9,786.2	8/1/94	3,913.4
7/5/99	11,193.7	4/1/99	10,789.0	8/1/95	4,610.6
7/12/99	11,209.8	5/1/99	10,559.7	8/1/96	5,616.2
7/19/99	10,911.0	6/1/99	10,970.8	8/1/97	7,622.4
7/26/99	10,655.1	7/1/99	10,655.1	8/1/98	7,539.1
8/2/99	10,714.0	8/1/99	10,829.3	8/1/99	10,829.3

3. Hamburger Heaven has hired a team of students from the local university to develop a forecasting tool for predicting weekly burger sales to assist in the purchasing of supplies. The assistant manager, who has taken a couple of college classes, has heard of exponential smoothing and suggests that the students try using it. He gives them the following data on sales for the past 16 weeks.

Week	1	2	3	4	5	6	7	8
Sales	3,500	3,700	3,400	3,900	4,100	3,500	3,600	4,200

Week	9	10	11	12	13	14	15	16
Sales	9,300	8,900	9,100	9,200	9,300	9,000	9,400	9,100

a. What happens if exponential smoothing (with no trend) is applied to these data in a conventional manner? Use a smoothing constant $\alpha = 0.3$.

b. Does it improve the forecast if we use exponential smoothing with a linear trend and smoothing constants $\alpha = \beta = 0.3$?

c. Suggest a modification of exponential smoothing that might make more sense for this situation.

4. Select-a-Model offers computer-generated photos of people posing with famous supermodels. You simply send in a photo of yourself, and the company sends back a photo of you skiing, or boating, or night clubbing, or whatever, with a model. Of course, Select-a-Model must pay the supermodels for the use of their images. To anticipate cash flows, the company wants to set up a forecasting system to predict sales. The following table gives monthly demand for the past two years for three of the top-selling models.

Month	Model 1	Model 2	Model 3
1	82	95	148
2	25	12	125
3	44	90	78
4	36	56	53
5	27	54	25
6	91	65	29
7	100	65	9
8	33	92	68
9	97	91	84
10	92	116	110
11	39	141	147
12	94	137	120
13	70	124	147
14	72	90	109
15	90	72	96
16	73	71	70
17	6	92	42
18	30	140	36
19	98	170	34
20	9	150	28
21	0	141	71
22	17	180	102
23	25	171	103
24	11	124	144

a. Plot the demand data for all three models, and suggest a forecasting model that might be suited to each.

b. Find suitable constants for model 1. How good a predictor is the resulting model?

c. Find suitable constants for model 2. How good a predictor is the resulting model?

d. Find suitable constants for model 3. How good a predictor is the resulting model?

5. Can-Do Canoe sells lightweight portable canoes. Quarterly demand for its most popular product family over the past three years has been as follows:

Year	1996				1997				1998			
Quarter	1	2	3	4	1	2	3	4	1	2	3	4
Demand	25	120	40	60	30	140	60	80	35	150	55	90

    a. Use an exponential smoothing model with smoothing constant $\alpha = 0.2$ to develop a forecast for these data. How does it fit? What is the resulting MSD?

    b. Use an exponential smoothing with a linear trend model with smoothing constants $\alpha = \beta = 0.2$ to develop a forecast for these data. How does it fit? What is the resulting MSD?

    c. Use the Winters method with smoothing constants $\alpha = \beta = \gamma = 0.2$ to develop a forecast for these data. How does it fit? What is the resulting MSD?

    d. Find smoothing constants that minimize MSD over the second two years of data. How does the resulting forecast fit the data in the third year?

    e. Find smoothing constants that minimize MSD over the third year of data. How much better does the model fit the data in the third year than that of part *d*? Which model, *d* or *e*, do you think is likely to better predict demand in year 4?

6. Suppose a plant produces 50 customized high-performance bicycles per day and maintains on average 10 days' worth of WIP in the system.

    a. What is the average cycle time (i.e., time from when an order is released to the plant until the bicycle is completed, ready to ship)?

    b. When would the *conveyor model* predict that the 400th bicycle will be completed?

    c. Suppose we currently have orders for 1,000 bicycles (i.e., including the orders for the 500 bicycles that have already been released to the plant) and a customer is inquiring about when we could deliver an order of 50 bicycles. Use the conveyor model to predict when this new order will be completed. If we have flexibility concerning the due date we quote to the customer, should we quote a date calculated earlier, later, or at the same time as that computed using the conveyor model? Why?

7. Marco, the manager of a contractor's supply store, is concerned about predicting demand for the DeWally 519 hammer drill in order to help plan for purchasing. He has brought in a team of MBAs, who have suggested using a moving-average or exponential smoothing method. However, Marco is not sure this is the right approach because, as he points out, sales of the drill are affected by price. Since the store periodically runs promotions during which the price is reduced, he thinks that price should be accounted for in the forecasting model. The following are price and sales data for the past 20 weeks.

Week	Price	Sales
1	199	25
2	199	27
3	199	24
4	179	35
5	199	21
6	199	26
7	199	29
8	199	28
9	199	32
10	169	48
11	169	45
12	199	30
13	199	38
14	199	37
15	199	38
16	199	39
17	179	45
18	199	40
19	199	39
20	199	42

a. Propose an alternative to a time series model for forecasting demand for the Dewally 519.
b. Use your method for the first n weeks of data to predict sales in week $n + 1$ for $n = 15, \ldots, 19$. How well does it work?
c. What does your model predict sales will be in week 21 if the price is $199? If the price is $179?

8. Suppose Clutch-o-Matic, Inc., has been approached by an automotive company to provide a particular model of clutch on a daily basis. The automotive company needs 1,000 clutches per day, but expects to divide this production among several suppliers. What the company wants from Clutch-o-Matic is a commitment to supply a specific number each day (i.e., a daily quota). Under the terms of the contract, failure to supply the quota will result in a financial penalty.

Clutch-o-Matic has a line it could dedicate to this customer and has computed that the line has a mean daily production of 250 clutches with a standard deviation of 50 clutches under single (eight-hour) shift production. A clutch sells for $200, of which $30 is profit. If overtime is used, union rules require at least two hours of overtime pay. The cost of worker pay, supervisor pay, utilities, etc., for running a typical overtime shift has been estimated at $6,200.

a. What is the profit-maximizing quota from the perspective of Clutch-o-Matic?
b. What is the average daily profit to Clutch-o-Matic if the quota is set at the level computed in *a*?
c. If the automotive company insists on 250 clutches per day, is it still profitable for Clutch-o-Matic? How much of a decrease in profit does this cause relative to the quota from *b*?
d. How might a quota-setting model like this one be used in the negotiation process between a supplier and its customers requesting JIT contracts?

14　Shop Floor Control

Even a journey of one thousand li begins with a single step.
Lao Tzu

14.1　Introduction

Shop floor control (SFC) is where planning meets parts. As such, it is the foundation of a production planning and control system. Because of its proximity to the actual manufacturing process, SFC is also a natural vehicle for collecting data for use in the other planning and control modules. A well-designed SFC module both controls the flow of material through the plant and makes the rest of the production planning system easier to design and manage.[1]

Despite its logical importance in a production planning hierarchy, SFC is frequently given little attention in practice. In part, this is because it is perceived, too narrowly, we think, as purely material flow control. This view makes it appear that once one has a good schedule in hand, the SFC function can be accomplished by routing slips attached to parts and giving the sequence of process centers to be visited; one simply works on parts in the order given by the schedule and then moves them according to the routing slips. As we will see here and in Chapter 15, even with an effective scheduling module, the control of material flow is frequently not so simple. No scheduling system can anticipate random disruptions, but the SFC module must accommodate them anyway. Furthermore, as we have already noted and will discuss further in this chapter, material flow control is simply too narrow a focus for SFC. When one includes the other functions that are appropriately included in SFC, this module assumes a critical function in the overall planning hierarchy.

There may be another reason for the lack of attention to SFC. A set of results from the operations management literature indicates that decisions affecting material flow are less important to plant performance than are decisions dealing with shaping the production environment. Krajewski et al. (1987) used simulation experiments to show

---

[1]We remind the reader that we are using the term *module* to include all the decision making, record keeping, and computation associated with a particular planning or control problem. So while the SFC module may make use of a computer program, it involves more than this. Indeed, some SFC modules may not even be computerized at all.

that the benefits from improving the production environment by reducing setup times, improving yields, and increasing worker flexibility were far larger than the benefits from switching to a kanban system from a reorder point or MRP system. On the basis of their study, they concluded that (1) reshaping the production environment was key to the Japanese success stories, and (2) if a firm improves the environment enough, it does not make much difference what type of production control system is used. In a somewhat narrower vein, Roderick et al. (1991) used simulation to show that the release rate had a far greater effect on performance than did work sequencing at individual machines. Their conclusion was that master production schedule (MPS) smoothing is likely to have a stronger beneficial effect than sophisticated dispatching techniques for controlling work within the line.

If one narrowly interprets SFC to mean dispatching or flow control between machines, then studies like these do indeed tend to minimize its importance. However, if one takes the broader view that SFC controls flow *and* establishes links between other functions, then the design of the SFC module serves to shape the production environment. For instance, the very decision to install a kanban system evinces a commitment to small-lot manufacture and setup reduction. Moreover, a pull system automatically governs the release rate into the factory, thereby achieving the key benefits identified by Roderick et al.

But is kanban (or something like it) *essential* to achieving these environmental improvements? Krajewski et al. imply that environmental improvements, such as setup reduction, could be just as effective without kanban, while JIT proponents contend that kanban is needed to apply the necessary pressure to force these improvements. Our view is closer to that of the JIT proponents; without an SFC module that promotes environmental improvements and, by means of data collection, documents their effectiveness, it is extremely difficult to identify areas of leverage and make changes stick. Thus, we will take the reshaping of the production environment as part and parcel of SFC module design.

On the basis of our discussions in Chapters 4, 10, and 13, we feel that the most effective (and manageable) production environment is that established by a pull system. Recall that the basic distinction between push and pull is that push systems schedule production, while pull systems authorize production. Fundamental to any pull mechanism for authorizing production is a WIP cap that limits the total inventory in a production line. In our terminology, a system cannot be termed *pull* if it does not establish a WIP cap. Complementing this defining feature are a host of other supporting characteristics of pull systems, including setup time reduction, worker cross-training, cellular layouts, quality at the source, and so on. The manner and extent to which these techniques can be used depend on the specific system. The objective for the SFC module is to make the actual production environment as close as possible to the ideal environments we examined in Chapters 4 and 10. At the same time, the SFC module should be relatively easy to use, integrate well with the other planning functions, and be flexible enough to accommodate changes the plant is likely to face. As we will see, because manufacturing settings differ greatly, the extent to which we can do this will vary widely, as will the nature of the appropriate SFC module.

Figure 14.1 illustrates the range of functions one can incorporate into the SFC module. At the center of these functions is **material flow control (MFC),** without which SFC would not be shop floor control. Material flow control is the mechanism by which we decide which jobs to release into the factory, which parts to work on at the individual workstations, and what material to move to and between workstations. Although SFC is sometimes narrowly interpreted to consist solely of material flow control, there are a

Figure 14.1

Potential functions of the SFC module

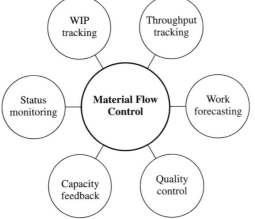

number of other functions that are integrally related to material flow control, and a good SFC module can provide platform for these.

WIP tracking, status monitoring, and **throughput tracking** deal with what is happening in the plant in real time. WIP tracking involves identifying the current location of parts in the line. Its implementation can be detailed and automated (e.g., through the use of optical scanners) or rough and manual (e.g., performed by log entries at specified points in the line). Status monitoring refers to surveillance of other parameters in the line besides WIP position, such as machine status (i.e., up or down) or staffing situation. Throughput tracking consists of measuring output from the line or plant against an established production quota and/or customer due dates, and it can be used to anticipate the need for overtime or staffing shifts.

Since the SFC module is the place where real-time control decisions are implemented, it is a natural place for monitoring these types of changes in real-time status of the line. If the SFC module is implemented on a computer, these data collection and display tasks are likely to share files used by the SFC module for material flow control. Even if material flow control is implemented as a manual system, it makes sense to think about monitoring the system in conjunction with controlling it, since this may have an impact on the way paperwork forms are devised. A specific mechanism for monitoring the system is **statistical throughput control (STC),** in which we track progress toward making the periodic production quota. We give details on STC in Section 14.5.1.

In addition to collecting information about real-time status, the SFC module is a useful place to collect and process some information pertaining to the future beyond real time. One possibility is the **real-time simulation** function, in which projections are made about the timing of arrival of specific parts at various points in the line. Chapter 13 addressed this function as an off-line activity. However, it is also possible to incorporate a version of the real-time simulation module directly into the SFC module. The basic mechanism is to use information about current WIP position, collected by the WIP tracking function, plus a model of material flow (e.g., based on the conveyor model) to predict when a particular job will reach a specific workstation. Being able to call up such information from the system can allow line personnel to anticipate and prepare for jobs.

A different function of the SFC module is the collection of data to update capacity estimates. This **capacity feedback** function is important for ensuring that the high-level planning modules are consistent with low-level execution, as we noted in Chapter 13. Since the SFC module governs the movement of materials through the plant, it is the natural place to measure output. By monitoring input over time we can estimate the actual capacity of a line or plant. We will discuss the details of how to do this in Section 14.5.2.

The fact that move points represent natural opportunities for quality assurance establishes a link between the SFC module and **quality control.** If the operator of a downstream workstation has the authority to refuse parts from an upstream workstation on the basis of inadequate quality, then the SFC module must recognize this disruption of a requested transaction. The material flow control function must realize that replacements for rejected parts are required or that rework will cause delays in part arrivals; the WIP tracking function must note that these parts did not move as anticipated; and the work forecasting function must consider the delay in order to make work projections. Furthermore, since quality problems must be noted for these control purposes, it is often convenient to use the system to keep a record of them. These records provide a link to a statistical process control (SPC) system for monitoring quality performance and identifying opportunities for improvement.

In the remainder of this chapter, we give

1. An overview of issues that must be resolved prior to designing an SFC module.
2. A discussion of CONWIP as the basis for an SFC module.
3. Extensions of CONWIP schemes.
4. Mechanisms for tracking production in order to measure progress toward quota in the short term, and collecting and validating capacity data for other planning modules in the long term.

14.2   General Considerations

One is naturally tempted to begin a discussion of the design of an SFC system by addressing questions about the control mechanism itself: Should work releases be controlled by computer? Should kanban cards be used? How do workers know which jobs to work on? And so on. However, even more basic questions should be addressed first. These deal with the general physical and logical environment in which the SFC system must operate.

To develop a reasonable perspective on the management implications of the SFC module, it is important to consider shop floor control from both a *design* and a *control* standpoint. Design issues deal with establishing a system within which to make decisions, while control issues treat the decisions themselves. For instance, choosing a work release mechanism is a design decision, while selecting parameters (e.g., WIP levels) for making the mechanism work is a control issue. We will begin by addressing relatively high-level design topics and will move progressively toward lower-level control topics throughout the chapter.

14.2.1   Gross Capacity Control

Production control systems work best in stable environments. When demand is steady, product mix is constant, and processes are well behaved, almost any type of system (e.g., reorder points, MRP, or kanban) can work well, as shown by the simulation studies of Krajewski et al. (1987). From a manufacturing perspective, we would like to set up production lines and run them at a nice, steady pace without disruptions. Indeed, to a large extent, this is precisely what JIT, with its emphasis on production smoothing and setup reduction, attempts to do. But efforts to create a smooth, easy production environment can conflict with the business objectives to make money, grow and maintain market share, and ensure long-term viability. Customer demand fluctuates, products emerge and decline, technological competition forces us to rely on new and unstable processes. Therefore, while we should look for opportunities to stabilize the environment, we must take care not to lose sight of higher-level objectives in our zeal to do this. We shouldn't forgo an opportunity to gain a strategic edge via a new technology simply because the old technology is more stable and easier to manage.

Even while we respond to market needs, there are things we can do to avoid unnecessary volatility in the plant. One way to stabilize the environment in which the SFC module must operate is to use gross capacity control to ensure that, when running, the lines are close to optimally loaded. The goal is to avoid drastic swings in line speed by controlling the amount of time the line, or part of it, is used. Specific options for gross capacity control include

1. *Varying the number of shifts.* For instance, three shifts per day may be used during periods of heavy demand, but only two shifts during periods of lighter demand. A plant can use this option to match capacity to seasonal fluctuations in demand. However, since it typically involves laying off and rehiring workers, it is only appropriate for accommodating persistent demand changes (e.g., months or more).

2. *Varying the number of days per week.* For instance, weekends can be used to meet surges of demand. Since weekend workers can be paid on overtime, a plant can use this approach on much shorter notice than it can use shift changes. Notice that we are talking here of *planned overtime,* where the weekends are scheduled in advance because of heavy demand. This is in contrast with *emergency overtime* used to make up quota shortfalls, as we discussed in Chapter 13.

3. *Varying the number of hours per day.* Another source of planned overtime is to lengthen the workday, for instance, from 8 to 10 hours.

4. *Varying staffing levels.* In manual operations, capacity can be augmented by adding workers (e.g., floating workers from another part of the plant, or temporary hires). In multimachine workstations, managers can alter capacity by changing the number of machines in use, possibly requiring staffing changes as well.

5. *Using outside vendors.* One way to maintain a steady loading on a plant or line is to divert work beyond a specified level to another firm. Ideally, this transfers at least part of the burden of demand variability to the vendor.[2]

As the term *gross* capacity control implies, these activities can only alter the effective capacity in a rough fashion. Shifts must be added whole and only infrequently removed. Weekend overtime may have to be added in specific amounts (e.g., a day or half-day) due to union rules or personnel policy. Options for varying capacity through floating workers are limited by worker skill levels and loadings in other portions of the plant. Adding and releasing temporary workers requires training and other expenses, which

---

[2]Of course, there is no guarantee that a vendor will be able to accommodate varying demand any better than the firm itself. Moreover, vendors who can are likely to charge for it. So while vendors can be useful, they are hardly a panacea.

Figure 14.2

Throughput as a function of WIP in a single-product line

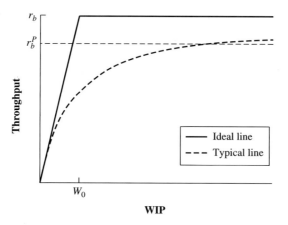

limits the flexibility of this option. Vendoring contracts may require minimum and/or maximum amounts of work to be sent to the vendor, so this approach may remove only part of the demand variability faced by the firm. Moreover, since finding and certifying vendors is a time-consuming process, vendor contracts are likely to persist over time.

Despite the limitations of the options discussed, it is important that they, or other methods, be used to match capacity to demand at least roughly. Huge variations in the workload of a line will induce tremendous variability throughout the line and will seriously degrade its performance. Kanban or CONWIP requires fairly steady rate-driven lines. We will discuss a pull alternative for lines that cannot achieve this type of stability via gross capacity control. However, no system can entirely mitigate the negative effects of highly variable demand.

14.2.2   Bottleneck Planning

In Part II we stressed that the rate of a line is ultimately determined by the bottleneck, or slowest, process. In the simple single-product, single-routing lines we considered in Chapter 7 to illustrate basic factory dynamics, the bottleneck process represents the maximum rate of the line. This rate is only achieved when the WIP in the line is allowed to become large,[3] as illustrated in Figure 14.2.

In lines where all parts follow the same routing and processing times are such that the same process is the slowest operation for all parts, the conveyor model is an accurate representation of reality and useful for analysis, as well as intuition. In such cases, the bottleneck plays a key role in the performance of the line and therefore should be given special attention by the SFC module. Because throughput is a direct function of the utilization of the bottleneck, it makes sense to trigger releases into the line according to the status of the bottleneck. Such "pull from the bottleneck" schemes can work well in some systems, and we will discuss them further.

In spite of the theoretical importance of bottlenecks, it has been our experience that few manufacturers can identify their bottleneck process with any degree of confidence. The reason is that few manufacturing environments closely resemble a single-product, single-routing line. Most systems involve multiple products with different processing times. As a result, the bottleneck machine for one product may not be the bottleneck for

---

[3]What is meant by *large,* of course, depends on the amount of variability in the line, as we noted in Chapter 9.

FIGURE 14.3

Routings with a shared resource

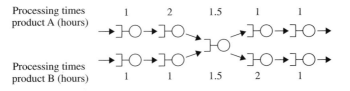

another product. This can cause the bottleneck to "float," depending on the product mix. Recall that Figure 10.9 illustrated this type of behavior with an example where machine 2 is the bottleneck for product A, machine 4 is the bottleneck for product B, and machine 3 is the bottleneck for a 50-50 mix of A and B.

Multiproduct systems also often involve different routings for different products. For instance, Figure 14.3 shows a two-routing system with a single shared workstation. Whether or not machine 3 is the bottleneck for product A depends on the volume of product B. Similarly, the bottleneck for product B depends on the volume of product A. Thus, the bottlenecks in this system can also float depending on product mix. Furthermore, if the two product lines are under separate management, the location of the bottleneck in each line may be outside the control of the line manager.

This discussion has two important implications for design of the SFC module:

1. *Stable bottlenecks are easier to manage.* A line with a distinct identifiable bottleneck is simpler to model (i.e., with the conveyor model) and control than a line with multiple moving bottlenecks. A manager can focus on the status of the bottleneck and think about the rest of the line almost exclusively in terms of its impact on the bottleneck (i.e., preventing starvation or blocking of the bottleneck). If we are fortunate enough to have a line with a distinct bottleneck, we should exploit this advantage with an SFC module that gives the bottleneck favorable treatment and provides accurate monitoring of its status.

2. *Bottlenecks can be designed.* Although some manufacturing systems have their bottleneck situation more or less determined by other considerations (e.g., the capacity of all key processes would be too expensive to change), we can often proactively influence the bottleneck. For instance, we can reduce the number of potential bottlenecks by adding capacity at some stations to ensure that they virtually never constrain throughput. This may make sense for stations where capacity is inexpensive.[4] Or interacting lines can be separated into cells; for example, the two lines in Figure 14.3 could be separated by adding an additional machine 3 (or dedicating machines to lines, if station 3 is a multimachine workstation). This type of cellular manufacturing has become increasingly popular in industry, in large part because small, simple cells are easier to manage than large, complex plants.

Although it is difficult to estimate accurately the cost benefits of simplifying bottleneck behavior, it is clear that there *are* costs associated with complexity. The simplest plant to manage is one with separate routings and distinct, steady bottlenecks. Any departures from this only serve to increase variability, congestion, and inefficiency. This does not mean that we should automatically add capacity until our plant resembles this ideal; only that we should consider the motivation for departures from it. If we are

---

[4]Note that the idea of deliberately adding capacity that will result in some resources being underutilized runs counter to the principle of line balancing. Economic justification of unbalancing the line requires taking a linewide perspective that considers variability, as we have stressed throughout this book.

plagued by a floating bottleneck that could be eliminated via inexpensive capacity, the addition deserves consideration. If interacting routings could be separated without large cost, we should look into it.

Moreover, line design and capacity allocation need not be plantwide to be effective. Sometimes great improvements can be achieved by assigning a few high-volume product families to separate, well-designed cells, leaving many low-volume families to an inefficient job shop portion of the plant. This "factory within a factory" idea has been promoted by various researchers and practitioners, most prominently Wickham Skinner (1974), as part of the **focused factory** philosophy. The main idea behind focused factories is that plants can do only a few things very well and therefore should be focused on a narrow range of products, processes, volumes, and markets. As we will see repeatedly throughout Part III, simplicity offers substantial benefits throughout the planning hierarchy, from low-level shop floor control to long-range strategic planning.

14.2.3 Span of Control

In Chapter 13, we discussed disaggregation of the production planning problem into smaller, more manageable units. We devoted most of that discussion to disaggregation along the time dimension, into short-, intermediate-, and long-range planning. But other dimensions can be important as well. In particular, in large plants it is essential to divide the plant by product or process in order to avoid overloading individual line managers.

Typically, a reasonable **span of control,** which usually refers to the number of employees under direct supervision of the manager, is on the order of 10 employees. A line with many more workers than this will probably require intermediate levels of management (foremen, lead technicians, multiple layers of line managers). Of course, 10 is only a rough rule of thumb; the appropriate number of employees under direct supervision of a manager will vary across plants. Strictly speaking, the term *span of control* should really refer to more than simply the number of subordinates, to consider the range of products or processes the manager must supervise.

For instance, printed-circuit board (PCB) manufacture involves, among other operations, a lamination process, in which copper and fiberglass sheets are pressed together, and a circuitize process, in which the copper sheets are etched to produce the desired circuitry. The technology, equipment, and logistics of the two processes are very different. Lamination is a batch process involving large mechanical presses, while circuitizing is a combination of a one-board-at-a-time process using optical expose machines and a conveyorized flow process involving chemical etching. These differences, along with physical separation, make it logical to assign different managers to the two processes.

How a line is broken up, for bottleneck design, span of control, or other considerations, is relevant to the configuration of the SFC module. Depending on the complexity of the line, managers may be able to coordinate movement of material through the portion of the line for which they are responsible, with very little assistance from the production control system. But the managers cannot coordinate activities outside their areas. Presumably, a higher-level manager has responsibility spanning disparate portions of the plant (and, of course, the plant manager has ultimate responsibility for the whole plant). However, at a real-time control level, these higher-level managers cannot force coordination. This must be done by the line managers, using information provided by the SFC module.

At a minimum, the SFC module must tell managers what parts are required by downstream workstations. If the module can also project what materials will be arriving at each station, so much the better, since this information enables the line managers to plan their activities in advance. The division of the line for management purposes provides a natural set of points in the line for reporting this information. How the line is divided may also affect the other functions of the SFC module listed in Figure 14.1. For purposes of accountability, it may be desirable to build in quality checks between workstations under separate management (e.g., the downstream station checks parts from an upstream station and refuses to accept them if they do not meet specifications). Under these conditions, the links between SFC and quality control must be made with this in mind.

14.3   CONWIP Configurations

As we observed in Chapter 5, JIT authors sometimes get carried away with the rhetoric of simplicity, making statements like "Kanban ... can be installed ... in 15 minutes, using a few containers and masking tape" (Schonberger 1990, p. 308). As any manager who has installed a pull system knows, getting a system that works well is *not* simple or easy. Manufacturing enterprises are complex, varied activities. Neither the high-level philosophical guidelines of "romantic JIT" nor the collection of techniques from "pragmatic JIT" can possibly provide ready-made solutions for individual manufacturing environments. With this in mind, we begin our review of possible SFC configurations. We start with the simplest possibilities, note where they will and won't work well, and move to more sophisticated methods for more complex environments. Since we cannot discuss every option in detail, our hope is that the range offered here will provide the reader with a mix-and-match starting point for choosing and developing SFC modules for specific applications.

14.3.1   Basic CONWIP

The simplest manufacturing environment from a management standpoint is the single-routing, single-family production line. If the following conditions hold, then this model is an accurate approximation of reality, and a basic CONWIP system (where releases are coordinated with completions to hold the WIP level in the line constant) will work well in the SFC module, for the reasons discussed in Chapter 10:

1. There are *constant routings* so that all parts traverse the same sequence of machines. Actually, if some parts contain a few extra operations (e.g., installation of a deluxe feature) that do not substantially alter flow time, we may be able to ignore this and still use basic CONWIP. However, if routings are conditional (e.g., jobs may be diverted to a rework line or sent out to a vendor), then we may not be able to treat the line as a single routing and will require more than basic CONWIP.

2. *Processing times are similar* so that all parts require roughly the same amount of time at each process center. This implies that the bottleneck will be stable. We do not require that the bottleneck be sharply defined (i.e., significantly slower than other machines), however.

3. There are *no significant setups* so that the time through the line for an individual job is not strongly affected by the sequence of jobs.

4. There are *no assemblies*, so we can view the progression of jobs as a linear flow. We will modify basic CONWIP to accommodate assemblies later.

Perhaps the simplest way to maintain the constant-WIP protocol is by means of physical cards or containers, as Figure 14.4 illustrates. Raw materials arrive to the line in standard containers but are only released into the line if there is an available CONWIP card. These cards can be laminated sheets of paper, metal or plastic tags, or the empty containers themselves. Since no routing or product information is required on the cards, they can be very simple. Provided that work is only released into the line with a card, and cards are faithfully recycled (they don't get trapped with a job diverted for rework or terminated by an engineering change order), the WIP in the line will remain constant at the level set by the number of CONWIP cards.

Even in this simple system there are SFC issues to resolve.

1. *Work backlog.* Because the CONWIP cards do not contain product information, a line manager or operator needs additional information to select jobs to release into the line. This is the task of the sequencing and scheduling module, which may use a simple earliest due date (EDD) sequence (because of the no-setup assumption) or a more involved batching routine (to achieve a rhythm by working on similar parts for extended periods). Once generated, the backlog can be communicated to the line in a variety of ways. The simplest consists of a piece of paper with a prioritized list of jobs. Whenever a CONWIP card is available, the next job for which raw materials are available is released into the line. Some situations may call for more sophisticated work backlog displays, for example, showing priorities or projected arrival times.

2. *Line discipline.* In general, a line should maintain a first-in-system first-out order. This means that, barring yield loss, rework problems, or passing at multimachine stations, the jobs will exit the line in the same order they were released. Since the CONWIP protocol keeps the line running at a steady pace, this makes it easy to predict when jobs—even those still on the work backlog—will be completed. However, if the CONWIP line is long, there may arise situations in which certain jobs require expediting. While we wish to discourage incautious use of expediting because it can dramatically increase variability in the line, it is unreasonable to expect the firm never to expedite. To minimize the resulting disruption, it may make sense to allow only two levels of priority and to establish specific **passing points.** The passing points are buffers or stock points in the line, typically between segments run as CONWIP loops, where "hot" jobs are allowed to pass "normal" jobs. The discipline of a workstation taking material from such a buffer is to take the first job from the **hot list,** if there is one, and, if not, the oldest job currently in the buffer. To allow passing only at designated points in the line makes it easier to build a model (the real-time simulation module) for predicting when jobs will exit the line. If many levels of priority and unrestricted passing are permitted, the variability or "churn" in the line can become acute, and it can be almost impossible to predict line behavior.

FIGURE 14.4

A CONWIP line using cards

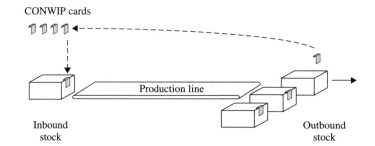

CONWIP cards

Production line

Inbound stock

Outbound stock

3. *Card counts.* To be effective, a CONWIP SFC module must fix a reasonable WIP level. As we noted in Chapter 13, setting card counts is a function that should be done infrequently (e.g., monthly or quarterly), not in real time with the work releases. If CONWIP is being implemented on an established line, the easiest approach for setting card counts is to begin with a count that fixes WIP at the historical level. After the line has stabilized, look at the workstations for persistent queues. If a station's queue virtually never empties, then reducing the card count will not have much effect on throughput and therefore should be done. Make periodic reviews of queue lengths to adjust card counts to accommodate physical changes (hopefully improvements) in the line. If CONWIP is being implemented on a new line, then a reasonable approach is to select the WIP level by choosing a reasonable and *feasible* cycle time CT and estimating the practical production rate of the line r_b^P. Then, using Little's law, set the WIP level as:

$$\text{WIP} = \text{CT} \times r_b^P$$

Assuming actual throughput is close to r_b^P, this will set WIP levels appropriately, provided that CT is actually feasible. Care must be taken not to get overly optimistic about cycle time, since it will lead to an underestimate of the required WIP and therefore a reduction in throughput.

4. *Card deficits.* If the card count is sufficiently large relative to variability in the line, rigidly adhering to the CONWIP protocol can work well. However, there are situations where we may be tempted to violate the constant-WIP release rule. Figure 14.5 illustrates one such situation, where a nonbottleneck machine downstream from the bottleneck is experiencing an unusually long failure, causing the bottleneck to starve for lack of cards. If the nonbottleneck machine is substantially faster than the bottleneck, then it will easily catch up once it is repaired. But in the meantime, we are losing valuable time at the bottleneck. One remedy for this situation is to run a **card deficit,** in which we release some jobs without CONWIP cards into the line. This will allow the bottleneck to resume work. Once the failure situation is resolved, we revert to the CONWIP rules and only allow releases with cards. The jobs without cards will eventually clear the line, and WIP will fall back to the target level. Another remedy for this type of problem is to pull from the bottleneck instead of the end of the line. We discuss this in Section 14.4.2.

5. *Work ahead.* One of the benefits of CONWIP that we identified in Chapter 10 is its ability to opportunistically work ahead of schedule when events permit. For instance, if the bottleneck is unusually fast or reliable this week, we may be able to do more work than we had planned. Assuming that the master production schedule is full, it probably makes sense to take advantage of our good fortune—up to a limit. While it almost certainly makes sense to start some of next week's jobs, it may not make sense to start jobs that are not due for months. If the MPS for a particular routing is not full, which is a real possibility in a plant with many routings, each of which is used sporadically, then we may want to establish a **work-ahead window.**

For instance, when authorized by the CONWIP mechanism, we may release the next job into the line, *provided that it is within n weeks of its due date.* Setting the limit *n* is an

FIGURE 14.5

*CONWIP card deficits in
failure situations*

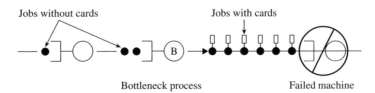

additional CONWIP design question, which is closely related to the concepts of frozen zones and time fences discussed in Chapter 3. Since jobs within the frozen zone of their due dates are not subject to change, it makes sense to allow CONWIP to work ahead on them. Jobs beyond the restricted frozen zone (or partially restricted time fences) are much riskier to work ahead on, since customer requirements for these jobs may change. Clearly, the choice of an appropriate work-ahead policy is strongly dependent on the manufacturing environment.

14.3.2   Tandem CONWIP Lines

Even if we satisfy the conditions for basic CONWIP to be applicable (constant routings, similar processing times, no significant setups, and no assemblies), we may not want to run the line as a single CONWIP loop. The reason is that span-of-control considerations may encourage us to decouple the line into more manageable parts. One way to do this is to control the line as several tandem CONWIP loops separated by WIP buffers. The WIP levels in the various loops are held constant at specified levels. The interloop buffers hold enough WIP to allow the loops to temporarily run at different speeds without affecting (blocking or starving) one another. This makes it easier for different managers to be in charge of the different loops. The extra WIP and cycle time introduced by the buffers also degrade efficiency. This is a tradeoff one must evaluate in light of the particular needs of the manufacturing system.

Figure 14.6 illustrates different CONWIP breakdowns of a single production line, ranging from treating the entire line as a single CONWIP loop to treating each workstation as a CONWIP loop. Notice that this last case, with each workstation as a loop, is identical to one-card kanban. In a sense, basic CONWIP and kanban are extremes in a continuum of CONWIP-based SFC configurations. The more CONWIP loops we break the line into, the closer its behavior will be to kanban. As we discussed in Chapter 10, kanban provides tighter control over the material flow through individual workstations and, if WIP levels are low enough, can promote communication between adjacent stations. However, because there are more WIP levels to set in kanban, it tends to be more complex to implement than basic CONWIP. Therefore, in addition to the efficiency/span-of-control tradeoff to consider in determining how many CONWIP loops to use to control a line, we should think about the complexity/communication tradeoff.

Another control issue that arises in a line controlled with multiple tandem CONWIP loops concerns when to release cards. The two options are (1) when jobs enter the interloop buffers or (2) when they leave them. If CONWIP cards remain attached to jobs in the buffer at the end of a loop, then the sum of the WIP in the line plus the WIP in the buffer will remain constant. Therefore, if WIP in the buffer reaches the level specified by the card count, then the loop will shut down until the downstream loop removes WIP from the buffer and releases some cards. As Figure 14.7 illustrates (in loops 1 and 3), this mechanism makes sense for nonbottleneck loops that are fast enough to keep pace with the overall line. If we did not link loop 1 to the pace of the line by leaving cards attached to jobs in the buffer, it could run far ahead of other loops, swamping the system with WIP.

If one loop is a clearly defined bottleneck, however, we may want to decouple it from the rest of the line, in order to let it run as fast as it can (i.e., to work ahead). As loop 2 in Figure 14.7 illustrates, we accomplish this by releasing cards as soon as jobs exit the end of the line—before they enter the downstream buffer. This will let the loop run as fast as it can, subject to availability of WIP in the upstream buffer and subject to a WIP cap on the total amount of inventory that can be in the line at any point in time.

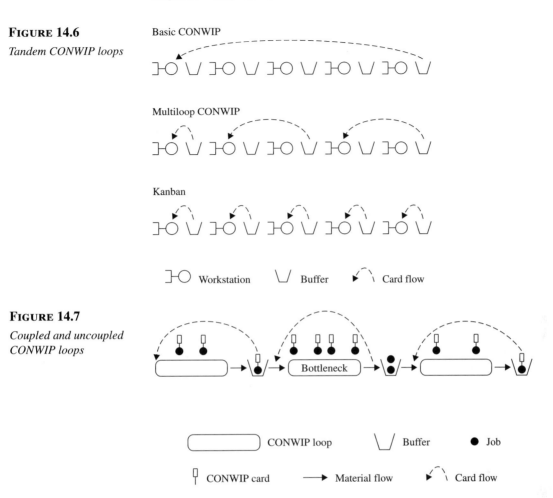

FIGURE 14.6

Tandem CONWIP loops

Basic CONWIP

Multiloop CONWIP

Kanban

⊐O Workstation   ⋁ Buffer   ⤷ Card flow

FIGURE 14.7

Coupled and uncoupled CONWIP loops

Bottleneck

⟨ ⟩ CONWIP loop   ⋁ Buffer   ● Job

CONWIP card   → Material flow   ⤷ Card flow

Of course, this means that the WIP in the downstream buffer can float without bound, but as long as the rest of the line is consistently faster than the bottleneck loop, the faster portion will catch up and therefore WIP will not grow too large. Of course, in the long run, all the CONWIP loops will run at the same speed, that set by the bottleneck loop.

14.3.3   Shared Resources

While it is certainly simplest from a logistics standpoint if machines are dedicated to routings—and this is precisely what is sometimes achieved by assigning a set of product families to manufacturing cells—other considerations sometimes make this impossible. For instance, if a certain very expensive machine is required by two different products with otherwise separate routings, it may not be economical to duplicate the machine in order to completely separate the routings. The result will be something like that illustrated previously in Figure 14.3. If several multiple resources are shared across many routings, the situation can become quite complex.

Shared resources complicate both control and prediction of CONWIP lines. Control is complicated at a shared resource because we must choose a job to work on from multiple incoming routings. If the shared resource is in the interior of a CONWIP loop, then the natural information to use for making this choice is the "age" of the incoming jobs. The proper choice is to work on jobs in FISFO (first-in-system first-out) order, because the time a job entered the line corresponds to the time of a downstream demand, as it is a pull system. Hence FISFO will coordinate production with demand.

If it is important to ensure that the shared resource works on jobs imminently needed downstream, then it may make sense to break the line into separate CONWIP loops before and after the shared resource, as Figure 14.8 illustrates. This figure shows two routings, for product families A and B, that share a common resource. Both routings are treated as CONWIP loops before and after the common resource. This provides the common resource with incoming parts in the upstream buffers, and with cards indicating downstream replenishment needs. Working on jobs whose cards have been waiting longest (provided there are appropriate materials in the incoming buffer) is a simple way to force the shared resource to work on parts most likely to be needed soon. If a machine setup is required to switch between families, then an additional rule about how many parts of one family to run before switching may be required.

Shared resources also complicate prediction. While the conveyor model can be quite accurate for estimating the exit times of jobs from a single CONWIP line, it is not nearly as accurate for a line with resources shared by other lines. The reason is that the outputs from one line can strongly depend on what is in the other lines. A simple way to adapt the conveyor model to approximate this situation is to preallocate capacity. For example, suppose two CONWIP lines, for product families A and B, share a common resource, where on average family A utilizes 60 percent of the time of this resource and family B utilizes 40 percent. Then we can roughly treat the line for family A by inflating the process times on the shared resource by dividing them by 0.6 to account for the fact that the resource devotes only 60 percent of its time to family A. Likewise, we treat the line for family B by dividing processing times on the shared resource by 0.4.

To illustrate this analysis in a little greater detail, suppose that the shared resource in Figure 14.8 requires one hour per job on routing A and two hours per job on routing B. If 60 percent of the jobs processed by this resource are from routing A and 40 percent are from B, then the fraction of processing hours (hours spent running product) that are devoted to A is given by

$$\frac{1 \times 0.6}{1 \times 0.6 + 2 \times 0.4} = 0.4286$$

Therefore, the fraction of processing hours devoted to B is $1 - 0.4286 = 0.5714$. The 42.86 percent number is very much like an *availability* caused by machine outages. In

FIGURE 14.8

Splitting a CONWIP loop at a shared resource

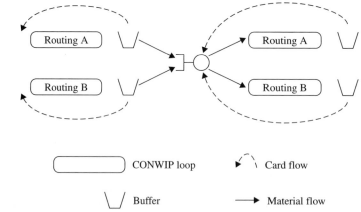

effect, the resource is available to A only 42.86 percent of the time. Thus, while the rate of the shared resource would be one job per hour if only A parts were run, it is reduced to 1×0.4286 job per hour due to the sharing with B. The average processing time is the inverse of this rate, or $1/0.4286 = 2.33$ hours per job. Similarly, the average processing of a B job is

$$\frac{2}{0.5714} = 3.50 \text{ hours per job}$$

Using these inflated processing times for the shared resource, we can now treat routings A and B as entirely separate CONWIP lines for the purposes of analysis. Of course, if the volumes on the two routings fluctuate greatly, then the output times will vary substantially above and below those predicted by the conveyor model. The effect will be very much the same as having highly variable (e.g., long infrequent, as opposed to short frequent) outage times on a resource in a CONWIP line. Therefore, if we use such a model to quote due dates, we have to add a larger inflation factor to compensate for this extra variability.

14.3.4   Multiple-Product Families

We now begin relaxing the assumptions needed to justify basic CONWIP by considering the situation where the line has multiple-product families. We still assume a simple flow line with constant routings and no assemblies, but now we allow different product families to have substantially different processing times and possibly sequence-dependent setups. Under these conditions, it may no longer be reasonable to fix the WIP level in a CONWIP loop by holding the number of units in the line constant. The reason is that the total workload in the line may vary greatly due to the difference in processing times across products. It may make more sense to adjust the WIP count for capacity.

One plausible measure of the work in the system would be hours of processing time at the bottleneck machine. Under this approach, if a unit of product A requires one hour on the bottleneck and B requires two hours, then when one unit of B departs the line, we allow two units of product A to enter (provided that it is next on the work backlog). As long as the location of the bottleneck is relatively insensitive to product mix, this mechanism will tend to maintain a stable workload at the bottleneck. If the bottleneck changes with mix (i.e., different products have different machines as their slowest resource), then computing a capacity-adjusted WIP level is more difficult. We could use total hours of processing time on all machines. However, we will probably need a higher WIP level than would be required for a system with a stable bottleneck, to compensate for the variability caused by the moving bottleneck. Furthermore, if the total processing times of different products do not vary much, this approach will not be much different from the simpler approach of counting WIP in physical units.

If we count WIP in capacity-adjusted standard units, it becomes more difficult to control the WIP level with a simple mechanism like cards. Instead of trying to attach multiple cards to jobs to reflect their differing complexity, it probably makes sense to use an electronic system for monitoring WIP level. Figure 14.9 illustrates an electronic **CONWIP controller,** which consists of a local-area network (LAN) with computers located at the front and back of the line. The computers monitor the adjusted WIP level and indicate when it falls below the target level (e.g., by changing an indicator light from red to green). When this happens, the operator of the first workstation selects the next job on the work backlog for which the necessary materials are available (displayed on the computer terminal as showing the due date, DD, part number, PN, and quantity to be

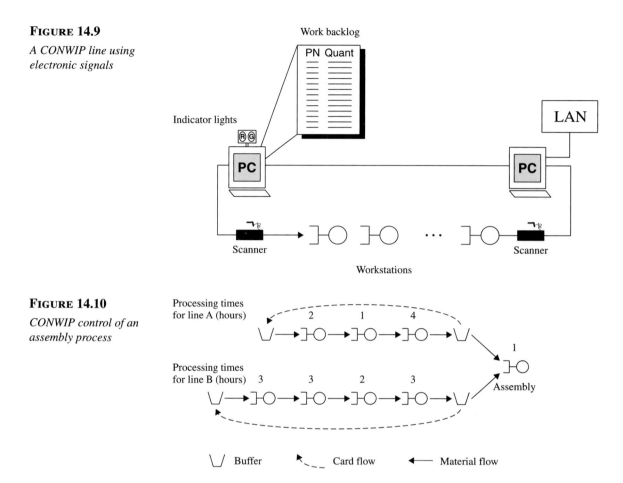

FIGURE 14.9

A CONWIP line using electronic signals

FIGURE 14.10

CONWIP control of an assembly process

released, (Quant)) and releases it into the line. This release is recorded by keyboard or optical scanner and is added to the capacity-adjusted WIP level. At the end of the line, job outputs are also recorded and subtracted from the WIP level. Exceptions, such as fallout due to yield loss, may also need to be recorded on one of the computer terminals.

14.3.5   CONWIP Assembly Lines

We now further extend the CONWIP concept to systems with assembly operations. Figure 14.10 illustrates the simple situation in which an assembly operation is fed by two fabrication lines. Each assembly requires one subcomponent from family A and one subcomponent from family B. The assembly operation cannot begin until both subcomponents are available. The two fabrication lines are controlled as CONWIP loops with fixed, but not necessarily identical, WIP levels. Each time an assembly operation is completed, a signal (e.g., CONWIP card or electronic signal) triggers a new release in each fabrication line. As long as a FISFO protocol is maintained in the fabrication lines, the final assembly sequence will be the same as the release sequence.

Notice that assembly completions need not trigger releases of subcomponents destined for the same assembly. If line A has a WIP level of 9 jobs and line B has a WIP level of 18 jobs, then the release authorized by the next completion into line A will be used 9 assemblies from now, while the release into line B will be used 18 assemblies from now. If the total process time for line B is longer than that for line A, this type of imbalance makes sense. In general, the longer line will require a larger WIP level (i.e., because of Little's law). Determining precise WIP levels is a bit trickier. Fortunately, performance is robust in WIP level, provided that the lines have sufficient WIP to prevent excessive starvation of the bottleneck.

To illustrate a mechanism for setting ballpark WIP levels in an assembly system, consider the data given in Figure 14.10. Notice that the systemwide bottleneck is machine 3 of line A. Hence, the bottleneck rate is $r_b = 0.25$ job per hour. If we look at the two lines, including assembly, as separate fabrication lines, we can use the critical WIP formula from Chapter 7 on each line. This shows that the WIP levels under ideal (i.e., perfectly deterministic) conditions need to be

$$W_0^A = r_b T_0^A = \tfrac{1}{4}(2 + 1 + 4 + 1) = \tfrac{8}{4} = 2$$
$$W_0^B = r_b T_0^B = \tfrac{1}{4}(3 + 3 + 2 + 3 + 1) = \tfrac{12}{4} = 3$$

to achieve full throughput. Of course, in reality, there will be variability in the line, so the WIP levels will need to be larger than this. How much larger depends on how much variability there is in the line.

For a line corresponding to the practical worst case discussed in Chapter 7, we can compute the WIP level required to achieve throughput equal to 90 percent of the bottleneck rate by setting the throughput expression equal to $0.9r_b$ and solving for the WIP level w:

$$\frac{w}{W_0 + w - 1} r_b = 0.9 r_b$$
$$\frac{w}{W_0 + w - 1} = 0.9$$
$$w = 0.9(W_0 + w - 1)$$
$$w = 9W_0 - 9 = 9(W_0 - 1)$$

Inflating W_0^A and W_0^B according to this formula yields

$$w^A = 9(2 - 1) = 9$$
$$w^B = 9(3 - 1) = 18$$

Unless the line is highly variable, these WIP levels are probably reasonable starting points, from which a process of adjustment can be initiated. If the processing times on all the machines are less variable than the practical worst case (i.e., they have coefficients of variation smaller than one), then the line may operate effectively with smaller WIP levels than this. If the processing times on some machines are more variable than the practical worst case (i.e., they have coefficients of variation larger than one, due to long failures or setups, for example), then even more WIP than this may be required to achieve a reasonable throughput rate.

14.4   Other Pull Mechanisms

We look upon CONWIP as the first option to be considered as an SFC platform. It is simple, predictable, and robust. Therefore, unless the manufacturing environment is such that it is inapplicable, or another approach is likely to produce substantially better performance, CONWIP is a good, safe choice. By using the flexibility we discussed above to split physical lines into multiple CONWIP loops, one can tailor CONWIP to the needs of a wide variety of environments. But there are situations in which a suitable SFC module, while still a pull system, is not what we would term CONWIP. We discuss some possibilities below.

14.4.1   Kanban

As we noted earlier, kanban can be viewed as tandem CONWIP loops carried to the extreme of having only a single machine in each loop. So from a CONWIP enthusiast's perspective, kanban is just a special case of CONWIP. However, Ohno's book contains a diagram of a kanban system that looks very much like a set of CONWIP loops feeding an assembly line. Therefore, the developers of kanban may well have considered CONWIP a form of kanban. As far as we are concerned, this distinction is a matter of semantics; kanban and CONWIP are obviously closely related. The important question concerns when to use kanban (single-station loops) instead of CONWIP (multistation loops).

Kanban offers two potential advantages over CONWIP:

1. By causing each station to pull from the upstream station, kanban may force better interstation communication. Although there may be other ways to promote the same communication, kanban makes it almost automatic.

2. By breaking the line at every station, kanban naturally provides a mechanism like that illustrated in Figure 14.8, for sharing a resource among different routings.

However, kanban also has the following potential disadvantages:

1. It is more complex than CONWIP, requiring specification of more WIP levels. (However, recall that pull systems are fairly insensitive to WIP level. Hence, the WIP levels in kanban need not be set precisely for the system to function well, and therefore this increase in complexity may not be a major obstacle to kanban in most settings.)

2. It induces a tighter pacing of the line, giving operators less flexibility for working ahead and placing considerable pressure on them to replenish buffers quickly.

3. The use of product-specific cards means that at least one standard container of each part number must be maintained at each station, to allow the downstream stations to pull what they need. This makes it impractical for systems with numerous part numbers.

4. It cannot accommodate a changing product mix (unless a great deal of WIP is loaded into the system) because the product-specific card counts rigidly govern the mix of WIP in the system.

5. It is impractical for small, infrequent orders (onesies and twosies). Either WIP would have to be left unused on the floor for long spans of time (i.e., between orders), or the system would be unresponsive to such orders because authorizations signaled by the kanban cards would have to propagate all the way to the beginning of the line to trigger new releases of WIP.

There is little one can do to alleviate the first two disadvantages; complexity and pressure are the price one pays for the additional local control of kanban. However, the remaining disadvantages are a function of product-specific cards and therefore can be mitigated by using **routing-specific cards** and a **work backlog.** Figure 14.11 shows a kanban system with different-color cards for different routings. When a standard

FIGURE 14.11

Kanban with route-specific cards and a work backlog

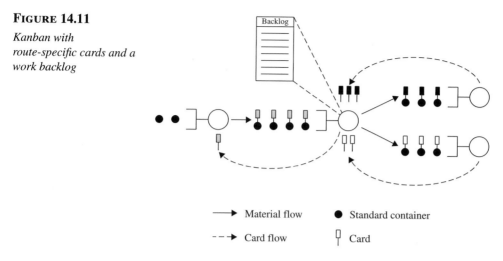

container is removed from the outbound stock point, the card authorizes production to replace it. The identity of the part that will be produced is determined by the work backlog, which must be established by the sequencing and scheduling module. If a part does not appear on the backlog for an extended period, then it will not be present in the line. The modification of route-specific (as opposed to part-specific) cards enables this approach to kanban to be used in systems with many part numbers.

On the basis of this discussion, it would appear that kanban is best suited to systems with many routings that share resources, especially if products and routings are frequently added and removed. If we are going to break the line into many CONWIP loops to make control of the shared resources easier, then moving all the way to kanban will not significantly change performance. Moreover, if a new routing converts a previously unshared resource to a shared resource, then a kanban configuration will already provide the desired break in the line.

On the other hand, if the various routings have few shared resources and new products tend to follow established routings, there would seem to be little incentive to incur the additional complexity of kanban. The system will probably function more simply and effectively under CONWIP, possibly broken into separate loops for span-of-control reasons, to give special treatment to a shared resource, or to feed buffers at assembly points.

14.4.2   Pull-from-the-Bottleneck Methods

Two problems that can arise with CONWIP (or kanban) in certain environments are the following:

1. *Bottleneck starvation* due to downstream machine failures. As we illustrated in Figure 14.5, we may want to allow releases beyond those authorized by cards to compensate for this situation.

2. *Premature releases* due to the requirement that the WIP level be held constant. Even if a part will not be needed for months, a CONWIP system may trigger its release because WIP in the loop has fallen below its target level. This can reduce flexibility for no good reason (e.g., engineering changes or changes in

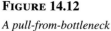

FIGURE 14.12

*A pull-from-bottleneck
system*

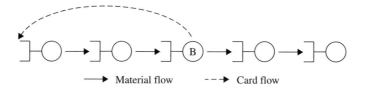

customer needs are much more difficult to accommodate once a job has been released to the floor).

We can modify CONWIP to address these situations. The basic idea is to devise a mechanism for enabling the bottleneck to work ahead, but at the same time provide a means of preventing it from working too far ahead. The techniques we will introduce are related to the technique termed **drum-buffer-rope (DBR)** developed by Goldratt (Goldratt and Fox 1986), although he presented DBR primarily as a scheduling methodology rather than an SFC mechanism.

We begin with the simplest version of the **pull-from-bottleneck (PFB)** strategy. Figure 14.12 shows such a system for a single line. This mechanism differs from CONWIP in that the WIP level is held constant in the machines up to and including the bottleneck, but is allowed to float freely past the bottleneck. Since machines downstream from the bottleneck are faster on average than the bottleneck, WIP will not usually build up in this portion of the line. However, if a failure in one of these machines causes a temporary buildup of WIP, it will not cause the bottleneck to shut down, as can occur under CONWIP if card deficits are not used. Therefore, a PFB approach may make sense as an alternative to card deficits in a line with a stable bottleneck. If the bottleneck shifts depending on product mix, then it is not clear where the pulling point should be located, and therefore one may be just as well off pulling from the end of the line (i.e., using regular CONWIP), possibly with a card deficit policy.

The simple PFB approach of Figure 14.12 can mitigate the bottleneck starvation problem associated with CONWIP, but does not address the issue of premature releases. When we are talking about a single line, we often speak as though the line will be kept running at close to full capacity. And this is frequently true in plants with few routings. But in plants with many routings (e.g., a plant tending toward a job shop configuration), some routings may not be used for substantial periods of time. For instance, we have seen plants with 5,000 distinct routings, only a relative few of which contained WIP at any given time. Clearly, under these conditions we do not want to maintain a constant WIP level along the routing, since this would result in releasing jobs that are not needed until far in the future.

Consider the situation illustrated in Figure 14.13, which shows four distinct product routings, three of which pass through the bottleneck. The goal of a PFB strategy is to ensure that jobs are released so that they arrive at the bottleneck a specified time before they are needed (i.e., so that waiting jobs will form a buffer in front of the bottleneck to prevent random variations from causing it to starve).

To make our approach precise, let

b_i = time required on bottleneck by job i on backlog. Note that jobs on different routings may have different processing times, and there may even be different families within same routing having different processing times.

ℓ_i = average time after release required for job i to reach bottleneck. Note that this time involves processing on nonbottleneck resources only. Since most

FIGURE 14.13

Routings in a job shop

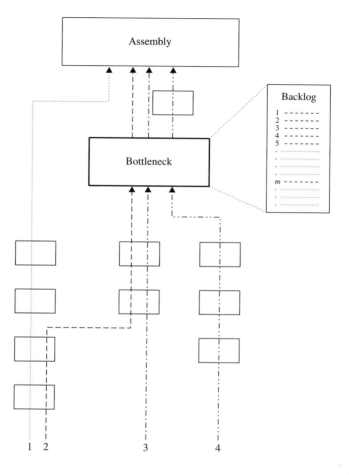

of queueing will occur at bottleneck, this time should be relatively constant for a given routing.[5] However, these times may differ substantially across routings.

$L =$ specified time for jobs to wait in buffer in front of bottleneck. This is a user-specified constant that depends on how much time protection is desired at bottleneck.

Now we can compute the amount of work at the bottleneck in the line by summing the b_i values. Suppose that the work backlog contains jobs in the sequence they will be worked on at the bottleneck, and suppose job 1 represents the current job being worked on at the bottleneck (where b_1 represents its remaining processing time). Then the amount of time until the bottleneck will be available to work on job j is

$$\sum_{i=1}^{j-1} b_i$$

Our goal is to release jobs on the backlog so that they will wait L time units in front of the bottleneck. Since job j takes ℓ_j on average to get to the bottleneck, we should

---

[5]This is in sharp contrast with MRP, which assumes constant lead times through the entire plant including the bottleneck. Because MRP does not maintain constant loadings on the plant, actual cycle times can vary greatly, making the constant-lead-time assumption very poor.

release job j whenever

$$\sum_{i=1}^{j-1} b_i \leq \ell_j + L$$

Therefore, if we track the number

$$\sum_{i=1}^{j-1} b_i - \ell_j - L \tag{14.1}$$

for every job on the backlog and release jobs when this quantity hits (or goes below) zero, we will maintain a constant workload on the bottleneck and jobs should arrive on average L time units before they are needed at the bottleneck. As long as L is large enough to prevent significant delays at the bottleneck, the actual work sequence at the bottleneck should be able to match the sequence on the work backlog reasonably well.

Notice that if the ℓ_j lead times differ for different routings, then the release sequence may be different from the sequence on the work backlog. All other things being equal, a job with a large ℓ_j will be released earlier than a job with a small ℓ_j, as one would expect, since its index in Equation (14.1) will go negative sooner. Furthermore, since the work backlog may have intervals during which no jobs along certain routings are required, this system may let WIP along some routings fall to zero at some points. Thus, while this mechanism induces a WIP cap it is not CONWIP in the sense of maintaining constant loadings along routings.

The PFB logic we have described so far is fine for routings 2, 3, and 4, but does not cover routing 1, which does not run through the bottleneck.[6] A sensible approach for this routing is to control it as a CONWIP loop. This will be effective as long as the need for parts from routing 1 is relatively stable. If the final assembly sequence contains intervals during which there is no need for routing 1 parts, then we might want to modify the CONWIP logic to include a requirement that the part be required within a certain time window (e.g., a week) before releasing it. Thus, we would release jobs when *both* the WIP level in routing 1 fell below its target level *and* the next part was needed within a specified time window.

14.4.3   Shop Floor Control and Scheduling

This last point about holding parts out until they are within a window of their due date makes it clear that there is potentially a strong link between the shop floor control module and the sequencing and scheduling module. If we have generated a schedule using the sequencing and scheduling module, then we can control individual routings by releasing jobs according to this schedule, *subject to a WIP cap*. That is, jobs will be released whenever the (capacity-adjusted) WIP along the routing is below the target level and a job is within a specified time window of its scheduled release date. If the schedule contains enough work to keep the routing fully loaded, this approach is equivalent to CONWIP. If there are gaps in the schedule for products along a routing, then the WIP level along that routing may fall below the target level, or even to zero.

A variety of scheduling systems could be used in conjunction with a WIP cap mechanism in this manner. We will discuss scheduling approaches based on the conveyor model that are particularly well suited to this purpose in Chapter 15. But one could also

---

[6]Observe that although routings 2 and 3 share nonbottleneck resources, we do not consider this in the release mechanism. As long as these shared resources are not close to being bottlenecks, this will probably work well. However, if these resources can become bottlenecks depending on the product mix, more complex scheduling and release methods may be required. We will discuss this in Chapter 15.

use something less ideal, such as MRP. The planned order releases generated by MRP represent a schedule. Instead of following these releases independently of what is going on in the factory, one could block releases along routings whose WIP levels are too high, and move up releases (up to a specified amount) along routings whose WIP levels are too low. The fixed-lead-time assumption of MRP will still tend to make the schedule inaccurate. But by forcing compliance with a WIP cap, this SFC approach will at least prevent the dreaded WIP explosion. The benefits of capping WIP in an MRP system were pointed out long ago in the MRP literature (Wight 1970), but mechanisms for actually achieving this have been rare in practice.

14.5   Production Tracking

As we mentioned, the SFC module is the point of contact with the real-time evolution of the plant. Therefore, it is the natural place to monitor plant behavior. We are interested in both the short term, where the concern is making schedule, and the long term, where the concern is collecting accurate data for planning purposes. Although individual plants may have a wide range of specific data requirements, we will restrict our attention to two generic issues: monitoring progress toward meeting our schedule in the short term, and tracking key capacity parameters for use in other planning modules in the long term.

14.5.1   Statistical Throughput Control

In the short term, the primary question concerns whether we are on track to make our scheduled commitments. If the line is running as a CONWIP loop with a specified production quota, then the question concerns whether we will make the quota by the end of the period (e.g., by the end of the day or week). If we are following a schedule for the routing, then this depends on whether we will be on schedule at the next overtime opportunity. If there is a good chance that we will be behind schedule, we may want to prepare for overtime (notify workers). Alternatively, if the SFC module can provide early enough warning that we are seriously behind schedule, we may be able to reallocate resources or take other corrective action to remedy the problem.

We can use techniques similar to those used in statistical process control (SPC) to answer the basic short-term production tracking questions. Because of the analogy with SPC, we refer to this function of the SFC module as **statistical throughput control (STC).** To see how STC works, we consider production in a CONWIP loop during a single production period. Common examples of periods are (1) an eight-hour shift (with a four-hour preventive maintenance period available for overtime), (2) first and second shifts (with third shift available for overtime), and (3) regular time on Monday through Friday (with Saturday and Sunday available for overtime).

We denote the beginning of the period as time 0 and the end of the regular time period as time R. At any intermediate point in time t, where $0 \leq t \leq R$, we must compare two pieces of information:

$n_t = $ *actual* cumulative production by line, possibly in capacity-adjusted units, in time interval $[0, t]$

$S_t = $ *scheduled* cumulative production for line for time interval $[0, t]$

First, note that since S_t represents *cumulative* scheduled production, it is always increasing in t. However, if we are measuring actual production at a point in the routing prior to an inspection point, at which yield fallout is possible, then n_t could potentially decrease. Second, note that if the line uses a detailed schedule, S_t may increase unevenly. However, if it uses a periodic production quota, without a detailed schedule, so that the target is to complete Q units of production by time R, then we assume that S_t is linear (i.e., constant) on the interval, so that

$$S_t = Q\frac{t}{R}$$

and hence $S_R = Q$. Figure 14.14 illustrates two possibilities for S_t.

Ideally, we would like actual production n_t to equal scheduled production S_t at every point in time between 0 and R. Of course, because of random variations in the plant, this will virtually never happen. Therefore, we are interested in characterizing how far ahead of or behind schedule we are. We could plot $n_t - S_t$ as a function of time t, to show this in units of production. When $n_t - S_t > 0$, we are ahead of schedule; when $n_t - S_t < 0$, we are behind it. However, the difference between n_t and S_t does not give direct information on how difficult it will be to make up a shortage or how much cushion is provided by an overage. Therefore, a more illuminating piece of information is the *probability of being on schedule by the end of the regular time period*, given how far we are ahead or behind now.

In Appendix 14A we derive an expression for this probability under the assumption that we can approximate the distribution of production during any interval of time by using the normal distribution. From a practical implementation standpoint, however, it is convenient to use the formula from Appendix 14A to precompute the overage levels (that is, $n_t - S_t$) that cause the probability of missing the quota to be any specified level α. If we know the mean and standard deviation of production during regular time (in capacity-adjusted units), denoted by μ and σ, this can be accomplished as follows.

Define x to be

$$x = -\frac{(\mu - Q)(R - t)}{R} - z_\alpha \sigma \sqrt{\frac{R - t}{R}} \tag{14.2}$$

where z_α is found from a standard normal table such that $\Phi(z_\alpha) = \alpha$. We show in Appendix 14A that if the overage level at time t is equal to x (that is, $n_t - S_t = x$), then the probability of missing the quota is exactly α. If $n_t - S_t > (<) x$, then the probability of missing quota is less than (greater than) α.

We can display this information in simple graphical form. Figure 14.15 plots the x values for specific probabilities of missing the quota. We have chosen to display these

FIGURE 14.14

Scheduled cumulative production functions, S_t

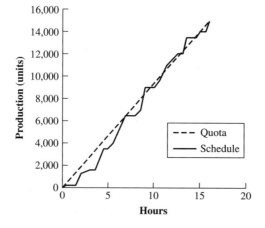

FIGURE 14.15

An STC chart when quota is equal to capacity

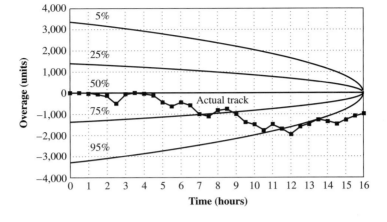

curves for probabilities of 5 percent, 25 percent, 50 percent, 75 percent, and 95 percent. In this example we are assuming a production quota, where regular time consists of two shifts, for a total of 16 hours, and historical data show that average production during 16 hours is 15,000 units and $\sigma = 2,000$ units. Quota is set equal to average capacity. That is, $S_t = Q_t/R$, where $Q = \mu = 15,000$. The curves in Figure 14.15 give an at-a-glance indication of how we stand relative to making the quota. For instance, if the overage level at time t (that is, $n_t - S_t$) lies exactly on the 75 percent curve, then the probability of missing the quota is 75 percent. On the basis of this information, the line manager may take action (e.g., shift workers) to speed things up. If $n_t - S_t$ rises above the 50 percent mark, this indicates that the action was successful. If it falls, say, below the 95 percent mark at time $t = 12$, then making the quota is getting increasingly improbable and perhaps it is time to announce overtime.

Notice that in Figure 14.15 the critical value (that is, x) for $\alpha = 0.5$ is always zero. The reason for this is that since the quota is set exactly equal to mean production, we always have a 50-50 chance of making it when we are exactly on time. The other critical values follow curved lines. For instance, the curve for $\alpha = 0.25$ indicates that we must be quite far ahead of scheduled production early in the regular time period to have only a 25 percent chance of missing the quota, but we must only be a little ahead of schedule near the end to have this same chance of missing the quota. The reason, of course, is that near the end of the period we do not have much of the quota remaining, and therefore less of a cushion is required to improve our chances of making it.

The Chapter 13 discussion on setting production quotas in pull systems pointed out that it may well be economically attractive to set the quota below mean regular time. When this is the case, we can still use Equation (14.2) to precompute the critical values for various probabilities of missing the quota. Figure 14.16 gives a graphical display of a case with a quota $Q = 14,000$ units, which is below mean regular time capacity $\mu = 15,000$ units. Notice that in this case, if we start out with no shortage or overage (that is, $n_0 - S_0 = 0$), then we begin with a greater than 50 percent chance of making the quota. This is because we have set the quota below the amount we can make on average during a regular time period. Since $Q < \mu$, on average we should be able to achieve a pace such that $n_i - S_t$ goes positive and continues to increase, that is, until the quota is reached and either production stops or we work ahead on the next period's quota. If something goes wrong, so that we fail to exceed the pace, then the position of the $n_t - S_t$ curve allows us to determine at a glance the probability of making the quota, given that we achieve historical average pace from time t until the end of regular time.

FIGURE 14.16

*An STC chart when the
quota is less than capacity*

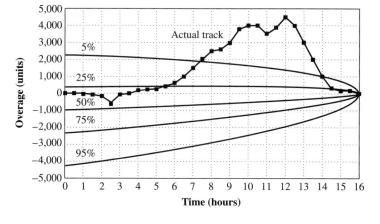

STC charts like those illustrated in Figures 14.15 and 14.16 can be generated by using Equation (14.2) and data on actual production (that is, n_t). The computer terminals of the CONWIP controller (see Figure 14.9) are a natural place to display these charts for CONWIP lines. STC charts can also be maintained and displayed at any critical resource in the plant.

STC charts can be useful even if n_t is not tracked in real time. For instance, if regular time consists of Monday through Friday and we only get readings on actual throughput at the end of each day, we could update the STC chart daily to indicate our chances for achieving the quota.

Finally, STC charts can be particularly useful at a critical resource that is shared by more than one routing. For instance, a system with two different circuit board lines running through a copper plating process could maintain separate STC charts for the two routings. Line managers could make decisions about which routing to work on from information about the quota status of the two routings. If line 1 is safely ahead of the quota, while line 2 is behind, then it makes sense to work on line 2 if incoming parts are available. Of course, we may need to use the information from the STC charts judiciously, to avoid rapid switches between lines if switching requires a significant setup.

14.5.2 Long-Range Capacity Tracking

In addition to providing short-term information to workers and managers, a production tracking system should provide input to other planning functions, such as aggregate and workforce planning and quota setting. The key data needed by these functions are the mean and standard deviation of regular time production of the plant in standard units of work. Since we are continually monitoring output via the SFC module, this is a reasonable place to collect this information.

In the following discussion, we assume that we can observe directly the amount of work (in capacity-adjusted standard units, if appropriate) completed during regular time. In a rigid quota system, in which work is stopped when the quota is achieved, even if this happens before the end of regular time, this procedure should *not* be used, since it will underestimate true regular time capacity. Instead, data should be collected on the mean and standard deviation of the *time to make quota,* which could be shorter or longer than the regular time period, and convert these to the mean and standard deviation of regular time production. The formulas for making this conversion are given in Spearman et al. (1989).

Since actual production during regular time is apt to fluctuate up and down due to random disturbances, it makes sense to smooth past data to produce estimates of the capacity parameters that are not inordinately sensitive to noise. The technique of exponential smoothing (Appendix 13A) is well suited to this task. We can use this method to take past observations of output to predict future capacity.

Let μ and σ represent the mean and standard deviation, respectively, of regular time production. These are the quantities we wish to estimate from past data. Let Y_n represent the nth observation of the amount produced during regular time, $\hat{\mu}(n)$ represent the nth smoothed estimate of regular time capacity, $\hat{T}(n)$ represent the nth smoothed trend, and α and β represent smoothing constants. We can iteratively compute $\hat{\mu}(n)$ and $\hat{T}(n)$ as

$$\hat{\mu}(n) = \alpha Y_n = (1 - \alpha)[\hat{\mu}(n - 1) + \hat{T}(n - 1)] \tag{14.3}$$

$$\hat{T}(n) = \beta[\hat{\mu}(n) - \hat{\mu}(n - 1)] + (1 - \beta)\hat{T}(n - 1) \tag{14.4}$$

At the end of each regular time period, we receive a new observation of output Y_n and can recompute our estimate of mean regular time capacity $\hat{\mu}(n)$. To start the method, we need estimates of $\hat{\mu}(0)$ and $\hat{T}(0)$. These can be reasonable guesses or statistical estimates based on historical data. Depending on the values of α and β, the effect of these initial values of $\hat{\mu}(0)$ and $\hat{T}(0)$ will "wash out" after a few actual observations.

Because we are making use of exponential smoothing with a trend, the system can also be used to chart improvement progress. The trend $\hat{T}(n)$ is a good indicator of capacity improvements. If positive, then average output is increasing. In a sell-all-you-can-make environment, higher mean capacity will justify higher production quotas and hence greater profits.

Recall that our computation of economic production quotas in Chapter 13 required the mean μ—*and* standard deviation σ—of regular time production. We can use exponential smoothing to track this parameter as well. Since variance is a much noisier statistic to track than the mean, it is more difficult to track trends explicitly. For this reason, we advocate using exponential smoothing with no trend.

Let Y_n represent the nth observation of the amount produced during regular time production, $\hat{\mu}(n)$ represent the nth estimate of mean regular time capacity, and γ denote a smoothing constant. Recall that the definition of variance of a random variable X is

$$\text{Var}(X) = E[(X - E[X])^2]$$

After the nth observation, we have estimated the mean of regular time capacity as $\hat{\mu}(n)$. Hence, we can make an estimate of the variance of regular time capacity after the nth observation as

$$\left[Y_n - \hat{\mu}(n)\right]^2$$

Since these estimates will be noisy, we smooth them with previous estimates to get

$$\hat{\sigma}^2(n) = \gamma \left[Y_n - \hat{\mu}(n)\right]^2 + (1 - \gamma)\hat{\sigma}^2(n - 1) \tag{14.5}$$

as our nth estimate of the variance of regular time production.

As usual with exponential smoothing, an estimate of $\hat{\sigma}^2(0)$ must be supplied to start the iteration. Thereafter, each new observation of regular time output yields a new estimate of the variance of regular time production. As we observed in Chapter 13, smaller variance enables us to set the quota closer to mean capacity and thereby yields greater profit. Therefore, a downward trend in $\hat{\sigma}^2(n)$ is a useful measure of an improving production system.

We now illustrate these calculations by means of the example described in Table 14.1. Regular time periods consist of Monday through Friday (two shifts per day), and we have collected 20 weeks of past data on weekly output. As a rough starting point we optimistically estimate capacity at 2,000 units per week, so we set $\hat{\mu}(0) = 2,000$. We have no evidence of a trend, so we set $\hat{T}(0) = 0$. We make a guess that the standard deviation of regular time production is around 100, so we set $\hat{\sigma}^2(0) = 100^2 = 10,000$. We will choose our smoothing constants to be

$$\alpha = 0.5$$
$$\beta = 0.2$$
$$\lambda = 0.4$$

Of course, as we discussed in Appendix 13A, choosing smoothing constants is something of an art, so trial and error on past data may be required to obtain reasonable values in actual practice.

Now we can start the smoothing process. Regular time production during the first period is 1,400 units, so using Equation (14.3), we compute our smoothed estimate of mean regular time capacity as

$$\hat{\mu}(1) = \alpha Y_1 + (1 - \alpha)[\hat{\mu}(0) + \hat{T}(0)]$$
$$= 0.5(1,400) + (1 - 0.5)(2,000 + 0)$$
$$= 1,700$$

TABLE 14.1   **Exponential Smoothing of Capacity Parameters**

n	Y_n	$\hat{\mu}(n)$	$\hat{T}(n)$	$\hat{\sigma}^2(n)$	$\hat{\sigma}(n)$
0	—	2,000.0	0.0	10,000.0	100.0
1	1,400	1,700.0	−60.0	42,000.0	204.9
2	1,302	1,471.0	−93.8	36,624.4	191.4
3	1,600	1,488.6	−71.5	26,938.6	164.1
4	2,100	1,758.5	−3.2	62,801.1	250.6
5	1,800	1,777.7	1.2	37,880.4	194.6
6	2,150	1,964.4	38.4	36,500.0	191.0
7	2,450	2,226.4	83.1	41,898.8	204.7
8	2,200	2,254.7	72.1	26,337.7	162.3
9	2,600	2,463.4	99.4	23,263.2	152.5
10	2,100	2,331.4	53.2	35,382.6	188.1
11	2,200	2,292.3	34.7	24,636.7	157.0
12	2,600	2,463.5	62.0	22,235.7	149.1
13	2,800	2,662.7	89.4	20,877.2	144.5
14	2,300	2,526.1	44.2	32,973.8	181.6
15	2,900	2,735.2	77.2	30,653.1	175.1
16	2,800	2,806.2	76.0	18,407.1	135.7
17	2,650	2,766.1	52.7	16,433.0	128.2
18	3,000	2,909.4	70.9	13,142.7	114.6
19	2,750	2,865.1	47.8	13,188.1	114.8
20	3,150	3,031.5	71.5	13,531.0	116.3

Similarly, we use Equation (14.4) to compute the smoothed trend as

$$\hat{T}(1) = \beta[\hat{\mu}(1) - \hat{\mu}(0)] + (1 - \beta)\hat{T}(0)$$
$$= 0.2(1,700 - 2,000) + (1 - 0.2)(0)$$
$$= -60$$

Finally, we use Equation (14.5) to compute the smoothed estimate of variance of regular time production as

$$\hat{\sigma}^2(1) = \gamma[Y_n - \hat{\mu}(1)]^2 + (1 - \gamma)\hat{\sigma}^2(0)$$
$$= 0.4(1,400 - 1,700)^2 = (1 - 0.4)(10,000)$$
$$= 42,000$$

Thus, the smoothed estimate of standard deviation of regular time production is $\hat{\sigma}(1) = \sqrt{42,000} = 204.9$.

We can continue in this manner to generate the numbers in Table 14.1. A convenient way to examine these data is to plot them graphically. Figure 14.17 compares the smoothed estimates with the actual values of regular time production. Notice that the smoothed estimate follows the upward trend of the data, but with less variability from period to period (it is called *smoothing,* after all). Furthermore, this graph makes it apparent that our initial estimate of regular time capacity of 2,000 units per week was somewhat high. To compensate, the smoothed estimate trends downward for the first few periods, until the actual upward trend forces it up again.

These trends can be directly observed in Figure 14.18, which plots the smoothed trend after each period. Because of the high initial estimate of $\hat{\mu}(0)$, this trend is initially negative. The eventual positive trend indicates that capacity is increasing in this plant, a sign that improvements are having an effect on the operation.

Finally, Figure 14.19 plots the smoothed estimate of the standard deviation of regular time production. This estimate appears to be constant or slightly decreasing. A decreasing estimate is an indication that plant improvements are reducing variability in output. Both this and the smoothed trend provide us with hard measures of continual improvement.

FIGURE 14.17

Exponential smoothing of mean regular time capacity

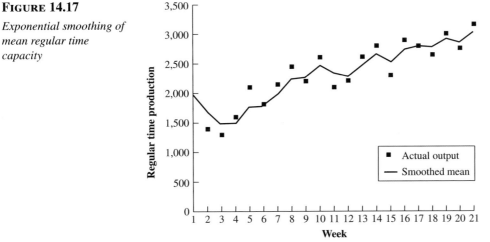

FIGURE 14.18

Exponential smoothing of trend in mean regular time capacity

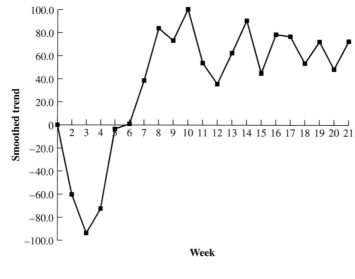

FIGURE 14.19

Exponential smoothing of variance of regular time capacity

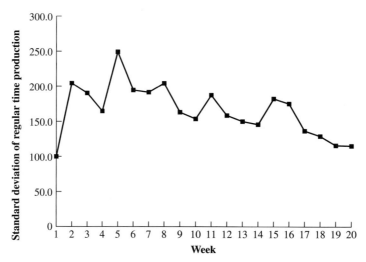

14.6 Conclusions

In this chapter, we have spent a good deal of time discussing the shop floor control (SFC) module of a production planning and control (PPC) system. We have stressed that a good SFC module can do a great deal more than simply govern the movement of material into and through the factory. As the lowest-level point of contact with the manufacturing process, SFC plays an important role in shaping the management problems that must be faced. A well-designed SFC module will establish a predictable, robust system with controls whose complexity is appropriate for the system's needs.

Because manufacturing systems are different, a uniform SFC module for all applications is impractical, if not impossible. A module that is sufficiently general to handle a broad range of situations is apt to be cumbersome for simple systems and ill suited for specific complex systems. More than any other module in the PPC hierarchy, the SFC module is a candidate for customization. It may make sense to make use of commercial bar coding, optical scanning, local area networks, statistical process control, and other technologies as components of an SFC module. However, there is no substitute for careful integration done with the capabilities and needs of the system in mind. It is our hope that the manufacturing professionals reading this book will provide such integration, using the basics, intuition, and synthesis skills they have acquired here and elsewhere.

Since we do not believe it is possible to provide a cookbook scheme for devising a suitable SFC module, we have taken the approach of starting with simple systems, highlighting key issues, and extending our approach to various more complex issues. Our basic scheme is to start with a simple set of CONWIP lines as the incumbent and ask why such a setup would not work. If it does work, as we believe it can in relatively uncomplicated flow shops, then this is the simplest, most robust solution. If not, then more complex schemes, such as that of pull-from-bottleneck (PFB), may be necessary. We hope that the variations on CONWIP we have offered are sufficient to spur the reader to think creatively of options for specific situations beyond those discussed here.

One last issue we have emphasized is that feedback is an *essential* feature of an effective production planning and control system. Unfortunately, many PPC systems evolve in a distributed fashion, with different groups responsible for different facets of the planning process. The result is that inconsistent data are used, communication between decision makers breaks down, and factionalism and finger pointing, instead of cooperation and coordination, become the standard response to problems. Furthermore, without a feedback mechanism, overly optimistic data (e.g., unrealistically high estimates of capacity) can persist in planning systems, causing them to be untrustworthy at best and downright humorous at worst. Statistical throughput control is one explicit mechanism for forcing needed feedback with regard to capacity data. Similar approaches can be devised to promote feedback on other key data, such as process yields, rework frequency, and learning curves for new products. The key is for management to be sensitive to the potential for inconsistency and to strive to make feedback systemic to the PPC hierarchy. Furthermore, to be effective, feedback mechanisms must be used in a spirit of problem solving, not one of blame fixing.

Although the SFC module performs some of the most lowly and mundane tasks in a manufacturing plant, it can play a critical role in the overall effectiveness of the system. A well-designed SFC module establishes a predictable environment upon which to build the rest of the planning hierarchy. Appropriate feedback mechanisms can collect useful data for such planning and can promote an environment of ongoing improvement. To recall our quote from the beginning of this chapter,

> *Even a journey of one thousand li begins with a single step.*
>
> Lao Tzu

The SFC module is not only the first step toward an effective production planning and control system, it is a very important step indeed.

Appendix 14A
Statistical Throughput Control

The basic quantity needed to address several short-term production tracking questions is the probability of making the quota by the end of regular time production, given that we know how much has been produced thus far. Since output from each line must be recorded in order to maintain a constant WIP level in the line, a CONWIP line will have the requisite data on hand to make this calculation.

To do this, we define the length of regular time production as R. We assume that production during this time, denoted by N_R, is normally distributed, with mean μ and standard deviation σ. We let N_t represent production, in standard units, during $[0, t]$, where $t \leq R$. We model N_t as continuous and normally distributed with mean $\mu t / R$ and variance $\sigma^2 t / R$. In general, the assumption that production is normal will often be good for all but small values of t. The assumption that the mean and variance of N_t are as given here is equivalent to assuming that production during nonoverlapping intervals is independent. Again, this is probably a good assumption except for very short intervals.

We are interested primarily in the process $N_t - S_t$, where S_t is the cumulative scheduled production up to time t. If we are using a periodic production quota, then $S_t = Qt/R$. The quantity $N_t - S_t$ represents the overage, or amount by which we are ahead of schedule, at time t. If this quantity is positive, we are ahead; if negative, we are behind. In an ideal system with constant production rates, this quantity would always be zero. In a real system, it will fluctuate, becoming positive and/or negative.

From our assumptions, it follows that $N_t - Qt/R$ is normally distributed with mean $(\mu - Q)t/R$ and variance $\sigma^2 t / R$. Likewise, N_{R-t} is normally distributed, with mean $\mu(R - t)/R$ and variance $\sigma^2(R - t)/R$. Hence, if at time t, $N_t = n_t$, where $n_t - Qt/R = x$ (we are x units ahead of schedule), then we will miss the quota by time R only if $N_{R-t} < Q - n_t$. Thus, the probability of missing the quota by time R given a current overage of x is given by

$$P(N_{R-t} \leq Q - n_t) = P\left(N_{R-t} \leq Q - x - \frac{Qt}{R}\right)$$

$$= P\left(N_{R-t} \leq \frac{Q(R-t)}{R} - x\right)$$

$$= \Phi\left[\frac{(Q - \mu)(R - t)/R - x}{\sigma\sqrt{(R-t)/R}}\right]$$

where $\Phi(\cdot)$ represents the standard normal distribution.

From a practical implementation standpoint, it is more convenient to precompute the overage levels that cause the probability of missing the quota to be any specified level α. These can be computed as follows:

$$\Phi\left[\frac{(Q - \mu)(R - t)/R - x}{\sigma\sqrt{(R-t)/R}}\right] = \alpha$$

which yields

$$x = -\frac{(\mu - Q)(R - t)}{R} - z_\alpha \sigma \sqrt{\frac{R - t}{R}}$$

where z_α is chosen such that $\Phi(z_\alpha) = \alpha$. This x is the overage at time t that results in a probability of missing the quota exactly equal to α, and is Equation (14.2), upon which our STC charts are based.

Study Questions

1. What is the motivation for limiting the span of control of a manager to a specified number of subordinates or manufacturing processes? What problems might this cause in coordinating the plant?

2. We have repeatedly mentioned that throughput is an increasing function of WIP. Therefore, we could conceivably vary the WIP level as a way of matching production to the demand rate. Why might this be a poor strategy in practice?

3. What factors might make kanban inappropriate for controlling material flow through a job shop, that is, a system with many, possibly changing, routings with fluctuating volumes?

4. Why might we want to violate the WIP cap imposed by CONWIP and run a card deficit when a machine downstream from the bottleneck fails? If we allow this, what additional discipline might we want to impose to prevent WIP explosions?

5. What are the advantages of breaking a long production line into tandem CONWIP loops? What are the disadvantages?

6. For each of the following situations, indicate whether you would be inclined to use CONWIP (C), kanban (K), PFB (P), or an individual system (I) for shop floor control.
   a. A flow line with a single-product family.
   b. A paced assembly line fed from inventory storage.
   c. A steel mill where casters feed hot strip mills (with slab storage in between), which feed cold rolling mills (with coil storage in between).
   d. A plant with several routings sharing some resources with significant setup times, and all routings are steadily loaded over time.
   e. A plant with many routings sharing some resources but where some routings are sporadically used.

7. What is meant by statistical throughput control, and how does it differ from statistical process control? Could one use SPC tools (i.e., control charts) for throughput tracking?

8. Why is the STC chart in Figure 14.15 symmetric, while the one in Figure 14.16 is asymmetric? What does this indicate about the effect of setting production quotas at or near average capacity?

9. Why might it make sense to use exponential smoothing with a linear trend to track mean capacity of a line? How could we judge whether exponential smoothing without a linear trend might work as well or better?

10. What uses are there for tracking the standard deviation of periodic output from a production line?

Problems

1. A circuit board manufacturing line contains an expose operation consisting of five parallel machines inside a clean room. Because of limited space, there is only room for five carts of WIP (boards) to buffer expose against upstream variability. Expose is fed by a coater line, which consists of a conveyor that loads boards at a rate of three per minute and requires roughly one hour to traverse (i.e., a job of 60 boards will require 20 minutes to load plus one hour for the last loaded board to arrive in the clean room at expose). Expose machines take roughly two hours to process jobs of 60 boards each. Current policy is that whenever the WIP inside the clean room reaches five jobs (in addition to the five jobs being worked on at the expose machines), the coater line is shut down for three hours. Both expose and the coater are subject to variability due to machine failures, materials shortages, operator unavailability, and so forth. When all this is factored into a capacity analysis, expose seems to be the bottleneck of the entire line.
   a. What problem might the current policy for controlling the coater present?
   b. What alternative would you suggest? Remember that expose is isolated from the rest of the line by virtue of being in a clean room and that because of this, the expose operators cannot see the beginning of the coater; nor can the coater loader easily see what is going on inside the clean room.
   c. How would your recommendation change if the capacity of expose were increased (say, by using floating labor to work through lunches) so that it was no longer the long-term bottleneck?

2. Consider a five-station line that processes two products, A and B. Station 3 is the bottleneck for both products. However, product A requires one hour per unit at the bottleneck, while

FIGURE 14.20

Pull-from-bottleneck production system

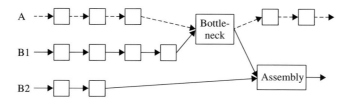

product B requires one-half hour. A modified CONWIP control policy is used under which the complexity-adjusted WIP is measured as the number of hours of work at the bottleneck. Hence, one unit of A counts as one unit of complexity-adjusted WIP, while one unit of B counts as one-half unit of complexity-adjusted WIP. The policy is to release the next job in the sequence whenever the complexity-adjusted WIP level falls to 10 or less.

a. Suppose the release sequence alternates between product A and B (that is, A-B-A-B-A-B-…). What will happen to the numbers of type A and type B jobs in the system over time?

b. Suppose the release sequence alternates between 10 units of A and 10 units of B. Now what happens to the numbers of type A and type B jobs in the system over time?

c. The JIT literature advocates a sequence like the one in *a.* Why? Why might some lines need to make use of a sequence like the one in *b*?

3. Consider the two-product system illustrated in Figure 14.20. Product A and component 1 of product B pass through the bottleneck operation. Components 1 and 2 of product B are assembled at the assembly operation. Type A jobs require one hour of processing at the bottleneck, while type B jobs require one and one-half hours. The lead time for type A jobs to reach the bottleneck from their release point is two hours. Component 1 of type B jobs takes four and one-half hours to react the bottleneck. The sequence of the next eight jobs to be processed at the bottleneck is as follows:

Job index	1	2	3	4	5	6	7	8
Job type	A	A	B	B	B	B	A	B

Jobs 1 through 6 have already been released but have not yet been completed at the bottleneck. Suppose that the system is controlled using the pull-from-the-bottleneck method described in Section 14.4.2, where the planned time at the bottleneck is $L = 4$ hours.

a. When should job 7 be released (i.e., now or after the completion of that job currently in the system)?

b. When should job 8 be released (i.e., now or after the completion of that job currently in the system)? Are jobs necessarily released in the order they will be processed at the bottleneck? Why or why not?

c. If we only check to see whether new jobs should be released when jobs are completed at the bottleneck, will jobs wait at the bottleneck more than, less than, or equal to the target time L? (*Hint:* What is the expected waiting time of job 8 at the bottleneck?) Could these be cases in which we would want to update the current workload at the bottleneck more frequently than at completion times of jobs?

d. Suppose that the lead time for component 2 of product B to reach assembly is one hour. If we want component 2 to wait for one and one-half hours on average at assembly, when should it be released relative to its corresponding component 1?

4. Consider a line that builds toasters runs five days per week, one shift per day (or 40 hours per week). A periodic quota of 2,500 toasters has been set. If this quota is not met by the end of work on Friday, overtime on the weekend is run to make up the difference. Historical data indicate that the capacity of the line is 2,800 toasters per week, with a standard deviation of 300 toasters.

    a. Suppose at hour 20 we have completed 1,000 toasters. Using the STC model, estimate the probability that the line will be able to make the quota by the end of the week.

    b. How many toasters must be completed by hour 20 to ensure a probability of 0.9 of making the quota?

    c. If the weekly quota is increased to 2,800 toasters per week, how does the answer to *b* change?

5. Output from the assembly line of a farm machinery manufacturer that produces combines has been as follows for the past 20 weeks:

Week	1	2	3	4	5	6	7	8	9	10
Output	22	21	24	30	25	25	33	40	36	39

Week	11	12	13	14	15	16	17	18	19	20
Output	50	55	44	48	55	47	61	58	55	60

    a. Use exponential smoothing with a linear trend and smoothing constants $\alpha = 0.4$ and $\beta = 0.2$ to track weekly output for weeks 2 to 20. Does there appear to be a positive trend to the data?

    b. Using mean square deviation (MSD) as your accuracy measure, can you find values of α and β that fit these data better than those given in *a*?

    c. Use exponential smoothing (without a linear trend) and a smoothing constant $\gamma = 0.2$ to track variance of weekly output for weeks 2 to 20. Does the variance seem to be increasing, decreasing, or constant?

15 PRODUCTION SCHEDULING

Let all things be done decently and in order.
I Corinthians

15.1 Goals of Production Scheduling

Virtually all manufacturing managers want on-time delivery, minimal work in process, short customer lead times, and maximum utilization of resources. Unfortunately, these goals conflict. It is much easier to finish jobs on time if resource utilization is low. Customer lead times can be made essentially zero if an enormous inventory is maintained. And so on. The goal of production scheduling is to strike a profitable balance among these conflicting objectives.

In this chapter we discuss various approaches to the scheduling problem. We begin with the standard measures used in scheduling and a review of traditional scheduling approaches. We then discuss why scheduling problems are so hard to solve and what implications this has for real-world systems. Next we develop practical scheduling approaches, first for the bottleneck resource and then for the entire plant. Finally, we discuss how to interface scheduling—which is push in concept—with a pull environment such as CONWIP.

15.1.1 Meeting Due Dates

A basic goal of production scheduling is to meet due dates. These typically come from one of two sources: directly from the customer or in the form of material requirements for other manufacturing processes.

In a make-to-order environment, customer due dates drive all other due dates. As we saw in Chapter 3, a set of customer requirements can be exploded according to the associated bills of material to generate the requirements for all lower-level parts and components.

In a make-to-stock environment there are no customer due dates, since all customer orders are expected to be filled immediately upon demand. Nevertheless, at some point, falling inventory triggers a demand on the manufacturing system. Demands generated in this fashion are just as real as actual customer orders since, if they are not met, customer demands will eventually go unfilled. These stock replenishment demands are exploded into demands for lower-level components in the same fashion as customer demands.

Several measures can be used to gauge due date performance, including these:

Service level (also known as simply **service**), typically used in make-to-order systems, is the fraction of orders filled on or before their due dates. Equivalently, it is the fraction of jobs whose cycle time is less than or equal to the planned lead time.

Fill rate is the make-to-stock equivalent of service level and is defined as the fraction of demands that are met from inventory, that is, without backorder.

Lateness is the difference between the order due date and the completion date. If we define d_j as the due date and c_j as the completion time of job j, the lateness of job j is given by $L_j = c_j - d_j$. Notice that lateness can be positive (indicating a late job) or negative (indicating an early job). Consequently, small *average* lateness has little meaning. It could mean that all jobs finished near their due dates, which is good; or it could mean that for every job that was very late there was one that was very early, which is bad. For lateness to be a useful measure, we must consider its *variance* as well as its *mean*. A small mean and variance of lateness indicates that most jobs finish on or near their due dates.

Tardiness is defined as the lateness of a job if it is late and zero otherwise. Thus, early jobs have zero tardiness. Consequently, *average* tardiness *is* a meaningful measure of customer due date performance.

These measures suggest several objectives that can be used to formulate scheduling problems. One that has become classic is to "minimize average tardiness." Of course, it is classic only in the production scheduling research literature, not in industry. As one might expect, "minimize lateness variance" has also seen very little use in industry.

Service level and fill rate *are* used in industry. This is probably because tardiness is difficult to track and because the measures of average tardiness and lateness variance are not intuitive. The percentage of on-time jobs is simpler to state than something like "the average number of days late, with early jobs counting as zero" or "the standard deviation of the difference between job due date and job completion date." However, service level and fill rate have obvious problems. Once a job is late, it counts against service *no matter how late it is*. Naive approaches can thus lead to ridiculous schedules that call for such things as never finishing late jobs or lying to customers. We present a due date quoting procedure in Section 15.3.2 that avoids these difficulties.

15.1.2   Maximizing Utilization

In industry, cost accounting encourages high machine utilization. Higher utilization of capital equipment means higher return on investment, provided of course that the equipment is utilized to increase revenue (i.e., to create products that are in demand). Otherwise, high utilization merely serves to increase inventory, not profits. High utilization makes the most sense when producing a commodity item to stock.

Factory physics also promotes high utilization, provided cycle times, quality, and service are not degraded excessively. However, recall that the Capacity Law implies that 100 percent utilization is impossible. How close to full utilization a line can run and still have reasonable WIP and cycle time depends on the level of variability. The more variability a line has, the lower utilization must be to compensate. Furthermore, as the practical worst case in Chapter 7 illustrated, *balanced* lines have more congestion than unbalanced ones, especially when variability is high. This implies that it may well be attractive not to have near 100 percent utilization of *all* resources in the line.

A measure that is closely related to utilization is **makespan,** which is defined as the time it takes to finish a fixed number of jobs. For this set of jobs, the production rate is the number of jobs divided by the makespan, and the utilization is the production rate divided by the capacity. Although makespan is not widely used in industry, it has seen frequent use in the theoretical scheduling research.

The decision of what target to use for utilization is a strategic one that belongs at the top of the in-plant planning hierarchy (Chapter 13). Because high-level decisions are made less frequently than low-level ones, utilization cannot be adjusted to facilitate production scheduling. Similarly, the level of variability in the line is a consequence of high-level decisions (e.g., capacity and process design decisions) that are also made much less frequently than are scheduling decisions. Thus, for the purposes of scheduling we can assume that utilization targets and variability levels are given. In most cases, the target utilization of the bottleneck resource will be high. The one important exception to this is a highly variable and customized demand process requiring an extremely quick response time (e.g., ambulances and fire engines). Such systems typically have very low utilization and are not well suited to scheduling. We will assume throughout, therefore, that the system is such that a fairly high bottleneck utilization is desirable.

15.1.3   Reducing WIP and Cycle Times

As we discussed in Part II, there are several motives for keeping cycle times short, including these:

1. *Better responsiveness to the customer.* If it takes less time to make a product, the lead time to the customer can be shortened.

2. *Maintaining flexibility.* Changing the list (backlog) of parts that are planned to start next is less disruptive than trying to change the set of jobs already in process. Since shorter cycle times allow for later releases, they enhance this type of flexibility.

3. *Improving quality.* Long cycle times typically imply long queues in the system, which in turn imply long delays between defect *creation* and defect *detection*. For this reason, short cycle times support good quality.

4. *Relying less on forecasts.* If cycle times are longer than customers are willing to wait, production must be done in anticipation of demand rather than in response to it. Given the lack of accuracy of most demand forecasts, it is extremely important to keep cycle times shorter than quoted lead times, whenever possible.

5. *Making better forecasts.* The more cycle times exceed customer lead times, the farther out the forecast must extend. Hence, even if cycle times cannot be reduced to the point where dependence on forecasting is eliminated, cycle time reduction can shorten the forecasting time horizon. This can greatly reduce forecasting errors.

Little's Law (CT = WIP/TH) implies that reducing cycle time and reducing WIP are equivalent, provided that throughput remains constant. However, the Variability Buffering Law implies that reducing WIP without reducing variability will cause throughput to decrease. Thus variability reduction is generally an important component of WIP and cycle time reduction programs.

Although WIP and cycle time may be virtually equivalent from a reduction policy standpoint, they are not equivalent from a measurement standpoint. WIP is often easier to measure, since one can count jobs, while cycle times require clocking jobs in and out of the system. Cycle times become even harder to measure in assembly operations. Consider an automobile, for instance. Does the cycle time start with the ordering of the components such as spark plugs and steel, or when the chassis starts down the assembly line? In such cases, it is more practical to use Little's Law to obtain an indirect measure of cycle time by measuring WIP (in dollars) over the system under consideration and dividing by throughput (in dollars per day).

15.2   Review of Scheduling Research

Scheduling as a practice is as old as manufacturing itself. Scheduling as a research discipline dates back to the scientific management movement in the early 1900s. But serious analysis of scheduling problems did not begin until the advent of the computer in the 1950s and 1960s. In this section, we review key results from the theory of scheduling.

15.2.1   MRP, MRP II, and ERP

As we discussed in Chapter 3, MRP was one of the earliest applications of computers to scheduling. However, the simplistic model of MRP undermines its effectiveness. The reasons, which we noted in Chapter 5, are as follows:

1. MRP assumes that lead times are attributes of parts, independent of the status of the shop. In essence, MRP assumes infinite capacity.
2. Since MRP uses only one lead time for offsetting and since late jobs are typically worse than excess inventory, there is strong incentive to inflate lead times in the system. This results in earlier releases, larger queues, and hence longer cycle times.

As we discussed in Part II, these problems prompted some scheduling researchers and practitioners to turn to enhancements in the form of MRP II and, more recently, ERP. Others rejected MRP altogether in favor of JIT. However, the majority of scheduling researchers focused on mathematical formulations in the field of operations research, as we discuss next.

15.2.2   Classic Scheduling

We refer to the set of problems in this section as *classic* scheduling problems because of their traditional role as targets of study in the operations research literature. For the most part, these problems are highly simplified and generic, which has limited their direct applicability to real situations. However, despite the fact that they are not classic from an applications perspective, they can offer some useful insights.

Most classical scheduling problems address one, two, or possibly three machines. Other common simplifying assumptions include these:

1. All jobs are available at the start of the problem (i.e., no jobs arrive after processing begins).
2. Process times are deterministic.
3. Process times do not depend on the schedule (i.e., there are no setups).
4. Machines never break down.
5. There is no preemption (i.e., once a job starts processing, it must finish).
6. There is no cancellation of jobs.

These assumptions serve to reduce the scheduling problem to manageable proportions, in some cases. One reason is that they allow us to restrict attention to simplified schedules, called sequences. In general, a **schedule** gives the anticipated start times of each job on each resource, while a **sequence** gives only the order in which the jobs are to be done. In some cases, such as the single-machine problem with all jobs available when processing begins, a simple sequence is sufficient. In more complex problems, separate sequences for different resources may be required. And in some problems a full-blown schedule is necessary to impart the needed instructions to the system. Not surprisingly, the more complex the form of the schedule that is sought, the more difficult it is to find it.

Some of the best-known problems that have been studied in the context of the assumptions discussed in the operations research literature are the following.

Minimizing average cycle time on a single machine. First, note that for the single-machine problem, the *total time* to complete all the jobs does not depend on the ordering—it is given by the sum of the processing times for the jobs. Hence an alternate criterion is needed. One candidate is the average cycle time (called **flow time** in the production scheduling literature), which can be shown to be minimized by processing jobs in order of their processing times, with the shortest job first and longest job last. This is called the **shortest process time (SPT)** sequencing rule. The primary insight from this result is that short jobs move through the shop more quickly than long jobs and therefore tend to reduce congestion.

Minimizing maximum lateness on a single machine. Another possible criterion is the maximum lateness that any job is late, which can be shown to be minimized by ordering the jobs according to their due dates, with the earliest due date first and the latest due date last. This is called the **earliest due date (EDD)** sequencing rule. The intuition behind this approach is that if it is *possible* to finish all the jobs on time, EDD sequencing will do so.

Minimizing average tardiness on a single machine. A third criterion for the single-machine problem is average tardiness. (Note that this is equivalent to total tardiness, since average tardiness is simply total tardiness divided by the number of jobs.) Unfortunately, there is no sequencing rule that is guaranteed to minimize this measure. Often EDD is a good heuristic, but its performance cannot be ensured, as we demonstrate in one of the exercises at the end of the chapter. Likewise, there is no sequencing rule that minimizes the variance of lateness. We will discuss the reasons why this scheduling problem and many others like it are particularly hard to solve.

Minimizing makespan on two machines. When the production process consists of two machines, the total time to finish all the jobs, the makespan, is no longer fixed. This is because certain sequences might induce idle time on the second machine as it waits for the first machine to finish a job. Johnson (1954) proposed an intuitive algorithm for finding the sequence that minimizes makespan for this problem, which can be stated as follows: Separate the jobs into two sets, A and B. Jobs in set A are those whose process time on the first machine is less than or equal to the process time on the second machine. Set B contains the remaining jobs. Jobs in set A go first and in the order of the shortest process time first. Then jobs in set B are appended in order of the longest process time first. The result is a sequence that minimizes the makespan over the two machines.

The insight behind Johnson's algorithm can be appreciated by noting that we want a short job in the first position because the second machine is idle until the first job finishes on the first machine. Similarly, we want a short job to be last since the first machine is idle while the second machine is finishing the last job. Hence, the algorithm implies that small jobs are better for reducing cycle times and increasing utilization.

Minimizing makespan in job shops. The problem of minimizing the time to complete n jobs with general routings through m machines (subject to all the assumptions previously discussed) is a well-known hard problem in the operations research literature. The reason for its difficulty is that the number of possible schedules to consider is enormous. Even for the modestly sized 10-job, 10-machine problem there are almost 4×10^{65} possible schedules (more atoms than there are in the earth). Because of this a 10-by-10 problem was not solved optimally until 1988 by using a mainframe computer and five hour of computing time (Carlier and Pinson 1988).

A standard approach to this type of problem is known as **branch and bound.** The basic idea is to define a **branch** by selecting a partial schedule and define **bounds** by computing a lower limit on the makespan that can be achieved with a schedule that includes this partial schedule. If the bound on a branch exceeds the makespan of the best (complete) schedule found so far, it is no longer considered. This is a method of **implicit enumeration,** which allows the algorithm to consider only a small subset of the possible schedules. Unfortunately, even a very small fraction of these can be an incredibly large number, and so branch and bound can be tediously slow. Indeed, as we will discuss, there is a body of theory that indicates that any exact algorithm for hard problems, like the job shop scheduling problem, will be slow. This makes nonexact *heuristic* approaches a virtual necessity. We will list a few of the many possible approaches in our discussion of the complexity of scheduling problems.

15.2.3   Dispatching

Scheduling is hard, both theoretically (as we will see) and practically speaking. A traditional alternative to scheduling all the jobs on all the machines is to simply **dispatch**—sort according to a specified order—as they arrive at machines. The simplest dispatching rule (and also the one that seems fairest when dealing with customers) is **first-in, first-out (FIFO).** The FIFO rule simply processes jobs in the order in which they arrive at a machine. However, simulation studies have shown that this rule tends not to work well in complex job shops. Alternatives that can work better are the SPT or EDD rules, which we discussed previously. In fact, these are often used in practice, as we noted in Chapter 3 in our discussion of shop floor control in ERP. Literally hundreds of different dispatching rules have been proposed by researchers as well as practitioners (see Blackstone, et al. 1982 for a survey).

All dispatching rules, however, are *myopic* in nature. By their very definition they consider only local and current conditions. Since the best choice of what to work on now at a given machine depends on future jobs as well as other machines, we cannot expect dispatching rules to work well all the time, and, in fact, they do not. But because the options for scheduling realistic systems are still very limited, dispatching continues to find extensive use in industry.

15.2.4   Why Scheduling Is Hard

We have noted several times that scheduling problems are hard. A branch of mathematics known as **computational complexity analysis** gives a formal means for evaluating just how hard they are. Although the mathematics of computational complexity is beyond our scope, we give a qualitative treatment of this topic in order to develop an appreciation of why some scheduling problems cannot be solved optimally. In these cases, we are forced to go from seeking the *best* solution to finding a *good* solution.

Problem Classes.   Mathematical problems can be divided into the following two classes according to their complexity:

1. **Class P problems** are problems that can be solved by algorithms whose computational time grows as a polynomial function of problem size.
2. **NP-hard problems** are problems for which there is no known polynomial algorithm, so that the time to find a solution grows exponentially (i.e., much more rapidly than a polynomial function) in problem size. Although it has not been definitively proved that there are no clever polynomial algorithms for solving NP-hard problems, many eminent mathematicians have tried and failed. At present, the preponderance of evidence indicates that efficient (polynomial) algorithms cannot be found for these problems.

Roughly speaking, class P problems are easy, while NP-hard problems are hard. Moreover, some NP-hard problems appear to be harder than others. For some, efficient algorithms have been shown empirically to produce good approximate solutions. Other NP-hard problems, including many scheduling problems, are even difficult to solve approximately with efficient algorithms.

To get a feel for what the technical terms **polynomial** and **exponential** mean, consider the single-machine sequencing problem with three jobs. How many ways are there to sequence three jobs? Any one of the three could be in the first position, which leaves two candidates for the second position, and only one for the last position. Therefore, the number of sequences or *permutations* is $3 \times 2 \times 1 = 6$. We write this as 3! and say "3 factorial." If we were looking for the best sequence with regard to some objective function for this problem, we would have to consider (explicitly or implicitly) six alternatives. Since the factorial function exhibits exponential growth, the number of alternatives we must search through, and therefore the amount of time required to find the optimal solution, also grows exponentially in problem size.

The reason this is important is that *any* polynomial function will *eventually* become dominated by *any* exponential function. For instance, the function $10,000n^{10}$ is a big polynomial, while the function $e^n/10,000$ appears small. Indeed, for small values of n, the polynomial function dominates the exponential. But at around $n = 60$ the exponential begins to dominate and by $n = 80$ has grown to be 50 million times larger than the polynomial function.

Returning to the single-machine problem with three jobs, we note that 3! does not seem very large. However, observe how quickly this function blows up: $3! = 6, 4! = 24$, $5! = 120, 6! = 720$, and so on. As the number of jobs to be sequenced becomes large, the number of possible sequences becomes quite ominous: $10! = 3,628,800$, $13! = 6,227,020,800$, and

$$25! = 15,511,210,043,330,985,984,000,000$$

To get an idea of how big this number is, we compare it to the national debt, which at the time of this writing had not yet reached $5 trillion. Nonetheless, suppose it were $5 trillion and we wanted to pay it in pennies. The 500 trillion pennies would cover almost one-quarter of the state of Texas. In comparison, 25! pennies would cover the *entire*

TABLE 15.1   Computer Times for Job Sequencing on a Slow Computer

Number of Jobs	Computer Time
5	0.12 millisec
6	0.72 millisec
7	5.04 millisec
8	40.32 millisec
9	0.36 sec
10	3.63 sec
11	39.92 sec
12	7.98 min
13	1.73 hr
14	24.22 hr
15	15.14 days
⋮	⋮
20	77,147 years

TABLE 15.2   Computer Times for Job Sequencing on a Computer 1,000 Times Faster

Number of Jobs	Computer Time
5	0.12 microsec
6	0.72 microsec
7	5.04 microsec
8	40.32 microsec
9	362.88 microsec
10	3.63 millisec
11	39.92 millisec
12	479.00 millisec
13	6.23 sec
14	87.18 sec
15	21.79 min
⋮	⋮
20	77.147 years

state of Texas—to a height of over *6,000 miles!* Now that's big. (Perhaps this is why mathematicians use the exclamation point to indicate the factorial function.)

Now let us relate these big numbers to computation times. Suppose we have a "slow" computer that can examine 1,000,000 sequences per second and we wish to build a scheduling system that has a response time of no longer than one minute. Assuming we must examine every possible sequence to find the optimum, how many jobs can we sequence optimally? Table 15.1 shows the computation times for various numbers of jobs and indicates that 11 jobs is the maximum we can sequence in less than one minute.

Now suppose we purchase a computer that runs 1,000 times faster than our old "slow" one (i.e., it can examine one billion sequences per second). Now how many jobs can be examined in less than one minute? From Table 15.2 we see that the maximum problem size we can solve only increases to 13 jobs (or 14 if we allow the maximum time to increase to one and one-half minutes). A 1,000 fold increase in computer speed only results in an 18 percent increase in size of the largest problem that can be solved in the specified time. The basic conclusion is that even big increases in computer speed do not dramatically increase our power to solve nonpolynomial problems.

For comparison, we now consider problems that do not grow exponentially. These are called **polynomial** problems because the time to solve them can be bounded a polynomial function of problem size (for example, n^2, n^3, etc., where n is a measure of problem size).

As a specific example, consider the job dispatching problem described in Section 15.2.3 and suppose we wish to dispatch jobs according to the SPT rule. This requires us to sort the jobs in front of the workstation according to process time.[1] There are well-known algorithms for sorting a list of elements whose computation time (i.e., number of steps) is proportional to $n \log n$, where n is the number of elements being

---

[1] Actually, in practice we would probably maintain the queue in sorted order, so we would not have to resort it each time a job arrived. This would make the problem even simpler than we indicate here.

TABLE 15.3 Computer Times for Job Sorting on the Slow Computer

Number of Jobs	Computer Time
10	3.6 sec
11	4.1 sec
12	4.7 sec
⋮	⋮
20	9.4 sec
30	16.1 sec
⋮	⋮
80	55.2 sec
85	59.5 sec
90	63.8 sec
⋮	⋮
100	72.6 sec
200	167.0 sec

TABLE 15.4 Computer Times for Job Sorting on a Computer 1,000 Times Faster

Number of Jobs	Computer Time
1,000	1.1 sec
2,000	2.4 sec
3,000	3.8 sec
⋮	⋮
10,000	14.5 sec
20,000	31.2 sec
30,000	48.7 sec
35,000	57.7 sec
36,000	59.5 sec
⋮	⋮
50,000	85.3 sec
100,000	181.4 sec
200,000	384.7 sec

sorted. This function is clearly bounded by n^2, a polynomial. Therefore, dispatching has polynomial complexity.

Suppose, just for the sake of comparison, that on the slow computer of the previous example it takes the same amount of time to sort 10 jobs as it does to examine 10! sequences (that is, 3.6 seconds). Table 15.3 reveals how the sorting times grow for lists of jobs longer than 10. Notice that we can sort 85 jobs and still remain below one minute (as compared to 11 jobs for the sequencing problem).

Even more interesting is what happens when we purchase the computer that works 1,000 times faster. Table 15.4 shows the computation times and reveals that we can go from sorting 85 jobs on the slow computer to sorting around 36,000 on the fast one. This represents an increase of over 400 percent, as compared to the 18 percent increase we observed for the sequencing problem. Evidently, we gain a lot from a faster computer for the "easy" (polynomial) sorting problem, but not much for the "hard" (exponential) sequencing problem.

Implications for Real Problems.   Because most real-world scheduling problems fall into the NP-hard category and tend to be large (e.g., involving hundreds of jobs and tens of machines), the above results have important consequences for manufacturing practice. Quite literally, they mean that it is impossible to solve many realistically sized scheduling problems optimally.[2]

Fortunately, the practical consequences are not quite so severe. Just because we cannot find the *best* solution does not mean that we cannot find a *good* one. In some ways, the nonpolynomial nature of the problem may even help, since it implies that there may

---

[2]A computer with as many bits as there are protons in the universe, running at the speed of light, for the age of the universe, would not have enough time to solve some of these problems. Therefore the word *impossible* is *not* an exaggeration.

be many candidates for a good solution. Reconsider the 25-job sequencing problem. If "good" solutions were extremely rare to the point that only one in a trillion of the possible solutions was good, there would still be more than 15 trillion good solutions. We can apply an approximate algorithm, called a **heuristic,** that has polynomial performance to search for one of these solutions. There are many types of heuristics, including such interestingly named techniques as *beam search, tabu search, simulated annealing,* and *genetic algorithms.* We will describe one of these (tabu search) in greater detail when we discuss bottleneck scheduling.

15.2.5   Good News and Bad News

We can draw a number of insights from this review of scheduling research that are useful to the design of a practical scheduling system.

The Bad News.   We begin with the negatives. First, unfortunately, most real-world problems violate the assumptions made in the classic scheduling theory literature in at least the following ways:

1. There are always more than two machines. Thus Johnson's minimizing makespan algorithm and its many variants are not directly useful.
2. Process times are not deterministic. In Part II we learned that randomness and variability contribute greatly to the congestion found in manufacturing systems. By ignoring this, scheduling theory may have overlooked something fundamental.
3. All jobs are *not* ready at the beginning of the problem. New jobs *do* arrive and continue arriving during the entire life of the plant. To pretend that this does not happen or to assume that we "clear out" the plant before starting new work is to deny a fundamental aspect of plant behavior.
4. Process times are frequently sequence-dependent. Often the number of setups performed depends on the sequence of the jobs. Jobs of like or similar parts can usually share a setup while dissimilar jobs cannot. This can be an important concern when scheduling the bottleneck process.

Second, real-world production scheduling problems are hard (in the NP-hard sense), which means

1. We cannot hope to find optimal solutions of many realistic-size scheduling problems.
2. Nonpolynomial approaches, like dispatching, may not work well.

The Good News.   Fortunately, there are also positives, especially when we realize that much of the scheduling research suffers from type III error: solving the *wrong* problem. The formalized scheduling problems addressed in the operations research literature are models, not reality. The constraints assumed in these models are not necessarily fixed in the real world since, to some extent, we can control the problem by controlling the environment. This is precisely what the Japanese did when they made a hard scheduling problem much easier by reducing setup times. When we think along these lines, the failures as well as the successes of the scheduling research literature can lead us to useful insights, including the following.

Due dates: We do have some control over due dates; after all, someone in the company sets or negotiates them. We do not have to take them as given, although this is exactly what some companies and most scheduling problem formulations do. Section 15.3.2 presents a procedure for quoting due dates that are both achievable and competitive.

Job splitting: The SPT results for a single machine suggest that small jobs clear out more quickly than large jobs. Similarly, the mechanics of Johnson's algorithm call for a sequence that has a small job at both the beginning and the end. Thus, it appears that small jobs will generally improve performance with regard to average cycle time and machine utilization. However, in Part II we also saw that small batches result in lost capacity due to an increased number of setups. Thus, if we can somehow have large *process* batches (i.e., many units processed between setups) and small *move* batches (i.e., the number accumulated before moving to the next process), we can have both short cycle times and high throughput. This concept of lot splitting, which was illustrated in Chapter 9, thus serves to make the system less sensitive to scheduling errors.

Feasible schedules: An *optimal* schedule is really only meaningful in a mathematical model. In practice what we need is a *good, feasible* one. This makes the scheduling problem much easier because there are so many more candidates for a good schedule than for an optimal schedule. Indeed, as current research is beginning to show, various heuristic procedures can be quite effective in generating reasonable schedules.

Focus on bottlenecks: Because bottleneck resources can dominate the behavior of a manufacturing system, it is typically most critical to schedule these resources well. Scheduling the bottleneck(s) separately and then propagating the schedule to nonbottleneck resources can break up a complex large-scale scheduling problem into simpler pieces. Moreover, by focusing on the bottleneck we can apply some of the insights from the single-machine scheduling literature.

Capacity: As with due dates, we have some control over capacity. We can use some capacity controls (e.g., overtime) on the same time frame as that used to schedule production. Others (e.g., equipment or workforce changes) require longer time horizons. Depending on how overtime is used, it can simplify the scheduling procedure by providing more options for resolving infeasibilities. Also, if longer-term capacity decisions are made with an eye toward their scheduling implications, these, too, can make scheduling easier. Chapter 16 discusses aggregate planning tools that can help facilitate this.

With these insights in mind, we now examine some basic scheduling scenarios in greater detail. The methods we offer are not meant as ready-to-use solutions—the range of scheduling environments is too broad to permit such a thing—but rather as building blocks for constructing reasonable solutions to real problems.

15.2.6   Practical Finite-Capacity Scheduling

In this section we discuss some representative scheduling approaches, called variously **advanced planning systems** and **finite-capacity scheduling,** available in commercial software systems. Since the problems they address are large and NP-hard, all these make use of heuristics and hence none produces an optimal schedule (regardless of what the marketing materials might suggest). Moreover, these scheduling applications are generally additions to the MRP (material requirements planning) module within the ERP (enterprise resources planning) framework. As such, they attempt to take the planned order releases of MRP and schedule them through the shop so as to meet due dates, reduce the number of setups, increase utilization, decrease WIP, and so on. Unfortunately, if the planned order releases generated by MRP represent an infeasible plan, no amount of rescheduling can make it feasible. This is a major shortcoming of such "bolt-on" applications.

Finite-capacity scheduling systems typically fall into two categories: simulation-based and optimization-based. However, many of the optimization-based methods also make use of simulation.

Simulation-Based Scheduling.   One way to avoid the NP-hard optimization problem is to simply ignore it. This can be done by developing a detailed and *deterministic* (i.e., no unpredictable variation in process times, no unscheduled outages, etc.) simulation model of the entire system. The model is then interfaced to the WIP tracking system of ERP to allow downloading of the current status of active jobs. Demand information is obtained from either the master production schedule module of ERP or another source. To generate a schedule, the model is run forward in time and records the arrival and departure of jobs at each station. Different schedules are generated by applying various **dispatching rules** at each station. These are evaluated according to selected performance measures to find the "best" schedule.

An advantage of the simulation approach is that it is easier to explain than most optimization-based methods. Since a simulator mimics the behavior of the actual system in an intuitive way, planners and operators alike can understand its logic. Another advantage is that it can quickly generate a variety of different schedules by simply changing dispatching rules and then reporting statistics such as machine utilization and the number of tardy jobs to the user. The user can choose from these the schedule that best fits his or her needs. For example, a custom job shop might be more interested in on-time delivery than in utilization, whereas a production system that uses extremely expensive equipment to make a commodity would be more interested in keeping utilization high.

However, there are also disadvantages. First, simulation requires an enormous amount of data that must be constantly maintained. Second, because the model does not account for variability, there can be large discrepancies between predicted and actual behavior. However, since virtually all finite-capacity scheduling procedures ignore variability, this problem is not limited to the simulation approach. The consequence is that to prevent error from piling up and completely invalidating the schedule over time it is important to regenerate the schedule frequently.

A third problem is that because there is no general understanding of when a given dispatching rule works well, finding an effective schedule is a trial-and-error process. Also, because dispatching rules are inherently myopic, it may be that no dispatching rule generates a good schedule.

Finally, the simulation approach, like the optimization approach, is generally used as an add-on to MRP. In a simulation-based scheduler, MRP release times are used to define the work that will be input into the model. However, if the MRP release schedule is inherently infeasible, simple dispatching cannot make it feasible. Something else— either capacity or demand—must change. But simulation-based scheduling methods are not well suited to suggesting ways to make an infeasible schedule feasible. For this an entirely different procedure is needed, as we discuss in Section 15.5.

Optimization-Based Scheduling.   Unlike classical optimization, optimization-based scheduling techniques use heuristic procedures for which there are few guarantees of performance. The difference between optimization-based and simulation-based scheduling techniques is that the former uses some sort of algorithm to actively search for a good schedule. We will provide a short overview of these techniques and refer the reader interested in more details to a book devoted to the subject by Morton and Pentico (1993).

There are a variety of ways to simplify a complex scheduling problem to facilitate a tractable heuristic. One approach is to use a simulation model, like the simulation-based methods discussed, and have the system search for parameters (e.g., dispatching rules) that maximize a specified objective function. However, since it only searches over a partial set of policies (e.g., those represented by dispatching rules), it is not a true optimization approach.

An approach that makes truer use of optimization is to reduce a line or shop scheduling problem to a single-machine scheduling problem by focusing on the bottleneck. We refer to heuristics that do this as "OPT-like" methods, since the package called "Optimized Production Technique" developed in the early 1980s by Eliyahu Goldratt and others was the first to popularize this approach. Although OPT was sold as a "black box" without specific details on the solution approach, it involved four basic stages:

1. Determine the bottleneck for the shop.
2. Propagate the due date requirements from the end of the line back to the bottleneck using a fixed lead time with a time buffer.
3. Schedule the bottleneck most effectively.
4. Propagate material requirements from the bottleneck backward to the front of the line using a fixed lead time to determine a release schedule.

Simons and Simpson (1997) described this procedure in greater detail, extending it to cases in which there are multiple bottlenecks and when parts visit a bottleneck more than once. Because they use an objective function that weights due date performance and utilization, OPT-like methods can be used to generate different types of schedules by adjusting the weights.

An entirely different optimization-based heuristic is **beam search,** which is a derivative of the branch-and-bound technique mentioned earlier. However, instead of checking each branch generated, beam search checks only relatively few branches that are selected according to some sort of "intelligent" criteria. Consequently, it runs much faster than branch-and-bound but cannot guarantee an optimal solution.

An entire class of optimization-based heuristics are those classed as **local search techniques,** which start with a given schedule and then search in the "neighborhood" of this schedule to find a better one. It turns out that "greedy" techniques, which always select the best nearby schedule, do not work well. This is because there are many schedules that are not very good overall but are best in a very small local neighborhood. A simple greedy method will usually end up with one of these and then quit.

Several methods have been proposed to avoid this problem. One of these is called **tabu search** because it makes the most recent schedules "taboo" for consideration, thereby preventing the search from getting stuck with a locally good but globally poor schedule. Consequently, the search will move away from a locally good schedule and, for awhile, may even get worse while searching for a better schedule. Another method for preventing local optima is use of **genetic algorithms** that consider the characteristics of several "parent" schedules to generate new ones and then allow only good "offspring" to survive and "reproduce" new schedules. Still another is **simulated annealing,** which selects candidate schedules in a manner that loosely mimics the gradual cooling of a metal to minimize stress. In simulated annealing, wildly random changes to the schedule can take place early in the process, where some improve the schedule and others make it worse. However, as time goes on, the schedule becomes less volatile (i.e., is "cooled") and the approach becomes more and more greedy. Of course, all local search methods "remember" the best schedule that has been found at any point, in case no better schedule can be found. We will contrast one of these techniques (tabu search) with the greedy method in Section 15.4 on bottleneck sequencing.

Optimization-based heuristics can be applied in many different ways to a variety of scheduling problems. Within a factory, the most common problem formulations are (1) minimizing some measure of tardiness, (2) maximizing resource utilization, and (3) some combination of these. We have seen that tardiness problems are extremely difficult even for one machine. Utilization (e.g., makespan) problems are a little easier. But they also become intractable when there are more than two machines. So developing effective heuristics is not simple. Pinedo and Chao (1999) give details on which methods work well in various settings and how they can be implemented effectively.

One problem with optimization-based scheduling is that many practical scheduling problems are not really optimization problems at all but, rather, are better characterized as *satisficing* problems. Most scheduling professionals would not consider a schedule that has several late jobs as optimal. This is because some constraints, such as due dates and capacity, are not *hard* constraints but are more of a "wish list." Although the scheduler would rather not add capacity, it could be done if required to meet a set of demands. Likewise, it might be possible to split jobs or postpone due dates if required to obtain a feasible schedule. It is better to have a schedule that is implementable than one that optimizes an abstract objective function but cannot possibly be accomplished.

As with simulation-based scheduling, optimization-based scheduling has found useful implementation despite its drawbacks. A number of firms have been successful in combining such software (some developed in-house) with MRP II systems to assist planners. Arguello (1994) provides an excellent survey of finite-capacity scheduling software (both optimization-based and simulation-based) used in the semiconductor industry. Since most of this software has also been applied in other industries, the survey is relevant to non-semiconductor practitioners as well.

15.3   Linking Planning and Scheduling

Within an enterprise resources planning system, the MRP module generates planned order releases based on fixed lead times and other simplifying assumptions. As has been discussed before, this often results in an infeasible schedule. Also, because finite-capacity scheduling is far from a mature technology, many of the advanced planning systems found in modern ERP systems are complex and cumbersome. The time required to generate a capacity-feasible schedule makes it impractical to do so with any kind of regularity.

These problems have led to the practice of treating material planning (e.g., MRP), capacity planning (e.g., capacity requirements planning (CRP)), and production execution (e.g., order release and dispatching) separately in terms of time, software, and personnel. For example, material requirements planning determines what materials are needed and provides a rudimentary schedule without considering capacity. Then the capacity planning function performs a check to see if the needed capacity exists. If not, either the user (e.g., by iterating CRP) or the system (e.g., by using some advanced planning systems) attempts to reschedule the releases. But because capacity was not considered when material requirements were set, the capacity planning problem may have been made unnecessarily difficult (indeed, impossible). The problem is further aggravated by the common practice of having one department (e.g., production control) generate the production plan (both materials and capacity) which is then handed off to a different department (manufacturing) to execute.

An important antidote to the planning/execution disconnect is cycle time reduction. If cycle times are short (e.g., the result of variability reduction and/or use of some sort of pull system), the short-term production planning function (i.e., committing to demands) can provide the production schedule.[3] However, before that can be done, the production planning and scheduling problem must be recast from one of *optimization,* subject to given constraints of capacity and demand, to one of *feasibility analysis,* to determine what must be done in order to have a practical production plan. This requires a procedure that analyzes both material and capacity requirements simultaneously. This can be done in theory with a large mathematical programming model. However, such formulations are usually slow and therefore prohibit making frequent feasibility checks as the situation evolves. We present a practical heuristic method that provides a quick feasibilty check in Section 15.5.2.

The remainder of this chapter focuses on issues central to the development of practical scheduling procedures. In the remainder of this section we consider techniques for making scheduling problems easier, namely, effective batching and due date quoting. Section 15.4 deals with bottleneck scheduling in the context of CONWIP lines. For more general situations, we provide a method that considers material and capacity simultaneously in Section 15.5. Finally, in Section 15.6 we show how to use scheduling (which is inherently "push" in nature) within a pull environment.

15.3.1   Optimal Batching

In Chapter 9 we observed that process batch sizes can have a tremendous impact on cycle time. Hence, batching can also have a major influence on scheduling. By choosing batch sizes wisely, to keep cycle times short, we can make it easier for a schedule to meet due dates. We now develop methods for determining batch sizes that minimize cycle time.

Optimal Serial Batches.   Figure 15.1 shows the relation between average cycle time and the serial batch size. With the formulas developed in Chapter 9, we could plot the total cycle time and find an optimal batch size for a single part at a station. However, this would be cumbersome and is of little value when we have multiple parts that interact with one another. So instead we derive a simple procedure that first finds the (approximately) optimal utilization of the station and then uses this to compute the serial batch size. We do this first for the case of a single part and then extend the approach to multiproduct systems.

Technical Note: Optimal Serial Process Batch Sizes

We first consider the case in which the product families are identical with respect to process and setup times and arrivals are Poisson. The problem is to find the serial batch size that minimizes total cycle time at a single station. This batch size should be good for the line if only one station has significant setups and tends to be the bottleneck.

Using the notation from Chapter 9, the effective process time for a batch is $t_e = s + kt$, and utilization is given by

$$u = \frac{r_a}{k}(s + kt)$$

---

[3]Long-term production planning, also known as aggregate planning, is used to set capacity levels, plan for workforce changes, etc. (see Chapter 16).

FIGURE 15.1

Average cycle time versus serial batch size

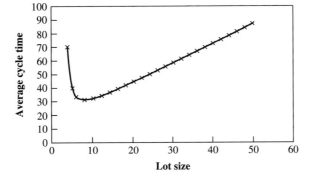

Now define the "utilization without setups" as $u_0 = r_a t$. A little algebra shows that the effective process time of a batch can be written

$$t_e = \frac{su}{u - u_0}$$

Since we are assuming Poisson arrivals (a good assumption if products arrive from a variety of sources), the arrival squared coefficient of variation (SCV) is $c_a^2 = 1$ and average cycle time is

$$\text{CT} = \left(\frac{1 + c_e^2}{2} \right) \left(\frac{u}{1 - u} \right) \frac{su}{u - u_0} + \frac{su}{u - u_0} \tag{15.1}$$

Written in this way, cycle time is a function of u only, instead of k and u. So minimizing cycle time boils down to finding the optimal station utilization. We do this by taking the derivative of (15.1) with respect to u, setting it equal to zero, and solving, which yields,

$$u^* = \frac{\alpha u_0 + \sqrt{\alpha^2 u_0^2 + [\alpha(1 + u_0) + 1]u_0}}{\alpha(1 + u_0) + 1} \tag{15.2}$$

where $\alpha = (1 + c_e^2)/2 - 1$. Note that in the special case where $c_e^2 = 1$ we have that $\alpha = 0$ and

$$u^* = \sqrt{u_0} \tag{15.3}$$

But even when c_e^2 is not equal to one, the value of u^* generally remains close to $\sqrt{u_0}$. For example, when $u_0 = 0.5$ and $c_e^2 = 15$, the difference is less than five percent. Moreover, the closer u_0 is to one (i.e., the higher the utilization of the system without setups), the smaller the difference between u^* and $\sqrt{u_0}$ for all c_e^2 (see Spearman and Kröckel 1999).

To obtain the batch size, recall that

$$u^* = \frac{r_a}{k^*} (s + k^* t) = \frac{r_a s}{k^*} + u_0$$

and solve for k^*.

The above analysis shows that a good approximation of the serial batch size that minimizes cycle time at a station is

$$k^* = \frac{r_a s}{u^* - u_0} \approx \frac{r_a s}{\sqrt{u_0} - u_0} \tag{15.4}$$

where $u_0 = r_a t$. We illustrate this with the following example.

Example: Optimal Serial Batching (Single Product)

Consider the serial batching example in Section 9.4 and shown in Figure 15.1. The utilization without considering setups u_0 is

$$u_0 = r_a t = (0.4 \text{ part/hour})(1 \text{ hour}) = 0.4$$

So, by Equation (15.3), optimal utilization is approximately

$$u^* = \sqrt{u_0} = \sqrt{0.4} = 0.6325$$

and by Equation (15.4) the optimal batch size is

$$k^* = \frac{r_a s}{u^* - u_0} = \frac{0.4(5)}{0.6325 - 0.4} = 8.6 \approx 9$$

From Figure 15.1, we see that this is indeed very close to the true optimum of eight. The difference in cycle time is less than one percent.

The insight that the optimal station utilization is very near to the square root of the utilization without setups is extremely robust. This allows it to be used as the basis for a serial batch-setting procedure in more general multiple-product family systems. We develop such an approach in the next technical note.

Technical Note—Optimal Serial Batches with Multiple Products

To model the multiproduct case we define the following:

n = number of products

i = index for products, $i = 1, \ldots, n$

r_{ai} = demand rate for product i (parts per hour)

t_i = mean time to process one part of product i (hours)

c_{ti}^2 = SCV of time to process one part of product i

s_i = mean time to perform setup when changing to product i (hours)

c_{si}^2 = SCV of time to perform setup when changing to product i

t_e = effective process time averaged over all products (hours)

c_e^2 = SCV of effective process time averaged over all products

$u_0 = \sum_i r_{ai} t_i$ = station utilization without setups

u = station utilization

k_i = lot size for product i

We can use the *VUT* equation to compute cycle time at the station as

$$CT = \left(\frac{Vu}{1 - u} + 1 \right) t_e \tag{15.5}$$

where $V = (1 + c_e^2)/2$. To use this, we must compute u, t_e and c_e from the individual part data. Utilization is given by

$$u = \sum_{i=1}^{n} \frac{r_{ai}}{k_i} (s_i + k_i t_i)$$

The effective process time is, in a sense, the "mean of the means." In other words, if the mean process time for a batch of i is $s_i + k_i t_i$ and the probability that the batch is for product i is π_i, then the effective process time is

$$t_e = \sum_{i=1}^{n} \pi_i (s_i + k_i t_i) \tag{15.6}$$

The probability that the batch is of a given product type is the ratio of that type's arrival rate to the total arrival rate

$$\pi_i = \frac{r_{ai}/k_i}{\sum_{j=1}^{n}(d_j/k_j)} \tag{15.7}$$

Using standard stochastic analysis, we compute the variance of the effective run time σ_e^2 as

$$\sigma_e^2 = \sum_{i=1}^{n} \pi_i(k_i c_{ti}^2 t_i^2 + c_{si}^2 s_i^2) + \left[\sum_{i=1}^{n} \pi_i(s_i + k_i t_i)^2 - t_e^2\right] \tag{15.8}$$

and hence the effective SCV is $c_e^2 = \sigma_e^2/t_e^2$.

Now, assuming as we did in the single-product case that $u^* = \sqrt{u_0}$ is a good approximation of the optimal utilization, the lot-sizing problem reduces to finding a set of k_i values that achieve u^* and keep c_e^2 and t_e small. From Equation (15.5) it is clear that this will lead to a small cycle time. Note that if all the values of $s_i + k_i t_i$, that is, all the average run lengths, were equal, the term in square brackets in Equation (15.8) would be zero. Thus, one way to keep both t_e and c_e^2 small is to minimize the average run length and to make all the run lengths the same. We can express this as the following optimization problem.

$$\text{Minimize} \qquad L$$
$$\text{Subject to:} \qquad s_i + k_i t_i \leq L \qquad \text{for } i = 1, \dots, n$$
$$\sum_{i=1}^{n} \frac{r_{ai}}{k_i}(s_i + k_i t_i) = u^*$$

The solution can be obtained from

$$s_i + k_i t_i = L$$
$$k_i = \frac{L - s_i}{t_i} \tag{15.9}$$

Then solve for L, using the constraint

$$\sum_{i=1}^{n} \frac{r_{ai}}{k_i}(s_i + k_i t_i) = u^*$$

$$\sum_{i=1}^{n} \frac{r_{ai} s_i}{k_i} = u^* - u_0$$

$$\sum_{i=1}^{n} \frac{r_{ai} s_i t_i}{L - s_i} = u^* - u_0$$

If the setup times are all close to the mean setup time, which we denote by $\bar{s}$, then we can solve for L as follows.

$$L = \frac{\sum_{i=1}^{n} r_{ai} s_i t_i}{u^* - u_0} + \bar{s} \tag{15.10}$$

Substituting this into Equation (15.9) yields approximately optimal batch sizes.

The above analysis shows that the serial batch size for product i that minimizes cycle time at a station with multiple products and setups is

$$k_i^* = \frac{L - s_i}{t_i} \tag{15.11}$$

where L is computed from Equation (15.10).

Example: Optimal Serial Batching (Multiple Products)

Consider an industrial process in which a blender blends three different products. Demand for each product is 15 blends per month and is controlled by an MRP system that uses a constant batch size for each product. Whenever the blender is switched from one product to another, a cleanup is required. Products A and B take four hours per blend and eight hours for cleanup. Product C requires eight hours per blend and 12 hours for cleanup. All process and setup times have a coefficient of variation of one-half. The blender is run two shifts per day, five days per week. With one hour lost for each shift and 52/12 weeks per month, this averages out to 303.33 hours per month.

In keeping with conventional wisdom (e.g., the EOQ model) that longer changeovers should have larger batch sizes, the firm is currently using batch sizes of 20 blends for products A and B and 30 blends for product C. The average cycle time through the process is currently around 32 shop days. But could they do better?

Converting demand to units of hours yields $r_{ai} = 15/303.33 = 0.0495$ blend per hour for all three products. The utilization without setups is therefore

$$u_0 = 0.0495(4 + 4 + 8) = 0.7912$$

Hence, the optimal utilization is $u^* = \sqrt{u_0} = \sqrt{0.7912} = 0.8895$.

The average setup time is $\bar{s} = (8 + 8 + 12)/3 = 9.33$ hours, so the sum needed in Equation (15.10) is

$$\sum_{i=1}^{3} r_{ai} s_i t_i = 0.0495[8(4) + 8(4) + 12(8)] = 7.912$$

and hence

$$L = \frac{7.912}{0.8895 - 0.7912} + 9.33 = 89.82$$

With this we can compute the approximately optimal batch sizes as follows.

$$k_A = k_B = \frac{L - s_A}{t_A} = \frac{89.82 - 8}{4} = 20.46 \approx 20$$

$$k_C = \frac{L - s_C}{t_C} = \frac{89.82 - 12}{8} = 9.73 \approx 10$$

Using these batch sizes results in an average cycle time of 20.28 days, a decrease of over 36 percent. Doing a complete search over all possible batch sizes shows that this is indeed the optimal solution.

Note that the batch size for part C is *smaller* than that for A and B. EOQ logic, which was developed assuming separable products, suggests that C should have a larger batch size because it has a longer setup time. But to keep the run lengths equal across products, we need to reduce the batch size of C.

Optimal Parallel Batches.   A machine with parallel batching is a true batch machine, such as a heat treat oven in a machine shop or a copper plater in a circuit-board plant. In these cases, the process time is the same regardless of how many parts are processed at once (the batch size).

In parallel batching situations, the basic tradeoff is between effective capacity utilization, for which we want large batches, and minimal wait-to-batch time, for which we want small batches. If the machine is a bottleneck, it is often best to use the largest batch possible (size of the batch operation). In nonbottlenecks, it can be best (in terms of cycle time) to process a partial batch. The following technical note describes a procedure for determining the optimal parallel batch size at a single station.

Technical Note—Optimal Parallel Batches

To find a batch size that minimizes cycle time at a parallel batch operation, it is convenient to find the best utilization and then translate this to a batch size, as we did in the case of serial process batching.

To do this, we make use of the following notation:

r_a = arrival rate (parts per hour)

c_a = coefficient of variation (CV) of interarrival times

t = time to process a batch (hours)

c_e = effective CV for processing time of a batch

B = maximum batch size (number of parts that can fit into process)

$u_m = r_a t$ = utilization resulting from batch size of one

u = station utilization

k = parallel batch size

Note that utilization is given by $u = r_a/(k/t)$, which must be less than one for the station to be stable. We can use $u_m = r_a t$ to rewrite this as $u = u_m/k$, which implies the batch size is $k = u_m/u$.

Recall, from Chapter 9, that the total time spent in a parallel batch operation includes wait-to-batch time (WTBT), queue time, and the time of the operation itself, which can be written

$$CT = WTBT + CT_q + t$$

$$= \frac{k-1}{2r_a} + \left(\frac{c_a^2/k + c_e^2}{2}\right)\left(\frac{u}{1-u}\right)t + t$$

$$= \frac{k-1}{2ku}t + \left(\frac{c_a^2/k + c_e^2}{2}\right)\left(\frac{u}{1-u}\right)t + t \qquad (15.12)$$

where the last equality follows from the fact that $r_a = uk/t$.

Substitution of $k = u_m/u$ allows us to rewrite Equation (15.12) as

$$CT = \left(\frac{u_m/u - 1}{2u_m} + \frac{c_a^2 u/u_m + c_e^2}{2}\frac{u}{1-u} + 1\right)t \qquad (15.13)$$

Unfortunately, minimizing CT with respect to utilization does not yield a simple expression. So to approximate, we will let $\beta = c_a^2 u/u_m$ and assume that this term can be treated as a constant. Our justification for this is that when k is large, u/u_m will be small, which will make β negligible. This reduces the expression for cycle time to

$$CT \approx \left(\frac{1}{2u} - \frac{1}{2u_m} + \frac{\beta + c_e^2}{2}\frac{u}{1-u} + 1\right)t$$

$$= \left(\frac{y(u)}{2} - \frac{1}{2u_m} + 1\right)t \qquad (15.14)$$

where

$$y(u) = \frac{1}{u} + \frac{\beta + c_e^2 u}{1-u}$$

Minimizing Equation (15.14) is equivalent to minimizing $y(u)$ with respect to u, which is fairly easy. Taking the derivative of $y(u)$ with respect to u, setting it equal to zero, and solving yields

$$u^* = \frac{1}{1 + \sqrt{\beta + c_e^2}} \tag{15.15}$$

If, as we suggested it might be, β is close to zero, then the optimal utilization reduces to

$$u^* \approx \frac{1}{1 + c_e} \tag{15.16}$$

When c_e^2 is not too small, dropping the $\beta = c_a^2 u / u_m$ term does not have a large impact and equation (15.16) is a fairly good approximation. However, when c_e^2 is small, dropping this term significantly changes the problem. Indeed, when $c_e^2 = 0$, equation (15.16) suggests that the optimal utilization is equal to one! Of course, we know that this is not reasonable, since if there is any variability in the arrival process, the queue will blow up.

So, to go back and reintroduce the β term, we substitute the approximate expression for u^* from equation (15.16) into $c_a^2 u / u_m$, so that

$$\beta = \frac{c_a^2}{u_m(1 + c_e)}$$

and

$$u^* = \frac{1}{1 + \sqrt{c_a^2/[u_m(1 + c_e)] + c_e^2}} \tag{15.17}$$

Once we have the optimal utilization u^*, we can easily find the optimal batch size k^* from $k = u_m/u$.

Thus, we have that the process batch size that minimizes cycle time at a parallel batch station is

$$k^* = \frac{u_m}{u^*} \tag{15.18}$$

where $u_m = r_a t$ and u^* is computed using Equation (15.17). To obtain an integer batch size, we will use the convention of rounding *up* the value from Equation (15.18). This will tend to offset some of the error introduced by the approximations made in the technical note.

In addition to a computational tool, Equations (15.17) and (15.16) yield some qualitative insight. They indicate that the more variability we have at the station, the less utilization it can handle. Specifically, as c_e or c_a increases, the optimal utilization of the system decreases. This is a consequence of the factory physics results on variability and utilization, which showed that these two factors combine to degrade performance. Hence, when we are optimizing performance, we must offset more variability via less utilization.

We illustrate the use of the formula for parallel batch sizing in the following example.

Example: Optimal Parallel Batching

Reconsider the burn-in operation discussed in Section 9.4, in which a facility tests medical diagnostic units in an operation that turns the units on and runs them in a temperature-controlled room for 24 hours regardless of how many units are being burned in. The burn-in room can hold 100 units at a time, and units arrive to burn in at a rate of one per hour (24 per day). Figure 9.6 plots cycle time versus batch size for this example and shows that cycle time is minimized at a batch size of 32, which achieves a cycle time of 42.88 hours.

Now consider the situation using the above optimal batch-sizing formulas. The arrival rate is $r_a = 1$ per hour and arrivals are Poisson, so $c_a = 1$. The process time is

$t = 24$ hours, and it has variability such that $c_e = 0.25$. So for stability we require a batch size $k > u_m = r_a t = 24$, which implies that the minimum batch size is 25.

However, if we use a batch size of 25, we get

$$u = \frac{r_a}{k/t} = \frac{1}{25/24} = 0.96$$

$$\text{WTBT} = \frac{k-1}{2r_a} = \frac{25-1}{2(1)} = 12 \text{ hours}$$

$$\text{CT}_q = \left(\frac{c_a^2/k + c_e^2}{2}\right)\left(\frac{u}{1-u}\right)t$$

$$= \left(\frac{1/25 + 0.25^2}{2}\right)\left(\frac{0.96}{1-0.96}\right)24 = 29.52 \text{ hours}$$

Hence, the average cycle time through the heat treat operation will be

$$\text{CT} = \text{WTBT} + \text{CT}_q + t = 12 + 29.52 + 24 = 65.52 \text{ hours}$$

Now consider the other extreme and let $k = 100$, the size of the burn-in room.

$$u = \frac{r_a}{k/t} = \frac{1}{100/24} = 0.24$$

$$\text{WTBT} = \frac{k-1}{2r_a} = \frac{100-1}{2(1)} = 49.5 \text{ hours}$$

$$\text{CT}_q = \left(\frac{c_a^2/k + c_e^2}{2}\right)\left(\frac{u}{1-u}\right)t$$

$$= \left(\frac{1/100 + 0.25^2}{2}\right)\left(\frac{0.24}{1-0.24}\right)24 = 0.27 \text{ hour}$$

So the average cycle time through the heat treat operation will be

$$\text{CT} = \text{WTBT} + \text{CT}_q + t = 49.5 + 0.27 + 24 = 73.77 \text{ hours}$$

Now to find the optimal batch size, we first compute the optimal utilization.

$$u^* = \frac{1}{1 + \sqrt{c_a^2/[u_m(1+c_e)] + c_e^2}}$$

$$= \frac{1}{1 + \sqrt{1/[24(1+0.25)] + 0.25^2}}$$

Then we use Equation (15.18) to compute

$$k^* = \frac{u_m}{u^*} = \frac{24}{0.7636} = 31.43 \approx 32$$

Note that this is exactly the optimal batch size we observed in Figure 9.6. Furthermore, the minimum batch size yields a cycle time that is 53 percent higher than the optimum, while the maximum batch size yields one that is 72 percent greater than optimal. Clearly, batching can have a significant impact on cycle times in parallel batch operations.

FIGURE 15.2

*Schematic of method for
quoting lead times*

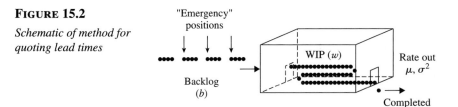

15.3.2   Due Date Quoting

Variability reduction (Chapter 9), pull production (Chapter 10), and efficient lot-sizing methods (previously described) all make a production system easier to schedule. Another technique for simplifying scheduling is due date quoting. Since scheduling problems that involve due dates are extremely hard, while the due date–setting problem can be relatively easy, this would seem worthwhile. Of course, in the real world, implementation is more than a matter of mathematics. Developing a due date–quoting system may involve a much more difficult problem—getting manufacturing and salespeople to talk to one another.

In addition to personnel issues, the difficulty of the due date–quoting problem depends on the manufacturing environment. To be able to specify reasonable due dates, we must be able to predict when jobs will be completed given a specified schedule of releases. If the environment is so complex that this is difficult, then due date quoting will also be difficult. However, if we simplify the environment in a way that makes it more predictable, then due date quoting can be made straightforward.

Quoting Due Dates for a CONWIP Line.   One of the most predictable manufacturing systems is the CONWIP line. As we noted previously, CONWIP behavior can be characterized via the conveyor model. This enables us to develop a simple procedure for quoting due dates.

Consider a CONWIP line that maintains w standard units[4] of WIP and whose output in each period (e.g., shift, day) is steady with mean μ and variance σ^2. Suppose a customer places an order that represents c standard units of work, and we are free to specify a due date. To balance responsiveness with dependability, we want to quote the earliest due date that ensures a service level (probability of on-time delivery) of s. Of course, the due date that will achieve this depends on how much work is ahead of the new order. This in turn depends on how customer orders are sequenced. One possibility is that jobs are processed in first-come, first-serve order, in which case we let b represent the current backlog (i.e., number of standard jobs that have been accepted but not yet released to the line). Alternatively, "emergency slots" for high-priority jobs could be maintained (see Figure 15.2) by quoting due dates for some lower-priority jobs as if there were "placeholder" jobs already ahead of them. In this case, we define b to represent the units of work until the first emergency slot.

In either case, the customer order will be filled after $m = w + b + c$ standard units of work are output by the line. Hence the problem of finding the earliest due date that guarantees a service level of s is equivalent to finding the time within which we are s percent certain of being able to complete m standard units of work. We derive an expression for this time in the following technical note.

---

[4]A standard unit of WIP is one that requires a certain amount of time at the bottleneck of the line. Thus, CONWIP maintains a constant workload in the line, as measured by time on the bottleneck.

Technical Note—Due Date Quoting for a CONWIP Line

Let X_t be a random variable representing the amount of work (in standard units) completed in period t. Assume that $X_t, t = 1, 2, \ldots$, are independent and normally distributed with mean μ and variance σ^2. To guarantee completion by time ℓ with probability s, the following must be true:

$$P\left\{\sum_{t=1}^{\ell} X_t \leq m\right\} = 1 - s$$

Note that since the means and variances of independent random variables are additive, the amount of work completed by time ℓ is given by

$$\sum_{t=1}^{\ell} X_t \sim N(\ell\mu, \ \ell\sigma^2)$$

That is, it is normally distributed with mean $\ell\mu$ and variance $\ell\sigma^2$. Hence,

$$P\left\{Z \leq \frac{m - \ell\mu}{\sqrt{\ell}\sigma}\right\} = 1 - s$$

where Z is the standard 0–1 normal random variable.

Therefore,

$$\frac{m - \ell\mu}{\sqrt{\ell}\sigma} = z_{1-s} \tag{15.19}$$

where z_{1-s} is obtained from a standard normal table.

We can rewrite Equation (15.19) as

$$\ell^2\mu^2 - (2\mu m + z_{1-s}^2\sigma^2)\ell + m^2 = 0 \tag{15.20}$$

which can be solved by using the quadratic equation. There are two roots to this equation; as long as $s \geq 0.5$, the larger one should always be used. This yields Equation (15.21).

The minimum quoted lead time for a new job consisting of c standard units that is sequenced behind a backlog of b standard units in a CONWIP line with a WIP level of w necessary to guarantee a service level of s is given by

$$\ell = \frac{m}{\mu} + \frac{z_{1-s}^2\sigma^2\left[1 + \sqrt{4\mu m/(z_{1-s}^2\sigma^2) + 1}\right]}{2\mu^2} \tag{15.21}$$

where $m = w + b + c$.

A possible criticism of the above method is that it is premised on service. Hence, a job that is one day late is considered just as bad as one that is one year late. A measure that better tracks performance from a customer perspective is tardiness. Fortunately, it turns out that quoting each job with the same service level also yields the minimum expected quoted lead time subject to a constraint on average tardiness (see Spearman and Zhang 1999).

Furthermore, to simplify implementation with little loss in performance, Equation (15.21) can be replaced by

$$\ell = \frac{m}{\mu} + \text{planned inventory time} \tag{15.22}$$

where planned inventory time can be adjusted by trial and error to achieve acceptable service (see Hopp and Roof 1998).

FIGURE 15.3

Quoted lead times versus the backlog

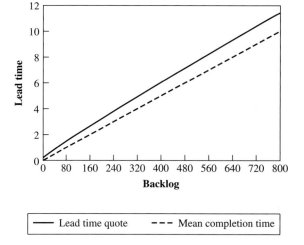

Example: Due Date Quoting

Suppose we have a CONWIP line that maintains 320 standard units of WIP and has an average output of 80 units per day with a standard deviation of 15 units. The line receives a high-priority order representing 20 standard units, and the first available emergency slot on the backlog is 100 jobs from the start of the line. We want to quote a due date with a service level of 99 percent.

To use Equation (15.21), we observe that $\mu = 80$, $\sigma^2 = 225$ (or, 15^2), $w = 320$, $b = 100$, and $c = 20$, so that $m = 440$. The value for $z_{1-s} = z_{0.01} = -2.33$ is found in a standard normal table. Thus,

$$
\ell = \frac{m}{\mu} + \frac{z_s^2 \sigma^2 \left[1 + \sqrt{4\mu m/(z_s^2 \sigma^2) + 1} \right]}{2\mu^2}
$$

$$
= \frac{440}{80} + \frac{(-2.33^2)(225)\left\{ 1 + \sqrt{4(80)(440)/[(-2.33)^2(225)] + 1} \right\}}{2(80^2)}
$$

$$
= 6.62
$$

and so we quote seven days to the customer.

Notice that the mean time to complete the order is $m/\mu = 440/80 = 5.5$ days. The additional one and one-half days represent **safety lead time** used as a buffer against the variability in the production process.

Figure 15.3 shows the lead time quotes as a function of total backlog m. The dashed line shows the mean completion time m/μ, which is what would be quoted if there were no variance in the production rate. The difference between the solid and dotted lines is the safety lead time, which we note increases in the backlog level. The reason is that the more work that must be completed to fill an order, the greater the variability in the completion time, and hence the higher the required safety lead time.

In an environment with multiple CONWIP routings, a similar set of computations would be performed for each routing in the plant. The only data needed are the first two moments of the production rate for the routing, the current WIP level (a constant under CONWIP), and the current status of the backlog. These data should be maintained in a central location accessible to both sales and manufacturing. Sales needs the information to quote due dates; manufacturing needs it to determine what to start next. Manufacturing can also track production against a backlog established by sales (e.g., the statistical throughput control procedure described in Chapter 14). The overall result will be due dates that are competitive, achievable, and consistent with manufacturing parameters.

15.4   Bottleneck Scheduling

A main conclusion of the scheduling research literature is that scheduling problems, particularly realistically sized ones, are very difficult. So it is common to simplify the problem by breaking it down into smaller pieces. One way to do this is by scheduling the bottleneck process by itself and then propagating that schedule to nonbottleneck stations. This is particularly effective in simple flow lines. However, bottleneck scheduling can also be an important component in more complex scheduling situations as well.

A major reason why restricting attention to the bottleneck can simplify the scheduling problem is that it reduces a multimachine problem to a single-machine problem. Recall from our discussion of scheduling research that simple sequences, as opposed to detailed schedules, are often sufficient for single-machine problems. Since a *schedule* presents information about when each job is to be run on each machine while a *sequence* only presents the order of processing the jobs, it is easier to compute a sequence. Furthermore, because schedules become increasingly inaccurate with time, sequences can be more robust in practice.

The scheduling problem can be further simplified if the manufacturing environment is made up of CONWIP lines. As we know (Chapter 13), a CONWIP line can be characterized as a conveyor with rate r_b^P (the practical production rate) and transit time T_0^P (minimum practical lead time). Since the parameters r_0^P and T_0^P are adjusted to include variability effects such as failures, variable process times, and setups, and because safety capacity (overtime) is used to ensure that the line achieves its target rate each period (day, week, or whatever), the deterministic conveyor model is a good approximation of the stochastic production system. Thus, by focusing on the bottleneck in a CONWIP line, we effectively reduce a very hard multistation stochastic scheduling problem to a much easier single-station deterministic scheduling problem. Also, since we use first-in-system first-out (FISFO) dispatching at each station, it is a trivial matter to propagate the bottleneck sequence to the other stations—simply use the same sequence at all stations. This sequence is the **CONWIP backlog** to which we have referred in previous chapters. In this section, we discuss how to generate this backlog.

15.4.1   CONWIP Lines Without Setups

We begin by considering the simplest case of CONWIP lines—those in which setups do not play a role in scheduling. This could be because there are no significant setups between any part changes. Alternatively, it could be because setups are done periodically (e.g., for cleaning or maintenance) but do not depend on the work sequence. Sequencing a single CONWIP line without setups is just like scheduling the single machine with due dates that we discussed earlier and hence can be done with the earliest due date (EDD) rule. Results from scheduling theory show that the EDD sequence will finish all the jobs on time if it is possible to do so. Of course, what this really means is that jobs will finish on time in the *planned* schedule. We cannot know in advance if this will occur, since it depends on random events. But starting with a feasible plan gives us a much better chance at good performance in practice than does starting with an infeasible plan.

A slightly more complex situation is one in which two or more CONWIP lines share one or more workstations. Figure 15.4 shows such a situation in which (1) two CONWIP lines share a machine that also happens to be the bottleneck and (2) the lines produce

FIGURE 15.4

Two CONWIP lines sharing a common process center

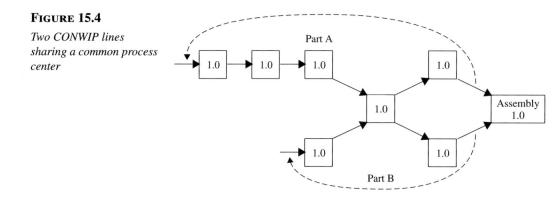

components for an assembly operation. We consider this case because it starkly illustrates the issues involved. However, the scheduling is fundamentally the same as scheduling a system with the lines feeding separate finished goods inventory (FGI) buffers instead of assembly.

In both cases, we should sequence releases into the individual lines according to the EDD rule and use this sequence at all nonshared stations, just as we did for the separate CONWIP line case. This leaves the question of what sequence to use at the shared stations.

One might intuitively think that using first-in-first-out (FIFO) would work well. However, if there is variability in the process times, then, for example, eventually a string of A jobs will arrive at the shared resource before the matching B jobs. Using FIFO will therefore only create a queue of unmatched parts at the assembly operation. In extreme cases, this could actually cause the bottleneck to starve for work since so much WIP is tied up at assembly.

A better alternative is first-in-system-first-out (FISFO) dispatching at the shared resource. Under this rule, jobs are sequenced according to when they entered the system (i.e., the times their CONWIP cards authorized their release). Since the CONWIP cards authorize releases for matching parts (i.e., one A and one B) at assembly at the same time, this rule serves to sequence the shared machine according to the assembly sequence. Hence it serves to synchronize arrivals to assembly as closely as possible. Of course, when there are no B jobs to work on at the shared machine (due to an unusually long process time upstream, perhaps) it will process only A jobs. But as soon as it receives B jobs to work on, it will.

15.4.2   Single CONWIP Lines with Setups

The situation becomes more difficult when we consider a CONWIP line with setups at the bottleneck. Indeed, even determining whether a sequence exists that will satisfy all the due dates is to answer an NP-complete question.

To illustrate the difficulty of this problem and to suggest a solution approach, we consider the set of 16 jobs shown in Table 15.5. Each job takes one hour to complete, not including a setup. Setups take four hours and occur if we go from any job family to any other. The jobs in Table 15.5 are arranged in earliest due date order. As we see, EDD does not appear very effective here, since it results in 10 setups and 12 tardy jobs for an average tardiness of 10.4. To find a better solution, we clearly do not want to evaluate every possibility, since there are $16! = 2 \times 10^{13}$ possible sequences. Instead we seek a heuristic that gives a good solution.

TABLE 15.5   EDD Sequence

Job Number	Family	Due Date	Completion Time	Lateness
1	1	5	5	0
2	1	6	6	0
3	1	10	7	−3
4	2	13	12	−1
5	1	15	17	2
6	2	15	22	7
7	1	22	27	5
8	2	22	32	10
9	1	23	37	14
10	3	29	42	13
11	2	30	47	17
12	2	31	48	17
13	3	32	53	21
14	3	32	54	22
15	3	33	55	22
16	3	40	56	16

One possible approach is known as a **greedy algorithm.** Each step of a greedy algorithm considers all simple alternatives (i.e., pairwise interchanges of jobs in the sequence) and selects the one that improves the schedule the most. This is why it is called greedy. The number of possible interchanges (120 in this case) is much smaller than the total number of sequences, and hence this algorithm will find a solution quickly. The question of course is, How good will the solution be? We consider this below.

Checking the total tardiness for every possible exchange between two jobs in the sequence reveals that the biggest decrease is achieved by putting job 4 after job 5. As shown in Table 15.6, this eliminates two setups (going from family 1 to family 2 and back again). The average tardiness is now 5.0 with eight setups.

We repeat the procedure in the second step of the algorithm. This time, the biggest reduction in total tardiness results from moving job 7 after job 8. Again, this eliminates two setups by grouping like families together. The average tardiness falls to 1.2 with six setups. The third step moves job 10 after job 12, which eliminates one setup and reduces the average tardiness to one-half. The resulting sequence is shown in Table 15.7.

At this point, no further single exchanges can reduce total tardiness. Thus the greedy algorithm terminates with a sequence that produces three tardy jobs. The question now is, Could we have done better?

The answer, as shown in Table 15.8, which gives a feasible sequence, is yes. But must we evaluate all 16! possible sequences to find it? Mathematically speaking, we must. However, practically speaking, we can often find a better (even feasible) sequence by using a slightly more clever approach than the simple greedy algorithm.

To develop such a procedure, we observe that the problem with greedy algorithms is that they can quickly converge to a **local optimum**—a solution that is better than any other adjacent solutions, but not as good as a nonadjacent solution. Since the greedy algorithm considered only adjacent moves (pairwise interchanges), it is vulnerable to getting stuck at a local optimum. This is particularly likely because NP-hard problems

TABLE 15.6   Sequence after First Swap in Greedy Algorithm

Job Number	Family	Due Date	Completion Time	Lateness
1	1	5	5	0
2	1	6	6	0
3	1	10	7	−3
5	1	15	8	−7
4	2	13	13	0
6	2	15	14	−1
7	1	22	19	−3
8	2	22	24	2
9	1	23	29	6
10	3	29	34	5
11	2	30	39	9
12	2	31	40	9
13	3	32	45	13
14	3	32	46	14
15	3	33	47	14
16	3	40	48	8

TABLE 15.7   Final Configuration Produced by Greedy Algorithm

Job Number	Family	Due Date	Completion Time	Lateness
1	1	5	5	0
2	1	6	6	0
3	1	10	7	−3
5	1	15	8	−7
4	2	13	13	0
6	2	15	14	−1
8	2	22	15	−7
7	1	22	20	−2
9	1	23	21	−2
11	2	30	26	−4
12	2	31	27	−4
10	3	29	32	3
13	3	32	33	1
14	3	32	34	2
15	3	33	35	2
16	3	40	36	−4

like this one tend to have many local optima. What we need, therefore, is a mechanism that will force the algorithm away from a local optimum in order to see if there are better sequences farther away.

TABLE 15.8   A Feasible Sequence

Job Number	Family	Due Date	Completion Time	Lateness
1	1	5	5	0
2	1	6	6	0
3	1	10	7	−3
5	1	15	8	−7
4	2	13	13	0
6	2	15	14	−1
8	2	22	15	−7
11	2	30	16	−14
12	2	31	17	−14
7	1	22	22	0
9	1	23	23	0
10	3	29	28	−1
13	3	32	29	−3
14	3	32	30	−2
15	3	33	31	−2
16	3	40	32	−8

One way to do this is to prohibit (make "taboo") certain recently considered moves. This approach is called **tabu search** (see Glover 1990), and the list of recent (and now forbidden) moves is called a **tabu list.** In practice, there are many ways to characterize moves. One obvious (albeit inefficient) choice is the entire sequence. In this case, certain sequences would become tabu once they were evaluated. But because there are so *many* sequences, the tabu list would need to be very long to be effective. Another, more efficient but less precise, option is the location of the job in the sequence. Thus, the move placing job 4 after job 5 (as we did in our first move) would become tabu once it was considered the first time. But because we need only prohibit this move temporarily in order to prevent the algorithm from settling into a local minimum, the length of the tabu list is limited. Once a tabu move has been on the list long enough, it is discarded and can then be considered again.

The tabu search can be further refined by not considering moves that we know cannot make things better. For example, in the above problem we know that making the sequence anything but EDD *within* a family (i.e., between setups) will only make things worse. For example, we would never consider moving job 2 after job 1 since these are of the same family and job 1 has a due date that is earlier than that for job 2. This type of consideration can limit the number of moves that must be considered and therefore can speed the algorithm.

Although tabu search is simple in principle, its implementation can become complicated (see Woodruff and Spearman 1992 for a more detailed discussion). Also, there are many other heuristic approaches that can be applied to sequencing and scheduling problems. Researchers are continuing to evolve new methods and evaluate which work best for given problems. For more discussion on heuristic scheduling methods, see Morton and Pentico (1994) and Pinedo (1995).

15.4.3   Bottleneck Scheduling Results

An important conclusion of this section is that scheduling need not be as hopeless as a narrow interpretation of the complexity results from scheduling theory might suggest. By simplifying the environment (e.g., with CONWIP lines) and using well-chosen heuristics, managers can achieve reasonably effective scheduling procedures.

In pull systems, such as CONWIP lines, simple sequences are sufficient, since the timing of releases is controlled by progress of the system. If there are no setups, an EDD sequence is an appropriate choice for a single CONWIP line. It is also suitable for systems of CONWIP lines with shared resources, as long as there are no significant setups and the FISFO dispatching rule is used at the shared resources. If there are significant setups, then a simple sequence is still sufficient for CONWIP lines, but not an EDD one. However, practical heuristics, such as tabu search, can be used to find good solutions for this case.

15.5   Diagnostic Scheduling

Unfortunately, not all scheduling situations are amenable to simple bottleneck sequencing. In some systems, the identity of the bottleneck shifts, due to changes in the product mix—when different products have different process times on the machines—or capacities change frequently, perhaps as a result of a fluctuating labor force. In some factories, extremely complicated routings do not allow use of CONWIP or any other pull system. In still others, WIP in the system is reassigned to different customers in response to a constantly changing demand profile.

A glib suggestion for dealing with these situations is to get rid of them. In some systems where this is possible, it may be the most sensible course of action. However, in others it may actually be infeasible physically or economically. In such cases, most firms turn to some variant of MRP. In concept, MRP can be applied to almost any manufacturing environment. However, as we noted in Chapters 3 and 5, the basic MRP model is flawed because of its underlying assumptions, particularly that of infinite capacity. In response, production researchers and software vendors have devoted increasing attention to finite-capacity schedulers. As stated earlier, this approach is often too little, too late since it relies on the MRP release schedule as input. The goal of this section is to maintain the structure of the ERP hierarchy while removing the defect in the MRP scheduling model.

In the real world, effective scheduling is more than a matter of finding good solutions to mathematical problems. Two important considerations are the following:

1. *Models depend on data, which must be estimated.* A common parameter required by many scheduling models is a tardiness cost, which is used to make a tradeoff between customer service and inventory costs. However, almost no one we have encountered in industry is comfortable with specifying such a cost in advance of seeing its effect on the schedule.

2. *Many intangibles are not addressed by models.* Special customer considerations, changing shop floor conditions, evolving relationships with suppliers and subcontractors, and so forth make completely automatic scheduling all but impossible. Consequently, most scheduling professionals with whom we have spoken feel that an effective scheduling system must allow for human intervention. To make effective use of human intelligence, such a system should evaluate the *feasibility* (not optimality) of a given schedule and, if it is infeasible, suggest changes. Suggestions might include adding capacity via overtime, temporary workers, or subcontracting; pushing out due dates of certain jobs,

and splitting large jobs. Human judgment is required to choose wisely among these options, in order to address such questions as, Which customers will tolerate a late or partial shipment? Which parts can be subcontracted now? Which groups of workers can and cannot be asked to work overtime?

Neither optimization-based nor simulation-based approaches are well suited to evaluating candidate schedules and offering improvement alternatives. Perhaps because of this, a survey of scheduling software found no systems with more than trivial diagnostic capability (Arguello 1994).

In contrast, the ERP paradigm is intended to develop *and evaluate* production schedules. The master production schedule (MPS) provides the demand; material requirements planning (MRP) nets demand, determines material requirements, and offsets them to provide a release schedule; and capacity requirements planning checks the schedule for feasibility. As a planning framework, this is ideally suited to real-world production control. However, as we discussed earlier, the basic model in MRP is too simple to accurately represent what happens in the plant. Similarly CRP is an inaccurate check on MRP because it suffers from the same modeling flaw (fixed lead times) as MRP. Even if CRP were an accurate check on schedule feasibility, it does not offer useful diagnostics on how to correct infeasibilities.

Thus, our goal is to provide a scheduling process that preserves the appropriate ERP framework but eliminates the modeling flaws of MRP. In this section, we discuss how and why infeasibilities arise and then offer a procedure for detecting them and suggesting corrective measures.

15.5.1   Types of Schedule Infeasibility

There are two basic types of schedule infeasibility. **WIP infeasibility** is caused by inappropriate positioning of WIP. If there is insufficient WIP in the system to facilitate fulfillment of near term due dates, then the schedule will be infeasible regardless of the capacity. The only way to remedy a WIP infeasibility is to postpone (push out) demand. **Capacity infeasibility** is caused by having insufficient capacity. Capacity infeasibilities can be remedied by either pushing out demand or adding capacity.

Example:

We illustrate the types and effects of schedule infeasibility by considering a line with a demonstrated capacity of $r_b^P = 100$ units per day and a practical minimum process time of $T_0^P = 3$ days. Thus, by Little's Law, the average WIP level will be 300 units. Currently, there are 95 jobs that are expected to finish at the end of day 1; 90 that should finish by the end of day 2; and 115 that have just started. Of these last 115 jobs, 100 will finish at the end of day 3. The remaining 15 will finish on day 4 due to the capacity constraint. The demands, which start out low but increase to above capacity, are given in Table 15.9.

First observe that total demand for the first three days is 280 jobs, while there are 300 units of WIP and capacity (each job is one unit). Demand for the next 12 days is 1,190 units, while there is capacity to produce 1,200 over this interval plus 20 units of current WIP left over after filling demand for the first three days. Thus, from a quick aggregate perspective, meeting demand appears feasible.

However, when we look more closely, a problem becomes apparent. At the end of the first day the line will output 95 units to meet a demand of 90 units, which leaves five units of finished goods inventory (FGI). After the second day 90 additional units will be

TABLE 15.9 Demand for Diagnostics Example

Day from Start	Amount Due
1	90
2	100
3	90
4	80
5	70
6	130
7	120
8	110
9	110
10	110
11	100
12	90
13	90
14	90
15	90

output, but demand for that day is 100. Even after the five units of FGI left over from day 1 are used, this results in a deficit of five units. At the end of the third day 100 units are output to meet demand of 90 units, resulting in an excess of 10 units. This can cover the deficit from day 2, but only if we are willing to be a day late on delivery.

The reason for the deficit in day 2 is that there is not enough WIP in the system within two days of completion to cover demand during the first two days. While total demand for days 1 and 2 is $90 + 100 = 190$ units, there are only $95 + 90 = 185$ units of WIP that can be output by the end of day 2. Hence, a five-unit deficit will occur no matter how much capacity the line has. This is an example of a WIP infeasibility. Note that because it does not involve capacity, MRP can detect this type of infeasibility.

Looking at the demands beyond day 3, we see that there are other problems as well. Figure 15.5 shows the maximum cumulative production for the line relative to the cumulative demand for the line. Whenever maximum cumulative production falls below cumulative demand, the schedule is infeasible. The surplus line, whose scale is on the right, is the difference between the maximum cumulative production and the cumulative demand. Negative values indicate infeasibility. This curve first becomes negative in day 2—the infeasibility caused by insufficient WIP in the line. After that, the line can produce more than demand, and the surplus curve becomes positive. It becomes negative again on day 8 when demand begins to exceed capacity and stays negative until day 14 when the line finally catches back up.

The infeasibility in day 8 is different from that in day 2 because it *is* a function of capacity. While no amount of extra capacity could enable the line to meet demand in day 2, production of an additional 25 units of output sometime before day 8 would allow it to meet demand on that day. Hence the infeasibility that occurred on day 8 is an example of a capacity infeasibility. Because MRP and CRP are based on an infinite-capacity model, they cannot detect this type of infeasibility.

FIGURE 15.5

Demand versus available production and WIP

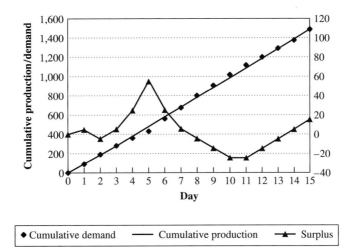

FIGURE 15.6

Demand versus available production and WIP after capacity increases

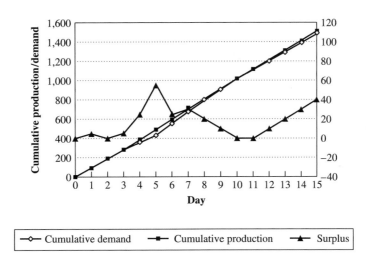

The two different types of infeasibilities require different remedies. Since adding capacity will not help a WIP infeasibility, the only solution is to push out due dates. For example, if five units of the 100 units due in day 2 could be pushed out to day 3, that portion of the schedule would become feasible.

Capacity infeasibilities can be remedied in two ways: by adding capacity or by pushing out due dates. For instance, if overtime were used on day 8 to produce 25 units of output, the schedule would be feasible. However, this will also increase the surplus by the end of the planning horizon (see Figure 15.6). Alternately, if 30 units of the 130 units demanded on day 6 are moved to days 12, 13, and 14 (10 each), the schedule also becomes feasible (see Figure 15.7). This results in a smaller surplus at the end of the planning horizon than occurs under the overtime alternative, since no capacity is added. Of course, in an actual scheduling situation we would have to correct these surpluses; the approach of the next section gives a procedure for doing this.

FIGURE 15.7

Demand versus available production and WIP after pushing out demand

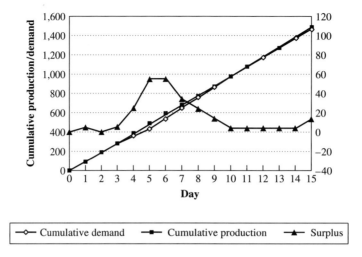

15.5.2 Capacitated Material Requirements Planning—MRP-C

A procedure designed to detect and remedy scheduling infeasibilities is **capacitated material requirements planning (MRP-C)** (see Tardif 1995 for details). MRP-C is similar to MRP except that it explicitly considers capacity. As such, it replaces MRP in the MRP II planning hierarchy.

The basic structure of MRP-C derives from the hierarchical nature of production planning. As we saw in Chapter 13, decision variables in high-level problems are often constraints in lower-level problems. For example, aggregate planning may treat capacity variables (e.g., overtime) as variables. Demand management may treat customer due dates as variables (if due date quoting is used). But scheduling frequently treats both capacities and due dates as constraints. As we have seen, these constraints result in some of the hardest problems in the production research literature (e.g., the minimum makespan job shop problem).

To make the scheduling problem tractable, MRP-C begins by seeking a low-level schedule that satisfies all due dates without violating capacity and builds as little inventory as possible before it is needed. We define **build-ahead** as inventory that is made earlier than the actual demand. The objective of MRP-C is to find a minimum build-ahead feasible schedule. If it cannot find such a schedule, then MRP-C highlights the causes of the infeasibility, enabling the planner to make changes in due dates, capacities, or both as appropriate to the specific situation.

The algorithm used in MRP-C is based on the conveyor model to characterize the behavior of each process (machine, line segment, or line, depending on the level of detail required) in the system. This requires estimates of the following two parameters:

1. **Minimum practical lead time** is denoted by T_0^P and represents the time to go through the line with no queueing. This should include any common delays such as waiting to move and minor adjustments and so will usually be larger than the raw process time T_0 used in Chapter 7.

2. **Practical production rate** is denoted by $r_b^P(t)$ and represents the realistic capacity of the line in time period t. If $r_b^P(t)$ is constant for all t, then because utilization of the bottleneck must be less than 100 percent, $r_b^P(t)$ must be smaller than the bottleneck rate r_b used in Chapter 7. However, if $r_b^P(t)$ varies (e.g., due to scheduled overtime), then $r_b^P(t)$ may exceed r_b for some t. However, on average it must be smaller than the long-run bottleneck rate.

By using only two values, MRP-C captures the basic relationships between WIP and cycle time for an entire process without the burden of a full-blown simulation or detailed capacity knowledge of each station.

MRP-C consists of two phases. The first compares near-term demands against WIP already in the line and available capacity to see if there are any infeasibilities. Once all infeasibilities have been addressed, the second phase works backward in time to determine the minimum releases into the line required to meet demand. We describe the mechanics of phase I of MRP-C in the following technical note.

Technical Note—MRP-C (Phase I)

The procedures divides time into periods, which could represent shifts, days, or weeks, depending on the level of resolution desired, and makes use of the following notation:

$$T = \text{planning horizon (last period) of problem}$$

$t = $ time index for periods, $t = 0, 1, \ldots, T$ (anything occurring in period 0 is before current period)

$T_0^P = $ minimum practical lead time of process under consideration

$\ell = $ lead time to obtain raw material

$r_b^P(t) = $ production rate (capacity) of process in period t

$D(t) = $ demand due at time t, that is, the master production schedule

$a(t) = $ scheduled receipts (arrivals) of raw material in period t

$w(t) = $ "timed-available" WIP (TAWIP) in line that is t periods away from completion, defined for $t = 0, 1, \ldots, T_P$. Note that $w(0)$ represents WIP already completed, which could be finished goods inventory or raw material for another process, while $w(t)$ represents WIP that is t periods from completion.

$\hat{w}(t) = $ capacity-adjusted timed-available WIP (CATAWIP) that takes into consideration the amount of capacity available in period t

$c(t) = $ carryover WIP at t

$I(t) = $ projected-on-hand FGI at time t

$N(t) = $ net FGI requirements for period t

The first phase computes the net WIP requirements $N(t)$ as follows:

1. Determine TAWIP, which can be in the form of existing WIP in the line or scheduled receipts. We do this by setting

$$w(t) = \begin{cases} \text{existing WIP} & \text{for } 1 \leq t \leq T_0^P \\ a(t - T_0^P) & \text{for } T_0^P < t \leq T_0^P + \ell \\ \infty & \text{for } t > T_0^P + \ell \end{cases}$$

For periods within the practical minimum process time, $w(t)$ is equal to the existing WIP in the line. In the previous example where $T_0^P = 3$ days, $w(1)$ has been in the line for two days and so is one day away from completion, $w(2)$ has been in for one day and therefore requires two more days for completion, and so on. For values of t beyond T_0^P but less than the time to obtain raw material, the timed-available WIP is equal to the arrivals of raw material received T_0^P periods before. For periods that are farther out than the raw material lead time (ℓ) plus the process time (T_0^P), the value is set to infinity since these materials can be ordered within their lead time.

2. Compute CATAWIP. We do this by starting with $c(0) = 0$ and computing

$$\hat{w}(t) = \min \{r_b^P(t), w(t) + c(t-1)\}$$
$$c(t) = w(t) + c(t-1) - \hat{w}(t)$$

for $t = 1, 2, \ldots, T$. This step accounts for the fact that no more than r_b^P units of WIP available in period t can actually be completed in period t, due to constrained capacity. So the $\hat{w}(t)$ values represent how much production can be done in each period running at full capacity. If more WIP is available than capacity in period t, then it is carried over as $c(t)$ and becomes available in period $t + 1$.

3. Compute projected on-hand FGI. We do this by starting with $I(0)$ equal to the initial finished goods inventory and computing

$$I(t) = I(t - 1) + \hat{w}(t) - D(t)$$

for $t = 1, 2, \ldots, T$. Using the maximum available capacity/raw materials, this step computes the ending net FGI in each period. If this value ever becomes negative, then it means that there is not sufficient WIP and/or capacity to meet demand.

4. Compute the net requirements. We do this by computing

$$N(t) = \max \{0, \min \{-I(t), D(t)\}\}$$

for $t = 1, 2, \ldots, T$. If $I(t)$ is greater than zero, there are no net requirements because there is sufficient inventory to cover the gross requirements. If $I(t)$ is negative but $I(t - 1) \geq 0$, then the net requirement is equal to the absolute value of $I(t)$. If $I(t)$ and $I(t - 1)$ are both negative, then $N(t)$ will be equal to demand for the period. Note that this is exactly analogous to the netting calculation in regular MRP.

If $N(t) > 0$ and $c(t) < N(t)$, then the schedule is WIP-infeasible and the only remedy is to move out $N(t) - c(t)$ units of demand. If $N(t) > 0$ and $c(t) \geq N(t)$, then the problem is a capacity infeasibility, which can be remedied either by moving out demand or by adding capacity.

5. After any change is made (e.g., moving out a due date), all values must be recomputed.

The MRP-C procedure detailed above appears complex, but is actually very straightforward to implement in a spreadsheet. The following example gives an illustration.

Example:
Applying the MRP-C procedure to the data of the previous example generates the results shown in Table 15.10. The WIP infeasibility of five units in period 2 is indicated by the fact that $N(2) = 5$. The only way to address this problem is to reduce demand in period 2 from 100 to 95 and then to move it into period 3 by increasing demand from 90 to 95. The fact that $N(t)$ reaches 25 for $t = 10, 11$ indicates a shortage of 25 units of capacity. One way to address this problem is to add enough overtime to produce 25 more units in period 8 (which we do in Table 15.11). Otherwise, if no extra capacity is available, we could have postponed the production of 25 units to later in the schedule by pushing back due dates. The projected on-hand figure indicates periods with additional capacity and/or WIP that could accept extra demand. The final schedule is shown in Table 15.11.

At this point, we know that a feasible schedule exists. However, the master production schedule generated is not a good schedule since it has periods of demand that exceed capacity. Thus, some build-ahead of inventory must be done. The second phase of MRP-C uses the constraints of capacity and WIP provided by the first phase to compute a schedule that is feasible and produces a minimum of build-ahead inventory. This is done by computing the schedule from the last period and working backward in time. The procedure is given in the following technical note.

TABLE 15.10   Feasibility Calculations

Period t	Demand D(t)	TAWIP w(t)	Capacity $r_b^P(t)$	CATAWIP $\hat{w}(t)$	Carryover c(t)	Projected on Hand I(t)	Net Requirements N(t)
0					0	0	
1	90	95	100	95	0	5	0
2	100	90	100	90	0	−5	5
3	90	115	100	100	15	5	0
4	80	∞	100	100	∞	25	0
5	70	∞	100	100	∞	55	0
6	130	∞	100	100	∞	25	0
7	120	∞	100	100	∞	5	0
8	110	∞	100	100	∞	−5	5
9	110	∞	100	100	∞	−15	15
10	110	∞	100	100	∞	−25	25
11	100	∞	100	100	∞	−25	25
12	90	∞	100	100	∞	−15	15
13	90	∞	100	100	∞	−5	5
14	90	∞	100	100	∞	5	0
15	90	∞	100	100	∞	15	0

TABLE 15.11   Final Feasible Master Production Schedule

Period t	Demand D(t)	TAWIP w(t)	Capacity $r_b^P(t)$	CATAWIP $\hat{w}(t)$	Carryover c(t)	Projected on Hand I(t)	Net Requirements N(t)
0					0	0	
1	90	95	100	95	0	5	0
2	95	90	100	90	0	0	0
3	90	115	100	100	15	10	0
4	85	∞	100	100	∞	25	0
5	70	∞	100	100	∞	55	0
6	130	∞	100	100	∞	25	0
7	120	∞	100	100	∞	5	0
8	110	∞	125	125	∞	20	0
9	110	∞	100	100	∞	10	0
10	110	∞	100	100	∞	0	0
11	100	∞	100	100	∞	0	0
12	90	∞	100	100	∞	10	0
13	90	∞	100	100	∞	20	0
14	90	∞	100	100	∞	30	0
15	90	∞	100	100	∞	40	0

Technical Note—MRP-C (Phase II)

To describe the MRP-C procedure for converting a schedule of (feasible) demands to a schedule of releases (starts), we make use of the following notation:

$D(t)$ = demand due at time t, that is, master production schedule

$I(t)$ = projected on-hand FGI at time t

$N(t)$ = net FGI requirements for period t

$\hat{w}(t)$ = CATAWIP available in period t

$X(t)$ = production quantity in period t

$Y(t)$ = amount of build-ahead inventory in period t, which represents production in period t intended to fill demand in periods beyond t

$S(t)$ = release quantities ("starts") in period t

The basic procedure is to first compute net demand by subtracting finished goods inventory in much the same way as MRP. Then available production in each period is given by the capacity-adjusted timed-available WIP (CATAWIP). Since this includes WIP in the line, we do not net it out as we would do in MRP. With this, the procedure computes production, build-ahead, and starts for each period.

The specific steps are as follows:

1. *Netting.* We first compute net requirements in the standard (MRP) way.

   a. Initialize variables:

   $$I(0) = \text{initial finished goods inventory}$$
   $$N(0) = 0$$

   b. For each period, beginning with period 1 and working to period T, we compute the projected on-hand inventory and the net requirements as follows.

   $$I(t) = I(t-1) + N(t-1) - D(t)$$
   $$N(t) = \max\{0, \min\{D(t), -I(t)\}$$

2. *Scheduling.* The scheduling procedure is done from the last (T) period, working backward in time.

   a. Initialize variables.

   $$D(T+1) = 0$$
   $$X(T+1) = 0$$
   $$Y(T+1) = \text{desired ending FGI level}$$

   b. For each period t, starting with T and working down to period 1, compute

   $$Y(t) = Y(t+1) + D(t+1) - X(t+1)$$
   $$X(t) = \min\{\hat{w}(t), D(t) + Y(t)\}$$

   c. The equation $Y(0) = Y(1) + D(1) - X(1)$ provides an easy capacity check. This value should be zero if all the infeasibilities were addressed in phase I. If not, the schedule is infeasible and phase I needs to be redone correctly.

   d. Assuming there are no remaining schedule infeasibilities, we compute the schedule of production starts by offsetting the production quantities by the minimum practical lead time as follows:

   $$S(t) = X(t + T_0^P) \qquad \text{for } t = 1, 2, \ldots, T - T_0^P$$

The MRP-C scheduling procedure computes the amount of build-ahead from the end of the time horizon T backward. The level of build-ahead in period T is the desired level of inventory at the end of the planning horizon. One would generally set this to zero, unless there were some exceptional reason to plan to finish the planning horizon with excess inventory. At each period, output will be either the capacity or the total demand (net demand plus build-ahead), whichever is less. This is intuitive since production *cannot* exceed the maximum rate of the line and *should not* exceed demand (including build-ahead).

If the build-ahead for period 0 is positive, the schedule is infeasible. The amount of build-ahead in period 0 indicates the amount of additional finished inventory needed at $t = 0$ to make the schedule feasible. However, if phase I has addressed all the capacity and WIP infeasibilities, $Y(0)$ will be zero. Indeed, this is the entire point of phase I.

The final output of the MRP-C procedure is a list of production starts that will meet all the (possibly revised) due dates within capacity and material constraints while producing a minimum of build-ahead inventory.

Example:

We continue with our example from phase I and apply the second phase of MRP-C. This generates the results in Table 15.12. Note that the schedule calls for production to be as high as possible, being limited by WIP in the first two periods, and then limited by capacity thereafter, until period 12. At this point, production decreases to 90 units, which is below CATAWIP but is sufficient to keep up with demand.

Notice that while MRP-C does the dirty work of finding infeasibilities and identifying possible actions for remedying them, it leaves the sensitive judgments concerning increasing capacity (whether, how, where) and delaying jobs (which ones, how much) up to the user. As such, MRP-C encourages appropriate use of the respective talents of humans and computers in the scheduling process.

TABLE 15.12   Final Production Schedule

Period t	Demand $D(t)$	Projected on Hand $I(t)$	Net Requirements $N(t)$	CATAWIP $\hat{w}(t)$	Build-Ahead $Y(t)$	Production $X(t)$	Starts $S(t)$
0		0	0		0		
1	90	−90	90	95	5	95	100
2	100	−95	95	90	0	90	100
3	90	−95	95	100	5	100	100
4	80	−80	80	100	25	100	100
5	70	−70	70	100	55	100	125
6	130	−130	130	100	25	100	100
7	120	−120	120	100	5	100	100
8	110	−110	110	125	20	125	100
9	110	−110	110	100	10	100	90
10	110	−110	110	100	0	100	90
11	100	−100	100	100	0	100	90
12	90	−90	90	100	0	90	90
13	90	−90	90	100	0	90	
14	90	−90	90	100	0	90	
15	90	−90	90	100	0	90	

15.5.3   Extending MRP-C to More General Environments

The preceding described how to use the MRP-C procedure to schedule one process (workstation, line, or line segment) represented by the conveyor model. The real power of MRP-C is that it can be extended to multistage systems with more than a single product.

For a serial line, this extension is simple. The production starts into a downstream station represent the demands upon the upstream station that feeds it. Thus, we can simply apply MRP-C by starting at the last station and working backward to the front of the line. Likewise, the time-adjusted WIP (TAWIP) levels will be generated by the production of the upstream process.

If there are assembly stations, then production starts must be translated to demands upon each of the stations feeding them. This is exactly analogous to the bill-of-material explosion concept of MRP, except applied to routings. Otherwise the MRP-C procedure remains unchanged.

In systems where multiple routings (i.e., producing different products) pass through a single station, we must combine the individual demands (i.e., production starts at downstream stations) to form *aggregate* demand. Since the different products may have different processing times at the shared resource, it is important that the MRP-C calculations be done in units of time instead of product. That is, capacity, demand, WIP, and so forth should all be measured in hours. This is similar in spirit to the idea of maintaining a constant amount of work rather than a constant number of units in a CONWIP line with multiple products, which we discussed in Chapter 14.

In systems with multiple products, things get a bit more complex because we must choose a method for breaking ties when more than one product requires build-ahead in the same period. The wrong choice can schedule early production of a product with little or no available WIP instead of another product that has plentiful WIP. This can cause a WIP infeasibility when the next stage is scheduled. Several clever means for breaking ties have been proposed by Tardif (1995), who also addresses other practical implementation issues.

15.5.4   Practical Issues

The MRP-C approach has two clear advantages over MRP: (1) It uses a more accurate model that explicitly considers capacity, and (2) it provides the planner with useful diagnostics. However, there are some problems.

First, MRP-C relies on a heuristic and therefore cannot be guaranteed to find a feasible schedule if one exists. (However, if it finds a feasible schedule, this schedule *is* truly feasible.) Although certain cases of MRP-C can make use of an exact algorithm, this is much slower (see Tardif 1995). In essence, the approach discussed above sacrifices accuracy for speed. Given that it is intended for use in an iterative, "decision support" mode, the additional speed is probably worth the small sacrifice in accuracy. Moreover, any errors produced by MRP-C will make the schedule more conservative. That is, MRP-C may require more adjustments than the minimum necessary to achieve feasibility. Hence, schedules will be "more feasible" than they really need to be and will thus have a better chance of being successfully executed.

Second, MRP-C, like virtually all scheduling approaches, implies a *push* philosophy (i.e., it sets release *times*). As we discussed in Chapter 10, this makes it subject to all the drawbacks of push systems. Fortunately, one can integrate MRP-C (and indeed any push system, including MRP) into a pull environment and obtain many of the efficiency, predictability, and robustness benefits associated with pull. We describe how this can be done in the following section.

15.6   Production Scheduling in a Pull Environment

Recall the definitions of push and pull production control. A push system *schedules* releases into the line based on due dates, while a pull system *authorizes* releases into the line based on operating conditions. Push systems control release rates (and thereby throughput) and measure WIP to see if the rates are too large or too small. Pull systems do the opposite. They control WIP and measure completions to determine whether production is adequate. Since WIP control is less sensitive than release control, pull systems are more robust to errors than are push systems. Also, since pull systems directly control WIP, they avoid WIP explosions and the associated overtime vicious cycle often observed in push systems. Finally, pull systems have the ability to work ahead for short periods, allowing them to exploit periods of better-than-average production.

For these reasons, we want to maintain the benefits of pull systems to whatever extent possible. The question is, How can it be done in an environment that requires a detailed schedule? In this section we discuss the link between scheduling and pull production.

15.6.1   Schedule Planning, Pull Execution

Even the best schedule is only a plan of what should happen, not a guarantee of what will happen. By necessity, schedules are prepared relatively infrequently compared to shop floor activity; the schedule may be regenerated weekly, while material flow, machine failures, and so forth happen in real time. Hence, they cannot help but become outdated, sometimes very rapidly. Therefore we should treat the schedule as a set of suggestions, not a set of requirements, concerning the order and timing of releases into the system.

A pull system is an ideal mechanism for linking releases to real-time status information. When the line is already congested with WIP, so that further releases will only increase congestion without making jobs finish sooner, a pull system will prevent releases. When the line runs faster than expected and has capacity for more work, a pull system will draw it in. Fortunately, using a pull system in concert with a schedule is not at all difficult.

To illustrate how this would work, suppose we have a CONWIP system in place for each routing and make use of MRP-C to generate a schedule for the overall system. Note that there is an important link between MRP-C and CONWIP: the conveyor model. Thus, if the parameters are correct, MRP-C will generate a set of release times that are very close to the times that the CONWIP system generates authorizations (pull signals) for the releases. Of course, variability will always prevent a perfect match, but on average actual performance will be consistent with the planned schedule.

When production falls behind schedule, we can catch up if there is a capacity cushion (e.g., a makeup time at the end of each shift or day) available. If no such cushion is available, we must adjust the schedule at the next regeneration. When production outpaces the schedule, we can allow it to work ahead, by allowing the line to pull in more than was planned. A simple rule comparing the current date and time with the date and time of the next release can keep the CONWIP line from working too far ahead. In this way, the CONWIP system can take advantage of the "good" production days without getting too far from schedule.

When we cannot rely on a capacity cushion to make up for lags in production (e.g., we are running the line as fast as we can), we can supplement the CONWIP control system with the statistical throughput control (STC) procedure described in Chapter 13. This provides a means for detecting when production is out of control relative to the schedule. When this occurs, either the system or the MRP-C parameters need adjustment. Which to adjust may pose an important management decision. Reducing MRP-C capacity parameters may be tantamount to admitting that corporate goals are not achievable. However, increasing capacity may require investment in equipment, staff, increased subcontracting costs, or consulting.

15.6.2   Using CONWIP with MRP

Nothing in the previous discussion about using CONWIP in conjunction with a schedule absolutely requires that the schedule be generated with MRP-C. Of course, since MRP-C considers capacity using the same conveyor model that underlies CONWIP, we would expect it to work well. But we can certainly use CONWIP with *any* scheduling system, including MRP. We would do this by using the MRP-generated list of planned order releases, sorted by routing, as the work backlogs for each CONWIP line. The CONWIP system then determines when jobs actually get pulled into the system.

As with MRP-C, we can employ a capacity cushion, work ahead, and track against schedule. The primary difference is that the underlying model of MRP and CONWIP are *not* consistent. Consequently, MRP is more likely to generate inconsistent planned order release schedules than is MRP-C. This can be mitigated, somewhat, by employing good master production scheduling techniques and by debugging the process using bottom-up replanning.

15.7   Conclusions

Production problems are notoriously difficult, both because they involve many conflicting goals and because the underlying mathematics can get very complex. Considerable scheduling research has produced formalized measures of the complexity of scheduling problems and has generated some good insights. However, it has not yielded good solutions to practical scheduling situations.

Because scheduling is difficult, an important insight of our discussion is that it is frequently possible to avoid hard problems by solving different ones. One example is to replace a system of exogenously generated due dates with a systematic means for quoting them. Another is to separate the problem of keeping cycle times short (solve by using small jobs) from the problem of keeping capacities high (solve by sequencing like jobs together for fewer setups). Given an appropriately formulated problem, good heuristics for identifying feasible (not optimal) schedules are becoming available.

An important recent trend in scheduling research and software development is toward finite-capacity scheduling. By overcoming the fundamental flaw in MRP, these models have the potential to make the MRP II hierarchy much more effective in practice. However, to provide flexibility for accommodating intangibles, an effective approach to finite-capacity scheduling is for the system to evaluate schedule feasibility and generate diagnostics about infeasibilities. A procedure designed to do this is capacitated material requirements planning—MRP-C.

Finally, although scheduling is essentially a push philosophy, it is possible to use a schedule in concert with a pull system. The basic idea is to use the schedule to plan work releases and the pull system to execute them. This offers the planning benefits of a scheduling system along with the environmental benefits of a pull system.

Study Questions

1. What are some goals of production scheduling? How do these conflict?
2. How does reducing cycle time support several of the above goals?
3. What motivates maximizing utilization? What motivates not maximizing utilization?
4. Why is average tardiness a better measure than average lateness?
5. What are some drawbacks of using service level as the only measure of due date performance?
6. For each of the assumptions of classic scheduling theory, give an example of when it might be valid. Give an example of when each is not valid.
7. Why do people use dispatching rules instead of finding an optimal schedule?
8. What dispatching rule minimizes average cycle time for a deterministic single machine? What rule minimizes maximum tardiness? How can one easily check to see if a schedule exists for which there are no tardy jobs?
9. Provide an argument that no matter how sophisticated the dispatching rule, it cannot solve the problem of minimizing average tardiness.
10. What is some evidence that there are some scheduling problems for which no polynomial algorithm exists?
11. Address the following comment: "Well, maybe today's computers are too slow to solve the job shop scheduling problem, but new parallel processing technology will speed them up to the point where computer time should not be an obstacle to solving it in the near future."
12. What higher-level planning problems are related to the production scheduling problem? What are the variables and constraints in the high-level problems? What are the variables and constraints in the lower-level scheduling problem? How are the problems linked?
13. How well do you think the policy of planning with a schedule and executing with a pull system should work using MRP-C and CONWIP? Why? How well should it work using MRP and kanban? Why?

Problems

1. Consider the following three jobs to be processed on a single machine:

Job Number	Process Time	Due Date
1	4	2
2	2	3
3	1	4

Enumerate all possible sequences and compute the average cycle time, total tardiness, and maximum lateness for each. Which sequence works best for each measure? Identify it as EDD, SPT, or something else.

2. You are in charge of the shearing and pressing operations in a job shop. When you arrived this morning, there were seven jobs with the following processing times.

	Processing Time	
Job	Shear	Press
1	6	3
2	2	9
3	5	3
4	1	8
5	7	1
6	4	5
7	9	6

    a. What is the makespan under the SPT dispatching rule?

    b. What sequence yields the minimum makespan?

    c. What is this makespan?

3. Your boss knows factory physics and insists on reducing average cycle time to help keep jobs on time and reduce congestion. For this reason, your personal performance evaluation is based on the average cycle time of the jobs through your process center. However, your boss also knows that late jobs are *extremely bad,* and she will fire you if you produce a schedule that includes any late jobs. The jobs listed below are staged in your process center for the first shift. Sequence them such that your evaluation will be the best it can be without getting you fired.

	Job				
	J_1	J_2	J_3	J_4	J_5
Processing time	6	2	4	9	3
Due date	33	13	6	23	31

4. Suppose daily production of a CONWIP line is nearly normally distributed with a mean of 250 pieces and a standard deviation of 50 pieces. The WIP level of the CONWIP line is 1,250 pieces. Currently there is a backlog of 1,400 pieces with an "emergency position" 150 pieces out. A new order for 100 pieces arrives.

    a. Quote a lead time with 95 percent confidence if the new order is placed at the end of the backlog and if it is placed in the emergency position.

    b. Quote a lead time with 99 percent confidence if the new order is placed at the end of the backlog and if it is placed in the emergency position.

5. Consider the jobs on the next page. Process times for all jobs are one hour. Changeovers between families require four hours. Thus, the completion time for job 1 is 5, for job 2 is 6, for job 3 is 11, and so on.

Job	Family Code	Due Date
1	1	5
2	1	6
3	2	12
4	2	13
5	1	13
6	1	19
7	1	20
8	2	20
9	2	26
10	1	28

    a. Compute the total tardiness of the sequence.
    b. How many possible sequences are there?
    c. Find a sequence with no tardiness.

6. The Hickory Flat Sawmill (HFS) makes four kinds of lumber in one mill. Orders come from a variety of lumber companies to a central warehouse. Whenever the warehouse hits the reorder point, an order is placed to HFS. Pappy Red, the sawmill manager, has set the lot sizes to be run on the mill based on historical demands and common sense. The smallest amount made is a lot of 1,000 board-feet (1 kbf). The time it takes to process a lot depends on the product, but the time does not vary more than 25 percent from the mean. The changeover time can be quite long depending on how long it takes to get the mill producing good product again. The shortest time that anyone can remember is two hours. Once it took all day (eight hours). Most of the time it takes around four hours. Demand data and run rates are given in Table 15.13. The mill runs productively eight hours per day, five days per week (assume 4.33 weeks per month).

    The lot sizes are 50 of the knotty 1 × 10, 34 for the clear 1 × 4, 45 for the clear 1 × 6, and 40 for the rough plank. Lots are run on a first-come, first-served basis as they arrive from the warehouse. Currently the average response time is nearly three weeks (14.3 working days). The distributor has told HFS that HFS needs to get this down to two weeks in order to continue being a supplier.

    a. Compute the effective SCV c_e^2 for the mill. What portion of c_e^2 is due to the term in square brackets in Equation (15.8)? What can you do to reduce it?
    b. Verify the 14.3-working-day cycle time.
    c. What can you do to reduce cycle times without investing in any more equipment or physical process improvements?

TABLE 15.13   Data for the Sawmill Problem

Parameter	Knotty 1 × 10	Clear 1 × 4	Clear 1 × 6	Rough Plank
Demand (kbf/mo)	50	170	45	80
One lot time (hour)	0.2000	0.4000	0.6000	0.1000

7. Single parts arrive to a furnace at a rate of 100 per hour with exponential times between arrivals. The furnace time is three hours with essentially no variability. It can hold 500 parts. Find the batch size that minimizes total cycle time at the furnace.

8. Consider a serial line composed of three workstations. The first workstation has a production rate of 100 units per day and a minimum practical lead time T_0^P of three days. The second has a rate of 90 units per day and $T_0^P = 4$ days; and the third has a rate of 100 and $T_0^P = 3$ days. Lead time for raw material is one day, and there are currently 100 units on hand.

   Currently there are 450 units of finished goods, 95 units ready to go into finished goods on the first day, 95 on the second, and 100 on the third; all from the last station. The middle station has 35 units completed and ready to move to the last station and 90 units ready to come out in each of the next four days. The first station has no WIP completed, 95 units that will finish on the first day, zero units that will finish the second day, and 100 units that will finish the third day.

   The demand for the line is given in the table below.

Day from Start	Amount Due
1	80
2	80
3	80
4	80
5	80
6	130
7	150
8	180
9	220
10	240
11	210
12	150
13	90
14	80
15	80

   Develop a feasible schedule that minimizes the amount of inventory required. If it is infeasible, adjust demands by moving them out. However, all demand must be met within 17 days.

16　Aggregate and Workforce Planning

And I remember misinformation followed us like a plague,
Nobody knew from time to time if the plans were changed.

Paul Simon

16.1　Introduction

A variety of manufacturing management decisions require information about what a plant will produce over the next year or two. Exampes include the following:

1. *Staffing*. Recruiting and training new workers is a time-consuming process. Management needs a long-term production plan to decide how many and what type of workers to add and when to bring them on-line in order to meet production needs. Conversely, eliminating workers is costly and painful, but sometimes necessary. Anticipating reductions via a long-term plan makes it possible to use natural attrition, or other gentler methods, in place of layoffs to achieve at least part of the reductions.

2. *Procurement*. Contracts with suppliers are frequently set up well in advance of placing actual orders. For example, a firm might need an opportunity to "certify" the subcontractor for quality and other performance measures. Additionally, some procurement lead times are long (e.g., for high-technology components they may be six months or more). Therefore, decisions regarding contracts and long-lead-time orders must be made on the basis of a long-term production plan.

3. *Subcontracting*. Management must arrange contracts with subcontractors to manufacture entire components or to perform specific operations well in advance of actually sending out orders. Determining what types of subcontracting to use requires long-term projections of production requirements and a plan for in-house capacity modifications.

4. *Marketing*. Marketing personnel should make decisions on which products to promote on the basis of both a demand forecast *and* knowledge of which products have tight capacity and which do not. A long-term production plan incorporating planned capacity changes is needed for this.

The module in which we address the important question of what will be produced and when it will be produced over the long range is the **aggregate planning (AP)** module. As Figure 13.2 illustrated, the AP module occupies a central position in the production planning and control (PPC) hierarchy. The reason, or course, is that so many important decisions, such as those listed, depend on a long-term production plan.

Precisely because so many different decisions hinge on the long-range production plan, many different formulations of the AP module are possible. Which formulation is appropriate depends on what decision is being addressed. A model for determining the time of staffing additions may be very different from a model for deciding which products should be manufactured by outside subcontractors. Yet a different model might make sense if we want to address both issues simultaneously.

The staffing problem is of sufficient importance to warrant its own module in the hierarchy of Figure 13.2, the **workforce planning (WP)** module. Although high-level workforce planning (projections of total staffing increases or decreases, institution of training policies) can be done using only a rough estimate of future production based on the demand forecast, low-level staffing decisions (timing of hires or layoffs, scheduling usage of temporary hires, scheduling training) are often based on the more detailed production information contained in the aggregate plan. In the context of the PPC hierarchy in Figure 13.2, we can think of the AP module as either refining the output of the WP module or working in concert with the WP module. In any case, they are closely related. We highlight this relationship by treating aggregate planning and workforce planning together in this chapter.

As we mentioned in Chapter 13, linear programming is a particularly useful tool for formulating and solving many of the problems commonly faced in the aggregate planning and workforce planning modules. In this chapter, we will formulate several typical AP/WP problems as linear programs (LPs). We will also demonstrate the use of linear programming (LP) as a solution tool in various examples. Our goal is not so much to provide specific solutions to particular AP programs, but rather to illustrate general problem-solving approaches. The reader should be able to combine and extend our solutions to cover situations not directly addressed here.

Finally, while this chapter will not make an LP expert out of the reader, we do hope that he or she will become aware of how and where LP can be used in solving AP problems. If managers can recognize that particular problems are well suited to LP, they can easily obtain the technical support (consultants, internal experts) for carrying out the analysis and implementation. Unfortunately, far too few practicing managers make this connection; as a result, many are hammering away at problems that are well suited to linear programming with manual spreadsheets and other ad hoc approaches.

16.2   Basic Aggregate Planning

We start with a discussion of simple aggregate planning situations and work our way up to more complex cases. Throughout the chapter, we assume that we have a **demand forecast** available to us. This forecast is generated by the forecasting module and gives estimates of periodic demand over the **planning horizon.** Typically, periods are given in months, although further into the future they can represent longer intervals. For instance, periods 1 to 12 might represent the next 12 months, while periods 13 to 16 might represent the four quarters following these 12 months. A typical planning horizon for an AP module is one to three years.

16.2.1   A Simple Model

Our first scenario represents the simplest possible AP module. We consider this case not because it leads to a practical model, but because it illustrates the basic issues, provides a basis for considering more realistic situations, and showcases how linear programming can support the aggregate planning process. Although our discussion does not presume any background in linear programming, the reader interested in how and why LP works is advised to consult Appendix 16A, which provides an elementary overview of this important technique.

For modeling purposes, we consider the situation where there is only a single product, and the entire plant can be treated as a single resource. In every period, we have a demand forecast and a capacity constraint. For simplicity, we assume that demands represent customer orders that are due at the end of the period, and we neglect randomness and yield loss.

It is obvious under these simplifying assumptions that if demand is less than capacity in every period, the optimal solution is to simply produce amounts equal to demand in every period. This solution will meet all demand just-in-time and therefore will not build up any inventory between periods. However, if demand exceeds capacity in some periods, then we must work ahead (i.e., produce more than we need in some previous period). If demand cannot be met even by working ahead, we want our model to tell us this. To model this situation in the form of a linear program, we introduce the following notation:

t = an index of time periods, where $t = 1, \ldots, \bar{t}$, so $\bar{t}$ is planning horizon for problem

d_t = demand in period t, in physical units, standard containers, or some other appropriate quantity (assumed due at end of period)

c_t = capacity in period t, in same units used for d_t

r = profit per unit of product sold (not including inventory-carrying cost)

h = cost to hold one unit of inventory for one period

X_t = quantity produced during period t (assumed available to satisfy demand at end of period t)

S_t = quantity sold during period t (we assume that units produced in t are available for sale in t and thereafter)

I_t = inventory at end of period t (after demand has been met); we assume I_0 is given as data

In this notation, X_t, S_t, and I_t are **decision variables.** That is, the computer program solving the LP is free to choose their values so as to minimize the objective, provided the constraints are satisfied. The other variables—d_t, c_t, r, h—are **constants,** which must be estimated for the actual system and supplied as data. Throughout this chapter, we use the convention of representing variables with capital letters and constants with lowercase letters.

We can represent the problem of maximizing net profit minus inventory carrying cost subject to capacity and demand constraints as

$$\text{Maximize} \qquad \sum_{t=1}^{\bar{t}} r S_t - h I_t \qquad\qquad (16.1)$$

Subject to:

$$S_t \le d_t \qquad\qquad t = 1, \ldots, \bar{t} \qquad\qquad (16.2)$$
$$X_t \le c_t \qquad\qquad t = 1, \ldots, \bar{t} \qquad\qquad (16.3)$$
$$I_t = I_{t-1} + X_t - S_t \qquad t = 1, \ldots, \bar{t} \qquad\qquad (16.4)$$
$$X_t, S_t, I_t \ge 0 \qquad\qquad t = 1, \ldots, \bar{t} \qquad\qquad (16.5)$$

The objective function computes net profit by multiplying unit profit r by sales S_t in each period t, and subtracting the inventory carrying cost h times remaining inventory I_t at the end of period t, and summing over all periods in the planning horizon. Constraints (16.2) limit sales to demand. If possible, the computer will make all these constraints tight, since increasing the S_t values increases the objective function. The only reason that these constraints will not be tight in the optimal solution is that capacity constraints (16.3) will not permit it.[1] Constraints (16.4), which are of a form common to almost all multiperiod aggregate planning models, are known as **balance constraints.** Physically, all they represent is conservation of material; the inventory at the end of period $t (I_t)$ is equal to the inventory at the end of period $t - 1 (I_{t-1})$ plus what was produced during period $t (X_t)$ minus the amount sold in period t (S_t). These constraints are what force the computer to choose values for X_t, S_t, and I_t that are consistent with our verbal definitions of them. Constraints (16.5) are simple nonnegativity constraints, which rule out negative production or inventory levels. Many, but not all, computer packages for solving LPs automatically force decision variables to be nonnegative unless the user specifies otherwise.

16.2.2   An LP Example

To make the above formulation concrete and to illustrate the mechanics of solving it via linear programming, we now consider a simple example. The Excel spreadsheet shown in Figure 16.1 contains the unit profit r of \$10, the one-period unit holding cost h of \$1, the initial inventory I_0 of 0, and capacity and demand data c_t and d_t for the next six months. We will make use of the rest of the spreadsheet in Figure 16.1 momentarily. For now, we can express LP (16.1)–(16.5) for this specific case as

Maximize $10(S_1 + S_2 + S_3 + S_4 + S_5 + S_6) - 1(I_1 + I_2 + I_3 + I_4 + I_5 + I_6)$ (16.6)

Subject to:

Demand constraints

$S_1 \leq 80$ (16.7)

$S_2 \leq 100$ (16.8)

$S_3 \leq 120$ (16.9)

$S_4 \leq 140$ (16.10)

$S_5 \leq 90$ (16.11)

$S_6 \leq 140$ (16.12)

Capacity constraints

$X_1 \leq 100$ (16.13)

$X_2 \leq 100$ (16.14)

$X_3 \leq 100$ (16.15)

$X_4 \leq 120$ (16.16)

$X_5 \leq 120$ (16.17)

$X_6 \leq 120$ (16.18)

---

[1]If we want to consider demand as inviolable, we could remove constraints (16.2) and replace S_t with d_t in the objective and constraints (16.4). The problem with this, however, is that if demand is capacity-infeasible, the computer will just come back with a message saying "infeasible," which doesn't tell us why. The formulation here will be feasible regardless of demand; it simply won't make sales equal to demand if there is not enough capacity, and thus we will know what demand we are incapable of meeting from the solution.

FIGURE 16.1

Input spreadsheet for linear programming example

	A	B	C	D	E	F	G	H
1	Constants:							
2	r	10						
3	h	1						
4	I_O	0						
5	t	1	2	3	4	5	6	Total
6	c_t	100	100	100	120	120	120	660
7	d_t	80	100	120	140	90	140	670
8								
9	Variables:							
10	t	1	2	3	4	5	6	Total
11	X_t	0	0	0	0	0	0	0
12	S_t	0	0	0	0	0	0	0
13	I_t	0	0	0	0	0	0	0
14								
15	Objective:							
16	Net Profit:	$0		r*(S_1+S_2+S_3+S_4+S_5+S_6) - h*(I_1+I_2+I_3+I_4+I_5+I_6)				
17								
18	Constraints:							
19	S_1	0	<=	80	d_1			
20	S_2	0	<=	100	d_2			
21	S_3	0	<=	120	d_3			
22	S_4	0	<=	140	d_4			
23	S_5	0	<=	90	d_5			
24	S_6	0	<=	140	d_6			
25	X_1	0	<=	100	c_1			
26	X_2	0	<=	100	c_2			
27	X_3	0	<=	100	c_3			
28	X_4	0	<=	120	c_4			
29	X_5	0	<=	120	c_5			
30	X_6	0	<=	120	c_6			
31	I_1-I_0-X_1+S_1	0	=	0				
32	I_2-I_1-X_2+S_2	0	=	0				
33	I_3-I_2-X_3+S_3	0	=	0				
34	I_4-I_3-X_4+S_4	0	=	0				
35	I_5-I_4-X_5+S_5	0	=	0				
36	I_6-I_5-X_6+S_6	0	=	0				
37					Note: X_t, S_t and I_t must be >= 0			

Inventory balance constraints

$$I_1 - X_1 + S_1 = 0 \quad\quad (16.19)$$

$$I_2 - I_1 - X_2 + S_2 = 0 \quad\quad (16.20)$$

$$I_3 - I_2 - X_3 + S_3 = 0 \quad\quad (16.21)$$

$$I_4 - I_3 - X_4 + S_4 = 0 \quad\quad (16.22)$$

$$I_5 - I_4 - X_5 + S_5 = 0 \quad\quad (16.23)$$

$$I_6 - I_5 - X_6 + S_6 = 0 \quad\quad (16.24)$$

Nonnegativity constraints

$$X_1, X_2, X_3, X_4, X_5, X_6 \geq 0 \quad\quad (16.25)$$

$$S_1, S_2, S_3, S_4, S_5, S_6 \geq 0 \quad\quad (16.26)$$

$$I_1, I_2, I_3, I_4, I_5, I_6 \geq 0 \quad\quad (16.27)$$

Some linear programming packages allow entry of a problem formulation in a format almost identical to (16.6) to (16.27) via a text editor. While this is certainly convenient for very small problems, it can become prohibitively tedious for large ones. Because of this, there is considerable work going on in the OM research community to develop **modeling languages** that provide user-friendly interfaces for describing large-scale optimization problems (see Fourer, Gay, and Kernighan 1993 for an excellent example of a modeling language). Conveniently for us, LP is becoming so prevalent that our spreadsheet package, Microsoft Excel, has an LP solver built right into it. We can represent and solve formulation (16.6) to (16.27) right in the spreadsheet shown in Figure 16.1. The following technical note provides details on how to do this.

Technical Note—Using the Excel LP Solver

Although the reader should consult the Excel documentation for details about the release in use, we will provide a brief overview of the LP solver in Excel 5.0. The first step is to establish cells for the decision variables (B11:G13 in Figure 16.1). We have initially entered zeros for these, but we can set them to be anything we like; thus, we could start by setting $X_t = d_t$, which would be closer to an optimal solution than zeros. The spreadsheet is a good place to play what-if games with the data. However, eventually we will turn over the problem of finding optimal values for the decision variables to the LP solver. Notice that for convenience we have also entered a column that totals X_t, S_t, and I_t. For example, cell H11 contains a formula to sum cells B11:G11. This allows us to write the objective function more compactly.

Once we have specified decision variables, we construct an objective function in cell B16. We do this by writing a formula that multiplies r (cell B2) by total sales (cell H12) and then subtracts the product of h (cell B3) and total inventory (cell H13). Since all the decision variables are zero at present, this formula also returns a zero; that is, the net profit on no production with no initial inventory is zero.

Next we need to specify the constraints (16.7) to (16.27). To do this, we need to develop formulas that compute the left-hand side of each constraint. For constraints (16.7) to (16.18) we really do not need to do this, since the left-hand sides are only X_t and S_t and we already have cells for these in the variables portion of the spreadsheet. However, for clarity, we will copy them to cells B19:B30. We will not do the same for the nonnegativity constraints (16.25) to (16.27), since it is a simple matter to choose all the decision variables and force them to be greater than or equal to zero in the Excel Solver menu. Constraints (16.19) to (16.24) require us to do work, since the left-hand sides are formulas of multiple variables. For instance, cell B31 contains a formula to compute $I_1 - I_0 - X_1 + S_1$ (that is, B13 − B4 − B11 + B12). We have given these cells names to remind us of what they represent, although any names could be used, since they are not necessary for the computation. We have also copied the values of the right-hand sides of the constraints into cells D19:D36 and labeled them in column E for clarity. This is not strictly necessary, but does make it easier to specify constraints in the Excel Solver, since whole blocks of constraints can be specified (for example, B19:B30 ≤ D19:D30). The equality and inequality symbols in column C are also unnecessary, but make the formulation easier to read.

To use the Excel LP Solver, we choose **Formula/Solver** from the menu. In the dialog box that comes up (see Figure 16.2), we specify the cells containing the objective, choose to maximize or minimize, and specify the cells containing decision variables (this can be done by pointing with the mouse). Then we add constraints by choosing **Add** from the constraints section of the form. Another dialog box (see Figure 16.3) comes up in which we fill in the cell containing the left-hand side of the constraint, choose the relationship (≥, ≤, or =), and fill in the right-hand side.

Note that the actual constraint is not shown explicitly in the spreadsheet; it is entered only in the **Solver** menu. However, the right-hand side of the constraint can be another cell in the spreadsheet or a constant. By specifying a range of cells for the right-hand side and a constant for the left-hand side, we can add a whole set of constraints in a single command. For instance, the range B11:G13 represents all the decision variables, so if we use this range as the left-hand side, a ≥ symbol, and a zero for the right-hand side, we will represent all the nonnegativity constraints (16.25) to (16.27). By choosing the **Add** button after each constraint we enter, we can add all the model constraints. When we are done, we choose the **OK** button, which returns us to the original form. We have the option to edit or delete constraints at any time.

Finally, before running the model, we must tell Excel that we want it to use the LP solution algorithm.[2] We do this by choosing the **Options** button to bring up another dialog box (see Figure 16.4) and choosing the **Assume Linear Model** option. This form also allows us to limit the time the model will run and to specify certain tolerances. If the model does not

---

[2]Excel can also solve nonlinear optimization problems and will apply the nonlinear algorithm as a default. Since LP is *much* more efficient, we definitely want to choose it as long as our model meets the requirements. All the formulations in this chapter are linear and therefore can use LP.

FIGURE 16.2

Specification of objectives and constraints in Excel

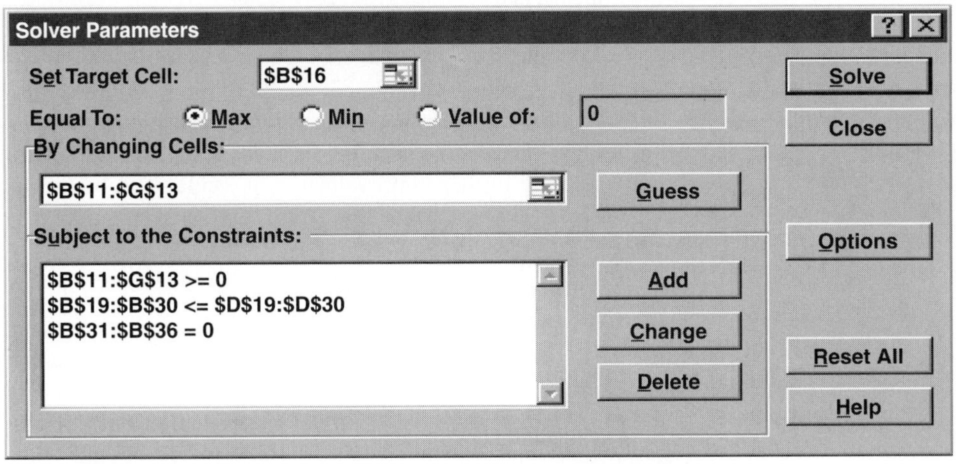

FIGURE 16.3

Add constraint dialog box in Excel

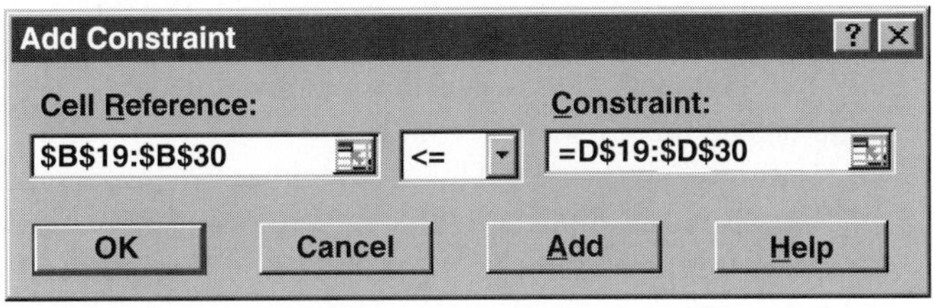

FIGURE 16.4

Setting Excel to use linear programming

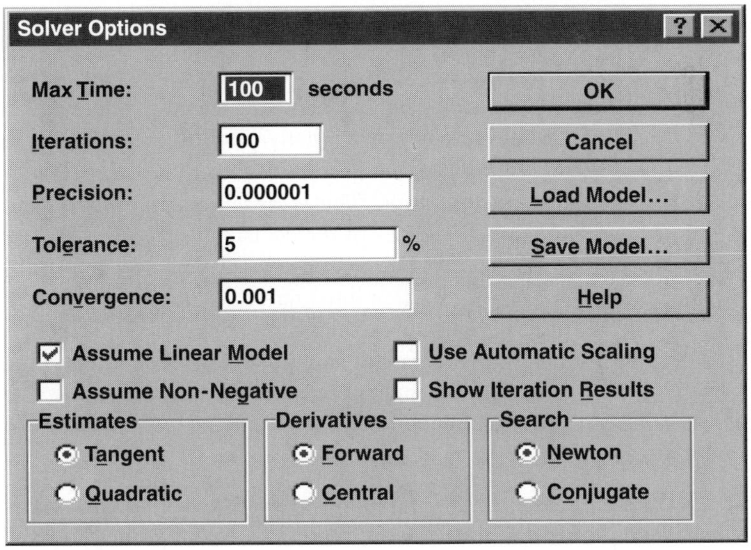

converge to an answer, the most likely reason is an error in one of the constraints. However, sometimes increasing the search time or reducing tolerances will fix the problem when the solver cannot find a solution. The reader should consult the Excel manual for more detailed documentation on this and other features, as well as information on upgrades that may have occurred since this writing. Choosing the **OK** button returns us to the original form.

Once we have done all this, we are ready to run the model by choosing the **Solve** button. The program will pause to set up the problem in the proper format and then will go through a sequence of trial solutions (although not for long in such a small problem as this).

Basically, LP works by first finding a feasible solution—one that satisfies all the constraints—and then generating a succession of new solutions, each better than the last. When no further improvement is possible, it stops and the solution is optimal: It maximizes or minimizes the objective function. Appendix 16A provides background on how this process works.

The algorithm will stop with one of three answers:

1. *Could not find a feasible solution.* This probably means that the problem is infeasible; that is, there is no solution that satisfies all the constraints. This could be due to a typing error (e.g., a plus sign was incorrectly typed as a minus sign) or a real infeasibility (e.g., it is not possible to meet demand with capacity). Notice that by clever formulation, one can avoid having the algorithm terminate with this depressing message when real infeasibilities exist. For instance, in formulation (16.6) to (16.27), we did not force sales to be equal to demand. Since cumulative demand exceeds cumulative capacity, it is obvious that this would not have been feasible. By setting separate sales and production variables, we let the computer tell us where demand cannot be met. Many variations on this trick are possible.

2. *Does not converge.* This means either that the algorithm could not find an optimal solution within the allotted time (so increasing the time or decreasing the tolerances under the **Options** menu might help) or that the algorithm is able to continue finding better and better solutions indefinitely. This second possibility can occur when the problem is **unbounded:** The objective can be driven to infinity by letting some variables grow positive or negative without bound. Usually this is the result of a failure to properly constrain a decision variable. For instance, in the above model, if we forgot to specify that all decision variables must be nonnegative, then the model will be able to make the objective arbitrarily large by choosing negative values of $I_t, t = 1, \ldots, 6$. Of course, we do not generate revenue via negative inventory levels, so it is important that nonnegativity constraints be included to rule out this nonsensical behavior.[3]

3. *Found a solution.* This is the outcome we want. When it occurs, the program will write the optimal values of the decision variables, objective value, and constraints into the spreadsheet. Figure 16.5 shows the spreadsheet as modified by the LP algorithm. The program also offers three reports—Answer, Sensitivity, and Limits—which write information about the solution into other spreadsheets. For instance, highlighting the Answer report generates a spreadsheet with the information shown in Figures 16.6 and 16.7. Figure 16.8 contains some of the information contained in the report generated by choosing Sensitivity.

Now that we have generated a solution, let us interpret it. Both Figure 16.5—the final spreadsheet—and Figure 16.6 show the optimal decision variables. From these we see that it is not optimal to produce at full capacity in every period. Specifically, the solution calls for producing only 110 units in month 5 when capacity is 120. This might seem odd given that demand exceeds capacity. However, if we look more carefully, we see that cumulative demand for periods 1 to 4 is 440 units, while cumulative capacity

---

[3]We will show how to modify the formulation to allow for backordering, which is like allowing negative inventory positions, without this inappropriately affecting the objective function, later in this chapter.

FIGURE 16.5

Output spreadsheet for LP example

	A	B	C	D	E	F	G	H
1	Constants:							
2	r	10						
3	h	1						
4	I_0	0						
5	t	1	2	3	4	5	6	Total
6	c_t	100	100	100	120	120	120	660
7	d_t	80	100	120	140	90	140	670
8								
9	Variables:							
10	t	1	2	3	4	5	6	Total
11	X_t	100	100	100	120	110	120	650
12	S_t	80	100	120	120	90	140	650
13	I_t	20	20	0	0	20	0	60
14								
15	Objective:							
16	Net Profit:	$6,440			r*(S_1+S_2+S_3+S_4+S_5+S_6) - h*(I_1+I_2+I_3+I_4+I_5+I_6)			
17								
18	Constraints:							
19	S_1	80	<=	80	d_1			
20	S_2	100	<=	100	d_2			
21	S_3	120	<=	120	d_3			
22	S_4	120	<=	140	d_4			
23	S_5	90	<=	90	d_5			
24	S_6	140	<=	140	d_6			
25	X_1	100	<=	100	c_1			
26	X_2	100	<=	100	c_2			
27	X_3	100	<=	100	c_3			
28	X_4	120	<=	120	c_4			
29	X_5	110	<=	120	c_5			
30	X_6	120	<=	120	c_6			
31	I_1-I_0-X_1+S_1	0	=	0				
32	I_2-I_1-X_2+S_2	0	=	0				
33	I_3-I_2-X_3+S_3	0	=	0				
34	I_4-I_3-X_4+S_4	0	=	0				
35	I_5-I_4-X_5+S_5	0	=	0				
36	I_6-I_5-X_6+S_6	0	=	0				
37					Note: X_t, S_t and I_t must be >= 0			

for those periods is only 420 units. Thus, even when we run flat out for the first four months, we will fall short of meeting demand by 20 units. Demand in the final two months is only 230 units, while capacity is 240 units. Since our model does not permit backordering, it does not make sense to produce more than 230 units in months 5 and 6. Any extra units cannot be used to make up a previous shortfall.

Figure 16.7 gives more details on the constraints by showing which ones are **binding** or **tight** (i.e., equal to the right-hand side) and which ones are **nonbinding** or **slack,** and by how much. Most interesting are the constraints on sales, given in (16.7) to (16.12), and capacity, in (16.13) to (16.18). As we have already noted, the capacity constraint on X_5 is nonbinding. Since we only produce 110 units in month 5 and have capacity for 120, this constraint is slack by 10 units. This means that if we changed this constraint by a little (e.g., reduced capacity in month 5 from 120 to 119 units), it would not change the optimal solution at all.

In this same vein, all sales constraints are tight except that for S_4. Since sales are limited to 140, but optimal sales are 120, this constraint has slackness of 20 units. Again, if we were to change this sales constraint by a little (e.g., limit sales to 141 units), the optimal solution would remain the same.

In contrast with these slack constraints, consider a binding constraint. For instance, consider the capacity constraint on X_1, which is the seventh one shown in Figure 16.7. Since the model chooses production equal to capacity in month 1, this constraint is tight. If we were to change this constraint by increasing or decreasing capacity, the solution would change. If we **relax** the constraint by increasing capacity, say, to 101 units, then we will be able to satisfy an additional unit of demand and therefore the net profit will

FIGURE 16.6

Optimal values report for LP example

Microsoft Excel 5.0 Answer Report
Worksheet: [BASICAP.XLS]Figure 16.5
Report Created: 5/15/95 12:22

Target Cell (Max)

Cell	Name	Original Value	Final Value
$B$16	Net Profit	0	6440

Adjustable Cells

Cell	Name	Original Value	Final Value
$B$12	S_1	0	80
$C$12	S_2	0	100
$D$12	S_3	0	120
$E$12	S_4	0	120
$F$12	S_5	0	90
$G$12	S_6	0	140
$B$11	X_1	0	100
$C$11	X_2	0	100
$D$11	X_3	0	100
$E$11	X_4	0	120
$F$11	X_5	0	110
$G$11	X_6	0	120
$B$13	I_1	0	20
$C$13	I_2	0	20
$D$13	I_3	0	0
$E$13	I_4	0	0
$F$13	I_5	0	20
$G$13	I_6	0	0

FIGURE 16.7

Optimal constraint status for LP example

Microsoft Excel 5.0 Answer Report
Worksheet: [BASICAP.XLS]Figure 16.5
Report Created: 5/15/95 12:22

Constraints

Cell	Name	Cell Value	Formula	Status	Slack
$B$19	S_1	80	$B$19<=$D$19	Binding	0
$B$20	S_2	100	$B$20<=$D$20	Binding	0
$B$21	S_3	120	$B$21<=$D$21	Binding	0
$B$22	S_4	120	$B$22<=$D$22	Not Binding	20
$B$23	S_5	90	$B$23<=$D$23	Binding	0
$B$24	S_6	140	$B$24<=$D$24	Binding	0
$B$25	X_1	100	$B$25<=$D$25	Binding	0
$B$26	X_2	100	$B$26<=$D$26	Binding	0
$B$27	X_3	100	$B$27<=$D$27	Binding	0
$B$28	X_4	120	$B$28<=$D$28	Binding	0
$B$29	X_5	110	$B$29<=$D$29	Not Binding	10
$B$30	X_6	120	$B$30<=$D$30	Binding	0
$B$31	I_1-I_0-X_1+S_1	0	$B$31=0	Binding	0
$B$32	I_2-I_1-X_2+S_2	0	$B$32=0	Binding	0
$B$33	I_3-I_2-X_3+S_3	0	$B$33=0	Binding	0
$B$34	I_4-I_3-X_4+S_4	0	$B$34=0	Binding	0
$B$35	I_5-I_4-X_5+S_5	0	$B$35=0	Binding	0
$B$36	I_6-I_5-X_6+S_6	0	$B$36=0	Binding	0
$B$12	S_1	80	$B$12>=0	Not Binding	80
$C$12	S_2	100	$C$12>=0	Not Binding	100
$D$12	S_3	120	$D$12>=0	Not Binding	120
$E$12	S_4	120	$E$12>=0	Not Binding	120
$F$12	S_5	90	$F$12>=0	Not Binding	90
$G$12	S_6	140	$G$12>=0	Not Binding	140
$B$11	X_1	100	$B$11>=0	Not Binding	100
$C$11	X_2	100	$C$11>=0	Not Binding	100
$D$11	X_3	100	$D$11>=0	Not Binding	100
$E$11	X_4	120	$E$11>=0	Not Binding	120
$F$11	X_5	110	$F$11>=0	Not Binding	110
$G$11	X_6	120	$G$11>=0	Not Binding	120
$B$13	I_1	20	$B$13>=0	Not Binding	20
$C$13	I_2	20	$C$13>=0	Not Binding	20
$D$13	I_3	0	$D$13>=0	Binding	0
$E$13	I_4	0	$E$13>=0	Binding	0
$F$13	I_5	20	$F$13>=0	Not Binding	20
$G$13	I_6	0	$G$13>=0	Binding	0

increase. Since we will produce the extra item in month 1, hold it for three months to month 4 at a cost of $1 per month, and then sell it for $10, the overall increase in the objective from this change will be $10 - 3 = $7. Conversely, if we **tighten** the constraint by decreasing capacity, say to 99 units, then we will only be able to carry 19 units from month 1 to month 3 and will therefore lose one unit of demand in month 3. The loss in net profit from this unit will be $8 ($10 - $2 for two months' holding).

The sensitivity data generated by the LP algorithm shown in Figure 16.8 gives us more direct information on the sensitivity of the final solution to changes in the constraints. This report has a line for every constraint in the model and reports three important pieces of information:[4]

1. The **shadow price** represents the amount the optimal objective will be increased by a unit increase in the right-hand side of the constraint.

2. The **allowable increase** represents the amount by which the right-hand side can be increased before the shadow price no longer applies.

3. The **allowable decrease** represents the amount by which the right-hand side can be decreased before the shadow price no longer applies.

Appendix 16A gives a geometric explanation of how these numbers are computed.

---

[4]The report also contains sensitivity information about the coefficients in the objective function. See Appendix 16A for a discussion of this.

FIGURE 16.8

Sensitivity analysis for LP example

Microsoft Excel 5.0 Sensitivity Report
Worksheet: [BASICAP.XLS]Figure 16.5
Report Created: 5/15/95 12:22

Changing Cells

Cell	Name	Final Value	Reduced Cost	Objective Coefficient	Allowable Increase	Allowable Decrease
$B$12	S_1	80	0	10	1E+30	3
$C$12	S_2	100	0	10	1E+30	2
$D$12	S_3	120	0	10	1E+30	1
$E$12	S_4	120	0	10	1	7
$F$12	S_5	90	0	10	1E+30	10
$G$12	S_6	140	0	10	1E+30	9
$B$11	X_1	100	0	0	1E+30	7
$C$11	X_2	100	0	0	1E+30	8
$D$11	X_3	100	0	0	1E+30	9
$E$11	X_4	120	0	0	1E+30	10
$F$11	X_5	110	0	0	1	9
$G$11	X_6	120	0	0	1E+30	1
$B$13	I_1	20	0	-1	3	7
$C$13	I_2	20	0	-1	2	7
$D$13	I_3	0	0	-1	1	7
$E$13	I_4	0	-11	-1	11	1E+30
$F$13	I_5	20	0	-1	1	9
$G$13	I_6	0	-2	-1	2	1E+30

Constraints

Cell	Name	Final Value	Shadow Price	Constraint R.H. Side	Allowable Increase	Allowable Decrease
$B$19	S_1	80	3	80	0	20
$B$20	S_2	100	2	100	0	20
$B$21	S_3	120	1	120	0	20
$B$22	S_4	120	0	140	1E+30	20
$B$23	S_5	90	10	90	10	90
$B$24	S_6	140	9	140	10	20
$B$25	X_1	100	7	100	20	0
$B$26	X_2	100	8	100	20	0
$B$27	X_3	100	9	100	20	0
$B$28	X_4	120	10	120	20	120
$B$29	X_5	110	0	120	1E+30	10
$B$30	X_6	120	1	120	20	10
$B$31	I_1-I_0-X_1+S_1	0	7	0	20	0
$B$32	I_2-I_1-X_2+S_2	0	8	0	20	0
$B$33	I_3-I_2-X_3+S_3	0	9	0	20	0
$B$34	I_4-I_3-X_4+S_4	0	10	0	20	120
$B$35	I_5-I_4-X_5+S_5	0	0	0	110	10
$B$36	I_6-I_5-X_6+S_6	0	1	0	20	10

To see how these data are interpreted, consider the information in Figure 16.8 on the seventh line of the constraint section for the capacity constraint $X_1 \leq 100$. The shadow price is $7, which means that if the constraint is changed to $X_1 \leq 101$, net profit will increase by $7, precisely as we computed above. The allowable increase is 20 units, which means that each unit capacity increase in period 1 up to a total of 20 units increases net profit by $7. Therefore, an increase in capacity from 100 to 120 will increase net profit by $20 \times 7 = \$140$. Above 20 units, we will have satisfied all the lost demand in month 4, and therefore further increases will not improve profit. Thus, this constraint will become nonbinding once the right-hand side exceeds 120. Notice that the allowable decrease is zero for this constraint. What this means is that the shadow price of $7 is not valid for decreases in the right-hand side. As we computed above, the decrease in net profit from a unit decrease in the capacity in month 1 is $8. In general, we can only determine the impact of changes outside the allowable increase or decrease range by actually changing the constraints and rerunning the LP solver.

The above examples are illustrative of the following general behavior of linear programming models:

1. Changing the right-hand sides of nonbinding constraints by a small amount does not affect the optimal solution. The shadow price of a nonbinding constraint is always zero.

2. Increasing the right-hand side of a binding constraint will increase the objective by an amount equal to the shadow price times the size of the increase, provided that the increase is smaller than the allowable increase.

3. Decreasing the right-hand side of a binding constraint will decrease the objective by an amount equal to the shadow price times the size of the decrease, provided that the decrease is smaller than the allowable decrease.

4. Changes in the right-hand sides beyond the allowable increase or decrease range have an indeterminate effect and must be evaluated by resolving the modified model.

5. All these sensitivity results apply to changes in *one right-hand side variable at a time*. If multiple changes are made, the effects are not necessarily additive. Generally, multiple-variable sensitivity analysis must be done by resolving the model under the multiple changes.

16.3   Product Mix Planning

Now that we have set up the basic framework for formulating and solving aggregate planning problems, we can examine some commonly encountered situations. The first realistic aggregate planning issue we will consider is that of product mix planning. To do this, we need to extend the model of the previous section to consider multiple products explicitly. As mentioned previously, allowing multiple products raises the possibility of a "floating bottleneck." That is, if the different products require different amounts of processing time on the various workstations, then the workstation that is most heavily loaded during a period may well depend on the mix of products run during that period. If flexibility in the mix is possible, we can use the AP module to adjust the mix in accordance with available capacity. And if the mix is essentially fixed, we can use the AP module to identify bottlenecks.

16.3.1   Basic Model

We start with a direct extension of the previous single-product model in which demands are assumed fixed and the objective is to minimize the inventory carrying cost of meeting these demands. To do this, we introduce the following notation:

i = an index of product, $i = 1, \ldots, m$, so m represents total number of products

j = an index of workstation, $j = 1, \ldots, n$, so n represents total number of workstations

t = an index of period, $t = 1, \ldots, \bar{t}$, so $\bar{t}$ represents planning horizon

$\bar{d}_{it}$ = maximum demand for product i in period t

$\underline{d}_{it}$ = minimum sales[5] allows of product i in period t

a_{ij} = time required on workstation j to produce one unit of product i

c_{jt} = capacity of workstation j in period t in units consistent with those used to define a_{ij}

r_i = net profit from one unit of product i

h_i = cost[6] to hold one unit of product i for one period t

---

[5]This might represent firm commitments that we do not want the computer program to violate.

[6]It is common to set h_i equal to the raw materials cost of product i times a one-period interest rate to represent the opportunity cost of the money tied up in inventory; but it may make sense to use higher values to penalize inventory that causes long, uncompetitive cycle times.

X_{it} = amount of product i produced in period t

S_{it} = amount of product i sold in period t

I_{it} = inventory of product i at end of period t (I_{i0} is given as data)

Again, X_{it}, S_{it}, and I_{it} are decision variables, while the other symbols are constants representing input data. We can give a linear program formulation of the problem to maximize net profit minus inventory carrying cost subject to upper and lower bounds on sales and capacity constraints as

$$\text{Maximize} \qquad \sum_{t=1}^{\bar{t}} \sum_{i=1}^{m} r_i S_{it} - h_i I_{it} \tag{16.28}$$

Subject to:

$$\underline{d}_{it} \le S_{it} \le \bar{d}_{it} \qquad \text{for all } i, t \tag{16.29}$$

$$\sum_{i=1}^{m} a_{ij} X_{it} \le c_{jt} \qquad \text{for all } j, t \tag{16.30}$$

$$I_{it} = I_{it-1} + X_{it} - S_{it} \qquad \text{for all } i, t \tag{16.31}$$

$$X_{it}, S_{it}, I_{it} \ge 0 \qquad \text{for all } i, t \tag{16.32}$$

In comparison to the previous single-product model, we have adjusted constraints (16.29) to include lower, as well as upper, bounds on sales. For instance, the firm may have long-term contracts that obligate it to produce certain minimum amounts of certain products. Conversely, the market for some products may be limited. To maximize profit, the computer has incentive to set production so that all these constraints will be tight at their upper limits. However, this may not be possible due to capacity constraints (16.30). Notice that unlike in the previous formulation, we now have capacity constraints for each workstation in each period. By noting which of these constraints are tight, we can identify those resources that limit production. Constraints (16.31) are the multiproduct version of the balance equations, and constraints (16.32) are the usual nonnegativity constraints.

We can use LP (16.28)–(16.32) to obtain several pieces of information, including

1. **Demand feasibility.** We can determine whether a set of demands is capacity-feasible. If the constraint $S_{it} \le \bar{d}_{it}$ is tight, then the upper bound on demand $\bar{d}_{it}$ is feasible. If not, then it is capacity-infeasible. If demands given by the lower bounds on demand $\underline{d}_{it}$ are capacity-infeasible, then the computer program will return a "could not find a feasible solution" message and the user must make changes (e.g., reduce demands or increase capacity) in order to get a solution.

2. **Bottleneck locations.** Constraints (16.30) restrict production on each workstation in each period. By noting which of these constraints are binding, we can determine which workstations limit capacity in which periods. A workstation that is consistently binding in many periods is a clear bottleneck and requires close management attention.

3. **Product mix.** If we are unable, for capacity reasons, to attain all the upper bounds on demand, then the computer will reduce sales below their maximum for some products. It will try to maximize revenue by producing those products with high net profit, but because of the capacity constraints, this is not a simple matter, as we will see in the following example.

16.3.2   A Simple Example

Let us consider a simple product mix example that shows why one needs a formal optimization method instead of a simpler ad hoc approach for these problems. We simplify matters by assuming a planning horizon of only one period. While this is certainly not a realistic assumption in general, in situations where we know in advance that we will never carry inventory from one period to the next, solving separate one-period problems for each period *will* yield the optimal solution. For example, if demands and cost coefficients are constant from period to period, then there is no incentive to build up inventory and therefore this will be the case.

Consider a situation in which a firm produces two products, which we will call products 1 and 2. Table 16.1 gives descriptive data for these two products. In addition to the direct raw material costs associated with each product, we assume a $5,000 per week fixed cost for labor and capital. Furthermore, there are 2,400 minutes (five days per week, eight hours per day) of time available on workstations A to D. We assume that all these data are identical from week to week. Therefore, there is no reason to build inventory in one week to sell in a subsequent week. (If we can meet maximum demand this week with this week's production, then the same thing is possible next week.) Thus, we can restrict our attention to a single week, and the only issue is the appropriate amount of each product to produce.

A Cost Approach.   Let us begin by looking at this problem from a simple cost standpoint. Net profit per unit of product 1 sold is $45 ($90 − 45), while net profit per unit of product 2 sold is $60 ($100 − 40). This would seem to indicate that we should emphasize production of product 2. Ideally, we would like to produce 50 units of product 2 to meet maximum demand, but we must check the capacity of the four workstations to make sure this is possible. Since workstation B requires the most time to make a unit of product 2 (30 minutes) among the four workstations, this is the potential constraint. Producing 50 units of product 2 on workstation B will require

$$30 \text{ minutes per unit} \times 50 \text{ units} = 1,500 \text{ minutes}$$

This is less than the available 2,400 minutes on workstation B, so producing 50 units of product 2 is feasible.

Now we need to determine how many units of product 1 we can produce with the leftover capacity. The unused time on workstations A to D after subtracting the time to

---

TABLE 16.1   Input Data for Single-Period AP Example

Product	1	2
Selling price	$90	$100
Raw material cost	$45	$40
Maximum weekly sales	100	50
Minutes per unit on workstation A	15	10
Minutes per unit on workstation B	15	30
Minutes per unit on workstation C	15	5
Minutes per unit on workstation D	15	5

make 50 units of product 2 we compute as

$$2,400 - 10(50) = 1,900 \text{ minutes on workstation A}$$
$$2,400 - 30(50) = 900 \text{ minutes on workstation B}$$
$$2,400 - 5(50) = 2,150 \text{ minutes on workstation C}$$
$$2,400 - 5(50) = 2,150 \text{ minutes on workstation D}$$

Since one unit of product 1 requires 15 minutes of time on each of the four workstations, we can compute the maximum possible production of product 1 at each workstation by dividing the unused time by 15. Since workstation B has the least remaining time, it is the potential bottleneck. The maximum production of product 1 on workstation B (after subtracting the time to produce 50 units of product 2) is

$$\frac{900}{15} = 60$$

Thus, even though we can sell 100 units of product 1, we only have capacity for 60.

The weekly profit from making 60 units of product 1 and 50 units of product 2 is

$$\$45 \times 60 + \$60 \times 50 - \$5,000 = \$700$$

Is this the best we can do?

A Bottleneck Approach.   The preceding analysis is entirely premised on costs and considers capacity only as an afterthought. A better method might be to look at cost *and* capacity, by computing a ratio representing *profit per minute of bottleneck time used* for each product. This requires that we first identify the bottleneck, which we do by computing the minutes required on each workstation to satisfy maximum demand and seeing which machine is most overloaded.[7] This yields

$$15(100) + 10(50) = 2,000 \text{ minutes on workstation A}$$
$$15(100) + 30(50) = 3,000 \text{ minutes on workstation B}$$
$$15(100) + 5(50) = 1,750 \text{ minutes on workstation C}$$
$$15(100) + 5(50) = 1,750 \text{ minutes on workstation D}$$

Only workstation B requires more than the available 2,400 minutes, so we designate it the bottleneck. Hence, we would like to make the most profitable use of our time on workstation B. To determine which of the two products does this, we compute the ratio of net profit to minutes on workstation B as

$$\frac{\$45}{15} = \$3 \text{ per minute spent processing product 1}$$

$$\frac{\$60}{30} = \$2 \text{ per minute spent processing product 2}$$

This calculation indicates the reverse of our previous cost analysis. Each minute spent processing product 1 on workstation B nets us $3, as opposed to only $2 per minute spent on product 2. Therefore, we should emphasize production of product 1, not product 2. If we produce 100 units of product 1 (the maximum amount allowed by the demand constraint), then since all workstations require 15 min per unit of one, the unused time on each workstation is

$$2,400 - 15(100) = 900 \text{ minutes}$$

---

[7]The alert reader should be suspicious at this point, since we know that the identity of the "bottleneck" can depend on the product mix in a multiproduct case.

Then since workstation B is the slowest operation for producing product 2, this is what limits the amount we can produce. Each unit of product 2 requires 30 minutes on B; thus, we can produce

$$\frac{900}{30} = 30$$

units of product 2. The net profit from producing 100 units of product 1 and 30 units of product 2 is

$$\$45 \times 100 + \$60 \times 30 - \$5{,}000 = \$1{,}300$$

This is clearly better than the $700 we got from using our original analysis and, it turns out, is the best we can do. But will this method always work?

A Linear Programming Approach.   To answer the question of whether the previous "bottleneck ratio" method will always determine the optimal product mix, we consider a slightly modified version of the previous example, with data shown in Table 16.2. The only changes in these data relative to the previous example are that the processing time of product 2 on workstation B has been increased from 30 to 35 minutes and the processing times for products 1 and 2 on workstation D have been increased from 15 and 5 to 25 and 14, respectively.

To execute our ratio-based approach on this modified problem, we first check for the bottleneck by computing the minutes required on each workstation to meet maximum demand levels:

$$15(100) + 10(50) = 2{,}000 \text{ minutes on workstation A}$$

$$15(100) + 35(50) = 3{,}250 \text{ minutes on workstation B}$$

$$15(100) + 5(50) = 1{,}750 \text{ minutes on workstation C}$$

$$25(100) + 14(50) = 3{,}200 \text{ minutes on workstation D}$$

Workstation B is still the most heavily loaded resource, but now workstation D also exceeds the available 2,400 minutes.

If we designate workstation B as the bottleneck, then the ratio of net profit to minute of time on the bottleneck is

$$\frac{\$45}{15} = \$3.00 \text{ per minute spent processing product 1}$$

$$\frac{\$60}{35} = \$1.71 \text{ per minute spent processing product 2}$$

TABLE 16.2   Input Data for Modified Single-Period AP Example

Product	1	2
Selling price	$90	$100
Raw material cost	$45	$40
Maximum weekly sales	100	50
Minutes per unit on workstation A	15	10
Minutes per unit on workstation B	15	35
Minutes per unit on workstation C	15	5
Minutes per unit on workstation D	25	14

which, as before, indicates that we should produce as much product 1 as possible. However, now it is workstation D that is slowest for product 1. The maximum amount that can be produced on D in 2,400 minutes is

$$\frac{2,400}{25} = 96$$

Since 96 units of product 1 use up all available time on workstation D, we cannot produce any product 2. The net profit from this mix, therefore, is

$$\$45 \times 96 - \$5,000 = -\$680$$

This doesn't look very good—we are losing money. Moreover, while we used workstation B as our bottleneck for the purpose of computing our ratios, it was workstation D that determined how much product we could produce. Therefore, perhaps we should have designated workstation D as our bottleneck. If we do this, the ratio of net profit to minute of time on the bottleneck is

$$\frac{\$45}{25} = \$1.80 \text{ per minute spent processing product 1}$$

$$\frac{\$60}{14} = \$4.29 \text{ per minute spent processing product 2}$$

This indicates that it is more profitable to emphasize production of product 2. Since workstation B is slowest for product 2, we check its capacity to see how much product 2 we can produce, and we find

$$\frac{2,400}{35} = 68.57$$

Since this is greater than maximum demand, we should produce the maximum amount of product 2, which is 50 units. Now we compute the unused time on each machine as

$$2,400 - 10(50) = 1,900 \text{ minutes on workstation A}$$
$$2,400 - 35(50) = 650 \text{ minutes on workstation B}$$
$$2,400 - 5(50) = 2,150 \text{ minutes on workstation C}$$
$$2,400 - 14(50) = 1,700 \text{ minutes on workstation D}$$

Dividing the unused time by the minutes required to produce one unit of product 1 on each workstation gives us the maximum production of product 1 on each to be

$$\frac{1,900}{15} = 126.67 \text{ units on workstation A}$$

$$\frac{650}{15} = 43.33 \text{ units on workstation B}$$

$$\frac{2,150}{15} = 143.33 \text{ units on workstation C}$$

$$\frac{1,700}{25} = 68 \text{ units on workstation D}$$

Thus, workstation B limits production of product 1 to 43 units, so total net profit for this solution is

$$\$45 \times 43 + \$60 \times 50 - \$5,000 = -\$65$$

This is better, but we are still losing money. Is this the best we can do?

Finally, let's bring out our big gun (not really that big, since it is included in popular spreadsheet programs) and solve the problem with a linear programming package.

Letting X_1 (X_2) represent the quantity of product 1 (2) produced, we formulate a linear programming model to maximize profit subject to the demand and capacity constraints as

$$\text{Maximize} \quad 45X_1 + 60X_2 - 5{,}000 \tag{16.33}$$

Subject to:

$$X_1 \leq 100 \tag{16.34}$$

$$X_2 \leq 50 \tag{16.35}$$

$$15X_1 + 10X_2 \leq 2{,}400 \tag{16.36}$$

$$15X_1 + 35X_2 \leq 2{,}400 \tag{16.37}$$

$$15X_1 + 5X_2 \leq 2{,}400 \tag{16.38}$$

$$25X_1 + 14X_2 \leq 2{,}400 \tag{16.39}$$

Problem (16.33)–16.39) is trivial for any LP package. Ours (Excel) reports the solution to this problem to be

$$\text{Optimal objective} = \$557.94$$

$$X_1^* = 75.79$$

$$X_2^* = 36.09$$

Even if we round this solution down (which will certainly still be capacity-feasible, since we are reducing production amounts) to integer values

$$X_1^* = 75$$

$$X_2^* = 36$$

we get an objective of

$$\$45 \times 75 + \$60 \times 36 - \$5{,}000 = \$535$$

So making as much product 1 as possible and making as much product 2 as possible both result in negative profit. But making a *mix* of the two products generates positive profit!

The moral of this exercise is that even simple product mix problems can be subtle. No trick that chooses a dominant product or identifies the bottleneck before knowing the product mix can find the optimal solution in general. While such tricks can work for specific problems, they can result in extremely bad solutions in others. The only method guaranteed to solve these problems optimally is an exact algorithm such as those used in linear programming packages. Given the speed, power, and user-friendliness of modern LP packages, one should have a very good reason to forsake LP for an approximate method.

16.3.3   Extensions to the Basic Model

A host of variations on the basic problem given in formulation (16.28)–(16.32) are possible. We discuss a few of these next; the reader is asked to think of others in the problems at chapter's end.

Other Resource Constraints.   Formulation (16.28)–(16.32) contains capacity constraints for the workstations, but not for other resources, such as people, raw materials, and transport devices. In some systems, these may be important determinants of overall capacity and therefore should be included in the AP module.

Generically, if we let

$b_{ij} =$ units of resource j required per unit of product i

$k_{jt} =$ number of units of resource j available in period t

$X_{it} =$ amount of product i produced in period t

we can express the capacity constraint on resource j in period t as

$$\sum_{i=1}^{m} b_{ij} X_{it} \leq k_{jt} \tag{16.40}$$

Notice that b_{ij} and k_{jt} are the nonworkstation analogs to a_{ij} and c_{jt} in formulation (16.28)–(16.32).

As a specific example, suppose an inspector must check products 1, 2, and 3, which require 1, 2, and 1.5 hours, respectively, per unit to inspect. If the inspector is available a total of 160 hours per month, then the constraint on this person's time in month t can be represented as

$$X_{1t} + 2X_{2t} + 1.5X_{3t} \leq 160$$

If this constraint is binding in the optimal solution, it means that inspector time is a bottleneck and perhaps something should be reorganized to remove this bottleneck. (The plant could provide help for the inspector, simplify the inspection procedure to speed it up, or use quality-at-the-source inspections by the workstation operators to eliminate the need for the extra inspection step.)

As a second example, suppose a firm makes four different models of circuit board, all of which require one unit of a particular component. The component contains leading-edge technology and is in short supply. If k_t represents the total number of these components that can be made available in period t, then the constraint represented by component availability in each period t can be expressed as

$$X_{1t} + X_{2t} + X_{3t} + X_{4t} \leq k_t$$

Many other resource constraints can be represented in analogous fashion.

Utilization Matching.   As our discussion so far shows, it is straightforward to model capacity constraints in LP formulations of AP problems. However, we must be careful about how we use these constraints in actual practice, for two reasons.

1. *Low-level complexity.* An AP module will necessarily gloss over details that can cause inefficiency in the short term. For instance, in the product mix example of the previous section, we assumed that it was possible to run the four machines 2,400 minutes per week. However, from our factory physics discussions of Part II, we know that it is virtually impossible to avoid some idle time on machines. Any source of randomness (machine failures, setups, errors in the scheduling process, etc.) can diminish utilization. While we cannot incorporate these directly in the AP model, we can account for their aggregate effect on utilization.

2. *Production control decisions.* As we noted in Chapter 13, it may be economically attractive to set the production quota below full average capacity, in order to achieve predictable customer service without excessive overtime costs. If the quota-setting module indicates that we should run at less than full utilization, we should include this fact in the aggregate planning module in order to maintain consistency.

These considerations may make it attractive to plan for production levels below full capacity. Although the decision of how close to capacity to run can be tricky, the mechanics of reducing capacity in the AP model are simple. If the c_{jt} parameters represent practical estimates of realistic full capacity of workstation j in period t, adjusted for setups, worker breaks, machine failures, and other reasonable detractors, then we can simply deflate capacity by multiplying these by a constant factor. For instance, if either historical experience or the quote-setting module indicates that it is reasonable to run at a fraction q of full capacity, then we can replace constraints (16.30) in LP (16.28)–(16.32) by

$$\sum_{i=t}^{m} a_{ij} X_{it} \leq q c_{jt} \qquad \text{for all } j, t$$

The result will be that a binding capacity constraint will occur whenever a workstation is loaded to $100q$ percent of capacity in a period.

Backorders. In LP (16.28)–(16.32), we forced inventory to remain positive at all times. Implicitly, we were assuming that demands had to be met from inventory or lost; no backlogging of unmet demand was allowed. However, in many realistic situations, demand is not lost when not met on time. Customers expect to receive their orders even if they are late. Moreover, it is important to remember that aggregate planning is a long-term planning function. Just because the model says a particular order will be late, that does not mean that this must be so in practice. If the model predicts that an order due nine months from now will be backlogged, there may be ample time to renegotiate the due date. For that matter, the demand may really be only a forecast, to which a firm customer due date has not yet been attached. With this in mind, it makes sense to think of the aggregate planning module as a tool for reconciling projected demands with available capacity. By using it to identify problems that are far in the future, we can address them while there is still time to do something about them.

We can easily modify LP (16.28)–(16.32) to permit backordering as follows:

$$\text{Maximize} \qquad \sum_{t=1}^{\bar{t}} r_i S_{it} - h_i I_{it}^+ - \pi_{it}^- \qquad (16.41)$$

Subject to:

$$\underline{d}_{it} \leq S_{it} \leq \bar{d}_{it} \qquad \text{for all } i, t \qquad (16.42)$$

$$\sum_{i=1}^{m} a_{ij} X_{it} \leq c_{jt} \qquad \text{for all } j, t \qquad (16.43)$$

$$I_{it} = I_{it-1} + X_{it} - S_{it} \qquad \text{for all } i, t \qquad (16.44)$$

$$I_{it} = I_{it}^+ - I_{it}^- \qquad \text{for all } i, t \qquad (16.45)$$

$$X_{it}, S_{it}, I_{it}^+, I_{it}^- \geq 0 \qquad \text{for all } i, t \qquad (16.46)$$

The main change was to redefine the inventory variable I_{it} as the difference $I_{it}^+ - I_{it}^-$, where I_{it}^+ represents the inventory of product i carried from period t to $t + 1$ and I_{it}^- represents the number of backorders carried from period t to $t + 1$. Both I_{it}^+ and I_{it}^- must be nonnegative. However, I_{it} can be either positive or negative, and so we refer to it as the **inventory position** of product i in period t. A positive inventory position indicates on-hand inventory, while a negative inventory position indicates outstanding backorders. The coefficient π_i is the backorder analog to the holding cost h_i and represents the penalty to carry one unit of product i on backorder for one period of time. Because both I_{it}^-

and I_{it}^+ appear in the objective with negative coefficients, the LP solver will never make both of them positive for the same period. This simply means that we won't both carry inventory and incur a backorder penalty in the same period.

In terms of modeling, the most troublesome parameters in this formulation are the backorder penalty coefficients π_i. What is the cost of being late by one period on one unit of product i? For that matter, why should the lateness penalty be linear in the number of periods late or the number of units that are late? Clearly, asking someone in the organization for these numbers is out of the question. Therefore, one should view this type of model as a tool for generating various long-term production plans. By increasing or decreasing the π_i coefficients relative to the h_i coefficients, the analyst can increase or decrease the relative penalty associated with backlogging. High π_i values tend to force the model to build up inventory to meet surges in demand, while low π_i values tend to allow the model to be late on satisfying some demands that occur during peak periods. By generating both types of plans, the user can get an idea of what options are feasible and select among them.

To accomplish this, we need not get overly fine with the selection of cost coefficients. We could set them with the simple equations

$$h_i = \alpha p_i \tag{16.47}$$

$$\pi_i = \beta \tag{16.48}$$

where α represents the one-period interest rate, suitably inflated to penalize uncompetitive cycle times caused by excess inventory, and p_i represents the raw materials cost of one unit of product i, so that αp_i represents the interest lost on the money tied up by holding one unit of product i in inventory. Analogously, β represents a (somewhat artificial) cost per period of delay on any product. The assumption here is that the true cost of being late (expediting costs, lost customer goodwill, lost future orders, etc.) is independent of the cost or price of the product. If Equations (16.47) and (16.48) are valid, then the user can fix α and generate many different production plans by varying the single parameter β.

Overtime.   The previous representations of capacity assume each workstation is available a fixed amount of time in each period. Of course, in many systems there is the possibility of increasing the time via the use of overtime. Although we will treat overtime in greater detail in our upcoming discussion of workforce planning, it makes sense to note quickly that it is a simple matter to represent the option of overtime in a product mix model, even when labor is not being considered explicitly.

To do this, let

l'_j = cost of one hour of overtime at workstation j; a cost parameter

O_{jt} = overtime at workstation j in period t in hours; a decision variable

We can modify LP (16.41)–(16.46) to allow overtime at each workstation as follows:

$$\text{Maximize} \quad \sum_{t=1}^{\bar{t}} \{ r_i S_{it} - h_i I_{it}^+ - \pi_i I_{it}^- - \sum_{j=1}^{n} l'_j O_{jt} \} \tag{16.49}$$

Subject to:

$$\underline{d}_{it} \le S_{it} \le \bar{d}_{it} \qquad \text{for all } i, t \tag{16.50}$$

$$\sum_{i=1}^{m} a_{ij} X_{it} \le c_{jt} + O_{jt} \qquad \text{for all } j, t \tag{16.51}$$

$$I_{it} = I_{it-1} + X_{it} - S_{it} \qquad \text{for all } i, t \tag{16.52}$$

$$I_{it} = I_{it}^+ - I_{it}^- \qquad \text{for all } i, t \qquad (16.53)$$

$$X_{it}, S_{it}, I_{it}^+, I_{it}^- O_{jt} \geq 0 \qquad \text{for all } i, j, t \qquad (16.54)$$

The two changes we have made to LP (16.41)–(16.46) were to

1. Subtract the cost of overtime at stations $1, \ldots, n$, which is $\sum_{t=1}^{\bar{t}} \sum_{j=1}^{n} l_j' O_{jt}$, from the objective function.

2. Add the hours of overtime scheduled at station j in period t, denoted by O_{jt}, to the capacity of this resource c_{jt} in constraints (16.51).

It is natural to include both backlogging and overtime in the same model, since these are both ways of addressing capacity problems. In LP (16.49)–(16.54), the computer has the option of being late in meeting demand (backlogging) or increasing capacity via overtime. The specific combination it chooses depends on the relative cost of back-ordering (π_i) and overtime (l_j'). By varying these cost coefficients, the user can generate a range of production plans.

Yield Loss. In systems where product is scrapped at various points in the line due to quality problems, we must release extra material into the system to compensate for these losses. The result is that workstations upstream from points of yield loss are more heavily utilized than if there were no yield loss (because they must produce the extra material that will ultimately be scrapped). Therefore, to assess accurately the feasibility of a particular demand profile relative to capacity, we must consider yield loss in the aggregate planning module in systems where scrap is an issue.

We illustrate the basic effect of yield loss in Figure 16.9. In this simple line, α, β, and γ represent the fraction of product that is lost to scrap at workstations A, B, and C, respectively. If we require d units of product to come out of station C, then, on average, we will have to release $d/(1-\gamma)$ units into station C. To get $d/(1-\gamma)$ units out of station B, we will have to release $d/[(1-\beta)(1-\gamma)]$ units into B on average. Finally, to get the needed $d/[(1-\beta)(1-\gamma)]$ out of B, we will have to release $d/[(1-\alpha)(1-\beta)(1-\gamma)]$ units into A.

We can generalize the specific example of Figure 16.9 by defining

$y_{ij} = $ cumulative yield from station j onward (including station j) for product i

If we want to get d units of product i out of the end of the line on average, then we must release

$$\frac{d}{y_{ij}} \qquad (16.55)$$

units of i into station j. These values can easily be computed in the manner used for the example in Figure 16.9 and updated in a spreadsheet or database as a function of the estimated yield loss at each station.

Using Equation (16.55) to adjust the production amounts X_{it} in the manner illustrated in Figure 16.9, we can modify the LP formulation (16.28)–(16.32) to consider

FIGURE 16.9

Yield loss in a three-station line

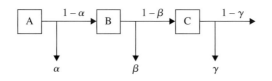

yield loss as follows:

$$\text{Maximize} \qquad \sum_{t=1}^{\bar{t}} r_i S_{it} - h_i I_{it} \qquad\qquad (16.56)$$

Subject to:

$$\underline{d}_{it} \le S_{it} \le \bar{d}_{it} \qquad \text{for all } i, t \qquad (16.57)$$

$$\sum_{i=1}^{m} \frac{a_{ij} X_{it}}{y_{ij}} \le c_{jt} \qquad \text{for all } j, t \qquad (16.58)$$

$$I_{it} = I_{it-1} + X_{it} - S_{it} \qquad \text{for all } i, t \qquad (16.59)$$

$$X_{it}, S_{it}, I_{it} \ge 0 \qquad \text{for all } i, t \qquad (16.60)$$

As one would expect, the net effect of this change is to reduce the effective capacity of workstations, particularly those at the beginning of the line. By altering the y_{ij} values (or better yet, the individual yields that make up the y_{ij} values), the planner can get a feel for the sensitivity of the system to improvements in yields. Again as one would intuitively expect, the impact of reducing the scrap rate toward the end of the line is frequently much larger than that of reducing scrap toward the beginning of the line. Obviously, scrapping product late in the process is very costly and should be avoided wherever possible. If better process control and quality assurance in the front of the line can reduce scrap later, this is probably a sound policy. An aggregate planning module like that given in LP (16.56)–(16.60) is one way to get a sense of the economic and logistic impact of such a policy.

16.4   Workforce Planning

In systems where the workload is subject to variation, due to either a changing workforce size or overtime load, it may make sense to consider the aggregate planning (AP) and workforce planning (WP) modules in tandem. Questions of how and when to resize the labor pool or whether to use overtime instead of workforce additions can be posed in the context of a linear programming formulation to support both modules.

16.4.1   An LP Model

To illustrate how an LP model can help address the workforce-resizing and overtime allocation questions, we will consider a simple single-product model. In systems where product routings and processing times are either almost identical, so that products can be aggregated into a single product, or entirely separate, so that routings can be analyzed separately, the single-product model can be reasonable. In a system where bottleneck identification is complicated by different processing times and interconnected routings, a planner would most likely need an explicit multiproduct model. This involves a straightforward integration of a product mix model, like those we discussed earlier, with a workforce-planning model like that presented next.

We introduce the following notation, paralleling that which we have used up to now, with a few additions to address the workforce issues.

$j = $ an index of workstation, $j = 1, \ldots, n$, so n represents total number of workstations

$t = $ an index of period, $t = 1, \ldots, \bar{t}$, so $\bar{t}$ represents planning horizon

$$\bar{d}_t = \text{maximum demand in period } t$$

$$\underline{d}_t = \text{minimum sales allowed in period } t$$

$$a_j = \text{time required on workstation } j \text{ to produce one unit of product}$$

$$b = \text{number of worker-hours required to produce one unit of product}$$

$$c_{jt} = \text{capacity of workstation } j \text{ in period } t$$

$$r = \text{net profit per unit of product sold}$$

$$h = \text{cost to hold one unit of product for one period}$$

$$l = \text{cost of regular time in dollars per worker-hour}$$

$$l' = \text{cost of overtime in dollars per worker-hour}$$

$$e = \text{cost to increase workforce by one worker-hour per period}$$

$$e' = \text{cost to decrease workforce by one worker-hour per period}$$

$$X_t = \text{amount produced in period } t$$

$$S_t = \text{amount sold in period } t$$

$$I_t = \text{inventory at end of } t \; (I_0 \text{ is given as data})$$

$$W_t = \text{workforce in period } t \text{ in worker-hours of regular time}$$
$$(W_0 \text{ is given as data})$$

$$H_t = \text{increase (hires) in workforce from period } t - 1 \text{ to } t \text{ in worker-hours}$$

$$F_t = \text{decrease (fires) in workforce from period } t - 1 \text{ to } t \text{ in worker-hours}$$

$$O_t = \text{overtime in period } t \text{ in hours}$$

We now have several new parameters and decision variables for representing the workforce considerations. First, we need b, the labor content of one unit of product, in order to relate workforce requirements to production needs. Once the model has used this parameter to determine the number of labor hours required in a given month, it has two options for meeting this requirement. Either it can schedule overtime, using the variable O_t and incurring cost at rate l'_t, or it can resize the workforce, using variables H_t and F_t and incurring a cost of e (e') for every worker added (laid off).

To model this planning problem as an LP, we will need to make the assumption that the cost of worker additions or deletions is linear in the number of workers added or deleted; that is, it costs twice as much to add (delete) two workers as it does to add (delete) one. Here we are assuming that e is an estimate of the hiring, training, outfitting, and lost productivity costs associated with bringing on a new worker. Similarly, e' represents the severance pay, unemployment costs, and so on associated with letting a worker go.

Of course, in reality, these workforce-related costs may not be linear. The training cost per worker may be less for a group than for an individual, since a single instructor can train many workers for roughly the same cost as a single one. On the other hand, the plant disruption and productivity falloff from introducing many new workers may be much more severe than those from introducing a single worker. Although one can use more sophisticated models to consider such sources of nonlinearity, we will stick with an LP model, keeping in mind that we are capturing general effects rather than elaborate details. Given that the AP and WP modules are used for long-term general planning purposes and rely on speculative forecasted data (e.g., of future demand), this is probably a reasonable choice for most applications.

We can write the LP formulation of the problem to maximize net profit, including labor, overtime, holding, and hiring/firing costs, subject to constraints on sales and capacity, as

$$\text{Maximize} \quad \sum_{t=1}^{\bar{t}} \{r S_t - h I_t - l W_t - l' O_t - e H_t - e' F_t\} \tag{16.61}$$

Subject to:

$$\underline{d}_t \le S_t \le \bar{d}_t \qquad\qquad \text{for all } t \quad (16.62)$$

$$a_j X_t \le c_{jt} \qquad\qquad \text{for all } j, t \ \ (16.63)$$

$$I_t = I_{t-1} + X_t - S_t \qquad\qquad \text{for all } t \quad (16.64)$$

$$W_t = W_{t-1} + H_t - F_t \qquad\qquad \text{for all } t \quad (16.65)$$

$$b X_t \le W_t + O_t \qquad\qquad \text{for all } t \quad (16.66)$$

$$X_t, S_t, I_t, O_t, W_t, H_t, F_t \ge 0 \qquad\qquad \text{for all } t \quad (16.67)$$

The objective function in formulation (16.61) computes profit as the difference between net revenue and inventory carrying costs, wages (regular and overtime), and workforce increase/decrease costs. Constraints (16.62) are the usual bounds on sales. Constraints (16.63) are capacity constraints for each workstation. Constraints (16.64) are the usual inventory balance equations. Constraints (16.65) and (16.66) are new to this formulation. Constraints (16.65) define the variables W_t, $t = 1, \ldots, \bar{t}$, to represent the size of the workforce in period t in units of worker-hours. Constraints (16.66) constrain the worker-hours required to produce X_t, given by $b X_t$, to be less than or equal to the sum of regular time plus overtime, namely, $W_t + O_t$. Finally, constraints (16.67) ensure that production, sales, inventory, overtime, workforce size, and labor increases/decreases are all nonnegative. The fact that $I_t \ge 0$ implies no backlogging, but we could easily modify this model to account for backlogging in a manner like that used in LP (16.41)–(16.46).

16.4.2 A Combined AP/WP Example

To make LP (16.61)–(16.67) concrete and to give a flavor for the manner in which modeling, analysis, and decision making interact, we consider the example presented in the spreadsheet of Figure 16.10. This represents an AP problem for a single product with unit net revenue of $1,000 over a 12-month planning horizon. We assume that each worker works 168 hours per month and that there are 15 workers in the system at the beginning of the planning horizon. Hence, the total number of labor hours available at the start of the problem is

$$W_0 = 15 \times 168 = 2{,}520$$

There is no inventory in the system at the start, so $I_0 = 0$.

The cost parameters are estimated as follows. Monthly holding cost is $10 per unit. Regular time labor (with benefits) costs $35 per hour. Overtime is paid at time-and-a-half, which is equal to $52.50 per hour. It costs roughly $2,500 to hire and train a new worker. Since this worker will account for 168 hours per month, the cost in terms of dollars per worker-hour is

$$\frac{\$2{,}500}{168} = \$14.88 \approx \$15 \text{ per hour}$$

Since this number is only a rough approximation, we will round to an even $15. Similarly, we estimate the cost to lay off a worker to be about $1,500, so the cost per hour of

FIGURE 16.10

Initial spreadsheet for workforce planning example

	A	B	C	D	E	F	G	H	I	J	K	L	M
1	Parameters:												
2	r	1000											
3	h	10											
4	l	35											
5	l'	52.5											
6	e	15											
7	e'	9											
8	b	12											
9	I_0	0											
10	W_0	2520											
11	t	1	2	3	4	5	6	7	8	9	10	11	12
12	d_t	200	220	230	300	400	450	320	180	170	170	160	180
13													
14	Decision Variables:												
15	t	1	2	3	4	5	6	7	8	9	10	11	12
16	Xt	0.00	0.00	0.00	0.00	0.00	0.00	0.00	0.00	0.00	0.00	0.00	0.00
17	Wt	0.00	0.00	0.00	0.00	0.00	0.00	0.00	0.00	0.00	0.00	0.00	0.00
18	Ht	0.00	0.00	0.00	0.00	0.00	0.00	0.00	0.00	0.00	0.00	0.00	0.00
19	Ft	0.00	0.00	0.00	0.00	0.00	0.00	0.00	0.00	0.00	0.00	0.00	0.00
20	It	0.00	0.00	0.00	0.00	0.00	0.00	0.00	0.00	0.00	0.00	0.00	0.00
21	Ot	0.00	0.00	0.00	0.00	0.00	0.00	0.00	0.00	0.00	0.00	0.00	0.00
22													
23	Objective:												
24	Profit:	$2,980,600.00											
25													
26	Constraints:												
27	I1-I0-X1	0.00	=	-200	-d_1								
28	I2-I1-X2	0.00	=	-220	-d_2								
29	I3-I2-X3	0.00	=	-230	-d_3								
30	I4-I3-X4	0.00	=	-300	-d_4								
31	I5-I4-X5	0.00	=	-400	-d_5								
32	I6-I5-X6	0.00	=	-450	-d_6								
33	I7-I6-X7	0.00	=	-320	-d_7								
34	I8-I7-X8	0.00	=	-180	-d_8								
35	I9-I8-X9	0.00	=	-170	-d_9								
36	I10-I9-X10	0.00	=	-170	-d_10								
37	I11-I10-X11	0.00	=	-160	-d_11								
38	I12-I11-X12	0.00	=	-180	-d_12								
39	W1-W0-H1+F1	-2520.00	=	0									
40	W2-W1-H2+F2	0.00	=	0									
41	W3-W2-H3+F3	0.00	=	0									
42	W4-W3-H4+F4	0.00	=	0									
43	W5-W4-H5+F5	0.00	=	0									
44	W6-W5-H6+F6	0.00	=	0									
45	W7-W6-H7+F7	0.00	=	0									
46	W8-W7-H8+F8	0.00	=	0									
47	W9-W8-H9+F9	0.00	=	0									
48	W10-W9-H10+F10	0.00	=	0									
49	W11-W10-H11+F11	0.00	=	0									
50	W12-W11-H12+F12	0.00	=	0									
51	bX1-W1-O1	0.00	<=	0									
52	bX2-W2-O2	0.00	<=	0									
53	bX3-W3-O3	0.00	<=	0									
54	bX4-W4-O4	0.00	<=	0									
55	bX5-W5-O5	0.00	<=	0									
56	bX6-W6-O6	0.00	<=	0									
57	bX7-W7-O7	0.00	<=	0									
58	bX8-W8-O8	0.00	<=	0									
59	bX9-W9-O9	0.00	<=	0									
60	bX10-W10-O10	0.00	<=	0									
61	bX11-W11-O11	0.00	<=	0									
62	bX12-W12-O12	0.00	<=	0									
63		*Note: All decision variables must be >= 0*											

reduction in the monthly workforce is

$$\frac{\$1,500}{168} = \$8.93 \approx \$9 \text{ per hour}$$

Again, we will use the rounded value of $9, since data are rough.

Notice that the projected demands (d_t) in the spreadsheet have a seasonal pattern to them, building to a peak in months 5 and 6, and tapering off thereafter. We will assume that backordering is not an option and that demands must be met, so the main issue will be how to do this.

Let us begin by expressing LP (16.61)–(16.67) in concrete terms for this problem. Because we are assuming that demands are met, we set $S_t = d_t$, which eliminates the need for separate sales variables S_t and sales constraints (16.62). Furthermore, to keep things simple, we will assume that the only capacity constraints are those posed by labor (i.e., it requires 12 hours of labor to produce each unit of product). No other machine or resource constraints need be considered. Thus we can omit constraints (16.63). Under these assumptions, the resulting LP formulation is

$$\text{Maximize} \quad 1{,}000(d_1 + \cdots + d_{12}) - 10(I_1 + \cdots + I_{12})$$
$$-35(W_1 + \cdots + W_{12}) - 52.5(O_1 + \cdots + O_{12})$$
$$-15(H_1 + \cdots + H_{12}) - 9(F_1 + \cdots + F_{12}) \tag{16.68}$$

Subject to:

$$I_1 - I_0 - X_1 = -d_1 \tag{16.69}$$
$$I_2 - I_1 - X_2 = -d_2 \tag{16.70}$$
$$I_3 - I_2 - X_3 = -d_3 \tag{16.71}$$
$$I_4 - I_3 - X_4 = -d_4 \tag{16.72}$$
$$I_5 - I_4 - X_5 = -d_5 \tag{16.73}$$
$$I_6 - I_5 - X_6 = -d_6 \tag{16.74}$$
$$I_7 - I_6 - X_7 = -d_7 \tag{16.75}$$
$$I_8 - I_7 - X_8 = -d_8 \tag{16.76}$$
$$I_9 - I_8 - X_9 = -d_9 \tag{16.77}$$
$$I_{10} - I_9 - X_{10} = -d_{10} \tag{16.78}$$
$$I_{11} - I_{10} - X_{11} = -d_{11} \tag{16.79}$$
$$I_{12} - I_{11} - X_{12} = -d_{12} \tag{16.80}$$
$$W_1 - H_1 + F_1 = 2{,}520 \tag{16.81}$$
$$W_2 - W_1 - H_2 + F_2 = 0 \tag{16.82}$$
$$W_3 - W_2 - H_3 + F_3 = 0 \tag{16.83}$$
$$W_4 - W_3 - H_4 + F_4 = 0 \tag{16.84}$$
$$W_5 - W_4 - H_5 + F_5 = 0 \tag{16.85}$$
$$W_6 - W_5 - H_6 + F_6 = 0 \tag{16.86}$$
$$W_7 - W_6 - H_7 + F_7 = 0 \tag{16.87}$$
$$W_8 - W_7 - H_8 + F_8 = 0 \tag{16.88}$$
$$W_9 - W_8 - H_9 + F_9 = 0 \tag{16.89}$$
$$W_{10} - W_9 - H_{10} + F_{10} = 0 \tag{16.90}$$
$$W_{11} - W_{10} - H_{11} + F_{11} = 0 \tag{16.91}$$
$$W_{12} - W_{11} - H_{12} + F_{12} = 0 \tag{16.92}$$
$$12X_1 - W_1 - O_1 \leq 0 \tag{16.93}$$
$$12X_2 - W_2 - O_2 \leq 0 \tag{16.94}$$
$$12X_3 - W_3 - O_3 \leq 0 \tag{16.95}$$

$$12X_4 - W_4 - O_4 \leq 0 \qquad (16.96)$$

$$12X_5 - W_5 - O_5 \leq 0 \qquad (16.97)$$

$$12X_6 - W_6 - O_6 \leq 0 \qquad (16.98)$$

$$12X_7 - W_7 - O_7 \leq 0 \qquad (16.99)$$

$$12X_8 - W_8 - O_8 \leq 0 \qquad (16.100)$$

$$12X_9 - W_9 - O_9 \leq 0 \qquad (16.101)$$

$$12X_{10} - W_{10} - O_{10} \leq 0 \qquad (16.102)$$

$$12X_{11} - W_{11} - O_{11} \leq 0 \qquad (16.103)$$

$$12X_{12} - W_{12} - O_{12} \leq 0 \qquad (16.104)$$

$$X_t, I_t, O_t, W_t, H_t, F_t \geq 0 \quad t = 1, \ldots, 12 \qquad (16.105)$$

Objective (16.68) is identical to objective (16.61), except that the S_t variables have been replaced with d_t constants.[8] Constraints (16.69)–(16.80) are the usual balance constraints. For instance, constraint (16.69) simply states that

$$I_1 = I_0 + X_1 - d_1$$

That is, inventory at the end of month 1 equals inventory at the end of month 0 (i.e., the beginning of the problem) plus production during month 1, minus sales (demand) in month 1. We have arranged these constraints so that all decision variables are on the left-hand side of the equality and constants (d_t) are on the right-hand side. This is often a convenient modeling convention, as we will see in our analysis.

Constraints (16.81) to (16.92) are the labor balance equations given in constraints (16.65) of our general formulation. For instance, constraint (16.81) represents the relation

$$W_1 = W_0 + H_1 - F_1$$

so that the workforce at the end of month 1 (in units of worker-hours) is equal to the workforce at the end of month 0, plus any additions in month 1, minus any subtractions in month 1.

Constraints (16.93) to (16.104) ensure that the labor content of the production plan does not exceed available labor, which can include overtime. For instance, constraint (16.93) can be written as

$$12X_1 \leq W_1 + O_1$$

In the spreadsheet shown in Figure 16.10, we have entered the decision variables X_t, W_t, H_t, F_t, I_t, and O_t into cells B16:M21. Using these variables and the various coefficients from the top of the spreadsheet, we express objective (16.68) as a formula in cell B24. Notice that this formula reports a value equal to the unit profit times total demand, or

$$1,000(200 + 220 + 230 + 300 + 400 + 450 + 320$$
$$+ 180 + 170 + 170 + 160 + 180) = \$2,980,000$$

because all other terms in the objective are zero when the decision variables are set at zero.

We enter formulas for the left-hand sides of constraints (16.69) to (16.80) in cells B27:B38, the left-hand sides of constraints (16.81) to (16.92) in cells B39:B50, and the

---

[8] Since the d_t values are fixed, the first term in the objective function is not a function of our decision variables and could be left out without affecting the solution. We have kept it in so that our model reports a sensible profit function.

left-hand sides of constraints (16.93) to (16.104) in cells B51:B62. Notice that many of these constraints are not satisfied when all decision variables are equal to zero. This is hardly surprising, since we cannot expect to earn revenues from sales of product we have not made.

A convenient aspect of using a spreadsheet for solving LP models is that it provides us with a mechanism for playing with the model to gain insight into its behavior. For instance, in the spreadsheet of Figure 16.11 we try a **chase solution** where we set production equal to demand ($X_t = d_t$) and leave $W_t = W_0$ in every period. Although this satisfies the inventory balance constraints in cells B27:B38, and the workforce balance constraints in cells B39:B50, it violates the labor content constraints in cells B52:B57. The reason, of course, is that the current workforce is not sufficient to meet demand without using overtime. We could try adding overtime by adjusting the O_t variables in cells B21:M21. However, searching around for an optimal solution can be difficult, particularly in large models. Therefore, we will let the LP solver in the software do the work for us.

Using the procedure we described earlier, we specify constraints (16.69) to (16.105) in our model and turn it loose. The result is the spreadsheet in Figure 16.12. Based on the costs we chose, it turns out to be optimal not to use any overtime. (Overtime costs $52.5 - 35 = 15.50$ per hour each month, while hiring a new worker costs only \$15 per hour as a one-time cost.) Instead, the model adds 1,114.29 hours to the workforce, which represents

$$\frac{1,114.29}{168} = 6.6$$

new workers. After the peak season of months 4 to 7, the solution calls for a reduction of $1,474.29 + 120 = 1,594.29$ hours, which implies laying off

$$\frac{1,594.29}{168} = 9.5$$

workers. Additionally, the solution involves building in excess of demand in months 1 to 4 and using this inventory to meet peak demand in months 5 to 7. The net profit resulting from this solution is \$1,687,337.14.

From a management standpoint, the planned layoffs in months 8 and 9 might be a problem. Although we have specified penalties for these layoffs, these penalties are highly speculative and may not accurately consider the long-term effects of hiring and firing on worker morale, productivity, and the firm's ability to recruit good people. Thus, it probably makes sense to carry our analysis further.

One approach we might consider would be to allow the model to hire but not fire workers. We can easily do this by eliminating the F_t variables or, since this requires fairly extensive changes in the spreadsheet, specifying additional constraints of the form

$$F_t = 0 \quad t = 1, \dots, 12$$

Rerunning the model with these additional constraints produces the spreadsheet in Figure 16.13. As we expect, this solution does not include any layoffs. Somewhat surprising, however, is the fact that it does not involve any new hires either (that is, $H_t = 0$ for every period). Instead of increasing the workforce size, the model has chosen to use overtime in months 3 to 7. Evidently, if we cannot fire workers, it is uneconomical to hire additional people.

However, when one looks more closely at the solution in Figure 16.13, a problem becomes evident. Overtime is too high. For instance, month 6 has more hours of overtime than hours of regular time! This means that our workforce of 15 people has

FIGURE 16.11

Infeasible "chase" solution

	A	B	C	D	E	F	G	H	I	J	K	L	M
1	Parameters:												
2	r	1000											
3	h	10											
4	l	35											
5	l′	52.5											
6	e	15											
7	e′	9											
8	b	12											
9	I_0	0											
10	W_0	2520											
11	t	1	2	3	4	5	6	7	8	9	10	11	12
12	d_t	200	220	230	300	400	450	320	180	170	170	160	180
13													
14	Decision Variables:												
15	t	1	2	3	4	5	6	7	8	9	10	11	12
16	Xt	200.00	220.00	230.00	300.00	400.00	450.00	320.00	180.00	170.00	170.00	160.00	180.00
17	Wt	2520.00	2520.00	2520.00	2520.00	2520.00	2520.00	2520.00	2520.00	2520.00	2520.00	2520.00	2520.00
18	Ht	0.00	0.00	0.00	0.00	0.00	0.00	0.00	0.00	0.00	0.00	0.00	0.00
19	Ft	0.00	0.00	0.00	0.00	0.00	0.00	0.00	0.00	0.00	0.00	0.00	0.00
20	It	0.00	0.00	0.00	0.00	0.00	0.00	0.00	0.00	0.00	0.00	0.00	0.00
21	Ot	10.00	0.00	0.00	0.00	0.00	0.00	0.00	0.00	0.00	0.00	0.00	0.00
22													
23	Objective:												
24	Profit:	$1,921,600.00											
25													
26	Constraints:												
27	I1-I0-X1	-200.00	=	-200	-d_1								
28	I2-I1-X2	-220.00	=	-220	-d_2								
29	I3-I2-X3	-230.00	=	-230	-d_3								
30	I4-I3-X4	-300.00	=	-300	-d_4								
31	I5-I4-X5	-400.00	=	-400	-d_5								
32	I6-I5-X6	-450.00	=	-450	-d_6								
33	I7-I6-X7	-320.00	=	-320	-d_7								
34	I8-I7-X8	-180.00	=	-180	-d_8								
35	I9-I8-X9	-170.00	=	-170	-d_9								
36	I10-I9-X10	-170.00	=	-170	-d_10								
37	I11-I10-X11	-160.00	=	-160	-d_11								
38	I12-I11-X12	-180.00	=	-180	-d_12								
39	W1-W0-H1+F1	0.00	=	0									
40	W2-W1-H2+F2	0.00	=	0									
41	W3-W2-H3+F3	0.00	=	0									
42	W4-W3-H4+F4	0.00	=	0									
43	W5-W4-H5+F5	0.00	=	0									
44	W6-W5-H6+F6	0.00	=	0									
45	W7-W6-H7+F7	0.00	=	0									
46	W8-W7-H8+F8	0.00	=	0									
47	W9-W8-H9+F9	0.00	=	0									
48	W10-W9-H10+F10	0.00	=	0									
49	W11-W10-H11+F11	0.00	=	0									
50	W12-W11-H12+F12	0.00	=	0									
51	bX1-W1-O1	-120.00	<=	0									
52	bX2-W2-O2	120.00	<=	0									
53	bX3-W3-O3	240.00	<=	0									
54	bX4-W4-O4	1080.00	<=	0									
55	bX5-W5-O5	2280.00	<=	0									
56	bX6-W6-O6	2880.00	<=	0									
57	bX7-W7-O7	1320.00	<=	0									
58	bX8-W8-O8	-360.00	<=	0									
59	bX9-W9-O9	-480.00	<=	0									
60	bX10-W10-O10	-480.00	<=	0									
61	bX11-W11-O11	-600.00	<=	0									
62	bX12-W12-O12	-360.00	<=	0									
63		Note: All decision variables must be >= 0											

2,880/15 = 192 hours of overtime in the month, or about 48 hours per week per worker. This is obviously excessive.

One way to eliminate this overtime problem is to add some more constraints. For instance, we might specify that overtime is not to exceed 20 percent of regular time. This would correspond to the entire workforce working an average of one full day of

FIGURE 16.12

LP optimal solution

	A	B	C	D	E	F	G	H	I	J	K	L	M
1	Parameters:												
2	r	1000											
3	h	10											
4	l	35											
5	l′	52.5											
6	e	15											
7	e′	9											
8	b	12											
9	I_0	0											
10	W_0	2520											
11	t	1	2	3	4	5	6	7	8	9	10	11	12
12	d_t	200	220	230	300	400	450	320	180	170	170	160	180
13													
14	Decision Variables:												
15	t	1	2	3	4	5	6	7	8	9	10	11	12
16	Xt	302.86	302.86	302.86	302.86	302.86	302.86	302.86	180.00	170.00	170.00	170.00	170.00
17	Wt	3634.29	3634.29	3634.29	3634.29	3634.29	3634.29	3634.29	2160.00	2040.00	2040.00	2040.00	2040.00
18	Ht	1114.29	0.00	0.00	0.00	0.00	0.00	0.00	0.00	0.00	0.00	0.00	0.00
19	Ft	0.00	0.00	0.00	0.00	0.00	0.00	0.00	1474.29	120.00	0.00	0.00	0.00
20	It	102.86	185.71	258.57	261.43	164.29	17.14	0.00	0.00	0.00	0.00	10.00	0.00
21	Ot	0.00	0.00	0.00	0.00	0.00	0.00	0.00	0.00	0.00	0.00	0.00	0.00
22													
23	Objective:												
24	Profit:	$1,687,337.14											
25													
26	Constraints:												
27	I1-I0-X1	-200.00	=	-200	-d_1								
28	I2-I1-X2	-220.00	=	-220	-d_2								
29	I3-I2-X3	-230.00	=	-230	-d_3								
30	I4-I3-X4	-300.00	=	-300	-d_4								
31	I5-I4-X5	-400.00	=	-400	-d_5								
32	I6-I5-X6	-450.00	=	-450	-d_6								
33	I7-I6-X7	-320.00	=	-320	-d_7								
34	I8-I7-X8	-180.00	=	-180	-d_8								
35	I9-I8-X9	-170.00	=	-170	-d_9								
36	I10-I9-X10	-170.00	=	-170	-d_10								
37	I11-I10-X11	-160.00	=	-160	-d_11								
38	I12-I11-X12	-180.00	=	-180	-d_12								
39	W1-W0-H1+F1	0.00	=	0									
40	W2-W1-H2+F2	0.00	=	0									
41	W3-W2-H3+F3	0.00	=	0									
42	W4-W3-H4+F4	0.00	=	0									
43	W5-W4-H5+F5	0.00	=	0									
44	W6-W5-H6+F6	0.00	=	0									
45	W7-W6-H7+F7	0.00	=	0									
46	W8-W7-H8+F8	0.00	=	0									
47	W9-W8-H9+F9	0.00	=	0									
48	W10-W9-H10+F10	0.00	=	0									
49	W11-W10-H11+F11	0.00	=	0									
50	W12-W11-H12+F12	0.00	=	0									
51	bX1-W1-O1	0.00	<=	0									
52	bX2-W2-O2	0.00	<=	0									
53	bX3-W3-O3	0.00	<=	0									
54	bX4-W4-O4	0.00	<=	0									
55	bX5-W5-O5	0.00	<=	0									
56	bX6-W6-O6	0.00	<=	0									
57	bX7-W7-O7	0.00	<=	0									
58	bX8-W8-O8	0.00	<=	0									
59	bX9-W9-O9	0.00	<=	0									
60	bX10-W10-O10	0.00	<=	0									
61	bX11-W11-O11	0.00	<=	0									
62	bX12-W12-O12	0.00	<=	0									
63		*Note: All decision variables must be >= 0*											

overtime per week in addition to the normal five-day workweek. We could do this by adding constraints of the form

$$O_t \leq 0.2W_t \qquad t = 1, \ldots, 12 \qquad (16.106)$$

Doing this to the spreadsheet of Figure 16.13 and resolving results in the spreadsheet

FIGURE 16.13

Optimal solution when $F_t = 0$

	A	B	C	D	E	F	G	H	I	J	K	L	M
1	Parameters:												
2	r	1000											
3	h	10											
4	l	35											
5	l′	52.5											
6	e	15											
7	e′	9											
8	b	12											
9	I_0	0											
10	W_0	2520											
11	t	1	2	3	4	5	6	7	8	9	10	11	12
12	d_t	200	220	230	300	400	450	320	180	170	170	160	180
13													
14	Decision Variables:												
15	t	1	2	3	4	5	6	7	8	9	10	11	12
16	Xt	210.00	210.00	230.00	300.00	400.00	450.00	320.00	180.00	170.00	170.00	160.00	180.00
17	Wt	2520.00	2520.00	2520.00	2520.00	2520.00	2520.00	2520.00	2520.00	2520.00	2520.00	2520.00	2520.00
18	Ht	0.00	0.00	0.00	0.00	0.00	0.00	0.00	0.00	0.00	0.00	0.00	0.00
19	Ft	0.00	0.00	0.00	0.00	0.00	0.00	0.00	0.00	0.00	0.00	0.00	0.00
20	It	10.00	0.00	0.00	0.00	0.00	0.00	0.00	0.00	0.00	0.00	0.00	0.00
21	Ot	0.00	0.00	240.00	1080.00	2280.00	2880.00	1320.00	0.00	0.00	0.00	0.00	0.00
22													
23	Objective:												
24	Profit:	$1,512,000.00											
25													
26	Constraints:												
27	I1-I0-X1	-200.00	=	-200	-d_1								
28	I2-I1-X2	-220.00	=	-220	-d_2								
29	I3-I2-X3	-230.00	=	-230	-d_3								
30	I4-I3-X4	-300.00	=	-300	-d_4								
31	I5-I4-X5	-400.00	=	-400	-d_5								
32	I6-I5-X6	-450.00	=	-450	-d_6								
33	I7-I6-X7	-320.00	=	-320	-d_7								
34	I8-I7-X8	-180.00	=	-180	-d_8								
35	I9-I8-X9	-170.00	=	-170	-d_9								
36	I10-I9-X10	-170.00	=	-170	-d_10								
37	I11-I10-X11	-160.00	=	-160	-d_11								
38	I12-I11-X12	-180.00	=	-180	-d_12								
39	W1-W0-H1+F1	0.00	=	0									
40	W2-W1-H2+F2	0.00	=	0									
41	W3-W2-H3+F3	0.00	=	0									
42	W4-W3-H4+F4	0.00	=	0									
43	W5-W4-H5+F5	0.00	=	0									
44	W6-W5-H6+F6	0.00	=	0									
45	W7-W6-H7+F7	0.00	=	0									
46	W8-W7-H8+F8	0.00	=	0									
47	W9-W8-H9+F9	0.00	=	0									
48	W10-W9-H10+F10	0.00	=	0									
49	W11-W10-H11+F11	0.00	=	0									
50	W12-W11-H12+F12	0.00	=	0									
51	bX1-W1-O1	0.00	<=	0									
52	bX2-W2-O2	0.00	<=	0									
53	bX3-W3-O3	0.00	<=	0									
54	bX4-W4-O4	0.00	<=	0									
55	bX5-W5-O5	0.00	<=	0									
56	bX6-W6-O6	0.00	<=	0									
57	bX7-W7-O7	0.00	<=	0									
58	bX8-W8-O8	-360.00	<=	0									
59	bX9-W9-O9	-480.00	<=	0									
60	bX10-W10-O10	-480.00	<=	0									
61	bX11-W11-O11	-600.00	<=	0									
62	bX12-W12-O12	-360.00	<=	0									
63		*Note: All decision variables must be >= 0*											

shown in Figure 16.14. The overtime limits have forced the model to resort to hiring. Since layoffs are still not allowed, the model hires only 508.57 hours worth of workers, or

$$\frac{508.57}{168} = 3$$

FIGURE 16.14

Optimal solution when $F_t = 0$ and $O_t \leq 0.2 W_t$

	A	B	C	D	E	F	G	H	I	J	K	L	M
1	Parameters:												
2	r	1000											
3	h	10											
4	l	35											
5	l'	52.5											
6	e	15											
7	e'	9											
8	b	12											
9	I_0	0											
10	W_0	2520											
11	t	1	2	3	4	5	6	7	8	9	10	11	12
12	d_t	200	220	230	300	400	450	320	180	170	170	160	180
13													
14	Decision Variables:												
15	t	1	2	3	4	5	6	7	8	9	10	11	12
16	Xt	302.86	302.86	302.86	302.86	302.86	302.86	302.86	180.00	170.00	170.00	160.00	180.00
17	Wt	3028.57	3028.57	3028.57	3028.57	3028.57	3028.57	3028.57	3028.57	3028.57	3028.57	3028.57	3028.57
18	Ht	508.57	0.00	0.00	0.00	0.00	0.00	0.00	0.00	0.00	0.00	0.00	0.00
19	Ft	0.00	0.00	0.00	0.00	0.00	0.00	0.00	0.00	0.00	0.00	0.00	0.00
20	It	102.86	185.71	258.57	261.43	164.29	17.14	0.00	0.00	0.00	0.00	0.00	0.00
21	Ot	605.71	605.71	605.71	605.71	605.71	605.71	605.71	0.00	0.00	0.00	0.00	0.00
22													
23	Objective:												
24	Profit:	$1,467,871.43											
25													
26	Constraints:												
27	I1-I0-X1	-200.00	=	-200	-d_1								
28	I2-I1-X2	-220.00	=	-220	-d_2								
29	I3-I2-X3	-230.00	=	-230	-d_3								
30	I4-I3-X4	-300.00	=	-300	-d_4								
31	I5-I4-X5	-400.00	=	-400	-d_5								
32	I6-I5-X6	-450.00	=	-450	-d_6								
33	I7-I6-X7	-320.00	=	-320	-d_7								
34	I8-I7-X8	-180.00	=	-180	-d_8								
35	I9-I8-X9	-170.00	=	-170	-d_9								
36	I10-I9-X10	-170.00	=	-170	-d_10								
37	I11-I10-X11	-160.00	=	-160	-d_11								
38	I12-I11-X12	-180.00	=	-180	-d_12								
39	W1-W0-H1+F1	0.00	=	0									
40	W2-W1-H2+F2	0.00	=	0									
41	W3-W2-H3+F3	0.00	=	0									
42	W4-W3-H4+F4	0.00	=	0									
43	W5-W4-H5+F5	0.00	=	0									
44	W6-W5-H6+F6	0.00	=	0									
45	W7-W6-H7+F7	0.00	=	0									
46	W8-W7-H8+F8	0.00	=	0									
47	W9-W8-H9+F9	0.00	=	0									
48	W10-W9-H10+F10	0.00	=	0									
49	W11-W10-H11+F11	0.00	=	0									
50	W12-W11-H12+F12	0.00	=	0									
51	bX1-W1-O1	0.00	<=	0									
52	bX2-W2-O2	0.00	<=	0									
53	bX3-W3-O3	0.00	<=	0									
54	bX4-W4-O4	0.00	<=	0									
55	bX5-W5-O5	0.00	<=	0									
56	bX6-W6-O6	0.00	<=	0									
57	bX7-W7-O7	0.00	<=	0									
58	bX8-W8-O8	-868.57	<=	0									
59	bX9-W9-O9	-988.57	<=	0									
60	bX10-W10-O10	-988.57	<=	0									
61	bX11-W11-O11	-1108.57	<=	0									
62	bX12-W12-O12	-868.57	<=	0									
63		*Note: All decision variables must be >= 0*											

new workers, as opposed to the 6.6 workers hired in the original solution in Figure 16.12. To attain the necessary production, the solution uses overtime in months 1 to 7. Notice that the amount of overtime used in these months is exactly 20 percent of regular time work hours, that is,

$$3,028.57 \times 0.2 = 605.71$$

What this means is that new constraints (16.106) are binding for periods 1 to 7, which we would be told explicitly if we printed out the sensitivity analysis reports generated by the LP solver. This implies that if it is possible to work more overtime in any of these months, we can improve the solution.

Notice that the net profit in the model of the spreadsheet shown in Figure 16.14 is $1,467,871.43, which is a 13 percent decrease over the original optimal solution of $1,687,337.14 in Figure 16.12. At first glance, it may appear that the policies of no layoffs and limits on overtime are expensive. On the other hand, it may really be telling us that our original estimates of the costs of hiring and firing were too low. If we were to increase these costs to represent, for example, long-term disruptions caused by labor changes, the optimal solution might be very much like the one arrived at in Figure 16.14.

16.4.3   Modeling Insights

In addition to providing a detailed example of a workforce formulation in LP (16.61)–(16.67), we hope that our discussion has helped the reader appreciate the following aspects of using an optimization model as the basis for an AP or WP module.

1. *Multiple modeling approaches.* There are often many ways to model a given problem, none of which is "correct" in any absolute sense. The key is to use cost coefficients and constraints to represent the main issues in a sensible way. In this example, we could have generated solutions without layoffs by either increasing the layoff penalty or placing constraints on the layoffs. Both approaches would achieve the same qualitative conclusions.

2. *Iterative model development.* Modeling and analysis almost never proceed in an ideal fashion in which the model is formulated, solved, and interpreted in a single pass. Often the solution from one version of the model suggests an alternate model. For instance, we had no way of knowing that eliminating layoffs would cause excessive overtime in the solution. We didn't know we would need constraints on the level of overtime until we saw the spreadsheet output of Figure 16.13.

16.5   Conclusions

In this chapter, we have given an overview of the issues involved in aggregate and workforce planning. A key observation behind our approach is that, because the aggregate planning and workforce planning modules use long time horizons, precise data and intricate modeling detail are impractical or impossible. We must recognize that the production or workforce plans that these modules generate will be adjusted as time evolves. The lower levels in the PPC hierarchy must handle the nuts-and-bolts challenge of converting the plans to action. The keys to a good AP module are to keep the focus on long-term planning (i.e., avoiding putting too many short-term control details in the model) and to provide links for consistency with other levels in the hierarchy. Some of the issues related to consistency were discussed in Chapter 13. Here, we close with some general observations about the aggregate and workforce planning functions:

1. *No single AP or WP module is right for every situation.* As the examples in this chapter show, aggregate and workforce planning can incorporate many different decision problems. A good AP or WP module is one that is tailored to address the specific issues faced by the firm.

2. *Simplicity promotes understanding*. Although it is desirable to address different issues in the AP/WP module, it is even more important to keep the model understandable. In general, these modules are used to generate candidate production and workforce plans, which will be examined, combined, and altered manually before being published as "The Plan." To generate a spectrum of plans (and explain them to others), the user must be able to trace changes in the model to changes in the plan. Because of this, it makes sense to start with as simple a formulation as possible. Additional detail (e.g., constraints) can be added later.

3. *Linear programming is a useful AP/WP tool*. The long planning horizon used for aggregate and workforce planning justifies ignoring many production details; therefore, capacity checks, sales restrictions, and inventory balances can be expressed as linear constraints. As long as we are willing to approximate actual costs with linear functions, an LP solver is a very efficient method for solving many problems related to the AP and WP modules. Because we are working with speculative long-range data, it generally does not make sense to use anything more sophisticated than LP (e.g., nonlinear or integer programming) in most aggregate and workforce planning situations.

4. *Robustness matters more than precision*. No matter how accurate the data and how sophisticated the model, the plan generated by the AP or WP module will never be followed exactly. The actual production sequence will be affected by unforeseen events that could not possibly have been factored into the module. This means that the mark of a good long-range production plan is that it enables us to do a reasonably good job even in the face of such contingencies. To find such a plan, the user of the AP module must be able to examine the consequences of various scenarios. This is another reason to keep the model reasonably simple.

APPENDIX 16A
LINEAR PROGRAMMING

Linear programming is a powerful mathematical tool for solving constrained optimization problems. The name derives from the fact that LP was first applied to find optimal schedules or "programs" of resource allocation. Hence, although LP generally does involve using a computer program, it does not entail programming on the part of the user in the sense of writing code.

In this appendix, we provide enough background to give the user of an LP package a basic idea of what the software is doing. Readers interested in more details should consult one of the many good texts on the subject (e.g., Eppen and Gould 1988 for an application-oriented overview, Murty 1983 for more technical coverage).

Formulation

The first step in using linear programming is to formulate a practical problem in mathematical terms. There are three basic choices we must make to do this:

1. **Decision variables** are quantities under our control. Typical examples for aggregate planning and workforce planning applications of LP are production quantities, number of workers to hire, and levels of inventory to hold.

2. **Objective function** is what we want to maximize or minimize. In most AP/WP applications, this is typically either to maximize profit or minimize cost. Beyond simply stating the objective, however, we must specify it in terms of the decision variables we have defined.

3. **Constraints** are restrictions on our choices of the decision variables. Typical examples for AP/WP applications include capacity constraints, raw materials limitations, restrictions on how fast we can add workers due to limitations on training capacity, and restrictions on physical flow (e.g., inventory levels as a direct result of how much we produce/procure and how much we sell).

When one is formulating an LP, it is often useful to try to specify the necessary inputs in the order in which they are listed. However, in realistic problems, one virtually never gets the "right" formulation in a single pass. The example in Section 16.4.2 illustrates some of the changes that may be required as a model evolves.

To describe the process of formulating an LP, let us consider the problem presented in Table 16.2. We begin by selecting decision variables. Since there are only two products and because demand and capacity are assumed stationary over time, the only decisions to make concern how much of each product to produce per week. Thus, we let X_1 and X_2 represent the weekly production quantities of products 1 and 2, respectively.

Next, we choose to maximize profit as our objective function. Since product 1 sells for $90 but costs $45 in raw material, its net profit is $45 per unit.[9] Similarly, product 2 sells for $100 but costs $40 in raw material, so its net unit profit is $60. Thus, weekly profit will be

$$45X_1 + 60X_2 - \text{weekly labor costs } - \text{weekly overhead costs}$$

But since we assume that labor and overhead costs are not affected by the choice of X_1 and X_2, we can use the following as our objective function for the LP model:

$$\text{Maximize} \quad 45X_1 + 60X_2$$

Finally, we need to specify constraints. If we could produce as much of products 1 and 2 as we wanted, we could drive the above objective function, and hence weekly profit, to infinity. This is not possible because of limitations on demand and capacity.

The demand constraints are easy. Since we can sell at most 100 units per week of product 1 and 50 units per week of product 2, our decision variables X_1 and X_2 must satisfy

$$X_1 \leq 100$$

$$X_2 \leq 50$$

The capacity constraints are a little more work. Since there are four machines, which run at most 2,400 minutes per week, we must ensure that our production quantities do not violate this constraint on each machine. Consider workstation A. Each unit of product 1 we produce requires 15 minutes on this workstation, while each unit of product 2 we produce requires 10 minutes. Hence, the total number of minutes of time required on workstation A to produce X_1 units of product 1 and X_2 units of product 2 is[10]

$$15X_1 + 10X_2$$

so the capacity constraint for workstation A is

$$15X_1 + 10X_2 \leq 2,400$$

Proceeding analogously for workstations B, C, and D, we can write the other capacity constraints as follows:

$$15X_1 + 35X_2 \leq 2,400 \quad \text{workstation B}$$

$$15X_1 + 5X_2 \leq 2,400 \quad \text{workstation C}$$

$$25X_1 + 14X_2 \leq 2,400 \quad \text{workstation D}$$

---

[9]Note that we are neglecting labor and overhead costs in our estimates of unit profit. This is reasonable if these costs are not affected by the choice of production quantities, that is, if we won't change the size of the workforce or the number of machines in the shop.

[10]Note that this constraint does not address such detailed considerations as setup times that depend on the sequence of products run on workstation A or whether full utilization of workstation A is possible given the WIP in the system. But as we discussed in Chapter 13, these issues are addressed at a lower level in the production planning and control hierarchy (e.g., in the sequencing and scheduling module).

We have now completely defined the following LP model of our optimization problem:

$$\text{Maximize} \quad 45X_1 + 60X_2 \tag{16.107}$$

Subject to:

$$X_1 \leq 100 \tag{16.108}$$

$$X_2 \leq 50 \tag{16.109}$$

$$15X_1 + 10X_2 \leq 2,400 \tag{16.110}$$

$$15X_1 + 35X_2 \leq 2,400 \tag{16.111}$$

$$15X_1 + 5X_2 \leq 2,400 \tag{16.112}$$

$$25X_1 + 14X_2 \leq 2,400 \tag{16.113}$$

Some LP packages allow the user to enter the problem in a form almost identical to that shown in formulation (16.107)–(16.113). Spreadsheet programs generally require the decision variables to be entered into cells and the constraints specified in terms of these cells. More sophisticated LP solvers allow the user to specify blocks of similar constraints in a concise form, which can substantially reduce modeling time for large problems.

Finally, with regard to formulation, we point out that we have not stated explicitly the constraints that X_1 and X_2 be nonnegative. Of course, they must be, since negative production quantities make no sense. In many LP packages, decision variables are assumed to be nonnegative unless the user specifies otherwise. In other packages, the user must include the nonnegativity constraints explicitly. This is something to beware of when using LP software.

Solution

To get a general idea of how an LP package works, let us consider the above formulation from a mathematical perspective. First, note that any pair of X_1 and X_2 that satisfies

$$15X_1 + 35X_2 \leq 2,400 \quad \text{workstation B}$$

will also satisfy

$$15X_1 + 10X_2 \leq 2,400 \quad \text{workstation A}$$

$$15X_1 + 5X_2 \leq 2,400 \quad \text{workstation C}$$

because these differ only by having smaller coefficients for X_2. This means that the constraints for workstations A and C are redundant. Leaving them out will not affect the solution. In general, it does not hurt anything to have redundant constraints in an LP formulation. But to make our graphical illustration of how LP works as clear as possible, we will omit constraints (16.110) and (16.112) from here on.

Figure 16.15 illustrates problem (16.107)–(16.113) in graphical form, where X_1 is plotted on the horizontal axis and X_2 is plotted on the vertical axis. The shaded area is the **feasible region,** consisting of all the pairs of X_1 and X_2 that satisfy the constraints. For instance, the demand constraints (16.108) and (16.109) simply state that X_1 cannot be larger than 100, and X_2 cannot be larger than 50. The capacity constraints are graphed by noting that, with a bit of algebra, we can write constraints (16.111) and (16.113) as

$$X_2 \leq -\left(\frac{15}{35}\right)X_1 + \frac{2,400}{35} = -0.429X_1 + 68.57 \tag{16.114}$$

$$X_2 \leq -\left(\frac{25}{14}\right)X_1 + \frac{2,400}{14} = -1.786X_1 + 171.43 \tag{16.115}$$

If we replace the inequalities with equality signs in Equations (16.114) and (16.115), then these are simply equations of straight lines. Figure 16.15 plots these lines. The set of X_1 and X_2 points that satisfy these constraints is all the points lying below both of these lines. The points marked by the shaded area are those satisfying all the demand, capacity, and nonnegativity constraints. This type of feasible region defined by linear constraints is known as a **polyhedron.**

FIGURE 16.15

Feasible region for LP example

FIGURE 16.16

Solution to LP example

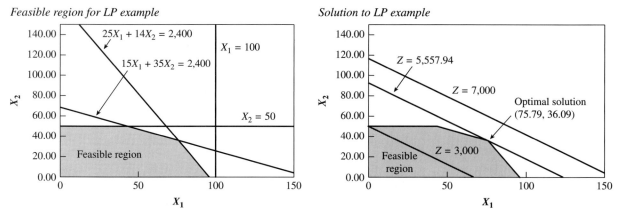

Now that we have characterized the feasible region, we turn to the objective. Let Z represent the value of the objective (i.e., net profit achieved by producing quantities X_1 and X_2). From objective (16.107), X_1 and X_2 are related to Z by

$$45X_1 + 60X_2 = Z \tag{16.116}$$

We can write this in the usual form for a straight line as

$$X_2 = \left(\frac{-45}{60}\right)X_1 + \frac{Z}{60} = -0.75X_1 + \frac{Z}{60} \tag{16.117}$$

Figure 16.16 illustrates Equation (16.117) for $Z = 3,000$, 5,557.94, and 7,000. Notice that for $Z = 3,000$, the line passes through the feasible region, leaving some points above it. Hence, we can feasibly increase profit (that is, Z). For $Z = 7,000$ the line lies entirely above the feasible region. Hence, $Z = 7,000$ is not feasible. For $Z = 5,557.94$, the objective function just touches the feasible region at a single point, the point ($X_1 = 75.79$, $X_2 = 36.09$). This is the **optimal solution.** Values of Z above 5,557.74 are infeasible, values below it are suboptimal. The optimal product mix, therefore, is to produce 75.79 (or 75, rounded to an integer value) units of product 1 and 36.09 (rounded to 36) units of product 2.

We can think of finding the solution to an LP by steadily increasing the objective value (Z), moving the objective function up and to the right, until it is just about to leave the feasible region. Because the feasible region is a polyhedron whose sides are made up of linear constraints, the last point of contact between the objective function and the feasible region will be a corner, or **extreme point,** of the feasible region.[11] This observation allows the optimization algorithm to ignore the infinitely many points inside the feasible region and search for a solution among the finite set of extreme points. The **simplex algorithm,** developed in the 1940s and still widely used, works in just this way, proceeding around the outside of the polyhedron, trying extreme points until an optimal one is found. Other, more modern algorithms use different schemes to find the optimal point, but will still converge to an extreme-point solution.

Sensitivity Analysis

The fact that the optimal solution to an LP lies at an extreme point enables us to perform useful sensitivity analysis on the optimal solution. The principal sensitivity information available to us falls into the following three categories.

---

[11]Actually, it is possible that the optimal objective function lies right along a flat spot connecting two extreme points of the polyhedron. When this occurs, there are many pairs of X_1 and X_2 that attain the optimal value of Z, and the solution is called **degenerate.** Even in this case, however, an extreme point (actually, at least two extreme points) will be among the optimal solutions.

1. **Coefficients in the objective function.** For instance, if we were to change the unit profit for product 1 from $45 to $60, then the equation for the objective function would change from Equation (16.117) to

$$X_2 = \left(-\frac{60}{60}\right) X_1 + \frac{Z}{60} = -X_1 + \frac{Z}{60} \tag{16.118}$$

so the slope changes form -0.75 to -1; that is, it gets steeper. Figure 16.17 illustrates the effect. Under this change, the optimal solution remains ($X_1 = 75.79$, $X_2 = 36.09$). Note, however that while the decision variables remain the same, the objective function does not. When the unit profit for product 1 increases to $60, the profit becomes

$$60(75.79) + 60(36.09) = \$6,712.80$$

The optimal decision variables remain unchanged until the coefficient of X_1 in the objective function reaches 107.14. When this happens, the slope becomes so steep that the point where the objective function just touches the feasible region moves to the extreme point ($X_1 = 96$, $X_2 = 0$). Geometrically, the objective function "rocked around" to a new extreme point. Economically, the profit from product 1 reached a point where it became optimal to produce all product 1 and no product 2.

In general, LP packages will report a range for each coefficient in the objective function for which the optimal solution (in terms of the decision variables) remains unchanged. Note that these ranges are valid only for one-at-a-time changes. If two or more coefficients are changed, the effect is more difficult to characterize. One has to rerun the model with multiple coefficient changes to get a feel for their effect.

2. **Coefficients in the constraints.** If the number of minutes required on workstation B by product 1 is changed from 15 to 20, then the equation defined by the capacity constraint for workstation B changes from Equation (16.114) to

$$X_2 \leq -\left(\frac{20}{35}\right) X_1 + \frac{2,400}{35} = -0.571X_1 + 68.57 \tag{16.119}$$

so the slope changes from -0.429 to -0.571; again, it becomes steeper. In a manner analogous to that described above for coefficients in the objective function, LP packages can determine how much a given coefficient can change before it ceases to define the optimal extreme point. However, because changing the coefficients in the constraints moves the extreme points themselves, the optimal decision variables will also change. For this reason, most LP packages do not report this sensitivity data, but rather make use of this product as part of a **parametric programming** option to quickly generate new solutions for specified changes in the constraint coefficients.

3. **Right-hand side coefficients.** Probably the most useful sensitivity information provided by LP models is for the right-hand side variables in the constraints. For instance, in formulation (16.107)–(16.113), if we run 100 minutes of overtime per week on machine B, then its right-hand

FIGURE 16.17

Effect of changing objective coefficients in LP example

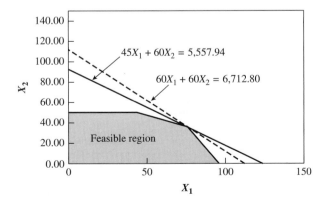

side will increase from 2,400 to 2,500. Since this is something we might want to consider, we would like to be able to determine its effect. We do this differently for two types of constraints:

a. **Slack constraints** are constraints that do not define the optimal extreme point. The capacity constraints for workstations A and C are slack, since we determined right at the outset that they could not affect the solution. The constraint $X_2 \leq 50$ is also slack, as can be seen in Figures 16.15 and 16.16, although we did not know this until we solved the problem.

Small changes in slack constraints do not change the optimal decision variables or objective value at all. If we change the demand constraint on product 2 to $X_2 \leq 49$, it still won't affect the optimal solution. Indeed, not until we reduce the constraint to $X_2 \leq 36.09$ will it have any effect. Likewise, increasing the right-hand side of this constraint (above 50) will not affect the solution. Thus, for a slack constraint, the LP package tells us how far we can vary the right-hand side without changing the solution. These are referred to as the **allowable increase** and **allowable decrease** of the right-hand side coefficients.

b. **Tight constraints** are constraints that define the optimal extreme point. Changing them changes the extreme point, and hence the optimal solution. For instance, the constraint that the number of hours per week on workstation B not exceed 2,400, that is,

$$15X_1 + 35X_2 \leq 2{,}400$$

is a tight constraint in Figures 16.15 and 16.16. If we increase or decrease the right-hand side, the optimal solution will change. However, if the changes are small enough, then the optimal extreme point will still be defined by the same constraints (i.e., the time on workstations B and D). Because of this, we are able to compute the following:

> **Shadow prices** are the amount by which the objective increases per unit increase in the right-hand side of a constraint. Since slack constraints do not affect the optimal solution, changing their right-hand sides has no effect, and hence their shadow prices are always zero. Tight constraints, however, generally have nonzero shadow prices. For instance, the shadow price for the constraint on workstation B is 1.31. (Any LP solver will automatically compute this value.) This means that the objective will increase by $1.31 for every extra minute per week on the workstation. So if we can work 2,500 minutes per week on workstation B, instead of 2,400, the objective will increase by $100 \times 1.31 = \$131$.
>
> **Maximum allowable increase/decrease** gives the range over which the shadow prices are valid. If we change a right-hand side by more than the maximum allowable increase or decrease, then the set of constraints that define the optimal extreme point may change, and hence the shadow price may also change. For example, as Figure 16.18 shows, if we increase the right-hand side of the constraint on workstation B from 2,400 to 2,770, the constraint moves to the very edge of the feasible region defined by $25X_1 + 14X_2 \leq 2{,}400$ (machine D) and $X_2 \leq 50$. Any further increases in the right-hand side will cause this constraint to become slack. Hence, the shadow price is $1.31 up to a maximum allowable

FIGURE 16.18

Feasible region when RHS of constraint of workstation B is increased to 2,770

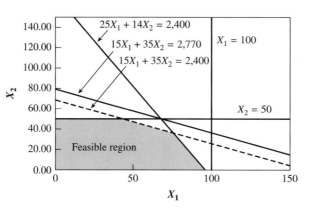

increase of 370 (that is, $2,770 - 2,400$). In this example, the shadow price is zero for changes above the maximum allowable increase. This is not always the case, however, so in general we must resolve the LP to determine the shadow prices beyond the maximum allowable increase or decrease.

Study Questions

1. Although the technology for solving aggregate planning models (linear programming) is well established and AP modules are widely available in commercial systems (e.g., MRP II systems), aggregate planning does not occupy a central place in the planning function of many firms. Why do you think this is true? What difficulties in modeling, interpreting, and implementing AP models might be contributing to this?

2. Why does it make sense to consider workforce planning and aggregate planning simultaneously in many situations?

3. What is the difference between a **chase** production plan and a **level** production plan, with respect to the amount of inventory carried and the fluctuation in output quantity over time? How do the production plans generated by an LP model relate to these two types of plan?

4. In a basic LP formulation of the product mix aggregate planning problem, what information is provided by the following?
   a. The optimal decision variables.
   b. The optimal objective function.
   c. Identification of which constraints are tight and which are slack.
   d. Shadow prices for the right-hand sides of the constraints.

Problems

1. Suppose a plant can supplement its capacity by subcontracting part of or all the production of certain parts.
   a. Show how to modify LP (16.28)–(16.32) to include this option, where we define

      V_{it} = units of product i received from a subcontractor in period t
      k_{it} = premium paid for subcontracting product i in period t (i.e., cost above variable cost of making it in-house)
      $\underline{v}_{it}$ = minimum amount of product i that must be purchased in period t (e.g., specified as part of long-term contract with supplier)
      $\bar{v}_{it}$ = maximum amount of product i that can be purchased in period t (e.g., due to capacity constraints on supplier, as specified in long-term contract)

   b. How would you modify the formulation in part a if the contract with a supplier stipulated only that total purchases of product i over the time horizon must be at least $\underline{v}_i$?
   c. How would you modify the formulation in part a if the supplier contract, instead of specifying $\underline{v}$ and $\bar{v}$, stipulated that the firm specify a base amount of product i, to be purchased every month, and that the maximum purchase in a given month can exceed the base amount by no more than 20 percent?
   d. What role might models like those in parts a to c play in the process of negotiating contracts with suppliers?

2. Show how to modify LP (16.49)–(16.54) to represent the case where overtime on all the workstations must be scheduled simultaneously (i.e., if one resource runs overtime, all resources run overtime). Describe how you would handle the case where, in general, different workstations can have different amounts of overtime, but two workstations, say A and B, must always be scheduled for overtime together.

3. Show how to modify LP (16.61)–(16.67) of the workforce planning problem to accommodate multiple products.

4. You have just been made corporate vice president in charge of manufacturing for an automotive components company and are directly in charge of assigning products to plants. Among many other products, the firm makes automotive batteries in three grades: heavy-duty, standard, and economy. The unit net profits and maximum daily demand for these products are given in the first table below. The firm has three locations where the batteries can be produced. The maximum assembly capacities, for any mix of battery grades, are given in the second table below. The number of batteries that can be produced at a location is limited by the amount of suitably formulated lead the location can produce. The lead requirements for each grade of battery and the maximum lead production for each location are also given in the following tables.

Product	Unit Profit ($/battery)	Maximum Demand (batteries/day)	Lead Requirements (lbs/battery)
Heavy-duty	12	700	21
Standard	10	900	17
Economy	7	450	14

Plant Location	Assembly Capacity (batteries/day)	Maximum Lead Production (lbs/day)
1	550	10,000
2	750	7,000
3	225	4,200

 a. Formulate a linear program that allocates production of the three grades among the three locations in a manner that maximizes profit.
 b. Suppose company policy requires that the fraction of capacity (units scheduled/assembly capacity) be the same at all locations. Show how to modify your LP to incorporate this constraint.
 c. Suppose company policy dictates that at least 50 percent of the batteries produced must be heavy-duty. Show how to modify your LP to incorporate this constraint.

5. Youohimga, Inc., makes a variety of computer storage devices, which can be divided into two main families that we call A and B. All devices in family A have the same routing and similar processing requirements at each workstation; similarly for family B. There are a total of 10 machines used to produce the two families, where the routings for A and B have some workstations in common (i.e., shared) but also contain unique (unshared) workstations.

 Because Youohimga does not always have sufficient capacity to meet demand, especially during the peak demand period (i.e., the months near the start of the school year in September), in the past it has contracted out production of some of its products to vendors (i.e., the vendors manufacture devices that are shipped out under Youohimga's label). This year, Youohimga has decided to use a systematic aggregate planning process to determine vendoring needs and a long-term production plan.

 a. Using the following notation

   X_{it} = units of family i (i = A, B) produced in month t ($t = 1, \ldots, 24$) and available to meet demand in month t

V_{it} = units of family i purchased from vendor in month t and available to meet demand in month t

I_{it} = finished goods inventory of family i at end of month t

d_{it} = units of family i demanded (and shipped) during month t

c_{jt} = hours available on work center j ($j = 1, \ldots, 10$) in month t

a_{ij} = hours required at work center j per unit of family i

v_i = premium (i.e., extra cost) per unit of family i that is vendored instead of being produced in-house

h_i = holding cost to carry one unit of family i in inventory from one month to the next

formulate a linear program that minimizes the cost (holding plus vendoring premium) over a two-year (24-month) planning horizon of meeting monthly demand (i.e., no backorders are permitted). You may assume that vendor capacity for both families is unlimited and that there is no inventory of either family on hand at the beginning of the planning horizon.

b. Which of the following factors might make sense to examine in the aggregate planning model to help formulate a sensible vendoring strategy?

- Altering machine capacities
- Sequencing and scheduling
- Varying size of workforce
- Alternate shop floor control mechanisms
- Vendoring individual operations rather than complete products
- All the above

c. Suppose you run the model in part a and it suggests vendoring 50 percent of the total demand for family A and 50 percent of the demand for B. Vendoring 100 percent of A and 0 percent of B is capacity-feasible, but results in a higher cost in the model. Could the 100–0 plan be preferable to the 50–50 plan in practice? If so, explain why.

6. Mr. B. O'Problem of Rancid Industries must decide on a production strategy for two top-secret products, which for security reasons we will call X and Y. The questions concern (1) whether to produce these products at all and (2) how much of each to produce. Both products can be produced on a single machine, and there are three brands of machine that can be leased for this purpose. However, because of availability problems, Rancid can lease at most one of each brand of machine. Thus, O'Problem must also decide which, if any, of the machines to lease. The relevant machine and product data are given below:

Machine	Hours to Produce One Unit of X	Hours to Produce One Unit of Y	Weekly Capacity (hours)	Weekly Lease + Operating Cost ($)
Brand 1	0.5	1.2	80	20,000
Brand 2	0.4	1.2	80	22,000
Brand 3	0.6	0.8	80	18,000

Product	Maximum Demand (units/week)	Net Unit Profit ($/unit)
X	200	150
Y	100	225

a. Letting X_{ij} represent the number of units of product i produced per week on machine j (for example, X_{A1} is the number of units of A produced on the brand 1 machine), formulate an LP to maximize weekly profit (including leasing cost) subject to the capacity and demand constraints. (*Hint:* Observe that the leasing/operating cost for a particular machine is only incurred if that machine is used and that this cost is fixed for any nonzero production level. Carefully define 0–1 integer variables to represent the all-or-nothing aspects of this decision.)

b. Suppose that the suppliers of brand 1 machines and brand 2 machines are feuding and will not service the same company. Show how to modify your formulation to ensure that Rancid leases either brand 1 or brand 2 or neither, but not both.

7. All-Balsa, Inc., produces two models of bookcases, for which the relevant data are summarized as follows:

	Bookcase 1	**Bookcase 2**
Selling price	$15	$8
Labor required	0.75 hr/unit	0.5 hr/unit
Bottleneck machine time required	1.5 hr/unit	0.8 hr/unit
Raw material required	2 bf/unit	1 bf/unit

$$P1 = \text{units of bookcase 1 produced per week}$$
$$P2 = \text{units of bookcase 2 produced per week}$$
$$OT = \text{hours of overtime used per week}$$
$$RM = \text{board-feet of raw material purchased per week}$$
$$A1 = \text{dollars per week spent on advertising bookcase 1}$$
$$A2 = \text{dollars per week spent on advertising bookcase 2}$$

Each week, up to 400 board feet (bf) of raw material is available at a cost of $1.50/bf. The company employs four workers, who work 40 hours per week for a total regular time labor supply of 160 hours per week. They work regardless of production volumes, so their salaries are treated as a fixed cost. Workers can be asked to work overtime and are paid $6 per hour for overtime work. There are 320 hours per week available on the bottleneck machine.

In the absence of advertising, 50 units per week of bookcase 1 and 60 units per week of bookcase 2 will be demanded. Advertising can be used to stimulate demand for each product. Experience shows that each dollar spent on advertising bookcase 1 increases demand for bookcase 1 by 10 units, while each dollar spent on advertising bookcase 2 increases demand for bookcase 2 by 15 units. At most, $100 per week can be spent on advertising.

An LP formulation and solution of the problem to determine how much of each product to produce each week, how much raw material to buy, how much overtime to use, and how much advertising to buy are given below. Answer the following on the basis of this output.

```
MAX     15 P1 + 8 P2 - 6 OT - 1.5 RM - A1 - A2
SUBJECT TO
        2)    P1 - 10 A1 <=    50
        3)    P2 - 15 A2 <=    60
        4)    0.75 P1 + 0.5 P2 - OT <=    160
        5)    2 P1 + P2 - RM <=    0
        6)    RM <=    400
        7)    A1 + A2 <=    100
        8)    1.5 P1 + 0.8 P2 <=    320
END
```

```
                    OBJECTIVE FUNCTION VALUE

           1)    2427.66700

     VARIABLE          VALUE          REDUCED COST
          P1       160.000000              .000000
          P2        80.000000              .000000
          OT          .000000             2.133334
          RM       400.000000              .000000
          A1        11.000000              .000000
          A2         1.333333              .000000

         ROW    SLACK OR SURPLUS    DUAL PRICES
          2)          .000000          .100000
          3)          .000000          .066667
          4)          .000000         3.866666
          5)          .000000         6.000000
          6)          .000000         4.500000
          7)        87.666660          .000000
          8)        16.000000          .000000

     NO. ITERATIONS=       5

     RANGES IN WHICH THE BASIS IS UNCHANGED:

                              OBJ COEFFICIENT RANGES
     VARIABLE          CURRENT       ALLOWABLE      ALLOWABLE
                        COEF         INCREASE       DECREASE
          P1       15.000000          .966667        .533333
          P2        8.000000          .266667        .483333
          OT       -6.000000         2.133334        INFINITY
          RM       -1.500000         INFINITY        4.500000
          A1       -1.000000         1.000000        5.333335
          A2       -1.000000         1.000000        7.249999

                              RIGHT-HAND SIDE RANGES
         ROW          CURRENT       ALLOWABLE      ALLOWABLE
                        RHS          INCREASE       DECREASE
           2       50.000000       110.000000      876.666600
           3       60.000000        20.000000     1315.000000
           4      160.000000        27.500000        2.500000
           5         .000000         6.666667       55.000000
           6      400.000000         6.666667       55.000000
           7      100.000000         INFINITY       87.666660
           8      320.000000         INFINITY       16.000000
```

a. If overtime costs only $4 per hour (and all other parameters remained unchanged), how much overtime should All-Balsa use?

b. If each unit of bookcase 1 sold for $15.50 (and all other parameters are unchanged), what will the optimal profit per week be—or can you not tell without resolving the LP?

c. What is the most All-Balsa should be willing to pay for another unit of raw material?

d. If each worker were required (as part of the regular workweek) to work 45 hours per week (and all other parameters remained unchanged), what would the company's profit be?

e. If each unit of bookcase 2 sold for $10 (and all other parameters remained unchanged), what would be the optimal quantity of bookcase 2 to produce—or can you not tell without resolving the LP?

    f. Reconsider the All-Balsa problem formulation and suppose that instead of having 400 bf of raw material available at $1.50/bf, All-Balsa faces a two-tier pricing scheme such that the first 200 bf/week costs $2.00/bf, but any amount above 200 bf/week up to a limit of an additional 300 bf/week costs $$p$/bf. (*Note: p is a constant,* not a variable, and we cannot purchase the $$p$/bf raw material unless we first purchase 200 bf of the $2.00 raw material.) To modify the LP to compute an "optimal" production/advertising policy, we define

    RM1 $=$ bf of raw material purchased at $2.00/bf

    RM2 $=$ bf of raw material purchased at $$p$/bf

    To formulate an appropriate LP to represent this new pricing scheme, we first replace 1.5RM in the objective function by 2RM1 $+$ pRM2.

    i. If $p > 2$, what other changes in the previous LP make it properly reflect the new pricing scheme?

    ii. If $p < 2$, what other changes in the previous LP make it properly reflect the new pricing scheme?

8. Consider a production line with four workstations, labeled $j = 1, 2, 3,$ and 4, in tandem (all products flow through all four machines in order). Three different products, labeled $i = $ A, B, and C, are produced on the line. The hours required on each workstation for each product and the net profit per unit sold (r_i) are given as follows:

i	1	2	3	4	r_i
A	2.4	1.1	0.8	3.0	$50
B	2.0	2.2	1.2	2.1	$65
C	0.9	0.9	1.0	2.5	$70

(Column header j spans columns 1–4.)

The number of hours available (c_{jt}) and the upper and lower limits on demand ($\bar{d}_{it}$ and $\underline{d}_{it}$) for each product over the next four quarters are as follows:

t	1	2	3	4
c_{1t}	640	640	1,280	1,280
c_{2t}	640	640	640	640
c_{3t}	1,920	1,920	1,920	1,920
c_{4t}	1,280	1,280	1,280	2,560
$\bar{d}_{At}$	100	50	50	75
$\underline{d}_{At}$	0	0	0	0
$\bar{d}_{Bt}$	100	100	100	100
$\underline{d}_{Bt}$	20	20	20	25
$\bar{d}_{Ct}$	300	250	250	400
$\underline{d}_{Ct}$	0	0	0	50

    a. Suppose we use a quarterly holding cost of $5 and a quarterly backorder cost of $10 per item on all products and allow backordering. Formulate an LP to maximize profit minus holding and backorder costs subject to the constraints on workstation capacity and minimum/maximum sales.

b. Using the LP solver of your choice, solve your formulation in part *a*. Which constraints are binding in your solution?

c. Suppose that there is an inspect operation immediately after station 2 (which has plenty of capacity and therefore does not need to be modeled as an extra resource) and 20 percent of the parts (regardless of product type) are recycled back through stations 1 and 2. Show how to modify your formulation in part *a* to model this.

9. A manufacturer of high-voltage switches projects demand (in units) for the upcoming year to be as follows.

Jan	1,000	Jul	3,200
Feb	1,000	Aug	2,000
Mar	1,000	Sep	1,000
Apr	2,000	Oct	900
May	2,400	Nov	800
Jun	2,500	Dec	800

The plant runs 160 hours per month and produces at an average rate of 10 switches per hour. Unit profit per switch sold is $50, and the estimated cost to hold a switch in inventory for one month is $5. There is no inventory at the start of the year. Overtime can be used at a cost of $300 per hour.

a. Compute the inventory-holding and overtime cost of a chase production strategy (i.e., producing the amount demanded in each month).

b. Compute the inventory holding and overtime cost of a level production strategy (i.e., producing the same amount each month). If the monthly production quantity is set equal to average monthly demand, how much inventory will be left at the end of the year?

c. Compute a production strategy by solving a linear program to maximize profit (i.e., net sales revenue minus inventory carrying cost minus overtime cost). Is the amount of overtime in the plan reasonable? If not, what changes to the LP model could be made to generate a more reasonable solution?

d. How does the solution change if the inventory carrying cost is reduced to $3 per unit per month? If overtime costs are reduced to $200 per hour? Given that these costs are approximate, what do these results imply about the production plan?

10. Reconsider Problem 2 of Chapter 6 in which a manufacturer produced three models of vacuum cleaner on a three-station production line.

a. Use linear programming to compute a monthly production plan that maximizes monthly profit, and compare it to the profit resulting from the current plan given in Chapter 6 and those suggested by the labor hours and ABA cost accounting calculations.

b. Could this LP solution have been arrived at by rank-ordering the products according to profitability by a cost accounting scheme? What does this say about the effectiveness of using accounting methods to plan production schedules?

17   SUPPLY CHAIN MANAGEMENT

*One's work may be finished some day,
but one's education never.*

Alexandre Dumas

17.1   Introduction

A major theme of this book is the central role of inventory in the operational behavior of a production system. We began with a historical review of inventory control and its relationship to production control in Part I. In Part II, we deepened our understanding of the interaction between inventory (WIP, in particular) and other performance measures, such as throughput and cycle time. Now in Part III we are ready to combine our historical and factory physics insights to address the practical problem of managing inventories in a manufacturing system. Our objective is to improve inventory *efficiency* throughout the system. That is, we do not simply seek to reduce inventories; we seek to ensure that the purpose of inventories is met with minimal dollar investment. In modern parlance, this overall systemwide coordination of inventory stocks and flows is known as **supply chain management.**

For purposes of our discussions here, we divide inventories in a supply chain into four categories:

1. **Raw materials** are components, subassemblies, or materials that are purchased from outside the plant and used in the fabrication/assembly processes inside the plant.

2. **Work in process (WIP)** includes all unfinished parts or products that have been released to a production line.

3. **Finished goods inventory (FGI)** is finished product that has not been sold.

4. **Spare parts** are components that are used to maintain or repair production equipment.

The reasons for holding each of these types of inventory, and therefore the options for improving efficiency, are different. Hence, we treat each category separately in the following discussions.

17.2   Reasons for Holding Inventory

17.2.1   Raw Materials

If we could receive raw materials from suppliers in literal just-in-time fashion (i.e., exactly when needed by the production system), we would not need to carry any raw materials inventories. Since this is never possible in practice, all manufacturing systems carry stocks of raw materials. There are three main factors that influence the size of these stocks.

1. **Batching.** Quantity discounts from suppliers, limited capacity of the plant's purchasing function (e.g., a limit on the number of purchase orders that can be placed and tracked), and economies of scale in deliveries provide incentive to order raw materials in bulk.[1] We refer to inventory that addresses batching considerations as **cycle stock,** since it represents stock held between ordering cycles.

2. **Variability.** When production gets ahead of schedule, supplier deliveries get behind schedule, or quality problems cause excessive scrap loss, the line will shut down for lack of materials if extra stock is not available. This extra stock can be planned for directly as a **safety stock** (i.e., by ordering so that expected stock levels remain above the safety level) or be the consequence of a **safety lead time** (i.e., order materials so that they arrive before needed and therefore wait in raw materials inventory). In either case, we refer to inventory carried as protection against variability as safety stock.

3. **Obsolescence.** Changes in demand or design can render some materials no longer needed, so some inventory in manufacturing systems does not address either of the above purposes. This inventory, which we term **obsolete inventory,** may have been ordered as cycle or safety stock, but is now essentially useless and must be disposed of and written off as quickly as financial reporting considerations will permit.

To recognize these reasons for carrying raw materials inventories is useful in identifying improved management policies. However, one should remember that they are not strictly separate. For instance, as we pointed out in Chapter 2, safety stock *and* cycle stock provide protection against variability (i.e., because if we order in very large batches, then we reduce the frequency with which inventory levels fall to the point where a stockout is possible). Also, the level of obsolete inventory is clearly affected by the levels of cycle and safety stock (i.e., if we order in large batches or carry large safety stocks, then we risk having large amounts of inventory become obsolete due to system changes). Appreciating these interactions can also help us devise raw materials management policies.

17.2.2   Work in Process

Despite the JIT goal of zero inventories, we can never operate a manufacturing system with zero WIP since, as we saw in Part II, zero WIP implies zero throughput. In Chapter 7, we derived a **critical WIP** level that represents the smallest WIP level required by

---

[1]These factors are precisely those that motivated the fixed order cost in the EOQ model presented in Chapter 2. The EOQ model balances this fixed cost against inventory carrying costs to determine an economic order quantity.

a line to achieve full throughput under the best conditions. Under realistic conditions, actual WIP levels frequently exceed the critical WIP level by a large amount (e.g., often 20 to 30 times). This WIP will be in one of five states:

1. **Queueing** if it is waiting for a resource (person, machine, or transport device).
2. **Processing** if it is being worked on by a resource.
3. **Waiting for batch** if it has to wait for other jobs to arrive in order to form a batch. This batch may serve to fill a bulk manufacturing operation (e.g., heat treat, in which a roomful of jobs is subjected to a burn-in operation simultaneously) or a move operation (e.g., when jobs are moved only in full pallets). Note that once the process or move batch has been formed, any additional waiting time for the resource (e.g., for the heat treater or the forklift to become available) is classed as queueing time.
4. **Moving** if it is actually being transported between resources.
5. **Waiting to match** if it consists of components waiting at an assembly operation for their counterparts to arrive so that an assembly can occur. Once the entire "kit" of parts has arrived, any additional waiting time for the assembly resource is defined as queueing time.

To use the above classification in a WIP management/reduction program, two observations are needed. First, as illustrated in Figure 17.1, in most manufacturing systems the fraction of WIP that is actually processing or moving is small (e.g., less than 10 percent; see Bradt 1983 for empirical documentation). The majority of WIP is in queue, waiting for batch, or waiting to match. Clearly, a WIP reduction program must address these latter categories to be successful.

Second, queueing WIP, wait-for-batch WIP, and wait-to-match WIP are the result of different causes. As we saw in Part II, the principal causes of queueing are high utilization and variability (both flow variability and process variability). Wait-for-batch WIP is clearly caused by batching for process or transport; the larger the batch size, the more WIP required. Wait-to-match WIP is caused by lack of synchronization in the arrival of parts to the assembly process, some of which is due to simple flow variability and some of which can be caused by the production control process. These differences imply that the different types of WIP are amenable to different management policies, as we will discuss later.

FIGURE 17.1

Typical breakdown of WIP in a manufacturing system

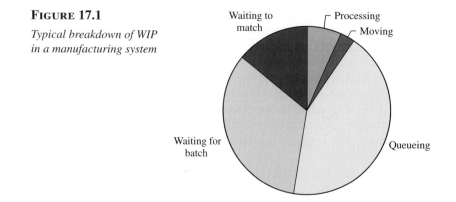

17.2.3 Finished Goods Inventory

If we could ship everything we produced directly to customers as soon as processing was complete, there would be no need for FGI. Although some manufacturing systems (e.g., heavily loaded job shops that make custom products) can almost achieve this, many cannot. There are five basic reasons for carrying FGI.

1. **Customer responsiveness.** To provide delivery lead times that are shorter than manufacturing cycle times, many firms make use of a **make-to-stock** (instead of a **make-to-order**) policy. For example, many products, such as building materials (e.g., roofing shingles, lumber), standard electrical components (e.g., resistors, capacitors), and basic food products (e.g., baking soda, corn oil) are **commodity** products. As such, their price and specifications (e.g., quality) are set by the market. The only competitive issue, then, is delivery. For this reason, such products are frequently produced to stock. The amount of FGI needed to support a given make-to-stock system depends on the variability of customer demand and the desired level of customer service.

An approach that combines the effectiveness of make-to-stock and make-to-order procedures is **assemble-to-order.** This procedure produces components to stock and then assembles these components to order. In the terminology of Chapter 10, make-to-order places the **push/pull interface** at raw materials, make-to-stock is places it at finished goods, while assemble-to-order places it somewhere in between. The result is faster response than the traditional make-to-order approach with less inventory than a make-to-stock policy.

2. **Batch production.** If, for whatever reason, production occurs in prespecified quantities (batches), then output will sometimes not match customer orders and any excess will go into finished goods inventory. For example, a steel mill that runs 250-ton batches (in order to efficiently utilize the casting furnace) but has customer orders averaging 50 tons will frequently have to place remnants of batches of various grades of steel into FGI.

3. **Forecast errors.** When jobs are released without firm customer orders, either to replenish stock in a make-to-stock system or to meet anticipated orders in a make-to-order system, product will inevitably be built that does not sell as anticipated. This excess will wind up in FGI.

4. **Production variability.** In a make-to-order system where orders cannot be shipped early (or have a limit on how early they can be shipped), variability in production *timing* will sometimes result in product that will have to reside in FGI while awaiting shipment. In either a make-to-order or a make-to-stock system, variability in production *quantity* (e.g., due to random yield loss) can result in overproduction relative to demand (e.g., if we "overinflate" to compensate for the yield loss). Again, the excess will go into FGI.

5. **Seasonality.** One approach to dealing with demand that varies with season (e.g., lawnmowers, snowblowers, room air conditioners) is to build inventory during the off season to meet peak demand. This **built-ahead inventory** will become part of FGI.

Notice that the factors motivating finished goods inventory interact. For instance, whenever we build FGI to provide short lead times or to cover seasonal demand we increase exposure of the system to forecasting errors. Because of this, it is important to view FGI holistically. Only by doing this can we consider basic structural changes that may offer significant potential. For instance, maybe the system should really be run in make-to-order instead of make-to-stock fashion; maybe excess capacity or seasonal labor should be used instead of built-ahead inventory to address seasonal demand, or maybe the push-pull interface should be relocated (e.g., to use an assemble-to-order strategy). We will return to these options in our discussion of improvement strategies.

17.2.4   Spare Parts

Spare parts are not used as direct inputs to finished products, but they do support the production process by keeping the machines running. In many systems the dollar value of inventory involved is not large, but the consequences of shortfalls can be severe (e.g., the entire line can be shut down for lack of a critical part). In some systems (e.g., a contract service operation that supports repairs in a nationwide network of machines), however, the dollar value of spare parts inventories can be substantial. In either case, the primary reasons for stocking spare parts are

1. **Service.** The main objective of any spare parts system is to support a maintenance and repair process. If repair personnel must wait for a part (e.g., from a central storage site or an outside supplier), then the time to complete a repair can be dramatically lengthened. All other things being equal, achieving higher service (i.e., avoidance of delay due to an out-of-stock part) requires a higher level of spare parts inventory.
2. **Purchasing/production lead times.** If spare parts could be purchased or produced instantly, there would be no need to stock them. Unfortunately, this is virtually never the case; so to provide the desired service, we must carry spare parts inventories. In general, the longer the lead time to obtain a part, the more stock we will have to carry.
3. **Batch replenishment.** If there are economies of scale in replenishing spare parts (e.g., quantity discounts on a purchased part or a large fixed cost to produce a part), then it may make sense to purchase them in bulk. Of course, a larger replenishment batch implies a higher average inventory level.

In theory, spare parts inventory systems are not much different from FGI systems. In both, we stock parts, possibly in batches, to satisfy an uncertain demand process with some level of service. Because of this similarity, it may well be possible to use similar tools for controlling spare parts and FGI. However, it is important to recognize the difference between the roles played by the two types of inventory. For instance, it may be reasonable to set a fill rate of 90 percent for FGI, based on industry benchmarking, say. But a 90 percent fill rate for spare parts may be far too low when one considers the logistical and financial consequences of causing a long machine outage by stocking out on a critical part. Thus, while we might use similar models to address the two types of inventory, we must carefully consider the costs and objectives involved in order to set appropriate parameters for the models.

Having reviewed the reasons for holding different types of inventory, we now review techniques for improving the efficiency (i.e., attaining the same benefits with a smaller overall investment) of each type of inventory.

17.3   Managing Raw Materials

As noted above, the objective in managing raw materials is to have them available when needed by the production process without carrying any more inventory than necessary. Some strategies can enhance our ability to do this for all parts. Others are economically viable for only certain classes of parts. Therefore, our basic strategy is one of "divide and conquer," in which we apply different approaches to different classes of raw material. In the following sections we present some overall improvement strategies, a classification scheme, and focused control policies geared to specific part classes.

17.3.1   Visibility Improvements

Obviously, we can do a better job of purchasing raw materials if we know what parts are needed than if we must guess. Unfortunately, manufacturing cycle times and purchasing lead times are frequently long enough to require us to purchase at least some of the materials before we have firm customer orders. In the short term, we may have no option other than to maintain safety stocks of raw materials to buffer against purchasing mistakes. In the long term, however, we can improve the situation via the following policies:

1. **Improve forecasting.** If forecasts of future demand are truly horrible, better projections may be possible through the use of systematic forecasting techniques (see Appendix 13A). However, such methods cannot get around the first law of forecasting—*forecasts are always wrong*. Thus, there are limits to the improvements possible through forecasting.

2. **Reduce cycle times.** Reduced manufacturing cycle times imply that jobs can be released closer to their due dates. Hence, purchased parts can be ordered later, when customer demands are firmer. In systems with long cycle times, cycle time reduction can improve forecasts much more than use of sophisticated forecasting techniques can. We discuss specific techniques for cycle time (and WIP) reduction in Section 17.4.

3. **Improve scheduling.** If scheduling is poor, then projected use of purchased parts may be very different from actual use. For instance, a schedule generated with an infinite-capacity MRP model may project much earlier completion of jobs than actually will occur. This will result in purchased parts arriving well before they are actually used and hence will cause raw materials inventories to be inflated. A good finite-capacity scheduler will generate more realistic schedules and thus will enable purchased parts to be brought in closer to when they are used.

17.3.2   ABC Classification

In most manufacturing systems, a small fraction of the purchased parts represent a large fraction of the purchasing expenditures.[2] To have maximum impact, therefore, management attention should be focused most closely on these parts. To accomplish this, many manufacturing firms use some sort of **ABC classification** for purchased parts and materials. In a typical definition of ABC categories, we rank-order the purchased parts according to the annual dollar value spent on each, and we define

A parts: the first 5 to 10 percent of the parts, accounting for 75 to 80 percent of total annual expenditures.

---

[2]This is an example of **Pareto's law,** commonly known as the "80-20 rule," named for Italian economist Vilfredo Pareto (1848–1923) who observed that a large fraction of wealth tends to be concentrated in a small fraction of the population.

B parts: the next 10 to 15 percent of the parts, accounting for 10 to 15 percent of total annual expenditures.

C parts: the bottom 80 percent or so of the parts, accounting for only 10 percent or so of total annual expenditures.

Because their number is relatively small and their cost is high, it makes sense to use sophisticated, time-consuming methods to tightly coordinate the arrival of A parts with their use by the production process. Such efforts are generally not warranted for C parts, since the cost of holding small excess quantities of inventory is not large. The B parts are in-between, so they deserve more attention than the C parts, but not as much as the A parts. Approaches may vary from system to system, but the main point of ABC classification remains the same: Inventories of different classes of parts should be treated differently.

We discuss some suitable techniques and where each is applicable in the following sections.

17.3.3   Just-in-Time

Very expensive A parts, for which holding inventory is costly, and extremely bulky parts (e.g., packaging materials), for which holding inventory is inconvenient, are good candidates for tight inventory control. The way to maintain the absolute minimum level of inventory of a part is to coordinate deliveries with use in the production process. This is precisely the idea behind just-in-time (JIT).

A typical JIT contract with a supplier calls for frequent deliveries (e.g., weekly, daily, or even more often, depending on the system) in small quantities closely matched to what is required by the production schedule. Since production schedules are prone to change, most JIT contracts allow adjustment of the order quantities almost up to the delivery time (although most contracts also specify limits on the amount of change allowed).

To give suppliers a reasonable chance of meeting delivery requirements, well-managed JIT procurement systems provide visibility of the production schedule to suppliers. The primary goal is to alert suppliers as quickly as possible to any changes in the schedule. But such visibility can have other benefits. It can eliminate the need for purchase orders. For instance, a contract with a supplier of automotive brakes might call for it to look at the final assembly schedule and deliver the proper brakes to support it. The system could go even further and eliminate invoices for the brakes by simply counting the number of automobiles produced and sending payment to the supplier for them. (The implicit, and reasonable, assumption is that every automobile has a set of brakes.)

In concept, JIT contracts with suppliers are very attractive. However, in order for them to work, suppliers must be reliable, with regard to both delivery timing and quality. If a shipment is late or defective, then the entire line may be stopped for lack of parts. Because of this, firms that rely extensively on JIT deliveries of raw materials generally institute some kind of **vendor certification** program. Good vendor certification programs involve both reviews of supplier procedures and efforts to help vendors improve their systems.

Because close supervision and cultivation of suppliers is a prerequisite for JIT deliveries of raw materials, this approach may not be a feasible option for smaller firms. A firm whose purchases compose a very small fraction of a supplier's business may simply lack the clout to persuade the supplier to deliver parts on a JIT basis. While the current trend toward responsiveness (e.g., as embodied in buzzwords such as *time-based competition*, *total cycle time*, *short-cycle manufacturing*) may be increasing the number of suppliers who are willing to offer JIT deliveries to firms other than their largest customers, true JIT contracts are still largely unavailable to the typical small firm. Thus, they must seek other approaches to managing expensive raw materials inventories.

17.3.4   Setting Safety Stock/Lead Times for Purchased Components

Even if a firm cannot or will not use JIT deliveries for expensive A parts, it still makes sense to link purchases of these parts closely to the production schedule (instead of, say, ordering infrequently in large batches and supplying the line from an amply stocked materials crib). In MRP language, this means that expensive parts should be ordered on a **lot-for-lot** basis. For example, if we plan to produce 1,000 high-resolution monitors *n* weeks from now, we should order 1,000 cathode-ray tubes to arrive some fixed safety lead time in advance of the schedule.[3]

Notice that this approach is different from JIT because we are ordering parts against a *planned* schedule, rather than having them delivered in synchronization with *actual* production. But if true JIT is not possible, this may be the best we can do. Of course, if (when) the schedule changes, production of the desired amounts may be impossible due to lack of appropriate raw materials. This implies that short delivery lead times are less difficult to work with than long ones, because purchases will be made closer to due dates, when the schedule consists more of firm orders and less of speculative forecasts. In the long run, a higher-priced supplier with short lead times may be more economical than a lower-priced one with long lead times.

As we noted in Chapter 12 in the context of supplier quality, management of purchased parts is extremely important in assembly systems with many parts. There we pointed out that if we purchase 10 parts with sufficient safety lead times such that each has a service level of 95 percent, then the probability of having all 10 parts arrive in time to meet the schedule is $0.95^{10} = 0.5987$, which represents very poor service. Assembly systems with many purchased parts require extremely high service for each part in order to meet schedules reliably. For instance, for all 10 parts to be available to meet the schedule 95 percent of the time requires that each part have a service level of $0.95^{1/10} = 0.9949$.

Finally, note that it is not necessary to set the same service level for every A part that is ordered on a lot-for-lot basis. If one part is particularly expensive, it might make sense to set its service relatively low (say, 96 percent) and the other service levels higher (say, 99.9 percent) to compensate. If we let S_j represent the service level chosen for the *j*th part and there are *n* parts in total, then we can ensure 95 percent compliance with the schedule provided we choose the S_j values such that

$$S_1 \cdot S_2 \cdots S_n = 0.95$$

A formal method for choosing service levels to meet an overall service level with minimal average investment in inventory is described in Hopp and Spearman (1993a).

17.3.5   Setting Order Frequencies for Purchased Components

The above JIT and lot-for-lot purchasing schemes are reasonable options for expensive A parts, and they might also work for intermediate B parts, but are generally not appropriate

---

[3]If yield loss is a problem, we may also need to maintain a planned level of safety stock.

590 Part III   Principles in Practice

for inexpensive C parts. It doesn't make sense to order screws, washers, two-cent resistors, etc., to be delivered in tight synchronization with the production schedule. The increased risk of an outage and the extra purchasing and material handling costs simply cannot be justified by reductions in inventory investment.

The problem of managing inexpensive purchased parts can be thought of in terms of **lot sizing.** The essential economic tradeoff is between inventory investment and purchasing cost. Recall that this is precisely the tradeoff addressed by the economic order quantity (EOQ) model. Indeed, we could directly apply the single-product model presented in Section 2.2, provided we are willing to ignore part interactions. That is, if we let

N = total number of distinct part numbers in system

D_j = demand rate (units per year) for part j

c_j = unit production cost of part j

A = fixed cost to place an order for any part

h_j = cost to hold one unit of part j for one year

Q_j = size of order or lot size for part j (decision variable)

we can compute the lot size for part j using the standard EOQ formula:

$$Q_j^* = \sqrt{\frac{2AD_j}{h_j}} \tag{17.1}$$

The most difficult input to estimate in this formula is the fixed order cost,[4] A. Ideally, this should reflect those costs that are incurred each time an order is placed. These could include actual shipping costs, purchasing agent time spent to process and follow up on the order, time required to receive the order, and so on. Overhead costs (e.g., maintenance of a purchasing department) should not be included in A_j.

A potential problem with the above approach is that it does not consider interactions between parts, which can occur when (1) parts share common delivery systems and (2) we consider the overall capacity of the purchasing department. For instance, if different parts can share common delivery trucks, then there is an incentive to order parts at the same time, when possible. In Chapter 2, we mentioned the powers-of-two replenishment policy as one way to accomplish this. Given the robustness of the EOQ cost function and the roughness of the input data, a reasonable approach to the multipart purchasing problem is to simply use the EOQ formula to compute an optimal order interval for each part (that is, D_j/Q_j^*) and then round to the nearest power of two of some convenient base ordering cycle. For instance, if weekly orders are practical, then round the EOQ interval to the nearest value in the set: 1 week, 2 weeks, 4 weeks, 8 weeks, etc.

To consider the overall capacity of the purchasing function, we could approach the problem as one of minimizing the total inventory holding cost for all parts subject to the constraint that the *average* order frequency not exceed some specified constant F. Since the total number of purchase orders placed per year is equal to the average order frequency per item multiplied by N, this formulation is equivalent to minimizing the total investment in inventory subject to the constraint that the total number of annual purchase orders not exceed NF. We have found it easier to think in terms of average order frequency, however, and therefore we state the problem in this way.

---

[4]Recall that in Part I we criticized the fixed-order-cost assumption for *production* systems because it frequently acts as a proxy for a capacity constraint, which changes over time and cannot be determined in advance of the schedule. However, for *purchasing* systems, capacity may not be a consideration, and therefore a fixed order cost is a much more plausible modeling assumption.

To formulate a mathematical model, we recall that if the order quantity for part j is Q_j, then the average inventory of part j (in units) is $Q_j/2$, and hence the annual holding cost is $h_i Q_i/2$. The order frequency of part j is D_j/Q_j. Therefore, total holding cost is $\sum_{j=1}^{N} h_j Q_j/2$, and the average order frequency is $1/N, \sum_{i=j}^{N} D_j/Q_j$. Thus, we can express the problem to minimize total holding cost subject to an average order frequency of no more than F as

$$\text{Minimum} \qquad \frac{\sum_{j=1}^{N} h_j Q_j}{2} \qquad (17.2)$$

$$\text{Subject to:} \qquad \frac{1}{N} \sum_{j=1}^{N} \frac{D_j}{Q_j} \leq F \qquad (17.3)$$

Notice that if we replace holding cost h_j by unit cost c_j, then the problem becomes one of minimizing total inventory *investment* subject to a constraint on average order frequency. Some decision makers find it easier to think in terms of inventory investment rather than holding cost. However, the two are equivalent (i.e., result in the same lot sizes) if $h_j = ic_j$, where i is an interest rate. So the decision of whether to use holding cost or inventory investment as the objective is generally just a matter of taste.

This formulation is an example of a **nonlinear programming problem.** The standard technique for solving such problems is the **method of Lagrange,** which converts a constrained optimization problem to an unconstrained one by attaching a penalty to violation of the constraint and incorporating it into the objective (Bazaraa and Shetty 1979). While this sounds complex, it really boils down to finding a fixed setup cost for (17.1) that causes constraint (17.3) to be satisfied. We do this by an iterative search method like the following.

Algorithm (Multiproduct EOQ Model)

> **Step 0.** Pick an initial value for A.
>
> **Step 1.** Use A in Equation (17.1) to compute the lot sizes Q_j for all $j = 1, \ldots, N$.
>
> **Step 2.** Compute the resulting order frequency:
>
> $$F(A) = \frac{1}{N} \sum_{j=1}^{N} \frac{D_j}{Q_j}$$
>
> **Step 3.** If $F(A) = F$, stop.[5] Else,
> > If $F(A) < F$, decrease A
> > If $F(A) > F$, increase A
> > and go to step 1.

The increases and decreases in A can be made by trial and error, or some more sophisticated search technique, such as interval bisection.[6] As long as the method we use takes smaller and smaller steps when we near the optimum, the procedure will eventually converge.

---

[5] Since $F(A)$ is a continuous number, it will never equal F exactly. So we typically stop when $F(A)$ is within some small prespecified tolerance of F.

[6] Basically, bisection starts with two points for A, an upper bound that is too high (i.e., causes $F(A) < F$) and a lower bound that is too low (i.e., causes $F(A) > F$), and tries the midpoint between them. If it is too high, then the midpoint replaces the upper bound; if it is too low, it replaces the lower bound. The gap between the lower and upper bounds will steadily decrease. When it is sufficiently small (i.e., below some specified tolerance), we stop.

At the end of this procedure, we will have the optimal order quantities Q_j^*, $j = 1, \ldots, N$. We also get the appropriate fixed order cost A. An alternate interpretation of this cost is the *decrease in total inventory holding cost per unit decrease in the average order frequency.* If we knew how much we were willing to pay in annual holding cost to decrease the average order frequency by one order per item per year, then we could immediately use this value in Equation (17.1) to compute the optimal order quantities. If, as is often the case, this is a difficult number to come up with, we can run the above algorithm for a variety of values of F and plot the optimal holding cost (or inventory investment, if we use c_j in place of h_j) versus average order frequency. Such a curve would represent the multiproduct analog to Figure 2.3 for the single-product case.

We could directly implement the optimal lot sizes Q_j, $j = 1, \ldots, N$, computed via the above procedure. However, if there are savings to ordering parts simultaneously, it may make sense to round the order intervals associated with these lot sizes to powers of two. We do this by noting that the reorder interval for part i is given by

$$T_j^* = \frac{Q_j^*}{D_j}$$

If we round the T_j^* values to the nearest power of two, then, as we discussed in Chapter 2, orders of different parts will tend to "line up." Of course, this rounding will affect both inventory and average order frequency. If we round the T_j^* values to T_j' values, then our order quantities become

$$Q_j' = T_j' D_j$$

Hence, the actual inventory holding cost will be

$$\frac{\sum_{i=j}^{N} c_j Q_j'}{2}$$

and the actual average order frequency will be

$$\frac{1}{N} \sum_{i=j}^{N} \frac{D_j}{Q_j'}$$

If the increase in inventory investment relative to the optimum is too great, or if the average order frequency is too much larger than the target level F, then the benefits from power-of-two rounding may not justify their costs. If the difference between the actual solution and the optimum is slight, then such rounding is probably worthwhile.

Example:

To illustrate the above procedure, we consider a very simple four-part example with data given in Table 17.1. The objective is to minimize average inventory investment subject to an average annual order frequency of $F = 12$ (i.e., once per month). Note that since the objective is average inventory *investment,* we use a holding cost rate equal to the unit cost $h_j = c_j$.

Table 17.2 summarizes the output of the above procedure applied to this example. The rightmost column in this table gives average inventory investment for each set of order quantities, which is calculated as

$$\frac{\sum_{i=j}^{N} c_j Q_j}{2}$$

To initiate the procedure, we begin with $A = 1$. As shown in Table 17.2, this results in an average order frequency of 96.85, which is much too high. Therefore, A must

TABLE 17.1   Input Data for Multipart Lot Size Example

Part j	D_j	c_j
1	1,000	100
2	1,000	10
3	100	100
4	100	10

TABLE 17.2   Calculations for Multipart Lot Size Example

Iteration	A	$Q_1(A)$	$Q_2(A)$	$Q_3(A)$	$Q_4(A)$	$F(A)$	Inventory Investment ($)
1	1.000	4.47	14.14	1.41	4.47	96.85	387.39
2	100.000	44.72	141.42	14.14	44.72	9.68	3,873.89
3	50.000	31.62	100.00	10.00	31.62	13.70	2,739.25
4	75.000	38.73	122.47	12.25	38.73	11.18	3,354.89
5	62.500	35.36	111.80	11.18	35.36	12.25	3,062.58
6	68.750	37.08	117.26	11.73	37.08	11.68	3,212.06
7	65.625	36.23	114.56	11.46	36.23	11.96	3,138.21
8	64.065	35.80	113.19	11.32	35.80	12.10	3,100.68
9	64.845	36.01	113.88	11.39	36.01	12.03	3,119.50
10	65.235	36.12	114.22	11.42	36.12	11.99	3,128.87
11	65.040	36.07	114.05	11.41	36.07	12.01	3,124.19
12	65.138	36.09	114.14	11.41	36.09	12.00	3,126.53

be increased. So we try $A = 100$. As we would expect, since we are penalizing frequent orders heavily, this results in much higher order quantities, and an average order frequency falls to 9.68. Since this is too low, we now have A bracketed. We know that the optimal value of A (the one that achieves an order frequency of 12) is between 1 and 100. So we try $A = 50$. Since this results in an order frequency of 13.70, it is too low. So we try $A = 75$. This decreases the order frequency to 11.18. Proceeding in this manner, the procedure eventually converges to the desired order frequency. Note that all the calculations involved are easily handled in a spreadsheet, provided that the number of parts is not too large. Indeed, it is a simple matter to use Goal Seek or Solver in Excel to search out the proper value of A.

The last line in Table 17.2 gives us the result from the multipart lot-sizing procedure. These numbers tell us that the optimal lot sizes for parts 1, 2, 3, and 4 are 36.09, 114.14, 11.41, and 36.09, respectively. Notice that the lot size of part 2 is larger than that of part 1, and the lot size of part 4 is larger than that of part 3. This is because part 2 is less costly than part 1 and part 4 is less costly than part 3. Intuitively, optimal lot size is decreasing in cost.

Furthermore, the lot size of part 1 is larger than the lot size of part 3, even though their costs are the same. This is because the demand is greater for part 1. The same

FIGURE 17.2

Inventory investment versus order frequency for multipart example

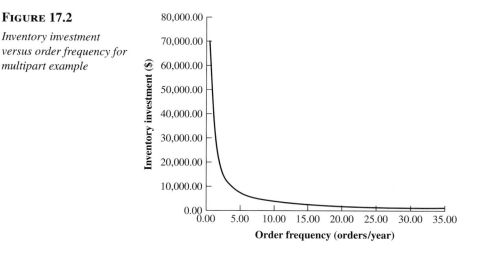

relationship holds between parts 2 and 4. As we would expect, lot size is increasing in demand rate.

Finally, notice that parts 1 and 4 have the same lot size. This is because

$$\frac{D_1}{c_1} = \frac{D_4}{c_4}$$

From expression (17.1), it is apparent that lot size depends on D_j and h_j (and hence c_j) only through their ratio.

The output from the procedure also tells us that $A = 65.138$. This gives us an estimate of the cost (in inventory investment) of changing the average order frequency. Increasing the order frequency by one (to 13 per year) would decrease inventory investment by \$65.14, while decreasing it by one (to 11 per year) would increase inventory investment by \$65.14. However, we must note that these costs are only approximate, since the true cost function is nonlinear. In reality, increasing the order frequency by one will save less than \$65.14, while decreasing it by one will cost more than \$65.14. However, it does give the user a rough idea of the inventory value of more frequent orders.

The resulting value of A also serves as a reality check on our original choice of order frequency target. If the actual cost of placing an order is less (more) than \$65.14, then we should have chosen an order frequency larger (smaller) than 12 times per year. The point is that if we have some idea of what A and F should be, but aren't completely certain about either, then we will get a better solution by cross-checking them against each other and adjusting until both are reasonable.

We can be more exact about the tradeoff between inventory investment and order frequency. Notice that if we keep track of the inventory investment, as we did in Table 17.2, then each choice of A gives us an inventory investment/order frequency pair. Hence, by varying A over a sufficiently wide range, we can generate a graph of inventory investment versus average order frequency. We do this in Figure 17.2. Notice that the inventory investment falls very rapidly as we increase the number of orders per year from zero to five. However, increasing the order frequency above this, and particularly above 10 per year, has a much smaller effect. This type of *diminishing returns* is exactly analogous to the behavior of the single-product model shown in Figure 2.3.

Last, if there are economies to joint orders, we might want to round our order intervals to powers of two. To do this, we first compute the order intervals:

$$T_1^* = \frac{Q_1^*}{D_1} = \frac{36.09}{1,000} = 0.03609 \text{ year} = 13.17 \text{ days}$$

$$T_2^* = \frac{Q_2^*}{D_2} = \frac{114.14}{1,000} = 0.11414 \text{ year} = 41.66 \text{ days}$$

$$T_3^* = \frac{Q_3^*}{D_3} = \frac{11.41}{100} = 0.11414 \text{ year} = 41.66 \text{ days}$$

$$T_4^* = \frac{Q_4^*}{D_4} = \frac{36.09}{100} = 0.3609 \text{ year} = 131.73 \text{ days}$$

Using days as our base time unit, we choose T_1' to be the closest power of two to 13.17, namely, $2^4 = 16$. We choose T_2' and T_3' as the closest power of two to 41.66, which is $2^5 = 32$. And we set T_4' equal to the closest power of two to 131.73, which is $2^7 = 128$. These order intervals translate to order quantities as follows:

$$Q_1' = \frac{D_1 T_1'}{365} = 1,000 \times \frac{16}{365} = 43.84 \text{ units}$$

$$Q_2' = \frac{D_2 T_2'}{365} = 1,000 \times \frac{32}{365} = 87.67 \text{ units}$$

$$Q_3' = \frac{D_3 T_3'}{365} = 100 \times \frac{32}{365} = 8.77 \text{ units}$$

$$Q_4' = \frac{D_4 T_4'}{365} = 100 \times \frac{128}{365} = 35.07 \text{ units}$$

Substituting these into the expressions for inventory investment and order frequency yields

$$\text{Inventory investment} = \frac{\sum_{j=1}^{4} c_j Q_j'}{2} = \$3,243.84$$

$$\text{Average order frequency} = \frac{1}{4} \sum_{j=1}^{4} \frac{D_j}{Q_j'} = 12.12$$

Since we presumably save some effort by combining orders due to the power of two order intervals, it may be acceptable to have a slightly higher average order frequency than the originally desired level of 12. Notice, however, that the inventory investment increases from \$3,126.53 to \$3,243.84. This increased cost must be offset by the benefits of joint replenishment (e.g., fewer separate purchase orders to issue, truck sharing) for the powers of two policy to be worthwhile.

17.4   Managing WIP

The first thing to note about managing WIP is that Little's law

$$\text{CT} = \frac{\text{WIP}}{\text{TH}}$$

implies that for fixed throughput, reducing WIP and reducing cycle time are directly linked. Therefore, the measures we will suggest to increase the efficiency of WIP are *precisely* the same as those one would use to reduce cycle times.

The second important point concerning WIP management is that, as we pointed out earlier, the bulk of work-in-process in most production systems (i.e., disconnected flow lines) is in queue (caused by variability and high utilization), waiting for batch (caused by batching), or waiting to match (caused by lack of synchronization). Thus, WIP reduction programs should be directed at (judiciously) lowering utilization, smoothing out variability, reducing batching, or improving synchronization.

In the following sections, we review techniques for reducing WIP in queue, waiting to move, and waiting to match.

17.4.1 Reducing Queueing

Recall that for a single-machine workstation, with mean processing time t_e, coefficient of variation of processing times c_e, coefficient of variation of arrivals c_a, and utilization u, cycle time can be approximated by

$$\text{CT} \approx \left(\frac{c_a^2 + c_e^2}{2}\right)\left(\frac{u}{1-u}\right) t_e + t_e \tag{17.4}$$

so by Little's law and the fact that $u = r_a t_e$, where r_a is the average arrival rate to the workstation,

$$\text{WIP} = \text{CT} \cdot r_a \approx \left(\frac{c_a^2 + c_e^2}{2}\right)\left(\frac{u}{1-u}\right) u + u \tag{17.5}$$

Thus, to reduce WIP and CT at the workstation, we can reduce the variability of arrivals to the station (c_a^2), the effective variability of the processing times at the station (c_e^2), or utilization (u).

Generic options for achieving these include the following:

1. **Equipment changes/additions.** The simplest way to increase capacity, and hence reduce utilization, of a station is to replace machines with faster models or augment the current machines with additional parallel capacity. While hardly imaginative, this option can be effective. However, to choose good equipment additions, we must consider the purchase cost, the effect on the capacity and variability at the station, and downstream (flow) variability effects. We discuss a framework for this in Chapter 18.

2. **Pull systems.** As we saw in Chapter 10, a pull system will achieve the same level of throughput with a lower average WIP level. The reason is that the releases to the line are coordinated with the status of the line (i.e., work is allowed to enter the line only when there is space for it). This is something like reducing c_a to the front of the line, but not quite. What pull systems really do is to tie releases to the line to completion of work within the line. Most importantly, they establish a WIP cap, which prevents the WIP level in a line from exceeding a specified quantity. Thus, pull systems can *mandate* a WIP reduction. The challenge is to achieve the WIP reduction without a loss in throughput. This requires making some of the other variability reduction or capacity enhancement changes suggested here.

3. **Finite-capacity scheduling.** If releases to the line are made without adequate attention to capacity (e.g., as in MRP), then WIP explosions at bottleneck resources are possible. As Chapter 15 described, a finite-capacity scheduling system can help regulate releases in accordance with system capacity. Although this does not tie releases to production quite as strongly as a pull system (a pull system links releases to *actual* production, while a finite-capacity scheduler links releases to *expected* production), finite-capacity schedulers can substantially reduce WIP by preventing systematic overreleasing to the line. Ideally, one should supplement a finite-capacity scheduling

system with a pull system, in order to keep the system under control when conditions depart from the schedule.

4. **Setup reduction.** All other things being equal, reducing setups will increase effective capacity, and therefore reduce utilization, of a workstation. However, typically when we reduce setups, we run smaller lots and hence perform more setups. Even if the increase in the number of setups completely offsets the capacity increase, as we discussed in Part II, shorter, more frequent setups will decrease effective variability at the workstation (c_e). This will serve to reduce queueing at the workstation and downstream (i.e., because flow variability will also be reduced). Moreover, as we noted earlier, if we can produce smaller batches, we will have less need to store excess production as finished goods inventory.

5. **Improved reliability/maintainability.** Increasing either the mean time to failure or the mean time to repair increases the availability of a machine and hence augments its capacity. In addition, decreasing the mean time to repair can significantly reduce the effective variability of the machine (c_e). Thus, these types of improvement can reduce queueing at a workstation and, by lowering downstream flow variability, also reduce queueing at subsequent stations.

6. **Enhanced quality.** As we noted in Chapter 12, reducing either rework or yield loss can substantially increase capacity and reduce effective variability. Because of this, quality improvement efforts can be major components of a WIP/cycle time reduction program.

7. **Floating work.** Cross-trained workers who can move to where capacity is required can increase the effective capacity of the line. Cross-training also tends to give workers a more global picture of the line and gets more brains thinking about the problems faced at each station in the line. In manual assembly systems, paced or unpaced, the effects of floating work can be achieved by designating certain tasks as "shared." For example, a particular component might be assigned to be attached by either worker A (upstream) or worker B (downstream). Whenever worker A is keeping up with the line, she will attach the shared component. However, if worker A gets behind (e.g., a quality glitch slows her down), then she can pass the component to worker B for him to attach. In general, floating work schemes only work effectively if the incentive system encourages cooperation toward a linewide goal (e.g., throughput).

Finally, we make the same point we made with regard to ABC classification of purchased parts: *Not all WIP need be treated equally.* It may make perfect sense to stratify parts by volume. High volume parts could be assigned to lines with few part families, and hence few setups, where the steadiness of flow facilitates use of a highly efficient pull system. Low volume parts could be produced in a job shop environment, so that high flexibility purchased at the cost of low efficiency would only affect a minor portion of the overall business. This type of **focused factory** strategy can greatly simplify management of a factory with many different parts.

17.4.2 Reducing Wait-for-Batch WIP

Batching for process reasons may be unavoidable (e.g., a batch burn-in operation that requires 24 hours may only be able to provide sufficient capacity when large batches are processed together). Batching for move reasons is another matter. Anything that enables jobs to move from one workstation to the next in smaller batches, and hence with less waiting, will clearly reduce WIP and cycle time. Specific approaches for doing this include these:

1. **Lot splitting.** Remember that **process lots** and **move lots** do not have to be the same. Even if long setup times at a workstation that processes jobs one at a time necessitate large batches for capacity reasons, there is no need to wait until the batch is complete before moving some of the jobs to the next workstation. For instance, a machining center that produces crankshafts in lots of 10,000 (i.e., before setting up to produce a different type of crankshaft) might send them to the subsequent finishing process in lots of 100. In theory, the crankshafts could even be moved one at a time from machining to finishing. The limiting factor is the amount of time required to move the material.

2. **Flow-oriented layout.** More frequent moves can be facilitated by the plant layout. One of the advantages of a cellular layout is that workstations are in close proximity so that material can move easily between them. Material handling systems (e.g., conveyors, AGVs) can also facilitate small lot transfer between workstations, even if they are not physically close to one another.

3. **Cart sharing.** In workstations with multiple parallel machines producing identical product, sharing incoming and/or outgoing carts (or whatever containers are used to move jobs between workstations) can reduce the amount of WIP waiting before and after the workstation. For instance, Figure 17.3 shows 12 machines filling different numbers of outgoing carts (we have not explicitly represented incoming carts). On average, the number of completed parts waiting to be moved to the next workstation in the system with one outgoing cart will be one-twelfth that in the system with 12 outgoing carts. Notice, however, that this assumes that the machine operators spend the same amount of time moving completed parts to the carts in both systems. If, because of geography, operators must walk farther to bring parts to the single shared cart than to put them

FIGURE 17.3

Cart-sharing arrangements

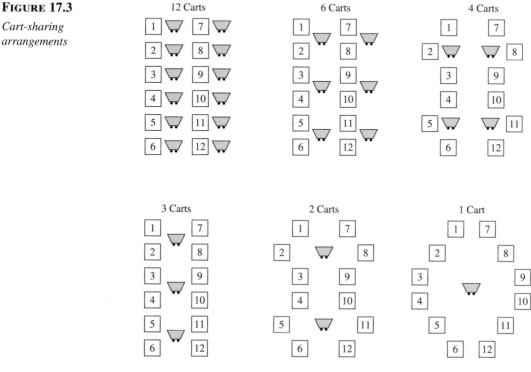

on individual carts in the 12-cart system, then cart sharing can lengthen the effective processing times. Depending on the system, the cycle time reduction from cart sharing might offset that from the capacity decrease. However, in general, cart sharing typically makes sense only where the time and inconvenience are slight. This consideration might make the three- or four-cart arrangement the most practical option for the 12-machine workstation in Figure 17.3.

17.4.3   Reducing Wait-to-Match WIP

At assembly stations, all subcomponents must be available in order for the assembly operation to occur. We have already discussed the problem of managing purchased parts feeding an assembly process in this chapter and in Chapter 12, so we will only consider the situation where subcomponents are produced on different fabrication lines within the plant.

Ideally, we would like to release work orders for the various subcomponents and process them in the fabrication lines so that they arrive at assembly at exactly the same time, in close coordination with the final assembly schedule. Variability generally makes this impossible, but there are things we can do to improve synchronization:

1. **Pull system.** As we know from Chapter 14, a pull system, and a CONWIP system in particular, will naturally synchronize releases into the fabrication lines with final assembly. If fabrication lines are of different length (i.e., in terms of the time required to traverse them), then different WIP levels (card counts) will be needed. This will mean that releases into the fabrication lines at the same time will not necessarily correspond to the same finished product. However, if the WIP levels in the fabrication lines are set appropriately, subcomponent arrivals to assembly will be synchronized.

2. **Common work backlog.** The above CONWIP scheme for coordinating releases with final assembly will only synchronize arrival of subcomponents to assembly if the release sequence is not scrambled in the fabrication lines. If, for instance, local dispatching rules such as shortest processing time (SPT) are used at individual workstations, then jobs can pass one another and synchronization will be lost. Even if we use first-in, first-out (FIFO) at the workstations in the fabrication lines, passing is still possible at multimachine stations. Thus, the way to maintain synchronization with the final assembly schedule is to follow a common **work backlog** at each workstation in the fabrication lines. This backlog simply lists the jobs in order of the final assembly sequence. As long as the fabrication workstations process jobs in the order specified by the backlog, the jobs will arrive synchronized to assembly. If the backlog must be routinely violated (e.g., because of batching or quality problems), then a buffer of WIP will have to be maintained in front of assembly to avoid stoppages due to "out-of-sync" arrivals.

3. **Balanced batching.** If one fabrication line uses large process lots because of a long setup, it may be unable to coordinate with the final assembly schedule. There are three ways to deal with this problem. (1) Produce well ahead of the final assembly schedule on this fabrication line, and maintain a substantial buffer between this line and final assembly. (2) Generate the final assembly schedule in accordance with the batching requirements of the fabrication line. (3) Reduce setup times or augment capacity in the fabrication line so that smaller lots become feasible and it can be synchronized with the desired final assembly schedule. The first two are short-term options; the third may require more time to implement.

17.5   Managing FGI

Finished goods inventory acts as a buffer between production and demand. As we noted earlier, such a buffer may be needed to (1) insulate customers from manufacturing cycle time, perhaps to provide "instant" delivery, (2) absorb variability in either the production or demand processes, or (3) level out capacity loading (e.g., due to seasonality). These imply that anything that links production and demand processes more closely will allow less FGI to be carried. Options for doing this include the following:

1. **Improved forecasting.** While we don't want to raise unrealistic expectations for a forecasting panacea, it is certainly the case that forecasting errors can inflate FGI. If better techniques for forecasting demand, like the time series methods of Chapter 13, can reduce the discrepancies between production and demand, then FGI will be reduced. Despite this fact, there are limits to our ability to predict the future, and so the other options below may be more promising in most systems.

2. **Dynamic lead time quoting.** Many systems quote fixed lead times to customers. However, because plant loading varies over time, actual manufacturing cycle times also vary over time. Therefore, if we set the fixed lead time such that the fraction of time we can deliver within this time is reasonably high, then a high percentage of jobs will finish early. If early delivery is not permitted, these jobs will wait in FGI. We can eliminate this problem by dynamically quoting customer lead times that are sensitive to plant loading.

For example, we worked with a manufacturer of metal cabinets that published 10-week fixed lead times in its product catalog. If it had used a dynamic lead time quoting system, customers who placed orders when the plant was almost empty might have received a two-week lead time, while customers who placed orders when the plant was backed up with work might have received a 12-week lead time. Overall, lead times would be shorter on average, and less product would have to wait in FGI for shipment to attain the same on-time delivery performance.

3. **Cycle time reduction.** A very effective way to reduce forecasting errors is to rely less on forecasting. If cycle time (including the entire value-added chain consisting of time to enter orders, code orders, engineer orders, schedule orders, manufacture products, deliver products, etc.) can be reduced, then work releases can be made closer to their due dates. Since forecasts tend to grow worse with distance into the future, later releases have the effect of making the master production schedule more reliable. If cycle times become short enough, then all releases can be made in conjunction with firm customer orders and therefore FGI due to forecasting errors can be eliminated altogether. Happily, all the WIP reduction techniques listed earlier are also cycle time reduction techniques (Little's law) and therefore are well suited to this purpose.

4. **Cycle time variability reduction.** Chapter 12 pointed out that if we want to guarantee a certain level of service, the lead time quoted to a customer is affected by both the average cycle time and the standard deviation of cycle time (see Figures 12.9 and 12.10). The more variability in cycle times, the more safety lead time we must build into our quotes to ensure a high percentage of on-time deliveries. Higher safety lead times imply that product will spend more time waiting in FGI, unless early delivery is permitted. Fortunately, many of the things we can do to reduce average cycle time (reduce setups, improve reliability/maintainability, implement pull mechanisms, reduce rework and scrap) also serve to reduce cycle time variance.

5. **Late customization.** Even if it is necessary to carry inventory in order to provide short customer lead times, it may not be necessary to carry the inventory in the form of FGI. In some cases, it may be possible to stock the product in semifinished form and assemble or customize to order. Semi-finished inventory is more flexible, provided it

can be used to produce more than one finished product, which makes it possible to carry less total inventory.

For example, a manufacturer of faucet fixtures might offer 20 different models made up of all combinations of five bases and four handle styles. By stocking the bases and handles, the manufacturer need maintain only nine different items in stock, instead of 20. Because of variability pooling, it is easier to forecast demand for the nine parts than for the 20 finished products, and hence less total stock will be required.

As another example, an appliance manufacturer might produce a family of electric mixers that differ according to accessories (a dough hook might or might not be included), retail outlet (labels and packaging might indicate a store brand), and market destination (instructions might be in different languages). By stocking generic families of mixers, distinguished by color of plastic parts, say, the manufacturer could quickly label and package mixers to supply demand for many different finished products. Under this strategy, forecasts would only have to be accurate at the family level, so FGI due to forecasting errors could be considerably reduced.

The potential drawbacks to this type of strategy are that (1) customer lead time is not reduced as much as if FGI is stocked in finished form, which could present a problem if the competition stocks at the FGI level, and (2) storage of semi-finished products can be difficult; for example, dirt and breakage might be a problem if mixers are not boxed.

The ability to store product at the semifinished level can also be a function of product design. For instance, the manufacturer of institutional cabinetry mentioned earlier had 10-week lead times in large part because of its large product line with each product built from scratch (i.e., sheet metal). A competitor was able to offer four-week lead times by offering a smaller product line built around a small set of standard modules (stocked) with different paint colors, face-framing options, and features (faucets, electrical hookups, glass doors, etc.) to allow them to meet customers' needs. Because customers were typically architects who were also frequently behind schedule, responsiveness was highly valued in this market, and the competitor was clearly gaining the upper hand as a result of the shrewd product design strategy.

6. **Balancing labor, capacity, and inventory.** In many markets, product is produced during periods of low demand and held as FGI to meet demand during peak periods. While this may be the best option in some cases, it is by no means the only way to address the problem of seasonal demand. An alternative approach may be to vary the size of the workforce, either by using temporary workers during the peak season or by pairing the product with one with an offset peak (e.g., lawnmowers with snowblowers) and transferring workers between lines. Another—heretical, to most traditional managers—option is to maintain enough excess capacity to meet peak demand without building inventory. When the costs of carrying FGI, obsolescence, and poor customer service due to forecasting errors are considered, it is possible that these other options may be more economical than building large stores of FGI. At the very least, it may make sense to use a combination of approaches, such as a limited inventory buildup, coupled with some excess capacity and some floating labor.

17.6   Managing Spare Parts

Managing spare parts is an important component of an overall maintenance policy, which can be a major determinant of operational efficiency in a manufacturing system. Because of its importance and complexity, a wide variety of spare parts practices are observed in industry (see Cohen, Zheng and Agrawal 1994 for a benchmark study). We will not attempt a survey of these practices. Instead, in this section, we establish a framework for evaluating spare parts inventories and build on the models from Chapter 2 to develop appropriate tools.

17.6.1   Stratifying Demand

There are two distinct types of spare parts, those used in scheduled **preventive maintenance** and those used in unscheduled **emergency repairs.** For instance, a filter may be used in a regular monthly maintenance procedure, while a fuse is replaced only when it fails. The two types of parts should be managed differently.

Scheduled maintenance represents a very predictable demand source. Indeed, if maintenance procedures are followed carefully, this demand may be much more stable than customer demand for finished products. Thus, standard MRP logic is probably applicable to these parts. That is, starting with projected demand, we net against current inventory (and scheduled receipts) and use a lot-sizing rule (lot for lot, fixed order quantity, etc.), to generate planned order receipts, and then back out according to purchasing lead times to generate purchase orders. If the parts are produced internally, we can substitute whatever scheduling procedure is used in place of the fixed purchase lead times to generate a production schedule. In either case, the stable predictable nature of the demand process makes these preventive maintenance parts relatively easy to manage.

Unscheduled emergency repairs are by definition unpredictable. Therefore, using MRP logic for these parts tends to work poorly. We address approaches for maintaining sufficient safety stock to support timely repair of equipment in the following section.

17.6.2   Stocking Spare Parts for Emergency Repairs

For spare parts whose demand is unpredictable, the challenge is to provide high service in a cost-efficient manner. Because demand is uncertain, the (Q, r) model we discussed in Chapter 2 is a potential tool for examining this tradeoff. To apply it, we must decide how to represent service in a multipart environment.

In spare parts systems, service is related to the availability of the machines being supported. Moreover, because a machine that is down for lack of a $2 fuse is just as unavailable as one that is down for lack of a $3,000 computer unit, it is often reasonable to assume that the cost of not having a part on hand is the same for all parts. Therefore, if we can specify either the backorder cost or the stockout cost, we can analyze the parts separately using one of the models of Section 2.4.3.

However, as we have noted before, backorder and stockout costs are often difficult to estimate. In the case of spare parts systems, the reason is that the cost of a part shortage depends on the cost of the machine outage caused by it, which in turn depends on the cost of customer delays caused by the outages. Because of this, it is frequently attractive to think of the problem in terms of a service constraint rather than a service cost. Fortunately, there is a close connection between the cost and constraint formulations.

To adapt the (Q, r) model to the multiproduct case, we make use of the same notation as in Section 2.4.3 with a subscript j to represent parameters for part j, $j = 1, \ldots, N$, so that

$$N = \text{total number of distinct part types in system}$$
$$D_j = \text{annual demand (units per year) for part } j$$
$$\ell_j = \text{replenishment lead time (days) for part } j$$

θ_j = expected demand during replenishment lead time for part j
($\theta_j = D_j \ell_j / 365$)

σ_j = standard deviation of demand during replenishment lead time for part j

$p_j(x)$ = probability of exactly x demands during replenishment lead time for part j (probability mass function)

$G_j(x) = \sum_{y=0}^{x} p_j(y)$, probability that demand for part j during replenishment lead time is less than or equal to x (cumulative distribution function)

A = setup or purchase order cost per replenishment for any part (dollars)

c_j = unit production or purchase cost of part j (dollars per unit)

h_j = annual unit holding cost for part j (dollars per unit per year)

k = cost per stockout for any part (dollars)

b = annual unit backorder cost for any part (dollars per unit of backorder per year). Note that failure to have inventory available to fill a demand is penalized by using either k_j or b_j but not both.

B = desired total backorder level

S = desired average service level

F = desired average order frequency

Q_j = order quantity for part j (decision variable)

r_j = reorder point for part j (decision variable)

$F_j(Q_j)$ = order frequency (replenishment orders per year) for part j as a function of Q_j

$S_j(Q_j, r_j)$ = fill rate (fraction of orders filled from stock) of part j as a function of Q_j and r_j

$B_j(Q_j, r_j)$ = average number of outstanding backorders for part j as a function of Q_j and r_j

$I_j(Q_j, r_j)$ = average on-hand inventory level (units) of part j as a function of Q_j and r_j

With this notation, we can represent the total cost in two ways. We develop both, along with their associated constraint formulations, below.

Backorder Model.   We begin by characterizing service by means of the average backorder level. We can formulate a cost function representing the sum of the setup plus backorder plus holding cost as

$$Y_b(\mathbf{Q}, \mathbf{r}) = \sum_{j=1}^{N} [A F_j(Q_j) + b B_j(Q_j, r_j) + h_j I_j(Q_j, r_j)] \qquad (17.6)$$

where $\mathbf{Q} = (Q_j, j = 1, \ldots, N)$ and $\mathbf{r} = (r_j, j = 1, \ldots, N)$ represent vectors of the order quantities and reorder points. Since the cost function Y_b is simply the sum of separate terms that depend on (Q_j, r_j) pairs, we can minimize it by minimizing the terms for each j separately. But we already did this in Chapter 2. Hence, using the same approximation we used there (i.e., approximating the (Q, r) backorder formula $B_j(Q_j, r_j)$ by the base stock backorder formula $B_j(r_j)$) leads to the same expressions for the optimal order quantities and reorder points:

$$Q_j^* = \sqrt{\frac{2AD_j}{h_j}} \tag{17.7}$$

$$G(r_j^*) = \frac{b}{b + h_j} \tag{17.8}$$

Note that these are the familiar EOQ and base stock formulas. Furthermore, if we assume that lead time demand for product j is normally distributed with mean θ_j and standard deviation σ_j, then we can simplify (17.8) to

$$r_j^* = \theta_j + z_j \sigma_j \tag{17.9}$$

where z_j is the value in the standard normal table such that $\Phi(z_j) = b/(b + h_j)$.

Note that these expressions for Q_i and r_i are sensitive to the differences between parts. For instance, all other things being equal, a high-cost part (which will have a higher h_j coefficient) will have both a smaller order quantity Q_j and reorder point r_j than will a low-cost part. In addition, as we would expect, Q_j and r_j are increasing in the demand rate[7] D_j. In the normal demand case, the reorder point r_j will also increase in the standard deviation of lead time demand provided that $z_j > 0$, which as we noted in Chapter 2 is true as long as $b > h_j$. Finally, we note that increasing the fixed order cost A increases all order quantities Q_j, and increasing the backorder cost b increases all reorder points r_j.

If we can specify reasonable values for the fixed setup (order) cost A and the unit backorder penalty b, we can use formulas (17.7) and (17.9) to compute stocking parameters for the multiproduct (Q, r) system. However, as we observed in Chapter 2, this is frequently difficult to do in practice. In production environments, A is often a proxy for capacity, since the motivation for producing in batches is to avoid capacity losses due to frequent setups. In purchasing environments where capacity is not a direct concern, estimating A directly is much easier. But even in this case, estimating the backorder cost b is problematic, since it involves placing a value on loss of customer goodwill and other intangibles. For this reason, it is often more intuitive to use a constrained model. When service is appropriately characterized by the total number of outstanding backorders (for all part types), then we can formulate the problem as:

> Minimize   Inventory holding cost
>
> Subject to: Average order frequency $\leq F$
>   Total backorder level $\leq B$

We can use an iterative procedure, like that we described for the multiproduct EOQ model earlier, to solve this constrained problem. The basic idea is to first adjust the fixed order cost A until the order frequency constraint is satisfied and then adjust the backorder cost b until the backorder level constraint is satisfied. Note that when we check to see whether a given set of (Q_j, r_j) values satisfies the backorder level constraint, we use the *exact* formula for computing backorder level, not the approximation we used to derive Equation (17.8). Also, because the backorder level $B_j(Q_j, r_j)$ depends on both Q_j and r_j, while the order frequency $F_j(Q_j) = D_j/Q_j$ depends only on Q_j, it is important to adjust A first and b second. We state the procedure formally on the next page.

---

[7]To see that r_j increases in D_j, note that increasing D_j increases θ_j and by Equation (17.9) we see that r_j increases in θ_j.

Algorithm (Multiproduct (Q, r) Backorder Model)

   Step 0. Pick initial values for A and b.

   Step 1. Use A in Equation (17.7) to compute the lot sizes Q_j for all $j = 1, \ldots, N$.

   Step 2. Compute the resulting order frequency

$$F(A) = \frac{1}{N} \sum_{j=1}^{N} \frac{D_j}{Q_j}$$

   Step 3. If $F(A) = F$, go to Step 4. Else,
      If $F(A) < F$, decrease A
      If $F(A) > F$, increase A
   and go to step 1.

   Step 4. Use b in Equation (17.9) to compute the reorder points r_j for all $j = 1, \ldots, N$.

   Step 5. Compute the resulting total backorder level

$$B(b) = \sum_{j=1}^{N} B_j(Q_j, r_j)$$

   Step 6. If $B(b) = B$, stop. Else,
      If $B(b) < B$, decrease b
      If $B(b) > B$, increase b
   and go to step 4.

Stockout Model.   If service is characterized better by average fill rate than by total backorder level, then we can formulate a cost function representing the sum of the setup plus stockout plus holding cost as

$$Y_s(\mathbf{Q}, \mathbf{r}) = \sum_{j=1}^{N} \left\{ A F_j(Q_j) + k[1 - S_j(Q_j, r_j)] + h_j I_j(Q_j, r_j) \right\} \quad (17.10)$$

where $\mathbf{Q} = (Q_j, j = 1, \ldots, N)$ and $\mathbf{r} = (r_j, j = 1, \ldots, N)$ represent vectors of the order quantities and reorder points. As with the backorder cost model, we can optimize this separately for each part j. Using the same approximation we used in Chapter 2 (i.e., that we can compute Q_j using the EOQ model and approximate the fill rate with the type II approximation $S_j(Q_j, r_j) \approx 1 - B_j(r_j)/Q_j$ and approximate the backorder level $B_j(Q_j, r_j)$ by the base stock backorder formula $B_j(r_j)$) leads to the same expressions for the optimal order quantities and reorder points:

$$Q_j^* = \sqrt{\frac{2 A D_j}{h_j}} \quad (17.11)$$

$$G(r_j^*) = \frac{k D_j}{k D_j + h_j Q_j} \quad (17.12)$$

If we further assume that lead time demand for product j is normally distributed with mean θ_j and standard deviation σ_j, then we can simplify Equation (17.12) to

$$r_j^* = \theta_j + z_j \sigma_j \quad (17.13)$$

where z_j is the value in the standard normal table such that $\Phi(z_j) = k D_j/(k D_j + h_j Q_j)$.

   As in the backorder model, these expressions for Q_i and r_i are sensitive to the differences between parts. Again, all other things being equal, a high-cost part will have both a smaller order quantity Q_j and reorder point r_j than will a low-cost part. Also, Q_j and r_j are again increasing in the demand rate D_j, and in the normal demand case, the reorder point r_j will increase in the standard deviation of lead time demand provided that $z_j > 0$. Finally, as we would expect, increasing the fixed order cost A increases all order quantities Q_j, and increasing the stockout cost k increases all reorder points r_j. A difference from the backorder model is that the r_j^* values depend on the Q_j values.

If we can specify reasonable values for the fixed setup (order) cost A and the unit stockout penalty k, we can use formulas (17.11) and (17.13) to compute stocking parameters for the multiproduct (Q, r) system. If, for the reasons discussed and in Chapter 2, we are not able to do this, we can use a constrained formulation. When service is appropriately characterized by the *average* fill rate, then we can formulate the problem as

> Minimize   Inventory holding cost
>
> Subject to: Average order frequency $\leq F$
>
> Average fill rate $\geq S$

We can use an analogous iterative procedure to that used above for the backorder model. As before, we make use of *exact* formulas for computing the fill rate in order to check the fill rate constraint. Again, it is important to adjust A to achieve the order frequency constraint before adjusting k to achieve the fill rate constraint. The formal procedure can be stated as follows:

Algorithm (Multiproduct (Q, r) Stockout Model)

Step 0. Pick initial values for A and k.

Step 1. Use A in Equation (17.11) to compute the lot sizes Q_j for all $j = 1, \ldots, N$.

Step 2. Compute the resulting order frequency

$$F(A) = \frac{1}{N} \sum_{j=1}^{N} \frac{D_j}{Q_j}$$

Step 3. If $F(A) = F$, go to step 4. Else,
> If $F(A) < F$, decrease A
> If $F(A) > F$, increase A
and go to step 1.

Step 4. Use k in Equation (17.13) to compute the reorder points r_j for all $j = 1, \ldots, N$.

Step 5. Compute the resulting total average fill rate

$$S(k) = \frac{\sum_{j=1}^{N} D_j S_j(Q_j, r_j)}{\sum_{j=1}^{N} D_j}$$

Step 6. If $S(k) = S$, stop. Else,
> If $S(k) < S$, increase k
> If $S(k) > S$, decrease k
and go to step 4.

Multiproduct (Q, r) Example.   To illustrate the use of the backorder and stockout models for the multiproduct (Q, r) problem, and the difference between them, we consider the example in Table 17.3. This table gives the unit cost c_j, annual demand D_j,

TABLE 17.3   Cost and Demand Data for Multipart (Q, r) Example

j	c_j ($/unit)	D_j (units/yr)	ℓ_j (days)	θ_j (units)	σ_j (units)
1	100	1,000	60	164.4	12.8
2	10	1,000	30	82.2	9.1
3	100	100	100	27.4	5.2
4	10	100	15	4.1	2.0

TABLE 17.4   Results of Multipart Stockout Model (Q, r) Calculations

j	Q_j (units)	$kD_j/(kD_j + h_j Q_j)$ (unitless)	r_j (units)	F_j (Order Freq.)	S_j (Fill Rate)	B_j (Backorder Level)	I_j (Inventory Level)($)
1	36.1	0.666	169.9	27.7	0.922	0.544	2,410.66
2	114.1	0.863	92.1	8.8	0.995	0.022	670.24
3	11.4	0.387	25.9	8.8	0.749	0.918	512.52
4	36.1	0.666	5.0	2.8	0.988	0.014	189.33
				12.0	0.950	1.497	3,782.75

replenishment lead time ℓ_j, and mean and standard deviation of lead time demand, θ_i and σ_i, respectively. Our objective is to minimize average inventory investment subject to constraints on average order frequency and either average fill rate or average backorder level. Note that since we are using inventory investment as our objective, we set the holding cost equal to unit cost: $h_j = c_j$.

First we address the problem of setting the order quantities Q_j. To do this, we assume a target average order frequency of $F = 12$ orders per year. Notice that the unit cost and annual demand data are identical to those in Table 17.1. Hence, we have already solved this problem because the portion of the multipart algorithms for computing Q_j is identical to the multipart EOQ algorithm. From our previous example, we know that choosing a fixed order cost of $A = 65.138$ yields Q_j values that achieve an average order frequency of 12 per year. These Q_j values are recorded in Tables 17.4 and 17.5.

This leaves only the problem of computing the reorder points r_j. We start by using the stockout model with a target average fill rate of $S = 0.95$. Using the above stockout model algorithm, we find that the penalty cost that makes the average fill rate equal 95 percent is $k = 7.213$. Table 17.4 reports the resulting critical ratios, reorder points, fill rates, backorder levels, and inventory levels for each part. It also computes the average fill rate (95 percent), the total backorder level (1.497 units), and the total inventory investment ($3,782.75).

Notice that the algorithm produces a very high fill rate (99.5 percent) for inexpensive, high-demand part 2, but a low fill rate (74.9 percent) for expensive, low-demand part 3. Intuitively, the algorithm is trying to achieve an average fill rate of 95 percent as cheaply as possible, so it makes service high where it can do so cheaply (i.e., where the unit cost is low) and where it has a big impact on the overall average (i.e., where annual demand is high).

TABLE 17.5   **Results of Multipart Backorder Model (Q, r) Calculations**

j	$b_j/(b_j + h)$ (unitless)	Q_j (units)	r_j (units)	F_j (Order Freq.)	S_j (Fill Rate)	B_j (Backorder Level)	I_j (Inventory Level)($)
1	0.538	36.1	165.6	27.7	0.875	0.974	2,024.77
2	0.921	114.1	95.0	8.8	0.997	0.010	698.76
3	0.538	11.4	27.9	8.8	0.840	0.511	671.85
4	0.921	36.1	7.0	2.8	0.998	0.002	209.10
				12.0	0.934	1.497	3,604.48

An alternative to characterizing service via fill rate is to use the backorder level instead. We can do this by using the backorder model algorithm to adjust the backorder cost b until the total backorder level achieves a specified target. To make a comparison of the stockout and backorder models, we take as our total backorder target the level that resulted from the stockout model, that is, $B = 1.497$ units.

Before going on, we pause to note that establishing a target backorder level is not always an easy thing to do. Unlike the fill rate, which is expressed in a unitless percentage, the total backorder level measures the average number of outstanding backorders at any time. Therefore, one cannot easily translate a backorder level from one system to another (e.g., an average backorder level of five might be horrendous service for a system with few parts and low demand and just fine for a system with many parts and high demand). One way to place the backorder level in a more intuitive context is to think of it in terms of the average wait a customer demand experiences as a result of backorders. If we let W represent the average wait of a demand and D represent the total number of demands per year, then by Little's law

$$B = D \times W$$

or
$$W = \frac{B}{D}$$

In this example, $D = 2,200$ units per year, so a backorder level of 1.497 units translates to

$$W = \frac{1.497}{2,200} = 6.8045 \times 10^{-4} \text{ years} = 5.96 \text{ hours}$$

This means that on average a part (any part, not just one that encounters a backorder situation) will experience 5.96 hours of delay due to lack of inventory. Of course, what this really means is that most parts will encounter no delay, while others will experience significantly longer than 5.96 hours. But looking at the average delay per part gives the decision maker a sense of how much disruption is implied by a given backorder level. Indeed, it is completely equivalent to use hours of delay as the performance target instead of backorder level in the algorithm—all we have to do is to divide by the demand rate and multiply by the number of hours in a year.

Now, supposing that the backorder level target of 1.497 is reasonable, we can use the backorder algorithm to find the backorder penalty that causes total backorders to achieve this level. It turns out that $b = 116.50$ does the trick. Table 17.5 reports the resulting critical ratios, reorder points, fill rates, backorder levels, and inventory levels for each part. It also computes the average fill rate (93.4 percent), the total backorder level (1.497 units), and the total inventory investment ($3,604.48).

Notice that the algorithm results in low backorder levels for inexpensive parts 2 and 4, but higher backorder levels for expensive parts 1 and 3. In addition, it tends to have higher backorder levels for higher-demand parts (i.e., part 1 is higher than part 3, and part 2 is higher than part 4) because higher demand produces more backorders when all other things are equal. As did the stockout model, the backorder model places the bulk of its inventory investment in the expensive, high-demand part 1.

But there are some key differences between the two solutions. Notice that while the total backorder levels are the same, as we forced them to be, the fill rates and inventory levels are different. The backorder model achieves a given backorder level with a smaller investment in inventory ($3,604.48 versus $3,782.75). But it does so at the price of a lower fill rate (93.4 percent versus 95 percent). If we had used the backorder model to adjust the backorder cost b to make the fill rate equal 95 percent, it would have resulted in a higher inventory investment than did the stockout model. The conclusion is that the stockout model finds a policy that efficiently uses inventory to achieve a given fill rate, while the backorder model finds a policy that efficiently uses inventory to achieve a given total backorder level. Thankfully, this is exactly what we would expect them to do. But since the two models articulate different tradeoffs, it is important that we choose the right one for a given situation. If fill rate is the right measure of service, the stockout model is appropriate. If backorder level (or time delay) is a better representation of service, then the backorder model makes more sense.

Finally, we observe that we can use either the stockout or the base stock model to generate a tradeoff curve between inventory investment and either fill rate or backorder level. We do this by simply varying the stockout cost k or the backorder cost b and plotting the resulting pairs of inventory investment and fill rate (or backorder level). Figure 17.4 depicts curves for the previous example for a variety of order frequencies. Note that, as we expect, inventory investment grows exponentially as we approach a 100 percent fill rate. Furthermore, we can see that the inventory reduction from adding an additional six replenishment orders per year diminishes as the number of orders increases. These curves represent **efficient frontiers,** since they represent the lowest inventory investment for each order frequency/fill rate pair. A manager can use a graph like this

FIGURE 17.4

Tradeoff between order frequency, fill rate, and inventory investment in multipart (Q, r) model

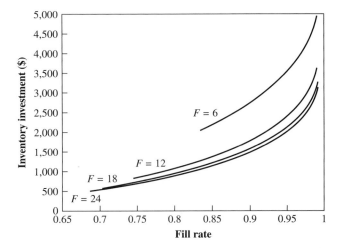

to get a feel for how much investment in inventory is required to achieve various service levels. With this information, he or she can choose a sensible fill rate target. A similar curve of inventory investment versus fill rate could be generated by using the backorder model.

17.7   Multiechelon Supply Chains

Many supply chains, including those for spare parts, involve multiple levels as well as multiple parts. For instance, a retailer might stock inventory in regional warehouses, which supply individual outlets, which in turn supply customers. Alternatively, an equipment manufacturer that offers service contracts on its products may stock spare parts in a main distribution center, which supplies regional facilities, which in turn provide parts to maintain customer equipment. Because of variability pooling, stocking inventory in a central location, such as a warehouse or distribution center, allows holding less safety stock than holding separate inventories at individual demand sites. However, holding inventory in distributed fashion (e.g., at the retail outlets or service facilities) enables swifter response to demand because of geographic proximity. The basic challenge in multiechelon supply chains is to balance the efficiency of central inventories with the responsiveness of distributed inventories so as to provide high system performance without excessive investment in inventory. Research indicates that doing this by directly applying single-level approaches to multilevel problems can work poorly (Hausman and Erkip 1994, Muckstadt and Thomas 1980). This motivates us to give multiechelon systems special treatment.

The complexity and variety of multiechelon supply chains make them very challenging from an analysis standpoint. Serious study of such systems dates back to the classical work of Clark and Scarf (1960) and continues today (see Federgruen 1993, Axsäter 1993, Nahmias and Smith 1992 for excellent surveys and Schwartz 1981 for an anthology on the subject). More modern studies place multiechelon inventory management in the context of supply chain management (see, e.g., Lee and Billington 1992; Fisher 1997; Simchi-Levi, Kaminsky, and Simchi-Levi 1999). Since it is not possible for us to give anything close to a comprehensive treatment here, we will focus instead on defining the issues and indicating how some of the earlier single-level results can be adapted to the multilevel setting.

17.7.1   System Configurations

The defining feature of a multiechelon supply chain is that lower-level locations are supplied by higher-level locations. However, within this framework there are many possible variations, and, if we allow transshipment between locations at the same level (e.g., regional warehouses can supply one another), then the very definition of a level becomes hazy. In short, multiechelon systems can be extremely complex.

For the purposes of our discussion, we will concentrate primarily on **arborescent** systems, in which each inventory location is supplied by a single source (see Figure 17.5). In particular, we will consider the two-level arborescent system in which a single central warehouse (depot, distribution center) supplies multiple retail outlets (facilities, demand sites). We do this because (1) such systems are common in practice; (2) good approximate models of their behavior exist (see Deuermeyer and Schwartz 1981, Sherbrooke 1992, Svoronos and Zipkin 1988); and (3) approaches to the two-level problem can be used as building blocks for developing approaches to more complex multilevel systems.

FIGURE 17.5

Arborescent multiechelon supply chains

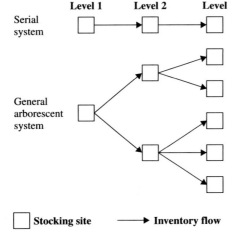

Before we move on to analysis, however, it is important to point out that the system configuration itself is a decision variable. Just because a system is currently configured using a three-level arborescent structure does not mean that this must always be the case. Indeed, determining the number of inventory levels, the locations of warehouses, and the policies for interconnecting them can be among the most important logistics decisions a firm can make about its distribution system. Even though these systems present challenging problems, it is better to address them openly than to miss significant opportunities because the status quo is viewed as immovable.

As an example of this type of rethinking the system configuration, we offer the case of an equipment manufacturer with whom we are familiar. This firm offered service contracts on its equipment (e.g., a guarantee of a maximum number of hours of downtime per month) and stocked spare parts to support the maintenance process. These parts were stocked at three levels: at a main distribution center, at regional facilities, and at customer sites (for customers whose service contracts specified it). Virtually all shipments from the distribution center to facilities were made via overnight mail (except for one facility that was close enough to the distribution center for the maintenance personnel to physically pick up parts needed for repairs). Maintenance personnel replenished on-site inventories from the facilities. Roughly one-half of the total inventory in the system was held at the distribution center, with the remainder in the field (i.e., at facilities and sites).

This configuration raises an obvious question. Why stock parts at a distribution center at all?[8] A facility can receive a part overnight equally well from the distribution center or from another facility. (Indeed, we discovered that the facility managers had an informal system of getting parts from one another via overnight mail when the distribution center was out of stock.) Thus, it might be possible for the distribution center to divide its inventories among the facilities. This would place the inventory geographically closer to the demand sites and therefore make it less likely that customers with broken machines would have to wait overnight for a crucial part. Moreover, if a facility lacked a part, it could still get it overnight, from another facility instead of the distribution center, provided that some facility in the system had the part in stock. The distribution center would cease to be a physical stocking site and would become the logical purchasing agent (i.e., to order parts from vendors or to be manufactured internally) and coordinating

---

[8]We are indebted to Professor Yehuda Bassok for pointing out this "obvious" question to us.

mechanism (i.e., by maintaining the information system that kept track of the location of the inventory in the system). The net result would be that for the same total amount of inventory in the system, customers would receive better repair service. This kind of bold reconfiguration might well offer greater overall benefits than detailed optimization of the existing system.

17.7.2   Performance Measures

To make design decisions or develop a model, it is essential that desired system performance be specified in concrete terms. A host of measures can be used, including these:

1. **Fill rate** is the fraction of demands that are met out of stock. This could apply at any level in the system. It is important to remember, however, that a measure applied to higher levels (e.g., the central warehouse) is only a means to an end. It is the performance of the low levels that actually service customers that determines the ultimate performance of the system.

2. **Backorder level** is the average number of orders waiting to be filled. This measure applies to systems where backordering occurs (e.g., spare parts systems, where a demand must eventually be filled whether or not the part is in stock at the time of the demand). As we noted earlier, backorder level is closely related to the average backorder delay, since we can apply Little's law to conclude

$$\text{Average backorder delay} = \frac{\text{Average backorder level}}{\text{Average demand rate}}$$

For instance, if a particular part has an annual usage of 100 parts per year and the average backorder level is one part, then the average delay seen by a part (any part, not just those that get backordered) is $\frac{1}{100}$ year, or 3.65 days.

3. **Lost sales** is the number of potential orders lost due to stockout. This measure applies to systems in which customers go elsewhere rather than wait for a backordered item (e.g., retail outlets). If every demand that encounters a stockout situation is lost, then the expected lost sales per year is related to fill rate by

$$\text{Lost sales} = (1 - \text{Fill rate}) \times \text{Average demand rate}$$

For instance, if the fill rate for a given part is 95 percent and annual demand is 100 parts per year, then $(1 - 0.95)(100) = 5$ parts per year will be lost due to stockout.

4. **Probability of delay** is the likelihood that an activity (e.g., a machine repair, shipment of a multipart customer order) will be delayed for lack of inventory. This measure is often used in systems where high reliability is required (e.g., aircraft maintenance). In general, the probability of delay in a multipart, multilevel system is a function of the fill rates of the various parts, although depending on the manner in which parts are demanded together (e.g., used on the same repair or customer order), this dependence can be complex (see Sherbrooke 1992 for a more complete discussion).

From these discussions we conclude that fill rate and average backorder level are key measures, since the other measures can be computed from them. For this reason, the majority of mathematical models either use these measures directly or use cost functions that rely on them.

17.7.3   The Bullwhip Effect

An important issue that arises in multiechelon supply chains is that of **channel alignment.** This refers to the coordination of policies between the various levels and can involve information sharing, inventory control, and transportation, among other management decisions. Because there are so many possible decision variables, channel coordination is challenging even when a single firm controls all the levels in the supply chain. When the levels consist of different firms, the problem becomes even more daunting.

A natural response to the complexity of multiechelon supply chains is to treat the various levels independently. That is, allow each level to use local information to implement locally "optimal" policies. Indeed, when levels consist of separate firms, such a strategy is the traditional default. But while natural to implement, the approach of separating levels can lead to very poor performance of the overall supply chain. The most obvious consequence of poor channel coordination is inefficiency (i.e., inventory will be held in inefficient quantities and locations). But a more subtle, though equally damaging, consequence is the **bullwhip effect,** which refers to the amplification of demand fluctuations from the bottom of the supply chain to the top.

Figure 17.6 illustrates the bullwhip effect. Even though demand at the bottom of the supply chain (e.g., retail level) is relatively stable over time, it is quite volatile at the top level (e.g., manufacturer level). This phenomenon was observed by Forrester (1961) in case studies of industrial dynamics models. It was also noted in a behavioral context as part of the well-known Beer Game, developed at MIT in the 1960s (see Sterman 1989). More recently it has been observed in practice. For example, Procter & Gamble noted that retail demand for Pamper brand diapers was fairly stable, while distributor orders to the manufacturing plant were highly variable. Similar behavior has been observed in the demand for printers by Hewlett-Packard and for insulin produced by Eli Lilly. As we know, variability must be buffered—by inventory, capacity, or time. Hence, the bullwhip effect leads to negative consequences, such as excessive WIP, poor use of capacity, long customer backlogs, and expediting costs.

Given that the bullwhip effect is real, the key questions are, What causes it? and What can be done about it? Lee, Padmanabhan, and Whang (1997a, 1997b) classified the causes of the bullwhip effect into four categories. Following their structure, we will summarize these along with potential remedies.

Batching.   At the lowest level of the supply chain (e.g, the retail level) demand is often steady, or at least predictable, because purchases are made in small quantities.

FIGURE 17.6

Demand seen by different levels of the supply chain

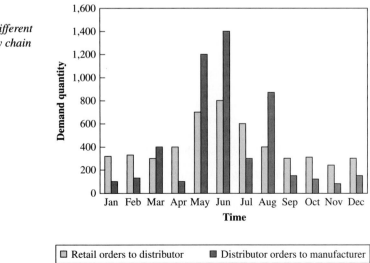

For instance, individual diabetics typically purchase small supplies of insulin, adequate to meet needs for a few weeks or months. Since the diabetics make their decisions independently, total retail demand is extremely level over time. This smoothness would be preserved throughout the supply chain if the retailer replenished its stock directly by placing lot-for-lot orders on the distributor, and the distributor did the same with its orders to the manufacturer. However, if retailers and distributors use some kind of lot-sizing rule (e.g., they follow a (Q, r) policy and hence wait until their requirements justify a replenishment order of size Q), then their demands will be much lumpier than those at the retail level. Furthermore, if there is synchronization among the decision makers at a given level (e.g., they all regenerate their MRP systems at the beginning of the month[9]), then this lumpiness will be even more exaggerated.

Since the amplification of demand variability is the result of batch ordering, policies that facilitate replenishment of stock in smaller quantities will reduce the bullwhip effect. Some options are to

1. *Reduce the cost of the replenishment order.* As we know from Chapter 2, one of the main reasons for ordering in bulk is the cost of placing a purchase order. One way to lower this cost is by using **electronic data interchange (EDI)** to reduce or eliminate purchase orders. By greatly reducing the amount of paperwork involved, such "paperless" ordering systems can facilitate more frequent replenishment in smaller quantities.

2. *Consolidate the orders to fill the trucks.* Another reason for ordering in bulk is the cost of transportation. It is not uncommon for wholesalers or distributors to set their order quantities equal to a full truckload. This is because the cost of shipping in full truckloads is significantly less than that for less-than-full truckloads. However, a truck need not necessarily be filled with the same product. So one way to reduce order quantities while retaining the full-truck cost advantage is to order multiple products from the same supplier. Alternatively, the replenishment process could be turned over to a third-party logistics company, which would consolidate loads from multiple suppliers and/or multiple customers. In either case, the result would facilitate more frequent deliveries.

Forecasting.   In supply chains where the levels are managed by independent decision makers (e.g., they consist of separate companies), demand forecasting can amplify order variability. To see how, suppose that the retailer sees a small spike in demand. Because orders must cover both anticipated demand and safety stock, this leads to an order spike that is larger than the demand spike. The distributor, who forecasts demand on the basis of retailer orders, sees this spike, adds its own safety stock to the anticipated demand, and passes on an even larger order spike to the manufacturer. The reverse situation happens when the retailer sees a dip in demand. Hence, demand volatility increases as we progress up the supply chain.

The basic reason that forecasting aggravates the bullwhip effect is that each level updates its forecast on the basis of the demand *it sees,* rather than on actual customer demand. Hence, policies that serve to consolidate demand forecasting will reduce the bullwhip effect. Some options include these:

1. *Share demand data.* A simple remedy for reducing the amplification effect of separate forecasting at multiple levels is to use a common set of demand data. In supply chains owned by a single firm, sharing demand data from the lowest level is conceptually

---

[9]The phenomenon of synchronized MRP systems causing total demand to spike at certain times is sometimes called the **MRP jitters.**

straightforward (although far from universally practiced). In supply chains involving multiple firms, it requires explicit cooperation. For example, IBM, Hewlett-Packard, and Apple all require sell-through data from their resellers as part of their contracts. In supply chains where the participants make use of EDI, information sharing is relatively simple in principle; the challenge is to achieve the necessary degree of partnering to make it happen.

2. *Vendor-managed inventory.* A more aggressive way to ensure that forecasting is done using low-level demand data is to have a single entity do it. In **vendor-managed inventory (VMI)** systems, the manufacturer controls resupply over the entire chain. For example, Proctor & Gamble controls inventories of Pampers all the way from its supplier (3M) to its customer (Wal-Mart). The fact that alliances using VMI can pool inventory across levels enables them to operate with substantially less inventory than is needed in uncoordinated supply chains.

3. *Lead time reduction.* The magnification effect of forecasting on orders is a function of the amount of safety stock a demand spike drives into the system. But as we saw in Chapter 2, safety stock increases with replenishment lead time. Hence, an obvious, but potentially significant, way to reduce demand volatility due to forecasting is through lead time reduction. Any of the efficiency improvements discussed in Section 17.4 for WIP/cycle time reduction could be practiced at the various levels to achieve this.

Pricing.   Another factor that can cause demand seen at higher levels of the supply chain to "clump up" into spikes is price discounting. Whenever a product's price is low, due to promotional pricing, customers tend to forward-buy (i.e., purchase in greater quantities than needed). When prices return to normal, customers consume the excess stock and hence order less than normal. The result is a volatile demand process.

Since it is price variation that drives demand volatility, the obvious remedy is to stabilize prices. Specific policies for supporting more stable prices are

1. *Everyday low pricing.* The most straightforward way to stabilize prices is to simply reduce or eliminate reliance on promotions using discounting. In the grocery industry, several manufacturers have established uniform wholesale pricing policies and have promoted them via a marketing campaign centered on "everyday low prices" or "value prices."

2. *Activity-based costing.* Traditional accounting systems may not show the costs of some practices resulting from promotional pricing, such as when regional discounts cause retailers to buy in bulk in one area and ship product to other areas for consumption. Activity-based costing (ABC) systems account for inventory, shipping, handling, etc., and hence are useful in justifying and implementing an everyday low-pricing strategy.

Gaming Behavior.   One final factor that contributes to the bullwhip effect is the manner in which customers use their orders in a gaming fashion. For instance, suppose a supplier allocates a product in short supply to customers in proportion to the quantities they have on order. Then customers have a clear incentive to exaggerate their orders in hope of getting more product. When supply catches up with demand, the customers will cancel the excess orders, leaving the supplier awash in inventory. This occurred more than once during the 1980s in the computer memory chip market, when shortages encouraged computer makers to order chips from several suppliers, buy from the first one to deliver, and cancel the remaining orders.

The fundamental issue here is that when gaming behavior is present, customer orders can provide very bad information to the supplier about actual demand. Alternatives for reducing the incentive to game orders include the following:

1. *Allocate shortages according to past sales.* If a supplier facing a product shortage allocates its supply on the basis of historical demand, rather than current orders, then customers do not have an incentive to exaggerate orders in shortage situations.

2. *Use more stringent time fencing.* Recall from Chapter 3 that frozen zones and time fences are tools used to place restrictions or penalties on customers for making changes in orders. If customers cannot freely cancel orders, then gaming strategies become more costly. Of course, a supplier must decide on a reasonable balance between responsive customer service and demand stabilization.

3. *Reduce lead time.* Another situation that can lead to gaming behavior occurs when products involve long-lead-time components. For example, we worked with a printed-circuit board (PCB) plant that supplied computer assembly (box) plants. To assemble the circuit boards, the PCB plant had to purchase both the raw cards and the components to be mounted on them. Some of the components had very long procurement lead times of a year or more. To encourage its customers to communicate demands early, the PCB plant had a series of time fences that restricted the changes in order quantity and type at various lead times prior to the requested due date. However, because the company knew that long-lead-time parts would be difficult to obtain if demands were increased, customers had strong incentive to overestimate their requirements. Sure enough, when we checked the data, we found that at each time fence requirements dropped significantly (e.g., if a time fence allowed a 15 percent reduction in order quantity without cost penalty, then many orders were decreased by exactly this amount when they reached that time fence). The result was to drive excess quantities of the long-lead-time parts into the PCB plant's inventory. One remedy, as suggested above, would be to restrict customers' ability to alter orders. For instance, if the PCB plant had a frozen zone longer than the lead time of all its components, such gaming behavior would not occur. But of course it is not reasonable to impose a one year frozen zone on customers. The alternative, therefore, is to work to reduce lead times of the components so that customers will have less incentive to try to trick the system into overordering for these parts.

Finally, we observe that a sweeping policy for reducing all the factors contributing to the bullwhip effect is to eliminate whole layers of the supply chain. This is precisely what Dell Computer did with its direct marketing system in which computers were sold by the manufacturer to the customer without the use of resellers. In addition to giving Dell access to direct customer demand data, it eliminated a whole level of inventory and hence cost. This strategy played a major part in making Dell one of the most successful companies in America during the 1990s.

17.7.4   An Approximation for a Two-Level System

We now turn to a specific supply chain problem by considering a two-echelon inventory system with a single warehouse that supplies a number of facilities, which in turn supply customer demands. Assume that both warehouse and facilities make use of continuous review inventory control policies, where the warehouse uses a (Q, r) policy and the facilities use base stock policies (i.e., they replenish stock one at a time, so in effect they use (Q, r) policies with $r = 1$). This type of system makes sense for a spare parts system, where speed of delivery is crucial and volumes are relatively small. Thus, facilities are likely to receive shipments of parts from the warehouse on a frequent basis, and one-at-a-time replenishment is a practical option. This assumption may be less appropriate for retail systems, where outlets are replenished less frequently and high volumes make bulk deliveries necessary. We refer the interested reader to Nahmias and Smith (1992) for details on modeling retail systems.

The one-at-a-time facility replenishment assumption implies that demands at the facilities are passed directly back to the warehouse. This means that if demand for each part at each facility is distributed according to the Poisson distribution, then total demand at the warehouse is also Poisson-distributed. (Recall that in Chapter 2 we observed that the Poisson distribution is often a reasonable modeling assumption for representing demand processes.) This allows us to take the following approach. First we analyze the warehouse using a single-level (Q, r) model, where we fix the service level (fill rate) and compute order quantities and reorder points for each part. Then we compute the expected number of backorders outstanding at any point for each part and use this to estimate the delay that an order from a facility will experience. With this, we approximate lead times seen by the facility as the expectation of the actual delivery time from the warehouse plus this delay. Then, using these modified lead times, we apply a base stock model to each facility to compute reorder points for each part.

To develop a model, we will make use of the following notation, which is analogous to that used for the multi-item (Q, r) model above, with additional subscripts m to indicate the facility:

N = total number of distinct part types in system

M = number of facilities serviced by warehouse

$D_j = \sum_{m=1}^{M} D_{jm}$, annual demand (units per year) for part j at warehouse

ℓ_j = replenishment lead time (in days) for part j to warehouse, assumed constant

θ_j = expected demand during replenishment lead time for part j $(\theta_j = D_j \ell_j / 365)$

$p_j(x)$ = probability of exactly x demands during replenishment lead time for part j at warehouse (probability mass function)

$G_j(x) = \sum_{y=0}^{x} p_j(y)$, probability that demand for part j at warehouse during replenishment lead time is less than or equal to x (cumulative distribution function)

W_j = expected time an order for part j waits at warehouse due to backordering

D_{jm} = annual demand (units per year) for part j at facility m

ℓ_{jm} = lead time (in days) for facility m to receive part j from warehouse, assumed constant

θ_{jm} = expected demand during replenishment lead time for part j $(\theta_j = D_j \ell_j / 365)$

$p_{jm}(x)$ = probability of exactly x demands during replenishment lead time for part j at facility m (probability mass function)

$G_{jm}(x) = \sum_{y=0}^{x} p_j(y)$, probability that demand for part j at facility m during replenishment lead time is less than or equal to x (cumulative distribution function)

L_{jm} = lead time (including backordering delay) for an order of part j from facility m to be filled by warehouse, a random variable

c_j = unit cost (dollars) of part j

Q_j = order quantity for part j at warehouse (decision variable)

r_j = reorder point for part j at warehouse (decision variable)

r_{jm} = reorder point for part j at facility m (decision variable)

$$R_{jm} = r_{jm} + 1, \text{ base stock level for part } j \text{ at facility } m \text{ (decision variable equivalent to } r_{jm})$$

$$F_j(Q_j) = \text{order frequency (replenishment orders per year) for part } j \text{ at warehouse as a function of } Q_j$$

$$S_j(Q_j, r_j) = \text{fill rate (fraction of orders filled from stock) of part } j \text{ at warehouse as a function of } Q_j \text{ and } r_j$$

$$B_j(Q_j, r_j) = \text{average number of outstanding backorders for part } j \text{ at warehouse as a function of } Q_j \text{ and } r_j$$

$$I_j(Q_j, r_j) = \text{average on-hand inventory level (in units) of part } j \text{ at warehouse as a function of } Q_j \text{ and } r_j$$

Warehouse Level.   We can solve the warehouse problem (i.e., compute Q_j and r_j for all parts) by using any of the approaches given earlier for the single-level problem. That is, we could use a cost model in which we specify a fixed order cost A and either a backorder cost b or a stockout cost k. Or we could use a constrained model in which we specify constraints on the average number of orders per year F and either the fill rate S or the average backorder level B. Typically, it makes more sense to use a model based on a backorder cost or constraint, rather than one based on fill rate, since the reason for holding inventory in the warehouse is to minimize delay seen by the facilities (and hence the customers).

Regardless of what model we use, we will wind up with a set of Q_j and r_j values, which can then be used to compute F_j, S_j, B_j, and I_j for all parts $j = 1, \ldots, N$ using the functions developed in Chapter 2. We will use these as inputs to the calculations at the facility level.

Facility Level.   Observe that the expected time (in days) an order from a facility waits at the warehouse due to backordering is

$$W_j = \frac{365 B_j(Q_j, r_j)}{D_j} \tag{17.14}$$

Notice that this is nothing more than an application of Little's law to the backorders (i.e., the wait is analogous to cycle time, the backorder level is analogous to WIP, and the demand rate is analogous to throughput). Hence we can estimate the mean effective lead time (in days) for part j to facility m as

$$E[L_{jm}] = \ell_{jm} + W_j \tag{17.15}$$

We could just act as though this mean lead time were a constant and use it in the base stock model to compute performance measures for the facilities. Indeed, researchers have shown that treating these lead times as if they were equal to their means (that is, L_j) can yield reasonable results (see Sherbrooke 1992). However, it is clear that L_{jm} is a random variable that could exhibit a great deal of variability. When an order from the facility to the warehouse finds stock available, $L_{jm} = \ell_{jm}$. But when an order finds the warehouse in a state of stockout, then L_{jm} could be much longer than this. Computing the exact distribution of the effective lead time seen by a facility is complicated (see de Kok 1993). But we can incorporate the effect of lead time variability in an approximate way.

Technical Note

To approximate the variance of the effective lead time of an order from a facility to the warehouse, suppose that there are only two possibilities: Either the order sees no delay and the lead time is ℓ_{jm}, or it does encounter a stockout delay and has lead time $\ell_{jm} + y$, where y is a deterministic delay. Since the probability of stockout is $1 - S_j$ (we will omit the dependence of S_j and B_j on Q_j and r_j for notational convenience), we know that

$$E[L_{jm}] = S_j \ell_{jm} + (1 - S_j)(\ell_{jm} + y) = \ell_{jm} + (1 - S_j)(y) \tag{17.16}$$

But in order for this to match Equation (17.15), we must have

$$y = \frac{W_j}{1 - S_j} \tag{17.17}$$

To calculate the variance of L_{jm}, we first compute

$$E[L_{jm}^2] = S_j \ell_{jm}^2 + (1 - S_j)(\ell_{jm} + y)^2 \tag{17.18}$$

and then

$$\begin{aligned} \text{Var}(L_{jm}) &= E[L_{jm}^2] - E[L_{jm}]^2 \\ &= S_j(1 - S_j)y^2 \\ &= \frac{S_j}{1 - S_j} W_j^2 \end{aligned} \tag{17.19}$$

The standard deviation of the effective lead time to the facility (in days) is therefore approximately equal to

$$\sigma(L_{jm}) = \sqrt{\frac{S_j}{1 - S_j}} W_j \tag{17.20}$$

We can use $E[L_{jm}]$ and $\sigma(L_{jm})$ in a base stock model for each part j at facility m to compute a base stock level R_{jm}.

Integrating Levels.   There are two issues to be addressed to coordinate the two levels: the model to use at the warehouse level and the parameters to use in the model. Once we have chosen these, the above method for modeling the facility level will adjust the base stock levels for facilities accordingly.

In a multiechelon spare parts supply chain, the most natural model for the warehouse level is the backorder model. The reason is that service to the customer is closely related to delay caused by part outages. Hence, the key measure of service at the warehouse is time delay, which we have seen is proportional to backorder level. Therefore, a logical choice of a warehouse model is the backorder (Q, r) model with a constraint on backorder level. We can use the previously described algorithm to compute the order quantities Q_j and reorder points r_j for the warehouse. Equivalently, we could use the backorder model with a backorder cost b instead of a constraint on backorder level. However, it is usually more intuitive to set a target backorder level (or time delay) constraint than it is to specify a backorder cost.

In other multiechelon supply chains, such as retail systems, customer service may be more appropriately measured by the fill rate. For instance, if orders that cannot be filled immediately at the warehouse are either lost or shunted to a (more expensive) third party, then fill rate makes perfect sense as the service measure at the warehouse. However, we would need to modify the model to account for lost sales or a different dependence of the lead times on the warehouse service level.

Once we have a model for the warehouse level, we need to specify its parameters. If we use the constrained backorder model, then the key decisions concern what to use for the order frequency target F and the target backorder level B. The order frequency target can be selected directly by considering the capacity of the warehouse procurement system and hence the number of replenishment orders that it can accommodate annually. Alternatively, we could specify a fixed cost of placing an order A and use this in the multipart EOQ formula (17.7) to compute order quantities.

Selecting the target backorder level is more difficult. How many backorders are allowable at the warehouse depends on what this does to performance at the facilities. Therefore, it is almost impossible to specify a backorder level target a priori. Instead, what we should do is to think of this backorder level target as a variable that we can adjust to seek the best overall system performance. Specifically, we solve the warehouse level using a given backorder level target. Then we solve the facility level so as to achieve the desired backorder level or fill rates at the facilities and observe the inventory holding cost (or investment). Finally we go back and try a different backorder level target at the warehouse and resolve both levels to see if the same performance at the facilities can be achieved with a lower inventory cost. Changing the backorder level target will alter the balance of inventory at the warehouse versus the facilities. The search for a backorder target that achieves the optimal balance can be automated within a spreadsheet or other optimization routine.

Example:

We conclude this section with a two-echelon example. Because our purpose is to highlight the relationship between levels, we will keep things simple by looking at only a single part.

Suppose the example we solved for Jack, the maintenance department manager (Chapter 2, Table 2.6), actually represents the warehouse in a two-echelon supply chain. Jack stocks spare parts at the warehouse in order to supply various regional facilities, which provide the parts for use in actual machine repair. Omitting the subscripts j because this is a single-part example, we see the key data for the warehouse are $D = 14$ parts per year, $Q = 4$, and $r = 3$. Recall that we computed the order quantity $Q = 4$ and reorder point $r = 3$ in Chapter 2 by using the backorder cost model (assuming a fixed setup cost of $A = \$15$ and a backorder cost of $b = \$100$). But we could have just as easily have used a constrained model with constraints on order frequency F and backorder level B.

Now let's extend this example by looking at a single facility with $D_m = 7$ (i.e., the facility accounts for one-half of the annual demand seen by the warehouse). From the calculations in Chapter 2, we know that $B(4, 3) = 0.0142$ unit, so the average time a replenishment order waits due to lack of inventory is

$$W = \frac{365 B(4, 3)}{D} = \frac{365(0.0142)}{14} = 0.3702 \text{ day}$$

Supposing that the actual delivery time to receive a part from the warehouse is one day, the expected lead time for a part is

$$E[L_m] = 1 + 0.3702 = 1.3702 \text{ days}$$

and hence expected demand during replenishment lead time to the facility is

$$\theta_m = \frac{1.3702 \times 7}{365} = 0.0263 \text{ unit}$$

Also from our previous calculations in Chapter 2, we know that the fill rate is $S(4, 3) = 0.965$. Hence, the standard deviation of replenishment lead time is

$$\sigma(L_m) = \sqrt{\frac{S}{1 - S}} W = \sqrt{\frac{0.965}{1 - 0.965}} (1.3702) = 1.944 \text{ days}$$

Assuming that demand at the facility level is Poisson, we can use Equation (2.58) to compute the standard deviation of lead time demand as

$$\sigma_m = \sqrt{\theta_m + \left(\frac{D_m}{365}\right)^2 \sigma(L_m)^2} = \sqrt{0.0263 + \left(\frac{7}{365}\right)^2 (1.944)^2} = 0.166 \text{ unit}$$

Note that in this example $\sigma_m = 0.166$ is very close to $\sqrt{\theta_m} = \sqrt{0.0263} = 0.162$. The reason is that the inflation factor in Equation (2.58) is relatively small. This implies that lead time demand is very close to Poisson. Hence, we can use the Poisson formulas to approximate the service that results from various base stock levels.[10] For instance, if we set the reorder point for the facility equal to $r_m = 0$, then the fill rate is given by

$$G_m(r_m) = \sum_{y=0}^{r_m} p(y) = p(0)$$

$$= \frac{\theta_m^0 e^{-\theta_m}}{0!} = e^{-0.0263}$$

$$= 0.974$$

If we increase the reorder point to $r_m = 1$, then service increases to 0.997. So, depending on the criticality of this part at the facility, it looks as if a reorder point of zero or one will be appropriate.

17.8   Conclusions

Inventory management is as old as manufacturing itself. Analytical approaches to inventory control date back to the scientific management era (i.e., the early 20th century) and are among the earliest examples of operations research/management science. Despite this, the field continues to evolve. Even techniques as old as the EOQ and (Q, r) models are experiencing breakthroughs (e.g., new algorithms and use in multiechelon supply chains). Thus, it appears that the final word on inventory and supply chain management is far from written. The models presented in this chapter provide reasonable approaches to some settings, but better methods and extensions to new settings will undoubtedly evolve. This means that inventory will be an area ripe for continual improvement and that manufacturing managers will need to continue learning new tricks in this important field.

In the meantime, the following tips are worth keeping in mind:

1. *Understand why inventory is being held.* Different types of inventory are held for different reasons, some conscious and others unconscious. Rigorously asking the question of why each type of inventory is held in a given system can reveal inefficiencies that are being taken for granted.

2. *Look for structural changes.* Fine-tuning a supply chain through the use of sophisticated models is fine. However, really big improvements are likely to require structural changes. For instance, changing from a strategy of stocking FGI to one of stocking semifinished product and producing to order might have a dramatic effect on total inventory investment. Similarly, eliminating the central warehouse and stocking all spare parts at regional facilities could produce a substantial improvement in customer service with no increase in inventory. The specific changes that are possible depend on

---

[10]Since the actual variability is slightly greater than the Poisson distribution, actual service will be slightly lower than predicted by the Poisson formulas.

the system. The key to identifying them is to take for granted as little of the status quo as possible.

3. *Use empirical evaluation procedures.* Any model is based on simplifying assumptions (e.g., steady state, Poisson demand), and input data are approximate at best. Thus, the best analysis can do is to help us find a reasonable policy (finding the "optimum" is out of the question) and examine tradeoffs. Given this, we should be careful to supplement analysis with empirical observation and feedback. Examples of parameters we should monitor include (1) service levels, to compare with those predicted by our models and to determine whether policy changes are needed; (2) minimum inventory levels and stockout frequency of stock in raw materials and FGI, to determine whether we are carrying insufficient or excessive safety stock; and (3) queue lengths and starvation time at key workstations, to detect excessive or insufficient WIP. Many other measures may make sense depending on the system. The important thing is to identify a few key measures and set up an adequate data collection and interpretation system for them.

4. *Cycle time reduction is crucial.* Little's law tells us that where there is WIP, there is cycle time. So WIP reduction and cycle time reduction are virtually synonymous. But even more importantly, reduced cycle times make it possible to rely less on distant forecasts in the purchase of components and the scheduling of work. The net result, therefore, is smaller raw materials and FGI levels, as well as less WIP.

5. *Coordinate levels in multiechelon supply chains.* Inventory management grows more complex when stock is held at multiple levels. In addition to managing each level efficiently, it is critical to make sure that performance at the separate levels supports overall system efficiency. The bullwhip effect is an important example of how myopic control of the separate levels can cause huge problems for the system as a whole. To avoid these, it is important to analyze the supply chain as a whole, rather than as separate parts, share common data (e.g., retail demand data) wherever possible, and streamline the supply chain to avoid unnecessary complexity.

6. *Coordinate incentive systems with objectives.* It is well and good to set up an inventory management system with specific performance goals in mind. However, any such system will rely on people to make it work. Therefore, if the reward structure does not support the system goals, it is unlikely to work. (Recall the personnel law: *people, not organizations, are self-optimizing.*) For example, we recently worked for a company with a multiechelon supply chain in which facilities were evaluated primarily in terms of customer service but, in the name of inventory efficiency, also had their inventory levels audited once per month. Predictably, facility managers had a tendency to hoard inventory (i.e., carry more than the recommended levels) all month. Right before the end-of-month audit, they would send the excess back to the distribution center. Once the audit was completed, they would order back up to their "excessive" levels. The effect was to destroy any balance between inventory and service. Clearly, no modeling or analysis effort could correct this problem. Only revising the facility evaluation procedure (e.g., by using ratings that combine service with inventory level, where inventory is measured continuously or randomly in units of dollars) could rationalize the facility inventory levels.

Discussion Point

Suppose a manufacturer of electric mixers sells virtually identical models to several retailers. The major differences between models are the boxes (which are printed with glossy pictures of the mixer and the house brand of the retail outlet) and the paper inserts (which include instructions and retailer-specific information). Demand is strongly seasonal (i.e., peaking around Christmas), so the firm follows a strategy of building inventory (FGI) in the off season. The problem is that while forecasts for total volumes are typically reasonable, the forecasts for individual retailers can be awful. The result is that the firm is frequently short of fast-selling models and awash in slow-moving ones. What general strategies might the firm consider to improve customer service and reduce FGI?

Study Questions

1. Why might the EOQ model be better-suited to purchased parts than to internally manufactured products?

2. How can cycle time reduction reduce raw materials, WIP, and FGI?

3. In general, WIP reduction techniques are also lead time reduction techniques, but the reverse is not always true. List some lead time reduction techniques that do not reduce WIP.

4. What causes large inventories of unmatched parts at an assembly operation? What measures might we consider to address such a situation?

5. What is the difference between type I and type II service? What is the rationale for using type I service in a (Q, r)-type model?

6. Why do we use approximations for fill rate and backorder level in the algorithms for computing Q and r, but check the constraints on these measures against the exact formulas?

7. Suggest appropriate performance measures for evaluating the efficiency of raw materials, WIP, FGI, and spare parts in a manufacturing system.

8. List some examples of arborescent multiechelon supply chains. Can you think of a system that has the reverse of the anborescent structure (i.e., so that many high-level sites supply a few middle-level sites, which in turn supply a single low-level site)?

9. What are the four main causes of the bullwhip effect in multiechelon supply chains? Which causes are likely to have the largest effect in each of the following systems?
   a. A consumer products distribution network, consisting of the manufacturing plant, regional warehouses, and retail outlets.
   b. A spare parts network, consisting of a main distribution center, regional facilities, and customer sites.
   c. A military supply network, consisting of a central warehouse, regional depots, and field usage sites.

10. List some supply chains in which holding the bulk of the stock at the demand level (e.g., at retail outlets) and making use of lateral transshipments might make sense.

11. What incentive or reward system changes might be required to effectively reconfigure a multiechelon supply chain to do away with the central warehouse and store all inventory at regional facilities?

Problems

1. CMW, a custom metalwork shop, makes a variety of products from three basic inputs—bar stock, sheet metal, and rivets—which are purchased in bars, sheets, and kits (boxes of 100), respectively. Projected use and cost of these raw materials for the upcoming year are as follows:

Part	Use (1,000 units/yr)	Cost ($/unit)
Bar stock	120	40
Sheet metal	400	20
Rivet kits	1000	0.5

The shop estimates that issuing a purchase order for any type of material costs $100 and uses an interest rate of 15 percent to calculate holding costs.

a. Assuming steady use throughout the year, estimate the purchasing plus holding cost if all products are purchased four times per year. What happens to cost if we purchase each product 12 times per year?

b. What are the "optimal" order frequencies if we use the EOQ model separately for each product? How many total purchase orders must be placed under this policy?

c. Use the EOQ model to compute order quantities for each part and adjust the fixed cost of placing an order until the *average* order frequency is 12 times per year. How does the holding cost compare to that in part *a* where all parts are ordered 12 times per year?

2. Rivethead Charlie is in charge of the raw materials crib at a facility that manufactures specialized camping gear. In one part of the crib, Charlie stocks connectors. These are not included on the bills of material for the end items, but instead are ordered according to Charlie's "two-bin" system. Under this system, Charlie maintains two bins for each type of connector that hold 1,000 units each. Whenever one bin of a connector becomes empty, Charlie opens up the second bin and orders a refill (that is, 1,000 units) to replenish the first bin. The two most common connectors are rivets, which are used at an average rate of 2,000 per month, and screws, which are used at an average rate of 500 per month. The replenishment lead time from the supplier is two weeks (one-half month), and the unit cost is $0.10 for both rivets and screws. You can assume that demand (use in the manufacturing process) is Poisson for both types of connector.

a. Note that Charlie is following a (Q, r) policy. What are Q and r for rivets and screws under his policy?

b. What are the average fill rate and inventory investment (total for both parts) under Charlie's policy?

c. A summer intern suggests that Charlie should use "days of supply" to set the sizes of the bins, rather than a fixed size of 1,000. What would be the (Q, r) policy that would result if Charlie used bins sized to hold a one month supply of parts? What are the average fill rate and inventory investment under this new policy?

d. Suppose Charlie uses a two-bin policy in which bins hold five weeks (1.25 months) of supply. What are Q and r for rivets and screws, and what are the average fill rate and inventory investment? What do the results of parts *c* and *d* say about the efficacy of using the days-of-supply approach to bin sizing? Is the intern's suggestion a good one?

e. What type of policy might be better than a two-bin policy, with or without the days-of-supply modification?

3. Stock-a-Lot maintains inventories of parts to support repairs of manufacturing equipment. For a subset of its parts, the expected use, unit cost, and replenishment lead time for the upcoming year are forecast as follows:

Part	Use (units/yr)	Cost ($/unit)	Lead Time (months)
1	5	1,000	1
2	10	100	2
3	5	200	6
4	20	1,000	1
5	50	50	3

   a. Find order quantities that make the average order frequency equal to five times per year, by adjusting the fixed order cost and using the EOQ model.

   b. Using the order quantities from part *a*, compute the reorder points so that the fill rate is 95 percent for all parts; and compute the average inventory investment.

   c. Using the order quantities from part *a*, compute the reorder points that achieve an *average* fill rate of 95 percent, by adjusting the stockout cost in the stockout model algorithm.

   d. Compute the average backorder level resulting from the solution to part *c*. Using the backorder model algorithm and the order quantities from part *a*, find the reorder points that attain the same backorder level as part *c*. How does the total inventory investment compare to that from part *c*?

4. Reconsider the Stock-a-Lot problem, and suppose now that the warehouse supplies several regional facilities. Assume the warehouse is stocked according to the policy computed in part *c* of Problem 3. Consider a single facility supplied by the warehouse that has 12-hour actual delivery times and a demand rate for part 4 of 10 units per year. Compute the following for part 4.

   a. Find the expected number of outstanding backorders at the warehouse.

   b. Determine the expected effective lead time to the facility.

   c. Treating demand at the facility as Poisson, find the minimum base stock level for part 4 at the facility that achieves a target service level of 99 percent.

5. A&T, Inc., has a spare parts system that corresponds to the example depicted in Figure 17.4.

   a. A&T's current stocking policy has resulted in an average order frequency of $F = 12$, a fill rate of $S = 0.85$, and an inventory investment of $2,500. Comment on the quality of the policy. If you were to encounter a situation like this in practice, what system elements would you look at in the hope of making improvements?

   b. The president of A&T has demanded a system with a fill rate of $S = 0.95$ and inventory investment of no more than $1,000. What can you say about the feasibility of this demand? How could you respond to it?

6. Windsong, a novelty store that sells wind chimes and related items, stocks the popular "Old Ben" model. Sales are steady at a rate of one per day (365 per year), and demand can be regarded as Poisson. Windsong purchases Old Bens, along with other products from a supplier that makes daily deliveries. Hence, Windsong uses a base stock policy for its products.

   Suppose that the supplier has set its stocking policy such that the fill rate and average backorder level for Old Bens are 89.7 percent and 0.465 day, respectively. Replenishment lead time is seven days.

   a. What is the expected demand during replenishment lead time when delays by the supplier are taken into consideration?

   b. What is the standard deviation of lead time demand? Is it more or less variable than Poisson?

   c. If we assume demand is Poisson, what fill rate will result from a base stock policy with a reorder point of 10? Will the actual fill rate be higher or lower than this?

18 CAPACITY MANAGEMENT

You can't always get what you want.
No, you can't always get what you want.
But if you try sometimes, you just might find
You get what you need.

Rolling Stones

18.1 The Capacity-Setting Problem

Choices about how much and what type of capacity to install have a strong direct influence on a firm's bottom line. Additionally, because **capacity planning** is at the top of the plant planning hierarchy (see Figure 13.2), capacity decisions have a major impact on all other production planning issues (e.g., aggregate planning, demand management, sequencing and scheduling, shop floor control). In this chapter we invoke factory physics concepts to translate strategic capacity decisions into specific tactical terms. Our goal is to provide a framework for capacity planning that explicitly recognizes its impact on the overall plant management process.

18.1.1 Short-Term and Long-Term Capacity Setting

There are many times in the life cycle of a manufacturing facility when it makes sense to adjust capacity. Most often, the motivation is to accommodate a change in the total volume or the product mix of demand. In the short term, the facility can address demand changes through the use of overtime, addition or deletion of shifts, subcontracting, and workforce size changes. These policies were discussed in Chapter 16 in the context of aggregate planning; they are clearly options in capacity planning as well.

Some of these short-term options may also be viable as long-term policies. For instance, we could run three shifts or subcontract part of or all production on a semiper-manent basis. Of course, if we outsource manufacturing of a product to a vendor on a long-term basis, the vendor might eventually decide to sell it directly and become a competitor. Fortunately, however, there are barriers to entry that often prevent this. For example, nonmanufacturing factors such as rights to a recognizable brand name or possession of an effective delivery/service network can be critical. Even if eventual competition is not a serious risk, relying on vendors to manufacture parts or products makes them a significant partner in the quality management process, as we discussed in Chapter 12. Without measures to ensure vendor quality, the decision to outsource manufacturing can seriously hamper a firm's ability to control its destiny.

In the long term, we must go beyond these short-term options and consider permanent equipment, or "bricks and mortar" changes. These involve either major changes to an existing facility or construction of a new facility altogether. In some cases, a firm can permanently increase capacity by redesigning a product, using design for manufacture (DFM) approaches (see Turino 1992, Chapter 7 for a discussion). More frequently, however, the change must come from either adding machines or processing stations or making permanent changes in the productivity of existing equipment or procedures.

18.1.2   Strategic Capacity Planning

Before a firm can consider how much and what type of capacity to install, it must articulate a capacity strategy. Such a strategy hinges on decisions that are very close to the firm's core business plan. For instance, it may need to decide whether to enter a new market, whether to remain in an existing market, to lead or follow in the product innovation process, to make or outsource a product, what segment of the market to pursue, and many other questions. Taken together, these questions are tantamount to the fundamental strategic question of "What business are we in?" which lies beyond the scope of factory physics. The laws of physics can tell us how a particular physical system will behave but not what system we should be interested in. Similarly, the laws of manufacturing can help us design systems to attain specific objectives but cannot tell us what our objectives should be. Therefore, for the purposes of our discussion, we will assume that the above strategic decisions have been made and that the issue is how to evolve a capacity plan to support them.

Once we have decided that we need to add capacity, there are several issues to address:

1. *How much and when should capacity be added?* Should additions be made only when demand has already developed (when we are already losing sales), or in anticipation of future demand? If we don't anticipate demand, should we fill in the overcapacity periods by using short-term measures such as overtime or subcontracting? If we decide to anticipate demand, how far into the future should we try to cover? Adding large increments will satisfy demand farther into the future, will cause fewer construction disruptions, and can take advantage of economies of scale. However, large increments also imply poorer equipment utilization and greater exposure to risk. (What if the forecasted demand does not materialize?) The appropriate approach also depends on the production technology involved. For example, steel mills must generally add capacity in large units in the form of new furnaces or rolling mills, while a metalworking job shop can add small increments of capacity by adding individual machines. See Freidenfelds (1981) for an analysis of these issues.

2. *What type of capacity should be added?* The size of the capacity increment we can add also depends on the flexibility of the equipment we choose. If machines purchased now can be adapted to new products that will be introduced in the future, the risk of installing more capacity than currently needed is substantially less. In today's environment of rapid product change, product lifetimes are often less than the lifetimes of the production equipment; consequently, this type of flexibility has become a key

consideration in choosing new capacity. See Sethi and Sethi (1990) for a review of the different types of flexibility in manufacturing systems.

3. *Where should additional capacity be added?* Should we add capacity by expanding an existing facility, or should we build a new one? Although it is often more expensive to build a new facility than to expand an existing one, the new facility often affords new marketing and distribution efficiencies, for instance by being closer to either suppliers or customers. See Daskin (1995) for models of the facility location problem.

An important strategic concept is known as **production economies of scale.** The basic idea is that unit costs are typically (but not always) less for a large plant than for a small one. Hayes and Wheelwright (1984) discuss three different economies of scale: short-, intermediate-, and long-term.

Short-term economies of scale arise from the fact that in the very near term, many manufacturing costs are fixed. Although adjustable in the longer term, the production facility, its labor force, management, insurance cost, property taxes, and so on, for any given day, are all *fixed*. The cost of these does not depend on production volumes. Indeed, in the near term, the only true *variable* costs are material, some utilities, and some wear on machines. We can express cost per unit as

$$\text{Unit cost} = \frac{\text{Fixed cost} + \text{Variable cost}}{\text{Throughput}}$$

$$= \frac{\text{Fixed cost}}{\text{Throughput}} + \text{Variable unit cost}$$

Thus, in the short term, unit cost decreases as throughput increases.

Intermediate-term economies of scale depend on the run lengths used in production—the number of units of a product that are produced before the facility switches to another product. Given the changeover cost and run length of a particular product, unit cost can be expressed as

$$\text{Unit cost} = \frac{\text{Changeover cost}}{\text{Units per run}} + \text{Running cost per unit}$$

In this case, labor might or might not be fixed. Run lengths can be affected by setting up less frequently (facilitated through setup reduction), by dedicating equipment (so that some product families can be continually run without changing over), and by using specialized equipment (e.g., flexible manufacturing systems). Of course, some of these options can result in larger inventories, as we discussed in Part II.

Long-term economies of scale are functions of plant equipment itself. Economists have long noted that the cost of equipment tends to be proportional to its surface area, while capacity is more closely proportional to volume. To illustrate the implications of this, suppose the equipment is a cube with side length ℓ. Then we can express cost as

$$K = a_1 \ell^2$$

and capacity as

$$C = a_2 \ell^3$$

where a_1 and a_2 are proportionality constants. To express cost as a function of capacity, we solve for ℓ in terms of C, and we get $\ell = a_3 C^{1/3}$, with a_3 representing another constant; then we substitute into the cost expression. This yields

$$K(C) = a C^{2/3}$$

where, again, a is a proportionality constant.

For general (non-cube-shaped) equipment, cost as a function of the capacity can be approximated by

$$K(C) = aC^b$$

where b is typically between 0.6 and 1.

We can now express cost per unit as

$$\text{Unit cost} = \frac{K(C)}{C} = aC^{b-1}$$

Since b is usually less than one, this implies that unit cost tends to decrease with capacity. That is, large plants are more efficient than small ones.

In practice, economies of scale frequently do enable bigger plants to achieve lower unit costs, but not always. There can also be **diseconomies of scale** that cause the organization to lose efficiency as it becomes larger. One place this happens is in distribution. A small compact cell has less material handling than a large plant composed of many process centers. While process centers in the large plant may be more efficient than the single stations of which the cell is composed, jobs must also be moved greater distances. This increases material handling and cycle times. Also since large manufacturing plants typically serve larger areas than small ones, their freight costs are typically higher. In the case of bulky commodity products like bricks, the most profitable plant size may be quite small.

Another form of diseconomy of scale is due to bureaucratization. As the size of the operation increases, so does the required amount of supervision and support. To keep the span of control manageable, the large firm adds layers of management, which further decreases communication effectiveness. This can lead to compartmentalization and turf wars. If not managed carefully, such diseconomies can be very destructive.

Finally, larger plants naturally create more risk. Natural disasters such as earthquakes, fires, floods, and hurricanes will obviously have a greater negative impact on the company if they strike a single large plant than if they affect a single small facility among many. Similarly, poor management, strikes, and the like are more disruptive if the company capacity is concentrated than if it is distributed.

A natural question arises in this context: What is the optimal plant size? This question is largely one of strategy, which is beyond the scope of this book. Moreover, since it involves many firm-specific issues, a general-purpose answer is not possible. The above discussion gives a preliminary overview of the issues to be considered. More detailed treatments are available in the manufacturing strategy literature (e.g., see Hayes and Wheelwright 1984; Schmenner 1993).

In keeping with our focus on plant management, we will assume that the size of the facility has already been determined on the basis of strategic considerations. Thus, we will consider the problem of how to change capacity within a plant to attain a specified set of objectives. In particular, we examine two scenarios: building a new facility and changing an existing one.

18.1.3   Traditional and Modern Views of Capacity Management

To frame the capacity-planning problem at the plant level, it is useful to distinguish between the **traditional** and the **modern** views of the role of capacity (Suri and Treville 1993). The traditional view is based on the interpretation of manufacturing efficiency shown in the left portion of Figure 18.1. Here, the only question is whether there is enough capacity to meet a particular throughput target, and the answer is either yes or no. If utilization is below capacity, then production is feasible; otherwise, it is infeasible.

FIGURE 18.1

*Traditional versus modern
views of capacity planning*

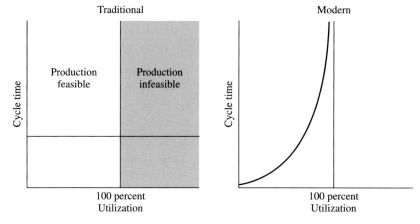

The modern view, which is more realistic and consistent with the principles of factory physics, holds that lead times and WIP levels grow continuously with increasing utilization; this is shown in the right side of Figure 18.1. In this view, there is no one point where production is infeasible. Instead a continuum of decreasing responsiveness occurs as capacity is utilized more heavily.

These two views imply very different approaches to the design of production lines. The traditional view suggests selecting a set of machines that have sufficient capacity, at the lowest possible cost. But doing this usually leads to problems when the line goes into production. We have encountered many plants with lines consisting of machines, each of which has rated capacity above the desired rate, but which consistently fall well short of their throughput targets. (The reader who has absorbed the factory physics principles of Part II should have a pretty clear idea of why such lines fail to meet throughput goals.)

The modern view affords a much richer interpretation of the capacity issue. Since capacity is more than a simple yes-or-no question, we must consider other measures of performance in addition to cost and throughput. WIP, mean cycle time, cycle time variance, and quality are all affected by capacity decisions. If we can state our objectives in terms of these measures, then we can formulate the capacity-planning problem very simply (solving it, however, is a different matter) as follows:

For a fixed budget, design the "best" facility possible.

This formulation is imprecise since what is "best" is difficult to define because we usually have more than one objective. For instance, is a line with low throughput and low cycle time better or worse than one with higher throughput and higher cycle time? As we discussed in Chapter 6, we get around the problem of dealing with multiple objectives by using the technique of **satisficing,** that is, by selecting one measure as the objective and fixing the remainder as constraints. In this way, the problem is divided into a **strategic** problem that defines one or more **tactical** problems. The strategic problem might be to choose how much capacity to have, how long cycle times should be, what types of capacity to use, what throughput is required, and so on. The tactical problem is then to minimize cost or some other quantity subject to the constraints imposed by the strategic problem. This approach of higher-level problems providing constraints for lower-level ones was discussed in Chapter 6.

One formulation would be to maximize throughput subject to a budget constraint and, possibly, constraints on WIP and cycle time. Another would be to minimize cycle time subject to constraints on budget and throughput. Still another would be to minimize cost subject to constraints on throughput, cycle time, and WIP. Which is best depends on the circumstances. If, on one hand, we are concerned with improving an existing line and have a fixed budget to spend, then the formulation to optimize something (maximize throughput or minimize cycle time) subject to a budget constraint makes perfect sense. If, on the other hand, we are designing a new line to achieve given performance specifications, then minimizing cost subject to constraints on things like throughput and cycle time is appropriate.

Regardless of the formulation chosen, we can use the resulting model to examine important tradeoffs. For instance, if we use a model to minimize cost subject to constraints on throughput and cycle time, we can vary the levels of the throughput and cycle time constraints to see how cost changes. The result will be curves of throughput versus cost and cycle time versus cost, both of which are useful in deciding whether our initial strategic specifications were reasonable.

In addition to focusing on the *optimality* of capacity decisions, we must be sensitive to their *robustness*. The requirements we specify today may be quite different from our requirements in the future. It is sometimes a good idea to spend a bit more money up front (e.g., on a capacity cushion, or on more expensive but more flexible equipment) to cover future contingencies. We can consider such options by examining various demand scenarios in the model. However, we must take care not to overbuild for the sake of robustness. One of the reasons that wafer fabrication facilities are enormously expensive is that they are designed in the hope of making almost anything that might be desired in the near future. Because technological uncertainty in semiconductor manufacturing is extremely high, this requires installing the very latest leading-edge (or "bleeding-edge") equipment.

For the remainder of this chapter, we will focus on the problem of minimizing the cost of installing or changing a line, subject to various performance constraints. We have chosen this particular formulation for the following reasons: (1) It is the most natural framework for considering the new line design problem, and (2) it is well adapted to generating cost-versus-performance tradeoff curves. However, one can easily analyze other formulations (e.g., to minimize cycle time subject to throughput and cost constraints) using the tools and techniques we present here.

18.2   Modeling and Analysis

We have relied heavily on models throughout this book, primarily because models force us to think carefully about the systems we are studying and help us develop intuition about how they behave. But at the practical level, without some form of model, either explicit or implicit, one cannot do analysis at all. Accounting, marketing, finance, quality control, and virtually all other business functions rely on models to interpret data, predict performance, and evaluate actions. Happily, the models upon which we rely to address the capacity-planning problem are largely the same as those we used in Part II to explain the concepts of factory physics. In particular, we use the queueing network representation of a manufacturing line to develop capacity analysis tools. Although we adhere to the basic formulas introduced in Part II, there is a large literature on these tools, and we refer the interested reader to Buzacott and Shanthikumar (1993), Suri et al. (1993), and Whitt (1983, 1993) for more details.

For clarity, we concentrate our analysis on a single line and regard the remainder of the production facility as fixed. We assume that the line has M workstations and that the "manufacturing recipe" is given—that is, the operations required at each station to produce the part or product are set in advance. We consider here only the case in which the line produces a single product, although we can accommodate the multiple-product case by attributing the variability due to different processing times of different products at the stations to the natural variability at the process centers (i.e., by inflating the coefficient of variation of the effective processing times). We number the stations $1, 2, \ldots, M$, where jobs arrive to station 1, which feeds them to station 2, which feeds them to station 3, and so on. In this discussion we do not consider rework or branching routings, although these can be accommodated by using more sophisticated versions of the queueing network models (see Suri et al. 1993).

For each station there are a number of different **technology options,** consisting of specific configurations of machines and/or operating policies, from which to select. These options might include different models of machines from various equipment vendors. They might also include a machine with and without a kit of field replacement parts, where the option with the replacement parts has shorter repair times but higher cost than the option without them. Notice that this definition makes identifying an appropriate set of technology options more than a matter of collecting data from equipment vendors. We must make use of our factory physics intuition from Part II to recognize options like field replacement parts that are potentially attractive. We assume here that a reasonable set of technology options can be generated and that cost, capacity, and variability parameters can be estimated for each option.

To keep the number of technology options and the analysis manageable, we assume that no mixing of machine types is allowed at multimachine stations. In other words, if the line requires three lathes and we have chosen the South Bend X-14 as our model, we will use three South Bend X-14s. We cannot use two South Bend X-14s and one Peoria P1000. This restriction is likely to be satisfied naturally in new lines, since we are unlikely to want to deal with two equipment vendors when we can deal with only one. In retrofit situations, it may not be literally satisfied, but is frequently not a major problem from a modeling perspective.

Each option at each station is described by five parameters:

t_e = mean effective process time for machine, including outages, setups, rework, and other routine disruptions

c_e = effective coefficient of variation (CV) for the machine, also considering outages, setups, rework, and other routine disruptions

m = number of (identical) machines at station

k = cost per machine

A = fixed cost of machine option

The total cost of installing the option is given by $A + km$. Thus, if it costs \$75,000 to install one machine and \$125,000 to install two machines, then $A = \$25,000$ and $k = \$50,000$. The idea here is to allow us to represent the costs of activities that need only be done once, regardless of the number of machines installed, such as modifying the electrical or ventilation systems or reinforcing the floor.

We described how to compute t_e and c_e^2 from more basic parameters in Chapter 8. Here we assume that these have already been computed for each option. However, it may be useful to examine the more basic parameters (MTTR, MTTF, etc.) to suggest other technology options.

To formulate constraints for the model, we assume that strategic decisions have been made regarding the overall performance of the line, which establish the following:

$$\text{TH} = \text{required throughput}$$

$$\text{CT} = \text{maximum total cycle time}$$

Then, using the above parameters and a description of the arrival process to the line, we compute the following for each station in the line:

$u(m)$ = utilization of station with m machines installed

$\text{CT}(m)$ = cycle time at station with m machines installed

c_a = CV of arrivals to station

c_d = CV of departures from station

The formulas for computing u and CT are familiar from Part II and can be expressed as

$$u(m) = \frac{r_a t_e}{m} \tag{18.1}$$

$$\text{CT}(m) = \left(\frac{c_a^2 + c_e^2}{2}\right)\left(\frac{u^{\sqrt{2(m+1)}-1}}{m(1-u)}\right) t_e + t_e \tag{18.2}$$

The squared coefficient of variation (SCV) of the arrivals c_a^2 is specified as a parameter for station 1, and for subsequent stations is equal to the SCV of the departures from the previous station. That is, letting $c_a^2(i)$ and $c_e^2(i)$ represent the SCV of the arrival times and effective processing times at station $i (i = 1, \ldots, M)$, respectively, we have

$$c_a^2(i) = \begin{cases} \text{a specified constant} & i = 1 \\ c_d^2(i-1) & i > 1 \end{cases} \tag{18.3}$$

where for $i = 1, \ldots, M$,

$$c_d^2(i) = 1 + [c_a^2(i-1) - 1][1 - u^2(m)] + \frac{u^2(m)}{\sqrt{m}}[c_e^2(i) - 1] \tag{18.4}$$

For a given equipment configuration (i.e., choice of technology option at each station) we use Equation (18.2) to compute $\text{CT}(m)$ and check the total cycle time constraint. If it is violated, we must consider more capacity or a lower variability option. The trick is to change the configuration in the most cost-effective fashion.

Before this can be done, however, we must have a starting point that has sufficient capacity. We call this a **capacity-feasible** solution and give an example of how to find it below.

18.2.1   Example: A Minimum Cost, Capacity-Feasible Line

Consider a four-station line with a throughput target of two and one-half jobs per hour or 60 jobs per day (running three shifts per day). Suppose the SCV of arrivals to the line is equal to 1.0 (recall that we termed this the moderate-variability case in Part II). Thus, TH = 2.5 jobs per hour and $c_a^2 = 1.0$ for the first station. Set the target cycle time for the line at CT = 16. To begin, assume that only one type of machine is available at each station (although we are allowed to choose the number of machines to install at each station). Table 18.1 gives the data for the four stations.

First, we perform a capacity check to determine the minimum number of machines we need at each station. We do this by solving Equation (18.1) for the minimum value of m that keeps utilization below one, that is,

$$u(m) = \frac{r_a t_e}{m} \qquad m < 1$$

or

$$m > r_a t_e$$

TABLE 18.1   **Basic Data for a Line Design Problem**

Station	Fixed Cost ($000)	Unit Cost ($000)	t_e (hours)	c_e^2
1	225	100	1.50	1.00
2	150	155	0.78	1.00
3	200	90	1.10	3.14
4	250	130	1.60	0.10

TABLE 18.2   **The Minimum Cost, Capacity-Feasible Solution**

Station	Number of Machines	Utilization	Cost ($000)
1	4	0.94	625
2	2	0.98	460
3	3	0.92	470
4	5	0.80	900
Total			2,455

For the first station,

$$r_a t_e = 2.5 \text{ jobs/hour } \times 1.5 \text{ hours} = 3.75$$

which indicates we require at least four machines. Table 18.2 summarizes the other machine requirements and their corresponding utilization.

Note that for station 4,

$$r_a t_e = 2.5 \text{ jobs/hour } \times 1.6 \text{ hours} = 4.00$$

However, this would yield a utilization of exactly 1.0. Since the utilization law of factory physics stated that utilization must always be *strictly less than* 1.0, we must assign five machines to station 4, thereby lowering the utilization to 0.80.

Note that the solution in Table 18.2 is the least-cost configuration that has sufficient capacity. This is called the **minimum cost, capacity-feasible (MCCF)** configuration and in this case costs $2,455,000.

It is easy to extend this analysis to find the MCCF configuration when there is more than one technology option at each station. For each station we determine how many machines of each option are required to meet the capacity target and choose the option with the smallest total cost. Doing this for each station will result in an MCCF configuration for the line.

18.2.2   Forcing Cycle Time Compliance

Once we have a capacity-feasible configuration, we then check the cycle time, using Equations (18.2) and (18.4).

Station 1:

$$CT(4) = \left(\frac{1.0 + 1.0}{2}\right)\left(\frac{0.94^{\sqrt{2(4+1)}-1}}{4(1 - 0.94)}\right)1.5 + 1.5 = 6.72 \text{ hours}$$

$$c_d^2 = 1 + (1 - 1)(1 - 0.94^2) + \frac{0.94^2}{\sqrt{4}}(1 - 1) = 1.0$$

Station 2:

$$CT(2) = \left(\frac{1.0 + 1.0}{2}\right)\left(\frac{0.98^{\sqrt{2(2+1)}-1}}{2(1 - 0.98)}\right)0.78 + 0.78 = 15.82 \text{ hours}$$

$$c_d^2 = 1 + (1 - 1)(1 - 0.98^2) + \frac{0.98^2}{\sqrt{2}}(1 - 1) = 1.0$$

Station 3:

$$CT(3) = \left(\frac{1.0 + 3.14}{2}\right)\left(\frac{0.92^{\sqrt{2(3+1)}-1}}{3(1 - 0.92)}\right)1.1 + 1.1 = 8.87 \text{ hours}$$

$$c_d^2 = 1 + (1 - 1)(1 - 0.92^2) + \frac{0.92^2}{\sqrt{3}}(3.14 - 1) = 2.0$$

Station 4:

$$CT(5) = \left(\frac{2.0 + 0.1}{2}\right)\left(\frac{0.80^{\sqrt{2(5+1)}-1}}{5(1 - 0.80)}\right)1.6 + 1.6 = 2.59 \text{ hours}$$

The sum of these cycle times is 34 hours, which is significantly greater than the target of 16. Clearly, the line needs changes to obtain a design that complies with the strategic specifications.

There are three basic improvement alternatives: (1) Modify the existing machines, (2) change the machine options, or (3) add more machines. Chapter 9 described how to use factory physics principles to *diagnose* problems in a line. This approach could be used to determine the cause of long cycle times (e.g., long and infrequent outages) and therefore what machine modifications would be most effective. It might be worthwhile to spend money to reduce variability or speed up a machine rather than to purchase an additional one. Of course, if we are designing a new line, there are no "existing" tools, and hence alternative one is not available.

Altering machine options in the pursuit of shorter cycle times might entail purchasing a different and perhaps more expensive machine with better operating characteristics (e.g., faster rate or smaller process variability). Often, however, especially in high-tech situations, the number of distinct machine types is quite limited. In some cases there may be only a single equipment vendor available. When this is the case, most of the technology options that can be used to reduce cycle time are modifications of a given machine type. Modifications include speeding up the machine, reducing setup time, reducing MTTR, and so on.

The most obvious way to reduce excess cycle time is simply to purchase more machines. If capacity comes in small increments, this might well be the most economical approach.

Depending on the size of the required reduction in cycle time, the range of available technology options, and the cost and size of capacity increments, the best approach may consist of any number of combinations of these types of alternatives.

18.3   Modifying Existing Production Lines

We now offer a heuristic procedure for determining a least-cost configuration that meets the throughput and cycle time constraints. The heuristic starts with the MCCF configuration and then looks for the change that results in the "biggest bang for the buck" with respect to cycle time improvement.

To illustrate this approach, we reconsider the example of Table 18.1. Recall that the minimum cost, capacity-feasible configuration (Table 18.2) did not satisfy the cycle time constraint. Specifically, desired total cycle time was 16 hours, but the resulting total cycle time of the minimum cost configuration was 34 hours. We now consider how to bring the configuration into cycle time compliance in a cost-efficient fashion. Note that this is precisely the type of problem faced by firms trying to implement the methods of cycle time reduction or time-based competition in an existing facility.

To make the example more realistic, suppose we can modify as well as add machines at each station. In particular, suppose that by spending $10,000 per machine at the third station, we could alter long and infrequent random outages to shorter but more frequent ones with the same availability (recall the discussion in Chapter 8 that showed why this is desirable). We might be able to accomplish this by installing field replacement parts and/or doing more preventive maintenance. We assume here that this does not change t_e, but does reduce c_e^2 from 3.14 to 1.0. Using these cost and performance data, we can consider this variability reduction option as an alternative to adding machines.

Hence, these are the available options: At any station, we can add a machine; at station 3, we can either add a machine or reduce machine variability by changing the characteristics of the machine. For each alternative, we can compute the change in cycle time at the station and the change in cost.[1] A reasonable measure of the effectiveness of the change is the ratio of the change in cost to the change in cycle time. The "best single change" is that with the lowest ratio. We compute these ratios for each option in Table 18.3.

The first thing we notice from Table 18.3 is that no single change reduces total cycle time by enough to satisfy the cycle time constraint—we need an 18-hour reduction. The smallest ratio is obtained by modifying the machine at station 3 (by reducing repair time variability) with cycle time reduced by 4.49 hours at a cost of $30,000. This takes us down to 29.51 hours, still considerably longer than the 16 hours allotted. If we repeat the analysis, the minimum ratio occurs by adding a machine to station 2, which costs $155,000 and further reduces cycle time by 14.7 hours. This takes us down to 14.81 hours, which is within the 16-hour constraint.

Although we are not guaranteed that repeatedly choosing the best single change will bring us within the cycle time constraint at a minimum cost, this approach usually works well. In any case, it does yield a configuration that is throughput- and cycle time–feasible. For this example, the resulting solution is given in Table 18.4.

The total cost is $2,640,000, or $185,000 more than the MCCF configuration. In addition, notice that this line is not even close to balanced. Surprisingly, the most expensive station (number 4) has the lowest utilization. This is because both the fixed cost and the unit cost at station 4 are quite high, and because four machines at station 4 result in 100 percent utilization.

---

[1] We ignore what might happen downstream at this point, so our calculations are actually approximations of the change in cycle time for the entire line. It is easy enough to go back and check the line cycle time for a specific option, and for that matter it is not too hard to include downstream effects when estimating the effect of a single change. However, if we do this, we can only evaluate changes one at a time—the reduction in total cycle time from two options together is *not* necessarily the sum of the reductions from each separately.

TABLE 18.3   Cost and Cycle Time Impacts of Improvement Alternatives

Station	Current Number of Machines	Change	Cost Increase ($000)	CT Decrease (hours)	Ratio ($000/hour)
1	4	Add machine	100	4.63	21.61
2	2	Add machine	155	14.73	10.52
3	3	Add machine	90	7.20	12.49
3	3	Reduce variability	30	4.49	6.67
4	5	Add machine	130	0.71	183.10

TABLE 18.4   Capacity- and Cycle Time–Feasible Configuration

Station	Number of Machines	Utilization	Station Cost ($000)
1	4	0.94	625
2	3	0.65	615
3	3 (modified)	0.92	500
4	5	0.80	900
Total			2,640

18.4   Designing New Production Lines

The problem of designing a new line is different from that of modifying an existing one, in that there are typically many more options to consider. In a new line, we are not constrained by existing machines, facilities, or even structure. Indeed, we may have *so much* freedom that the problem becomes almost impossible to solve in an optimal fashion.

18.4.1   The Traditional Approach

In the 18th century, when the first factories were designed, a major consideration was how to arrange the various operations in order to run them from a single source of power—the waterwheel. Consequently, operations were arranged in linear fashion along the waterwheel shaft, each connected to a belt on a properly sized gear to obtain the required turning speed from the waterwheel. Today, it is not uncommon to find factories that follow this traditional design, their process centers laid out in straight lines within a rectangular facility.

We found this curious, since manufacturing plants have not relied on water power for 150 years, and we questioned several architectural engineers who design complex plants (e.g., wafer fabs) and manufacturing engineers who work in existing plants. We discerned that a typical procedure for designing new plants and new lines goes something like this:

1. Establish the basic size and shape of the new facility.
2. Determine where the support facilities (electricity, steam headers, process gases, etc.) should go to minimize the cost of the facility.

3. Determine where the workstations should go within the facility to minimize cost.

4. Determine the product flow.

Given this, the tendency toward linear layouts is not surprising. Since the design process *starts* with the size and shape of the facility, tradition exerts strong influence over the resulting design. But there are obvious problems with this scheme. The most serious is that little consideration is given to product flow until after most of the plant has been designed.

18.4.2   A Factory Physics Approach

A good alternate approach is to view the problem from a customer perspective. This makes it clear that the main purpose of a line or plant is to provide quality product in a timely and competitive fashion. A facility design process consistent with this goal, which is almost the reverse of the traditional approach, is the following:

1. The customer determines the product. Mixes, volumes, and cycle times are forecast.

2. The product(s) determine(s) the processes. For most products, there is a basic recipe of steps that must be done to produce a unit.

3. The processes determine a basic set of machines. Machine descriptions will start out very general and will acquire detail as the planning process evolves.

4. The machines determine the facilities needed to support them.

5. The facilities determine the overall structure and size of the plant.

Of course, if we were to literally follow this procedure, we could end up with a facility that is well equipped to make the product in the volumes desired but is too costly to build. Focusing solely on product flow in order to minimize cycle times may lead us to install multiple expensive machines when one would have done. For instance, in a wafer fab, the photolithography operation is typically one of the more expensive machines in the fab. Its facility requirements are enormous, and to make matters worse, the wafers must visit the operation for each layer (often 10 or more) applied during fabrication. A pure cycle time minimization perspective might suggest installing 10 sets of equipment at a tremendous cost. A pure cost minimization perspective would call for only one set of equipment. The "best" option can only be determined by considering photolithography in the context of the other operations and comparing relative costs of different configurations that meet performance targets.

As a result, it makes sense to approach the facility design problem from a combination of the traditional and factory physics perspectives. We start with an idea of the basic processes and layout of the factory. Using the basic layout, we install the process centers, sizing them to meet desired throughput and cycle time levels. If the resulting configuration results in too high a facility cost, we reconsider the basic layout. On the other hand, if cycle times are excessive, we consider installing more support facilities to improve process flows.

As part of the analysis, we might also want to do a Pareto analysis of the product mix to determine if a "factory within a factory" concept is applicable. If most of the volume is for a relatively small number of products, it may make sense to duplicate processes in the plant. One set, in a tight flow line configuration, is dedicated to the small number of products representing the large portion of throughput. The other is arranged in more of a job shop configuration that maximizes flexibility at the expense of lower utilization or

FIGURE 18.2

*Plot of total equipment
cost versus total cycle time*

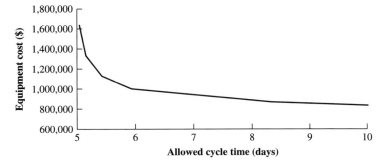

higher cycle times. Low utilization should be expected in this portion since the volumes are (by design) low.

Once we have settled on a basic layout, we turn to detailed selection of specific options and numbers of machines. A relatively simple procedure is to start with the MCCF configuration and then successively choose the best single change, as described, to bring the line into cycle time compliance. To be effective, we should include as many available technology options (i.e., including both purchasing additional machines and modifying machines and/or procedures on site) as we can without overwhelming the decision maker. We want to avoid overlooking an inexpensive modification that alleviates a performance problem and eliminates the need for additional expensive machines. Factory physics diagnostic procedures (Chapter 9) are useful in identifying promising options.

Of course, as we know, the performance requirements (e.g., throughput and cycle time targets) are themselves decision variables. Although we can specify plausible values to start the analysis, it makes sense to examine tradeoffs between cost and performance. For example, if we could shorten cycle times by five days at a cost of $100,000, we might well decide to do it. We can do this with our model by solving it for various values of the throughput or cycle time constraints in order to generate a cost-versus-performance curve. A typical plot of cost versus total cycle time is shown in Figure 18.2. While the model cannot specify which point on this curve is optimal, it does provide useful information to help the decision maker make a rational choice.

18.4.3 Other Facility Design Considerations

These discussions offer some perspective on how to incorporate cost, throughput, cycle time, and other factors into a customer-oriented facility design process. However, there is more to the facility design problem than we have dealt with here. Indeed, there exists a vast literature called, broadly, **plant layout** or **facilities planning,** which deals with topics ranging from the placement of various process centers to minimize product flow, to determining the number of employee parking spaces. This literature addresses the important issues of materials handling, physical plant layout, storage and warehousing, office planning, facility services, and developing and maintaining facilities plans. We suggest Tompkins and White (1984) as a good introduction to this field.

18.5   Capacity Allocation and Line Balancing

As the previous example illustrated, factory physics procedures for line design are unlikely to result in a balanced line. The reasons are as follows:

1. An unbalanced flow line with distinct bottleneck is easier to manage and exhibits better logistical behavior (i.e., has a characteristic curve closer to the best case) than a corresponding balanced line.
2. The cost of capacity is typically not the same at each station, so it is cheaper to maintain excess capacity at some stations than at others.
3. Capacity is frequently available only in discrete-size increments (e.g., we can buy one or two lathes, but not one and one-half), so it may be impossible to match capacity of a given station to a particular target.

When appropriate consideration is given to these factors, the optimal configuration of most flow lines will be an unbalanced line.

18.5.1   Paced Assembly Lines

Despite the arguments in favor of unbalanced lines, sometimes line balancing makes sense. Indeed, the line-of-balance (LOB) problem is a classic problem in industrial engineering. However, it is applicable only to **paced assembly lines,** not **flow lines.** In a flow line, stations are essentially independent. Each station operates at its own speed, so the bottleneck is the slowest station in the line. In a paced assembly line, parts flow through the line on a belt or chain that moves at a constant speed. The parts move through **zones** that usually contain one or more operators. The line is designed so that the operators will almost always be able to complete their task while the part is in their zone. If not, the line would be disrupted as workers tried to finish tasks in the next worker's zone. Hence, the bottleneck of a paced assembly line is not the slowest station in the line but the line-moving mechanism itself.

Additionally, capacity increments in a paced assembly line are usually much smaller than those in a flow line. In a paced assembly line, tasks are typically assigned to workers on the line and can be split into fine increments. For example, in a manual electronic assembly operation, each station "stuffs" circuit boards with a number of components. Since there are many components, the line can be balanced by adjusting the amount of stuffing done at each station. A discussion and an example technique for solving the LOB problem are given in Appendix 18A.

Another justification for a balanced assembly line is one of personnel management. No one likes to be in a situation in which they are constantly expected to do more than their peers for the same pay. Since most assembly lines are staffed by people (although some assembly lines use robots), the issue of fairness is an important one. In these cases a line in, which each station has nearly the same amount of work is desirable.

In contrast, in a flow line, the tasks depend more on the machines themselves and are therefore less easily divided. To increase capacity at a particular station, we must either add an additional machine to that station or speed up the existing ones. Unfortunately, the notion of a balanced line has become so ingrained that it is often applied when it is inappropriate. This and the desire to have high utilization are the reasons one frequently encounters nearly balanced flow lines.

18.5.2   Unbalancing Flow Lines

The previous reasons for unbalancing flow lines suggest that a process with small and inexpensive capacity increments should never be a bottleneck. Such a process can easily and inexpensively add small increments of capacity until it no longer causes problems due to insufficient capacity. On the other hand, a process for which capacity comes in large expensive blocks is a good choice to be the line bottleneck.

As an example, consider two different process centers in a circuit board plant: copper plate and manual inspect. The manual inspect operation occurs before the copper plate operation.[2] Copper plate utilizes a machine that involves a chemical bath along with enormous amounts of electricity. Each machine has a capacity of around 2,000 panels per day. Adding an additional machine at copper plate costs more than $2 million in machine and facility costs and requires a significant amount of floor space. Copper plate represents one of the largest and most expensive machines in the plant. In contrast, each of the stations in manual inspect requires one semiskilled operator, an illuminated magnifier, and a touch-up tool. Each station can inspect around 150 panels per day. None of these stations costs more than $100, and the floor space requirements are small.

If these were the only two stations in the line, the situation would be easy to analyze. If we designate the copper plater to be the bottleneck, then we can easily and inexpensively keep it from starving by adding capacity to the manual inspect operation. It is of little consequence that manual inspect is not fully utilized. On the contrary, to designate manual inspect as the bottleneck and to keep it from starving,[3] we would have to add a large and costly increment of capacity to the copper plate operation. Thus, it makes more sense to designate copper plate as the bottleneck and to manage it accordingly.

18.6   Conclusions

This chapter has focused primarily on applying the factory physics framework to the design of new production lines and improvement of existing ones with respect to capacity. Our main points can be summarized as follows:

1. *Capacity decisions have a strategic impact on the competitiveness of the manufacturing operation.* A capacity strategy has a strong direct effect on costs and many indirect effects on performance by influencing other planning and control problems, including aggregate planning, scheduling, and shop floor control. Decisions include how much, when, where, and what type of capacity to add. Other strategic issues involve various economies and diseconomies of scale.

2. *Factory physics formulas can provide the basis for line design and improvement procedures.* By allowing computation of throughput, cycle time, and WIP for a given configuration, these formulas enable us to frame the line design or improvement problem as one to minimize cost subject to specified throughput, cycle time, and/or WIP constraints. By varying the constraints, we can also generate cost-versus-performance constraints.

3. *Capacity additions and equipment or procedure modifications can be viable alternatives and/or complements to one another.* For instance, reducing repair times on an existing machine can sometimes have similar logistical effects as adding capacity to a station in the form of additional machines. All other things being equal, the value of procedural changes is typically greater than that of equipment additions, because the learning and discipline gained from improving a line can be translated to other lines, while simple capacity additions offer no such learning opportunities.

---

[2]The capacities, capabilities, and even the process description have been altered here from those in a circuit board plant in which the authors have consulted.

[3]Recall that in a CONWIP line, there really is no *front* to the line. Thus, workstations earlier in the line can be starved by later workstations if the pull signals (i.e., the CONWIP "cards") are not returned in a timely manner.

4. *Flow lines should generally be unbalanced.* Logistical and cost differences between stations make it sensible to configure flow lines to have different levels of utilization at the stations.

5. *Paced assembly lines should generally be balanced.* On paced assembly lines it is the pacing mechanism (e.g., the conveyor or chain) that is typically the bottleneck. To enable workers to complete their assigned tasks within the allotted pacing time, as well as to allocate work fairly, it makes sense to divide tasks among stations as evenly as possible, subject to precedence and discreteness requirements.

It is important to note that lines designed using factory physics procedures are likely to be more expensive than lines designed using a traditional minimum cost, capacity-feasible approach. However, they are also much more likely to do what they were designed to do. When one considers factors such as lost sales due to inability to meet throughput targets, loss of customer goodwill due to inability to meet cycle time targets, and the confusion that results in trying to operate a line that is in a constant state of chaos, the more expensive factory physics lines are likely to be much more profitable in the long run.

APPENDIX 18A
THE LINE-OF-BALANCE PROBLEM

Assigning tasks to stations on a paced assembly line should be done so that each station has nearly the same amount of work. There are two good reasons for this: to use labor efficiently and to avoid issues of fairness that result when one station must work much harder than another.

Assume there are n tasks to be performed on each piece moving through the line and the time to do the ith task is t_i. These tasks are assigned to k workstations where $k \leq n$. If t_0 is the time allowed for each station (i.e., the time for the conveyor to move through a workstation), then the rate of the line will be $r_b = 1/t_0$.

Since the tasks have random times, we need to make some allowance for variability. We define $c < t_0$ to be the maximum time allowed for task assignment. By requiring the sum of the mean task times to be less than or equal to c, we provide some extra time at each station to accommodate the inherent variability of the tasks. Note that $u = c/t_0$ is the maximum utilization of any station in the line and is always less than one.

In many texts dealing with the LOB problem, c is called the *cycle time*. However, since we use this term to refer to the time through an entire routing, we will refer to c as the **conveyor time** (i.e., because it is the time the conveyor allows at each station).

The objective of most line-of-balance algorithms is to minimize total idle time, which we write as

$$\text{Total idle time} = kc - \sum_{i=1}^{n} t_i$$

An equivalent measure is known as **balance delay**

$$b = \frac{kc - \sum_{i=1}^{n} t_i}{kc}$$

which represents the total fraction of idle time.

To further complicate matters, we must consider a number of other constraints. The most common are **precedence** constraints, which occur when certain tasks must be done before others. We will consider only precedence constraints, but refer the reader to Hax and Candea (1984, section 5.4) for a more complete discussion of the LOB problem and a survey of relevant literature.

It turns out that the LOB problem is very complex (i.e., NP-hard), so that optimal algorithms often require excessive amounts of computer time for realistically sized problems (e.g., with 100 tasks or more). For this reason, most commercial packages rely on heuristic methods.

We illustrate a heuristic LOB algorithm using a simple procedure that is similar to that of Kilbridge and Wester (1961) by using an example from Johnson and Montgomery (1974, p. 369). To do this, consider the nine tasks whose precedence relations are given in Figure 18.3. The times for these tasks and the number of successors are given in Table 18.5. Note that task 5 has the largest average performance time of 10. Thus, $c \geq 10$. Also note that the sum of the performance times is $\sum_i t_i = 48$.

To have zero idle time, the ratio $\sum_{i=1}^{n} t_i / c$ must be an integer. However, this does not guarantee zero idle time because the precedence constraints might prevent the required assignment of tasks to stations. Nonetheless, this fact and

$$\max_i \{t_i\} \leq c \leq \sum_{i=1}^{n} t_i$$

help to determine an appropriate value for c. If we factor $\sum_{i=1}^{n} t_i = 48$, we get

$$2 \times 2 \times 2 \times 2 \times 3 = 48$$

The combinations of these factors that are between 10 (the largest performance time) and 48 (the sum of the performance times) are

$$2 \times 2 \times 2 \times 2 \times 3 = 48$$
$$2 \times 2 \times 2 \times 3 = 24$$
$$2 \times 2 \times 2 \times 2 = 16$$
$$2 \times 2 \times 3 = 12$$

FIGURE 18.3

Precedence diagram for LOB example

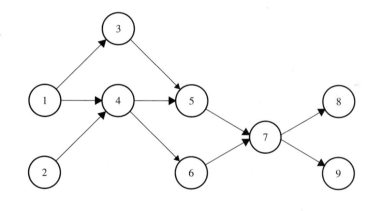

TABLE 18.5   Data for LOB Problem Example

Task Number	Average Performance Time	Number of Successors
1	5	7
2	3	6
3	6	4
4	8	5
5	10	3
6	7	3
7	1	2
8	5	0
9	3	0

So we *might* be able to achieve a perfectly balanced line (i.e., no idle time) with either $48/48 = 1$ station (obvious and not very useful), $48/24 = 2$ stations, $48/16 = 3$ stations, or $48/12 = 4$ stations. Let us consider the case with $c = 16$, the three-station case.[4]

To describe our procedure, define N to be the current station number, T the set of tasks assigned to the current station, A the time available to be assigned at the current station, and S the set of available tasks to be assigned, that is, those tasks whose precedence constraints have been satisfied and whose performance times fit within the remaining time. The algorithm then proceeds as follows:

Step 1. Set the current station number N to 1.

Step 2. Set the time available to c, $A \leftarrow c$, and $T = \phi$, indicating no assignments thus far.

Step 3. Determine the set of candidate tasks for assignment S. To be a candidate, two conditions must be satisfied:

1. All predecessors of the candidate must be scheduled, or equivalently, the candidate has no predecessors.
2. The performance time does not exceed the time available: $t_j \leq A$.

Step 4. Choose the task j from the set S, using the following two rules:

1. Choose the task that has the largest number of total successors.
2. Break ties by choosing the task with the longest performance time.

Place the task in T.

Step 5. Update the available time $A \leftarrow A - t_j$. Remove task j from set S.

Step 6. Repeat steps 3, 4, and 5 until no candidate tasks remain (i.e., set S is empty).

Step 7. If there are tasks remaining, increment the station number and go to step 2. Otherwise, stop.

To apply this algorithm to our example, we start with
$$N = 1 \quad A = 16 \quad S = \{1, 2\} \quad T = \phi$$
Set S contains tasks 1 and 2 only, since they are the only tasks without any predecessors. Since task 1 has the most successors, we assign it first to station 1. We now have
$$N = 1 \quad A = 11 \quad S = \{2, 3\} \quad T = \{1\}$$
Note that task 3 is now a candidate since its only precedence, task 1, has been scheduled. Since task 2 has the most successors and fits within the available time, we schedule it next.
$$N = 1 \quad A = 8 \quad S = \{3, 4\} \quad T = \{1, 2\}$$
Both tasks 3 and 4 are now candidates for the next slot. Here we see the importance (and arbitrariness) of the heuristic rules. Since our rule is to select the task with the most successors, we select task 4 which fits perfectly (using all eight time units remaining). If we had selected task 3, we would have had time remaining at the station after the task assignments. More sophisticated LOB algorithms would try all combinations of the tasks remaining and see if any are a perfect fit. This, of course, increases the amount of computer time required. The status of the algorithm is now
$$N = 1 \quad A = 0 \quad S = \phi \quad T = \{1, 2, 4\}$$
There are no candidate tasks because the time remaining is zero. We must now move on to schedule the second station. We reset $A = c$ and note that there are now two candidate tasks
$$N = 2 \quad A = 16 \quad S = \{3, 6\} \quad T = \phi$$
Task 3 has the greatest number of successors and so is scheduled first at station 2. The status is now
$$N = 2 \quad A = 10 \quad S = \{5, 6\} \quad T = \{3\}$$

---

[4]Of course, by choosing the value $c = 16$ we have established the throughput of the line. If we need greater throughput, we might be better off with $c = 12$, even though the line will not be perfectly balanced and even though there is more idle time. These issues are often not considered in LOB software.

Tasks 5 and 6 both have three successors. However, task 5 is the longest task and just fits in the time remaining. We finish station 2 with

$$N = 2 \quad A = 0 \quad S = \{6\} \quad T = \{3, 5\}$$

The remaining tasks all fit within the conveyor time c at station 3.

$$N = 3 \quad A = 0 \quad S = \phi \quad T = \{6, 7, 8, 9\}$$

The schedule is optimal with $b = 0$.

Note how many times during the algorithm that we got lucky when tasks "just fit" in the time remaining. This is not typical and, in fact, would not happen when $c = 12$ or $c = 24$. Most commercial algorithms try many different values of c and different tie-breaking rules within the procedure.

Study Questions

1. Why would anyone want to add capacity before demand has materialized? Why would anyone want to lag behind demand?

2. Why is the unit cost usually less expensive in a large plant than in a small one? What might cause this not to be true?

3. Why is the traditional view of capacity management inadequate? What law from factory physics speaks to this directly?

4. Consider this statement: For a fixed budget, design the "best" facility possible. Provide a more specific problem statement in terms of cost, cycle time, throughput, and so on.

5. Why is it appropriate to balance a paced assembly line but not a line of independent workstations? What is the bottleneck of a paced assembly line?

6. Consider the line-of-balance problem. Why should the conveyor time c be greater than the maximum time assigned at any station? What might happen if it were not?

7. What are some shortcomings of the traditional approach to designing factories in which we start with the size and shape of the plant, decide where the support facilities go, and then decide where to place the tools? What are some shortcomings of the factory physics approach?

Problems

1. You are charged with designing a three-station flow line that must achieve a target throughput of five jobs per hour and a total cycle time of three hours or less. Each station must consist of a single machine purchased from a vendor who will construct it to your specifications, any speed you desire. However, the price depends on the speed as follows:

$$K(i) = a(i) \left[\frac{1}{t_e(i)} \right]^{b(i)}$$

where $K(i)$ is the (total) equipment cost at station i; $t_e(i)$ is the effective process time of the machine at station i; and $a(i)$ and $b(i)$ are constants. Assume that the arrival coefficient of variation (CV) to the line is equal to one and that $c_e(i) = 1$ for $i = 1, 2, 3$ (i.e., the process CV for all machines is equal to one, regardless of the speed).

   a. Suppose that $a(i) = \$10,000$ and $b(i) = \frac{2}{3}$ for $i = 1, 2, 3$. Find the values of $t_e(i)$ for $i = 1, 2, 3$ that achieve target throughput and cycle time with minimum total equipment cost. (*Hint:* The *Solver* tool in Excel is very handy for this.) Is the result a balanced line? Explain why or why not.

   b. Suppose that $a(1) = \$1,000$, $a(2) = \$100,000$, $a(3) = \$10,000$, and $b(i) = \frac{2}{3}$ for $i = 1, 2, 3$. Find the values of $t_e(i)$ for $i = 1, 2, 3$ that achieve target throughput and cycle time with minimum total equipment cost. Is the result a balanced line? Explain why or why not.

   c. Suppose that everything is the same as in part *a* except that now $t_e(i)$ can only be chosen in multiples of 0.05 hour (0.05, 0.1, 0.15, etc.). Find the values of $t_e(i)$ for $i = 1, 2, 3$ that achieve target throughput and cycle time with minimum total equipment cost. Is the result a balanced line? Explain why or why not.

   d. What implications do the results of this simplified model have for designing realistic flow lines?

2. Table 18.6 gives the speeds (in pieces per hour), the CV, and the cost for a set of machines for a circuit board line. Jobs go through the line in totes that hold 50 panels each; this cannot be changed. The CVs represent the *effective* process times and thus include the effects of downtime, setups, and other common disruptions.

     The desired average cycle time through this workstation is one day. The maximum demand is 1,000 panels per day.

   a. What is the least-cost configuration that meets demand requirements?

   b. How many possible configurations are there?

   c. Find a good configuration.

3. *Challenge:* Consider the data in Table 18.1 along with the option of reducing the c_e^2 for station 3 as described in Section 18.3. Design a line with maximum throughput that has cycle times of not more than 16 hours and an equipment budget of no more than $2,800,000.

4. Assembling a computer monitor requires a chassis, two main circuit boards and components, a yoke, followed by a test. These are performed according to the following precedence requirements:

- The chassis must be put down first. This takes two minutes.
- Board 1 requires only a chassis. It takes three minutes.
- Components 1 require that board 1 be in place. Placing these components on the board takes three minutes.
- Board 2 requires that board 1 be in place. Board 2 takes four minutes to insert.
- Components 2 require that board 2 be in place. These take two minutes to insert.
- The yoke requires that all the boards and the components be in place and takes three minutes to install.
- Testing, naturally, requires that all the assembly be finished and takes five minutes to perform.

   a. Draw a precedence diagram of the assembly of a computer monitor.

   b. What is the minimum conveyor time that could possibly result in zero balance delay?

   c. If the expected utilization is 0.85, how many monitors will be produced per hour using the minimum conveyor time computed above?

   d. Assign the tasks to stations using the minimum conveyor time. What is the balance delay?

TABLE 18.6   Possible Machines to Purchase for Each Work Center

| Station | Possible Machines (Speed (pieces/hour), CV, Cost ($000)) | | | |
	Type 1	Type 2	Type 3	Type 4
MMOD	42, 2.0, $50	42, 1.0, $85	50, 2.0, $65	10, 2.0, $110.5
SIP	42, 2.0, $50	42, 1.0, $85	50, 2.0, $65	10, 2.0, $110.5
ROBOT	25, 1.0, $100	25, 0.7, $120	—	—
HDBLD	50, 0.75, $20	5.5, 0.75, $22	6, 0.75, $24	—

19 SYNTHESIS—PULLING IT ALL TOGETHER

This is not the end. It is not even the beginning of the end. But it is, perhaps, the end of the beginning.

Winston Churchill, November 10, 1942

19.1 The Strategic Importance of Details

We will be the first to admit that the treatment of manufacturing in this book has been technical. Manufacturing *is* technical. It would be nice if we could just do what feels right, get product out the door, and make a living. But there are fewer and fewer businesses in which this is possible. Under the pressure of intense global competition, manufacturing firms are *forced* to continually improve cost efficiency, product quality, and delivery responsiveness. Certainly a strategic vision is essential to foster an environment where this kind of performance is possible. But it is only through careful attention to technical detail that it can be achieved.

In the 1950s and 1960s America could afford to gloss over the details of manufacturing and concentrate on high-level marketing and finance issues. In the wake of World War II, American manufacturers did not need to worry about costs or defect levels that were a few percent too high. Customers had few alternatives and low expectations. In the 1980s and 1990s, however, consumers began to see high-quality, reasonably priced products from Japan, Germany, Korea, and many other places, and accordingly, they grew to expect more from American manufacturers. As a result, today even a relatively small gap in cost, quality, or customer service can drive a firm right out of a market.

The strategic value of details, however, goes well beyond their role in achieving small but important performance improvements. The most important reason that we need a deeper understanding of manufacturing systems is that the pace of technological change in recent years has made trial-and-error solutions almost useless. Henry Ford produced the Model T for an entire generation, so he could evolve systems and solutions by observing and tinkering with the production line. In contrast, the typical life span of a personal computer is less than two years, which means that modern PC manufacturers must set up the facilities, ramp up the volumes, attain the efficiencies needed to make a profit, achieve the level of predictability needed to ensure good customer service, and phase out the product, all in a very short time. Predicting and analyzing the behavior of a system *before* it is in place requires sound intuition and appropriate models, both of which are premised on an understanding of the technical details of manufacturing.

19.2   The Practical Matter of Implementation

Having the proper analysis tools is a key prerequisite for making significant improvements to a manufacturing system. But implementation is more than a matter of being right. An effective manufacturing manager must pull together a coherent plan and nurture it to fruition. This requires (1) addressing the *right* problem and (2) convincing others that it needs to be solved. The first is the subject of systems analysis, while the second deals with the human element of manufacturing management. Chapters 6 and 11 addressed these; but they are so central to the implementation process that we revisit them briefly here.

19.2.1   A Systems Perspective

The laws and formulas of factory physics can help identify areas of leverage, build intuition about why certain approaches work in certain environments, and evaluate and compare specific policies. But they cannot generate original ideas. The managers of a manufacturing system must determine *what* they want it to do before any tools can be applied to the question of *how* to do it. Therefore, to fully exploit the strategic potential of factory physics, it is important to use it in the larger problem-solving framework of systems analysis.

Recall from Chapter 6 that the essential aspects of systems analysis (as well as the modern variant of systems analysis, business process reengineering) are as follows:

1. *A systems view.* The problem is viewed in the context of a system of interacting subsystems. The emphasis is on taking a broad, holistic view of the problem, rather than a narrow, reductionist one.

2. *Means-ends analysis.* The objective is always specified first, and then alternatives are sought and evaluated in terms of this objective. For instance, a systems analysis project might use the objective "to deliver finished goods swiftly and conveniently to customers," but would *not* use the objective "to improve the efficiency of processing purchase orders." The latter is a "means-first" approach, which could rule out potentially attractive options—such as doing away with purchase orders under an entirely new procedure.

In systems analysis, objectives are typically organized into a hierarchy of objectives, which identifies the links between the fundamental objective and various lower-level objectives. This helps identify conflicting objectives (e.g., low inventory and high fill rate) and highlights lower-level objectives that support more than one higher-level objective (e.g., short cycle times allow for better manufacturing quality as well as better customer responsiveness).

3. *Creative alternative generation.* With the objective in mind, the systems approach seeks as broad a range of alternate policies as possible. For instance, to reduce manufacturing cycle time, we should go beyond simply considering how to speed up individual processes and think about basic causes of cycle time. Many formalized brainstorming techniques have been developed to encourage expansive thinking about nonobvious alternatives.

4. *Modeling and optimization.* To compare alternatives in terms of the objective, the project requires some kind of quantification. The modeling/optimization step for doing this may be as simple as computing costs for each alternative and choosing the cheapest one, or it may require analysis of a sophisticated mathematical model. The appropriate level of detail will vary depending on the complexity of the system and the magnitude of the potential impact.

5. *Iteration.* In every complex systems analysis project, the objective, alternatives, and model are revised repeatedly. This is because as we perform the analysis, we learn more about the system. In Chapter 6, we formalized this procedure as the "conjecture and refutation" process.

The systems analysis procedure helps focus attention on the correct problem (i.e., where major leverage exists), promotes insight into the system, and fosters a sense of teamwork toward the project. As such, it is a vital starting point and frame of reference for virtually any manufacturing improvement project.

19.2.2   Initiating Change

Systems analysis is valuable in generating and evaluating ideas. But no matter how good an idea is, it will never be implemented if it cannot be communicated. All the factory physics arguments in the world will not change a manufacturing organization unless the people in it are convinced of the need for change and know what they must do to bring it about.

Overcoming institutional momentum can be very difficult. As Machiavelli put it:

> There is nothing more difficult to take in hand, more perilous to conduct, or more uncertain in its success, than to take the lead in the introduction of a new order of things.

The amount of effort required to put through a program of change depends on the situation. If the manager of a production line has used her factory physics insight to recognize that reducing setups on a particular machine would reduce WIP and cycle time, and she has the authority to form a setup reduction team consisting of machine operators and staff engineers, then she should probably go ahead and do it. No hoopla, slogans, or revolutions are required to make small, incremental changes in the system. And while such changes will not remake the company, they can be important parts in the process of ongoing improvement.

Bigger changes, such as refocusing a plant as part of a time-based competition strategy, require much more institutional support. Radically reducing customer lead times by addressing the entire product delivery process—which involves sales, order entry, manufacturing, customer service, and possibly many other functions—demands the leadership of someone with sufficient clout to make the necessary changes. Depending on the system, this might be the plant manager, or if influence beyond the plant is needed (e.g., product development or component production), someone even higher, perhaps the vice president for manufacturing or chief operating officer. Once the leader has been assigned, it is critical for him/her to instigate the change *and* provide ongoing support for it. If the leader gives a few fiery speeches and then disappears, momentum for change will quickly evaporate.

An effective leader with the requisite authority can get people inspired to change, but cannot actually carry out the change. Systems analysis teams are typically needed to do the analysis and oversee the implementation required to actually reshape an organization. These teams can be configured and managed in many different ways (see Hayes, Wheelwright, and Clark 1988; Hammer and Champy 1993 for examples). We will not go into a great deal of depth about this, but we make the following observations about systems analysis teams:

1. Teams should *not* be committees. That is, they should be small enough to function aggressively. If the number of people on a team exceeds 10 or so, it becomes so difficult to get everyone together that the team becomes ineffective.

2. The team should consist of key people from the major functional areas affected by the change. For instance, a cycle time reduction effort should involve people from sales, manufacturing, production control, and so on. These people must be chosen to have a "big picture" attitude, so that they are not simply protecting their turf. Alternatively, they could be assigned 100 percent to the systems analysis team with the knowledge that after the team is dissolved, they will *not* go back to their previous position. The idea is to motivate people to think in terms of what is good for the overall system, not just for their part of it.

3. The team should include some outsiders, people not directly connected with the system under consideration. These could be people from elsewhere in the organization or independent consultants. The purpose of these outsiders is to act as provocateurs who will challenge assumptions and traditions. It is altogether too easy for a team of all insiders to mistake the way things are for the way things must be.

When supported by an influential leader and well-chosen analysis team, a systems analysis can be a powerful tool for bringing about dramatic change in an organization.

19.3   Focusing Teamwork

Often in modern manufacturing organizations, it is not the big failures that are most damaging, but rather the small successes. A highly visible failure that occurs when a firm attempts to push out the envelope of manufacturing practice is a noble effort and a valuable learning opportunity. In the right environment (one that does not punish people for taking good risks or become overly conservative in reaction to a failure), such failures are necessary and positive steps on the road of continual improvement.

In contrast, small safe projects that make tiny improvements can ensure their leaders of positive performance evaluations, but can steadily undermine the competitiveness of a firm. The reason is that they sap the resources of the organization. A firm that devotes too much energy to the easy marginal improvements is open prey to a competitor who aims higher. In this era of intense competition, the "all safe" strategy is almost a sure formula for failure.

This observation implies that a critical first step in setting up a systems analysis team is to focus the team on a problem of real importance. One way to do this is to make sure the original topic of a systems analysis study is sufficiently broad to allow the team to identify the major areas of leverage for themselves. As illustration we offer the example of a systems analysis in which the authors participated some years ago. At the inaugural workshop, the objective was stated as increasing the efficiency of the painting process. After listening to a great many details about the problems in painting, we asked about the motive for improving painting and learned that manufacturing cycle times were too long relative to the competition. But after we asked more questions, we were able to estimate that painting accounted for less than one day of a 10-week cycle time. Eventually, we discovered that the single major determinant of cycle time was the order entry process, which accounted for four weeks or more. Thus, although we eventually arrived at an appropriate focus for the study, we would have gotten there much more efficiently had the initial focus been on something broad like "remaining profitable in the face of faster competition," instead of the restrictive "improving painting efficiency."

19.3.1   Pareto's Law

A basic tool for sifting through a complex manufacturing system and picking out the most important aspects is **Pareto's law,** also known as the 80-20 rule. Pareto originally offered it as the law of economics that 80 percent of the wealth is owned by 20 percent of the people. Applied more generally, it states that a large fraction of any problem (or benefit) is caused by a small fraction of the constituents. For instance, a small percentage of part numbers accounts for the majority of demand, a small number of maintenance items accounts for the majority of the maintenance budget, a small number of customers accounts for both a large fraction of sales as well as complaints.

Pareto's law can be used as a management guide, suggesting the "important few" be given separate treatment from the "less important many." The few high-volume part numbers might be dedicated to efficient flow lines, while the many lower-volume part numbers are produced in a less efficient job shop environment. The few high-volume materials might be delivered in daily just-in-time fashion, while the many low-volume materials are purchased and stocked in bulk. The few machines accounting for a large fraction of downtime may have dedicated repair kits and specialized procedures, while the many machines causing less downtime are handled with routine maintenance procedures. The few big customers might be (probably will be) given preferential treatment relative to the many small customers. In each case, the idea is to allocate limited resources to the places where they will do the most good.

Pareto's law can also be used as a simplification tool. For instance, the routings in a manufacturing plant may seem like a hopelessly intricate mess when all part numbers are considered. But when only major families are considered, a much simpler pattern may emerge. Studying this simplified system is likely to be tractable and to lead to an understanding of the essential behavior of the overall system.

19.3.2   Factory Physics Laws

Once the system has been pared down to a manageable level using Pareto's law, the fundamental tools at the disposal of a systems analysis team are the laws of factory physics. First and foremost, these offer intuition about the way a manufacturing system will tend to behave. Additionally, they provide analytical methods that can be supplemented by many other modeling and analysis techniques as appropriate to the particular study.

The following is a summary of the key factory physics principles that have been introduced in this book.

Law (Little's Law):

$$\text{WIP} = \text{TH} \times \text{CT}$$

Law (Best-Case Performance):   *The minimum cycle time for a given WIP level w is given by*

$$\text{CT}_{\text{best}} = \begin{cases} T_0 & \text{if } w \leq W_0 \\ \dfrac{w}{r_b} & \text{otherwise} \end{cases}$$

The maximum throughput for a given WIP level w is given by

$$\text{TH}_{\text{best}} = \begin{cases} \dfrac{w}{T_0} & \text{if } w \leq W_0 \\ r_b & \text{otherwise} \end{cases}$$

Law (Worst-Case Performance): *The worst-case cycle time for a given WIP level w is given by*

$$\text{CT}_{\text{worst}} = wT_0$$

The worst-case throughput for a given WIP level w is given by

$$\text{TH}_{\text{worst}} = \frac{1}{T_0}$$

Definition (Practical Worst-Case Performance): *The practical worst-case (PWC) cycle time for a given WIP level w is given by*

$$\text{CT}_{\text{PWC}} = T_0 + \frac{w - 1}{r_b}$$

The PWC throughput for a given WIP level w is given by

$$\text{TH}_{\text{PWC}} = \frac{w}{W_0 + w - 1} r_b$$

Law (Labor Capacity): *The maximum capacity of a line staffed by n cross-trained operators with identical work rates is*

$$\text{TH}_{\text{max}} = \frac{n}{T_0}$$

Law (CONWIP with Flexible Labor): *In a CONWIP line with n identical workers and w jobs, where $w \geq n$, any policy that never idles workers when unblocked jobs are available will achieve a throughput level $\text{TH}(w)$ bounded by*

$$\text{TH}_{\text{CW}}(n) \leq \text{TH}(w) \leq \text{TH}_{\text{CW}}(w)$$

where $\text{TH}_{\text{CW}}(x)$ represents the throughput of a CONWIP line with all machines staffed by workers and x jobs in the system.

Law (Variability): *Increasing variability always degrades the performance of a production system.*

Corollary (Variability Placement): *In a line where releases are independent of completions, variability early in a routing increases cycle time more than equivalent variability later in the routing.*

Law (Variability Buffering): *Variability in a production system* will *be buffered by some combination of*

1. *Inventory*
2. *Capacity*
3. *Time*

Corollary (Buffer Flexibility): *Flexibility reduces the amount of variability buffering required in a production system.*

Law (Conservation of Material):   *In a stable system, over the long run, the rate out of a system will equal the rate in, less any yield loss, plus any parts production within the system.*

Law (Capacity):   *In steady state, all plants* will *release work at an average rate that is strictly less than the average capacity.*

Law (Utilization):   *If a station increases utilization without making any other changes, average WIP and cycle time will increase in a highly nonlinear fashion.*

Law (Process Batching):   *In stations with batch operations or with significant changeover times:*

1. *The minimum process batch size that yields a stable system may be greater than one.*
2. *As process batch size becomes large, cycle time grows proportionally with batch size.*
3. *Cycle time at the station will be minimized for some process batch size, which may be greater than one.*

Law (Move Batching):   *Cycle times over a segment of a routing are roughly proportional to the transfer batch sizes used over that segment, provided there is no waiting for the conveyance device.*

Law (Assembly Operations):   *The performance of an assembly station is degraded by increasing any of the following:*

1. *Number of components being assembled.*
2. *Variability of component arrivals.*
3. *Lack of coordination between component arrivals.*

Definition (Station Cycle Time):   *The average cycle time at a station is made up of the following components:*

$$\text{Cycle time} = \text{move time} + \text{queue time} + \text{setup time} + \text{process time}$$
$$+ \text{ wait-to-batch time} + \text{wait-in-batch time}$$
$$+ \text{ wait-to-match time}$$

Definition (Line Cycle Time):   *The average cycle time in a line is equal to the sum of the cycle times at the individual stations, less any time that overlaps two or more stations.*

Law (Rework):   *For a given throughput level, rework increases both the mean and standard deviation of the cycle time of a process.*

Law (Lead Time):   *The manufacturing lead time for a routing that yields a given service level is an increasing function of both the mean and standard deviation of the cycle time of the routing.*

Law (CONWIP Efficiency):   *For a given level of throughput, a push system will have more WIP on average than an equivalent CONWIP system.*

Law (CONWIP Robustness):   *A CONWIP system is more robust to errors in WIP level than a pure push system is to errors in release rate.*

Law (Self-Interest):   *People, not organizations, are self-optimizing.*

Law (Individuality):   *People are different.*

Law (Advocacy):   *For almost any program, there exists a champion who can make it work—at least for a while.*

Law (Burnout):   *People get burned out.*

Law (Responsibility):   *Responsibility without commensurate authority is demoralizing and counterproductive.*

19.4   A Factory Physics Parable

In this book we have introduced a host of widely varied concepts in order to develop the perspective, intuition, and tools for designing and improving manufacturing systems. To illustrate how many of these factory physics pieces might fit together in a systems analysis project to improve a specific system, we now consider a case study. The scenario is actually a composite of many different companies. Much of the data come from an excellent case by Bourland (1992). However, any lack of literary merit is entirely the responsibility of the authors.

19.4.1   Hitting the Trail

It was 6:20 on a Friday afternoon when Carol snapped her briefcase shut and stood up to go. Her one thought was, *Time to hit the trail!* She had been promised a week's vacation when she joined Texas Tool and Die as manager of manufacturing engineering four months ago. But every time she made plans, a plant crisis forced her to postpone. *Not this time. I've been wanting to go riding in west Texas for years.*

Before she could reach the door, the phone rang. *Not again!* She knew she shouldn't answer it, but her travel agent had said he might call with some last-minute schedule changes. So, gingerly, she picked up the phone.

"Carol Moura."

"Carol. Claude. Good thing you're still here. Milling is out of control again, and Bill wants us in his office *now.* I'll come by."

Carol clapped the phone into the receiver hard. *This will never end!* Not since her freshman year as an engineering student at Michigan State, far from her tight-knit family in Connecticut, had she felt so alone and depressed.

On the way to Bill's office, Claude Chadwick, a production manager, chattered on about the current situation, making sure to stress how critical Carol was to a solution. *Sure. All he wants is for someone to do his work so he can get out this weekend. Him and his marketing MBA. He doesn't care about the plant. It's just a stepping stone to bigger and better things. "Doing my time," he says. As if the plant is a prison.*

Carol's jaw tightened as she spied the sign on the office suite—William Whyskrak, Vice President of Manufacturing. *Bill Whyskrak! "Wiss-krek" he pronounces it. He's forever finding ways to make me look bad. Like that time in printing. First he tells me my cart-sharing idea for reducing cycle times is the stupidest thing he ever heard. Then he gives me a royal chewing out for going ahead with it. But when it worked, he takes all the credit. Worse, he tells Mr. Walker now he'd been trying to get me to do it for weeks and that I had been dragging my feet. Mr. Walker told him to "keep up the good work,"* but only smiled at me. *What did that mean? Well, I was looking for a job when I found this one."*

In his office, this time Carol doesn't even give Bill time to explain the latest crisis.

"Bill, I've postponed my vacation three times now. I deserve this time off. If I don't go now, I never will. See you in a week."

That wasn't so hard. On her way to the airport she began to forget the plant. It was early May, the flowers were gorgeous, the weather clear and cool. She let herself relax and started to enjoy the drive. *A week with nothing but my horse, sleeping bag, slicker, and hat to think about. My only problems will be food and water, and there's plenty of that on the wagon. It's going to be a good week.*

Carol spent the first three days on the trail trying not to think about the plant, and mostly succeeding. But on the morning of the fourth day, it forced its way into her consciousness. *What have I really accomplished in four months? A few small things and a lot of crisis management. But I haven't turned things around by a long shot. Bill has no faith in me. Maybe Mr. Walker doesn't either—I can never tell with him. Maybe I won't have a job when I get back. I was looking* hard *for a job when I found this one.*

Bob McAlister, the trail boss, broke her reverie by pulling up to ride alongside her. "Good thing that horse knows where to go."

"What do you mean?" So far, she had had little to do with Bob. He was usually busy making sure everyone's gear was right and had been quiet the rest of the time. Almost all he had said to her was, "Mornin' Ma'am." Even when he checked her saddle girth, all he did was pat the back end of her horse and tip his hat. Bob really seemed to fit the image of the silent cowboy.

"What I mean is that you're not *here.* You're back *there.* If you're going to spend good money to get away from there, why do you want to bring it here?"

"You're pretty smart," Carol admitted.

"You got to have a PhD in psychology to be a trail boss—state law, you know." Bob was the kind of Texan who liked to make outrageous statements with a straight face and see how long it took the non-Texans to catch on. "Trail ridin' takes brains. Your horse ain't gonna tell you he's goin' lame, and that mama cow over there ain't gonna e-mail you she's runnin' dry. It's clear that somethin's botherin' you. Why, you're twitchin' like a long-tailed cat in a room full of rockin' chairs."

Carol laughed. "You're right. I've been wondering if I'll have a job to go back to."

"Maybe I can help. I know you're some kind of big engineer at a plant. I'm no engineer, but you never know, comin' at it from a different angle, I might just see somethin'. Anyway, we got a long way to ride today, and we might as well talk a spell."

"All right, but I'm warning you, it's technical. We make parts and assemblies for aircraft. I'm responsible for making hubs. We get orders…"

Carol talked for 10 minutes before Bob interrupted, "I don't want to know all that. I'm a simple cowboy—just give me the basics. You're tryin' to take one piece of metal and turn it into a different piece, right?"

"Yes, but there are a lot of different pieces…"

"And after you do it, you want to sell the right number of the right piece of metal to the right customer, right?"

"Of course, but there are all kinds of…"

"And you need to do all this with the equipment you got in your plant right now, right?"

"Yes, but…"

"And you want to do it without keepin' your customers waitin' or havin' a lot of extra stock layin' around, right?"

"Yes, but it's a complicated plant. The issues are just not that simple!"

"Who said they were? But I know one thing."

"What's that?"

"Details may not be simple, but *principles* are!" Bob pulled out his canteen, took a drink and offered it to Carol.

Carol took a drink, wiped her mouth, and asked, "OK, what *are* the principles? I've taken every short course there is and have come to the conclusion that for every expert telling me to do one thing, there's another expert telling me to do something else."

"Well, I don't really know."

Carol rolled her eyes. "Great! Maybe I can get a job shoeing horses."

"Wouldn't recommend it. Too hard on your back. What I do know is that there *are* principles and the important ones ain't that hard. You know, like an apple fallin' from a tree. Sometimes the principle is just hidden. You can't see the forest for the trees—that is, if you got trees. Out here I guess it's the hill for the rocks." Bob surveyed the landscape and continued.

"Anyway, a couple years ago, the Extension Service sent out this young expert to make the local feed co-op more efficient." Bob nearly spat out the word *expert.* "By the time he was finished, the place was a mess. I was so mad, I stood up in a meetin' and said a ol' cowpoke like me could've done a better job. Durned if they didn't vote me president that year. Well, I had to do somethin' then. So, I went in, called a meetin' and asked a single question, just one: What in the world is it we're tryin' to do here?

"You should've seen the looks I got. They thought I was dumber than dirt. But when folks started answerin' the question, the place really heated up. We got somethin' like 20 different answers and almost a fight or two. But folks got the picture. Nobody had any idea what we were trying to do. So we sat down, agreed on some goals, and figured out ways to make 'em happen. Actually, it was pretty simple once we got started."

"But what were the principles?" Carol asked. But Bob wasn't looking at her. He was staring at one of the horses near the front of the line.

"Pardon, Ma'am, but it looks like we got a runaway. Talk to you later." Bob spurred his horse and took off after a galloping mare carrying a frightened boy.

Bob stopped the horse and returned the boy to his mother in short order. But his horse had lost a shoe. It stumbled on the way back to the group and threw Bob to the ground. His knee hit a rock and knocked a pin loose from an old rodeo injury. Jedidiah the cook took him to the first ranch house they came to and he was hurried to the hospital. The damage turned out not to be serious, but Bob wouldn't ride again for a month.

After the excitement had died down, Carol began to think about "principles." *If only my problems were that simple. But then, I don't think the co-op problem was all that simple, no matter what Bob says. After all, the "expert" wasn't able to solve it. Maybe most people's problems are just as hard as mine. Maybe everyone has to look for principles of some kind. Like the apple falling from a tree. That's physics. But I have a factory to manage…Wait a minute, what about that factory physics I learned about in B-school? Didn't that have principles that are supposed to be relevant to factories?*

For the rest of the trip, Carol continued to muse about using principles to figure out what was wrong with the plant and fix it. She soon realized she would need help. *Jane Snyder—she was just promoted to manager of marketing—she seems sharp. And Ed Burleson, the manufacturing engineer who came in with me, is a computer whiz. Both strike me as go-getters. What principles do they use? Maybe I can get them together and we can develop a plan. Of course, we can't spend much money. Bill would never go for that. But we could do pretty much whatever we want on the plant floor. No one really pays attention to that—until the end of the quarter—or when customers are screaming.*

I hear they're going to sell the plant. But if we can make the operation run better, we might just keep our jobs.

19.4.2   The Challenge

Texas Tool and Die, which was founded in the 1950s, makes components for the aircraft industry at a single plant near Fort Worth, Texas. Two years prior to Carol's arrival, TTD had been bought out by an investment group that hoped to improve operations and sell it for a profit. An immediate reorganization brought in Bill Whyskrak, a polished speaker with management experience in several industries, and his assistant Claude Chadwick. But despite the changes and a major influx of capital, profits had steadily declined in the face of increasingly stiff competition from firms with lower prices and better customer responsiveness.

The managing owner was a man named Sam Walker, who had started his career as a design engineer and had worked his way into management. Sam was convinced that they had to find ways to increase throughput (to lower unit costs so they would allow more competitive pricing) and to reduce cycle times (so they could offer competitive customer deliveries). He directed Bill to bring in more manufacturing talent—which led to the hiring of Carol Moura, a manufacturing engineering manager with 10 years of experience and an MBA in operations, and Ed Burleson, a manufacturing engineer with a BS in industrial engineering. Two months after Carol and Ed came on board, things had gotten so bad that some of the investors were at the point of wanting to sell the company, take their losses, and move on. Sam convinced the other owners to give the throughput enhancement and cycle time reduction efforts one more chance. The other owners agreed to six more months of operations, with the stipulation that no large capital expenditures be made.

19.4.3   The Lay of the Land

Historically, company policy had been to collect customer orders during the week and group them into jobs every Friday. In its product catalog, TTD promised delivery four weeks after the close of business on Friday. Unfortunately, the competition was offering three-week lead times and had been steadily reducing these each year. Worse, TTD had not been able to achieve even the four-week target with regularity. Average cycle time for some parts was well over eight weeks.

Although average demand was still high, it was variable, to the point that there were times when there was almost no demand for the week. Figure 19.1 shows the aggregate demand for the previous year. Table 19.1 gives projected demand for the next year for the four largest-selling products, which accounted for 90 percent of total demand, along with the lot size for each product. Demand for other products was met by production from a job shop separate from the part of the plant that produced hubs 1 through 4.

Several months before Carol and Ed had arrived, Bill and Claude had organized the main processes for producing hubs 1 to 4 into a cellular layout in an attempt to reduce cycle times by eliminating unnecessary material handling. The anticipated reduction had yet to materialize. The cell consisted of three benches (which served as preparation stations), four vertical lathes (VTL), one deburring station, four inspection stations, two mills, two drills, and one rework station. All machines were subject to occasional breakdown. Table 19.2 gives data gathered on mean times to failure and mean times to repair.

There were 14 workers in the cell, with three prep workers assigned to the benches, three repair operators assigned to the deburr and rework stations, three inspectors

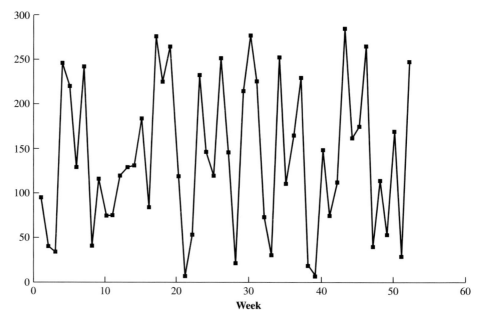

FIGURE 19.1

Total demand for previous year

TABLE 19.1   Average Demand and Lot Sizes

Part	Average Demand	Lot Size
Hub 1	2,100	40
Hub 2	1,700	30
Hub 3	2,000	44
Hub 4	1,500	30

assigned to the inspection stations, and five machinists assigned to the lathes, drills, and mills. Figure 19.2 shows the layout of the facility, along with the labor assignments. Due to breaks—scheduled and unscheduled—workers were generally considered available only 90 percent of the time.

The sequence of operations (routing) for hub 1 is shown in Figure 19.3. Run times, setup times, and labor times are given in Table 19.3. Because many of the operations were automated, the labor time for some operations was less than machine time, so it was possible for an operator to monitor multiple machines. The routings and process times for the other products were similar to those for hub 1.[1]

As Figure 19.3 shows, an average of 15 percent of the hub 1 parts were found to be defective at the inspection station. An average of two-thirds of these were sent to rework; the others were scrapped. In rework, an average of 20 percent were reworked without success and were eventually scrapped. The remaining 80 percent were reworked and sent back to inspect, where they might or might not be certified as good parts.

---

[1]The details of all the parts are not central to our story. The interested reader is referred to Bourland (1992) for other details of the case.

TABLE 19.2   Equipment Data

Equipment Group	Number in Group	Reliability		Labor Group Assigned
		MTTF (hour)	MTTR (hour)	
Bench	3	160	8	Prep
VTL	4	160	16	Machinist
Deburr	1	80	8	Repair
Inspect	4	40	8	Inspector
Repair	1	160	8	Repair
Mill	2	80	4	Machinist
Drill	2	160	4	Machinist

FIGURE 19.2

Cell layout

FIGURE 19.3

Operations and routings

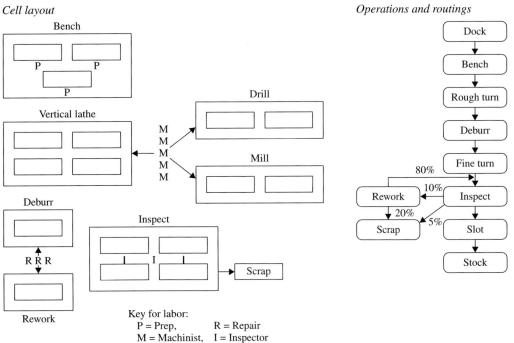

Each hub was composed of four to six mountings and a single sleeve. Each mounting was composed of two brackets and two bolts. The brackets, bolts, and sleeves were all purchased from outside suppliers. Since these parts were common to many assemblies, TTD tended to keep ample stocks of them. Table 19.4 gives the process times for the unpacking and inspection of the purchased parts. The assembly of the mounts, sleeves, and hubs took place in the assembly area, which seemed to have sufficient capacity and rarely failed to keep up with the cell.

TABLE 19.3   Operation Assignments and Process Times for Hub 1

Operation	Equipment	Time at Equipment		Labor Times	
		Setup Time (minute)	Run Time (minute/piece)	Setup Time (minute)	Run Time (minute/piece)
Bench	Bench	0	10	0	10
Rough turn	VTL	180	17	180	15
Deburr	Deburr	0	10	0	10
Finish turn	VTL	120	26	120	20
Inspect	Inspect	7	12	7	7
Rework	Rework	90	32	90	32
Slot	Mill	60	60	60	40

TABLE 19.4   Operation Assignments and Process Times for Purchased Parts

Operation	Equipment	Time at Equipment		Labor Times	
		Setup Time (minute)	Run Time (minute/piece)	Setup Time (minute)	Run Time (minute/piece)
Mounting					
Unpack	Bench	12	2	12	2
Inspect	Inspect	0	3	0	3
Bracket					
Unpack	Bench	12	0	12	0
Inspect	Inspect	10	0	4	0
Bolt					
Unpack	Bench	12	0	12	0
Inspect	Inspect	12	0	4	0
Sleeve					
Unpack	Bench	12	3	12	3
Inspect	Inspect	0	3	0	3

19.4.4   Teamwork to the Rescue

Carol returned from her vacation rested but anxious. There were seven progressively shrill calls from Bill Whyskrak on her voice mail. *Big surprise.* Before returning them, she called Jane Snyder and Ed Burleson—who both agreed that the plant was in big trouble—and asked them to meet her after work at the local watering hole. They agreed. Then she called Bill and endured another haranguing.

No sooner had she hung up than Claude slithered into her office with his version of the past week's disasters and bitter complaints about having to work all weekend. *About time!* When he had gone *(Finally!),* Carol moved the pile of unanswered mail to the side of her desk *(It'll keep one more day),* got out her old *Factory Physics* text *(Dusty but it still looks almost new),* and began looking for "principles." When it was time to go to the bar, she was ready.

Principles.   "What in the world is it that we're trying do do?" Carol asked as she, Jane, and Ed waited for the beer and nachos to arrive. After some discussion of basic concerns like "keep our jobs," the three agreed that two fundamental problems were driving costs up and revenues down: insufficient throughput and excessive cycle times. If they could make a significant difference in these, they believed TTD could be made profitable.

Carol had anticipated this and was armed with some principles from *Factory Physics.* She began by pointing out that Little's Law shows that throughput and cycle times are related:

Law (Little's Law):

$$WIP = TH \times CT$$

"Cool!" Ed observed. "If we can get throughput up to capacity and keep it there, then reducing WIP will reduce cycle time."

"Exactly!" Carol knew there was a reason she had asked Ed along. "Except that we have to be careful about aiming for capacity." She displayed her next factory physics law.

Law (Capacity): *In steady state, all plants* will *release work at an average rate that is strictly less than the average capacity.*

"Okay. That's what I meant, actually. Everyone knows that machines can't run all the time."

"Oh yeah?" Jane raised her eyebrows. "How many times have you heard Bill screaming for 100 percent utilization of the lathes? But if we're going to talk about principles, let's leave Bill out of it." Ignoring Ed's groan, Jane went on. "Carol, I'm wondering about that Little's Law. It looks like we can get the same throughput with small WIP and small cycle times or big WIP and big cycle times. It's pretty clear which category we fall into, but what's the difference?"

"I couldn't have set it up better myself." Carol smiled and presented her next law.

Law (Variability): *Increasing variability always degrades performance of a production system.*

"And I found one more that follows up on the variability theme."

Law (Variability Buffering): *Variability in a production system* will *be buffered by some combination of*

1. Inventory

2. Capacity

3. Time

"The book also refers to this as the pay-me-now-or-pay-me-later law," she said.

"Nice name," grinned Ed. "But what's it mean?"

"It means we have either too much variability or too much WIP. But if we keep WIP too low, we lose on throughput and so we have a capacity buffer," Carol explained.

"How could we be keeping WIP too low? I thought we had too much WIP."

"Whenever we turn off releases because WIP has gotten out of hand, we lose throughput."

"You mean like the week you were gone."

"Uh huh. But before we can even talk about a reasonable target throughput, we need to know what our capacity is."

"How do we do that?"

"You guys up for a walk? Let's go back to the plant," Carol suggested, as she picked up the check.

The scene at the manufacturing cell was all too familiar. The trio found WIP piled high in front of the bench operation, vertical lathes, and the milling machines. Things were so bad that the prep workers had just returned a load of materials to the storeroom to relieve the congestion. The machinists were complaining that they were being overworked again as the repair operators were "just sitting around." When questioned, an idle repair operator explained that his load was sporadic; he couldn't help it if he sometimes ran out of work to do.

"We've got our work cut out for us," said Ed as they walked out to the parking lot.

"But where do we start?" asked Jane.

Carol reached her car first and unlocked the door. "I suggest we listen to the machinists. Maybe they *are* overworked. I'm going to run some numbers. Let's talk about it tomorrow, okay Ed? Night, Jane."

"Night."

Capacity Analysis.   The next morning, Carol set up a spreadsheet and did a quick estimate of the utilization levels of the machinists and repair operators. She did this by calculating the total load generated by production needed to meet demand, including setups, at the current lot sizes. This showed that the average workload of the machinists was indeed higher than that of the repair operators. Ed determined that one repair operator could be moved into the machinist pool without compromising the ability of the repair operators to do their work. Fortunately, one of the operators had worked as a machinist, was bored with his repair job, and welcomed the move. Since no one could come up with a reason not to, Carol talked the foreman into making the switch that afternoon.

Cycle Time Analsyis.   What to do next was not so obvious. Carol's simple spreadsheet did not suggest any more easy labor reassignments, and no one could offer a clear idea of how variability was affecting the system. Almost for lack of anything else to do, Ed volunteered to develop a simulation of the facility. After a week of coding, debugging, and preliminary runs, he had a basic working model. He was pleased to be able to show Carol and Jane that his simulation predicted extremely long (indeed unstable) cycle times in the cell when staffed by three repair operators and five machinists. However, if one repair operator were reassigned, so that there would be two repair operators and six machinists, the simulated cycle times dropped to between four and seven weeks, with hub 1 having the longest.

"It looks like we did the right thing," he concluded with a grin. "Cycle times should be coming down soon."

And for a while the system really did seem to be improving. Two weeks after reclassifying the repair operator as a machinist, throughput was up noticeably. But cycle times were still well above the levels predicted by the simulation. The team was puzzled at the discrepancy and rechecked the process times on the machines. The times used in the simulation were found to be, if anything, longer than those observed in the actual system.

"It's not the rate data." Ed looked up from his keyboard. "What else could be making the cycle times so much longer than the model says they should be? Do we have any other data we could check?"

"Not many," Carol admitted. "But we do have these WIP sheets. What does the simulation say about WIP?"

"I don't know. I'll run it again and generate WIP-versus-time charts for the different equipment groups."

"Good. I'll make up the same charts from these sheets. Let's meet for coffee around four. I'll call Jane."

Four o'clock found the team members hunched over a cafeteria table, studying the two charts. They did not look anything alike. The simulation model predicted fairly modest increases and decreases in WIP, while the actual WIP charts showed huge "bubbles" of WIP that drifted through the plant.

"What's causing that?" Jane asked.

"Queueing," Carol answered.

"What's that equation for queue time again?" Jane reached for the no-longer-dusty copy of *Factory Physics.*

"Whoa!" Ed feigned falling out of his chair. "A marketing person asking for an equation!"

"Give me a break! Marketing *is* quantitative, you know. Here it is."

$$CT_q = \underbrace{\frac{c_a^2 + c_e^2}{2}}_{\text{Variability}} \times \overbrace{\frac{u}{1-u}}^{\text{Utilization}} \times \underbrace{t_e}_{\text{Process time}}$$

Jane studied the formula carefully and mused, "Hmmm. Since our process times are conservative, utilization must also be conservative, since the throughput is right."

"Wow! I guess you marketing types do know your way around an equation," said Carol, obviously impressed.

"So it must be in the variability numbers," Ed added swiftly, not wanting to be outdone in the technical analysis department.

"Which one?" Jane asked.

"Well, the c-sub-e number could be big, but not that big. And I don't see how the c-sub-a number can get very big either," Carol said with a puzzled look.

"What are c-sub-e and c-sub-a?" asked Jane.

"The c-sub-e is a measure of how variable the machine process times are, while the c-sub-a measures the variability of arrivals," Ed explained, a little relieved to have an opportunity to display his knowledge.

"What does it mean for arrivals to be variable?"

"If they don't come in one at a time, regularly, like clockwork, then they're variable."

"Well, of course they don't come in like that. We release jobs in week-long batches. It's part of our marketing strategy," Jane explained.

"Hello!" Ed grinned. "Maybe you better tell us more about that strategy."

"We publish a lead time to our customers. Any order we get during a given week will be delivered four weeks later. The close-out day is Friday. Orders are batched over the weekend and then sent to the floor on Monday. We've been doing it for years. Efficiency considerations, you know."

"Well, it might make things more efficient, but I'll bet it's driving the heck out of cycle time. No wonder we see all these WIP bubbles." Carol said and turned to Ed. "What c-sub-a do we have in the model?"

"For lack of a better number, we used one, the usual exponential assumption." Ed snuck a glance over at Jane to see if this technical talk was making her nervous. It wasn't.

"Probably way too low. My guess would be more like 10."

"It might even be worse," Jane added. "There's a lot of variability in our demand as well. Take a look at this."

The chart (Figure 19.1) showed that total weekly demand for the past 12 months averaged 146 pieces, but ranged between 6 and 284. Thus, while the capacity of the plant was around 160 parts per week, it was faced with a "feast or famine" situation. Clearly, this meant that in some weeks the plant was starved for work, while in others it was completely swamped.

Ed stood up. "I've got to change the way I model demand. I'll talk to you tomorrow."

Carol accompanied Jane back to her office. "Jane, what would happen if, instead of publishing a fixed lead time, we quoted delivery dates to our customers. And what if those dates were closer in than four weeks?"

"Well, getting lead times below four weeks would be great. The competition is killing us on that. And I guess most of our customers would probably like a quotation better—provided we deliver on time. But some customers have their MRP system loaded with our lead time. Could we have a fixed lead time for them?"

"I think so, at least most of the time. But when we're really busy, we may not be able to meet the fixed lead times."

"Actually, now that I think of it, that might not be so bad. Usually, when we're swamped, so are our competitors."

"Good point. The main thing, though, is that we'll be able to quote shorter lead times on average."

"Our customers will like that. What do we need to do?"

"It's called *due date quoting,* and we can do it for each of our product lines. This gives some details." Carol handed Jane the *Factory Physics* book. "See the chapter on scheduling."

"All right, I'll get on it."

The next morning, Ed was in Carol's office early.

"Got it! I changed the arrival processes, and the simulation matches on cycle times pretty well. Now what?"

"Now we get rid of those WIP bubbles."

"How?"

"Well, I think a pull system will smooth the workload. I'll work on that. You see if you can find ways to reduce process variability. Okay?"

"Sounds like a plan."

During the next month, Carol set up a CONWIP system in the cell. The mechanics were simple, basically consisting of nothing more than laminated cards to limit WIP and the standard work list to sequence releases. More challenging was breaking the tradition of bulk releases. Carol carefully involved the operators in the implementation process, and even shut down the cell for a two-hour "all hands" orientation meeting. (She thought Bill was going to burst a vein over that!) To the operators, CONWIP seemed almost obvious; after all, why release work into the cell until there is capacity to work on it? A couple of people in production control, who were responsible for running the MRP system that scheduled the bulk releases, initially raised some objections about having their schedules overridden by the CONWIP system. But Jane helped Carol win them over, by stressing the marketing value of shorter cycle times.

Meanwhile, Ed searched his simulation and the cell for large sources of variability in effective process times. At first, the process times seemed extremely regular, since processes were largely automated. Then he realized that he needed to consider the effect of downtimes that averaged from 4 to 16 hours on the various machines. Ed performed a Pareto analysis of previous failures and found that most of the maintenance calls were the result of a small set of problems. He and the maintenance superintendent developed efficient procedures for handling the most common problems and then documented them. Where appropriate, they also installed field-ready replacement kits. The result was that mean time to repair on all machines dropped to less than four hours. Although they would not have data to document it for months, the beneficial effects on the line were felt almost immediately.

After the blowup about Carol's CONWIP meeting, Bill mysteriously emerged as a convert to JIT. He gave Carol and Claude a popular JIT book and ordered Carol to install a kanban system in the cell and Claude to implement JIT deliveries of raw material. Carol ignored the book, but was careful to refer to her CONWIP system as a kanban system whenever she spoke to Bill. Luckily for her, Bill didn't have time to pay too much attention to what she was doing because of problems with Claude's policies.

With Bill's blessing, Claude changed from purchasing commonly used pieces of bar stock in one-month supplies to having daily deliveries from a local vendor. Raw material inventory dropped by 80 percent, but delivery charges went up dramatically as well. Bill stepped in and threatened to cancel the contract because of the higher delivery cost. The offended vendor responded by canceling the contract himself. The production schedule was badly scrambled, and production came to a virtual halt for almost two days before Sam Walker smoothed things over with the vendor and reestablished the supply.

Also at Bill's instigation, Claude began a plantwide setup reduction program that made use of single minute exchange of die (SMED) techniques Ed had developed previously for a specific machine. Because these techniques did not apply universally and because effort was spread over so many processes, Claude got off to a slow start. By mid-July, after almost two months' work, he had achieved significant setup reductions only in the labeling area. However, about the time Claude's program was beginning to stall, Ed became convinced from his ongoing simulation study that setup reduction was important on the VT lathe, drilling, and milling. He took over (unofficial) leadership of this part of the program, and by the end of August they had reduced the setup times of the VT lathe, drilling, and milling by 50 percent. With these and the other changes they had made, Ed's model predicted cycle times of 9 to 22 days, compared with the original 5 to 9 weeks.

At the next team meeting, Carol copied the basic cycle time equation from the increasingly ragged copy of *Factory Physics* to the board:

Definition (Station Cycle Time): *The average cycle time at a station is made up of the following components:*

Cycle time = Move time + queue time + setup time + process time
$\qquad\qquad$ + wait-to-batch time + wait-in-batch time + wait-to-match time

"The way I see it, CONWIP and due date quoting have brought queue times down by something like 80 percent. Process times and move times were never big. Wait-to-match time doesn't apply in the cell. So, the only remaining area to be addressed is wait-for-batch time." Carol sat down. "Ed, what move batch sizes are we using in the model?"

"The ones they use in the plant. They were computed using the square root formula. I think. Why?"

"So the batch sizes are the same for both move batches and process batches?"

"What do you mean by *move batch* and *process batch?*" Jane asked. "I've never heard anyone here use those terms."

"That could be our problem." Carol answered. "The process batch is how many parts we run between setups. The move batch is how many we move at once to the next operation. They don't have to be the same."

"Why didn't I think of that!" Ed began sliding his chair back. "Let me see what happens in the model if we leave our process batch sizes alone but make all the move batches in the cell equal to one."

"Wait. Let me get this straight," Jane jumped in before Ed could escape. "You mean, like for hub 1, we process 40 units before changing over to another hub but move them one at a time as soon as they're done?"

"Exactly!"

Carol was confident that she knew what Ed's simulation would show. Smaller move batches would result in shorter cycle times. But while she was waiting for him to estimate the size of the reduction, Carol began thinking about the process batch sizes. *Since we reduced setup times, we should be able to reduce batch sizes as well. But how much? That silly EOQ formula won't help because we have no idea what setup cost should be. Besides, the interaction between the batch sizes of the various hubs is probably complex. Wasn't there something in the scheduling chapter about optimal batch sizing to minimize cycle times?*

She picked up the phone to call Ed, but he walked in before she had a chance to dial.

"Good news! The cycle times should drop another 30 percent by simply making the move sizes equal to one. But I think we could do even better if we adjust the process batch sizes, so I started reading in Chapter 15 about..."

"Optimal process batch sizes! You're reading my mind. I was just calling you to suggest we fiddle with process batch sizes."

Carol and Ed spent a few hours building an optimal batch-sizing model. Using it along with some trial-and-error, they settled on the set of batch sizes shown in Table 19.5. The next morning Ed met with the shop superintendent, who readily agreed to the changes in process and move batch size. Congestion in the cell steadily declined. By the end of September, cycle times had fallen to between four and seven days.

19.4.5   How the Plant Was Won

October was judgment time. Sam Walker gave Bill responsibility for organizing an overview of the improvement program at a meeting of the owners. Bill told Carol and Claude that he'd handle the presentation himself. Carol made up some slides anyway, just in case. Claude did not.

TABLE 19.5   Recommended Batch Sizes and Resulting Cycle Times

Part	Recommended Batch Size	Predicted Cycle Time
Hub 1	10	6.7
Hub 2	15	3.4
Hub 3	20	5.6
Hub 4	15	3.7

Sam began the meeting with a brief overview of how much output had increased, cycle times had decreased, and customer relations had improved. He concluded with, "And now I'm going to ask Bill to tell us just what was done to make this good news possible. Bill?"

Bill was dressed to the nines and had slick color slides. A couple of owners even laughed at his introductory jokes. *He's going to pull this off! All the work we did, and we won't get a shred of credit.* Carol sighed as Bill moved into the core of his presentation.

"The key to our cycle time reduction program was recognizing what cycle time is." Bill put up his main slide, which showed:

Cycle time = Value-added time + non-value-added time

"Things like setup time, move time, unnecessary meeting time," Bill emphasized the last item with a glance at Carol, "are all waste. Or, as they say in Japan, *muda.* Eliminate *muda* and you'll reduce cycle times." Bill flipped up the next slide. "One of our most successful efforts was reducing setups through the use of SMED techniques. Take labeling for instance…"

"Wait a minute, Bill," Sam interrupted. "Why do we want to reduce setup times in labeling? We've got plenty of capacity there, and I've never seen much WIP in that area. What's the point?"

"Well, as I said, setups represent non-value-added time. They should be eliminated."

"Is that what you were doing last winter in printing? I recall that once you got Carol going, you eliminated a cart at each table and had the operators share a single cart. Seems to me like you added quite a bit of walking around. Isn't that non-value-added?"

Got Carol going! Carol's heart sank. *He thinks I'm in the way!*

"Well, er, it depends. In this case…," Bill's polished demeanor faltered just a bit. "Claude, didn't you want to say something about our lean manufacturing program to Mr. Walker?"

Carol watched the panic rise in Claude's face. *Well, at least I'm not the only one Bill makes look bad.* But Claude covered neatly.

"Well, I think it's pretty clear that the proof's in the pudding. As you can all see, Bill's program has really turned things around." Claude turned from Bill to Sam. "Regardless of what you call it. After all, we're here to run the plant, not name things."

Some of the owners nodded in agreement. Sam was noncommittal and quickly looked back to Bill. "Wasn't there more to the program than setup reduction?"

"Yes. You'll recall that we also implemented just-in-time deliveries."

"I remember," muttered Sam under his breath.

"And we installed a simple kanban system in the cell that increases efficiency by pulling parts between machines and…"

"Excuse me Bill," Sam interrupted again. "I've been down to the cell and I believe I've heard the operators referring to the new system as CONWIP, not kanban. Why is that?"

"Oh! Well,…, it's basically the same thing. Actually, Carol helped me quite a bit with that part, so maybe we should ask her."

Carol swallowed hard and walked up to the projector.

"*CONWIP* stands for constant work in process and is *not* quite the same thing as what most people mean by kanban…" Carol gathered steam as she spoke. She rolled through the importance of variability, the effects of batching, and even put up a few factory physics graphs. She showed plots of the progressively shorter cycle times predicted by the simulation model as improvements were incorporated. Her speech grew more rapid, her gestures more animated. Before she knew it, she had spoken for 20 minutes without a single interruption. She stopped and looked up anxiously for questions. The room was silent.

"Thank you, Ms. Moura." Sam had a sly smile on his face.

What can that mean? I must have talked too much, and I shouldn't have contradicted Bill. Now I've done it!

"Thank you all. This is a fine piece of work. Now, if you'll excuse us, I need to wrap up with the owners." Sam motioned them to the door.

As she filed out with Bill and Claude, Carol could hear the owners congratulating Sam. One was shaking his hand, and Sam was smiling broadly.

"I think that went well," said Bill as soon as they were in the hall. "Except for you boring them with your quantoid stuff, Carol. Kanban, CONWIP—nobody cares! But at least we're still in business."

"Yeah." Carol didn't want to join the post mortem with Bill and Claude. "I've got to take care of some things. See you."

Forty-five minutes later, back in her office, Carol was mechanically answering e-mail when the phone rang. It was Sam. They wanted her back in the conference room. Filled with dread, she went.

"Hello, Carol." Sam offered her a seat. "We've been working on a few changes of our own." He flipped on the overhead projector, revealing an organization chart. Carol hastily scanned it for her position. It was unfilled. *Oh no! Well, I did it this time. Now I am looking for a job! Me and my big mouth!*

One of the owners said, "Congratulations, Carol!"

Congratulations!? Why that sarcastic... Carol looked back at the screen. In the box labeled VP Manufacturing was her name. Next to it in the position of Manager, Manufacturing Engineering was the name of Edward Burleson. Jane Snyder was listed as VP Marketing for the division.

Sam read the question in her eyes. "We have already discussed matters with Mr. Whyskrak, and he and Mr. Chadwick have decided to leave the company to form their own concern."

Carol sped down the hall in search of Ed and Jane. This called for more than beer and nachos!

19.4.6   Epilogue

Carol was unpacking in her new office. She pulled out the battered copy of *Factory Physics,* with its dog-eared pages and broken spine, and placed it gently on the shelf. *This is about to fall apart. I need a new copy. I sure hope it's still in print.*

When she had emptied and disposed of the boxes, she began sifting through her mail. She spied a piece with a familiar name on it.

Whyskrak & Company
"We add value by eliminating waste."

Sounds good to me! She tossed the flyer into the waste paper basket.

Then she pulled an old card from her organizer and dialed the number. After a pause she said, "Bob? This is Carol Moura from Texas Tool and Die. Remember our discussion about principles?"

19.5   The Future

This book has focused on manufacturing management, within the scope of operations, and using factory physics as the unifying perspective. It is fitting that we close with an assessment of what factory physics is and what we can expect from it in the future.

1. *Factory physics is a start to a science of manufacturing.* We have argued that a science of manufacturing is needed to enable managers to judge which policies will be effective in their system and which will not. In the past 30 years or so, manufacturing has been besieged by one "revolution" after another—MRP, JIT, TQM, TBC (time-based competition), BPR (business process reengineering), SCM (supply chain management), and so on—each of which has undoubtedly contained useful insights. But because each presents only a specific perspective, generally sold in fire-breathing revolutionary rhetoric and justified primarily in terms of anecdotal evidence, the manufacturing manager has no basis on which to choose between them, combine features of different approaches, or develop a unique system adapted to the particular environment. Only a science that describes the critical behavior and interactions in a manufacturing system can provide the over arching understanding needed for this.

Our efforts in this book at the development of a science of manufacturing are far from complete. However, we feel that we have at least framed the problem in the correct context. While we have relied on mathematical formulas, we have *not* sought a "factory mathematics." Our focus has consistently been on the physical behavior of manufacturing systems; mathematics are simply the language for describing this behavior precisely. For example, the basic factory dynamics formulas of Chapter 7 were developed in response to the question, How do WIP, throughput, and cycle time depend on one another? By making various assumptions about the behavior of the plant (e.g., the best case, worst case, and practical worst case), we were able to develop formulas for the curves of throughput versus WIP and cycle time versus WIP. These relationships sharpened our insight into questions like why many plants have excessive WIP levels, why variability reductions can reduce cycle times, and how improvements in a production line can be characterized. However, these formulas are certainly not the final word on the WIP, throughput, and cycle time relationships. In Chapter 12, we returned to these curves and showed that when scrap loss is considered, throughput may eventually decrease in the WIP level—something that our cases in Chapter 7 did not allow.

Because manufacturing systems are complex and diverse, some systems undoubtedly exhibit types of behavior that we have not described in this book. Indeed, as we write this, considerable research is being devoted to describing many different production systems (see Askin and Standridge 1993; Buzacott and Shanthikumar 1993; and Graves, Rinnooy Kan, and Zipkin 1993 for good, up-to-date summaries). Thus, in the next few years, we can expect the range and depth of factory physics to expand significantly. Although advances in manufacturing science will never enable manufacturing management to become merely an analytical exercise, our hope is that it will become more like medicine (i.e., science-based, with a strong human element) and less like fashion (i.e., trendy, without guiding principles).

2. *Factory physics is a pedagogical framework for conveying;*
   a. *Basics*
   b. *Intuition*
   c. *Synthesis*

To give precise descriptions of factory behavior under various conditions, we need appropriate tools (e.g., statistics, queueing theory, reliability). In a factory physics framework, therefore, these become important not just for their own sake, but as building blocks for answering fundamental questions about how plants behave.

We have repeatedly stressed that sound intuition is perhaps the single most important skill of the manufacturing manager, enabling him or her to focus attention on the areas of greatest leverage. By describing the natural tendencies of manufacturing systems, factory physics provides a structure within which to build intuition. The manager who understands factory physics principles and can interpret empirical observations in terms of them will acquire insight into the behavior of a system far more rapidly than a manager without these skills.

We have also stressed that manufacturing systems are complex, multifaceted organizations involving many different processes, people, and machines, and multiple objectives. In such environments, the major opportunities for improvements often lie at the interfaces (e.g., between sales and manufacturing, or between product development and manufacturing). By providing a general description of the manufacturing system, factory physics gives us a means for evaluating the impacts of external changes on plant behavior. As such, it represents a linking mechanism between manufacturing and other business functions.

3. *Factory physics is a link between the* process *and* systems *views of manufacturing.* Manufacturing specialists tend to come in two varieties. One group focuses on the specific processes involved in manufacturing, such as robotics, surface finishing, grinding, injection molding. The other group (to which the authors belong) focuses on systems, such as scheduling, inventory control, production planning. Clearly, both sets of concerns are critical to effective operation of a plant. Unfortunately, members of each group are inclined to act as if their view of manufacturing were the only "correct" one. As a result, processes are chosen with little regard for systems impact, and systems are designed with little detailed consideration of processes. Factory physics uses process-oriented descriptors (e.g., mean time to failure, mean time to repair, setup time), condensed into logistics-oriented descriptors (e.g., mean and SCV of effective processing times), to estimate systems-oriented measures (throughput, WIP, cycle time). Thus, it provides a means for interpreting process changes in systems terms.

4. *Factory physics is a collection of tools for quantifying tradeoffs.* As we have seen, increasing capacity, reducing scrap, improving reliability and maintainability, reducing or externalizing setups, upgrading the quality of purchased parts, more frequent moves of smaller batches, and many other policies can have related logistical impacts. By combining the factory physics tools for evaluating these effects with estimates of costs, we can examine the relative attractiveness of each. Moreover, by using the plant-level measures provided by factory physics under different configurations, we can generate cost versus performance curves (e.g., throughput versus cost or cycle time versus cost) and determine strategically desirable targets.

Finally, from an impact standpoint, it is difficult to overstate the importance of factory physics. Roughly one-half of the U.S. economy (jobs, as well as GNP) still depends on manufacturing. Indeed, operational improvements in the manufacturing sector were instrumental in the productivity gains that drove the economic boom of the 1990s. But as competitiveness in the world of manufacturing continues to escalate, the ability to deliver diverse products with high quality, low cost, swift delivery, and reliable service is fast evolving from a recipe for success to a requirement for survival. In the past it was possible to develop effective manufacturing practices by trial and error. In the future there won't be time. Only by sustaining a rapid cycle of continual improvement through the use of principles to quickly develop practices that support strategy will firms be able to keep pace. In the 21st century, mastery of the concepts of factory physics will be as vital a core manufacturing competency as the concepts of mass production were in the 20th century.

TABLE Cumulative Probabilities of the Standard Normal Distribution

Entry is area $\Phi(z)$ under the standard normal curve from $-\infty$ to z.

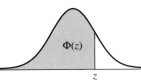

z	0.00	0.01	0.02	0.03	0.04	0.05	0.06	0.07	0.08	0.09
−3.4	0.0003	0.0003	0.0003	0.0003	0.0003	0.0003	0.0003	0.0003	0.0003	0.0002
−3.3	0.0005	0.0005	0.0005	0.0004	0.0004	0.0004	0.0004	0.0004	0.0004	0.0003
−3.2	0.0007	0.0007	0.0006	0.0006	0.0006	0.0006	0.0006	0.0005	0.0005	0.0005
−3.1	0.0010	0.0009	0.0009	0.0009	0.0008	0.0008	0.0008	0.0008	0.0007	0.0007
−3.0	0.0013	0.0013	0.0013	0.0012	0.0012	0.0011	0.0011	0.0011	0.0010	0.0010
−2.9	0.0019	0.0018	0.0017	0.0017	0.0016	0.0016	0.0015	0.0015	0.0014	0.0014
−2.8	0.0026	0.0025	0.0024	0.0023	0.0023	0.0022	0.0021	0.0021	0.0020	0.0019
−2.7	0.0035	0.0034	0.0033	0.0032	0.0031	0.0030	0.0029	0.0028	0.0027	0.0026
−2.6	0.0047	0.0045	0.0044	0.0043	0.0041	0.0040	0.0039	0.0038	0.0037	0.0036
−2.5	0.0062	0.0060	0.0059	0.0057	0.0055	0.0054	0.0052	0.0051	0.0049	0.0048
−2.4	0.0082	0.0080	0.0078	0.0075	0.0073	0.0071	0.0069	0.0068	0.0066	0.0064
−2.3	0.0107	0.0104	0.0102	0.0099	0.0096	0.0094	0.0091	0.0089	0.0087	0.0084
−2.2	0.0139	0.0136	0.0132	0.0129	0.0125	0.0122	0.0119	0.0116	0.0113	0.0110
−2.1	0.0179	0.0174	0.0170	0.0166	0.0162	0.0158	0.0154	0.0150	0.0146	0.0143
−2.0	0.0228	0.0222	0.0217	0.0212	0.0207	0.0202	0.0197	0.0192	0.0188	0.0183
−1.9	0.0287	0.0281	0.0274	0.0268	0.0262	0.0256	0.0250	0.0244	0.0239	0.0233
−1.8	0.0359	0.0352	0.0344	0.0336	0.0329	0.0322	0.0314	0.0307	0.0301	0.0294
−1.7	0.0446	0.0436	0.0427	0.0418	0.0409	0.0401	0.0392	0.0384	0.0375	0.0367
−1.6	0.0548	0.0537	0.0526	0.0516	0.0505	0.0495	0.0485	0.0475	0.0465	0.0455
−1.5	0.0668	0.0655	0.0643	0.0630	0.0618	0.0606	0.0594	0.0582	0.0571	0.0559
−1.4	0.0808	0.0793	0.0778	0.0764	0.0749	0.0735	0.0722	0.0708	0.0694	0.0681
−1.3	0.0968	0.0951	0.0934	0.0918	0.0901	0.0885	0.0869	0.0853	0.0838	0.0823
−1.2	0.1151	0.1131	0.1112	0.1093	0.1075	0.1056	0.1038	0.1020	0.1003	0.0985
−1.1	0.1357	0.1335	0.1314	0.1292	0.1271	0.1251	0.1230	0.1210	0.1190	0.1170
−1.0	0.1587	0.1562	0.1539	0.1515	0.1492	0.1469	0.1446	0.1423	0.1401	0.1379
−0.9	0.1841	0.1814	0.1788	0.1762	0.1736	0.1711	0.1685	0.1660	0.1635	0.1611
−0.8	0.2119	0.2090	0.2061	0.2033	0.2005	0.1977	0.1949	0.1922	0.1894	0.1867
−0.7	0.2420	0.2389	0.2358	0.2327	0.2296	0.2266	0.2236	0.2206	0.2177	0.2148
−0.6	0.2743	0.2709	0.2676	0.2643	0.2611	0.2578	0.2546	0.2514	0.2483	0.2451
−0.5	0.3085	0.3050	0.3015	0.2981	0.2946	0.2912	0.2877	0.2843	0.2810	0.2776
−0.4	0.3446	0.3409	0.3372	0.3336	0.3300	0.3264	0.3228	0.3192	0.3156	0.3121
−0.3	0.3821	0.3783	0.3745	0.3707	0.3669	0.3632	0.3594	0.3557	0.3520	0.3483
−0.2	0.4207	0.4168	0.4129	0.4090	0.4052	0.4013	0.3974	0.3936	0.3897	0.3859
−0.1	0.4602	0.4562	0.4522	0.4483	0.4443	0.4404	0.4364	0.4325	0.4286	0.4247
−0.0	0.5000	0.4960	0.4920	0.4880	0.4840	0.4801	0.4761	0.4721	0.4681	0.4641
0.0	0.5000	0.5040	0.5080	0.5120	0.5160	0.5199	0.5239	0.5279	0.5319	0.5359
0.1	0.5398	0.5438	0.5478	0.5517	0.5557	0.5596	0.5636	0.5675	0.5714	0.5753
0.2	0.5793	0.5832	0.5871	0.5910	0.5948	0.5987	0.6026	0.6064	0.6103	0.6141
0.3	0.6179	0.6217	0.6255	0.6293	0.6331	0.6368	0.6406	0.6443	0.6480	0.6517
0.4	0.6554	0.6591	0.6628	0.6664	0.6700	0.6736	0.6772	0.6808	0.6844	0.6879
0.5	0.6915	0.6950	0.6985	0.7019	0.7054	0.7088	0.7123	0.7157	0.7190	0.7224
0.6	0.7257	0.7291	0.7324	0.7357	0.7389	0.7422	0.7454	0.7486	0.7517	0.7549
0.7	0.7580	0.7611	0.7642	0.7673	0.7704	0.7734	0.7764	0.7794	0.7823	0.7852
0.8	0.7881	0.7910	0.7939	0.7967	0.7995	0.8023	0.8051	0.8078	0.8106	0.8133
0.9	0.8159	0.8186	0.8212	0.8238	0.8264	0.8289	0.8315	0.8340	0.8365	0.8389
1.0	0.8413	0.8438	0.8461	0.8485	0.8508	0.8531	0.8554	0.8577	0.8599	0.8621
1.1	0.8643	0.8665	0.8686	0.8708	0.8729	0.8749	0.8770	0.8790	0.8810	0.8830
1.2	0.8849	0.8869	0.8888	0.8907	0.8925	0.8944	0.8962	0.8980	0.8997	0.9015
1.3	0.9032	0.9049	0.9066	0.9082	0.9099	0.9115	0.9131	0.9147	0.9162	0.9177
1.4	0.9192	0.9207	0.9222	0.9236	0.9251	0.9265	0.9278	0.9292	0.9306	0.9319
1.5	0.9332	0.9345	0.9357	0.9370	0.9382	0.9394	0.9406	0.9418	0.9429	0.9441
1.6	0.9452	0.9463	0.9474	0.9484	0.9495	0.9505	0.9515	0.9525	0.9535	0.9545
1.7	0.9554	0.9564	0.9573	0.9582	0.9591	0.9599	0.9608	0.9616	0.9625	0.9633
1.8	0.9641	0.9649	0.9656	0.9664	0.9671	0.9678	0.9686	0.9693	0.9699	0.9706
1.9	0.9713	0.9719	0.9726	0.9732	0.9738	0.9744	0.9750	0.9756	0.9761	0.9767
2.0	0.9772	0.9778	0.9783	0.9788	0.9793	0.9798	0.9803	0.9808	0.9812	0.9817
2.1	0.9821	0.9826	0.9830	0.9834	0.9838	0.9842	0.9846	0.9850	0.9854	0.9857
2.2	0.9861	0.9864	0.9868	0.9871	0.9875	0.9878	0.9881	0.9884	0.9887	0.9890
2.3	0.9893	0.9896	0.9898	0.9901	0.9904	0.9906	0.9909	0.9911	0.9913	0.9916
2.4	0.9918	0.9920	0.9922	0.9925	0.9927	0.9929	0.9931	0.9932	0.9934	0.9936
2.5	0.9938	0.9940	0.9941	0.9943	0.9945	0.9946	0.9948	0.9949	0.9951	0.9952
2.6	0.9953	0.9955	0.9956	0.9957	0.9959	0.9960	0.9961	0.9962	0.9963	0.9964
2.7	0.9965	0.9966	0.9967	0.9968	0.9969	0.9970	0.9971	0.9972	0.9973	0.9974
2.8	0.9974	0.9975	0.9976	0.9977	0.9977	0.9978	0.9979	0.9979	0.9980	0.9981
2.9	0.9981	0.9982	0.9982	0.9983	0.9984	0.9984	0.9985	0.9985	0.9986	0.9986
3.0	0.9987	0.9987	0.9987	0.9988	0.9988	0.9989	0.9989	0.9989	0.9990	0.9990
3.1	0.9990	0.9991	0.9991	0.9991	0.9992	0.9992	0.9992	0.9992	0.9993	0.9993
3.2	0.9993	0.9993	0.9994	0.9994	0.9994	0.9994	0.9994	0.9995	0.9995	0.9995
3.3	0.9995	0.9995	0.9995	0.9996	0.9996	0.9996	0.9996	0.9996	0.9996	0.9997
3.4	0.9997	0.9997	0.9997	0.9997	0.9997	0.9997	0.9997	0.9997	0.9997	0.9998

Selected Percentiles

Cumulative probability $\Phi(z)$:	.90	.95	.975	.98	.99	.995	.999
z:	1.282	1.645	1.960	2.054	2.326	2.576	3.090

Ackoff, R. L. 1956. "The Development of Operations Research as a Science," *Operations Research* **4**: 256.

Aggarwal, S. C. 1985. "MRP, JIT, OPT, FMS? Making Sense of Production Operations Systems," *Harvard Business Review,* September–October, pp. 8–16.

Anderson, J., R. Schroeder, S. Tupy, and E. White. 1982. "Material Requirements Planning Systems: The State-of-the-Art," *Production and Inventory Management* **23**(4): 51–67.

Arguello, M. 1994. "Review of Scheduling Software," Technology Transfer 93091822A-XFR, SEMATECH, Austin, TX.

Askin, R. G., and C. R. Standridge. 1993. *Modeling and Analysis of Manufacturing Systems.* New York: Wiley.

A. W. Shaw Company. 1915. *The Library of Factory Management,* vols. 1–5. Chicago: A. W. Shaw.

Axsäter, S. 1993. "Continuous Review Policies for Multi-Level Inventory Systems with Stochastic Demand," in *Handbooks in Operations Research and Management Science, vol. 4: Logistics of Production and Inventory.* S. C. Graves, A. H. G. Rinnooy Kan, and P. H. Zipkin (eds.). New York: North-Holland.

Babbage, C. 1832. *On the Economy of Machinery and Manufactures.* London: Charles Knight. Reprint, Augustus M. Kelley, New York, 1963.

Bahl, H. C., L. P. Ritzman, and J. N. D. Gupta. 1987. "Determining Lot Sizes and Resource Requirements: A Review," *Operations Research* **35**(3): 329–345.

Baker, K. R. 1993. "Requirements Planning," in *Handbooks in Operations Research and Management Science, vol 4: Logistics of Production and Inventory,* S. C. Graves, A. H. G. Rinnooy Kan, and P. H. Zipkin (eds.). New York: North-Holland.

Baker, W. M. 1994. "Understanding Activity-Based Costing," *Industrial Management,* March/April, **36:** 28–30.

Barnard, C. I. 1938. *The Functions of the Executive.* Cambridge, MA: Harvard University Press.

Barnes, R. 1937. *Motion and Time Study.* New York: Wiley.

Bartholdi, J. J., and D. D. Eisenstein. 1996. "A Production Line that Balances Itself," *Operations Research* **44**(1): 21–34.

Baumol, W. J., S. Blackman, and E. N. Wolff. 1989. *Productivity and American Leadership: The Long View.* Cambridge, MA: MIT Press.

Bazaraa, M. S., and C. M. Shetty. 1979. *Nonlinear Programming: Theory and Algorithms.* New York: Wiley.

Benjaafar, S., and M. Sheikhzadeh. 1997. "Scheduling Policies, Batch Sizes, and Manufacturing Lead Times," *IIE Transactions* **29**(2): 159–166.

Blackburn, J. D. (ed.). 1991. *Time-Based Competition: The Next Battleground in American Manufacturing.* Homewood, IL: Irwin.

Blackstone, J. H., Jr., D. T. Phillips, and G. L. Hogg. 1982. "A State-of-the-art Survey of Dispatching Rules for Manufacturing Job Shop Operations," *International Journal of Production Research* **20**(1): 27–45.

Bonneville, J. H. 1925. *Elements of Business Finance.* Englewood Cliffs, NJ: Prentice-Hall.

Boorstein, D. J. 1958. *The Americans: The Colonial Experience.* New York: Random House.

———. 1965. *The Americans: The National Experience.* New York: Random House.

———. 1973. *The Americans: The Democratic Experience.* New York: Random House.

Boudette, N. 1999. "Europe's SAP Scrambles to Stem Big Glitches—Software Giant to Tighten Its Watch After Snafus at Whirlpool, Hershey," *The Wall Street Journal,* Nov 4.

Bourland, K. 1992. "Spartan Industries," Case Study, Amos Tuck School, Dartmouth College.

Box, G. E. P., and G. M. Jenkins. 1970. *Time Series Analysis, Forecasting and Control.* San Francisco: Holden-Day.

Bradt, L. J. 1983. "The Automated Factory: Myth or Reality," *Engineering: Cornell Quarterly* **3**(13).

Brown, R. G. 1967. *Decision Rules for Inventory Management.* New York: Holt, Rinehart and Winston.

Browne, J., J. Harhen, and J. Shivnan. 1988. *Production Management Systems.* Reading, MA: Addison-Wesley.

Bryant, K. L., and H. C. Dethloff. 1990. *A History of American Business.* Englewood Cliffs, NJ: Prentice-Hall.

Buzacott, J. A., and J. G. Shanthikumar. 1993. *Stochastic Models of Manufacturing Systems.* Englewood Cliffs, NJ: Prentice-Hall.

Carlier, J., and E. Pinson. 1988. "An Algorithm for Solving the Job-Shop Problem," *Management Science* **35:** 164–176.

Carnegie, A. 1920. *Autobiography of Andrew Carnegie.* Boston: Houghton Mifflin.

Cerveny, R. P., and L. W. Scott. 1989. "A Survey of MRP Implementation," *Production and Inventory Management* **30**(3): 31–34.

Chandler, Alfred D., Jr. 1977. *The Visible Hand: The Managerial Revolution in American Business.* Cambridge, MA: Belknap Press.

———. 1984. "The Emergence of Managerial Capitalism," *Business History Review* **58:** 473–503.

———. 1990. *Scale and Scope: The Dynamics of Industrial Capitalism.* Cambridge, MA: Harvard University Press.

Chandler, Alfred D., and S. Salsbury. 1971. *Pierre S. Du Pont and the Making of the Modern Corporation.* New York: Harper & Row.

Charney, C. 1991. *Time to Market: Reducing Product Lead Time.* Dearborn, MI: Society of Manufacturing Engineers.

Cherington, P. T. 1920. *The Elements of Marketing.* New York: Macmillan.

Churchman, C. W. 1968. *The Systems Approach.* New York: Dell.

Clark, A., and H. Scarf. 1960. "Optimal Policies for a Multi-Echelon Inventory Problem," *Management Science* **36:** 1329–1338.

Clark, K. B., R. H. Hayes, and C. Lorenz. 1985. *The Uneasy Alliance: Managing the Productivity-Technology Dilemma.* Boston: Harvard Business School Press.

Cohen, M. A., Y. S. Zheng, and V. Agrawal. 1994. "Service Parts Logistics Benchmark Study." Working paper, Wharton School, University of Pennsylvania, Philadelphia.

Cohen, S. S., and J. Zysman. 1987. *Manufacturing Matters: The Myth of the Post-Industrial Economy.* New York: Basic Books.

Consiglio, M. 1969. "Leonardo da Vinci: The First IE?" *Industrial Engineering* **1:** 71.

Copley, F. B. 1923. *Frederick W. Taylor: Father of Scientific Management.* New York: Harper and Brothers.

Cray, E. 1979. *Chrome Colossus: General Motors and Its Times.* New York: McGraw-Hill.

Crosby, P. B. 1979. *Quality Is Free: The Art of Making Quality Certain.* New York: McGraw-Hill.

———. 1984. *Quality Without Tears: The Art of Hassle-Free Management.* New York: McGraw-Hill.

Daskin, M. S. 1995. *Network and Discrete Location.* New York: Wiley.

Davidson, K. M. 1990. "Do Megamergers Make Sense?" in *Mergers, Acquisitions and Leveraged Buyouts,* R. L. Kuhn (ed.) Homewood, IL: Irwin.

de Kok, T. 1993. "Back-Order Lead Time Behavior in (S,Q)-Inventory Models with Compound Renewal Demand." Working paper, School of Technology Management, Eindhoven University of Technology, Eindhoven, The Netherlands.

Deleersnyder, J. L., T. J. Hodgson, R. E. King, P. J. O'Grady, and A. Savva. 1992. "Integrating Kanban Type Pull Systems and MRP Type Push Systems: Insights from a Markovian Model," *IIE Transactions* **24**(3): 43–56.

Deming, W. E. 1950a. *Some Theory of Sampling.* New York: Wiley.

———. 1950b. *Elementary Principles of the Statistical Control of Quality.* Tokyo: Union of Japanese Science and Engineering.

———. 1960. *Sample Design in Business Research.* New York: Wiley.

———. 1982. *Quality Productivity and Competitive Position.* Cambridge, MA: Massachusetts Institute of Technology, Center for Advanced Engineering Study.

———. 1986. *Out of the Crisis.* Cambridge, MA: MIT Press.

Dertouzos, M. L., R. K. Lester, and R. M. Solow. 1989. *Made in America: Regaining the Productive Edge.* Cambridge, MA: MIT Press.

Deuermeyer, B. 1994. Interoffice Memorandum on Undergraduate Curriculum in Industrial Engineering, Texas A&M University.

Deuermeyer, B., and L. B. Schwarz. 1981. "A Model for the Analysis of System Service Level in Warehouse/Retailer Distribution Systems: The Identical Retailer Case." In: *Multi-Level Production/Inventory Control Systems: Theory and Practice.* L. B. Schwarz (ed.). Amsterdam: North-Holland, 163–193.

DeVor, R., T. Chang, and J. Sutherland. 1992. *Statistical Quality Design and Control: Contemporary Concepts and Methods.* New York: Macmillan.

Drucker, P. F. 1954. *The Practice of Management.* New York: Harper & Row.

Dudek, R. A., S. S. Panwalkar, and M. L. Smith. 1992. "The Lessons of Flowshop Scheduling Research," *Operations Research* **40**(1): 7–13.

Duncan, W. J. 1989. *Great Ideas in Management: Lessons from the Founders and Foundations of Managerial Practice.* San Francisco: Jossey-Bass Publishers.

Edmondson, G., and A. Reinhardt. 1997. "Silicon Valley on the Rhine," *Business Week,* November 3, 162–166.

Einstein, A. 1950. Quoted by Lincoln Barnett, "The Meaning of Einstein's New Theory," *Life,* January 9, 22.

Emerson, H. P., and D. C. E. Naehring. 1984. *Origins of Industrial Engineering.* Norcross, GA: Institute of Industrial Engineers.

Erlenkotter, D. 1989. "An Early Classic Misplaced: Ford W. Harris's Economic Order Quantity Model of 1915," *Management Science* **35**(7): 898–900.

———. 1990. "Ford Whitman Harris and the Economic Order Quantity Model," *Operations Research* **38**(6): 937–946.

Fayol, H. 1916. *Administration industrielle et générale,* Paris: Dunod. In English, *General and Industrial Management.* (Constance Storrs, trans.) London: Sir Isaac Pitman and Sons, 1949.

Federgruen, A. 1993. "Centralized Planning Models for Multi-Echelon Inventory Systems under Uncertainty." In *Handbooks in Operations Research and Management Science, vol. 4: Logistics of Production and Inventory,* S. C. Graves, A. H. G. Rinnooy Kan, and P. H. Zipkin (eds.). New York: North-Holland.

Federgruen, A., and Y. Zheng. 1992. "The Joint Replenishment Problem with General Joint Cost Structures," *Operations Research* **40:** 384–403.

——— and ———. 1992. "An Efficient Algorithm for Computing an Optimal (r, Q) Policy in Continuous Review Stochastic Inventory Systems," *Operations Research* **40:** 808–813.

Federgruen, A., and P. Zipkin. 1984. "Computational Issues in an Infinite Horizon, Multi-Echelon Inventory Model," *Operations Research* **32:** 818–836.

Feigenbaum, A. V. 1956. "Total Quality Control," *Harvard Business Review,* November.

———. 1961. *Total Quality Control: Engineering and Management.* New York: McGraw-Hill.

Feitzinger, E., and H. L. Lee. 1997. "Mass Customization at Hewlett-Packard: The Power of Postponement." *Harvard Business Review,* January–February, 116–121.

Fish, J. C. L. 1915. *Engineering Economics: First Principles.* New York: McGraw-Hill.

Fisher, M. L. 1997. "What Is the Right Supply Chain for Your Product?" *Harvard Business Review,* March–April, 105–116.

Flink, J. J. 1970. *America Adopts the Automobile, 1895–1910.* Cambridge, MA: MIT Press.

Follett, M. P. 1942. *Dynamic Administration: The Collected Papers of Mary Parker Follett.* H. C. Metcalf, and L. Urwick (eds.). New York: Harper.

Ford, H. 1926. *Today and Tomorrow.* New York: Doubleday. Reprint, Productivity Press, 1988.

Fordyce, J. M, and F. M. Webster. 1984. "The Wagner-Whitin Algorithm Made Simple." *Production and Inventory Management* **25**(2): 21–30.

Forrester, J. 1961. *Industrial Dynamics.* New York: MIT Press and Wiley.

Fourer, R., D. M. Gay, and B.W. Kernighan. 1993. *AMPL: A Modeling Language for Mathematical Programming.* San Francisco: Scientific Press.

Fox, R. E. 1980. "Keys to Successful Materials Management Systems: A Contrast Between Japan, Europe and the U.S." *23rd Annual Conference Proceedings,* APICS, 440–444.

Freidenfelds, J. 1981. *Capacity Extension: Simple Models and Applications.* Amsterdam: North-Holland.

Galbraith, J. K. 1958. *The Affluent Society.* Boston: Houghton Mifflin.

Garvin, D. 1988. *Managing Quality: The Strategic and Competitive Edge.* New York: Free Press.

Gilbreth, F. B. 1911. *Motion Study.* New York: Van Nostrand.

Gilbreth, F. B., and E. G. Gilbreth Carey. 1949. *Cheaper by the Dozen.* New York: T. Y. Crowell.

Gilbreth, L. M. 1914. *The Psychology of Management.* New York: Sturgis and Walton. Reprinted in W. R. Spriegel, and C. E. Myers (eds.). *The Writings of the Gilbreths.* Homewood IL: Irwin, 1953.

Glover, F. 1990. "Tabu Search: A Tutorial." *Interfaces* **20**(4): 79–94.

Goldratt, E. M., and J. Cox. 1984. *The Goal: A Process of Ongoing Improvement.* Croton-on-the-Hudson, NY: North River Press.

Goldratt, E. M., and R. E. Fox. 1986. *The Race.* Croton-on-the-Hudson, NY: North River Press.

Gordon, R. A., and J. E. Howell. 1959. *Higher Education for Business.* New York: Columbia University Press.

Gould, L. 1985. "Computers Run the Factory." *Electronics Week,* March 25.

Grant, E. L. 1930. *Principles of Engineering Economy.* New York: Ronald Press.

Grant, E. L., and R. Leavenworth. 1946. *Statistical Quality Control.* Milwaukee, WI: American Society for Quality Control.

Graves, S. C., A. H. G. Rinnooy Kan, and P. H. Zipkin (eds.). 1993. *Handbooks in Operations Research and Management Science, vol. 4: Logistics of Production and Inventory.* New York: North-Holland.

Gross, D., and C. Harris. 1985. *Fundamentals of Queueing Theory.* 2d ed. New York: Wiley.

Hackman, S. T., and R.C. Leachman. 1989. "A General Framework for Modeling Production," *Management Science* **35**(4): 478–495.

Hadley, G., and T. M. Whitin. 1963. *Analysis of Inventory Systems.* Englewood Cliffs, NJ: Prentice-Hall.

Hall, R. W. 1981. *Driving the Productivity Machine: Production Planning and Control in Japan.* Falls Church, VA: American Production and Inventory Control Society, Inc.

———. 1983. *Zero Inventories.* Homewood, IL: Dow Jones-Irwin.

Hammer, M., and J. Champy. 1993. *Reengineering the Corporation.* New York: HarperCollins.

Harris, F. W. 1913. "How Many Parts to Make at Once." *Factory: The Magazine of Management* **10**(2): 135–136, 152. Also reprinted in *Operations Research* **38**(6): 947–950, 1990.

Hausman, W. H., and N. K. Erkip. 1994. "Multi-Echelon vs. Single-Echelon Inventory Control Policies for Low-Demand Items." *Management Science* **40:** 597–602.

Hax, A. C., and D. Candea. 1984. *Production and Inventory Management.* Englewood Cliffs, NJ: Prentice-Hall.

Hayes, R. 1981. "Why Japanese Factories Work." *Harvard Business Review,* July–August, pp. 57–66.

Hayes, R., and S. Wheelwright. 1984. *Restoring Our Competitive Edge: Competing through Manufacturing.* New York: Wiley.

Hayes, R., S. Wheelwright, and K. Clark. 1988. *Dynamic Manufacturing: Creating the Learning Organization.* New York: Free Press.

Hitomi, K. 1979. *Manufacturing Systems Engineering.* London: Taylor and Francis.

Hodge, A. C., and J. O. McKinsey. 1921. *Principles of Accounting.* Chicago: University of Chicago Press.

Hopp, W. J., and M. L. Roof. 1998. "Quoting Manufacturing Due Dates Subject to a Service Level Constraint," Technical Report, Department of Industrial Engineering, Northwestern University, Evanston, IL.

Hopp, W. J., and M. L. Spearman. 1991. "Throughput of a Constant Work in Process Manufacturing Line Subject to Failures." *International Journal of Production Research* **29**(3): 635–655.

Hopp, W. J., and M. L. Spearman. 1993. "Setting Safety Leadtimes for Purchased Components in Assembly Systems." *IIE Transactions* **25**(2): 2–11.

Hopp, W. J., and M. L. Spearman, and I. Duenyas. 1993. "Economic Production Quotas for Pull Manufacturing Systems." *IIE Transactions* **25**(2): 71–79.

Hopp, W. J., and M. L. Spearman, and D. L. Woodruff. 1990. "Practical Strategies for Lead Time Reduction." *Manufacturing Review* **3**(2): 78–84.

Industrial Engineering. 1991. "Competition in Manufacturing Leads to MRP II." *Industrial Engineering,* **23**(7): 10–13.

Inman, R. A., and S. Mehra. 1990. "The Transferability of Just-in-Time Concepts to American Small Businesses." *Interfaces* **20:** 30–37, March–April.

Inman, R. R. 1993. "Inventory Is the *Flower* of All Evil." *Production and Inventory Management Journal* **34**(4): 41–45.

Jackson, P. L., W. L. Maxwell, and J. A. Muckstadt. 1985. "The Joint Replenishment Problem With a Powers of Two Restriction." *IIE Transactions* **17:** 25–32.

Jacobs, F. R. 1984. "OPT Uncovered: Many Production Planning and Scheduling Concepts Can Be Applied With or Without the Software." *Industrial Engineering,* October, 32–41.

Johnson, H. T., and R. S. Kaplan. 1987. *Relevance Lost: The Rise and Fall of Management Accounting.* Cambridge, MA: Harvard Business School Press.

Johnson, L. A., and D. C. Montgomery. 1974. *Operations Research in Production Planning, Scheduling, and Inventory Control.* New York: Wiley.

Johnson, S. M. 1954. "Optimal Two- and Three-Stage Production Schedules with Setup Times Included," *Naval Research Logistics Quarterly* **1:** 61–68.

Juran, J. M. 1964. *Managerial Breakthrough.* New York: McGraw-Hill.

———(ed.). 1988. *Juran's Quality Control Handbook,* 4th ed., F.M. Gryna (assoc. ed.). New York: McGraw-Hill.

———. 1989. *Juran on Leadership for Quality: An Executive Handbook.* New York: Free Press.

———. 1992. *Juran on Quality by Design: The New Steps for Planning Quality into Goods and Services.* New York: Free Press.

Kakar, S. 1970. *Frederick Taylor: A Study in Personality and Innovation.* Cambridge, MA: MIT Press.

Kanet, J. J. 1984. "Inventory Planning at Black & Decker." *Production and Inventory Management* **25**(3): 62–74.

———. 1988. "MRP 96: Time to Rethink Manufacturing Logistics." *Production and Inventory Management* **29**(2): 57–61.

Kaplan, R. S. 1986. "Must CIM Be Justified by Faith Alone?" *Harvard Business Review,* March–April, 87–95.

Karmarkar, U. S. 1987. "Lot Sizes, Lead Times and In-Process Inventories." *Management Science* **33**(3): 409–423.

———. 1989. "Getting Control of Just-in-Time," *Harvard Business Review,* September–October, 122–131.

Kearns, D. T., and D. A. Nadler. 1992. *Prophets in the Dark: How Xerox Reinvented Itself and Beat Back the Japanese.* New York: HarperCollins.

Kellermann, A. L., F. P. Rivara, N. B. Rushforth, J. G. Banton, D. T. Reay, J. T. Francisco, A. B. Locci, J. Prodzinski, B. B. Hackman, and G. Somes. 1993. "Gun Ownership as a Risk Factor for Homicide in the Home." *New England Journal of Medicine,* **329**(15): 1084–1091.

Kilbridge, M. D., and L. Wester. 1961. "A Heuristic Method of Assembly Line Balancing," *Journal of Industrial Engineering.* **12**(4): 292–298.

Klein, J. A. 1989. "The Human Costs of Manufacturing Reform." *Harvard Business Review,* March–April, pp. 60–66.

Kleinrock, L. 1975. *Queueing Systems,* vol. I: *Theory.* New York: Wiley.

Krajewski, L. J., B. E. King, L. P. Ritzman, and D. S. Wong. 1987. "Kanban, MRP, and Shaping the Manufacturing Environment." *Management Science* **33**(1): 39–57.

Kuhn, T. S. 1970. *The Structure of Scientific Revolutions.* Chicago: University of Chicago Press.

LaForge, R., and V. Sturr. 1986. "MRP Practices in a Random Sample of Manufacturing Firms," *Production and Inventory Management* **28**(3): 129–137.

Lamm, R. D. 1988. "Crisis: The Uncompetitive Society." In *Global Competitiveness.* M. K. Starr (ed.). New York: Norton.

Lee, H. L., and C. Billington. 1992. "Managing Supply Chain Inventory: Pitfalls and Opportunities," *Sloan Management Review* **33:** 65–73.

———. 1995. "The Evolution of Supply-Chain-Management Models and Practice at Hewlett-Packard." *Interfaces* **25**(5): 42–63.

Lee, H. L., C. Billington, and B. Carter. 1993. "Hewlett-Packard Gains Control of Inventory and Service through Design for Localization." *Interfaces* **23**(4): 1–20.

Lee, H. L., V. Padmanabhan, and S. Whang. 1997a. "The Bullwhip Effect in Supply Chains." *Sloan Management Review* **38**(3): 93–102.

———. 1997b. "Information Distortion in a Supply Chain: The Bullwhip Effect." *Management Science* **43**(4): 546–558.

Little, J. D. C. 1992. "Tautologies, Models and Theories: Can We Find 'Laws' of Manufacturing?" *IIE Transactions* **24:** 7–13.

Lough, W. H. 1920. *Business Finance.* New York: Ronald Press.

Lundrigan, R. 1986. "What's This Thing Called OPT?" *Production and Inventory Management* **27**(2): 2–12.

Maddison, A. 1984. "Comparative Analysis of the Productivity Situation in the Advanced Capitalist Countries." In *International Comparisons of Productivity and Causes of the Slowdown.* J. W. Kendrick (ed.). Cambridge, MA: Ballinger.

Majone, G. 1985. "Systems Analysis: A Genetic Approach." In *Handbook of Systems Analysis: Overview of Uses, Procedures, Applications, and Practice,* chapter 2. Hugh J. Misner and Edward S. Quade (eds.). New York: Elsevier.

Marion, J. B. 1970. *Classical Dynamics of Particles and Systems,* 2d ed. New York: Academic Press, 266.

Martino, J. P. 1983. *Technological Forecasting for Decision Making,* 2d ed. New York: North-Holland.

Maslow, A. 1954. *Motivation and Personality.* New York: Harper.

Mayo, E. 1933. *The Human Problems of an Industrial Civilization.* New York: Macmillan.

———. 1945. *The Social Problems of an Industrial Civilization.* Cambridge, MA: Division of Research, Graduate School of Business Administration, Harvard University.

McClain, J. O., and L. J. Thomas. 1985. *Operations Management: Production of Goods and Services,* 2d ed. Englewood Cliffs, NJ: Prentice-Hall.

McCloskey, J. F. 1987a. "The Beginnings of Operations Research 1934–1941." *Operations Research* **35**(1): 143–152.

———. 1987b. "British Operational Research in World War II." *Operations Research* **35**(3): 453–470.

———. 1987c. "U.S. Operations Research in World War II." *Operations Research* **35**(6): 910–925.

McGregor, D. 1960. *The Human Side of Enterprise.* New York: McGraw-Hill.

Michel, R. 1997. "Reinvention Reigns: ERP Vendors Redefine Value, Planning, and Elevate Customer Service." *Manufacturing Systems,* July, 28.

Micklethwait, J., and A. Woolridge. 1996. *The Witch Doctors.* New York: Random House.

Miller, J. G., and T. E. Vollmann. 1985. "The Hidden Factory." *Harvard Business Review,* September–October, 142–150.

Miser, H. J., and E. S. Quade (eds.). 1985. *Handbook of Systems Analysis: Overview of Uses, Procedures, Applications, and Practice.* New York: North-Holland.

——— (eds.). 1988. *Handbook of Systems Analysis: Craft Issues and Procedural Choices.* New York: North-Holland.

Mitchell, W. N. 1931. *Production Management.* Chicago: University of Chicago Press.

Monden, Y. 1983. *Toyota Production System: Practical Approach to Production Management.* Norcross, GA: Industrial Engineering and Management Press.

Montgomery, D. C. 1991. *Introduction to Statistical Quality Control,* 2d ed. New York: Wiley.

Morton, T. E., and D. W. Pentico. 1993. *Heuristic Scheduling Systems with Applications to Production Systems and Project Management.* New York: Wiley.

Muckstadt, J. A., and L. J. Thomas. 1980. "Are Multi-Echelon Inventory Methods Worth Implementing in Systems with Low Demand Rates?" *Management Science* **26:** 483–494.

Muhs, W. F., C. D. Wrege, and A. Murtuza. 1981. "Extracts from Chordal's Letters: Pre-Taylor Shop Management." *Proceedings of the Academy of Management,* 41st annual meeting, San Diego, CA.

Mumford, L. 1943. *Technics and Civilization.* New York: Harcourt and Brace.

Munsterberg, H. 1913. *Psychology and Industrial Efficiency.* Boston: Houghton Mifflin.

Myers, F. S. 1990. "Japan's Henry Ford." *Scientific American* **262**(5): 98.

Nahmias, S. 1993. *Production and Operations Analysis.* 2d ed. Homewood, IL: Irwin.

Nahmias, S., and S. Smith. 1992. "Mathematical Models of Retailer Inventory Systems: A Review." In *Perspectives in Operations Management: Essays in Honor of Elwood S. Buffa.* R. K. Sarin (ed.). Boston: Kluwer.

Nellemann, D., and L. Smith. 1982. "Just-in-Time" vs. Just-in-Case Production/Inventory Systems Concepts Borrowed Back from Japan." *Production and Inventory Management,* second quarter, 12–21.

Nelson, D. 1990. *Frederick W. Taylor and the Rise of Scientific Management.* Madison: University of Wisconsin Press.

Niebel, B. 1993. *Motion and Time Study,* 9th ed. Homewood, IL: Irwin.

Ohno, T. 1988. *Toyota Production System: Beyond Large-Scale Production.* Cambridge, MA: Productivity Press (translation of *Toyota seisan hoshiki,* Tokyo: Diamond, 1978).

Ohno, T., and S. Mito. 1988. *Just-in-Time for Today and Tomorrow.* Cambridge, MA: Productivity Press (translation of *Naze hitsuyeo na mono o hitsuyeo na bun dake hitsuyeo na toki ni teikyeo shinai no ka,* Tokyo: Diamond, 1986).

Orlicky, J. 1975. *Material Requirements Planning: The New Way of Life in Production and Inventory Management.* New York: McGraw-Hill.

Parker, K. 1997. "The Great Trek Begins: Mid-sized Manufacturers Migrate to Client/Server Enterprise Systems," *Manufacturing Systems,* January.

Peterson, R., and E. A. Silver. 1985. *Decision Systems for Inventory Management and Production Planning.* 2d ed., New York: Wiley.

Pierson, F. C., et al. 1959. *The Education of American Businessmen.* New York: McGraw-Hill.

Pinedo, M. 1995. *Scheduling: Theory, Algorithms, and Systems.* Englewood Cliffs, NJ: Prentice-Hall.

Pinedo, M., and X. Chao. 1999. *Operations Scheduling: With Applications in Manufacturing and Services.* Boston: Irwin/McGraw-Hill.

Plossl, G. W. 1985. *Production and Inventory Control,* 2d ed. Englewood Cliffs, NJ: Prentice-Hall.

Pollard, H. R. 1974. *Developments in Management Thought.* London: Heinemann.

Polya, G. 1954. *Patterns of Plausible Inference.* Princeton, NJ: Princeton University Press.

Popper, K. 1963. *Conjectures and Refutations.* London: Routledge & Kegan Paul Ltd.

Rao, A. 1989. "A Survey of MRP-II Software Suppliers' Trends in Support of Just-in-Time." *Production and Inventory Management,* third quarter, 14–17.

Ravenscraft, D. J., and F. M. Scherer. 1987. *Mergers, Sell-Offs, and Economic Efficiency.* Washington: Brookings Institute.

Raymond, F. E. 1931. *Quantity and Economy in Manufacture.* New York: McGraw-Hill.

Roderick, L. M., D. T. Phillips, and G. L. Hogg, 1991. "A Comparison of Order Release Strategies in Production Control Systems." *International Journal of Production Research* **30**(2): 1991.

Roethlisberger, F. J., and W. J. Dickson. 1939. *Management and the Worker.* Cambridge, MA: Harvard University Press.

Roundy, R. 1985. "98% Effective Integer Ratio Lot-Sizing for One Warehouse Multi-Retailer Systems." *Management Science* **31**: 1416–1430.

———. 1986. "98% Effective Lot-Sizing Rule for Multi-Product, Multi-Stage Production Inventory Systems." *Mathematics of Operations Research* **11**: 699–727.

Sage, A. P. 1992. *Systems Engineering.* New York: Wiley.

Sanderson, R. J., J. A. Cambell, and J. D. Meyer. 1982. *Industrial Robots, A Summary and Forecast for Manufacturing Managers.* Lake Geneva, WI: Tech Tran Corporation.

Scherer, F. M., and D. Ross. 1990. *Industrial Market Structure and Economic Performance,* 3d ed. Boston: Houghton Mifflin.

Schmenner, R. W. 1993. *Production/Operations Management: From the Inside Out,* 5th ed. New York: Macmillian.

Schonberger, R. J. 1982. *Japanese Manufacturing Techniques: Nine Hidden Lessons in Simplicity.* New York: Free Press.

———. 1986. *World Class Manufacturing: The Lessons of Simplicity Applied.* New York: Free Press.

———. 1990. *Building a Chain of Customers: Linking Business Functions to Create a World Class Company.* New York: Free Press.

Schroeder, R., J. Anderson, S. Tupy, and E. White. 1981. "A Study of MRP Benefits and Costs." *Journal of Operations Management* **2**(1): 1–9.

Schumacher, B. G. 1986. *On the Origin and Nature of Management.* Norman, OK: Eugnosis.

Schwarz, L. B. (ed.). 1981. *Multi-Level Production/Inventory Control Systems: Theory and Practice.* Amsterdam: North-Holland.

———. 1998. "A New Teaching Paradigm: The Information/Control/Buffer Portfolio." *Production and Operations Management* **7**(2): 125–131, summer.

Scott, W. D. 1913. *Increasing Human Efficiency in Business.* New York: Macmillan.

Sethi, K. S., S. P. Sethi. 1990. "Flexibility in Manufacturing: A Survey," *International Journal of Flexible Manufacturing Systems* **2**, 289–328.

Shafritz, J. M., and J. S. Ott. 1992. *Classics of Organization Theory,* 3d ed., Pacific Grove, CA: Brooks/Cole Publishing Company.

Sherbrooke, C. C. 1992. *Optimal Inventory Modeling of Systems: Multi-Echelon Techniques.* New York: Wiley.

Shewhart, W. A. 1931. *Economic Control of Quality of Manufactured Product.* New York: Van Nostrand.

Shingo, S. 1985. *A Revolution in Manufacturing: The SMED System.* Cambridge, MA: Productivity Press.

———. 1986. *Zero Quality Control: Source Inspection and the Poka-Yoke System.* Cambridge, MA: Productivity Press.

———. 1989. *A Study of the Toyota Production System from an Industrial Engineering Viewpoint.* Cambridge, MA: Productivity Press.

———. 1990. *Modern Approaches to Manufacturing Improvement: The Shingo System.* A. Robinson (ed.). Cambridge, MA: Productivity Press.

Silver, A., D. Pyke, and R. Peterson. 1998. *Inventory Management and Production Planning and Scheduling.* New York: Wiley.

Simchi-Levi, D., P. Kaminsky, and E. Simchi-Levi. 1999. *Designing and Managing the Supply Chain: Concepts, Strategies and Cases.* Burr Ridge, IL: Irwin/McGraw-Hill.

Simons, Jr., J. V., and W. P. Simpson III. 1997. "An Exposition of Multiple Constraint Scheduling as Implemented in the Goal System (Formerly Disaster)." *Production and Operations Management* **6**(1): 3–22.

Singer, C., E. Flomyard, A. Hall, and T. Williams. 1958. *A History of Technology.* Oxford: Clarendon Press.

Skinner, W. 1969. "Manufacturing—The Missing Link in Corporate Strategy." *Harvard Business Review,* May/June, 156.

——. 1974. "The Focused Factory." *Harvard Business Review,* May–June, 113-121.

——. 1985. *Manufacturing: The Formidable Competitive Weapon.* New York: Wiley.

——. 1985b. "The Taming of Lions: How Manufacturing Leadership Evolved, 1780–1984." In K. B. Clark, R. H. Hayes, and C. Lorenz, *The Uneasy Alliance: Managing the Productivity-Technology Dilemma,* Boston: Harvard University Press.

——. 1986. "The Productivity Paradox." *Harvard Business Review,* July–August, 55–59.

——. 1988. "What Matters to Manufacturing." *Harvard Business Review,* January–February, 10–16.

Smith, A. 1776. *An Inquiry into the Nature and Causes of the Wealth of Nations.* Chicago: Great Books of the Western World, vol. 39, *Encyclopaedia Britannica,* 1952.

Spearman, M. L. 1991. "An Analytic Congestion Model for Closed Production Systems with IFR Processing Times," *Management Science* **37**(8): 1015–1029.

Spearman, M. L., W.J. Hopp, and D. L. Woodruff. 1989. "A Hierarchical Control Architecture for CONWIP Production Systems." *Journal of Manufacturing and Operations Management* **2:** 147–171.

Spearman, M. L., and S. Kröckel. 1999. "Batch Sizing to Minimize Flow Times in a Multi-Product System with Significant Changeover Times." Technical Report. Atlanta: Georgia Institute of Technology.

Spearman, M. L., D. L. Woodruff, and W. J. Hopp. 1989. "CONWIP: A Pull Alternative to Kanban." *International Journal of Production Research* **28**(5): 879–894.

Spearman, M. L., and M. A. Zazanis. 1992. "Push and Pull Production Systems: Issues and Comparisions." *Operations Research* **40**(3): 521–532.

Spearman, M. L., and R. Q. Zhang. 1999. "Optimal Lead Time Policies." *Management Science* **45**(2): 290–295.

Spriegel, W. R., and C. E. Myers (eds.). 1953. *The Writings of the Gilbreths.* Homewood, IL: Irwin.

Stalk, G., and T. M. Hout. 1990. *Competing Against Time: How Time-Based Competition Is Reshaping Global Markets.* New York: Free Press.

Stedman, C. 1999. "Survey: ERP Costs More Than Measurable ROI," *Computerworld,* April 5.

Sterman, J. D. 1989. "Modeling Managerial Behavior: Misperceptions of Feedback in a Dynamic Decision Making Experiment." *Management Science* **35**(3): 321–339.

Stover, John, F. 1961. *American Railroads.* Chicago: University of Chicago Press.

Suri, R. 1998. *Quick Response Manufacturing: A Companywide Approach to Reducing Leadtimes.* Portland, OR: Productivity Press.

Suri, R., and S. de Treville. 1992. "Time Is Money." *OR/MS Today,* October.

——. 1993. "Rapid Modeling: The Use of Queueing Models to Support Time-Based Competitive Manufacturing." In *Operations Research in Production Planning and Control.* G. Fandel, T. Gulledge, and A. Jones (eds.). New York: Springer-Verlag.

Suri, R., J. L. Sanders, and M. Kamanth. 1993. "Performance Evaluation of Production Networks." In *Handbooks in Operations Research and Management Science, vol 4: Logistics of Production and Inventory.* S. C. Graves, A. H. G. Rinnooy Kan, and P. H. Zipkin (eds.). New York: North-Holland.

Svoronos, A., and P. Zipkin. 1988. "Estimating the Performance of Multi-Level Inventory Systems." *Operations Research* **36:** 57–72.

Tardif, V. 1995. "Detecting Scheduling Infeasibilities in Multi-Stage, Finite Capacity, Production Environments," Ph.D. dissertation, Northwestern University, Evanston, IL.

Taft, E. W. 1918. "Formulas for Exact and Approximate Evaluation—Handling Cost of Jigs and Interest Charges of Product Manufactured Included." *The Iron Age* **101:** 1410–1412.

Taylor, A. 1997. "How Toyota Defies Gravity." *Fortune,* December 8, 100–108.

Taylor, F. W. 1903. "Shop Management." *Transactions of the ASME* **24:** 1337–1480.

———. 1911. *The Principles of Scientific Management.* New York: Harper & Row.

Thomas, P. R. 1990. *Competitiveness Through Total Cycle Time: An Overview for CEO's.* New York: McGraw-Hill.

———. 1991. *Getting Competitive: Middle Managers and the Cycle Time Ethic.* New York: McGraw-Hill.

Thompkins, J. A., and J. A. White. 1984. *Facilities Planning.* New York: Wiley.

Thompson, J. R., and J. Koronacki. 1992. *Statistical Process Control for Quality.* New York: Chapman & Hall.

Thompson, M. B. 1992. "Why Finite Capacity?" *APICS—The Performance Advantage,* June, 50–54.

Towne, H. R. 1886. "The Engineer as an Economist." *ASME Transactions* **7:** 428–432.

Turino, J. 1992. *Managing Concurrent Engineering Buying Time to Market.* New York: Van Nostrand Reinhold.

U.S. Department of Commerce. 1972. *Statistical Abstract of the United States.* 93d annual edition, Bureau of the Census.

———. 1977. *Statistical Abstract of the United States.* Economics and Statistics Administration, Bureau of the Census, Table 664, 842 and 758.

Ure, A. 1835. *The Philosophy of Manufactures: Or an Exposition of the Scientific, Moral and Commercial Economy of the Factory System of Great Britain.* London: Charles Knight. Reprint, Augustus M. Kelley, New York, 1967.

Urwick, L. 1947. *The Elements of Administration.* London: Pitman.

Vollmann, T. E., W. L. Berry, and D. C. Whybark. 1992. *Manufacturing Planning and Control Systems,* 3d ed., Burr Ridge, IL: Irwin.

Wack, P. 1985. "Scenarios: Uncharted Waters Ahead." *Harvard Business Review,* September–October, 73–89.

Wagner, H. M., and T. M. Whitin. 1958. "Dynamic Version of the Economic Lot Size Model." *Management Science* **5**(1): 89–96.

Waring, S. P. 1991. *Taylorism Transformed: Scientific Management Theory Since 1945.* Chapel Hill: University of North Carolina Press.

Wellington, A. M. 1877. *The Economic Theory of the Location of Railways.* New York: Wiley.

Wheelwright, S. 1981. "Japan—Where Operations Really Are Strategic." *Harvard Business Review,* July–August, 67–74.

Whiteside, D., and J. Arbose. 1984. "Unsnarling Industrial Production: Why Top Management Is Starting to Care," *International Management,* March, 20–26.

Whitin, T. M. 1953. *The Theory of Inventory Management.* Princeton, NJ: Princeton University Press.

Whitt, W. 1983. "The Queueing Network Analyzer." *Bell System Technology Journal* **62**(9): 2779–2815.

———. 1993. "Approximating the GI/G/m Queue." *Production and Operations Management,* **2**(2): 114–161.

Wight, O. 1970. "Input/Output Control: A Real Handle on Lead Time." *Production and Inventory Management Journal* **11**(3): 9–31.

———. 1974. *Production and Inventory Management in the Computer Age.* Boston: Cahners Books.

———. 1981. *MRP II: Unlocking America's Productivity Potential.* Boston: CBI Publishing.

Wilson, B. 1984. *Systems: Concepts, Methodologies, and Applications.* New York: Wiley.

Wilson, R. H. 1934. "A Scientific Routine for Stock Control." *Harvard Business Review* **13**(1): 116–128.

Winters, P. 1960. "Forecasting Sales by Exponentially Weighted Moving Averages." *Management Science* **6:** 324–342.

Woodruff, D., and M. Spearman. 1992. "Sequencing and Batching for Two Classes of Jobs with Deadlines and Setup Times," *Journal of Production and Operations Management,* **1:** 87–102.

Wrege, C. D., and R. G. Greenwood. 1991. *Frederick W. Taylor—The Father of Scientific Management: Myth and Reality.* Homewood, IL: Irwin.

Wren, D. 1987. *The Evolution of Management Thought.* 3d ed. New York: Wiley

Yates, R. 1992. "On the Road with the 'Messiah of Management' as He Tries to Do for His Country What He Did for Japan." *Chicago Tribune,* February 16, Section 10, 16.

Zais, A. 1986. "IBM Reigns in Dynamic MRP II Marketplace." *Computerworld,* January 27.

Zipkin, P. H. 1986. "Inventory Service-Level Measures: Convexity and Approximation." *Management Science* **32:** 975–981.

———. 1991. "Does Manufacturing Need a JIT Revolution?" *Harvard Business Review,* January–February, 40–50.

———. 1999. *Foundations of Inventory Management.* New York: McGraw-Hill.

Notation

General Conventions:

- A subscript "a" indicates a parameter that describes interarrival times to a station. For example, t_a represents the average time between arrivals to a station or line.
- A subscript "e" indicates a parameter that describes "effective" process times at a station. For example, t_e represents the average process time at a station including detractors such as downtime, setups, yield loss, etc.
- A parameter followed by (i) indicates that the parameter applies to station i, as in $TH(i)$, $CT(i)$, $t_e(i)$, $c_e(i)$, and so on.
- A superscript $*$ indicates a parameter that describes an "ideal" system without detractors. For example, r_b^* and T_0^* are the bottleneck rate and raw process time for a line with no downtime, setups, yield loss, or other inefficiencies.
- A superscript "P" indicates a parameter that describes a "practical" system. For example, r_b^P and T_0^P are the bottleneck rate and raw process time for a line operating under realistic conditions.

Mathematical Symbols:

CV   coefficient of variation of a random variable, which is the standard deviation divided by the mean.

c_0 CV of natural (no detractors) process time at a station.

c_a CV of the time between arrivals to a station.

c_e CV of effective process time at a station.

c_d CV of the time between departures from a station.

CT_q average queue time at a station. For single machine stations: $CT_q = \left(\frac{c_a^2 + c_e^2}{2}\right)\left(\frac{u}{1-u}\right)t_e$.

CT   cycle time, which is measured as the average time from when a job is released into a station or line to when it exits. (Where ambiguity is possible cycle time at station i is written as $CT(i)$.) Note that $CT = CT_q + t_e$.

FGI   finished goods inventory. For end items, FGI represents the store of final product waiting to be shipped to customers. For components, FGI can also represent "crib" inventory, which is stock in an intermediate location such as before an assembly operation.

LT   lead time, a management constant indicating the time allotted for production of a part on a given routing.